VOLUME FOUR HUNDRED AND EIGHTY-FIVE

METHODS IN ENZYMOLOGY

Constitutive Activity in Receptors and Other Proteins, Part B

METHODS IN ENZYMOLOGY

Editors-in-Chief

JOHN N. ABELSON AND MELVIN I. SIMON

Division of Biology
California Institute of Technology
Pasadena, California

Founding Editors

SIDNEY P. COLOWICK AND NATHAN O. KAPLAN

VOLUME FOUR HUNDRED AND EIGHTY-FIVE

METHODS IN ENZYMOLOGY

Constitutive Activity in Receptors and Other Proteins, Part B

EDITED BY

P. MICHAEL CONN

*Divisions of Reproductive Sciences and Neuroscience (ONPRC)
Departments of Pharmacology and Physiology,
Cell and Developmental Biology, and Obstetrics
and Gynecology (OHSU),
Beaverton, OR, USA*

[redacted]

ELSEVIER

AMSTERDAM • BOSTON • HEIDELBERG • LONDON
NEW YORK • OXFORD • PARIS • SAN DIEGO
SAN FRANCISCO • SINGAPORE • SYDNEY • TOKYO
Academic Press is an imprint of Elsevier

Academic Press is an imprint of Elsevier
525 B Street, Suite 1900, San Diego, CA 92101-4495, USA
30 Corporate Drive, Suite 400, Burlington, MA 01803, USA
32 Jamestown Road, London NW1 7BY, UK

First edition 2010

ISBN: 978-0-12-381296-4
ISSN: 0076-6879

Printed and bound in United States of America
10 11 12 10 9 8 7 6 5 4 3 2 1

Contents

Contributors

Hiroshi Akazawa
Department of Cardiovascular Medicine, Osaka University Graduate School of Medicine, Yamadaoka, Suita, Osaka, Japan

Noelle C. Anastasio
Center for Addiction Research, and Department of Pharmacology and Toxicology, University of Texas Medical Branch, Galveston, Texas, USA

Remko A. Bakker
CardioMetabolic Diseases Research, Boehringer Ingelheim Pharma GmbH & Co. KG, Birkendorferstrasse 65, Biberach an der Riss, Germany

Thomas J. Baranski
Department of Medicine, and Department of Developmental Biology, Washington University Medical School, St. Louis, Missouri, USA

Gabriel Barreda-Gómez
Department of Pharmacology, Faculty of Medicine, University of the Basque Country, Vizcaya, Spain

Annette G. Beck-Sickinger
Institute of Biochemistry, Leipzig University, Brüderstrasse, Leipzig, Germany

Jeffrey M. Becker
The University of Tennessee, Department of Microbiology, Knoxville, Tennessee, USA

Bruce R. Bianchi
Neuroscience Research, Global Pharmaceutical Research and Development, Abbott Laboratories, Abbott Park, Illinois, USA

Joël Bockaert
Centre National de la Recherche Scientifique, Institut de Génomique Fonctionnelle; Institut National de la Santé et de la Recherche Médicale; Université Montpellier 1; and Université Montpellier 2, Montpellier, France

William Bourguet
INSERM U554 and CNRS UMR5048, Centre de Biochimie Structurale, Universités Montpellier 1 & 2, Montpellier, France

L. J. Bridge
Centre for Mathematical Medicine and Biology, School of Mathematical Sciences, University of Nottingham, United Kingdom

Paola Casarosa
Respiratory Diseases Research, Boehringer Ingelheim Pharma GmbH & Co. KG, Birkendorferstrasse 65, Biberach an der Riss, Germany

Jun Chen
Neuroscience Research, Global Pharmaceutical Research and Development, Abbott Laboratories, Abbott Park, Illinois, USA

Chang K. Choi
Michigan Technological University, Department of Mechanical Engineering-Engineering Mechanics, Houghton, Michigan, USA

Constance Chollet
Institute of Biochemistry, Leipzig University, Brüderstrasse, Leipzig, Germany

Claude Cochet
INSERM, U873, CEA, iRTSV/LTS, Université Joseph Fourier, Grenoble, France

P. Michael Conn
Divisions of Reproductive Sciences and Neuroscience (ONPRC), and Departments of Pharmacology and Physiology, Cell and Developmental Biology, and Obstetrics and Gynecology (OHSU), Beaverton, Oregon, USA

Susanna Cotecchia
Department of General and Environmental Physiology, University of Bari, Italy, and Department of Pharmacology and Toxicology, University of Lausanne, Switzerland

Kathryn A. Cunningham
Center for Addiction Research, and Department of Pharmacology and Toxicology, University of Texas Medical Branch, Galveston, Texas, USA

Angel R. de Lera
Departamento de Química Orgánica, Facultad de Química, Universidade de Vigo, Vigo, Spain

Andria L. Del Tredici
Trianode, Inc., San Diego, California, USA

Mabrouka Doghman
Institut de Pharmacologie Moléculaire et Cellulaire, CNRS UMR 6097, Université de Nice Sophia Antipolis, Valbonne, France

Mercedes Dosil
Centro de Investigación del Cáncer and Instituto de Biología Molecular y Celular del Cáncer, CSIC-University of Salamanca, Campus Unamuno, Salamanca, Spain

Sylvia Els
Institute of Biochemistry, Leipzig University, Brüderstrasse, Leipzig, Germany

Michael A. Farrar
Center for Immunology, Masonic Cancer Center, and Department of Laboratory Medicine and Pathology, University of Minnesota, Minneapolis, USA

Stephanie Federico
Dipartimento di Scienze Farmaceutiche, Università degli Studi di Trieste, Piazzale Europa, Trieste, Italy

Odile Filhol
INSERM, U873, CEA, iRTSV/LTS, Université Joseph Fourier, Grenoble, France

Tung M. Fong
Department of Pharmacology, Forest Research Institute, Harborside Financial Center, Jersey City, New Jersey, USA

Jadwiga Furmaniak
FIRS Laboratories, RSR Ltd, Parc Ty Glas, Llanishen, Cardiff, United Kingdom

Guerrini Gabriella
Dipartimento di Scienze Farmaceutiche, Laboratorio di Progettazione, Sintesi e Studio di Eterocicli Biologicamente attivi (HeteroBioLab) Università degli Studi di Firenze, Via U. Schiff, Polo Scientifico, Sesto Fiorentino, Firenze, Italy

Line Garneau
Department of Physiology, Groupe d'étude des protéines membranaires, Université de Montréal, Montreal, Canada

Marvin C. Gershengorn
Clinical Endocrinology Branch, National Institute of Diabetes and Digestive and Kidney Diseases, National Institutes of Health, Bethesda, Maryland, USA

Ciciani Giovanna
Dipartimento di Scienze Farmaceutiche, Laboratorio di Progettazione, Sintesi e Studio di Eterocicli Biologicamente attivi (HeteroBioLab) Università degli Studi di Firenze, Via U. Schiff, Polo Scientifico, Sesto Fiorentino, Firenze, Italy

M. Teresa Giralt
Department of Pharmacology, Faculty of Medicine, University of the Basque Country, Vizcaya, Spain

Sujatha Gopalakrishnan
Advanced Technology, Global Pharmaceutical Research and Development, Abbott Laboratories, Abbott Park, Illinois, USA

Hinrich Gronemeyer
Department of Cancer Biology, Institut de Génétique et de Biologie Moléculaire et Cellulaire (IGBMC), C. U. de Strasbourg, France

Bonnie J. Hanson
Cell Systems Division, Discovery Assays and Services, Life Technologies Corp., Madison, Wisconsin, USA

Christian Höcht
Cátedra de Farmacología, Instituto de Fisiopatología y Bioquímica Clínica, Cátedra de Química Medicinal, Facultad de Farmacia y Bioquímica, Universidad de Buenos Aires, CONICET, Junín 256, Buenos Aires, Argentina

Peter Hodder
Scripps Research Institute Molecular Screening Center and Molecular Therapeutics, The Scripps Research Institute, Scripps Florida, Jupiter, Florida, USA

Jo Ann Janovick
Divisions of Reproductive Sciences and Neuroscience, (ONPRC), Beaverton, Oregon, USA

Holger Jaeschke
Department for Internal Medicine, Neurology and Dermatolgy; Clinic for Endocrinology and Nephrology, University of Leipzig, Leipzig, Germany

Tobias Kiechle
Respiratory Diseases Research, Boehringer Ingelheim Pharma GmbH & Co. KG, Birkendorferstrasse 65, Biberach an der Riss, Germany

Heejung Kim
The University of Tennessee, Department of Microbiology, Knoxville, Tennessee, USA

Hélène Klein
Department of Physiology, Groupe d'étude des protéines membranaires, Université de Montréal, Montreal, Canada

Issei Komuro
Department of Cardiovascular Medicine, Osaka University Graduate School of Medicine, Yamadaoka, Suita, Osaka, Japan

Masahiro Kono
Department of Ophthalmology, Medical University of South Carolina, Charleston, South Carolina, USA

James B. Konopka
Department of Molecular Genetics and Microbiology, State University of New York, Stony Brook, New York, USA

Marc R. Lake
Advanced Technology, Global Pharmaceutical Research and Development, Abbott Laboratories, Abbott Park, Illinois, USA

Enzo Lalli
Institut de Pharmacologie Moléculaire et Cellulaire, CNRS UMR 6097, Université de Nice Sophia Antipolis, Valbonne, France

Maria Fe Lanfranco
Center for Addiction Research, and Department of Pharmacology and Toxicology, University of Texas Medical Branch, Galveston, Texas, USA

Béatrice Laudet
INSERM, U873, CEA, iRTSV/LTS, Université Joseph Fourier, Grenoble, France

Paul H. Lee
Lead Discovery, Amgen, Inc., Thousand Oaks, California, USA

John Locklear
Advanced Technology, Global Pharmaceutical Research and Development, Abbott Laboratories, Abbott Park, Illinois, USA

Franck Madoux
Scripps Research Institute Molecular Screening Center and Molecular Therapeutics, The Scripps Research Institute, Scripps Florida, Jupiter, Florida, USA

Philippe Marin
Centre National de la Recherche Scientifique, Institut de Génomique Fonctionnelle; Institut National de la Santé et de la Recherche Médicale; Université Montpellier 1 and 2, Montpellier, France

Tom Masi
The University of Tennessee, Department of Microbiology, Knoxville, Tennessee, USA

Shin-ichiro Miura
Department of Cardiology, Fukuoka University School of Medicine, Nanakuma, Jonan-ku, Fukuoka, Japan

Federico Monczor
Cátedra de Farmacología, Instituto de Fisiopatología y Bioquímica Clínica, Cátedra de Química Medicinal, Facultad de Farmacia y Bioquímica, Universidad de Buenos Aires, CONICET, Junín 956, Buenos Aires, Argentina

Stefano Moro
Molecular Modeling Section (MMS), Dipartimento di Scienze Farmaceutiche, Università di Padova, via Marzolo, Padova, Italy

Virginie Moucadel
INSERM, U873, CEA, iRTSV/LTS, Université Joseph Fourier, Grenoble, France

Sandra Mueller
Department for Internal Medicine, Neurology and Dermatolgy; Clinic for Endocrinology and Nephrology, University of Leipzig, Leipzig, Germany

Ricardo Núñez Miguel
FIRS Laboratories, RSR Ltd, Parc Ty Glas, Llanishen, Cardiff, United Kingdom

Susanne Neumann
Clinical Endocrinology Branch, National Institute of Diabetes and Digestive and Kidney Diseases, National Institutes of Health, Bethesda, Maryland, USA

Gregory V. Nikiforovich
MolLife Design LLC, St. Louis, Missouri, USA

Silvia Paoletta
Molecular Modeling Section (MMS), Dipartimento di Scienze Farmaceutiche, Università di Padova, via Marzolo, Padova, Italy

Lucie Parent
Department of Physiology, Groupe d'étude des protéines membranaires, Université de Montréal, Montreal, Canada

Giljun Park
The University of Tennessee, Department of Microbiology, Knoxville, Tennessee, USA

Ralf Paschke
Department for Internal Medicine, Neurology and Dermatolgy; Clinic for Endocrinology and Nephrology, University of Leipzig, Leipzig, Germany

Fabrice Piu
Otonomy Inc., San Diego, California, USA

Renaud Prudent
INSERM, U873, CEA, iRTSV/LTS, Université Joseph Fourier, Grenoble, France

Bruce M. Raaka
Clinical Endocrinology Branch, National Institute of Diabetes and Digestive and Kidney Diseases, National Institutes of Health, Bethesda, Maryland, USA

Bernard Rees Smith
FIRS Laboratories, RSR Ltd, Parc Ty Glas, Llanishen, Cardiff, United Kingdom

Regina M. Reilly
Neuroscience Research, Global Pharmaceutical Research and Development, Abbott Laboratories, Abbott Park, Illinois, USA

Rafael Rodríguez-Puertas
Department of Pharmacology, Faculty of Medicine, University of the Basque Country, Vizcaya, Spain

Jane Sanders
FIRS Laboratories, RSR Ltd, Parc Ty Glas, Llanishen, Cardiff, United Kingdom

Céline F. Sautel
INSERM, U873, CEA, iRTSV/LTS, Université Joseph Fourier, Grenoble, France

Rémy Sauvé
Department of Physiology, Groupe d'étude des protéines membranaires, Université de Montréal, Montreal, Canada

Erich H. Schneider
Laboratory of Molecular Immunology, NIAID/NIH, Bethesda, Maryland, USA

Roland Seifert
Institute of Pharmacology, Medical School of Hannover, Hannover, Germany

Mathieu Seimandi
Centre National de la Recherche Scientifique, Institut de Génomique Fonctionnelle; Institut National de la Santé et de la Recherche Médicale; Université Montpellier 1 and 2, Montpellier, France

Patricia K. Seitz
Center for Addiction Research, and Department of Pharmacology and Toxicology, University of Texas Medical Branch, Galveston, Texas, USA

Giampiero Spalluto
Dipartimento di Scienze Farmaceutiche, Università degli Studi di Trieste, Piazzale Europa, Trieste, Italy

Tim E. Sparer
The University of Tennessee, Department of Microbiology, Knoxville, Tennessee, USA

Alan C. Spivey
Department of Chemistry, Imperial College, London, United Kingdom

Carlos A. Taira
Cátedra de Farmacología, Instituto de Fisiopatología y Bioquímica Clínica, Cátedra de Química Medicinal, Facultad de Farmacia y Bioquímica, Universidad de Buenos Aires, CONICET, Junín 956, Buenos Aires, Argentina

Chih-Chung Tseng
Chemical Synthesis Laboratory@Biopolis, Institute of Chemical and Engineering Sciences, Singapore

Noritaka Yasuda
Department of Cardiovascular Science and Medicine, Chiba University Graduate School of Medicine, Inohana, Chuo-ku, Chiba, Japan

Preface

The observation that mutant (and sometimes wild type) receptors, ion channels, and enzymes exist in states of constitutive activation has provided insight into the etiology of disease and the mechanism of protein function in ligand recognition, effector coupling, ion conductance, and catalysis. The observation that constitutive activity is a surprisingly common event has supported the view that biologically active molecules exist in both inactive and active states. Moreover, the observation that many drugs already at market are actually inverse agonists (agents that inhibit constitutive activity) makes understanding constitutive activity important for therapeutic drug design.

This volume provides descriptions of methods used to assess the mechanism of protein function and pharmacological tools and methodological approaches to the analysis of constitutive activity. The authors explain how these methods are able to provide important biological insights.

Authors were selected on the basis of research contributions in the area about which they have written and on the basis of their ability to describe their methodological contribution in a clear and reproducible way. They have been encouraged to make use of graphics, to do comparisons with other methods, and to provide tricks and approaches not revealed in prior publications that make it possible to adapt methods to other systems.

The editor expresses appreciation to the contributors for providing their contributions in a timely fashion, to the senior editors for guidance, and to the staff at Academic Press for helpful input.

JUNE, 2010 P. MICHAEL CONN

Methods in Enzymology

Volume I. Preparation and Assay of Enzymes
Edited by Sidney P. Colowick and Nathan O. Kaplan

Volume II. Preparation and Assay of Enzymes
Edited by Sidney P. Colowick and Nathan O. Kaplan

Volume III. Preparation and Assay of Substrates
Edited by Sidney P. Colowick and Nathan O. Kaplan

Volume IV. Special Techniques for the Enzymologist
Edited by Sidney P. Colowick and Nathan O. Kaplan

Volume V. Preparation and Assay of Enzymes
Edited by Sidney P. Colowick and Nathan O. Kaplan

Volume VI. Preparation and Assay of Enzymes *(Continued)*
Preparation and Assay of Substrates
Special Techniques
Edited by Sidney P. Colowick and Nathan O. Kaplan

Volume VII. Cumulative Subject Index
Edited by Sidney P. Colowick and Nathan O. Kaplan

Volume VIII. Complex Carbohydrates
Edited by Elizabeth F. Neufeld and Victor Ginsburg

Volume IX. Carbohydrate Metabolism
Edited by Willis A. Wood

Volume X. Oxidation and Phosphorylation
Edited by Ronald W. Estabrook and Maynard E. Pullman

Volume XI. Enzyme Structure
Edited by C. H. W. Hirs

Volume XII. Nucleic Acids (Parts A and B)
Edited by Lawrence Grossman and Kivie Moldave

Volume XIII. Citric Acid Cycle
Edited by J. M. Lowenstein

Volume XIV. Lipids
Edited by J. M. Lowenstein

Volume XV. Steroids and Terpenoids
Edited by Raymond B. Clayton

Volume XVI. Fast Reactions
Edited by Kenneth Kustin

Volume XVII. Metabolism of Amino Acids and Amines (Parts A and B)
Edited by Herbert Tabor and Celia White Tabor

Volume XVIII. Vitamins and Coenzymes (Parts A, B, and C)
Edited by Donald B. McCormick and Lemuel D. Wright

Volume XIX. Proteolytic Enzymes
Edited by Gertrude E. Perlmann and Laszlo Lorand

Volume XX. Nucleic Acids and Protein Synthesis (Part C)
Edited by Kivie Moldave and Lawrence Grossman

Volume XXI. Nucleic Acids (Part D)
Edited by Lawrence Grossman and Kivie Moldave

Volume XXII. Enzyme Purification and Related Techniques
Edited by William B. Jakoby

Volume XXIII. Photosynthesis (Part A)
Edited by Anthony San Pietro

Volume XXIV. Photosynthesis and Nitrogen Fixation (Part B)
Edited by Anthony San Pietro

Volume XXV. Enzyme Structure (Part B)
Edited by C. H. W. Hirs and Serge N. Timasheff

Volume XXVI. Enzyme Structure (Part C)
Edited by C. H. W. Hirs and Serge N. Timasheff

Volume XXVII. Enzyme Structure (Part D)
Edited by C. H. W. Hirs and Serge N. Timasheff

Volume XXVIII. Complex Carbohydrates (Part B)
Edited by Victor Ginsburg

Volume XXIX. Nucleic Acids and Protein Synthesis (Part E)
Edited by Lawrence Grossman and Kivie Moldave

Volume XXX. Nucleic Acids and Protein Synthesis (Part F)
Edited by Kivie Moldave and Lawrence Grossman

Volume XXXI. Biomembranes (Part A)
Edited by Sidney Fleischer and Lester Packer

Volume XXXII. Biomembranes (Part B)
Edited by Sidney Fleischer and Lester Packer

Volume XXXIII. Cumulative Subject Index Volumes I-XXX
Edited by Martha G. Dennis and Edward A. Dennis

Volume XXXIV. Affinity Techniques (Enzyme Purification: Part B)
Edited by William B. Jakoby and Meir Wilchek

Volume XXXV. Lipids (Part B)
Edited by John M. Lowenstein

Volume XXXVI. Hormone Action (Part A: Steroid Hormones)
Edited by Bert W. O'Malley and Joel G. Hardman

Volume XXXVII. Hormone Action (Part B: Peptide Hormones)
Edited by Bert W. O'Malley and Joel G. Hardman

Volume XXXVIII. Hormone Action (Part C: Cyclic Nucleotides)
Edited by Joel G. Hardman and Bert W. O'Malley

Volume XXXIX. Hormone Action (Part D: Isolated Cells, Tissues, and Organ Systems)
Edited by Joel G. Hardman and Bert W. O'Malley

Volume XL. Hormone Action (Part E: Nuclear Structure and Function)
Edited by Bert W. O'Malley and Joel G. Hardman

Volume XLI. Carbohydrate Metabolism (Part B)
Edited by W. A. Wood

Volume XLII. Carbohydrate Metabolism (Part C)
Edited by W. A. Wood

Volume XLIII. Antibiotics
Edited by John H. Hash

Volume XLIV. Immobilized Enzymes
Edited by Klaus Mosbach

Volume XLV. Proteolytic Enzymes (Part B)
Edited by Laszlo Lorand

Volume XLVI. Affinity Labeling
Edited by William B. Jakoby and Meir Wilchek

Volume XLVII. Enzyme Structure (Part E)
Edited by C. H. W. Hirs and Serge N. Timasheff

Volume XLVIII. Enzyme Structure (Part F)
Edited by C. H. W. Hirs and Serge N. Timasheff

Volume XLIX. Enzyme Structure (Part G)
Edited by C. H. W. Hirs and Serge N. Timasheff

Volume L. Complex Carbohydrates (Part C)
Edited by Victor Ginsburg

Volume LI. Purine and Pyrimidine Nucleotide Metabolism
Edited by Patricia A. Hoffee and Mary Ellen Jones

Volume LII. Biomembranes (Part C: Biological Oxidations)
Edited by Sidney Fleischer and Lester Packer

VOLUME LIII. Biomembranes (Part D: Biological Oxidations)
Edited by SIDNEY FLEISCHER AND LESTER PACKER

VOLUME LIV. Biomembranes (Part E: Biological Oxidations)
Edited by SIDNEY FLEISCHER AND LESTER PACKER

VOLUME LV. Biomembranes (Part F: Bioenergetics)
Edited by SIDNEY FLEISCHER AND LESTER PACKER

VOLUME LVI. Biomembranes (Part G: Bioenergetics)
Edited by SIDNEY FLEISCHER AND LESTER PACKER

VOLUME LVII. Bioluminescence and Chemiluminescence
Edited by MARLENE A. DELUCA

VOLUME LVIII. Cell Culture
Edited by WILLIAM B. JAKOBY AND IRA PASTAN

VOLUME LIX. Nucleic Acids and Protein Synthesis (Part G)
Edited by KIVIE MOLDAVE AND LAWRENCE GROSSMAN

VOLUME LX. Nucleic Acids and Protein Synthesis (Part H)
Edited by KIVIE MOLDAVE AND LAWRENCE GROSSMAN

VOLUME 61. Enzyme Structure (Part H)
Edited by C. H. W. HIRS AND SERGE N. TIMASHEFF

VOLUME 62. Vitamins and Coenzymes (Part D)
Edited by DONALD B. MCCORMICK AND LEMUEL D. WRIGHT

VOLUME 63. Enzyme Kinetics and Mechanism (Part A: Initial Rate and Inhibitor Methods)
Edited by DANIEL L. PURICH

VOLUME 64. Enzyme Kinetics and Mechanism (Part B: Isotopic Probes and Complex Enzyme Systems)
Edited by DANIEL L. PURICH

VOLUME 65. Nucleic Acids (Part I)
Edited by LAWRENCE GROSSMAN AND KIVIE MOLDAVE

VOLUME 66. Vitamins and Coenzymes (Part E)
Edited by DONALD B. MCCORMICK AND LEMUEL D. WRIGHT

VOLUME 67. Vitamins and Coenzymes (Part F)
Edited by DONALD B. MCCORMICK AND LEMUEL D. WRIGHT

VOLUME 68. Recombinant DNA
Edited by RAY WU

VOLUME 69. Photosynthesis and Nitrogen Fixation (Part C)
Edited by ANTHONY SAN PIETRO

VOLUME 70. Immunochemical Techniques (Part A)
Edited by HELEN VAN VUNAKIS AND JOHN J. LANGONE

Volume 89. Carbohydrate Metabolism (Part D)
Edited by Willis A. Wood

Volume 90. Carbohydrate Metabolism (Part E)
Edited by Willis A. Wood

Volume 91. Enzyme Structure (Part I)
Edited by C. H. W. Hirs and Serge N. Timasheff

Volume 92. Immunochemical Techniques (Part E: Monoclonal Antibodies and General Immunoassay Methods)
Edited by John J. Langone and Helen Van Vunakis

Volume 93. Immunochemical Techniques (Part F: Conventional Antibodies, Fc Receptors, and Cytotoxicity)
Edited by John J. Langone and Helen Van Vunakis

Volume 94. Polyamines
Edited by Herbert Tabor and Celia White Tabor

Volume 95. Cumulative Subject Index Volumes 61–74, 76–80
Edited by Edward A. Dennis and Martha G. Dennis

Volume 96. Biomembranes [Part J: Membrane Biogenesis: Assembly and Targeting (General Methods; Eukaryotes)]
Edited by Sidney Fleischer and Becca Fleischer

Volume 97. Biomembranes [Part K: Membrane Biogenesis: Assembly and Targeting (Prokaryotes, Mitochondria, and Chloroplasts)]
Edited by Sidney Fleischer and Becca Fleischer

Volume 98. Biomembranes (Part L: Membrane Biogenesis: Processing and Recycling)
Edited by Sidney Fleischer and Becca Fleischer

Volume 99. Hormone Action (Part F: Protein Kinases)
Edited by Jackie D. Corbin and Joel G. Hardman

Volume 100. Recombinant DNA (Part B)
Edited by Ray Wu, Lawrence Grossman, and Kivie Moldave

Volume 101. Recombinant DNA (Part C)
Edited by Ray Wu, Lawrence Grossman, and Kivie Moldave

Volume 102. Hormone Action (Part G: Calmodulin and Calcium-Binding Proteins)
Edited by Anthony R. Means and Bert W. O'Malley

Volume 103. Hormone Action (Part H: Neuroendocrine Peptides)
Edited by P. Michael Conn

Volume 104. Enzyme Purification and Related Techniques (Part C)
Edited by William B. Jakoby

VOLUME 105. Oxygen Radicals in Biological Systems
Edited by LESTER PACKER

VOLUME 106. Posttranslational Modifications (Part A)
Edited by FINN WOLD AND KIVIE MOLDAVE

VOLUME 107. Posttranslational Modifications (Part B)
Edited by FINN WOLD AND KIVIE MOLDAVE

VOLUME 108. Immunochemical Techniques (Part G: Separation and Characterization of Lymphoid Cells)
Edited by GIOVANNI DI SABATO, JOHN J. LANGONE, AND HELEN VAN VUNAKIS

VOLUME 109. Hormone Action (Part I: Peptide Hormones)
Edited by LUTZ BIRNBAUMER AND BERT W. O'MALLEY

VOLUME 110. Steroids and Isoprenoids (Part A)
Edited by JOHN H. LAW AND HANS C. RILLING

VOLUME 111. Steroids and Isoprenoids (Part B)
Edited by JOHN H. LAW AND HANS C. RILLING

VOLUME 112. Drug and Enzyme Targeting (Part A)
Edited by KENNETH J. WIDDER AND RALPH GREEN

VOLUME 113. Glutamate, Glutamine, Glutathione, and Related Compounds
Edited by ALTON MEISTER

VOLUME 114. Diffraction Methods for Biological Macromolecules (Part A)
Edited by HAROLD W. WYCKOFF, C. H. W. HIRS, AND SERGE N. TIMASHEFF

VOLUME 115. Diffraction Methods for Biological Macromolecules (Part B)
Edited by HAROLD W. WYCKOFF, C. H. W. HIRS, AND SERGE N. TIMASHEFF

VOLUME 116. Immunochemical Techniques
(Part H: Effectors and Mediators of Lymphoid Cell Functions)
Edited by GIOVANNI DI SABATO, JOHN J. LANGONE, AND HELEN VAN VUNAKIS

VOLUME 117. Enzyme Structure (Part J)
Edited by C. H. W. HIRS AND SERGE N. TIMASHEFF

VOLUME 118. Plant Molecular Biology
Edited by ARTHUR WEISSBACH AND HERBERT WEISSBACH

VOLUME 119. Interferons (Part C)
Edited by SIDNEY PESTKA

VOLUME 120. Cumulative Subject Index Volumes 81–94, 96–101

VOLUME 121. Immunochemical Techniques (Part I: Hybridoma Technology and Monoclonal Antibodies)
Edited by JOHN J. LANGONE AND HELEN VAN VUNAKIS

VOLUME 122. Vitamins and Coenzymes (Part G)
Edited by FRANK CHYTIL AND DONALD B. MCCORMICK

Volume 123. Vitamins and Coenzymes (Part H)
Edited by Frank Chytil and Donald B. McCormick

Volume 124. Hormone Action (Part J: Neuroendocrine Peptides)
Edited by P. Michael Conn

Volume 125. Biomembranes (Part M: Transport in Bacteria, Mitochondria, and Chloroplasts: General Approaches and Transport Systems)
Edited by Sidney Fleischer and Becca Fleischer

Volume 126. Biomembranes (Part N: Transport in Bacteria, Mitochondria, and Chloroplasts: Protonmotive Force)
Edited by Sidney Fleischer and Becca Fleischer

Volume 127. Biomembranes (Part O: Protons and Water: Structure and Translocation)
Edited by Lester Packer

Volume 128. Plasma Lipoproteins (Part A: Preparation, Structure, and Molecular Biology)
Edited by Jere P. Segrest and John J. Albers

Volume 129. Plasma Lipoproteins (Part B: Characterization, Cell Biology, and Metabolism)
Edited by John J. Albers and Jere P. Segrest

Volume 130. Enzyme Structure (Part K)
Edited by C. H. W. Hirs and Serge N. Timasheff

Volume 131. Enzyme Structure (Part L)
Edited by C. H. W. Hirs and Serge N. Timasheff

Volume 132. Immunochemical Techniques (Part J: Phagocytosis and Cell-Mediated Cytotoxicity)
Edited by Giovanni Di Sabato and Johannes Everse

Volume 133. Bioluminescence and Chemiluminescence (Part B)
Edited by Marlene DeLuca and William D. McElroy

Volume 134. Structural and Contractile Proteins (Part C: The Contractile Apparatus and the Cytoskeleton)
Edited by Richard B. Vallee

Volume 135. Immobilized Enzymes and Cells (Part B)
Edited by Klaus Mosbach

Volume 136. Immobilized Enzymes and Cells (Part C)
Edited by Klaus Mosbach

Volume 137. Immobilized Enzymes and Cells (Part D)
Edited by Klaus Mosbach

Volume 138. Complex Carbohydrates (Part E)
Edited by Victor Ginsburg

Volume 157. Biomembranes (Part Q: ATP-Driven Pumps and Related Transport: Calcium, Proton, and Potassium Pumps)
Edited by Sidney Fleischer and Becca Fleischer

Volume 158. Metalloproteins (Part A)
Edited by James F. Riordan and Bert L. Vallee

Volume 159. Initiation and Termination of Cyclic Nucleotide Action
Edited by Jackie D. Corbin and Roger A. Johnson

Volume 160. Biomass (Part A: Cellulose and Hemicellulose)
Edited by Willis A. Wood and Scott T. Kellogg

Volume 161. Biomass (Part B: Lignin, Pectin, and Chitin)
Edited by Willis A. Wood and Scott T. Kellogg

Volume 162. Immunochemical Techniques (Part L: Chemotaxis and Inflammation)
Edited by Giovanni Di Sabato

Volume 163. Immunochemical Techniques (Part M: Chemotaxis and Inflammation)
Edited by Giovanni Di Sabato

Volume 164. Ribosomes
Edited by Harry F. Noller, Jr., and Kivie Moldave

Volume 165. Microbial Toxins: Tools for Enzymology
Edited by Sidney Harshman

Volume 166. Branched-Chain Amino Acids
Edited by Robert Harris and John R. Sokatch

Volume 167. Cyanobacteria
Edited by Lester Packer and Alexander N. Glazer

Volume 168. Hormone Action (Part K: Neuroendocrine Peptides)
Edited by P. Michael Conn

Volume 169. Platelets: Receptors, Adhesion, Secretion (Part A)
Edited by Jacek Hawiger

Volume 170. Nucleosomes
Edited by Paul M. Wassarman and Roger D. Kornberg

Volume 171. Biomembranes (Part R: Transport Theory: Cells and Model Membranes)
Edited by Sidney Fleischer and Becca Fleischer

Volume 172. Biomembranes (Part S: Transport: Membrane Isolation and Characterization)
Edited by Sidney Fleischer and Becca Fleischer

Volume 173. Biomembranes [Part T: Cellular and Subcellular Transport: Eukaryotic (Nonepithelial) Cells]
Edited by Sidney Fleischer and Becca Fleischer

Volume 174. Biomembranes [Part U: Cellular and Subcellular Transport: Eukaryotic (Nonepithelial) Cells]
Edited by Sidney Fleischer and Becca Fleischer

Volume 175. Cumulative Subject Index Volumes 135–139, 141–167

Volume 176. Nuclear Magnetic Resonance (Part A: Spectral Techniques and Dynamics)
Edited by Norman J. Oppenheimer and Thomas L. James

Volume 177. Nuclear Magnetic Resonance (Part B: Structure and Mechanism)
Edited by Norman J. Oppenheimer and Thomas L. James

Volume 178. Antibodies, Antigens, and Molecular Mimicry
Edited by John J. Langone

Volume 179. Complex Carbohydrates (Part F)
Edited by Victor Ginsburg

Volume 180. RNA Processing (Part A: General Methods)
Edited by James E. Dahlberg and John N. Abelson

Volume 181. RNA Processing (Part B: Specific Methods)
Edited by James E. Dahlberg and John N. Abelson

Volume 182. Guide to Protein Purification
Edited by Murray P. Deutscher

Volume 183. Molecular Evolution: Computer Analysis of Protein and Nucleic Acid Sequences
Edited by Russell F. Doolittle

Volume 184. Avidin-Biotin Technology
Edited by Meir Wilchek and Edward A. Bayer

Volume 185. Gene Expression Technology
Edited by David V. Goeddel

Volume 186. Oxygen Radicals in Biological Systems (Part B: Oxygen Radicals and Antioxidants)
Edited by Lester Packer and Alexander N. Glazer

Volume 187. Arachidonate Related Lipid Mediators
Edited by Robert C. Murphy and Frank A. Fitzpatrick

Volume 188. Hydrocarbons and Methylotrophy
Edited by Mary E. Lidstrom

Volume 189. Retinoids (Part A: Molecular and Metabolic Aspects)
Edited by Lester Packer

VOLUME 190. Retinoids (Part B: Cell Differentiation and Clinical Applications)
Edited by LESTER PACKER

VOLUME 191. Biomembranes (Part V: Cellular and Subcellular Transport: Epithelial Cells)
Edited by SIDNEY FLEISCHER AND BECCA FLEISCHER

VOLUME 192. Biomembranes (Part W: Cellular and Subcellular Transport: Epithelial Cells)
Edited by SIDNEY FLEISCHER AND BECCA FLEISCHER

VOLUME 193. Mass Spectrometry
Edited by JAMES A. MCCLOSKEY

VOLUME 194. Guide to Yeast Genetics and Molecular Biology
Edited by CHRISTINE GUTHRIE AND GERALD R. FINK

VOLUME 195. Adenylyl Cyclase, G Proteins, and Guanylyl Cyclase
Edited by ROGER A. JOHNSON AND JACKIE D. CORBIN

VOLUME 196. Molecular Motors and the Cytoskeleton
Edited by RICHARD B. VALLEE

VOLUME 197. Phospholipases
Edited by EDWARD A. DENNIS

VOLUME 198. Peptide Growth Factors (Part C)
Edited by DAVID BARNES, J. P. MATHER, AND GORDON H. SATO

VOLUME 199. Cumulative Subject Index Volumes 168–174, 176–194

VOLUME 200. Protein Phosphorylation (Part A: Protein Kinases: Assays, Purification, Antibodies, Functional Analysis, Cloning, and Expression)
Edited by TONY HUNTER AND BARTHOLOMEW M. SEFTON

VOLUME 201. Protein Phosphorylation (Part B: Analysis of Protein Phosphorylation, Protein Kinase Inhibitors, and Protein Phosphatases)
Edited by TONY HUNTER AND BARTHOLOMEW M. SEFTON

VOLUME 202. Molecular Design and Modeling: Concepts and Applications (Part A: Proteins, Peptides, and Enzymes)
Edited by JOHN J. LANGONE

VOLUME 203. Molecular Design and Modeling: Concepts and Applications (Part B: Antibodies and Antigens, Nucleic Acids, Polysaccharides, and Drugs)
Edited by JOHN J. LANGONE

VOLUME 204. Bacterial Genetic Systems
Edited by JEFFREY H. MILLER

VOLUME 205. Metallobiochemistry (Part B: Metallothionein and Related Molecules)
Edited by JAMES F. RIORDAN AND BERT L. VALLEE

VOLUME 206. Cytochrome P450
Edited by MICHAEL R. WATERMAN AND ERIC F. JOHNSON

VOLUME 207. Ion Channels
Edited by BERNARDO RUDY AND LINDA E. IVERSON

VOLUME 208. Protein–DNA Interactions
Edited by ROBERT T. SAUER

VOLUME 209. Phospholipid Biosynthesis
Edited by EDWARD A. DENNIS AND DENNIS E. VANCE

VOLUME 210. Numerical Computer Methods
Edited by LUDWIG BRAND AND MICHAEL L. JOHNSON

VOLUME 211. DNA Structures (Part A: Synthesis and Physical Analysis of DNA)
Edited by DAVID M. J. LILLEY AND JAMES E. DAHLBERG

VOLUME 212. DNA Structures (Part B: Chemical and Electrophoretic Analysis of DNA)
Edited by DAVID M. J. LILLEY AND JAMES E. DAHLBERG

VOLUME 213. Carotenoids (Part A: Chemistry, Separation, Quantitation, and Antioxidation)
Edited by LESTER PACKER

VOLUME 214. Carotenoids (Part B: Metabolism, Genetics, and Biosynthesis)
Edited by LESTER PACKER

VOLUME 215. Platelets: Receptors, Adhesion, Secretion (Part B)
Edited by JACEK J. HAWIGER

VOLUME 216. Recombinant DNA (Part G)
Edited by RAY WU

VOLUME 217. Recombinant DNA (Part H)
Edited by RAY WU

VOLUME 218. Recombinant DNA (Part I)
Edited by RAY WU

VOLUME 219. Reconstitution of Intracellular Transport
Edited by JAMES E. ROTHMAN

VOLUME 220. Membrane Fusion Techniques (Part A)
Edited by NEJAT DÜZGÜNEŞ

VOLUME 221. Membrane Fusion Techniques (Part B)
Edited by NEJAT DÜZGÜNEŞ

VOLUME 222. Proteolytic Enzymes in Coagulation, Fibrinolysis, and Complement Activation (Part A: Mammalian Blood Coagulation Factors and Inhibitors)
Edited by LASZLO LORAND AND KENNETH G. MANN

Volume 223. Proteolytic Enzymes in Coagulation, Fibrinolysis, and Complement Activation (Part B: Complement Activation, Fibrinolysis, and Nonmammalian Blood Coagulation Factors)
Edited by Laszlo Lorand and Kenneth G. Mann

Volume 224. Molecular Evolution: Producing the Biochemical Data
Edited by Elizabeth Anne Zimmer, Thomas J. White, Rebecca L. Cann, and Allan C. Wilson

Volume 225. Guide to Techniques in Mouse Development
Edited by Paul M. Wassarman and Melvin L. DePamphilis

Volume 226. Metallobiochemistry (Part C: Spectroscopic and Physical Methods for Probing Metal Ion Environments in Metalloenzymes and Metalloproteins)
Edited by James F. Riordan and Bert L. Vallee

Volume 227. Metallobiochemistry (Part D: Physical and Spectroscopic Methods for Probing Metal Ion Environments in Metalloproteins)
Edited by James F. Riordan and Bert L. Vallee

Volume 228. Aqueous Two-Phase Systems
Edited by Harry Walter and Göte Johansson

Volume 229. Cumulative Subject Index Volumes 195–198, 200–227

Volume 230. Guide to Techniques in Glycobiology
Edited by William J. Lennarz and Gerald W. Hart

Volume 231. Hemoglobins (Part B: Biochemical and Analytical Methods)
Edited by Johannes Everse, Kim D. Vandegriff, and Robert M. Winslow

Volume 232. Hemoglobins (Part C: Biophysical Methods)
Edited by Johannes Everse, Kim D. Vandegriff, and Robert M. Winslow

Volume 233. Oxygen Radicals in Biological Systems (Part C)
Edited by Lester Packer

Volume 234. Oxygen Radicals in Biological Systems (Part D)
Edited by Lester Packer

Volume 235. Bacterial Pathogenesis (Part A: Identification and Regulation of Virulence Factors)
Edited by Virginia L. Clark and Patrik M. Bavoil

Volume 236. Bacterial Pathogenesis (Part B: Integration of Pathogenic Bacteria with Host Cells)
Edited by Virginia L. Clark and Patrik M. Bavoil

Volume 237. Heterotrimeric G Proteins
Edited by Ravi Iyengar

Volume 238. Heterotrimeric G-Protein Effectors
Edited by Ravi Iyengar

VOLUME 239. Nuclear Magnetic Resonance (Part C)
Edited by THOMAS L. JAMES AND NORMAN J. OPPENHEIMER

VOLUME 240. Numerical Computer Methods (Part B)
Edited by MICHAEL L. JOHNSON AND LUDWIG BRAND

VOLUME 241. Retroviral Proteases
Edited by LAWRENCE C. KUO AND JULES A. SHAFER

VOLUME 242. Neoglycoconjugates (Part A)
Edited by Y. C. LEE AND REIKO T. LEE

VOLUME 243. Inorganic Microbial Sulfur Metabolism
Edited by HARRY D. PECK, JR., AND JEAN LEGALL

VOLUME 244. Proteolytic Enzymes: Serine and Cysteine Peptidases
Edited by ALAN J. BARRETT

VOLUME 245. Extracellular Matrix Components
Edited by E. RUOSLAHTI AND E. ENGVALL

VOLUME 246. Biochemical Spectroscopy
Edited by KENNETH SAUER

VOLUME 247. Neoglycoconjugates (Part B: Biomedical Applications)
Edited by Y. C. LEE AND REIKO T. LEE

VOLUME 248. Proteolytic Enzymes: Aspartic and Metallo Peptidases
Edited by ALAN J. BARRETT

VOLUME 249. Enzyme Kinetics and Mechanism (Part D: Developments in Enzyme Dynamics)
Edited by DANIEL L. PURICH

VOLUME 250. Lipid Modifications of Proteins
Edited by PATRICK J. CASEY AND JANICE E. BUSS

VOLUME 251. Biothiols (Part A: Monothiols and Dithiols, Protein Thiols, and Thiyl Radicals)
Edited by LESTER PACKER

VOLUME 252. Biothiols (Part B: Glutathione and Thioredoxin; Thiols in Signal Transduction and Gene Regulation)
Edited by LESTER PACKER

VOLUME 253. Adhesion of Microbial Pathogens
Edited by RON J. DOYLE AND ITZHAK OFEK

VOLUME 254. Oncogene Techniques
Edited by PETER K. VOGT AND INDER M. VERMA

VOLUME 255. Small GTPases and Their Regulators (Part A: Ras Family)
Edited by W. E. BALCH, CHANNING J. DER, AND ALAN HALL

VOLUME 293. Ion Channels (Part B)
Edited by P. MICHAEL CONN

VOLUME 294. Ion Channels (Part C)
Edited by P. MICHAEL CONN

VOLUME 295. Energetics of Biological Macromolecules (Part B)
Edited by GARY K. ACKERS AND MICHAEL L. JOHNSON

VOLUME 296. Neurotransmitter Transporters
Edited by SUSAN G. AMARA

VOLUME 297. Photosynthesis: Molecular Biology of Energy Capture
Edited by LEE MCINTOSH

VOLUME 298. Molecular Motors and the Cytoskeleton (Part B)
Edited by RICHARD B. VALLEE

VOLUME 299. Oxidants and Antioxidants (Part A)
Edited by LESTER PACKER

VOLUME 300. Oxidants and Antioxidants (Part B)
Edited by LESTER PACKER

VOLUME 301. Nitric Oxide: Biological and Antioxidant Activities (Part C)
Edited by LESTER PACKER

VOLUME 302. Green Fluorescent Protein
Edited by P. MICHAEL CONN

VOLUME 303. cDNA Preparation and Display
Edited by SHERMAN M. WEISSMAN

VOLUME 304. Chromatin
Edited by PAUL M. WASSARMAN AND ALAN P. WOLFFE

VOLUME 305. Bioluminescence and Chemiluminescence (Part C)
Edited by THOMAS O. BALDWIN AND MIRIAM M. ZIEGLER

VOLUME 306. Expression of Recombinant Genes in Eukaryotic Systems
Edited by JOSEPH C. GLORIOSO AND MARTIN C. SCHMIDT

VOLUME 307. Confocal Microscopy
Edited by P. MICHAEL CONN

VOLUME 308. Enzyme Kinetics and Mechanism (Part E: Energetics of Enzyme Catalysis)
Edited by DANIEL L. PURICH AND VERN L. SCHRAMM

VOLUME 309. Amyloid, Prions, and Other Protein Aggregates
Edited by RONALD WETZEL

VOLUME 310. Biofilms
Edited by RON J. DOYLE

VOLUME 345. G Protein Pathways (Part C: Effector Mechanisms)
Edited by RAVI IYENGAR AND JOHN D. HILDEBRANDT

VOLUME 346. Gene Therapy Methods
Edited by M. IAN PHILLIPS

VOLUME 347. Protein Sensors and Reactive Oxygen Species (Part A: Selenoproteins and Thioredoxin)
Edited by HELMUT SIES AND LESTER PACKER

VOLUME 348. Protein Sensors and Reactive Oxygen Species (Part B: Thiol Enzymes and Proteins)
Edited by HELMUT SIES AND LESTER PACKER

VOLUME 349. Superoxide Dismutase
Edited by LESTER PACKER

VOLUME 350. Guide to Yeast Genetics and Molecular and Cell Biology (Part B)
Edited by CHRISTINE GUTHRIE AND GERALD R. FINK

VOLUME 351. Guide to Yeast Genetics and Molecular and Cell Biology (Part C)
Edited by CHRISTINE GUTHRIE AND GERALD R. FINK

VOLUME 352. Redox Cell Biology and Genetics (Part A)
Edited by CHANDAN K. SEN AND LESTER PACKER

VOLUME 353. Redox Cell Biology and Genetics (Part B)
Edited by CHANDAN K. SEN AND LESTER PACKER

VOLUME 354. Enzyme Kinetics and Mechanisms (Part F: Detection and Characterization of Enzyme Reaction Intermediates)
Edited by DANIEL L. PURICH

VOLUME 355. Cumulative Subject Index Volumes 321–354

VOLUME 356. Laser Capture Microscopy and Microdissection
Edited by P. MICHAEL CONN

VOLUME 357. Cytochrome P450, Part C
Edited by ERIC F. JOHNSON AND MICHAEL R. WATERMAN

VOLUME 358. Bacterial Pathogenesis (Part C: Identification, Regulation, and Function of Virulence Factors)
Edited by VIRGINIA L. CLARK AND PATRIK M. BAVOIL

VOLUME 359. Nitric Oxide (Part D)
Edited by ENRIQUE CADENAS AND LESTER PACKER

VOLUME 360. Biophotonics (Part A)
Edited by GERARD MARRIOTT AND IAN PARKER

VOLUME 361. Biophotonics (Part B)
Edited by GERARD MARRIOTT AND IAN PARKER

Volume 362. Recognition of Carbohydrates in Biological Systems (Part A)
Edited by Yuan C. Lee and Reiko T. Lee

Volume 363. Recognition of Carbohydrates in Biological Systems (Part B)
Edited by Yuan C. Lee and Reiko T. Lee

Volume 364. Nuclear Receptors
Edited by David W. Russell and David J. Mangelsdorf

Volume 365. Differentiation of Embryonic Stem Cells
Edited by Paul M. Wassauman and Gordon M. Keller

Volume 366. Protein Phosphatases
Edited by Susanne Klumpp and Josef Krieglstein

Volume 367. Liposomes (Part A)
Edited by Nejat Düzgüneş

Volume 368. Macromolecular Crystallography (Part C)
Edited by Charles W. Carter, Jr., and Robert M. Sweet

Volume 369. Combinational Chemistry (Part B)
Edited by Guillermo A. Morales and Barry A. Bunin

Volume 370. RNA Polymerases and Associated Factors (Part C)
Edited by Sankar L. Adhya and Susan Garges

Volume 371. RNA Polymerases and Associated Factors (Part D)
Edited by Sankar L. Adhya and Susan Garges

Volume 372. Liposomes (Part B)
Edited by Nejat Düzgüneş

Volume 373. Liposomes (Part C)
Edited by Nejat Düzgüneş

Volume 374. Macromolecular Crystallography (Part D)
Edited by Charles W. Carter, Jr., and Robert W. Sweet

Volume 375. Chromatin and Chromatin Remodeling Enzymes (Part A)
Edited by C. David Allis and Carl Wu

Volume 376. Chromatin and Chromatin Remodeling Enzymes (Part B)
Edited by C. David Allis and Carl Wu

Volume 377. Chromatin and Chromatin Remodeling Enzymes (Part C)
Edited by C. David Allis and Carl Wu

Volume 378. Quinones and Quinone Enzymes (Part A)
Edited by Helmut Sies and Lester Packer

Volume 379. Energetics of Biological Macromolecules (Part D)
Edited by Jo M. Holt, Michael L. Johnson, and Gary K. Ackers

Volume 380. Energetics of Biological Macromolecules (Part E)
Edited by Jo M. Holt, Michael L. Johnson, and Gary K. Ackers

VOLUME 381. Oxygen Sensing
Edited by CHANDAN K. SEN AND GREGG L. SEMENZA

VOLUME 382. Quinones and Quinone Enzymes (Part B)
Edited by HELMUT SIES AND LESTER PACKER

VOLUME 383. Numerical Computer Methods (Part D)
Edited by LUDWIG BRAND AND MICHAEL L. JOHNSON

VOLUME 384. Numerical Computer Methods (Part E)
Edited by LUDWIG BRAND AND MICHAEL L. JOHNSON

VOLUME 385. Imaging in Biological Research (Part A)
Edited by P. MICHAEL CONN

VOLUME 386. Imaging in Biological Research (Part B)
Edited by P. MICHAEL CONN

VOLUME 387. Liposomes (Part D)
Edited by NEJAT DÜZGÜNEŞ

VOLUME 388. Protein Engineering
Edited by DAN E. ROBERTSON AND JOSEPH P. NOEL

VOLUME 389. Regulators of G-Protein Signaling (Part A)
Edited by DAVID P. SIDEROVSKI

VOLUME 390. Regulators of G-Protein Signaling (Part B)
Edited by DAVID P. SIDEROVSKI

VOLUME 391. Liposomes (Part E)
Edited by NEJAT DÜZGÜNEŞ

VOLUME 392. RNA Interference
Edited by ENGELKE ROSSI

VOLUME 393. Circadian Rhythms
Edited by MICHAEL W. YOUNG

VOLUME 394. Nuclear Magnetic Resonance of Biological Macromolecules (Part C)
Edited by THOMAS L. JAMES

VOLUME 395. Producing the Biochemical Data (Part B)
Edited by ELIZABETH A. ZIMMER AND ERIC H. ROALSON

VOLUME 396. Nitric Oxide (Part E)
Edited by LESTER PACKER AND ENRIQUE CADENAS

VOLUME 397. Environmental Microbiology
Edited by JARED R. LEADBETTER

VOLUME 398. Ubiquitin and Protein Degradation (Part A)
Edited by RAYMOND J. DESHAIES

VOLUME 433. Lipidomics and Bioactive Lipids: Specialized Analytical Methods and Lipids in Disease
Edited by H. ALEX BROWN

VOLUME 434. Lipidomics and Bioactive Lipids: Lipids and Cell Signaling
Edited by H. ALEX BROWN

VOLUME 435. Oxygen Biology and Hypoxia
Edited by HELMUT SIES AND BERNHARD BRÜNE

VOLUME 436. Globins and Other Nitric Oxide-Reactive Protiens (Part A)
Edited by ROBERT K. POOLE

VOLUME 437. Globins and Other Nitric Oxide-Reactive Protiens (Part B)
Edited by ROBERT K. POOLE

VOLUME 438. Small GTPases in Disease (Part A)
Edited by WILLIAM E. BALCH, CHANNING J. DER, AND ALAN HALL

VOLUME 439. Small GTPases in Disease (Part B)
Edited by WILLIAM E. BALCH, CHANNING J. DER, AND ALAN HALL

VOLUME 440. Nitric Oxide, Part F Oxidative and Nitrosative Stress in Redox Regulation of Cell Signaling
Edited by ENRIQUE CADENAS AND LESTER PACKER

VOLUME 441. Nitric Oxide, Part G Oxidative and Nitrosative Stress in Redox Regulation of Cell Signaling
Edited by ENRIQUE CADENAS AND LESTER PACKER

VOLUME 442. Programmed Cell Death, General Principles for Studying Cell Death (Part A)
Edited by ROYA KHOSRAVI-FAR, ZAHRA ZAKERI, RICHARD A. LOCKSHIN, AND MAURO PIACENTINI

VOLUME 443. Angiogenesis: *In Vitro* Systems
Edited by DAVID A. CHERESH

VOLUME 444. Angiogenesis: *In Vivo* Systems (Part A)
Edited by DAVID A. CHERESH

VOLUME 445. Angiogenesis: *In Vivo* Systems (Part B)
Edited by DAVID A. CHERESH

VOLUME 446. Programmed Cell Death, The Biology and Therapeutic Implications of Cell Death (Part B)
Edited by ROYA KHOSRAVI-FAR, ZAHRA ZAKERI, RICHARD A. LOCKSHIN, AND MAURO PIACENTINI

VOLUME 447. RNA Turnover in Bacteria, Archaea and Organelles
Edited by LYNNE E. MAQUAT AND CECILIA M. ARRAIANO

Volume 448. RNA Turnover in Eukaryotes: Nucleases, Pathways and Analysis of mRNA Decay
Edited by Lynne E. Maquat and Megerditch Kiledjian

Volume 449. RNA Turnover in Eukaryotes: Analysis of Specialized and Quality Control RNA Decay Pathways
Edited by Lynne E. Maquat and Megerditch Kiledjian

Volume 450. Fluorescence Spectroscopy
Edited by Ludwig Brand and Michael L. Johnson

Volume 451. Autophagy: Lower Eukaryotes and Non-Mammalian Systems (Part A)
Edited by Daniel J. Klionsky

Volume 452. Autophagy in Mammalian Systems (Part B)
Edited by Daniel J. Klionsky

Volume 453. Autophagy in Disease and Clinical Applications (Part C)
Edited by Daniel J. Klionsky

Volume 454. Computer Methods (Part A)
Edited by Michael L. Johnson and Ludwig Brand

Volume 455. Biothermodynamics (Part A)
Edited by Michael L. Johnson, Jo M. Holt, and Gary K. Ackers (Retired)

Volume 456. Mitochondrial Function, Part A: Mitochondrial Electron Transport Complexes and Reactive Oxygen Species
Edited by William S. Allison and Immo E. Scheffler

Volume 457. Mitochondrial Function, Part B: Mitochondrial Protein Kinases, Protein Phosphatases and Mitochondrial Diseases
Edited by William S. Allison and Anne N. Murphy

Volume 458. Complex Enzymes in Microbial Natural Product Biosynthesis, Part A: Overview Articles and Peptides
Edited by David A. Hopwood

Volume 459. Complex Enzymes in Microbial Natural Product Biosynthesis, Part B: Polyketides, Aminocoumarins and Carbohydrates
Edited by David A. Hopwood

Volume 460. Chemokines, Part A
Edited by Tracy M. Handel and Damon J. Hamel

Volume 461. Chemokines, Part B
Edited by Tracy M. Handel and Damon J. Hamel

Volume 462. Non-Natural Amino Acids
Edited by Tom W. Muir and John N. Abelson

Volume 463. Guide to Protein Purification, 2nd Edition
Edited by Richard R. Burgess and Murray P. Deutscher

Volume 464. Liposomes, Part F
Edited by Nejat Düzgüneş

Volume 465. Liposomes, Part G
Edited by Nejat Düzgüneş

Volume 466. Biothermodynamics, Part B
Edited by Michael L. Johnson, Gary K. Ackers, and Jo M. Holt

Volume 467. Computer Methods Part B
Edited by Michael L. Johnson and Ludwig Brand

Volume 468. Biophysical, Chemical, and Functional Probes of RNA Structure, Interactions and Folding: Part A
Edited by Daniel Herschlag

Volume 469. Biophysical, Chemical, and Functional Probes of RNA Structure, Interactions and Folding: Part B
Edited by Daniel Herschlag

Volume 470. Guide to Yeast Genetics: Functional Genomics, Proteomics, and Other Systems Analysis, 2nd Edition
Edited by Gerald Fink, Jonathan Weissman, and Christine Guthrie

Volume 471. Two-Component Signaling Systems, Part C
Edited by Melvin I. Simon, Brian R. Crane, and Alexandrine Crane

Volume 472. Single Molecule Tools, Part A: Fluorescence Based Approaches
Edited by Nils G. Walter

Volume 473. Thiol Redox Transitions in Cell Signaling, Part A Chemistry and Biochemistry of Low Molecular Weight and Protein Thiols
Edited by Enrique Cadenas and Lester Packer

Volume 474. Thiol Redox Transitions in Cell Signaling, Part B Cellular Localization and Signaling
Edited by Enrique Cadenas and Lester Packer

Volume 475. Single Molecule Tools, Part B: Super-Resolution, Particle Tracking, Multiparameter, and Force Based Methods
Edited by Nils G. Walter

Volume 476. Guide to Techniques in Mouse Development, Part A Mice, Embryos, and Cells, 2nd Edition
Edited by Paul M. Wassarman and Philippe M. Soriano

Volume 477. Guide to Techniques in Mouse Development, Part B Mouse Molecular Genetics, 2nd Edition
Edited by Paul M. Wassarman and Philippe M. Soriano

Volume 478. Glycomics
Edited by Minoru Fukuda

SECTION ONE

INVERSE AGONISM AND INVERSE AGONISTS

CHAPTER ONE

Identification and Characterization of Steroidogenic Factor-1 Inverse Agonists

Mabrouka Doghman,*[,1] Franck Madoux,†[,1] Peter Hodder,† *and* Enzo Lalli*

Contents

Abstract

The transcription factor Steroidogenic Factor-1 (Ad4BP/SF-1; NR5A1 according to the standard nomenclature) has an essential role in adrenogonadal development. Furthermore, SF-1 is amplified and overexpressed in most cases of adrenocortical tumor occurring in children; studies performed in transgenic mice have shown that an increased SF-1 dosage triggers tumor formation in the adrenal cortex. For these reasons, drugs interfering with SF-1 action would represent a promising tool to be added to the current pharmacological

* Institut de Pharmacologie Moléculaire et Cellulaire, CNRS UMR 6097, Université de Nice Sophia Antipolis, Valbonne, France
† Scripps Research Institute Molecular Screening Center and Molecular Therapeutics, The Scripps Research Institute, Scripps Florida, Jupiter, Florida, USA
1 These authors contributed equally to this work

Methods in Enzymology, Volume 485
ISSN 0076-6879, DOI: 10.1016/S0076-6879(10)85001-3

protocols in the therapy of adrenocortical cancer. Here, we describe the methods how isoquinolinone compounds inhibiting the constitutive transcriptional activity of SF-1 (SF-1 inverse agonists) were identified and characterized. These compounds have the attributes to inhibit the increase in proliferation triggered by an augmented SF-1 dosage in adrenocortical tumor cells and to reduce their steroid production. This latter property may also reveal beneficial for drugs used in the therapy of adrenocortical tumors to alleviate symptoms of virilization and Cushing often associated with tumor burden.

1. Introduction

The nuclear receptor Steroidogenic Factor-1 (Ad4BP/SF-1; NR5A1 according to the standard nomenclature) is a transcription factor playing an essential role in adrenal and gonadal development and function (reviewed in Hoivik *et al.*, 2010; Lalli, 2010). Moreover, recent studies have shown that an increase of SF-1 expression, a frequent finding in adrenocortical tumors (Almeida *et al.*, 2010; Pianovski *et al.*, 2006), may have an important role in the pathogenesis of these neoplasms (Doghman *et al.*, 2007; Lalli, 2010).

SF-1 has the capacity of binding as a monomer to nuclear receptor half sites on DNA (Wilson *et al.*, 1993) and to activate transcription. In adrenocortical cells, SF-1 has been shown to regulate transcription of its target genes by increased association with their promoters and coactivator recruitment in a cAMP-dependent fashion (Winnay and Hammer, 2006). SF-1 transcriptional activity can be modulated by (1) posttranscriptional modifications, namely phosphorylation by different kinases at Ser203 (Hammer *et al.*, 1999; Lewis *et al.*, 2008) and sumoylation, affecting SF-1 subnuclear localization and DNA-binding activity (Campbell *et al.*, 2008; Chen *et al.*, 2004; Komatsu *et al.*, 2004; Lee *et al.*, 2005), and (2) interaction with positive and negative cofactors (reviewed in Hoivik *et al.*, 2010). For a long time, SF-1 has been considered to be an orphan receptor. However, recent structural studies have shown that its transcriptional activity can be regulated by phospholipid ligands binding inside the pocket present in the C-terminal domain of the protein, which bears homology with the ligand-binding domain (LBD) of classical nuclear hormone receptors (Krylova *et al.*, 2005; Li *et al.*, 2005; Wang *et al.*, 2005). The importance of phospholipids in the regulation of SF-1 transcriptional activity is shown by the impaired function of mutants in residues lining the LBD pocket and contacting the phospholipids. However, it remains to be established whether phospholipid binding may affect SF-1 transcriptional activity in the physiological setting.

Small molecules able to modulate SF-1 transcriptional activity have recently been described (Fig. 1.1). A first report described the

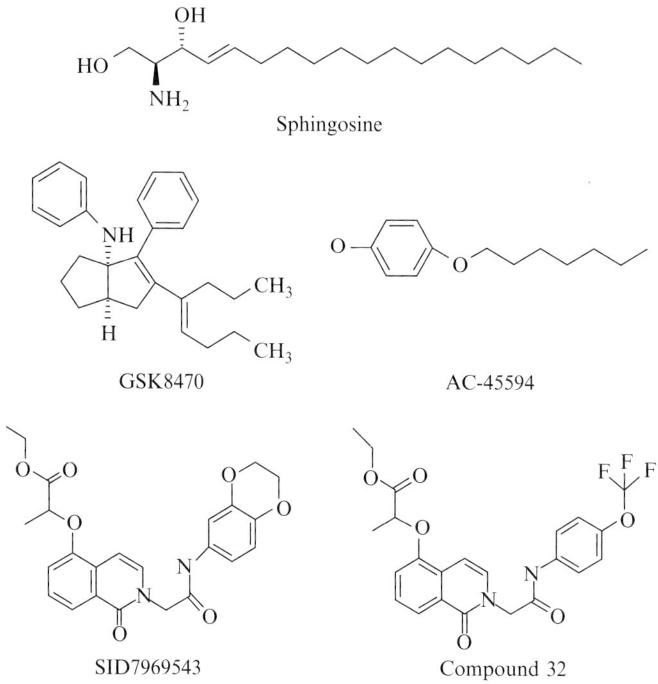

Figure 1.1 Structure of compounds modulating SF-1 transcriptional activity.

characterization of GSK8470, an amino *cis*-bicyclo[3.3.0]-oct-2-ene identified in a high-throughput screen using a FRET-based assay, which is able to displace endogenous lipids from the ligand-binding pocket of SF-1 and of its closely related nuclear receptor LRH-1 (NR5A2). This compound has been reported to modestly increase transactivation by LRH-1 of the SHP promoter (an LRH-1 target gene) linked to luciferase and to increase the levels of SHP mRNA in cultured hepatocytes (Whitby *et al.*, 2006). Conversely, the natural compound sphingosine was reported to be bound to SF-1 in adrenocortical cells and to inhibit activation of the CYP17 promoter in a dosage-dependent manner (Urs *et al.*, 2006). However, this sphingolipid had no effect on SF-1-dependent gene expression in other contexts (Doghman *et al.*, 2007).

Two distinct classes of small molecules capable of inhibiting SF-1 transactivation have been described. These compounds act as inverse agonists for SF-1 because of the constitutive activity of this transcription factor in transfection assays. The first class of compounds consists of alkyloxyphenols developed after the identification of the 4-(heptyloxy)phenol (AC-45594) lead in a functional cell-based assay (Del Tredici *et al.*, 2008). AC-45594 and analogs were reported to inhibit transcriptional activation of

SF-1-responsive promoters by SF-1 and steroidogenic gene expression in human H295 cells. The second class of compounds consists of isoquinolinones identified through a screening of chemicals able to modulate transcriptional activity of a Gal4–SF-1 chimera (Madoux *et al.*, 2008).

This chapter describes the characterization of isoquinolinones as SF-1 inverse agonists using an uHTS (ultra-high throughput screening) luciferase assay and the methods how their effects as inhibitors of SF-1-dependent proliferation and steroid production were studied in adrenocortical tumor cells.

2. Characterization of Inverse Agonists of SF-1 by uHTS

2.1. Transient transfection assays as tools to identify compounds able to modulate SF-1 transcriptional activity

The strategy for characterization of SF-1 inverse agonists involved the screening of the activity of a chemical library on SF-1 transactivation capacity, using an assay employing transient cotransfection of a Gal4–SF-1 chimera together with a Gal4-responsive luciferase promoter in mammalian cells (Fig. 1.2). A Gal4–RORA chimera was used as a control of specificity. The advantage of transient transfection methods to study transcription factor activity resides in the possibility of fine tuning the transcription factor/reporter ratio and flexibility to adjust luminescence levels of the assay.

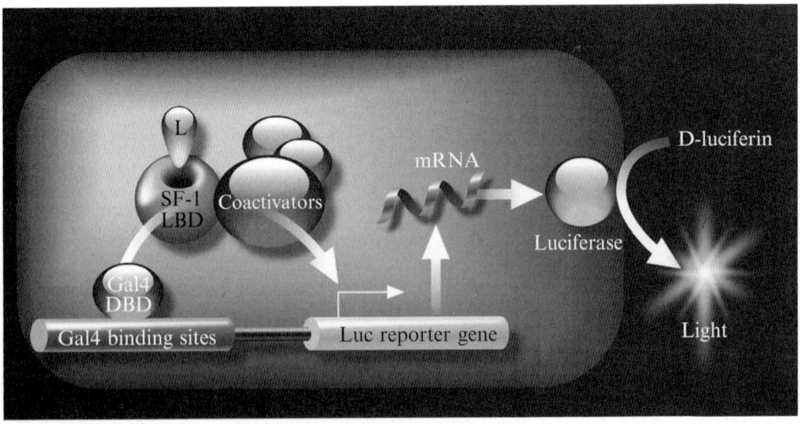

Figure 1.2 The Gal4 system. Mammalian cells are cotransfected with a reporter plasmid harboring Gal4 binding sites and another plasmid to express a Gal4 DBD–SF-1 LBD fusion protein. This protein binds to the Gal4 binding sites, interacts with coactivator complexes and increases the expression of the luciferase reporter gene. This enzyme generates light from the D-luciferin substrate. L, ligand. (See Color Insert.)

2.1.1. Plasmids construction

The first step required cloning of the human SF-1 and the mouse RORA LBDs in-frame with the yeast Gal4 DNA-binding domain (DBD) to yield a chimeric transcription factor able to bind Gal4 binding sites. The yeast Gal4 transcription factor binds specifically through its DBD to a 17-mer DNA sequence that is absent in the genome of higher eukaryotes. For this reason, reporter genes harboring Gal4 binding sites have very low background expression when transfected in mammalian cells (Kakidani and Ptashne, 1988). Primers GATCGGATCCCCGGAGCCTTATGCCAGCCC (forward) and GATCTCTAGATCAAGTCTGCTTGGCTTGCAGCAT TTCGATGAG (reverse), harboring a *Bam*HI and a *Xba*I site, respectively (underlined sequence), were used to amplify the SF-1 LBD (aa. 198–462) cDNA using an expressed sequence tag (EST) clone (Invitrogen) as a template. The RORA LBD (aa. 266–523) cDNA was generated by PCR using primers GCCGCCCCCGGGCCGAACTAGAACACCTTGCCC (forward) and TATATAAAGCTTTCCTTACCCATCGATTTGCAT GG (reverse), harboring a *Xma*I and a *Hin*dIII site, respectively (underlined sequence), and a mouse liver cDNA library (Clontech) as a template. PCR-amplified SF-1 and RORA LBD were cloned into the pFA-CMV (Stratagene) expression vector using the appropriate sites in the vector polylinker. pFA-CMV is a commercial vector apt for cloning cDNAs encoding protein domains of interest in frame with the yeast Gal4 DBD (1–147).

2.1.2. Cell culture and transient transfection

CHO (Chinese hamster ovary) cells of the K1 subtype (American Type Culture Collection) were used to express the Gal4–LBD chimeras by transient transfection. These cells proliferate rapidly and can be easily grown in large quantities. An example of the optimization results using the protocol below is shown in Fig. 1.3.

2.1.2.1. Required materials

- Growth medium: Ham's F-12 (Invitrogen) supplemented with 10% (v/v) fetal bovine serum (Gemini Bio-products) and 1% (v/v) penicillin/streptomycin/neomycin mix (Invitrogen).
- Culture conditions: 37 °C, 5% CO_2, 95% relative humidity, using 20 mL of complete medium in T-175 flasks (Corning).
- Subculture: Split when cells are at 70–80% confluency, approximately 1:4–1:8 (1–2 x 104 cells/cm^2). Trypsinize using 0.25% trypsin–EDTA solution (Invitrogen).
- Plasmids: The expression vectors encoding Gal4 DBD fusions described above and the reporter vector pG5*luc* (Promega), harboring five Gal4 binding sites upstream of the adenovirus major late promoter and the luciferase gene. All plasmids were prepared on a large scale using a commercial silica column method (Qiagen).

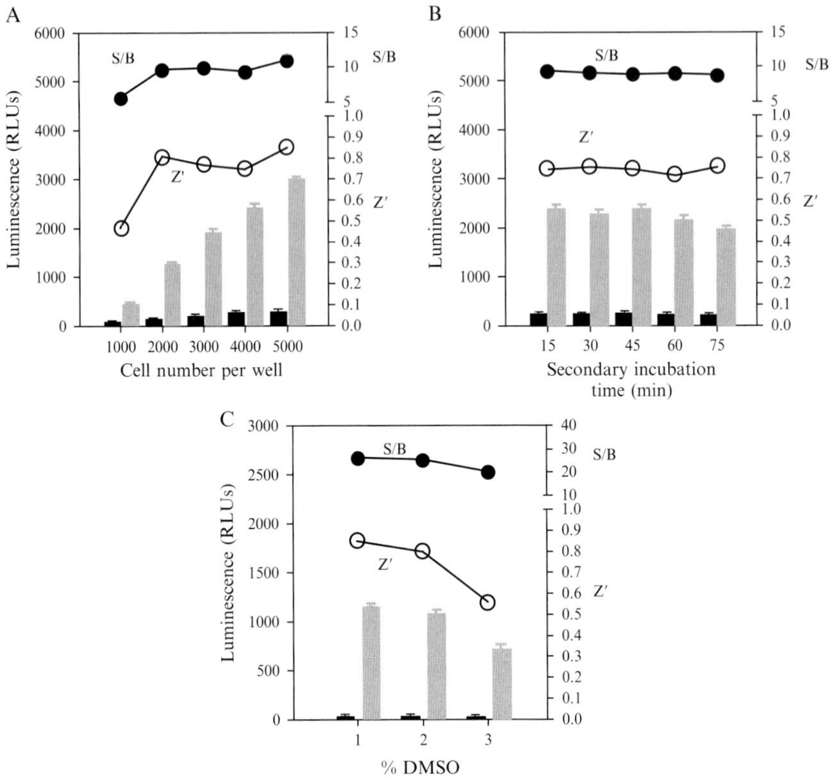

Figure 1.3 Assay optimization in the 1536-well plate format. (A) Cell seeding density optimization. CHO cells were transfected with the reporter plasmid pG5luc alone (black bars, low control) or cotransfected with pG5luc and pGal4 DBD–SF-1 LBD (gray bars, high control). Cells were seeded at different densities ranging from 1000 to 5000 cells per well. Z′ (○) and signal-to-background ratio (•) calculated based on relative luminescence unit (RLU) values between the high and low controls are shown for each tested cell density (n=24 for each control type). (B) Secondary incubation time optimization. After a 20 h incubation time, plates containing cells seeded at 4000 cells per well were read at 15 min intervals up to 75 min. Note the steady raw RLU values and assay statistics over the course of the examined timeline. (C) DMSO tolerance assessment. Assay statistics of Gal4-SF1 assay plates treated with increasing DMSO concentrations are reported. For all three bar graph of this figure, error bars represent the standard deviation (n = 4).

2.1.2.2. Methods

2.1.2.2.1. One day prior to transfection Cells were rinsed with PBS and dissociated using trypsin–EDTA solution. 6×10^6 CHO-K1 cells were seeded in each T-175 flask in 20 mL complete growth medium and incubated overnight.

2.1.2.2.2. On the day of transfection Quantities and volumes of reagents are intended for transfection of each T-125 flask.

1. In a sterile, plastic 12 × 75 mm tube (Becton Dickinson), the TransIT-CHO Reagent (Mirus Bioproducts; 54 μL) was added dropwise into 1.2 mL of serum-free Ham's F-12 medium and mixed thoroughly by vortexing.
2. Incubation at room temperature was performed for 20 min.
3. 250 ng of pGal4$_{DBD}$–SF-1$_{LBD}$ or 125 ng of pGal4$_{DBD}$–RORA$_{LBD}$ together with 9 μg of pG5*luc* and 8.75 μg of empty pcDNA3.1 plasmid (Invitrogen) were added to the diluted TransIT-CHO reagent and mixed by gentle pipetting. As a negative control for the assay, "−NR" cells were transfected with the pG5*luc* and empty pcDNA3.1 plasmids alone, as opposed to "+NR" cells, which were cotransfected with the Gal4-LBD encoding plasmids.
4. Incubation at room temperature was performed for 20 min.
5. TransIT-CHO Mojo reagent (Mirus Bioproducts; 9 μL) was added to the mixture and mixed by gentle pipetting.
6. Incubation at room temperature was performed for 20 min.
7. The mixture was added dropwise to CHO-K1 cells seeded the day before in a T-175 flask in 20 mL complete growth medium. Flasks were gently rocked from side to side to distribute the DNA–carrier complexes evenly. Flasks were then placed back in the incubator at 37 °C, 5% CO_2, 95% relative humidity. Four hours after transfection, cells were trypsinized and suspended to a concentration of 8 × 10^5 cells/mL in complete growth medium.

2.1.3. uHTS luciferase assay

2.1.3.1. Required materials

- White solid-bottomed 1536-well plates (Greiner)
- Bottle Valve liquid dispenser (GNF/Kalypsys)
- Chemical library (from the NIH Molecular Library Screening Center Network, http://mli.nih.gov/mli/). In the version of the library used, 64,908 compounds are dissolved in DMSO at the concentration of 1 m*M*.
- 1536-well head PinTool unit (GNF/Kalypsys)
- SteadyLite HTS reagent (PerkinElmer)
- ViewLux microplate imager (PerkinElmer)

2.1.3.2. Methods

1. 5 μL of the cell suspension were dispensed into each well (4000 cells/well) of 1536-well plates using a Bottle Valve liquid dispenser. 52 plates were used for the SF-1 screening. Cells transfected only with empty

expression plasmids (−NR) were seeded in the first two columns (starting from the left) of each 1536-well plate and designated as Low Control, while the other 46 columns were seeded with cells transfected with the plasmid encoding pGal4$_{DBD}$–SF-1$_{LBD}$ (+NR). A parallel screening using the same format was performed with cells transfected with pGal4$_{DBD}$–RORA$_{LBD}$.

2. One hour after seeding, 50 nL of each compound (44 columns) or DMSO alone (2 columns) were added in singlicate to each well containing +NR cells using a PinTool unit. Cells treated with DMSO were designated as High Control. The final nominal test concentration of the chemicals was 10 μ*M*, in a final DMSO concentration of 1%.
3. Incubation at 37 °C, 5% CO_2, 95% relative humidity was continued for 20 h.
4. Plates were taken out of the incubator and equilibrated at room temperature for 20 min.
5. 5 μL/well of the SteadyLite HTS reagent were added to each well using a Bottle Valve liquid dispenser. After a 15-min incubation, light emission was measured for 30 s using the ViewLux instrument. This method enabled screening of about 11,500 compounds per hour.
6. Activity of each compound was calculated on a plate-per-plate basis using Eq. (1.1):

$$\%\ \text{inhibition} = 100 \times \left(1 - \frac{\text{Test well} - \text{Median Low control}}{\text{Median High Control} - \text{Median Low Control}}\right) \tag{1.1}$$

where High Control represents +NR cells treated with DMSO ($n = 24$) and Low Control –NR cells treated with DMSO ($n = 24$). A stepwise protocol of the above methods is presented in Table 1.1.

2.1.4. Hit identification and selection

A summary of HTS hit selection is presented in Table 1.2. Compound activity data were uploaded and analyzed in the Scripps HTS database (MDL Information Systems). The parameter termed Z' factor measures the applicability of a method to HTS and takes into account, at the same time, the dynamic range of the method and data variability (Zhang *et al.*, 1999). This statistic factor is defined as in Eq. (1.2):

$$Z' = \frac{(\text{AVG}_{\text{max}} - 3\text{SD}_{\text{max}}/\sqrt{n}) - (\text{AVG}_{\text{min}} - 3\text{SD}_{\text{min}}/\sqrt{n})}{\text{AVG}_{\text{max}} - \text{AVG}_{\text{min}}} \tag{1.2}$$

Table 1.1 SF-1 assay protocol in 1536-well plate format

Order	Step	Condition	Comments
1	Cell preparation	Transfection	4 h incubation
2	Cell dispensing	5 μL/well	4000 cells/well
3	Primary incubation time	1 h	Room temperature
4	Compound addition	50 nL/well	Test concentration: 10 μ*M* final DMSO concentration: 1%
5	Secondary incubation time	20 h	37 °C, 95% relative humidity, 5% CO_2
6	Luciferase detection reagent addition	5 μL/well	–
7	Tertiary incubation time	20 min	Room temperature
8	Luminescence acquisition	30 s/plate	–

A Z′ factor value higher than 0.5 is considered adequate for high-throughput procedures. In the case of the SF-1 screening, the Z′ factor value was 0.72, while it was equal to 0.67 for the RORA screening.

For determination of active compounds, an algorithm was used that takes into account the mean percentage inhibition of all chemicals tested and three times their standard deviation (Hodder *et al.*, 2003). The sum of these values ($\mu + 3\sigma$) was used to establish a cutoff beyond which any compound exhibiting greater inhibition was considered active. For the SF-1 campaign, the cutoff was calculated at 47.96% inhibition. This way, 359 primary hits were identified. The majority of primary hits identified in the SF-1 campaign showed comparable inhibition values in both the SF-1 and RORA assays. Many of these hits corresponded to known cytotoxic drugs (e.g., doxorubicin, vinblastine sulfate) or to compounds known to produce artifacts in luciferase-based assays (e.g., analogs of 3-phenoxy-methyl-benzoic acid).

2.1.5. Dose-response activity studies of selected primary hits

New DMSO dilutions of the compounds representing the primary SF-1 hits were prepared starting from concentrated stocks available at the National Institutes of Health's Molecular Libraries Small Molecule Repository. 10-point, 1:3 serial dilutions starting from nominal 10 m*M* solution were prepared using the Biomek FX automated liquid handler (Beckman Coulter). Titration experiments of the inhibition of these compounds against $pGal4_{DBD}$–$SF\text{-}1_{LBD}$, $pGal4_{DBD}$–$RORA_{LBD}$ and the unrelated transcriptional activator $pGal4_{DBD}$–VP16, as a control for their specificity of action, were performed in triplicate using 1536-well plates and the method described above. For each chemical, mean values of their percentage

Table 1.2 uHTS campaign summary and results

Step	Screen type	Target	PubChem AID[a]	Test compounds number	Selection criteria	Selected compounds number	Assay statistics	
							Z'	S/B
1	Primary screen	SF-1	525	64907	%Inh>47.96%[b]	359	0.72 ± 0.06	53 ± 7
2	Parallel titration	SF-1	600	359	SF-1 IC_{50}< 10 μM and RORA IC_{50}> 10-fold SF-1 IC_{50}	2	0.63 ± 0.08	38 ± 2
		RORA	599	359			0.60 ± 0.05	72 ± 4

[a] PubChem AIDs are accessible on-line at http://www.ncbi.nlm.nih.gov/sites/entrez?db=pcassay&term=xxxx, where xxxx represents the PubChem AID number listed in the table.

[b] The primary screen hit-cutoff was calculated as the average percent activation of all test compounds plus three times the standard deviation.

inhibition of nuclear receptor activity at each dilution were plotted against compound concentration. IC_{50} values were calculated by a four-parameter equation describing a sigmoidal curve, using the Assay Explorer software (MDL Information Systems). In cases where the highest compound concentration tested (99 μM) did not result in $>50\%$ inhibition or where no curve fit was achieved, the IC_{50} was determined manually depending on the observed inhibition at the individual concentrations. Compounds with IC_{50} values of greater than 10 μM were considered inactive, while compounds with IC_{50} equal or less than 10 μM are considered active.

Compounds selected for further investigation fulfilled the following criteria:

- IC_{50} value in the SF-1 assay <1 μM;
- IC_{50} SF-1 assay $\leq 10\times$ IC_{50} RORA assay.

Two isoquinolinone derivatives, SID7969543 and SID7970631, met these requirements. IC_{50} values in the $pGal4_{DBD}$–$SF\text{-}1_{LBD}$ assay for the two compounds were 760 and 255 nM, respectively, while they were higher than 30 μM in both the $pGal4_{DBD}$–$RORA_{LBD}$ and the $pGal4_{DBD}$–VP16 assays.

2.1.6. SF-1 response element assay

In this assay, the inhibitory action of the selected isoquinolinone compounds was confirmed on the activity of full-length SF-1 on a luciferase reporter under the control of a synthetic, multimerized SF-1 binding site. Specificity of action of each compound was also tested on the related LRH-1 nuclear receptor.

2.1.6.1. Required materials

- Cells: human embryonic kidney (HEK) cells of the 293T subtype (American Type Culture Collection).
- Growth medium: Dulbecco's modified Eagle medium (Invitrogen) supplemented with 10% (v/v) fetal bovine serum (Gemini Bio-products) and 1% (v/v) penicillin/streptomycin//glutamine mix (Invitrogen).
- Culture conditions: 37 °C, 5% CO_2, 95% relative humidity, using 20 mL of complete medium in T-175 flasks (Corning).
- Subculture: Split when cells are at 70–80% confluency, approximately 1:4–1:8 (1–2×10^4 cells/cm^2). Trypsinize using 0.25% trypsin–EDTA solution (Invitrogen).
- Plasmids: pCMV-SF-1 and pCMV LRH-1 expression vectors (Open Biosystems); p5xSFRE, harboring five copies of a consensus SF-1 response element cloned upstream a luciferase reporter gene (a kind gift of Dr. Donald McDonnell, Duke University Medical Center, Durham, NC). All plasmids were prepared on a large scale using a commercial silica column method (Qiagen).

- Chemicals: an 11-point, 1:3 serial dilution of each compound starting at 0.4 m*M* (40× of final concentration in the assay) was prepared in PBS—5% DMSO using the PlateMate Plus liquid handling robot (Matrix Technologies).

2.1.6.2. *Methods*

2.1.6.2.1. One day prior to transfection Cells were rinsed with PBS and dissociated using trypsin–EDTA solution. $5x10^3$ HEK 293T cells were seeded into the wells of a 384-well plate in a volume of 39 μL.

2.1.6.2.2. On the day of transfection

1. In a sterile, plastic 12 × 75 mm tube (Becton Dickinson), 0.15 μL/sample of Fugene 6 transfection reagent (Roche) was diluted into 1 μL/sample of serum-free Dulbecco's modified Eagle medium and mixed thoroughly by vortexing.
2. Incubation at room temperature was performed for 5 min.
3. 25 ng of p5xSFRE and 25 ng of either expression vector were added to the diluted Fugene 6 reagent (3:1 ratio) and mixed by vortexing.
4. Incubation at room temperature was performed for 15 min.
5. Complexes were added to cells.
6. Plates were incubated at 37 °C, 5% CO_2, 95% relative humidity.
7. 24 h after transfection, cells were treated with 1 μL of each point of the compound 40× serial dilution ($n = 6$ for each dilution point). Final DMSO concentration in each well was 0.125%.
8. Plates were incubated at 37 °C, 5% CO_2, 95% relative humidity.
9. 24 h later, plates were allowed to equilibrate for 15 min at room temperature and luciferase activity was assayed by adding 25 μL/well BriteLite reagent (Perkin Elmer), incubating 2 min and then reading luminescence using an EnVision Multilabel Plate Reader (Perkin Elmer).

In the full-length SF-1 assay, IC_{50} values for SID7969543 and SID7970631 were 30 and 16 n*M*, respectively, while they had no significant inhibition on LRH-1 transcriptional activity.

2.2. Cytotoxicity of isoquinolinone inverse agonists for SF-1

An essential parameter in the initial characterization of the pharmacological properties of SF-1 inverse agonists is their cytotoxicity.

2.2.1. Cytotoxicity assay

This assay measures cell viability using a luminescence method based on light emission by luciferase proportional to the ATP content of cells.

2.2.1.1. *Required materials*

- White solid-bottomed 1536-well plates (Greiner)
- Growth medium for CHO-K1 cells: Ham's F-12 (Invitrogen) supplemented with 10% (v/v) fetal bovine serum (Gemini Bio-products) and 1% (v/v) penicillin/streptomycin//neomycin mix (Invitrogen)
- CellTiter-Glo reagent (Promega)
- ViewLux microplate imager (PerkinElmer)
- Chemicals: a 10-point, 1:3 serial dilution in DMSO of each compound starting at 10 m*M*
- 1536-well head PinTool unit (GNF/Kalypsys)

2.2.1.2. *Method*

1. Plate CHO-K1 cells at a density of 500/well in 1536-well plates in a volume of 5 μL complete growth medium.
2. Add 50 nL/well of each dilution of each compound or DMSO control ($n = 24$) using the PinTool unit.
3. Incubate plates at 37 °C, 5% CO_2, 95% relative humidity for 20 h.
4. Add 5 μL/well of CellTiter-Glo reagent.
5. Incubate at room temperature for 10 min to allow for stabilization of the light signal.
6. Record luminescence.
7. Express results in compound-treated samples as percentage of luminescence signal in DMSO-treated samples.

Using this method, no cytotoxicity could be detected for the SID7969543 compound, while SID7970631 displayed a half-maximal cytotoxic effect at concentrations higher than 30 μ*M*.

2.3. Structure–activity relationship of isoquinolinone inverse agonists for SF-1

An *in silico* structure–activity relationship study was performed by retrieving results of compounds present in the HTS screening and containing the isoquinolinone scaffold. A compound harboring a substitution of the branched methyl on the phenoxy group by a hydrogen atom was less active. Compounds similar to SID7970631 harboring substitutions of the dioxolane moiety also exhibited reduced SF-1 inhibition activity (Madoux *et al.*, 2008).

Furthermore, the structures of SID7969543 and SID7970631 were used as starting points for the development of SF-1 small molecule probes with increased potency. This study yielded compounds (termed #31 and #32 in Roth *et al.*, 2008) displaying SF-1 inhibitor activity in the 100–200 n*M* range (a three- to sixfold improvement over SID7969543/SID7970631; Fig. 1.4),

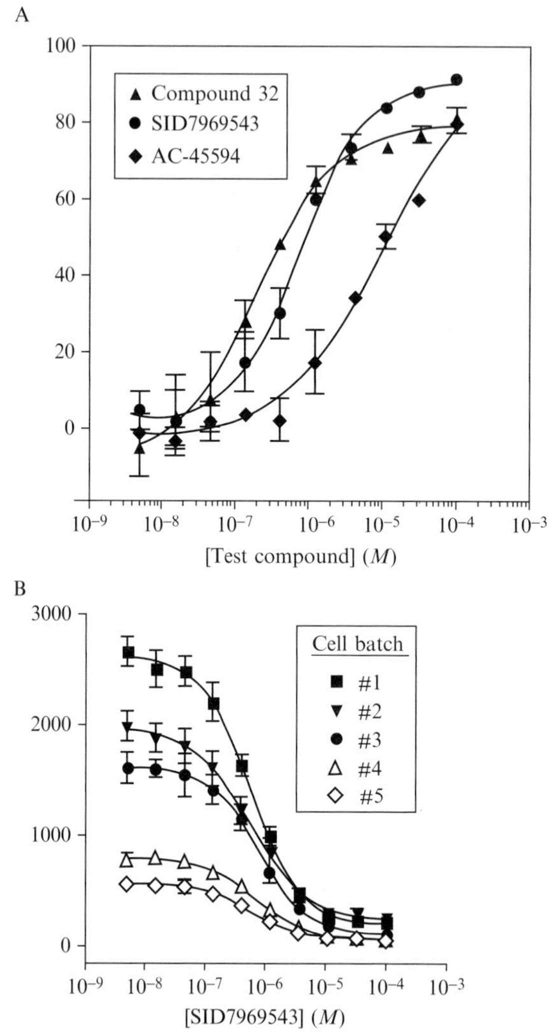

	Cell batch				
	#1	#2	#3	#4	#5
Hillslope	−1.065	−0.9279	−1.113	−1.005	−1.089
EC50	6.027×10^{7}	5.774×10^{7}	8.178×10^{7}	7.397×10^{7}	6.598×10^{7}

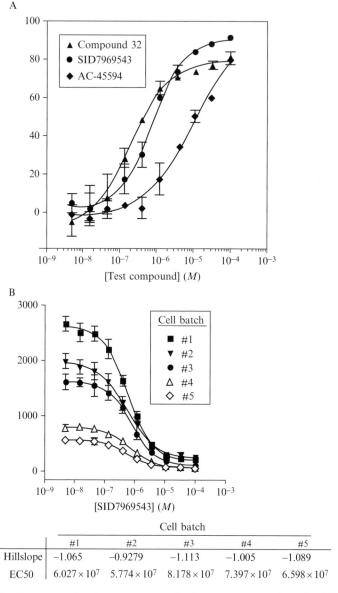

Figure 1.4 Representative dose-responses of reference compounds in the Gal4–SF-1 assay. (A) Normalized titration curves of AC-45594 (◆) and the isoquinolinones SID7969543 (•) and Compound 32 (▲) in the Gal4–SF1 assay. Error bars represent the standard deviation of three separate experiments. (B) Raw data titration curves of SID7969543 measured with different transfection batches. Note that the variability of the raw RLU data does not affect the pharmacology, as the calculated hill slope and IC_{50} values show a coefficient of variation below 15%. Error bars represent the standard deviation ($n = 16$).

inactive in the RORA and VP-16 assays and devoid of cellular toxicity, as measured in the luminescence assay described before (Roth *et al.*, 2008).

3. Effect of SF-1 Inverse Agonists on Adrenocortical Tumor Cell Proliferation and Steroid Production

Previous studies have shown that an increased SF-1 dosage has an important role in the pathogenesis of childhood adrenocortical tumors, where the gene encoding SF-1 is frequently amplified and the protein overexpressed (Almeida *et al.*, 2010; Doghman *et al.*, 2007; Figueiredo *et al.*, 2008; Pianovski *et al.*, 2006). Transgenic mice overexpressing SF-1 in their adrenal gland develop tumors expressing gonadal markers, which are probably derived from common adrenogonadal precursor cells situated in the subcapsular zone of the murine adrenal cortex (Doghman *et al.*, 2007; Lalli, 2010). Given the restricted tissue expression pattern of SF-1, compounds specifically inhibiting its transcriptional activity represent potentially useful drugs against adrenocortical tumors that will probably display limited side effects for other tissues.

3.1. *In vitro* assay of the antiproliferative activity of SF-1 inverse agonists for adrenocortical tumor cells

For this assay, we have used the H295R-derived adrenocortical tumor cell line (H295R/TR SF-1) that we have developed, where SF-1 dosage can be modulated in a tetracycline-dependent fashion. We have previously shown that proliferation of this cell line is modulated by SF-1 in a dosage-dependent fashion. This effect relies on an intact SF-1 transcriptional activity (Doghman *et al.*, 2007). As a control for the specificity of the compounds, we used the SW-13 cell line, a poorly differentiated adrenocortical tumor cell line not expressing SF-1 (Doghman *et al.*, 2009).

3.1.1. Proliferation assay

3.1.1.1. Required materials

- Growth medium—H295R/TR SF-1 cells: DMEM/F-12 (Invitrogen) supplemented with 2% (v/v) NuSerum (Becton Dickinson), 1% ITS+ Premix (Becton Dickinson), 1% (v/v) penicillin/streptomycin mix (Invitrogen), blasticidin (5 μg/mL; Cayla), and zeocin (100 μg/mL; Cayla). SW-13 cells: DMEM/F-12 (Invitrogen) supplemented with 10% (v/v) fetal calf serum (Invitrogen) and 1% (v/v) penicillin/streptomycin mix (Invitrogen).

- Culture conditions: 37 °C, 5% CO_2, 95% relative humidity, using 15 mL of complete medium in T-75 flasks (Becton Dickinson).
- Subculture: Split when cells are at 70–80% confluency, approximately 1:3 to 1:4. Trypsinize using 0.25% trypsin–EDTA solution (Invitrogen).
- Inverse agonists for SF-1: stock solutions in DMSO (10 m*M*).
- Countess automated cell counting apparatus (Invitrogen).

3.1.1.2. Methods

1. Seed cells (H295R/TR SF-1 and SW-13) in 24-well plates in 1 mL complete medium/well at the density of 3×10^4 cells/well in duplicate.
2. Allow cells to attach overnight.
3. Add dilutions of the compounds of interest or DMSO as a control; add doxycycline (Dox, a stable tetracycline analogue; 1 μg/mL) to relevant wells containing H295R/TR SF-1 cells to increase SF-1 expression from their integrated transgene.
4. Incubate cells for 3 days.
5. Trypsinize and count cells using the automated cell counting apparatus.

Remarkably, isoquinolinones SID7969543, compound #31 and compound #32 significantly inhibited H295R/TR SF-1 proliferation only in conditions of increased SF-1 dosage (+Dox cells) starting from concentrations of 1 μ*M*, while they had no effect in conditions of basal SF-1 expression (−Dox) (Fig. 1.5). None of these compounds inhibited

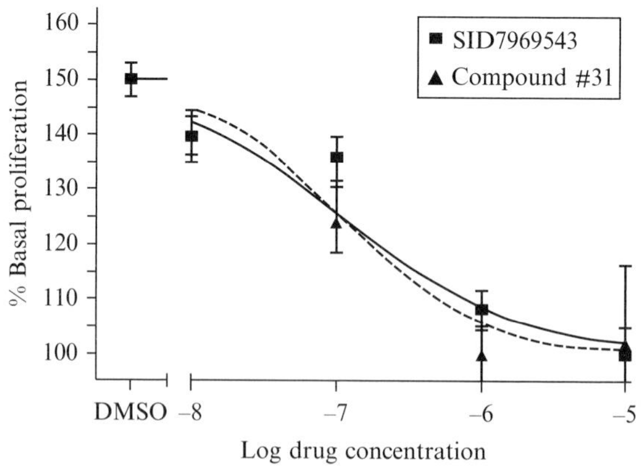

Figure 1.5 Inhibition of H295R/TR SF-1 human adrenocortical carcinoma cells proliferation by isoquinolinones SID7969543 (IsoQ A in Doghman *et al.*, 2009) (■) and compound #31 (▲). Cells were treated with 1 μg/mL doxycyline (Dox) to increase the expression of SF-1 and with the indicated concentrations of isoquinolinone compounds. Data is expressed as percentage of cell proliferation in basal culture conditions (no Dox; basal SF-1 expression).

proliferation of SF-1-negative SW-13 cells. Conversely, SID7970631 displayed a cytotoxic effect not only on H295R/TR SF-1 cells, but also on SW-13 cells. SF-1 inverse agonists of the alkyloxyphenol class (AC-45594 and octyloxyphenol; Del Tredici *et al.*, 2008) also displayed a nonspecific cytotoxic effect in this assay (Doghman *et al.*, 2009).

3.1.2. Cell cycle analysis

3.1.2.1. Required materials

- Propidium iodide and ribonuclease A (Sigma)
- FACScan instrument (Becton Dickinson)

3.1.2.2. Methods

1. Harvest cell by scraping in PBS; centrifuge them 5 min at 1200 rpm in Eppendorf tubes, then add while vortexing 500 μL 70% ethanol at 4 °C.
2. Incubate 15 min at room temperature; ethanol-fixed cells can be stored at 4 °C for at least 1 month.
3. Centrifuge cells 5 min at 1200 rpm. Resuspend them in 500 μL PBS containing ribonuclease A (50 μg/mL). Incubate 30 min at 37 °C.
4. Add propidium iodide at the final concentration of 50 μg/mL. Wait at least 15 min before analyzing cells on the FACScan.
5. Calculate cell cycle distribution using the instrument software.

Cell cycle analysis revealed an increase in the percentage of hypodiploid (apoptotic) cells after treatment with isoquinolinones #31 and #32.

3.2. Steroid hormone immunoassay

Since SF-1 is an essential regulator of the expression of the genes involved in steroid hormone production in adrenocortical cells (reviewed in Schimmer and White, 2010), it is important to evaluate the impact that inverse agonists for SF-1 have on steroid production in adrenocortical tumor cells, which may have clinical relevance. To this purpose, we measured aldosterone, cortisol, and DHEA-S levels in culture supernatants of H295R/TR SF-1 cells treated with isoquinolinone inverse agonists for SF-1.

3.2.1. Required materials

- Forskolin (Sigma); angiotensin II (Sigma).
- Tissue culture supernatants from H295R/TR SF-1 cells grown in basal conditions and stimulated with forskolin/angiotensin II.
- Immunoenzymatic assay kits for aldosterone, cortisol, and DHEA-S (Diagnostics Biochem Canada, Inc.).

3.2.2. Methods

1. Grow H295R/TR SF-1 cells in 24-well plates in basal conditions or under stimulation with forskolin (10 μg/mL) or angiotensin II (10 n*M*) for 4 days.
2. Collect supernatants. Harvest cells by scraping, centrifuge them 5 min at 1200 rpm in Eppendorf tubes, add 100 μL PBS and freeze–thaw them thrice in liquid nitrogen. Centrifuge 2 min at 12,000 × *g* to eliminate debris and transfer the extracts in new Eppendorf tubes. Measure protein content in extracts by Bradford assay using a BSA standard.
3. Assay aldosterone, cortisol, and DHEA-S concentration in tissue culture supernatants by enzyme-linked immunoassay, following the manufacturer's procedures.
4. Normalize steroid hormone concentration to protein content in each well.

Isoquinolinones SID7969543, #31 and #32 significantly inhibited angiotensin II-stimulated aldosterone and forskolin-stimulated cortisol productions in the presence of an increased SF-1 dosage. They also significantly inhibited forskolin-stimulated DHEA-S secretion in the presence of both basal and increased SF-1 dosage.

Table 1.3 summarizes the potency of different SF-1 inverse agonists measured in different assay formats, as reported in this review and elsewhere (Doghman *et al.*, 2009; Madoux *et al.*, 2008; Roth *et al.*, 2008).

4. Conclusion

Presented here is a set of assays and methodologies that facilitate identification and characterization of SF-1 inverse agonists. However, the mechanism of action of the isoquinolinone inverse agonists for SF-1 needs to be clarified. While it is likely that these compounds bind inside the SF-1 LBD, another possibility is that they may interfere with SF-1 interaction with transcriptional cofactors or alter its subcellular localization. A striking feature of the isoquinolinone inverse agonists for SF-1 is that they selectively interfere with the effects on cellular proliferation triggered by an increased dosage of SF-1 (Doghman *et al.*, 2009). These compounds may then represent an interesting option for the development of new tools to add to cytotoxic drugs and mitotane in the therapy of advanced adrenocortical carcinoma. Furthermore, inhibition of steroid production by SF-1 inverse agonists may be helpful to reduce clinical symptoms associated with these tumors due to hormone excess.

Table 1.3 SF-1 inverse agonists reported in the literature

Compound name	Method of identification	Library size	Reporter system	IC_{50} in the SF1 assay	Inhibition of adrenocortical carcinoma cell proliferation[a]	Original reference
AC-45594	HTS	~280,000	Beta-galactosidase	7000 nM[b]	Nonspecific	Del Tredici *et al.* (2008)
SID7969543	uHTS	~65,000	Luciferase	760 nM	Yes	Madoux *et al.* (2008)
Compound 32	Medicinal chemistry optimization of SID7969543	NA	Luciferase	200 nM	Yes	Roth *et al.* (2008)

[a] See Doghman *et al.* (2009) for more details.
[b] The original IC_{50} value reported in Del Tredici *et al.* (2008) in a R-SAT assay was 50–100 nM.

ACKNOWLEDGMENTS

The HTS methods described here were supported by the Molecular Libraries Screening Center Network (NIH grant # MH084512). Patrick Griffin, Juliana Conkright, and Gina Zastrow of the Scripps Florida are acknowledged for developing the SF-1 and LRH-1 response element assay methodology. The SF-1 and RORA vectors used for the HTS methods were provided by Orphagen Pharmaceuticals (NIH grant# MH077624). We thank F. Aguila for artwork. Research in E.L. laboratory is supported by Institut National du Cancer, Agence Nationale de la Recherche and FP7 ENS@T-CANCER.

REFERENCES

Almeida, M. Q., Soares, I. C., Ribeiro, T. C., Fragoso, M. C., Marins, L. V., Wakamatsu, A., Ressio, R. A., Nishi, M. Y., Jorge, A. A., Lerario, A. M., Alves, V. A., Mendonca, B. B., *et al.* (2010). Steroidogenic Factor 1 overexpression and gene amplification are more frequent in adrenocortical tumors from children than from adults. *J. Clin. Endocrinol. Metab.* **95,** 1458–1462.

Campbell, L. A., Faivre, E. J., Show, M. D., Ingraham, J. G., Flinders, J., Gross, J. D., and Ingraham, H. A. (2008). Decreased recognition of SUMO-sensitive target genes following modification of SF-1 (NR5A1). *Mol. Cell. Biol.* **28,** 7476–7486.

Chen, W. Y., Lee, W. C., Hsu, N. C., Huang, F., and Chung, B. C. (2004). SUMO modification of repression domains modulates function of nuclear receptor 5A1 (Steroidogenic Factor-1). *J. Biol. Chem.* **279,** 38730–38735.

Del Tredici, A. L., Andersen, C. B., Currier, E. A., Ohrmund, S. R., Fairbain, L. C., Lund, B. W., Nash, N., Olsson, R., and Piu, F. (2008). Identification of the first synthetic Steroidogenic Factor 1 inverse agonists: Pharmacological modulation of steroidogenic enzymes. *Mol. Pharmacol.* **73,** 900–908.

Doghman, M., Karpova, T., Rodrigues, G. A., Arhatte, M., De Moura, J., Cavalli, L. R., Virolle, V., Barbry, P., Zambetti, G. P., Figueiredo, B. C., Heckert, L. L., and Lalli, E. (2007). Increased Steroidogenic Factor-1 dosage triggers adrenocortical cell proliferation and cancer. *Mol. Endocrinol.* **21,** 2968–2987.

Doghman, M., Cazareth, J., Douguet, D., Madoux, F., Hodder, P., and Lalli, E. (2009). Inhibition of adrenocortical carcinoma cell proliferation by Steroidogenic Factor-1 inverse agonists. *J. Clin. Endocrinol. Metab.* **94,** 2178–2183.

Figueiredo, B. C., Cavalli, L. R., Pianovski, M. A., Lalli, E., Sandrini, R., Ribeiro, R. C., Zambetti, G., DeLacerda, L., Rodrigues, G. A., and Haddad, B. R. (2008). Amplification of the Steroidogenic Factor 1 gene in childhood adrenocortical tumors. *J. Clin. Endocrinol. Metab.* **90,** 615–619.

Hammer, G. D., Krylova, I., Zhang, Y., Darimont, B. D., Simpson, K., Weigel, N. L., and Ingraham, H. A. (1999). Phosphorylation of the nuclear receptor SF-1 modulates cofactor recruitment: Integration of hormone signaling in reproduction and stress. *Mol. Cell* **3,** 521–526.

Hodder, P., Cassaday, J., Peltier, R., Berry, K., Inglese, J., Feuston, B., Culberson, C., Bleicher, L., Cosford, N. D., Bayly, C., Suto, C., Varney, M., *et al.* (2003). Identification of metabotropic glutamate receptor antagonists using an automated high-throughput screening system. *Anal. Biochem.* **313,** 246–254.

Hoivik, E. A., Lewis, A. E., Aumo, L., and Bakke, M. (2010). Molecular aspects of Steroidogenic Factor 1 (SF-1). *Mol. Cell. Endocrinol.* **315,** 27–39.

Kakidani, H., and Ptashne, M. (1988). GAL4 activates gene expression in mammalian cells. *Cell* **52,** 161–167.

Komatsu, T., Mizusaki, H., Mukai, T., Ogawa, H., Baba, D., Shirakawa, M., Hatakeyama, S., Nakayama, K. I., Yamamoto, H., Kikuchi, A., and Morohashi, K.-I. (2004). Small ubiquitin-like modifier 1 (SUMO-1) modification of the synergy control motif of Ad4 binding protein/steroidogenic factor 1 (Ad4BP/SF-1) regulates synergistic transcription between Ad4BP/SF-1 and Sox9. *Mol. Endocrinol.* **18,** 2451–2462.

Krylova, I. N., Sablin, E. P., Moore, J., Xu, R. X., Waitt, G. M., MacKay, J. A., Juzumiene, D., Bynum, J. M., Madauss, K., Montana, V., Lebedeva, L., Suzawa, M., *et al.* (2005). Structural analyses reveal phosphatidyl inositols as ligands for the NR5 orphan receptors SF-1 and LRH-1. *Cell* **120,** 343–355.

Lalli, E. (2010). Adrenocortical development and cancer: Focus on SF-1. *J. Mol. Endocrinol.* **44,** 301–307.

Lee, M. B., Lebedeva, L. A., Suzawa, M., Wadekar, S. A., Desclozeaux, M., and Ingraham, H. A. (2005). The DEAD-box protein DP103 (Ddx20 or Gemin-3) represses orphan nuclear receptor activity via SUMO modification. *Mol. Cell. Biol.* **25,** 1879–1890.

Lewis, A. E., Rusten, M., Hoivik, E. A., Vikse, E. L., Hansson, M. L., Wallberg, A. E., and Bakke, M. (2008). Phosphorylation of Steroidogenic Factor 1 is mediated by cyclin-dependent kinase 7. *Mol. Endocrinol.* **22,** 91–104.

Li, Y., Choi, M., Cavey, G., Daugherty, J., Suino, K., Kovach, A., Bingham, N. C., Kliewer, S. A., and Xu, H. E. (2005). Crystallographic identification and functional characterization of phospholipids as ligands for the orphan nuclear receptor Steroidogenic Factor-1. *Mol. Cell* **17,** 491–502.

Madoux, F., Li, X., Chase, P., Zastrow, G., Cameron, M. D., Conkright, J. J., Griffin, P. R., Thacher, S., and Hodder, P. (2008). Potent, selective and cell penetrant inhibitors of SF-1 by functional ultra-high-throughput screening. *Mol. Pharmacol.* **73,** 1776–1784.

Pianovski, M. A., Cavalli, L. R., Figueiredo, B. C., Santos, S. C., Doghman, M., Ribeiro, R. C., Oliveira, A. G., Michalkiewicz, E., Rodrigues, G. A., Zambetti, G., Haddad, B. R., and Lalli, E. (2006). SF-1 overexpression in childhood adrenocortical tumours. *Eur. J. Cancer* **42,** 1040–1043.

Roth, J., Madoux, F., Hodder, P., and Roush, W. R. (2008). Synthesis of small molecule inhibitors of the orphan nuclear receptor steroidogenic factor-1 (NR5A1) based on isoquinolinone scaffolds. *Bioorg. Med. Chem. Lett.* **18,** 2628–2632.

Schimmer, B. P., and White, P. C. (2010). Minireview: Steroidogenic Factor 1: its roles in differentiation, development, and disease. *Mol. Endocrinol.* **24,** 1322–1337.

Urs, A. N., Dammer, E., and Sewer, M. B. (2006). Sphingosine regulates the transcription of CYP17 by binding to Steroidogenic Factor-1. *Endocrinology* **147,** 5249–5258.

Wang, W., Zhang, C., Marimuthu, A., Krupka, H. I., Tabrizizad, M., Shelloe, R., Mehra, U., Eng, K., Nguyen, H., Settachatgul, C., Powell, B., Milburn, M. V., *et al.* (2005). The crystal structures of human Steroidogenic Factor-1 and Liver Receptor Homologue-1. *Proc. Natl. Acad. Sci. USA* **102,** 7505–7510.

Whitby, R. J., Dixon, S., Maloney, P. R., Delerive, P., Goodwin, B. J., Parks, D. J., and Willson, T. M. (2006). Identification of small molecule agonists of the orphan nuclear receptors Liver Receptor Homolog-1 and Steroidogenic Factor-1. *J. Med. Chem.* **49,** 6652–6655.

Wilson, T. E., Fahrner, T. J., and Milbrandt, J. (1993). The orphan receptors NGFI-B and Steroidogenic Factor 1 establish monomer binding as a third paradigm of nuclear receptor-DNA interaction. *Mol. Cell. Biol.* **13,** 5794–5804.

Winnay, J. N., and Hammer, G. D. (2006). Adrenocorticotropic hormone-mediated signaling cascades coordinate a cyclic pattern of Steroidogenic Factor 1-dependent transcriptional activation. *Mol. Endocrinol.* **20,** 147–166.

Zhang, J. H., Chung, T. D., and Oldenburg, K. R. (1999). A simple statistical parameter for use in evaluation and validation of high throughput screening assays. *J. Biomol. Screen.* **4,** 67–73.

CHAPTER TWO

Assessment of Inverse Agonism for the Angiotensin II Type 1 Receptor

Hiroshi Akazawa,* Noritaka Yasuda,† Shin-ichiro Miura,‡ *and* Issei Komuro*

Contents

Abstract

The angiotensin II (AngII) type 1 (AT_1) receptor is a seven-transmembrane G-protein-coupled receptor that plays a regulatory role in the physiological and pathological processes of the cardiovascular system. AT_1 receptor inherently shows constitutive activity even in the absence of AngII, and it is activated not only by AngII but also by AngII-independent mechanisms. Especially, mechanical stress induces cardiac hypertrophy through activation of AT_1 receptor without the involvement of AngII. These AngII-independent activities of AT_1 receptor can be inhibited by inverse agonists, but not by neutral antagonists. In this chapter, we describe the methods used for biochemical assessment of

* Department of Cardiovascular Medicine, Osaka University Graduate School of Medicine, Yamadaoka, Suita, Osaka, Japan
† Department of Cardiovascular Science and Medicine, Chiba University Graduate School of Medicine, Inohana, Chuo-ku, Chiba, Japan
‡ Department of Cardiology, Fukuoka University School of Medicine, Nanakuma, Jonan-ku, Fukuoka, Japan

Methods in Enzymology, Volume 485
ISSN 0076-6879, DOI: 10.1016/S0076-6879(10)85002-5

inverse agonism of a ligand for AT_1 receptor. Their applications will improve our understanding of receptor activation and inactivation at a molecular level, and contribute to the development of AT_1 receptor blockers possessing superior therapeutic efficacy in cardiovascular diseases.

1. Introduction

The angiotensin II (AngII) type 1 (AT_1) receptor plays an important role in the physiological regulation of blood pressure and fluid–electrolyte balance. However, in addition to the homeostatic roles, AT_1 receptor participates in the progression of cardiovascular remodeling. For example, activation of AT_1 receptor stimulates multiple intracellular signaling pathways and enhances production of reactive oxygen species, which consequently evokes hypertrophic responses in cardiomyocytes (Hunyady and Catt, 2006; Kim and Iwao, 2000). During the past quarter of century, a growing body of evidence has accumulated indicating that AT_1 receptor blockers (ARBs) can prevent cardiac hypertrophy and reduce the morbidity and mortality in patients with heart failure (Jessup and Brozena, 2003).

The AT_1 receptor is a typical member of the G-protein-coupled receptor (GPCR) family, and is activated upon binding to AngII, the specific and endogenous agonist. In general, GPCRs are structurally flexible and instable, and show significant, albeit slight, levels of spontaneous activity in an agonist-independent manner (Leurs *et al.*, 1998; Milligan, 2003). The constitutive activity of AT_1 receptor is detectable when it is heterologously expressed in recombinant systems, and becomes manifest as a consequence of specific mutations (Balmforth *et al.*, 1997; Feng *et al.*, 1998; Groblewski *et al.*, 1997; Hunyady *et al.*, 2003; Noda *et al.*, 1996). Furthermore, we have recently demonstrated that AT_1 receptor is activated by mechanical stress independently of AngII (Yasuda *et al.*, 2008; Zou *et al.*, 2004). These observations have led to identification of inverse agonists for AT_1 receptor that are able to inhibit AngII-independent receptor activity in a dose-dependent manner (Bond and Ijzerman, 2006; Milligan, 2003; Strange, 2002). An inverse agonist stabilizes inactive conformation of the receptor and reduces constitutive activity of the receptor or the agonist-independent receptor activity.

In this section, we provide the detailed methods for assessing inverse agonism of a ligand for AT_1 receptor.

2. Protocol for Cell Culture and Transfection

To assess inverse agonism of a ligand for the AT_1 receptor, we need to assay the receptor activity unaffected by the agonistic action of AngII. For this purpose, we routinely use HEK293 or COS7 cells, because analysis by polymerase chain reaction with reverse transcription (RT-PCR) did not detect the transcript of the *angiotensinogen* gene (*ATG*) in these cells even after 50 cycles (Zou *et al.*, 2004). In addition, we observed that AngII did not activate extracellular signal-regulated protein kinases (ERKs) in these cells, but that forced expression of AT_1 receptor enabled these cells to activate ERKs in response to AngII (Zou *et al.*, 2004). We transfect HEK293 cells with expression vector for wild-type or mutant AT_1 receptor, and assess the constitutive activity or stretch-induced activation of AT_1 receptor without the involvement of AngII.

Required materials

- Cells of interest: HEK293 and COS7 cells can be obtained from the American Type Culture Collection (ATCC).
- DNA of interest: The mammalian expression vectors for either wild-type or N111G constitutively active AT_1 receptor (Miura *et al.*, 2006; Yasuda *et al.*, 2008).
- Growth medium: The growth medium for the cells contains Dubecco's Modified Eagle's Medium (DMEM) supplemented with 10% fetal calf serum (FCS) and penicillin/streptomycin.

1. Plate the cells at a field density of 5×10^3 cells/cm^2 in culture dishes with fresh growth medium.
2. At 24 h after plating, transfect the cells with empty vector or expression vector for wild-type or mutant AT_1 receptor. We use FuGENE 6 Transfection Reagent (Roche Diagnostics) for the transfection of plasmids.

3. Radioligand Assay for AT_1 Receptor

In this part, we describe the protocol for determining the maximum binding capacity (B_{max}) and dissociation constant (K_d) values of receptor binding by ^{125}I-[Sar^1, Ile^8] AngII-binding experiments under equilibrium conditions (Miura *et al.*, 1999).

3.1. Preparation of membrane-rich fractions

1. Remove the culture medium by aspiration, and wash the cells with ice-cold PBS $\times$ 1.
2. Homogenize the cells in lysis buffer (10 m*M* Tris–HCl pH 7.4, 5 m*M* EDTA, 5 m*M* EGTA, 5 μg/ml leupeptin, 5 μg/ml aprotinin, 10 μg/ml soybean trypsin inhibitor).
3. Centrifuge the homogenate at $800\times g$ for 10 min at 4 °C.

4. Transfer the supernatant to a new tube, and centrifuge at 50,000×*g* for 30 min at 4 °C.
5. Discard the supernatant, and resuspend the pellet in ice-cold lysis buffer.
6. Determine the protein concentration by bicinchoninic acid (BCA) method.

3.2. Radioligand receptor binding assay

1. Incubate 5 μg protein of membrane fraction with 10^{-10} *M* of ^{125}I-[Sar^1, Ile^8] AngII (Perkin Elmer) in binding buffer (75 m*M* Tris–HCl pH 7.4, 12.5 m*M* $MgCl_2$, 2m*M* EDTA) with a volume of 125 μl for 1 h at 22 °C.
2. Terminate binding reaction by adding excess volume (>3 ml) of ice-cold stop buffer (25 m*M* Tris–HCl, pH 7.4, 1 m*M* $MgCl_2$).
3. Separate the incubation mixture by vacuum filtration through Whatman GF/C glass filters (Disposable Filter Funnels Grade GF/C, GE Healthcare), presoaked with 0.1% polyethylenimine. Wash the filter with 3 ml of ice-cold stop buffer (25 m*M* Tris–HCl, pH 7.4, 1 m*M* $MgCl_2$).
4. Quantify the radioactivity trapped on the filters using an automatic γ-counter.
5. Carry out the competitive binding by using an increasing concentration (10^{-12}–10^{-6} *M*) of ^{125}I-[Sar^1, Ile^8] AngII (2200 Ci/mmol) and 10^{-6} *M* of unlabeled [Sar^1, Ile^8] AngII (LKT labs). Nonspecific binding of ^{125}I-[Sar^1, Ile^8] Ang II is determined in the presence of a high concentration (10^{-6} *M*) of unlabeled [Sar^1, Ile^8] AngII.
6. We analyze the binding kinetics, and calculate B_{max} and K_d with the LIGAND computer program (Elsevier-Biosoft).

4. Protocol for Cell Stretching

An inverse agonist for AT_1 receptor stabilizes inactive conformation of the receptor, and thereby inhibits stretch-induced receptor activation (Yasuda *et al.*, 2008; Zou *et al.*, 2004). Inverse agonism of a ligand is especially relevant to its ability to attenuate load-induced cardiac hypertrophy, because pressure overload by constricting the transverse aorta induced cardiac hypertrophy even in *angiotensinogen*-deficient mice as well as in wild-type mice, which was significantly inhibited by an inverse agonist for AT_1 receptor, candesartan (Zou *et al.*, 2004). In this part, we describe the protocols for applying mechanical stimulation by stretching the adherent cells cultured in elastic silicone dishes.

4.1. Preparation of collagen from rat-tail tendons

Based on our experiments, we recommend coating culture dishes with 0.01–0.1% collagen. Otherwise, cells will easily detach while applying stretch stimulation. Here we describe a protocol for extraction of intact collagen in large amounts from rat tails.

1. Animal experiments must comply with the law and regulations with regard to animal welfare legislation.
2. Soak rat tails in 95% ethanol for 15 min, and remove the skin to expose tendon bundles.
3. Rip off the bundles of white fibers using forceps from the distal tip of the tail.
4. Sterilize the ripped bundles by soaking them in 70% ethanol for 1 h.
5. Discard the ethanol, and air dry for 30 min.
6. Soak the air-dried bundles in a sterile beaker with 100–200 ml of 0.1% (vol/vol) acetic acid (glacial acetic acid), and keep this beaker well stirred on a magnetic stirrer for 24–48 h at 4 °C.
7. Transfer the solution (viscous) to sterile centrifuging tubes, and centrifuge the tubes at 15,000×*g* for 60 min at 4 °C.
8. Transfer the supernatant to 50 ml polypropylene tubes, and store at 4 °C.
9. Determine the protein concentration by BCA method.

4.2. Application of Mechanical Stretch to Cells

1. Coat flexible silicone rubber dishes (STB-CH-08, STREX Mechanical Cell Strain Instruments) with 0.01–0.1% collagen. The optimal concentration of collagens depends on the cells used in the experiments. We routinely coat silicone dishes with 0.01% collagen for COS7 cells, and 0.1% collagen for HEK293 cells. Dilute aqueous collagen solution in 0.1% (vol/vol) acetic acid (glacial acetic acid).
2. Place 4–6 silicone dishes in a 150-mm Petri dish, and pour 2 ml of collagen solution into each silicone dish.
3. Allow collagen solution in silicone dishes to be air-dried overnight in a laminar-flow clean bench under UV light.
4. Wash silicone dished in serum-free culture medium to remove surplus collagen solution.
5. Plate HEK293 or COS7 cells at a field density of 5×10^3 cells/cm^2, and culture in fresh growth medium for 24 h at 37 °C in 5% CO_2 incubator.
6. Starve cells under a serum-free condition for 24–48 h before stretch stimulation.
7. Pretreat the cells with ligands of interest for 30 min before stretch stimulation.

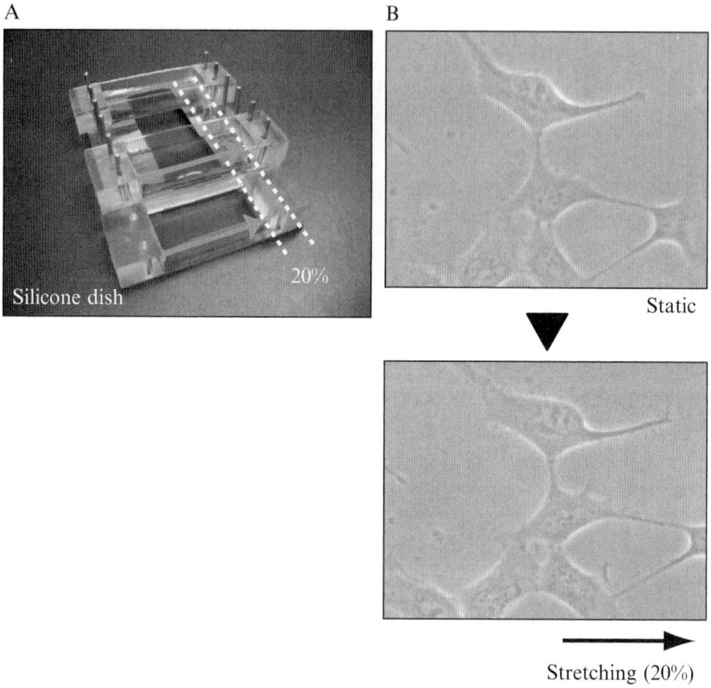

Figure 2.1 Passive stretch of cells cultured on extensible silicone dishes (A) Cells were plated on collagen-coated silicone rubber dishes, and silicone dishes were passively stretched longitudinally by 20%. (B) HEK293cells before and after stretching were shown as phase-contrast images. Figure is adopted from the work of Yasuda *et al.* (2008). (See Color Insert.)

8. Stretch the silicone dishes by attaching both ends of the dishes firmly to a fixed frame (STB-CH-08ST-XX, STREX Mechanical Cell Strain Instruments) to give a longitudinal stretch by 20% of the original length (Fig. 2.1). Maintain control dishes at static conditions with no application of stretching. Cells are kept stretched or left unstretched for 8 min.

5. Assessment of AT_1 Receptor Activation

Upon activation, AT_1 receptor stimulates diverse intracellular signaling pathways via $G_{q/11}$ protein coupling (Hunyady and Catt, 2006; Kim and Iwao, 2000). Especially, G protein coupling causes activation of phospholipase C, inositol phosphate (IP)-induced Ca^{2+} signal generation, activation of protein kinase C, and ERKs (Yamazaki *et al.*, 1995; Zou *et al.*, 1996).

In addition, AT_1 receptor activates several transcription factors, and thereby increases the expression levels of immediate early genes such as *c-fos*. Activation of AT_1 receptor also stimulates G-protein-independent signaling pathways such as the Jak/STAT pathway and the β-arrestin-mediated cascade (Hunyady and Turu, 2004). In this part, we provide the protocols for assaying AT_1 receptor activation in cultured cells.

5.1. Assay for total soluble IP production

Total soluble IP production can be extracted from the cells using the perchloric acid extraction method. Here, we show the protocol for assay for total IP production in HEK293 or COS7 cells transfected with wild-type or N111G constitutively active AT_1 receptor.

1. Plate the cells at a field density of 5×10^3 cells/cm^2 in 60-mm dishes, and culture in fresh growth medium at 37 °C in 5% CO_2 incubator.
2. At 24 h after plating, transfect the cells with empty vector or expression vector for wild-type or mutant AT_1 receptor.
3. Change culture medium 2 days after plasmid transfection and add 1.5 μl of 1 μCi/μl of [^{3}H]-myoinositol (TRK 883, GE healthcare) to 60-mm dish.
4. Incubate the cells for 20–24 h at 37 °C in 5% CO_2 incubator.
5. Wash the labeled cells with Hanks' Balanced Salt Solutions (HBSS) three times, and incubate the cells with HBSS containing 10 m*M* LiCl for 20 min at room temperature (RT).
6. Add ligands of interests, and continue incubation for another 45 min at 37 °C in 5% CO_2 incubator.
7. At the end of incubation, remove culture medium, and add 1 ml of ice-cold 0.4 *M* perchloric acid.
8. Freeze the cells at −80 °C for 10 min and scrape the cells by scraper after melting.
9. Transfer the scraped cells to a tube, and add 0.5 ml of 0.72 N KOH/0.6 *M* $KHCO_3$ to the tube.
10. After mixing, centrifuge the tube at 800×*g* for 10 min at RT.
11. In the meantime, resuspend the anion-exchange resin (AG1-X8, Bio-Rad) in 0.1 *M* formic acid, and pour the resin to a syringe column.
12. Apply the supernatant slowly to the column.
13. Wash with 2 ml of 0.1 *M* formic acid at four times, and elute three times with 1 ml of 1 *M* ammonium formate/0.1 *M* formic acid.
14. Collect each fraction in a scintillation vial containing 19 ml of scintillation cocktail (Ultima Gold MV, Perkin Elmer).
15. After mixing, count the vial using a standard scintillation counter.
16. Total IP production is expressed as the sum of radioactivity present in the three fractions.

5.2. Assay for phosphorylated levels of ERKs by Western blot analysis

A conventional method can be applied for protein extraction and membrane transfer. Here, we show the protocol for assay for phosphorylated ERKs after stretch stimulation in HEK293 transfected with wild-type AT_1 receptor. Since the basal ERKs activities in COS7 cells are relatively higher than those in HEK293 cells, we use HEK293 cells in assaying phosphorylated ERKs.

1. At the end of stretch stimulation, remove the culture medium by aspiration, wash the cells with ice-cold PBS × 1, and freeze the silicone dishes in liquid nitrogen.
2. Scrape the cells by scraper after melting, and transfer the scraped cells to a tube.
3. Homogenize the cells in lysis buffer (25 m*M* Tris–HCl pH 7.4, 25 m*M* NaCl, 0.5 m*M* EGTA, 10 m*M* $Na_4P_2O_7$, 1 m*M* Na_3VO_4, 10 m*M* NaF, 10 m*M* Okadaic acid, plus protease inhibitors [Complete mini, Roche Diagnostics]).
4. Determine the protein concentration by BCA method.
5. Fractionate total proteins (20 μg) by SDS-PAGE, and transfer the proteins to Hybond nitrocellulose membranes (GE Healthcare).
6. Block the blotted membranes with 10% skim milk in TBS buffer (50 mM Tris–HCl pH 7.6, 150 m*M* NaCl) containing 0.1% Tween 20 (TBS-T) for 1 h at RT.
7. Incubate the membranes with rabbit polyclonal antibody against phosphorylated ERKs (diluted 1:1000; Cell Signaling Technology, #9101) overnight at 4 °C. Wash the membrane in TBS-T for 3 × 5 min at RT.
8. Incubate with horseradish peroxidase (HRP)-conjugated goat anti-rabbit IgG antibody (diluted 1:5000; Jackson ImmunoResearch Laboratories, #111-035-003) for 1 h at RT. Wash the membrane in TBS-T for 3 × 5 min at RT.
9. Visualize the signals with the ECL detection kit (GE Healthcare).
10. Remove the primary and secondary antibodies from the membrane by incubating the membrane in stripping buffer (100 m*M* 2-mercaptoethanol, 2% SDS, 62.5 m*M* Tris–HCl pH 6.7) for 30 min at 50 °C. Wash the membrane in TBS-T for 2 × 10 min at RT.
11. Reprobe the membrane with rabbit polyclonal antibody against ERKs (diluted 1:2000; Invitrogen, #61-7400) overnight at 4 °C. Wash the membrane in TBS-T for 3 × 5 min a RT.
12. Incubate with HRP-conjugated goat anti-rabbit IgG antibody (diluted 1:5000; Jackson ImmunoResearch Laboratories, #111-035-003) for 1 h at RT. Wash the membrane in TBS-T for 3 × 5 min a RT.
13. Visualize the signals with the ECL detection kit (GE Healthcare).

5.3. Assay for transactivation of *c-fos* promoter by luciferase analysis

A conventional method can be applied for *c-fos* reporter gene assay. Here, we show the protocol for assay of *c-fos*-luciferase transactivation in HEK293 or COS7 cells transfected with wild-type or N111G constitutively active AT_1 receptor.

1. Plate the cells at a field density of 5×10^3 cells/cm^2 in 35-mm dishes, and culture in fresh growth medium at 37 °C in 5% CO_2 incubator.
2. At 24 h after plating, transfect the cells with the *c-fos* reporter plasmid (Tabuchi *et al.*, 1998) with empty vector or expression vector for wild-type or mutant AT_1 receptor. For each transfection, normalize the firefly luciferase activities with the *Renilla reniformis* luciferase activity by cotransfection of pRL-SV40 plasmid (Promega).
3. Start serum starvation at 6 h after transfection.
4. Measure luciferase activities at 18 h after the start of serum starvation with or without treatment with ligands of interest using the Dual-Luciferase Reporter Assay System (Promega).

6. Conclusion

We described the methods used for assessment of inverse agonism of a ligand for AT_1 receptor, which include biochemical assays of AngII-independent activity of the receptor. An inverse agonist of AT_1 receptor provides inhibitory effects on (i) basal activity of wild-type or N111G constitutively active AT_1 receptor and (ii) stretch-induced activation of AT_1 receptor. Both assays of AngII-independent receptor activity should be used in combination, because the conformation of AT_1 receptor during stretch-induced activation is different from that in the constitutively active AT_1 receptor, as revealed by studies using substituted cysteine accessibility mapping (Boucard *et al.*, 2003; Yasuda *et al.*, 2008). Transmembrane helix 7 undergoes a counterclockwise rotation and a shift into the ligand-binding pocket in response to mechanical stretch (Yasuda *et al.*, 2008), but it shifts apart from the ligand-binding pocket in the AT_1-N111G receptor (Boucard *et al.*, 2003). Therefore, inverse agonistic action of a ligand may differ according to the distinct processes of receptor activation (Qin *et al.*, 2009).

The inverse agonist activity of ARBs may have therapeutic benefits, because they inhibit both AngII-dependent and -independent receptor activation, and thus be a novel and important pharmacological parameter defining the beneficial effects on organ protection. The methods described here will make a contribution to development of a novel ARB with more potent inverse agonist activity and to enhance our understanding of the molecular basis for inverse agonism of a ligand.

ACKNOWLEDGMENTS

This work was supported in part by grants from the Japanese Ministry of Education, Science, Sports, and Culture, and Health and Labor Sciences Research Grants (to I. K. and H. A.); grants from Takeda Science Foundation, Astellas Foundation for Research on Metabolic Disorders, and Uehara Memorial Foundation (to H. A.).

REFERENCES

Balmforth, A. J., Lee, A. J., Warburton, P., Donnelly, D., and Ball, S. G. (1997). The conformational change responsible for AT1 receptor activation is dependent upon two juxtaposed asparagine residues on transmembrane helices III and VII. *J. Biol. Chem.* **272,** 4245–4251.

Bond, R. A., and Ijzerman, A. P. (2006). Recent developments in constitutive receptor activity and inverse agonism, and their potential for GPCR drug discovery. *Trends Pharmacol. Sci.* **27,** 92–96.

Boucard, A. A., Roy, M., Beaulieu, M. E., Lavigne, P., Escher, E., Guillemette, G., and Leduc, R. (2003). Constitutive activation of the angiotensin II type 1 receptor alters the spatial proximity of transmembrane 7 to the ligand-binding pocket. *J. Biol. Chem.* **278,** 36628–36636.

Feng, Y. H., Miura, S., Husain, A., and Karnik, S. S. (1998). Mechanism of constitutive activation of the AT1 receptor: Influence of the size of the agonist switch binding residue Asn(111). *Biochemistry* **37,** 15791–15798.

Groblewski, T., Maigret, B., Larguier, R., Lombard, C., Bonnafous, J. C., and Marie, J. (1997). Mutation of Asn111 in the third transmembrane domain of the AT1A angiotensin II receptor induces its constitutive activation. *J. Biol. Chem.* **272,** 1822–1826.

Hunyady, L., and Catt, K. J. (2006). Pleiotropic AT1 receptor signaling pathways mediating physiological and pathogenic actions of angiotensin II. *Mol. Endocrinol.* **20,** 953–970.

Hunyady, L., and Turu, G. (2004). The role of the AT1 angiotensin receptor in cardiac hypertrophy: angiotensin II receptor or stretch sensor? *Trends Endocrinol. Metab.* **15,** 405–408.

Hunyady, L., Vauquelin, G., and Vanderheyden, P. (2003). Agonist induction and conformational selection during activation of a G-protein-coupled receptor. *Trends Pharmacol. Sci.* **24,** 81–86.

Jessup, M., and Brozena, S. (2003). Heart failure. *N. Engl. J. Med.* **348,** 2007–2018.

Kim, S., and Iwao, H. (2000). Molecular and cellular mechanisms of angiotensin II-mediated cardiovascular and renal diseases. *Pharmacol. Rev.* **52,** 11–34.

Leurs, R., Smit, M. J., Alewijnse, A. E., and Timmerman, H. (1998). Agonist-independent regulation of constitutively active G-protein-coupled receptors. *Trends Biochem. Sci.* **23,** 418–422.

Milligan, G. (2003). Constitutive activity and inverse agonists of G protein-coupled receptors: A current perspective. *Mol. Pharmacol.* **64,** 1271–1276.

Miura, S., Feng, Y. H., Husain, A., and Karnik, S. S. (1999). Role of aromaticity of agonist switches of angiotensin II in the activation of the AT1 receptor. *J. Biol. Chem.* **274,** 7103–7110.

Miura, S., Fujino, M., Hanzawa, H., Kiya, Y., Imaizumi, S., Matsuo, Y., Tomita, S., Uehara, Y., Karnik, S. S., Yanagisawa, H., Koike, H., Komuro, I., *et al.* (2006). Molecular mechanism underlying inverse agonist of angiotensin II type 1 receptor. *J. Biol. Chem.* **281,** 19288–19295.

Noda, K., Feng, Y. H., Liu, X. P., Saad, Y., Husain, A., and Karnik, S. S. (1996). The active state of the AT1 angiotensin receptor is generated by angiotensin II induction. *Biochemistry* **35,** 16435–16442.

Qin, Y., Yasuda, N., Akazawa, H., Ito, K., Kudo, Y., Liao, C. H., Yamamoto, R., Miura, S., Saku, K., and Komuro, I. (2009). Multivalent ligand–receptor interactions elicit inverse agonist activity of AT(1) receptor blockers against stretch-induced AT(1) receptor activation. *Hypertens. Res.* **32,** 875–883.

Strange, P. G. (2002). Mechanisms of inverse agonism at G-protein-coupled receptors. *Trends Pharmacol. Sci.* **23,** 89–95.

Tabuchi, A., Sano, K., Nakaoka, R., Nakatani, C., and Tsuda, M. (1998). Inducibility of BDNF gene promoter I detected by calcium-phosphate-mediated DNA transfection is confined to neuronal but not to glial cells. *Biochem. Biophys. Res. Commun.* **253,** 818–823.

Yamazaki, T., Komuro, I., Kudoh, S., Zou, Y., Shiojima, I., Mizuno, T., Takano, H., Hiroi, Y., Ueki, K., and Tobe, K. (1995). Angiotensin II partly mediates mechanical stress-induced cardiac hypertrophy. *Circ. Res.* **77,** 258–265.

Yasuda, N., Miura, S., Akazawa, H., Tanaka, T., Qin, Y., Kiya, Y., Imaizumi, S., Fujino, M., Ito, K., Zou, Y., Fukuhara, S., Kunimoto, S., *et al.* (2008). Conformational switch of angiotensin II type 1 receptor underlying mechanical stress-induced activation. *EMBO Rep.* **9,** 179–186.

Zou, Y., Komuro, I., Yamazaki, T., Aikawa, R., Kudoh, S., Shiojima, I., Hiroi, Y., Mizuno, T., and Yazaki, Y. (1996). Protein kinase C, but not tyrosine kinases or Ras, plays a critical role in angiotensin II-induced activation of Raf-1 kinase and extracellular signal-regulated protein kinases in cardiac myocytes. *J. Biol. Chem.* **271,** 33592–33597.

Zou, Y., Akazawa, H., Qin, Y., Sano, M., Takano, H., Minamino, T., Makita, N., Iwanaga, K., Zhu, W., Kudoh, S., Toko, H., Tamura, K., *et al.* (2004). Mechanical stress activates angiotensin II type 1 receptor without the involvement of angiotensin II. *Nat. Cell Biol.* **6,** 499–506.

CHAPTER THREE

Measurement of Inverse Agonism in β-Adrenoceptors

Carlos A. Taira, Federico Monczor, *and* Christian Höcht

Contents

Abstract

Increasing numbers of compounds, previously classified as antagonists, were shown to inhibit this spontaneous or constitutive receptor activity, instead of leave it unaffected as expected for a formal antagonist. In addition, some other antagonists did not have any effect by themselves, but prevented the inhibition of constitutive activity induced by thought-to-be antagonists. These thought-to-be antagonists with negative efficacy are now known as "inverse agonists."

Inverse agonism at βAR has been evidenced for both subtypes in wild-type GPCRs systems and in engineered systems with high constitutive activity. It is important to mention that native systems are of particular importance for analyzing the *in vivo* relevance of constitutive activity because these systems have physiological expression levels of target receptors. Studies of inverse agonism of

Cátedra de Farmacología, Instituto de Fisiopatología y Bioquímica Clínica, Cátedra de Química Medicinal, Facultad de Farmacia y Bioquímica, Universidad de Buenos Aires, CONICET, Junín 956, Buenos Aires, Argentina

Methods in Enzymology, Volume 485
ISSN 0076-6879, DOI: 10.1016/S0076-6879(10)85003-7

β blockers in physiological setting have also evidenced that pathophysiological conditions can affect pharmacodynamic properties of these ligands.

To date, hundreds of clinically well-known drugs have been tested and classified for this property. Prominent examples include the beta-blockers propranolol, alprenolol, pindolol, and timolol used for treating hypertension, angina pectoris, and arrhythmia that act on the β_2ARs, metoprolol, and bisoprolol used for treating hypertension, coronary heart disease, and arrhythmias by acting on β_1ARs. Inverse agonists seem to be useful in the treatment of chronic disease characterized by harmful effects resulting from β_1AR and β_2AR overactivation, such as heart failure and asthma, respectively.

1. Introduction: Basal Spontaneous Receptor Activity. The Rise of the Concept of Inverse Agonism

Many activities run through effector molecules in the natural state of resting cells. Channels tunnel molecules, pumps pump ions, and cotransporters shovel nutrients to the enzyme catalyzed furnace. At resting conditions, these activities can be thought as "basal activity" of the various effector molecules.

In this context, the receptor molecules were mostly thought of as quiescent until the moment they are stimulated by an agonist. This view, perhaps due to its simplicity, dominated pharmacology until about two decades ago. In a way, such a view is more obvious in enzymology and transport physiology, since substrates or transported molecules in many types of experiments must be present in order to allow measurement of function. However, the above dogmatism about receptors seems to have vanished with the realization that even in the absence of agonists, many systems display "spontaneous activity" (Costa and Herz, 1989), while mutated receptors might be or become "constitutively active" (Kenakin, 1995; Milligan and Bond, 1997), and could explain the behavior of inverse agonists (Bond *et al.*, 1995; Kenakin, 1994).

Understandably, during this paradigm shift, the discoveries and recognition of spontaneous receptor activity briefly confused the definition of what agonists and antagonists should be, how to designate them, and furthermore, how to design experiments (Hoyer and Boddeke, 1993; Jenkinson, 1991; Kenakin, 1987). In those years, new notions and formulations were introduced, for example, concepts appeared such as "inverse agonism," "negative efficacy," and "negative intrinsic activity" as well as "negative antagonism" (Bond *et al.*, 1995; Costa *et al.*, 1992; Kenakin, 1994, 1995; Milligan *et al.*, 1995; Samama *et al.*, 1993). This confusion dissipated with the recognition of the two-state model (TSM) (Leff, 1995; Robertson *et al.*,

1994) and related reaction schemes developed later (Bindslev, 2004; Hall, 2000; Weiss *et al.*, 1996a–c). Increasing numbers of compounds, previously classified as antagonists, were shown to inhibit this spontaneous or constitutive receptor activity, instead of leave it unaffected (Kenakin, 2004) as expected for a formal antagonist. In addition, some other antagonists did not have any effect by themselves, but prevented the inhibition of constitutive activity induced by thought-to-be antagonists.

These thought-to-be antagonists with negative efficacy are now known as "inverse agonists," whereas compounds that antagonize the inhibitory effect of agonists and inverse agonists without an effect of their own are still signified as antagonists or even better as neutral antagonists (Chidiac, 2002; Kenakin, 2004; Milligan *et al.*, 1995; Strange, 2002). Classically, efficacy (whether positive, negative, or zero) is thought as a separate property unrelated to affinity. However in thermodynamic terms, this presents a paradox because the molecular forces that control affinity are the same as those that control efficacy (Kenakin, 2002). Considering this, it is not surprising that, when appropriately studied, 85% of the ligands formerly known as neutral antagonists were shown to possess negative efficacy (Kenakin, 2004). In the interest of this chapter, it is worth noting that many beta adrenergic ligands with clinical uses have been reclassified as inverse agonists according to this definition, but whether this negative efficacy is necessary to display the clinical effects, has not been established yet (Bosier and Hermans, 2007; Parra and Bond, 2007; Rodríguez-Puertas and Barreda-Gómez, 2006; Tao, 2008).

1.1. Ligand–receptor occupancy theoretical models and the mechanisms of inverse agonism

When we focus on receptor activity observed in functional studies, it can be either basal/spontaneous/constitutive, or agonist-induced. This is assumed independently of the model used to interpret the system. However, there can be hypothesized different spontaneous receptor conformations responsible for the basal ligand-independent activity, comprising a "receptor native ensemble." Hence, the uniformity of the concept of constitutive receptor activity is apparently challenged when the spontaneous receptor species (and therefore the source of the receptor basal activity) are explicitly modeled (Kenakin, 2002; Onaran and Costa, 1997).

For the TSM, receptors can spontaneously adopt solely two conformations, the resting or inactive state (R), and the active one (R*), to which the activity of the system in the absence of a ligand is formally attributed. However, when accessory proteins are included in the models, as in the case of ternary models where G-proteins are explicitly added, the native ensemble is modified. For the extended ternary complex model (ETCM) (Samama *et al.*, 1993), the native ensemble involves three distinct receptor forms including the inactive (R) and the active species (R*), but also an active G-protein-coupled receptor (GPCR) conformation that is considered responsible for basal activity (R*G).

Another advance from both TSMs is the cubic ternary complex model (CTCM) (Weiss *et al.*, 1996a–c). Considering the ETCM, this last model adds one more receptor species to the native ensemble, allowing receptor to couple to G-protein in an inactive form (RG).

Although the development of the ETCM was made necessary by experimental observations, the CTCM was originally proposed in an attempt to explore the mathematical and pharmacological implications that can be derived from permitting G-proteins to interact with receptors in their inactive and active forms, irrespectively. Thus, the CTCM was the culmination of a trend in increasing model complexity and statistical and thermodynamic completeness.

The three models and their relations are schematically presented in Fig. 3.1. The various models assume differences in the forms that receptors are able to adopt spontaneously. This implies that ligands that would bind to native receptor ensemble could stabilize different receptor conformations. The mechanisms by which inverse agonists are able to reduce receptor spontaneous activity vary according to the theoretical model used.

According to the law of mass action, when a situation affects the system established equilibriums, the receptor species amounts are redistributed. Consequently, for every model, inverse agonists exert their effect favoring

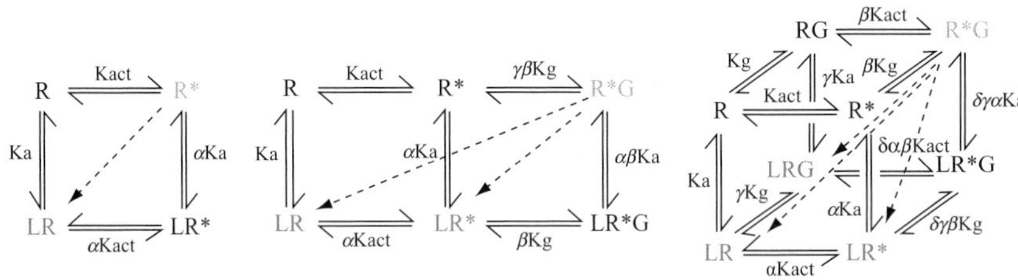

Figure 3.1 The schemes are intended to show the three more commonly used models of GPCR systems. The models of receptor activation describes quiescent receptor (R) in spontaneous equilibrium with an active state (R*). This activation of the receptor could be followed by binding of the active-state receptor to the G-protein (G). The cubic ternary complex model (right) implies the same, but also allows the liganded or unliganded inactive-state receptor (LR or R) to form a nonsignaling complex with the G-protein (LRG or RG, respectively). In green are shown receptors species responsible for constitutive activity, and in red those species that could be favored by inverse agonists diminishing receptor basal activity. It should be noted that is a trend in the complexity, that always the following scheme contains the previous, and that the liganded inactive receptor species are different for each model, thus differing the potential mechanisms of action accounting for inverse agonism. These different mechanisms are discussed in the main text.

the inactive receptor species at the expense of the active ones. However, since the array of considered species varies, the species subset favored by the ligand varies concomitantly. As will be further discussed, this not only affects the mechanism of action from a theoretical point of view, but also could have implications on desired or undesired effects of a drug.

As mentioned before, according to the TSM, it is interpreted that inverse agonists suppress the spontaneous activity of the receptors by stabilizing them in an inactive state. Considering the ETC model, inverse agonism can be achieved if the ligand stabilizes the R form of the receptor at the expense of the R*G form, thus suppressing basal activity (Fig. 3.1). They may act by binding to the R state of the receptor in preference to the R* state. Thus, as the ligand binds selectively to R instead of R*, the receptor species in the system will redistribute. If the system has R*G present (constitutive activity), then this species will be depleted as more receptor transforms into ligand-bound inactive R (LR), resulting in a decreased constitutive activity. There are several experiments that confirm this model, at least for the α2 and β2 adrenergic receptors (Samama *et al.*, 1993; Wade *et al.*, 2001).

Alternatively, ligands could bind to uncoupled states of the receptor (R and R*) in preference to the coupled state (R*G). This model experimentally lays on the sensitivity of the binding of inverse agonists to the effects of guanine nucleotide. This feature was described for 5-HT_{1A} and 5-HT_{2C} receptors, cardiac muscarinic receptors (M_2 receptors), dopamine D2 receptors and in the interest of this review for cardiac β adrenergic receptors (Bristow *et al.*, 1992; De Lean *et al.*, 1982; Hershberger *et al.*, 1990; Maack *et al.*, 2000; Martin *et al.*, 1984; McLoughlin and Strange, 2000; Westphal and Sanders-Bush, 1994; Yoshikawa *et al.*, 1996).

A third possibility may be that inverse agonists bias the receptor to an inactive conformation that exists in G-protein-coupled and uncoupled forms. Only the CTC model can account for this possibility, since it contains an inactive receptor conformation that can nevertheless couple to G-proteins (RG). It is worth noting that for the ETC model, a ligand with high affinity for receptor species coupled to G-protein necessarily elicits a response. In contrast, the CTC model allows a ligand with high affinity for the receptor form coupled to G-protein to behave as an antagonist or even as an inverse agonist. This point is a distinctive feature of the CTC model. This view can explain some striking results obtained for D2 dopamine and M3 muscarinic receptor ligands, which showed that an inverse agonist can exert its effects without promoting the expected receptor G-protein uncoupling (Dowling *et al.*, 2006; Wilson *et al.*, 2001).

Using this conceptual frame, it can be theoretically predicted, and then empirically proved, that an inverse agonist can stabilize a G-protein-coupled form of the receptor but inactive. Consequently, it can be inferred that if the G-protein is in a limiting quantity, the ligand will be able to interfere with the signaling of other unrelated GPCR that shares the signaling cascade. This effect can be interpreted in terms of a G-protein "molecular kidnapping" mediated by the inverse agonist bound receptor. This phenomenon could explain some observations made for ligands acting at H1 and H2 histamine receptors, at μ-opioid, and at CB1 cannabinoid receptors, that interfere on the signaling of other receptors (Bouaboula *et al.*, 1997; Brown and Pasternak, 1998; Fitzsimons *et al.*, 2004; Monczor *et al.*, 2003).

This unexpected interference of an inverse agonist on the signaling of a nontargeted receptor warrants the importance of the study of the mechanistic basis of action of ligands with efficacy (either positive or negative). Taking into account that this may be a generalized feature of several ligands with clinical uses, this understanding could help to rationalize the appearance of otherwise unexpected effects during or after a certain treatment. Regarding this, the current as well as potential uses and concerns of inverse agonists of the beta adrenergic system in clinical treatments will be further discussed.

2. β-Adrenoceptors: Main Features

The three subtypes of adrenoceptors (β_1AR, β_2AR, and β_3AR) are members of the GPCR super family (Dzmiri, 1999; Rozec and Gauthier, 2006; Taira *et al.*, 2008). Classical description of βARs stimulation includes a signaling cascade involving G stimulatory protein (Gs), adenylyl cyclase (AC), cAMP and protein kinase (PKA), and PKA-dependent protein phosphorylation (Dzmiri, 1999; Taira *et al.*, 2008; Xiao *et al.*, 2006).

The βAR subtypes are widely distributed on several tissues. Table 3.1 summarizes the anatomical localization of these adrenoceptor subtypes. The three subtypes of adrenoceptors are encoded by three distinct genes

Table 3.1 β-Adrenoceptors and main localizations

Subtype	Localization
β_1	Cardiac tissue
	Kidneys
	Adipocytes
β_2	Cardiac tissue
	Smooth muscle
	Skeletal muscle
β_3	Adipocytes
	Cardiac tissue
	Vascular and genitourinary smooth muscle

(β_1-adrenergic receptor [ADRB1], β_2-adrenergic receptor [ADRB2], and β_3-adrenergic receptor [ADRB3], all of them with polymorphic variants (Taylor and Bristow, 2004).

2.1. β_1-Adrenoceptors signaling pathway

Stimulation of β_1AR leads to production of cAMP and activation of PKA, which then phosphorylates several downstream target proteins including phospholamban, ryanodine receptors, L-type calcium channels, and cardiac troponins, resulting in positive inotropic, lusitropic, and chronotropic effects (Marian, 2006; Xiao *et al.*, 2004; Zhu *et al.*, 2005).

β_1AR stimulation also elicits a PKA-independent pathway. So, βAR stimulation activates ICa via a direct interaction of Gs with the calcium channel (Yatani and Brown, 1989), β_1AR modulation of cardiac excitation–contraction coupling invokes dual signaling pathways mediated by cAMP/PKA and calmodulin-dependent protein kinase II (CaMKII), respectively. Thus, persistent cardiac β_1AR stimulation changes the receptor signaling pathway form PKA to Ca^{2+}/CaMKII predominance, leading to myocyte apoptosis and maladaptive cardiac remodeling (Molenaar and Parsonage, 2005; Ostrom and Insel, 2004; Xiao *et al.*, 2004).

Phosphorylation and desensitization of the β_1AR is appreciated predominantly as a protective mechanism that decreases Gs-mediated signal transduction (Koch *et al.*, 2000). Interestingly, it was reported that GRK phosphorylation of β_1AR not only serves to reduce Gs/PKA-mediated signal transduction, but in parallel, serves to induce an antiapoptotic signal by mediating transactivation of the EGFR through a β-arrestin-dependent pathway (Noma *et al.*, 2007).

2.2. β_2-Adrenoceptor signaling pathway

β_2AR is linked by the Gs protein and AC, which increases cAMP, thus activating PKA, which affects calcium levels and reduces the efficiency of myosin light-chain kinase, causing smooth muscle relaxation (Giembycz and Newton, 2006; Johnson, 2006). This receptor also opens the K^+ channels by a cAMP/PKA mechanism (Giembycz and Newton, 2006). β_2AR can also couple to G inhibitory protein (Gi), that inhibits AC (Johnson, 2006; Xiao *et al.*, 2004).

β_2AR activates both Gs and Gi pathways in several smooth muscles and the mammalian heart, including humans (Daaka *et al.*, 1999; Dzmiri, 1999; Johnson, 2006; Xiao *et al.*, 2006; Zhu *et al.*, 2005). The Gi signaling pathways play a role as a cardioprotection against apoptosis of cardiomyocytes and attenuate the Gs-mediated inotropic effect. So, the cardioprotective effect of long β_2AR signaling is mediated by the Gi pathway which activates a cardiomyocyte survival pathway that involves G$\beta\gamma$, PI3K, and Akt (Chesley *et al.*, 2000; Zhu *et al.*, 2001).

It is well known that vascular β_2AR mediate vasorelaxation through an action on vascular smooth muscle cells via cAMP pathway. However, β_2AR-dependent vasorelaxation is mediated, at least in part, by endothelial nitric oxide (NO)-dependent processes (Ferro *et al.*, 1999). Iaccarino *et al.* (2002) demonstrated that the β_2ARs are expressed on endothelial cells and their stimulation causes endothelial nitric oxide synthase (eNOS) activation. It is also known that endothelial β_2AR regulates eNOS activity and consequently vascular tone, through means of PKB/Akt (Iaccarino *et al.*, 2004).

2.3. β_3-Adrenoceptor signaling pathway

In adipose tissue, stimulation of β_3AR increases cAMP production via classical activation of Gs and induces lipolysis (Arch *et al.*, 1984; Emorine *et al.*, 1989).

In the heart, β_3AR is coupled to Gi proteins (Gauthier *et al.*, 1996; Liggett *et al.*, 1993). Moreover, β_3AR can couple interchangeably to both Gs and Gi without a requirement for receptor phosphorylation (Liggett *et al.*, 1993; Soeder *et al.*, 1999).

Stimulation of cardiac β_3AR leads to a decrease in contractility via a NO release (Maffei *et al.*, 2007; Molenaar and Parsonage, 2005; Rozec and Gauthier, 2006). The vascular β_3AR stimulation induces vasodilation. In the rat thoracic aorta, endothelial β_3AR acts through activation of a NO synthase pathway and a subsequent increase in intracellular cGMP levels (Rozec and Gauthier, 2006) or involved several K^+ channels (Rautureau *et al.*, 2002).

It was found that β_3AR stimulation caused phosphorylation and activation of ERK1/2 in brown adipocytes (Lindquist and Rehnmark, 1998; Lindquist *et al.*, 2000; Shimizu *et al.*, 1997). Constitutive β_3AR coupling to Gi proteins serves both to restrain Gs-mediated activation of AC and to initiate the ERK1/2 MAP kinase cascade (Soeder *et al.*, 1999). β_3AR activates ERK by a mechanism that depends on a series of proline-rich motifs in its third intracellular domain and carboxyl terminus that are conserved among the mammalian homologues (Cao *et al.*, 2000). It is through these regions that the Src kinase is recruited to the β_3AR. This interaction triggers Src catalytic activity, which together are necessary steps in ERK activation (Cao *et al.*, 2000). One of the functional consequences of this β_3AR to ERK cascade in adipocytes is lipolysis wherein ERK functions together with PKA to produce maximal lipolytic capacity (Robidoux *et al.*, 2006). It was reported that a vimentin filament assembly is necessary for β_3AR-mediated ERK activation and lipolysis (Kumar *et al.*, 2007). On the other side, it was reported that β-adrenergic lipolysis, specifically β_3AR effect, which is realized via the AC/cAMP/PKA signaling cascade, involves NO production downstream of β_3AR/cAMP pathway (Canová *et al.*, 2006).

3. Methodological Aspects of the Assessment of Inverse Agonist Properties at βAR

Inverse agonism at βAR has been evidenced for both subtypes in wild-type GPCRs systems and in engineered systems with high constitutive activity. It is important to mention that native systems are of particular importance for analyzing the *in vivo* relevance of constitutive activity because these systems have physiological expression levels of target receptors.

3.1. Evaluation of inverse agonism of βAR ligands in engineered systems with high constitutive activity

Inverse agonism of beta blockers was described in different engineered systems with high constitutive activity due to expression of constitutively active mutant (CAM) of β_1AR or overexpression of wild-type βARs (Table 3.2).

Several CAMs of β_1AR have been designed by the replacement of a single aminoacid in different domains of the receptor protein. A point mutation at the second transmembrane domain of the β_1AR by the change of Asp104 to Ala104 increased the constitutively active in the second messenger assay (Ahmed *et al.*, 2006). By expressing either β_1AR wild-type or mutant gene into human embryonic kidney (HEK)-293 cells, the authors found a twofold increase in basal cAMP generation with the Asp104Ala variant with regards to the wild-type β_1AR receptor (Ahmed *et al.*, 2006). Moreover, while propranolol, atenolol, and carvedilol did not exert negative activity in HEK-293 cells expressing wild-type β_1AR, the three beta blockers decreased cAMP levels in a similar extent in Asp104Ala variant expressing cells (Ahmed *et al.*, 2006).

Lattion *et al.* (1999) also evaluated different CAMs of β_1AR by replacing Leucine 322 in the C-terminal portion of its third intracellular loop with seven different aminoacids. Expression of wild-type and CAM β_1ARs in HEK293 cells showed that the substitution of leucine by lysine enhanced maximal constitutive activation of adenylate cyclase. By using this CAM, it was shown that atenolol and propranolol act as partial agonists at both the recombinant human β_1AR and the L322K constitutive active variant. Conversely, betaxolol and ICI118,551 exert inverse agonist actions on the receptor-mediated activation of AC in HEK293 cells (Lattion *et al.*, 1999).

Constitutive active mutants of β_2AR were also obtained by point mutation or replacement of several aminoacids. Samama *et al.* (1993) first reported that the replacement of four amino acids of the third intracellular loop of the β_2AR by the corresponding residues of the α_{1B}AR led to

Table 3.2 Constitutive activity of β-adrenoceptors in engineered systems

Subtype of βAR	Type of engineered system	Tissue	Measured effect	Inverse agonists	Reference
β1AR	CAM Asp104Ala	HEK-293 cells	Intracellular cAMP levels	Carvedilol, propranolol, and atenolol	Ahmed *et al.* (2006)
β1AR	CAM L322K	HEK-293 cells	Intracellular cAMP levels	Betaxolol and ICI118,551	Lattion *et al.* (1999)
β1AR	Overexpression	COS-7 cells Transgenic mice	Intracellular cAMP levels Spontaneous beating rate of isolated right atria	CGP 20712A>Bisoprolol> Metoprolol Propranolol and carvedilol did not show inverse agonist activity	Engelhardt *et al.* (2001)
β2AR	CAM Replacement of four aminoacids of the third intracellular loop	CHO cells	Intracellular cAMP levels	–	Samama *et al.* (1993)
β2AR	CAM D130A D130N	COS-7 cells	Intracellular cAMP levels	–	Rasmussen *et al.* (1999)
β2AR	Overexpression	Transgenic mice	Cardiac parameters	ICI 118,551	Zhou *et al.* (1999)
β2AR	Overexpression	Sf9 cells	Intracellular cAMP levels	Timolol > propranolol > alprenolol > pindolol > labetalol > dichloroisoproterenol	Chidiac *et al.* (1994)

agonist-independent activation of adenylate cyclase. In another report, permanent expression of this CAM of β_2AR in Chinese hamster ovary (CHO) cell line have been associated with a greater desensitization of the receptor due to phosphorylation by recombinant βAR-specific kinase (βARK) with regards to wild-type expressing cells.

In addition, mutation of a highly conserved aspartic acid in the β_2AR also increased basal activation of this receptor. The change of aspartate at position 130 to asparagine (D130N) or to alanine (D130A) enhanced basal levels of cAMP accumulation compared with cells expressing the wild-type β_2AR (Rasmussen *et al.*, 1999).

Constitutive activity of human β_1 and β_2ARs were also evidenced by overexpression of receptor protein in cell lines. Briefly, transfection of the cDNAs of the human β_1 and β_2ARs into COS-7 cells induced a proportional increase in basal cAMP (Engelhardt *et al.*, 2001). Moreover, at comparable receptor level expression, increase in basal cAMP was about fivefold higher for the β_2- than for the β_1-subtype. In addition, transgenic mice overexpressing the human β_1AR have been used for the assessment of inverse agonism activity of different βAR blockers, including bisoprolol, metoprolol, and carvedilol (Engelhardt *et al.*, 2001). While bisoprolol and metoprolol reduced spontaneously beating of right atria from β_1AR transgenic mice showing inverse agonist activity, carvedilol slightly increase atrial frequency in this experimental conditions (Engelhardt *et al.*, 2001).

In another report, Zhou *et al.* (1999) have evaluated the impact of constitutive activity of β_2AR on cardiac parameters of transgenic mice with 200-fold overexpression of the adrenergic receptor. In this way, single murine cardiac myocytes were isolated from transgenic mice and their nontransgenic littermates and retrogradely perfused using the Langendorff method. By using confocal imaging, the authors compared the Ca^{2+} sparks and spatially resolved Ca^{2+} transients in single ventricular myocytes from transgenic and nontransgenic littermates (Zhou *et al.*, 1999). In addition, whole-cell voltage and clamp techniques were used to record L-type Ca^{2+} currents (ICa) and action potentials, respectively. Overexpression of β_2AR in cardiomyocytes increases the frequency and size of Ca^{2+} sparks when compared with nontransgenic littermates and this enhancement was blocked by treatment with the inverse agonist ICI 118,551 (Zhou *et al.*, 1999).

Agonist-independent activation of β_2ARs was also described in Sf9 cells by using the baculovirus expression systems. A proportional increase in cAMP production with respect to β_2AR expression level was found in this engineered system (Chidiac *et al.*, 1994). Moreover, the authors also compared the ability of different ligands to reduce constitutive activity yielding the following rank order of inverse agonism: timolol > or = propranolol > alprenolol > or = pindolol > labetalol > dichloroisoproterenol (Chidiac *et al.*, 1994).

3.2. Evaluation of inverse agonism of βAR ligands in native tissues

Most importantly, constitutive activity of βARs was also found in native tissues with physiological expression levels of target receptors (Table 3.3). In this regard, Mewes *et al.* (1993) have found that βARs are functionally active in the absence of agonist in isolated guinea pig and human cardiomyocytes. Using the patch-clamp technique in the single electrode mode for measuring whole-cell ICa, it was found that the β_1 selective antagonist atenolol induced a marked reduction of ICa in a concentration-dependent (Mewes *et al.*, 1993). Conversely, ICI 118,551, a β_2-selective antagonist, did not show activity in these experimental conditions. Taking together, these findings describe the existence of basal activity of myocardial β_1ARs but not of the β_2-subtype. A relevant methodological aspects of this study is the use of forskolin, an agent known to sensitize the AC signal transduction system, in order to evidence constitutive activity (Mewes *et al.*, 1993).

Basal activity of βARs has been reveled in membrane preparations of turkey erythrocytes by using a different experimental design. Briefly, the effects of different βARs ligands on native cAMP levels were studied by the hydrolysis-resistant GTP analogs, guanosine 5′-[gamma-thio]triphosphate and guanosine 5′-[β,γ-imino]triphosphate to increase in AC activity was studied in order to describe constitutive activity of βARs (Götze and Jakobs, 1994). As propranolol and pindolol completely prevented stimulation of AC by the GTP analog in a concentration-dependent, it could be suggested that in turkey erythrocyte membranes unoccupied beta-adrenoceptors can cause significant Gs protein and subsequent AC activation (Götze and Jakobs, 1994).

The existence of native constitutive activity of βARs is also supported by findings of functional experiments on isolated cardiac tissues. Varma *et al.* (1999a,b) assessed the negative inotropic effects of different βARs antagonists on electrically stimulated right atria, left atria, right ventricles, and left ventricular papillary muscles from reserpine-treated rats. Rats were pretreated with reserpine in order to reduce contamination of the preparation with endogenous catecholamines (Varma *et al.*, 1999a,b). An interesting aspect of the work of Varma *et al.* is the absence of methodological manipulations of the preparation in order to increase the basal activity and/or expression of myocardial βAR. Several findings are highly attractive, including the existence of a degree of constitutive activity of βAR in different myocardial tissues (Varma *et al.*, 1999a,b). In this regard, negative inotropic effect of βAR antagonists was most marked on the right atria with respect to left atria, right ventricles and left ventricular papillary muscles. On the other hand, the authors also found that the inverse agonist properties differ between βAR ligands establishing the following rank order: Propranolol $\sim$ ICI 118,551 $>$ timolol $\sim$ nadolol $\sim$ alprenolol metoprolol $\sim$ atenolol

Table 3.3 Constitutive activity of β-adrenoceptors in native tissues

Subtype of βAR	Tissue	Measured effect	Inverse agonists	Reference
βAR	Isolated guinea pig Human cardiomyocytes	Patch-clamp technique Whole-cell Ica	(-)-Propranolol ~ atenolol; ICI 118,551 did not show inverse agonist activity	Mewes *et al.* (1993)
βAR	Turkey erythrocytes	Intracellular cAMP levels	Propranolol and pindolol	Götze and Jakobs (1994)
βAR	Electrically stimulated right atria, left atria, right ventricles, and left ventricular papillary muscles	Inotropic response	Propranolol ~ ICI 118,551 > timolol ~ nadolol ~ alprenolol metoprolol ~ atenolol > acebutolol Pindolol acts as neutral antagonist	Varma *et al.* (1999a,b)
β1AR	HEK293 transfected cells	Fluorescent resonance energy transfer	Gly389 variant: bisoprolol, metoprolol, and carvedilol Asp289: carvedilol	Rochais *et al.* (2007)
β1AR	Human ventricular myocardium	Force of contraction	Metoprolol > bisoprolol = nebivolol > carvedilol	Maack *et al.* (2000)
β1AR	Human atrial myocardium	Force of contraction	Metoprolol > bisoprolol = nebivolol = carvedilol > bucindolol	Maack *et al.* (2001a,b)
β1AR	Rat isolated combined atria	Spontaneous beating rate of isolated right atria	Metoprolol, atenolol, and propranolol	Di Verniero *et al.* (2003, 2007, 2008), Höcht *et al.* (2004)

>acebutolol. Conversely, pindolol acts as neutral antagonist and may be an attractive drug for the antagonism of inverse agonist activity of other βAR ligands (Varma *et al.*, 1999a,b).

In an elegant study, Rochais *et al.* (2007) compared the inverse agonism properties of βAR ligands on different variants of β_1-adrenoceptors. The existence of allelic variant of β_1AR, which have an impact on the constitutive activity of the receptor and the sensitivity of the response to beta blocker treatment, is well known. Under native conditions, the Arg389 variant of the β_1AR has been associated with increased cAMP production with regards to the Gly389 variant (Rochais *et al.*, 2007). For the assessment of basal constitutive activity, the authors have developed a new technique based on fluorescent resonance energy transfer (FRET); this methodology allows the direct recording in real time of the conformational changes of the β_1AR protein that lead to its activation (Rochais *et al.*, 2007). More specifically, the degree of activation of β_1AR as assessed in living cells by the measurement of FRET between Cerulean, a mutant cyan fluorescent protein, and a yellow fluorescent protein inserted into the third intracellular loop of the β_1AR (Rochais *et al.*, 2007). Constitutive activity of both β_1AR variants was studied in HEK293 transfected cells using forskolin and the phosphodiesterase inhibitor IBMX. In this experimental conditions, maximal increase in cAMP levels was similar when comparing Arg389 and Gly389 variant transfected cells, suggesting that the two receptor variants showed similar native activity in terms of cAMP production (Rochais *et al.*, 2007).

Nevertheless, the most attractive finding of the work by Rochais *et al.* was the fact that β_1AR polymorphism has a different impact on inverse agonist properties of beta blockers used in the clinical practice for the treatment of heart failure (Rochais *et al.*, 2007). Briefly, bisoprolol, metoprolol, and carvedilol induced an increase in the FRET ratio suggesting inverse agonism activity. While the three blockers exerted varying degrees of inverse agonism on the Gly389 variant, only carvedilol displayed significant inverse agonist effect on the Arg389 variant of β_1AR (Rochais *et al.*, 2007).

Studies of inverse agonism of β blockers in physiological setting have also evidenced that pathophysiological conditions can affect pharmacodynamic properties of these ligands. Maack *et al.* (2000) studied inverse agonist activity in human ventricular myocardium of different commonly used β_1AR blockers. In preparations pretreated with forskolin, the authors established the following rank order of inverse agonist activity in this study: metoprolol > bisoprolol = nebivolol > carvedilol (Maack *et al.*, 2000). In addition, the same authors have demonstrated that desensitization of cardiac β_1AR in heart failure could alter the inverse agonist activity of βAR blockers and also established that β_1AR downregulation may reduce inverse agonist response (Maack *et al.*, 2000).

Hypertensive state and aging also can affect inverse agonist properties of βAR blockers. In isolated atria experiments, we found that the inverse agonist activity of metoprolol is reduced by ageing, but not by the hypertensive stage induced by aortic coarctation (Höcht *et al.*, 2004). More recently, metoprolol negative chronotropic inverse agonist activity was found to be blunted in isolated atria of spontaneously hypertensive rats (Di Verniero *et al.*, 2007). Moreover, a significant correlation between the ventricular weight/body weight ratio and the inverse agonist potency of metoprolol was demonstrated, suggesting a possible link between cardiac hypertrophy and the reduction in the inverse agonist activity of metoprolol (Di Verniero *et al.*, 2007).

4. Clinical Potential Uses of βAR Inverse Agonists

Before any therapeutic relevance is said to be a result of inverse agonism, it has been proposed that at least two criteria should be fulfilled; similarities or correlations between either normal or diseased tissues and the test system need to be found, and other simpler explanations must be excluded (Seifert and Wenzel-Seifert, 2002). It is now necessary to investigate the clinical and/or therapeutic relevance and applicability of inverse agonists.

Certain disease states may be only effectively treated with inverse agonists. There are instances where the pathological entity is a constitutively active GPCR, which produces physiological response in the absence of endogenous agonists. Mutations, which may be preserved in the germ line, have been shown to occur in GPCRs and result in constitutive receptor activity in patients with clinical syndromes.

In the interest of this chapter, there is a striking observation concerning two different polymorphisms in the β_1AR. The first one consists in the substitution of serine by glycine at residue 49 inducing receptor constitutive activity (Mason *et al.*, 1999). The long-term survival of patients with chronic heart failure was associated with the allelic distribution of this polymorphism. For example, whereas patients with the Ser49 genotype showed a mortality rate of 46% at 5 years, those either homozygous or heterozygous for the Gly49 variant showed a mortality rate of only 23% (Börjesson *et al.*, 2000).

The second polymorphism is the amino acid substitution of arginine by glycine at residue 389 (Arg389Gly) (Bruck *et al.*, 2005). It was reported that bucindolol behaved as an inverse agonist in ventricular strip preparations from heart failure patients homozygous for the Arg389 polymorphism, whereas it behaved as a neutral antagonist in Gly389 polymorphism carriers. Interestingly, when patients from the BEST (Beta-Blocker Evaluation of Survival Trial Investigators, 2001) study were stratified and compared for this genotype,

Arg389 homozygous patients treated with bucindolol had fewer numbers of hospitalizations and a higher probability of survival compared with Arg389 homozygous patients treated with placebo, whereas Gly389 carriers showed no improvement despite bucindolol treatment (Ligget *et al.*, 2006).

Hitherto only examples of desirable acute effects of inverse agonists were discussed; however, chronic effects of inverse agonist treatment should also be taken into account. In human heart failure β_1AR density is decreased, and as a consequence, there is a reduction in cardiac βARs functional responsiveness directly related to the severity of HF. Paradoxically, the most successful treatment for decreasing mortality in HF has been the chronic use of βARs blockers. Clinical studies have demonstrated ameliorated ventricular dysfunction in a dose- (Bristow *et al.*, 1996) and time-dependent manner (Hall *et al.*, 1995) in these patients. These effects could be attributed to inhibition of proliferative signaling (Katz *et al.*, 2003; Reiken *et al.*, 2003) and cardiomyocyte apoptosis (Pönicke *et al.*, 2003; Zaugg *et al.*, 2000) induced in the myocardium by compensatory elevated circulating catecholamines. Nevertheless, the beneficial effect of the antagonist is only observed after several weeks or months of βAR blockade, and there is no temporal correlation between the long-term changes in heart contractility and the improvements shown by the different studies. Thus, a compensatory mechanism, indirectly caused by the βARs blockade, appears to be one explanation able to simultaneously support a short-term detrimental effect with a chronic beneficial effect.

Remarkably, in a range of systems, sustained treatment with inverse agonists can produce substantially greater upregulation of receptor levels than antagonists. Several large-scale placebo-controlled clinical trials with carvedilol, metoprolol, and bisoprolol—β blockers with βARs inverse agonist properties—have demonstrated clinically relevant and statistically significant decreases in mortality and the number of hospitalizations in patients with New York Heart Association Class II, III, or IV heart failure (Hjalmarson *et al.*, 2000; CIBIS investigators, 1999; Lechat *et al.*, 1998; Packer *et al.*, 1996). Conversely, bucindolol, another β blocker reported as a neutral antagonist in this sample, did not achieve these endpoints.

When the heart failure analogy was applied to asthma, again only inverse agonists (this time at the β_2ARs) improved the airway hyperresponsiveness to methacholine in a murine model of the disease (Callaerts-Vegh *et al.*, 2004). The outcome of that study has led to a pilot clinical trial using the β inverse agonist nadolol; preliminary results suggest that nadolol may attenuate the hyperresponsiveness to methacholine in mild asthmatics (Hanania *et al.*, 2008).

Inverse agonistic properties of β-blockers may be therapeutically relevant in clinical practice (Parra and Bond, 2007). It is important to mention that excessive activation of cardiac βAR, either in response to agonist stimulation or by constitutive activity, induces deleterious effects, including cardiac hypertrophy, myocyte apoptosis, fibroblast hyperplasia, and arrhythmias (Metra *et al.*, 2004). It was demonstrated that inverse agonism at β_2AR could enhance coupling of this receptor to Gi-proteins, inhibiting hypertrophy and apoptosis of myocardial cells (Xiao *et al.*, 2003). In addition, it was found that constitutive activity of βAR results in pronounced agonist stimulation of these receptors (Milano *et al.*, 1994; Samama *et al.*, 1993). Therefore, inverse agonism at βAR could have beneficial effects in the clinical use of β-blockers.

In addition, excessive activation of cardiac βAR induces desensitization and loss of function of these receptor subtypes aggravating the loss of contractility of the failing heart. Therefore, restoration of β_1AR signaling is a therapeutic approach in the treatment of heart failure (Brodde, 2007). Considering that constitutive activity of GPCR induces receptor downregulation (Leurs *et al.*, 1998), it is expected that inverse agonists exert a greater effect in receptor upregulation with regards to neutral antagonists. It was found that β_1AR blockers with a high inverse agonist activity, such as bisoprolol and metoprolol (Maack *et al.*, 2001a,b), induced an increase in βAR density observed in human heart during long-term treatment with these β_1AR blockers (Brodde *et al.*, 1990; Gilbert *et al.*, 1996; Sigmund *et al.*, 1996). Conversely, carvedilol has been found to exert only weak inverse agonism (Maack *et al.*, 2000), and it did not increase cardiac βAR density in cardiac heart failure.

It is well known that chronic use of long-acting β_2AR agonists increased asthma morbidity and mortality due to receptor overactivation (Bond *et al.*, 2007). More recently, a possible therapeutic role of β_2AR inverse agonists in the treatment of asthma has been proposed. Preliminary studies have demonstrated that chronic treatment with β_2AR inverse agonists improved airway hyperresponsiveness to methacholine in a murine disease model (Callaerts-Vegh *et al.*, 2004). On the other side, the results of Lin *et al.* (2008) suggest that in the murine model of asthma, several compensatory changes associated with either increased bronchodilator signaling or decreased bronchoconstrictive signaling result from the chronic administration of certain β-blockers.

Taking together, inverse agonists seem to be useful in the treatment of chronic disease characterized by harmful effects resulting from β_1AR and β_2AR overactivation, such as heart failure and asthma, respectively.

REFERENCES

Ahmed, M., Muntasir, H. A., Hossain, M., *et al.* (2006). Beta-blockers show inverse agonism to a novel constitutively active mutant of beta1-adrenoceptor. *J. Pharmacol. Sci.* **102,** 167–172.

Arch, J. R., Ainsworth, A. T., Cawthorne, M. A., *et al.* (1984). Atypical β adrenoceptor on brown adipocytes as target for anti obesity drugs. *Nature* **309,** 163–165.

Beta-Blocker Evaluation of Survival Trial Investigators (2001). A trial of the beta-blocker bucindolol in patients with advanced chronic heart failure. *N. Engl. J. Med.* **344,** 1659–1667.

Bindslev, N. (2004). A homotropic two-state model and auto-antagonism. *BMC Pharmacol.* **4,** 11.

Bond, R. A., Leff, P., Johnson, T. D., *et al.* (1995). Physiological effects of inverse agonists in transgenic mice with myocardial overexpression of the beta 2-adrenoceptor. *Nature* **374,** 272–276.

Bond, R. A., Spina, D., Parra, S., and Page, C. P. (2007). Getting to the heart of asthma: Can "β blockers" be useful to treat asthma? *Pharmacol. Ther.* **115,** 360–374.

Börjesson, M., Magnusson, Y., Hjalmarson, A., and Andersson, B. (2000). A novel polymorphism in the gene coding for the beta(1)-adrenergic receptor associated with survival in patients with heart failure. *Eur. Heart J.* **21,** 1853–1858.

Bosier, B., and Hermans, E. (2007). Versatility of gpcr recognition by drugs: From biological implications to therapeutic relevance. *Trends Pharmacol. Sci.* **28,** 438–446.

Bouaboula, M., Perrachon, S., Milligan, L., *et al.* (1997). A selective inverse agonist for central cannabinoid receptor inhibits mitogen-activated protein kinase activation stimulated by insulin or insulin-like growth factor 1. Evidence for a new model of receptor/ligand interactions. *J. Biol. Chem.* **272,** 22330–22339.

Bristow, M. R., Larrabee, P., Minobe, W., *et al.* (1992). Receptor pharmacology of carvedilol in the human heart. *J. Cardiovasc. Pharmacol.* **19**(Suppl. 1), S68–S80.

Bristow, M. R., Gilbert, E. M., Abraham, W. T., *et al.* (1996). Carvedilol produces dose-related improvements in left ventricular function and survival in subjects with chronic heart failure mocha investigators. *Circulation* **94,** 2807–2816.

Brodde, O. E. (2007). β-adrenoceptor blocker treatment and the cardiac β-adrenoceptor-G-protein(s)-adenylyl cyclase system in chronic heart failure. *Naunyn Schmiedebergs Arch. Pharmacol.* **374,** 361–372.

Brodde, O. E., Daul, A., and Michel, M. C. (1990). Subtype-selective modulation of human β_1- and β_2-adrenoceptor function by β-adrenoceptor agonists and antagonists. *Clin. Physiol. Biochem.* **8,** 11–17.

Brown, G. P., and Pasternak, G. W. (1998). 3h-naloxone benzoylhydrazone binding in mor-1-transfected chinese hamster ovary cells: Evidence for G-protein-dependent antagonist binding. *J. Pharmacol. Exp. Ther.* **286,** 376–381.

Bruck, H., Leineweber, K., Temme, T., *et al.* (2005). The $Arg^{389}Gly$ β_1-adrenoceptor polymorphism and catecholamine effects on plasma-renin activity. *J. Am. Coll. Cardiol.* **46,** 2111–2115.

Callaerts-Vegh, Z., Evans, K. L., Dudekula, N., *et al.* (2004). Effects of acute and chronic administration of β-adrenoceptor ligands on airway function in a murine model of asthma. *Proc. Natl. Acad. Sci. USA* **101,** 4948–4953.

Canová, N. K., Lincová, D., Kmonícková, E., Kameníková, L., and Farghali, H. (2006). Nitric oxide production from rat adipocytes is modulated by β_3-adrenergic receptor agonists and is involved in a cyclic AMP-dependent lipolysis in adipocytes. *Nitric Oxide* **14,** 200–211.

Cao, W., Luttrell, L. M., Medvedev, A. V., *et al.* (2000). Direct binding of activated c-Src to the β_3 adrenergic receptor is required for MAP kinase activation. *J. Biol. Chem.* **275,** 38131–38134.

Chesley, A., Lundberg, M. S., Asai, T., *et al.* (2000). The β_2 adrenergic receptor delivers an antiapoptotic signal to cardiac myocytes through Gi-dependent coupling to phosphatidylinositol 3′-kinase. *Circ. Res.* **87,** 1172–1179.

Chidiac, P. (2002). Considerations in the evaluation of inverse agonism and protean agonism at G protein-coupled receptors. *Methods Enzymol.* **343,** 3–16.

Chidiac, P., Hebert, T. E., Valiquette, M., Dennis, M., and Bouvier, M. (1994). Inverse agonist activity of beta-adrenergic antagonists. *Mol. Pharmacol.* **45,** 490–499.

CIBIS investigators (1999). The cardiac insufficiency bisoprolol study II (CIBIS-II): A randomised trial. *Lancet* **353,** 9–13.

Costa, T., and Herz, A. (1989). Antagonists with negative intrinsic activity at delta opioid receptors coupled to GTP-binding proteins. *Proc. Natl. Acad. Sci. USA* **86,** 7321–7325.

Costa, T., Ogino, Y., Munson, P. J., Onaran, H. O., and Rodbard, D. (1992). Drug efficacy at guanine nucleotide-binding regulatory protein-linked receptors: Thermodynamic interpretation of negative antagonism and of receptor activity in the absence of ligand. *Mol. Pharmacol.* **41,** 549–560.

Daaka, Y., Luttrell, L. M., and Lefkowitz, R. J. (1999). Switching of the coupling of the beta2-adrenergic receptor to different G proteins by protein kinase A. *Nature* **390,** 88–91.

De Lean, A., Kilpatrick, B. F., and Caron, M. G. (1982). Dopamine receptor of the porcine anterior pituitary gland. Evidence for two affinity states discriminated by both agonists and antagonists. *Mol. Pharmacol.* **22,** 290–297.

Di Verniero, C., Höcht, C., Opezzo, J. A. W., and Taira, C. A. (2003). In vitro pharmacodynamic properties of beta adrenergic antagonists atenolol and propranolol in rats with aortic coarctation. *Rev. Argent. Cardiol.* **71,** 339–343.

Di Verniero, C., Höcht, C., Opezzo, J. A., and Taira, C. A. (2007). Changes in the in vitro pharmacodynamic properties of metoprolol in atria isolated from spontaneously hypertensive rats. *Clin. Exp. Pharmacol. Physiol.* **34,** 161–165.

Di Verniero, C. A., Silberman, E. A., Mayer, M. A., Opezzo, J. A., Taira, C. A., and Höcht, C. (2008). In vitro and in vivo pharmacodynamic properties of metoprolol in fructose-fed hypertensive rats. *J. Cardiovasc. Pharmacol.* **51,** 532–541.

Dowling, M. R., Willets, J. M., Budd, D. C., *et al.* (2006). A single point mutation (n514y) in the human m3 muscarinic acetylcholine receptor reveals differences in the properties of antagonists: Evidence for differential inverse agonism. *J. Pharmacol. Exp. Ther.* **317,** 1134–1142.

Dzmiri, N. (1999). Regulation of β adrenoceptor signaling in cardiac function and disease. *Pharmacol. Rev.* **51,** 465–501.

Emorine, L. J., Marullo, S., Breindt Sutren, M. M., Patey, G., and Tate, K. (1989). Molecular characterization of the human β_3 adrenergic receptor. *Science* **245,** 1118–1121.

Engelhardt, S., Grimmer, Y., Fan, G. H., and Lohse, M. J. (2001). Constitutive activity of the human beta(1)-adrenergic receptor in beta(1)-receptor transgenic mice. *Mol. Pharmacol.* **60,** 712–717.

Ferro, A., Queen, L. R., Priest, R. M., *et al.* (1999). Activation of nitric oxide synthase by β_2 adrenoceptors in human umbilical vein endothelium in vitro. *Br. J. Pharmacol.* **126,** 1872–1880.

Fitzsimons, C. P., Monczor, F., Fernández, N., Shayo, C., and Davio, C. (2004). Mepyramine, a histamine H1 receptor inverse agonist, binds preferentially to a g protein-coupled form of the receptor and sequesters g protein. *J. Biol. Chem.* **279,** 34431–34439.

Gauthier, C., Tavernier, G., Charpentier, F., Langin, D., and Le Marec, H. (1996). Functional β_3-adrenoceptor in the human heart. *J. Clin. Invest.* **98,** 556–562.

Giembycz, M. A., and Newton, R. (2006). Beyond the dogma: Novel β_2 adrenoceptor signalling in the airways. *Eur. Respir. J.* **27,** 1286–1306.

Gilbert, E. M., Abraham, W. T., Olsen, S., *et al.* (1996). Comparative hemodynamic, left ventricular functional, and antiadrenergic effects of chronic treatment with metoprolol versus carvedilol in the failing heart. *Circulation* **94,** 2817–2825.

Götze, K., and Jakobs, K. H. (1994). Unoccupied beta-adrenoceptor-induced adenylyl cyclase stimulation in turkey erythrocyte membranes. *Eur. J. Pharmacol.* **268,** 151–158.

Hall, D. A. (2000). Modeling the functional effects of allosteric modulators at pharmacological receptors: An extension of the two-state model of receptor activation. *Mol. Pharmacol.* **58,** 1412–1423.

Hall, S. A., Cigarroa, C. G., Marcoux, L., Risser, R. C., Grayburn, P. A., and Eichhorn, E. J. (1995). Time course of improvement in left ventricular function, mass and geometry in patients with congestive heart failure treated with beta-adrenergic blockade. *J. Am. Coll. Cardiol.* **25,** 1154–1161.

Hanania, N. A., Singh, S., El-Wali, R., *et al.* (2008). The safety and effects of the beta-blocker, nadolol, in mild asthma: An open-label pilot study. *Pulm. Pharmacol. Ther.* **21,** 134–141.

Hershberger, R. E., Wynn, J. R., Sundberg, L., and Bristow, M. R. (1990). Mechanism of action of bucindolol in human ventricular myocardium. *J. Cardiovasc. Pharmacol.* **15,** 959–967.

Hjalmarson, A., Goldstein, S., Fagerberg, B., *et al.* (2000). Effects of controlled-release metoprolol on total mortality, hospitalizations, and well-being in patients with heart failure: The metoprolol cr/xl randomized intervention trial in congestive heart failure (merit-hf). Merit-hf study group. *JAMA* **283,** 1295–1302.

Höcht, C., Di Verniero, C., Opezzo, J. A., and Taira, C. A. (2004). Pharmacokinetic-pharmacodynamic properties of metoprolol in chronic aortic coarctated rats. *Naunyn Schmiedebergs Arch. Pharmacol.* **370,** 1–8.

Hoyer, D., and Boddeke, H. W. (1993). Partial agonists, full agonists, antagonists: Dilemmas of definition. *Trends Pharmacol. Sci.* **14,** 270–275.

Iaccarino, G., Cipolletta, E., Fiorillo, A., *et al.* (2002). β_2 Adrenergic receptor gene delivery to the endothelium corrects impaired adrenergic vasorelaxation in hipertensión. *Circulation* **106,** 349–355.

Iaccarino, G., Ciccarelli, M., Sorriento, D., *et al.* (2004). AKT participates in endothelial dysfunction in hypertension. *Circulation* **109,** 2587–2593.

Jenkinson, D. H. (1991). How we describe competitive antagonists: Three questions of usage. *Trends Pharmacol. Sci.* **12,** 53–54.

Johnson, M. (2006). Molecular mechanisms of β_2 adrenergic receptor function, response, and regulation. *J. Allergy Clin. Inmunol.* **117,** 18–24.

Katz, A. M. (2003). Heart failure: A hemodynamic disorder complicated by maladaptive proliferative responses. *J. Cell. Mol. Med.* **7,** 1–10.

Kenakin, T. (1987). Agonists, partial agonists, antagonists, inverse agonists and agonist/antagonists? *Trends Pharmacol. Sci.* **8,** 423–426.

Kenakin, T. (1994). On the definition of efficacy. *Trends Pharmacol. Sci.* **15,** 408–409.

Kenakin, T. (1995). Agonist-receptor efficacy. i: Mechanisms of efficacy and receptor promiscuity. *Trends Pharmacol. Sci.* **16,** 188–192.

Kenakin, T. (2002). Efficacy at G-protein-coupled receptors. *Nat. Rev. Drug Discov.* **1,** 103–110.

Kenakin, T. (2004). Efficacy as a vector: The relative prevalence and paucity of inverse agonism. *Mol. Pharmacol.* **65,** 2–11.

Koch, W. J., Lefkowitz, R. J., and Rockman, H. A. (2000). Functional consequences of altering myocardial adrenergic receptor signaling. *Annu. Rev. Physiol.* **62,** 237–260.

Kumar, N., Robidoux, J., Daniel, W. D., Guzman, G., Floering, L. M., and Collins, S. (2007). Requirement of vimentin filament assembly for β_3 adrenergic receptor activation of ERK/MAP Kinase and lipolysis. *J. Biol. Chem.* **282,** 9244–9250.

Lattion, A., Abuin, L., Nenniger-Tosato, M., and Cotecchia, S. (1999). Constitutively active mutants of the beta1-adrenergic receptor. *FEBS Lett.* **457,** 302–306.

Lechat, P., Packer, M., Chalon, S., Cucherat, M., Arab, T., and Boissel, J. P. (1998). Clinical effects of beta-adrenergic blockade in chronic heart failure: A meta-analysis of double-blind, placebo-controlled, randomized trials. *Circulation* **98,** 1184–1191.

Leff, P. (1995). The two-state model of receptor activation. *Trends Pharmacol. Sci.* **16,** 89–97.

Leurs, R., Smit, M. J., Alewijnse, A. E., and Timmerman, H. (1998). Agonist-independent regulation of constitutively active G-protein-coupled receptors. *Trends Biochem. Sci.* **23,** 418–422.

Liggett, S. B., Freedman, N. J., Schwinn, D. A., and Lefkowitz, R. J. (1993). Structural basis for receptor subtype-specific regulation revealed by a chimeric β_3/β_2 adrenergic receptor. *PNAS* **90,** 3665–3669.

Liggett, S. B., Mialet-Perez, J., Thaneemit-Chen, S., *et al.* (2006). A polymorphism within a conserved beta(1)-adrenergic receptor motif alters cardiac function and beta-blocker response in human heart failure. *Proc. Natl. Acad. Sci. USA* **103,** 11288–11293.

Lin, R., Peng, H., Nguyen, L. P., *et al.* (2008). Changes in β_2-adrenoceptor and other signaling proteins produced by chronic administration of "beta-blockers" in a murine asthma. *Pulm. Pharmacol. Ther.* **21,** 115–124.

Lindquist, J. M., and Rehnmark, S. (1998). Ambient temperature regulation of apoptosis in brown adipose tissue. *J. Biol. Chem.* **273,** 30147–30156.

Lindquist, J. M., Fredriksson, J. M., Rehnmark, S., Cannon, B., and Nedergaard, J. (2000). β_3- and β_1 adrenegic Erk1/2 activation is Src- but not Gi-mediated in brown adipocytes. *J. Biol. Chem.* **275,** 22670–22677.

Maack, C., Cremers, B., Flesh, M., Höper, A., Südkamp, M., and Böhm, M. (2000). Different intrinsic activities of bucindolol, carvedilol and metoprolol in human failing myocardium. *Br. J. Pharmacol.* **130,** 1131–1139.

Maack, C., Tyroller, S., Schnabel, P., Cremers, B., Dabew, E., Südkamp, M., and Böhm, M. (2001a). Characterization of beta(1)-selectivity, adrenoceptor-G(s)-protein interaction and inverse agonism of nebivolol in human myocardium. *Br. J. Pharmacol.* **132,** 1817–1826.

Maack, C., Tyroller, S., Schnabel, P., *et al.* (2001b). Characterization of β_1-selectivity, adrenoceptor–Gs-protein interaction and inverse agonism of nebivolol in human myocardium. *Br. J. Pharmacol.* **132,** 1817–1826.

Maffei, A., Di Pardo, A., Carangi, R., *et al.* (2007). Nebivolol induces nitric oxide release in the heart through inducible nitric oxide synthase activation. *Hypertension* **50,** 652–656.

Marian, A. J. (2006). β-Adrenergic receptors signaling and heart failure in mice, rabbits and human. *J. Mol. Cell. Cardiol.* **41,** 11–13, (Editorial).

Martin, M. W., Smith, M. M., and Harden, T. K. (1984). Modulation of muscarinic cholinergic receptor affinity for antagonists in rat heart. *J. Pharmacol. Exp. Ther.* **230,** 424–430.

Mason, D. A., Moore, J. D., Green, S. A., and Liggett, S. B. (1999). A gain-of-function polymorphism in a G-protein coupling domain of the human beta1-adrenergic receptor. *J. Biol. Chem.* **274,** 12670–12674.

McLoughlin, D. J., and Strange, P. G. (2000). Mechanisms of agonism and inverse agonism at serotonin 5-ht1a receptors. *J. Neurochem.* **74,** 347–357.

Metra, M., Dei Cas, L., di Lenarda, A., and Poole-Wilson, P. (2004). β-blockers in heart failure: Are pharmacological differences clinically important? *Heart Fail. Rev.* **9,** 123–130.

Mewes, T., Dutz, S., Ravens, U., and Jakobs, K. H. (1993). Activation of calcium currents in cardiac myocytes by empty beta-adrenoceptors. *Circulation* **88,** 2916–2922.

Milano, C. A., Allen, L. F., Rockman, H. A., *et al.* (1994). Enhanced myocardial function in transgenic mice overexpressing the β_2-adrenergic receptor. *Science* **264,** 582–586.

Milligan, G., and Bond, R. A. (1997). Inverse agonism and the regulation of receptor number. *Trends Pharmacol. Sci.* **18,** 468–474.

Milligan, G., Bond, R. A., and Lee, M. (1995). Inverse agonism: Pharmacological curiosity or potential therapeutic strategy? *Trends Pharmacol. Sci.* **16,** 10–13.

Molenaar, P., and Parsonage, W. A. (2005). Fundamental considerations of β-adrenoceptor subtypes in human heart failure. *TIPS* **26,** 368–375.

Monczor, F., Fernandez, N., Legnazzi, B. L., Riveiro, M. E., Baldi, A., Shayo, C., and Davio, C. (2003). Tiotidine, a histamine h2 receptor inverse agonist that binds with high affinity to an inactive G-protein-coupled form of the receptor. Experimental support for the cubic ternary complex model. *Mol. Pharmacol.* **64,** 512–520.

Noma, T., Lemaire, A., Naga Prasad, S. V., *et al.* (2007). β-Arrestin-mediated β_1 adrenergic receptor transactivation of the EGFR confers cardioprotection. *J. Clin. Invest.* **117,** 2445–2458.

Onaran, H. O., and Costa, T. (1997). Agonist efficacy and allosteric models of receptor action. *Ann. N. Y. Acad. Sci.* **812,** 98–115.

Ostrom, R. S., and Insel, P. A. (2004). The evolving role of lipid rafts and caveolae in G protein-coupled receptor signaling: Implications for molecular pharmacology. *Br. J. Pharmacol.* **143,** 235–245.

Packer, M., Bristow, M. R., Cohn, J. N., Colucci, W. S., Fowler, M. B., Gilbert, E. M., and Shusterman, N. H. (1996). The effect of carvedilol on morbidity and mortality in patients with chronic heart failure. U.S. carvedilol heart failure study group. *N. Engl. J. Med.* **334,** 1349–1355.

Parra, S., and Bond, R. A. (2007). Inverse agonism: From curiosity to accepted dogma, but is it clinically relevant? *Curr. Opin. Pharmacol.* **7,** 146–150.

Pönicke, K., Heinroth-Hoffmann, I., and Brodde, O. (2003). Role of beta 1- and beta 2-adrenoceptors in hypertrophic and apoptotic effects of noradrenaline and adrenaline in adult rat ventricular cardiomyocytes. *Naunyn Schmiedebergs Arch. Pharmacol.* **367,** 592–599.

Rasmussen, S. G., Jensen, A. D., Liapakis, G., Ghanouni, P., Javitch, J. A., and Gether, U. (1999). Mutation of a highly conserved aspartic acid in the beta2 adrenergic receptor: Constitutive activation, structural instability, and conformational rearrangement of transmembrane segment 6. *Mol. Pharmacol.* **56**(1), 175–184.

Rautureau, Y., Toumaniantz, G., Serpillon, S., Jourdon, P., Trochu, J. N., and Gauthier, C. (2002). β_3-adrenoceptor in rat aorta: Molecular and biochemical characterization and signalling pathway. *Br. J. Pharmacol.* **137,** 153–161.

Reiken, S., Wehrens, X. H. T., Vest, J. A., Barbone, A., Klotz, S., Mancini, D., Burkhoff, D., and Marks, A. R. (2003). Beta-blockers restore calcium release channel function and improve cardiac muscle performance in human heart failure. *Circulation* **107,** 2459–2466.

Robertson, M. J., Dougall, I. G., Harper, D., McKechnie, K. C., and Leff, P. (1994). Agonist–antagonist interactions at angiotensin receptors: Application of a two-state receptor model. *Trends Pharmacol. Sci.* **15,** 364–369.

Robidoux, J., Kumar, N., Daniel, K. W., *et al.* (2006). Maximal β_3 adrenergic regulation of lipolysis involves Src and epidermal growth factor receptor-dependent ERK1/2 activation. *J. Biol. Chem.* **281,** 37794–37802.

Rochais, F., Vilardaga, J. P., Nikolaev, V. O., Bünemann, M., Lohse, M. J., and Engelhardt, S. (2007). Real-time optical recording of beta1-adrenergic receptor activation reveals supersensitivity of the Arg^{389} variant to carvedilol. *J. Clin. Invest.* **117,** 229–235.

Rodríguez-Puertas, R., and Barreda-Gómez, G. (2006). Development of new drugs that act through membrane receptors and involve an action of inverse agonism. *Recent Pat. CNS Drug Discov.* **1,** 207–217.

Rozec, B., and Gauthier, C. (2006). β_3-Adrenoceptors in the cardiovascular system: Putative roles in human pathologies. *Pharmacol. Ther.* **111,** 652–673.

Samama, P., Cotecchia, S., Costa, T., and Lefkowitz, R. J. (1993). A mutation-induced activated state of the beta 2-adrenergic receptor. Extending the ternary complex model. *J. Biol. Chem.* **268,** 4625–4636.

Seifert, R., and Wenzel-Seifert, K. (2002). Constitutive activity of G-protein-coupled receptors: Cause of disease and common property of wild-type receptors. *Naunyn Schmiedebergs Arch. Pharmacol.* **366,** 381–416.

Shimizu, Y., Tanishita, T., Minokoshi, Y., and Shimazu, T. (1997). Activation of mitogen-activated protein kinase by norepinephrine in brown adipocytes from rats. *Endocrinology* **138,** 248–253.

Sigmund, M., Jakob, H., Becker, H., *et al.* (1996). Effects of metoprolol on myocardial β-adrenoceptors and Gi-proteins in patients with congestive heart failure. *Eur. J. Clin. Pharmacol.* **51,** 127–132.

Soeder, K. J., Snedden, S. K., Cao, W., *et al.* (1999). The β_3 adrenergic receptor activates mitogen-activated protein kinase in adipocytes through a Gi-dependent mechanism. *J. Biol. Chem.* **274,** 12017–12022.

Strange, P. G. (2002). Mechanisms of inverse agonism at G-protein-coupled receptors. *Trends Pharmacol. Sci.* **23,** 89–95.

Taira, C. A., Carranza, A., Mayer, M., Di Verniero, C., Opezzo, J. A. W., and Höcht, C. (2008). Therapeutic implications of beta-adrenergic receptor pharmacodynamic properties. *Curr. Clin. Pharmacol.* **3,** 174–184.

Tao, Y. (2008). Constitutive activation of G protein-coupled receptors and diseases: Insights into mechanisms of activation and therapeutics. *Pharmacol. Ther.* **120,** 129–148.

Taylor, M. R., and Bristow, M. R. (2004). The emerging pharmacogenomics of the beta-adrenergic receptors. *Congest. Heart. Fail.* **10,** 281–288.

Varma, D. R., Shen, H., Deng, X. F., Peri, K. G., Chemtob, S., and Mulay, S. (1999a). Inverse agonist activities of β-adrenoceptor antagonists in rat myocardium. *Br. J. Pharmacol.* **127,** 895–902.

Varma, D. R., Shen, H., Deng, X. F., Peri, K. G., Chemtob, S., and Mulay, S. (1999b). Inverse agonist activities of β-adrenoceptor antagonists in rat myocardium. *Br. J. Pharmacol.* **127,** 895–902.

Wade, S. M., Lan, K., Moore, D. J., and Neubig, R. R. (2001). Inverse agonist activity at the alpha(2a)-adrenergic receptor. *Mol. Pharmacol.* **59,** 532–542.

Weiss, J. M., Morgan, P. H., Lutz, M. W., and Kenakin, T. P. (1996a). The cubic ternary complex receptor-occupancy model. iii. Resurrecting efficacy. *J. Theor. Biol.* **181,** 381–397.

Weiss, J. M., Morgan, P. H., Lutz, M. W., and Kenakin, T. P. (1996b). The cubic ternary complex receptor-occupancy model. ii. Understanding apparent affinity. *J. Theor. Biol.* **178,** 169–182.

Weiss, J. M., Morgan, P. H., Lutz, M. W., and Kenakin, T. P. (1996c). The cubic ternary complex receptor-occupancy model. i. Model description. *J. Theor. Biol.* **178,** 151–167.

Westphal, R. S., and Sanders-Bush, E. (1994). Reciprocal binding properties of 5-hydroxytryptamine type 2c receptor agonists and inverse agonists. *Mol. Pharmacol.* **46,** 937–942.

Wilson, J., Lin, H., Fu, D., Javitch, J. A., and Strange, P. G. (2001). Mechanisms of inverse agonism of antipsychotic drugs at the d(2) dopamine receptor: Use of a mutant d(2) dopamine receptor that adopts the activated conformation. *J. Neurochem.* **77,** 493–504.

Xiao, R. P., Zhu, W., Zheng, M., *et al.* (2004). Subtype specific β adrenoceptor signaling pathways in the heart and their potential clinical implications. *TIPS* **25,** 358–365.

Xiao, R. P., Zhu, W., Zheng, M., *et al.* (2006). Subtype specific α_1- and β adrenoceptor signaling in the heart. *TIPS* **27,** 330–337.

Yatani, A., and Brown, A. M. (1989). Rapid β-adrenergic modulation of cardiac calcium channel currents by a fast G protein pathway. *Science* **245,** 71–74.

Yoshikawa, T., Port, J. D., Asano, K., Chidiak, P., Bouvier, M., Dutcher, D., Roden, R. L., Minobe, W., Tremmel, K. D., and Bristow, M. R. (1996). Cardiac adrenergic receptor effects of carvedilol. *Eur. Heart J.* **17**(Suppl. B), 8–16.

Zaugg, M., Xu, W., Lucchinetti, E., Shafiq, S. A., Jamali, N. Z., and Siddiqui, M. A. (2000). Beta-adrenergic receptor subtypes differentially affect apoptosis in adult rat ventricular myocytes. *Circulation* **102,** 344–350.

Zhou, Y. Y., Song, L. S., Lakatta, E. G., Xiao, R. P., and Cheng, H. (1999). Constitutive beta2-adrenergic signalling enhances sarcoplasmic reticulum Ca2+ cycling to augment contraction in mouse heart. *J. Physiol.* **521,** 351–361.

Zhu, W. Z., Zheng, M., Lefkowitz, R. J., Koch, W. J., Kobilka, B., and Xiao, R. P. (2001). Dual modulation of cell survival and cell death by β_2 adrenergic signaling in adult mouse cardiac myocytes. *PNAS* **98,** 1607–1612.

Zhu, W., Zeng, X., Zheng, M., and Xiao, R. P. (2005). The enigma of β_2-adrenergic receptor Gi signaling in the heart: The good, the bad, and the ugly. *Circ. Res.* **97,** 507–509.

CHAPTER FOUR

Techniques for Studying Inverse Agonist Activity of Antidepressants at Recombinant Nonedited 5-HT$_{2C\text{-INI}}$ Receptor and Native Neuronal 5-HT$_{2C}$ Receptors

Mathieu Seimandi,[*,†,‡,§] Joël Bockaert,[*,†,‡,§] *and* Philippe Marin[*,†,‡,§]

Contents

* Centre National de la Recherche Scientifique, Institut de Génomique Fonctionnelle, Montpellier, France
† Institut National de la Santé et de la Recherche Médicale, Montpellier, France
‡ Université Montpellier 1, Montpellier, France
§ Université Montpellier 2, Montpellier, France

Methods in Enzymology, Volume 485
ISSN 0076-6879, DOI: 10.1016/S0076-6879(10)85004-9

Abstract

Serotonin (5-HT)$_{2C}$ receptors play a major role in the regulation of mood, and alteration of their functional status has been implicated in the etiology of affect disorders. Correspondingly, they represent an important target for various antidepressant categories, including tricyclics, tetracyclics, mCPP derivatives, specific serotonin reuptake inhibitors, and agomelatine, which exhibit medium to high affinities for 5-HT$_{2C}$ receptors and behave as antagonists. Antidepressant effects of 5-HT$_{2C}$ antagonists have been attributed to a disinhibition of mesocorticolimbic dopaminergic pathways, which exert a beneficial influence upon mood and cognitive functions altered in depression. However, recent experimental evidence revealed a prominent role of constitutive activity in the tonic inhibitory control of dopaminergic transmission exerted by 5-HT$_{2C}$ receptors in specific brain areas such as the nucleus accumbens. Accordingly, alteration in the constitutive activity of 5-HT$_{2C}$ receptors might participate in the induction of depressed states and drugs with inverse agonist properties should themselves be effective antidepressant agents and, possibly, more active than neutral antagonists. This highlights the relevance of systematically evaluating inverse agonist versus neutral antagonist activities of antidepressants acting at 5-HT$_{2C}$ receptors. Here, we provide a detailed description of a palette of cellular assays exploiting constitutive activity of 5-HT$_{2C}$ receptor expressed in heterologous cells (such as HEK-293 cells) toward Gq-operated signaling or their constitutive association with β-arrestins to evaluate inverse agonist activity of antidepressants. We also describe an approach allowing discrimination between inverse agonist and neutral antagonist activities of antidepressants at *native* constitutively active receptors expressed in cultured cortical neurons, based on previous findings indicating that prolonged treatments with inverse agonists, but not with neutral antagonists, induce functional 5-HT$_{2C}$ receptor-operated Ca^{2+} responses in neurons.

1. Introduction

Among the G-protein-coupled receptors (GPCRs) activated by serotonin (5-hydroxytryptamine, 5-HT), 5-HT$_{2C}$ receptors still raise particular interest in view of their implication in multiple physiological functions, including regulation of monoaminergic transmission, mood, motor behavior, appetite, sleep, and endocrine secretion (Berg *et al.*, 2008a; Bockaert *et al.*, 2006, 2010; Di Giovanni *et al.*, 2006; Giorgetti and Tecott, 2004; Millan, 2005). Alteration of their functional status has also been associated with the development of numerous pathological situations such as anxio-depressive states, schizophrenia, obesity, sleep disorders, and motor dysfunction. Accordingly, they represent important targets for a large number of psychoactive drugs including antidepressants, anxiolytics, "non-classical" antipsychotics, and appetite suppressants.

At the cellular level, 5-HT_{2C} receptors are known to primarily couple to $G\alpha_q$-phospholipase C (PLC) pathway in both heterologous cell lines as well as in authentic cellular context such as epithelial choroid plexus cells, which express the highest receptor density (Bockaert *et al.*, 2010; Marazziti *et al.*, 1999; Millan *et al.*, 2008). In addition, they activate PLA2, possibly via $G\alpha_{13}$ that also recruits a RhoA/PLD pathway (McGrew *et al.*, 2002; Millan *et al.*, 2008). Coupling to $G\alpha_{i3}$ has also been established, but only in recombinant systems (Cussac *et al.*, 2002). The 5-HT_{2C} receptor can activate the extracellular signal-regulated kinase (Erk)1,2 pathway independently of its coupling to heterotrimeric G proteins via a mechanism requiring physical association of receptor with β-arrestins and calmodulin (Labasque *et al.*, 2008). Additional mechanisms such as transactivation of tyrosine kinase receptors can also contribute to engagement of Erk by 5-HT_{2C} receptors, depending on the cell background and the agonist used to stimulate the receptor (Werry *et al.*, 2005, 2008).

The 5-HT_{2C} receptor is, to date, the only GPCR, whose mRNA was found to undergo posttranscriptional adenosine-to-inosine editing. Editing leads to the generation of multiple receptor isoforms (14 in human) with amino acid substitutions within the putative second intracellular loop (at positions 156, 158, and 160), ranging from the unedited (5-$HT_{2C\text{-}INI}$) to the fully edited (5-$HT_{2C\text{-}VGV}$) one and exhibiting different regional distributions (Burns *et al.*, 1997). Differentially edited receptors display different agonist potencies, patterns of G protein activation and ligand functional selectivity (Berg *et al.*, 2008b; Herrick-Davis *et al.*, 1999; Niswender *et al.*, 1999; Price *et al.*, 2001; Werry *et al.*, 2008). They also exhibit different levels of constitutive activity at Gq protein-dependent signaling, ranging from the highest for the nonedited 5-$HT_{2C\text{-}INI}$ receptor to intermediate for partially edited isoforms and negligible for the highly edited forms such as the full-edited one (5-$HT_{2C\text{-}VGV}$) (Herrick-Davis *et al.*, 1999; Niswender *et al.*, 1999). RNA editing also affects recruitment of β-arrestins by 5-HT_{2C} receptor and the ability of variants to spontaneously associate with β-arrestin is strongly correlated with their degree of constitutive activity toward the Gq-PLC effector pathway: the 5-$HT_{2C\text{-}INI}$ receptor binds to β-arrestins in an agonist-independent manner, a process resulting in constitutive receptor internalization and their predominant localization in intracellular compartments (Marion *et al.*, 2004). Moreover, constitutive interaction with β-arrestins is reversed by inverse agonist treatments, which promote receptor redistribution to the plasma membrane. Accordingly, prolonged treatments of cells expressing constitutively active receptors with inverse agonists increase receptor responsiveness, notwithstanding the paradoxical receptor downregulation induced by such treatments. In contrast, the 5-$HT_{2C\text{-}VGV}$ variant, which displays the lowest degree of constitutive activity toward Gq, does not spontaneously associate with β-arrestin, is mainly localized at the cell surface under basal conditions, and only undergoes endocytosis upon agonist stimulation (Marion *et al.*, 2004).

The first evidence of receptor constitutive activity *in vivo* came from microdialysis studies, which revealed a prominent role of constitutive activity in the tonic inhibition by 5-HT_{2C} receptors of mesocorticolimbic dopaminergic neurons and of dopamine release in the nucleus accumbens: SB206,553, a prototypic inverse agonist, was much more efficacious than SB242,084 (a neutral antagonist at PLC signaling) in elevating accumbal dopamine release and the effect of SB206,553 was abolished by concomitant injection of SB242,084 (Aloyo *et al.*, 2009; De Deurwaerdere *et al.*, 2004). Moreover, SB206,553-induced dopamine release was insensitive to the selective lesion of 5-HT neurons in Raphe nucleus by 5,7-dihydroxytryptamine neurotoxin. Inasmuch as mesocorticolimbic dopaminergic transmission exerts a positive influence upon mood, excessive signaling at constitutively active 5-HT_{2C} receptors inhibiting this pathway might participate in the induction of depressed states (Berg *et al.*, 2008c; Millan, 2005). Accordingly, constitutively active 5-HT_{2C} receptors might be an important target for antidepressant pharmacotherapy and 5-HT_{2C} inverse agonists should be more active than neutral antagonists. Although the clinical significance of 5-HT_{2C} inverse agonist concept remains to be established, the demonstration that tetracyclic antidepressants behave as inverse agonists at nonedited 5-$HT_{2C\text{-}INI}$ receptors underscores the need for systematic evaluation of inverse agonist activity of candidate antidepressants exhibiting some affinity for 5-HT_{2C} sites.

Classically, the level of constitutive activity expressed by a GPCR for a given pathway is estimated by measuring the change in basal response induced by expression of the receptor in heterologous cells. A major advantage of using heterologous systems for a first screen of inverse agonist activity is the possibility to assess specificity of drugs for the receptor by using nontransfected cells. We will hereby describe methods to evaluate inverse agonist activity of drugs at recombinant 5-HT_{2C} receptors transiently expressed in human embryonic kidney (HEK)-293 cells. HEK-293 cells are used in many laboratories to investigate GPCR-operated signal transduction, as it is possible to express recombinant receptors at moderate levels (80–300 fmol/mg protein, depending of the receptor), thereby limiting activation of nonphysiological signals. The degree of constitutive activity, and consequently the magnitude of inverse agonist effects, depends on receptor density, the proportion of receptor in the active conformation, and the efficiency of the active receptor conformation to activate the pathway of interest. The degree of constitutive activity is also pathway-dependent. As the 5-HT_{2C} receptor was found to exhibit higher level of constitutive activity toward the Gq-PLC pathway than toward the PLA2 pathway (Aloyo *et al.*, 2009), we will describe here methods to screen for inverse agonist activity of antidepressants toward constitutive receptor coupling to $G\alpha_q$ protein and production of inositol phosphates (IPs) in cells expressing nonedited 5-$HT_{2C\text{-}INI}$ receptor, the variant exhibiting the highest degree of constitutive activity. We will then provide a detailed description of methods exploiting the ability of prolonged treatments with inverse agonists to promote plasma membrane insertion of constitutively internalized 5-$HT_{2C\text{-}INI}$ receptors and, thus, to increase their responsiveness.

The level of constitutive activity of a given GPCR is also strongly dependent on the cell background, underscoring the importance of evaluating constitutive activity (and inverse agonism) in an authentic cellular context, an issue often challenged by the low density of native receptors. We have recently addressed this important question in primary cultures of mice cortical neurons which were found to express a complex profile of differentially edited 5-HT_{2C} receptor mRNAs, the majority of them encoding receptors known to exhibit constitutive activity (Chanrion *et al.*, 2008). 5-HT did not elicit detectable increases in cytosolic Ca^{2+} concentration in cortical neurons, consistent with constitutive desensitization (and internalization) of 5-HT_{2C} receptors expressed in these neurons. 5-HT_{2C} receptor-operated Ca^{2+} responses were only detected after prolonged exposure of neurons to inverse agonists (e.g., SB206,553) but not to neutral antagonists (e.g., SB242,084). Induction of functional responses mediated by 5-HT_{2C} probably was not related to an upregulation of $G\alpha_q$ protein and probably reflected plasma membrane insertion of constitutively internalized receptors. In the last section of this chapter, we will describe optimal procedures for detecting 5-HT_{2C} receptor-mediated Ca^{2+} responses in primary cultured neurons after prolonged treatments with inverse agonist, an approach providing a bridge to studies of inverse agonist activity of drugs *in vivo*.

2. Constitutive Activity Toward the $G\alpha_Q$-PLC Effector Pathway of 5-$HT_{2C\text{-}INI}$ Receptors Transiently Expressed in HEK-293 Cells

In this section, we provide the optimal conditions for detecting constitutive activation of $G\alpha_q$ protein and PLC in HEK-293 cells transiently expressing human 5-$HT_{2C\text{-}INI}$ receptor. Cells are transfected with cDNA encoding 5-$HT_{2C\text{-}INI}$ receptor using electroporation, a method compatible with large-scale studies such as pharmacological studies. A high transfection rate of over 50–60% can be achieved using electroporation, corresponding to receptor density of ~300 fmol/mg of protein (assessed 24 h after transfection), which is similar to that measured in the choroid plexus, the structure expressing the highest density of receptors (Chanrion *et al.*, 2008; Marazziti *et al.*, 1999). Nonetheless, no significant increase in basal IP production was detected in cells transfected with a plasmid encoding 5-$HT_{2C\text{-}INI}$ receptor alone (Fig. 4.1), probably due to an insufficient

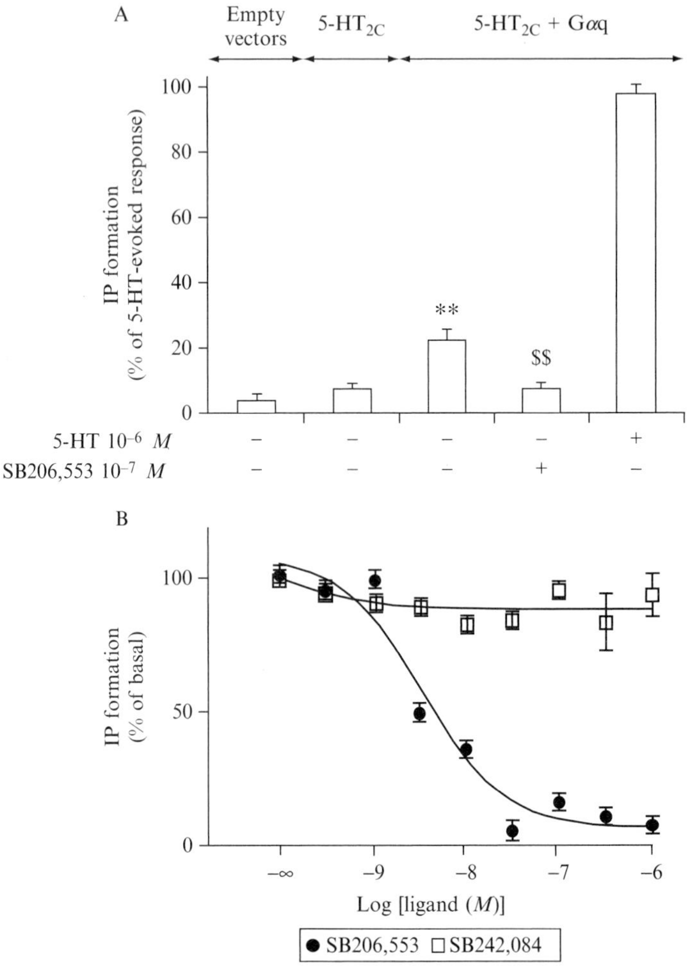

Figure 4.1 Constitutive activity of nonedited 5-HT$_{2C\text{-INI}}$ receptors expressed in HEK-293 cells cotransfected with Gα_q toward PLC effector pathway. (A) HEK-293 cells were transfected with either empty vectors or pRK5/c-Myc-5-HT$_{2C\text{-INI}}$ and/or pRK5/Gα_q vectors. Six hours after transfection, they were labeled with myo-[^{3}H] inositol and IP formation was measured after a 30-min exposure to the indicated treatments in the presence of 10 m*M* LiCl. Data, expressed as the percentage of 5-HT (1 μ*M*)-induced IP formation measured in cotransfected cells, are the mean ± SEM of triplicate determination in a representative experiment. ** $p < 0.01$ versus basal IP formation in cells transfected with empty vector. $$ $p < 0.01$ versus basal IP formation in cells cotransfected with pRK5/c-Myc-5-HT$_{2C\text{-INI}}$ and/or pRK5/Gα_q vectors. (B) HEK-293 cells cotransfected with pRK5/c-Myc-5-HT$_{2C\text{-INI}}$ and pRK5/Gα_q and labeled with myo-[^{3}H]inositol were exposed for 30 min with incremental concentrations of either SB206,553 or SB242,084 in the presence of 10 m*M* LiCl. Data, expressed in % of basal IP formation, are the mean ± SEM of triplicate determination in a representative experiment.

reservoir of endogenous $G\alpha_q$, compared to the overexpressed receptor. Cotransfection with a plasmid encoding $G\alpha_q$ is necessary to enhance constitutive activity up to measurable level. For instance, cotransfecting cells with equal amounts of plasmids encoding 5-$HT_{2C\text{-}INI}$ receptor and $G\alpha_q$ resulted in a eightfold increase in basal IP formation (Chanrion *et al.*, 2008). This basal activity, which represents ~20% of that induced by a maximally effective concentration of 5-HT (1 μM, Fig. 4.1A), is optimal to measure inverse agonist activity (Fig. 4.1B).

2.1. Cell cultures and transfection

2.1.1. Required devices and materials for electroporation

- Biorad Gene Pulser X cell (165-2661, Bio-Rad Laboratories, Hercules, CA)
- Electroporation cuvettes (CE-0004-50, Eurogentec, Seraing, Belgium)
- Electroporation buffer (K_2HPO_4, 50 m*M*; CH_3COOK, 20 m*M*; KOH, 20 m*M*, pH 7.4). The electroporation buffer is stored after filtration 5× concentrated.

Other reagents

- Cells of interest: HEK-293 cells can be obtained from the American type culture collection (ATCC).
- DNAs of interest: plasmids encoding Myc-tagged or Flag-tagged nonedited human 5-HT_{2C} receptor (e.g., pRK5/c-Myc-5-$HT_{2C\text{-}INI}$ or pCMV/Flag-5-$HT_{2C\text{-}INI}$, available on request) and $G\alpha_q$ (e.g., pRK5/$G\alpha_q$, available on request) and carrier DNA (e.g., pRK5 plasmid or PCMV/Flag plasmid). DNA of high quality (e.g., prepared using the plasmid Maxi Kit (12163, Qiagen, Courtaboeuf, France) is highly recommended to obtain optimized transfection rates.
- Growth medium: Grow cells in Dulbecco's modified Eagle's medium (DMEM) supplemented with 10% fetal calf serum (FCS, DE14-801F, Lonza, Verviers, Belgium) and antibiotics (penicillin/streptomycin, 5 IU/ml–5 mg/ml, Invitrogen, Carlsbad, CA). After transfection, replace FCS by *dialyzed* (i.e., 5-HT-free) FCS (DE14-810F, Lonza) in the culture medium. Antibiotics should also be removed at these stages.
- Trypsin 25% solution (15090-046, Invitrogen, Cergy Pontoise, France)
- PBS (Mg^{2+} and Ca^{2+}-free, BE17-715, Lonza)
- Poly-D-Lysine (354510, BD Biosciences)
- 150-mm culture plates (353025, BD Biosciences)
- 96-well culture plates (CLS3599, sigma-Aldrich)
- 1.5 ml Eppendorf tubes (96.7514.9.01, Treff Lab., Degersheim, Switzerland)
- 15 ml Falcon tubes (352096, BD Biosciences)

2.1.2. Cell transfection

One day prior to electroporation, seed HEK-293 cells in 150-mm plates in fresh growth medium so that the cells reach 60–80% confluence on the transfection day.

On the day of transfection, wash cells with PBS and dissociate them using prewarmed trypsin solution (2.5% in PBS). Stop the reaction by adding an equal volume of prewarmed growth medium. Centrifuge dissociated HEK-293 cells at 200×*g* for 5 min at room temperature (RT) to remove cell debris. Count the cells and resuspend them in electroporation buffer (10×10^6 cells/100 μl). Prepare cDNAs in 1.5 ml Eppendorf tube. Classically, each sample contains 1 μg of each coding cDNA and between 8 and 12 μg of carrier cDNA (in 10 μl). Add 190 μl of electroporation solution (5× electroporation buffer, 40 μl; $MgSO_4$ 1 *M*, 8 μl; H_2O 142 μl). Add 100 μl of cell suspension, gently shake up and down, and store samples for about 10 min at RT. For HEK-293 cells, pulse conditions of the electroporation device (Biorad Gene Pulser X cell) are as follows: 260 V and 1000 μF. Transfer samples (cDNAs + cells) in an electroporation cuvette, apply the pulse, and then immediately transfer electroporated cells into a Falcon tube with growth medium supplemented with dialyzed FCS.

Seed electroporated HEK-293 cells in either 150-mm or 96-well culture plates precoated with poly-D-lysine (10 μg/ml, 1 h). We recommended to seed electroporated cells at theoretical densities of 20×10^6 cells/150-mm plates and 0.25×10^6 cells per well in 96-well plates, to reach 60–80% confluence 24 h after seeding.

2.2. Antibody capture/scintillation proximity assay coupled to [^{35}S]GTPγS binding

This section describes an antibody capture assay coupled with a detection technique employing anti-IgG-coated scintillation proximity assay (SPA) beads to measure basal [^{35}S]GTPγS binding to $G\alpha_q$ protein in membranes of HEK-293 cells transiently expressing 5-$HT_{2C\text{-}INI}$ receptors.

2.2.1. Required materials and reagents

- Polytron homogenizer (Kinematica, Liittau-Liucerne, Switzerland)
- [^{35}S]GTPγS (1250 Ci/mmol, NEG030H00, PerkinElmer, Massachussetts)
- Polyclonal anti-$G\alpha_{q/11}$ antibody (sc392, Santa Cruz Biotechnology, Santa Cruz, CA)
- SPA beads coated with secondary antibody (RPNQ0299, GE Healthcare, Saclay, France)
- 96-well optiplates (6005290, PerkinElmer)
- Microplate scintillation counter (e.g., Microbeta, PerkinElmer)

2.2.2. Membrane preparation

Twenty-four hours after transfection, harvest HEK-293 cells grown in 100-mm culture dishes in PBS containing 33 m*M* glucose (PBS-glucose) and centrifuge them at 200×*g* for 5 min to eliminate cell debris. Then resuspend the cell pellet in PBS-glucose, homogenize them with a Polytron homogenizer, and centrifuge the homogenate at 20,000×*g* for 20 min at 4 °C. Resuspend the pellet (membrane fraction) in the assay buffer containing HEPES (20 m*M*, pH 7.4), $MgCl_2$ (50 m*M*), NaCl (150 m*M*), and GDP (0.1 m*M*) at a final protein concentration of 0.5–1 mg/ml.

2.2.3. Measurement of [^{35}S]GTPγS binding to $G\alpha_q$ protein

Incubate membranes (~20–30 μg/well) for 30 min at 37 °C with incremental concentrations of drugs to be tested in 96-well optiplates. We recommend to perform each determination in triplicate and to use SB206,553 (100 n*M*, Sigma-Aldrich, Saint Louis, MO) as a reference inverse agonist in each experiment. If basal response is modified by drugs, renew the experiment in presence of increasing concentrations of SB242,084 (neutral antagonist). Start the reaction by adding 0.2 n*M* [^{35}S]GTPγS to samples (final volume of 200 μl). At the end of the incubation period (60 min at RT), stop the reaction by solubilizing cell membranes with Nonidet P-40 (0.27%, v/v final) and gently shake the plates for an additional 30-min period. Add the rabbit anti-$G\alpha_{q/11}$ polyclonal antibody (1.74 μg/ml final) to each well for 1 h and then SPA beads coated with secondary anti-rabbit antibody (at the concentration recommended by the manufacturer). Incubate the plates for 3 h with gentle agitation and centrifuge them at 1300×*g*. Measure the radioactivity in each sample using a microplate scintillation counter.

2.3. Inositol phosphate production

In this section, we describe a miniaturized assay allowing measurement of constitutive IP formation in HEK-293 cells cultured in 96-well plates.

2.3.1. Required devices and materials

- Myo-[2-^{3}H(N)]-inositol (10–25 Ci/mmol, NET114A00, Perkin Elmer)
- Ion exchange chromatography DOWEX AG 1-X8 resin (140-1454, Bio-Rad)
- $MultiScreen_{HTS}$ Filter Plates (MSHVN 4550, Millipore, Billerica, MA)
- 96-well polyethylene terephtalate (PET) sample plates (1450-401, Perkin Elmer)
- Filtration device (UniVac 3 vacuum manifold, 7705-0102, Whatman International Ltd., Kent, UK)
- Microplate scintillation counter (Microbeta, Perkin Elmer)
- Scintillation reagent (1200-439, Optiphase Supermix, Perkin Elmer)

Other reagents

- HEPES buffer (NaCl, 150 m*M*; HEPES, 20 m*M*, pH 7.4; KCl, 4.2 m*M*; $MgCl_2$, 0.5 m*M*; $CaCl_2$, 1.8 m*M*; glucose, 33 m*M*)
- LiCl (10 56 79, Merck, Darmstadt, Germany)
- Formic acid (20318-297, Merck)
- Ammonium formate (16861, Arcos Organics, Gell, Belgium)
- Triton X-100 (17-1315-01, GE Healthcare)

2.3.2. Inositol phosphate assay

One day prior IP assay, seed HEK-293 cells cotransfected with plasmids encoding $G\alpha_q$ and $5\text{-}HT_{2C\text{-}INI}$ receptors in poly-D-lysine-coated 96-well plates (0.25×10^6 cells/well). Six hours after seeding, label cells overnight with 0.5 μCi/well myo-[^{3}H]inositol in culture medium containing dialyzed FCS. On the day of the measurement, wash the cells three times (10 min each) with HEPES buffer and incubate them in HEPES buffer supplemented with 10 m*M* LiCl (50 μl/well) for 10 min. Add drugs (prepared 2× concentrated in HEPES buffer plus LiCl) to be tested for 30 min (final reaction volume: 100 μl). Again, we recommend to perform each determination in triplicate and to use SB206,553 (100 n*M*) as a reference inverse agonist in each experiment. If basal response is modified by drugs, renew the experiment in presence of increasing concentrations of SB242,084 (neutral antagonist) to assess specificity of the drugs for $5\text{-}HT_{2C}$ receptors. Stop IP generation by replacing the incubation medium by 100 μl formic acid (0.1 *M*).

Deposit 100 μl of a suspension of DOWEX AG 1-X8 resin (250 g/l) in MultiScreen Filter plate wells and wash the resin twice with 200 μl water. Transfer the cell supernatants on the resin. After 5 min, elute nonretained material in a 96-well PET sample plate. Add 100 μl scintillant to each well, shake the plate for 15 min, and count the radioactivity (inositol uptake into cells) in the microplate scintillation counter. Wash the resin three times with 200 μl water and elute [^{3}H]IPs with 100 μl of a solution of 0.7 *M* ammonium formate/0.1 *M* formic acid in a second 96-well PET sample plate. Add 100 μl scintillant to each well, shake the plate for 15 min, and count the radioactivity. Solubilize cell membranes in 100 μl of Triton-X100 10%/NaOH 0.1 *M* (fraction containing [^{3}H]phosphatidyl inositol, PI), transfer solubilized material into a third 96-well PET sample plate, add 100 μl scintillant to each well, shake the plate for 15 min, and count the radioactivity incorporated into cell membranes. Express results as the amount of [^{3}H] IP produced in comparison with either radioactivity present in the Triton-X100 10%/NaOH 0.1 *M*-solubilized membrane fraction or radioactivity present in the first eluted sample ([^{3}H]-inositol uptake into cells).

3. Plasma Membrane Insertion of 5-HT$_{2C\text{-}INI}$ Receptors

This section takes advantage of the ability of inverse agonists to inhibit constitutive association of 5-HT$_{2C\text{-}INI}$ receptor with β-arrestins and thereby to inhibit constitutive receptor internalization (Marion *et al.*, 2004). Consequently, prolonged treatment of 5-HT$_{2C\text{-}INI}$ receptor expressing cells with inverse agonists (e.g., SB206,553 or the tetracyclic antidepressant mirtazapine), but not with neutral antagonist (e.g., SB242,084) induce a progressive insertion of constitutively internalized receptor to the plasma membrane, which can be visualized by immunofluorescence staining of epitope-tagged receptor combined with confocal microscopy and quantified by ELISA under nonpermeabilized conditions (Fig. 4.2) (Chanrion *et al.*, 2008). These assays can evaluate inverse agonist activity of drugs toward a process that does not directly rely on G-protein activation, contrasting with the methods described in the previous section.

3.1. Measurement of 5-HT$_{2C\text{-}INI}$ receptor cell surface expression in HEK-293 cells by ELISA

3.1.1. Devices and reagents for ELISA

- Wallac Victor2 Luminescence counter (Perkin Elmer)
- 96-well optiplates (6005290, PerkinElmer)

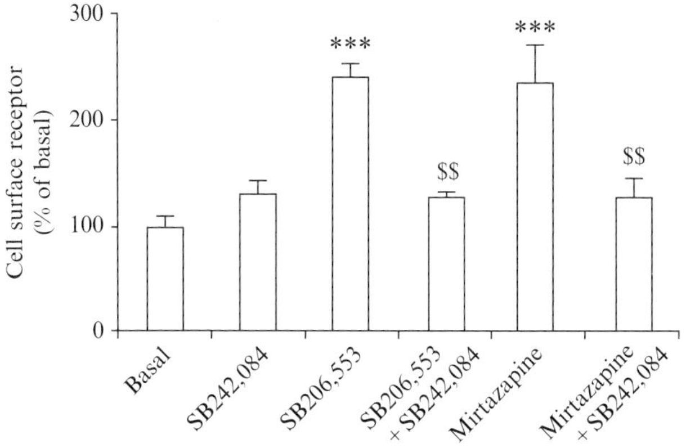

Figure 4.2 Increased cell surface localization of 5-HT$_{2C\text{-}INI}$ receptors expressed in HEK-293 cells following prolonged treatment to inverse agonists, HEK-293 cells transiently transfected with pCMV/Flag-5-HT$_{2C\text{-}INI}$ plasmid were exposed for 18 h to the indicated drugs (all at 1 μ*M*). Cellular surface 5-HT$_{2C\text{-}INI}$ receptors were determined by ELISA. Data, expressed in % of values measured in cells exposed to vehicle (Basal), are the mean ± SEM of triplicate determinations in a representative experiment. *** $p < 0.001$ versus Basal. $$$ $p < 0.001$ versus corresponding treatment in absence of SB242,084.

- Horseradish peroxidase-conjugated anti-Flag monoclonal antibody (1:5000, A8592, Sigma-Aldrich)
- Super signal ELISA Femto (37075, Pierce Biotechnology, Rockford, IL 61105)
- Paraformaldehyde (16% solution, 15710-S, Electron Microscopy Science, Hatfield, PA)

3.1.2. ELISA

One day prior the assay, seed HEK-293 cells transfected with the pCMV/Flag-5-$HT_{2C\text{-}INI}$ plasmid in poly-D-lysine-coated 96-well optiplates (0.25×10^6 cells/well). Cells should reach 80–90% confluence on the day of measurement. Six hours after transfection, add drugs to be tested to the culture medium for 18 h. As mentioned above, we recommend to compare the effect of drugs on receptor trafficking with those induced by prototypical inverse agonist and neutral antagonist (SB206,553 and SB242,084, respectively). At the end of the incubation period, wash cells with PBS and fix them with 4% paraformaldehyde for 20 min at RT. Then, wash them three times with PBS containing 0.1 *M* glycine. Do not permeabilize the cells in order to only quantify receptor cell surface expression. Incubate fixed cells in PBS containing 1% FCS for 30 min at RT and then with a horseradish peroxidase-conjugated anti-Flag monoclonal antibody (1:5000 in PBS containing 1% FCS) for 60 min. After five washes in PBS, add the chromogenic substrate (SuperSignal ELISA Femto) and measure immunoreactivity at 492 nm with a luminescence counter. It is highly recommended to perform control experiments omitting the primary antibody or using cells transfected with empty vectors to evaluate background signals. Remove background from values and normalize them with respect to the total amount of protein in each well.

3.2. Immunofluorescence staining of Flag-tagged 5-$HT_{2C\text{-}INI}$ receptor

3.2.1. Devices and materials for immunofluorescence

- Confocal laser scanning microscope (e.g., LSM 510 META Confocal System, Carl Zeiss MicroImaging GmbH, Göttingen, Germany)
- Glass coverslips (18-mm diameter, 01-115-80, Marienfeld, Lauda-Königshofen, Germany)
- Paraformaldehyde 16% solution (15710-S, Electron Microscopy Sciences)
- Rabbit anti-Flag antibody (F7425, Sigma-Aldrich)
- Cyanine 3-labelled goat anti-rabbit antibody (Invitrogen)
- Goat serum (S-1000, Vector Laboratories, Burlingam, CA)
- Glycine (G7126, Sigma-Aldrich)
- TritonX-100 (17-1315-01, GE Healthcare)
- Mowiol 4.88 (475904, Merck)

3.2.2. Cells preparation prior to immunofluorescence experiment

One day prior to immunofluorescence experiment, seed HEK-293 cells transfected with the pCMV/Flag-5-HT$_{2C\text{-}INI}$ plasmid on poly-D-lysine-coated glass coverslips placed in 12-well plates (0.2 × 10^6 cells/coverslip). As for ELISA, add drugs to be tested to the culture medium containing 10% dialyzed FCS 6 h after transfection.

3.2.3. Immunofluorescence staining and confocal microscopy

At the end of the 18-h incubation period, wash cells with PBS and fix them with 4% paraformaldehyde for 15 min at RT. Wash fixed cells three times with PBS containing 0.1 *M* glycine and permeabilize them with Triton X-100 (0.05% w/v in PBS) for 5 min. After two washes in PBS, incubate cells with PBS containing goat serum (10%, v/v) for 3 h at RT and then overnight at 4 °C with a rabbit anti-Flag antibody (1:1000 in PBS supplemented with 5% goat serum). After three washes in PBS containing 5% goat serum, incubate cells with the cyanine 3-labeled anti-rabbit antibody (1:2000 in PBS containing 5% goat serum). After three washes in PBS, mount dried coverslips on glass slides in Mowiol 4.88. Acquire series of optical sections with a step of 0.40 mm of immunostained cells using a confocal laser scanning microscope with a 63× oil immersion objective and collect images at 1024 × 1024 pixel resolution. Compare receptor distribution in cells exposed to drugs of interest with its localization in cells exposed for 18 h to either a prototypic inverse agonist (e.g., SB206,553) or a neutral antagonist (e.g., SB242,084).

4. Enhanced Responsiveness of Constitutively Active 5-HT$_{2C}$ Receptors Upon Prolonged Treatment with Inverse Agonists

Redistribution of 5-HT$_{2C\text{-}INI}$ receptor expressed in HEK-293 cells to the plasma membrane upon prolonged treatments with inverse agonists (e.g., SB206,553 or mirtazapine) results in a marked enhancement of 5-HT (1 μ*M*)-induced IP production (maximal response), relative to vehicle (Fig. 4.3). The same inverse agonist treatments result in induction of otherwise absent 5-HT$_{2C}$ receptor-operated Ca^{2+} responses in a subset of neurons in primary cortical cultures (Fig. 4.4) (Chanrion *et al.*, 2008). In this section, we will describe methods exploiting this enhanced receptor responsiveness as another index of inverse agonism at recombinant 5-HT$_{2C\text{-}INI}$ receptor expressed in HEK-293 cells and native receptors expressed in cultured cortical neurons.

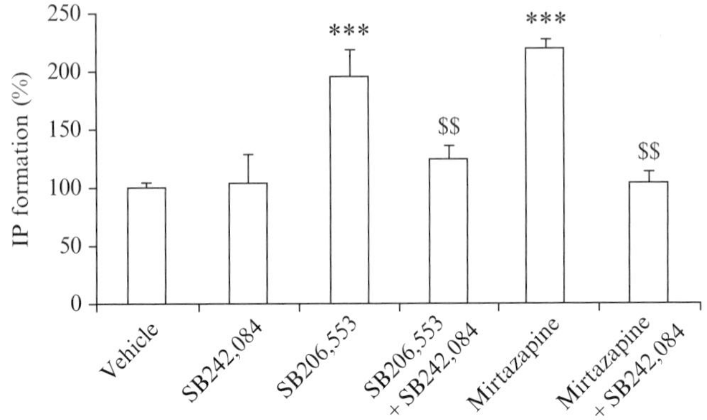

Figure 4.3 Increased responsiveness of 5-HT$_{2C\text{-INI}}$ receptors expressed in HEK-293 cells following prolonged treatment to inverse agonists, HEK-293 cells cotransfected with pRK5/c-Myc-5-HT$_{2C\text{-INI}}$ and pRK5/Gα_q plasmids were labelled with myo-[^{3}H]inositol and exposed to either vehicle, or SB206,553 (1 μM) or mirtazapine (1 μM) in the absence or presence of 1 μM SB242,084 for 18 h. Cells were then subjected to three 10-min washes in HEPES buffer, incubated for an additional 10-min period in HEPES buffer supplemented with 10 mM LiCl and exposed to 5-HT (1 μM) for 30 min in the presence of LiCl. Data, expressed as the percentage of 5-HT-induced IP formation measured in cells pretreated with vehicle, are the mean ± SEM of triplicate determination in a representative experiment. *** $p < 0.001$ versus 5-HT-induced formation in cells pretreated with vehicle. $$$ $p < 0.001$ versus corresponding condition in absence of SB242,084.

4.1. Responsiveness of 5-HT$_{2C\text{-INI}}$ receptor expressed in HEK-293 cells

The method described here is aimed at evaluating the impact of prolonged exposure to antidepressants on the ability of 5-HT (or any other agonist of 5-HT$_{2C}$ receptor) to increase IP production in HEK-293 cells transiently expressing 5-HT$_{2C\text{-INI}}$ receptor.

Required materials and reagents are identical to those described in Section 2.3.1. Six hours after transfection, incubate cotransfected HEK-293 cells with drugs to be tested and 0.5 μCi/well myo-[^{3}H]inositol for 18 h in culture medium containing dialyzed FCS. On the day of IP assay, wash the cells three times (10 min each) with HEPES buffer and incubate them in HEPES buffer supplemented with 10 mM LiCl for 10 min. Then expose cells for 30 min to incremental concentrations of 5-HT (from 1 nM to 1 μM). Stop the reaction and measure [^{3}H]IP formation as indicated in Section 2.3.2.

4.2. Induction of 5-HT$_{2C}$ receptor-operated Ca^{2+} responses in primary cultures of cortical neurons

Prolonged treatment of primary cultures of cortical neurons, which express a majority of constitutively active 5-HT$_{2C}$ receptors, induces 5-HT$_{2C}$ receptor-operated Ca^{2+} responses (which result from engagement of the

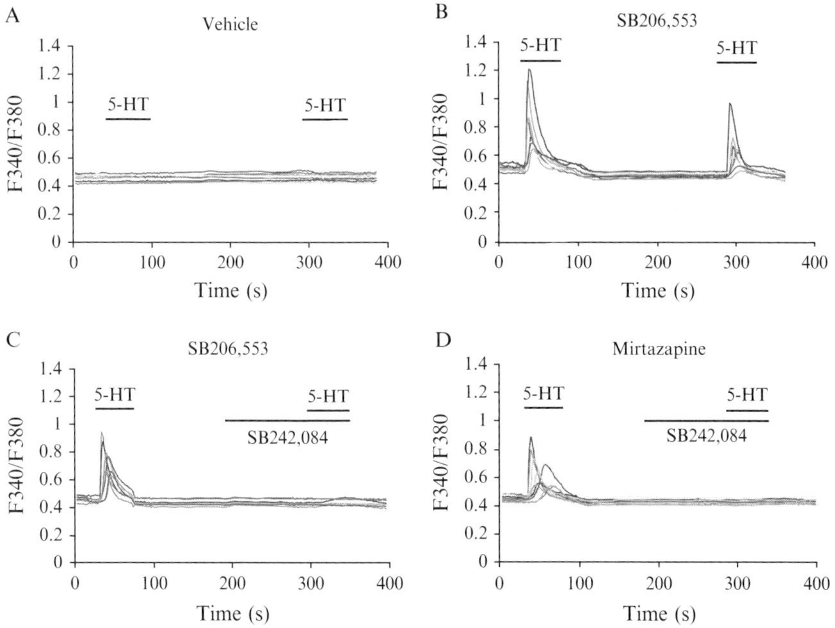

Figure 4.4 Induction of 5-HT$_{2C}$ receptor-operated Ca^{2+} responses in primary cultured cortical neurons upon prolonged exposure to inverse agonists. Cultured cortical neurons grown for 12 days in serum-free medium were exposed for 18 h to either Vehicle (A), or SB206,553 (100 n*M*, B and C) or mirtazapine (100 n*M*, D). Neurons were then exposed to 5-HT (1 μ*M*) for 1 min. To confirm that Ca^{2+} responses were mediated by 5-HT$_{2C}$ receptor, a second 5-HT challenge was performed in absence (A, B) or presence (C, D) of SB242,084 (1 μ*M*) after a 3-min washout. Variations in cytosolic Ca^{2+} concentration measured in a representative field are illustrated. For the clarity of the figures, only Ca^{2+} traces in neurons responding to 5-HT (14.6% and 10.9% following SB206,553 and mirtazapine treatments, respectively) are depicted.

Gq-PLC pathway) only in a subset of the neurons present in the culture (less than 15% of neurons respond to 5-HT following 18-h exposure to 100 n*M* SB206,553) (Chanrion *et al.*, 2008). Therefore, we hereby describe a Fura-2 Ca^{2+} imaging method allowing detection of 5-HT$_{2C}$ receptor-operated responses in *individual* neurons.

4.2.1. Devices and materials for Fura-2 imaging

- Inverted tissue culture microscope equipped for epifluorescence (e.g., IX70 microscope, Olympus France, Rungis) and a 20× water immersion objective
- Illumination system allowing rapid wavelength switching (e.g., DG4 filter wheel, Sutter Instrument Company, Novato, CA)

- CCD camera & acquisition software (e.g., CoolSNAP HQ2 camera, Roper Scientific, Ottobrunn, Germany)
- Fluorescence Ratio Imaging Software (e.g., MetaFluor Imaging System, Molecular Devices, Downingtown, PA)
- MultiChannel Perfusion System (Word Precision Instruments, Sarasota, FL)

Other Reagents

- Fura2 acetoxy methyl (AM) ester, cell permeant, special packaging (20–50 μg, F1221, Invitrogen)
- Pluronic F127 20% solution in DMSO (P-3000MP, Invitrogen)
- HEPES buffer (NaCl, 150 m*M*; HEPES, 20 m*M*, pH 7.4; KCl, 4.2 m*M*; $MgCl_2$, 0.5 m*M*; $CaCl_2$, 1.8 m*M*; glucose, 33 m*M*)

4.2.2. Material and reagents for primary cultures

- Lab-Tek II chambered coverglass Systems (155379, Nalge Nunc International, Napperville, IL)
- Pregnant Swiss mice (16 days pregnancy)
- Culture medium: The culture medium for primary cultures of neurons contains DMEM-F12 (1-1, 32500-035, Invitrogen) supplemented with glucose (33 m*M*), glutamine (2 m*M*), $NaHCO_3$ (13 m*M*), HEPES buffer (5 m*M*, pH 7.4), penicillin–streptomycin (5 IU/ml–5 mg/ml), and a mixture of salt and hormones containing transferrin (100 μg/ml, T1147, Sigma-Aldrich), insulin (25 μg/ml, I6634, Sigma-Aldrich), progesterone (20 n*M*, P8783, Sigma-Aldrich), putrescine (60 n*M*, P5780, Sigma-Aldrich), and Na_2SeO_3 (30 n*M*, S5261, Sigma-Aldrich). Cultures are maintained for 12 days at 37 °C in a humidified atmosphere containing 5% CO_2. At this stage, cultures were shown to contain at least 95% of neurons (Weiss *et al.*, 1986).
- FCS (DE14-801F, Lonza)
- Poly-D-Lysine (354210, BD Biosciences)
- Laminin (1 mg/ml solution, L2020, Sigma-Aldrich)
- PBS-glucose (Ca^{2+} and Mg^{2+}-free PBS supplemented with 33 m*M* glucose)

4.2.3. Preparation of primary cultured neurons

Recover the Lab-Tek II chamber slides with poly-D-lysine (10 μg/ml) for 1 h at 37 °C. After two washes in water, recover the slides with PBS containing 10% FCS and laminin (1 μg/ml) for 2 h at 37 °C. Decapitate mice embryos, remove the brain, and dissect cortices in PBS-glucose. Dissociate cortical cells mechanically using a fire-narrowed Pasteur pipette in PBS-glucose and centrifuge the cell suspension at $200 \times g$ for 5 min to remove cell debris. Resuspend the cell pellet rapidly in serum-free culture medium. After removing the last coating solution, seed cortical cells in Lab-Tek II chambers (1×10^6 cells/well in 2 ml culture medium).

4.2.4. Fura-2 Ca^{2+} imaging

Eighteen hours prior to Ca^{2+} recordings, add drugs to be tested (in comparison with reference inverse agonist and neutral antagonist) directly to the culture medium. Solubilize Fura-2 in Pluronic F127 (50 μg in 50 μl) and dilute solubilized Fura-2 in HEPES buffer (1 μg/ml final concentration). Wash the cells with HEPES buffer and load them with diluted Fura-2 solution for 45 min at 37 °C. We recommend adding drugs to be tested to the loading solution. Wash the cells three times for 10 min with HEPES buffer, mount the chamber on the stage of the microscope, check adequate loading of cells, and start recording of Fura-2 fluorescence (one image/s) at 340 and 380 nm excitation via the 20× water immersion objective for 30 s to establish a stable baseline Ca^{2+} measurement. Then perform two sequential 1-min applications of 5-HT (1 μ*M*, delivered with the superfusion device), separated by 3-min washout, as depicted in Fig. 4.4. This standard challenge interval protocol allows entire intracellular Ca^{2+} store refilling after the first 5-HT application. To assess implication of 5-HT_{2C} receptor in 5-HT-elicited Ca^{2+} increases, challenge cell with SB242,084 during the 3-min washout period before the second 5-HT exposure. Draw Ca^{2+} traces for all neurons responding to the first 5-HT challenge and calculate the percentage of responding neurons in each recorded field.

ACKNOWLEDGMENTS

The research relevant to this chapter was supported by grants from CNRS, INSERM, Servier Pharmaceuticals, l'Agence Nationale de la Recherche, and la Fondation pour la Recherche Médicale (contracts Equipe FRM 2005 and Equipe FRM 2009).

REFERENCES

Aloyo, V. J., Berg, K. A., Spampinato, U., Clarke, W. P., and Harvey, J. A. (2009). Current status of inverse agonism at serotonin2A (5-HT2A) and 5-HT2C receptors. *Pharmacol. Ther.* **121,** 160–173.

Berg, K. A., Clarke, W. P., Cunningham, K. A., and Spampinato, U. (2008a). Fine-tuning serotonin2c receptor function in the brain: Molecular and functional implications. *Neuropharmacology* **55,** 969–976.

Berg, K. A., Dunlop, J., Sanchez, T., Silva, M., and Clarke, W. P. (2008b). A conservative, single-amino acid substitution in the second cytoplasmic domain of the human Serotonin2C receptor alters both ligand-dependent and -independent receptor signaling. *J. Pharmacol. Exp. Ther.* **324,** 1084–1092.

Berg, K. A., Harvey, J. A., Spampinato, U., and Clarke, W. P. (2008c). Physiological and therapeutic relevance of constitutive activity of 5-HT 2A and 5-HT 2C receptors for the treatment of depression. *Prog. Brain Res.* **172,** 287–305.

Bockaert, J., Claeysen, S., Becamel, C., Dumuis, A., and Marin, P. (2006). Neuronal 5-HT metabotropic receptors: Fine-tuning of their structure, signaling, and roles in synaptic modulation. *Cell Tissue Res.* **326,** 553–572.

Bockaert, J., Claeysen, S., Dumuis, A., and Marin, P. (2010). Classification and signaling characteristics of 5-HT receptors. *In* "Handbook of the Behavioural Neurobiology of Serotonin," (C. P. Müller and B. L. Jacobs, eds.), pp. 123–138. Elsevier, New York.

Burns, C. M., Chu, H., Rueter, S. M., Hutchinson, L. K., Canton, H., Sanders-Bush, E., and Emeson, R. B. (1997). Regulation of serotonin-2C receptor G-protein coupling by RNA editing. *Nature* **387,** 303–308.

Chanrion, B., Mannoury la Cour, C., Gavarini, S., Seimandi, M., Vincent, L., Pujol, J. F., Bockaert, J., Marin, P., and Millan, M. J. (2008). Inverse agonist and neutral antagonist actions of antidepressants at recombinant and native 5-hydroxytryptamine2C receptors: Differential modulation of cell surface expression and signal transduction. *Mol. Pharmacol.* **73,** 748–757.

Cussac, D., Newman-Tancredi, A., Duqueyroix, D., Pasteau, V., and Millan, M. J. (2002). Differential activation of Gq/11 and Gi(3) proteins at 5-hydroxytryptamine(2C) receptors revealed by antibody capture assays: Influence of receptor reserve and relationship to agonist-directed trafficking. *Mol. Pharmacol.* **62,** 578–589.

De Deurwaerdere, P., Navailles, S., Berg, K. A., Clarke, W. P., and Spampinato, U. (2004). Constitutive activity of the serotonin2C receptor inhibits in vivo dopamine release in the rat striatum and nucleus accumbens. *J. Neurosci.* **24,** 3235–3241.

Di Giovanni, G., Di Matteo, V., Pierucci, M., Benigno, A., and Esposito, E. (2006). Central serotonin2C receptor: From physiology to pathology. *Curr. Top. Med. Chem.* **6,** 1909–1925.

Giorgetti, M., and Tecott, L. H. (2004). Contributions of 5-HT(2C) receptors to multiple actions of central serotonin systems. *Eur. J. Pharmacol.* **488,** 1–9.

Herrick-Davis, K., Grinde, E., and Niswender, C. M. (1999). Serotonin 5-HT2C receptor RNA editing alters receptor basal activity: Implications for serotonergic signal transduction. *J. Neurochem.* **73,** 1711–1717.

Labasque, M., Reiter, E., Becamel, C., Bockaert, J., and Marin, P. (2008). Physical interaction of calmodulin with the 5-hydroxytryptamine2C receptor C-terminus is essential for G protein-independent, arrestin-dependent receptor signaling. *Mol. Biol. Cell* **19,** 4640–4650.

Marazziti, D., Rossi, A., Giannaccini, G., Zavaglia, K. M., Dell'Osso, L., Lucacchini, A., and Cassano, G. B. (1999). Distribution and characterization of [3H]mesulergine binding in human brain postmortem. *Eur. Neuropsychopharmacol.* **10,** 21–26.

Marion, S., Weiner, D. M., and Caron, M. G. (2004). RNA editing induces variation in desensitization and trafficking of 5-hydroxytryptamine 2c receptor isoforms. *J. Biol. Chem.* **279,** 2945–2954.

McGrew, L., Chang, M. S., and Sanders-Bush, E. (2002). Phospholipase D activation by endogenous 5-hydroxytryptamine 2C receptors is mediated by Galpha13 and pertussis toxin-insensitive Gbetagamma subunits. *Mol. Pharmacol.* **62,** 1339–1343.

Millan, M. J. (2005). Serotonin 5-HT2C receptors as a target for the treatment of depressive and anxious states: Focus on novel therapeutic strategies. *Therapie* **60,** 441–460.

Millan, M. J., Marin, P., Bockaert, J., and la Cour, C. M. (2008). Signaling at G-protein-coupled serotonin receptors: Recent advances and future research directions. *Trends Pharmacol. Sci.* **29,** 454–464.

Niswender, C. M., Copeland, S. C., Herrick-Davis, K., Emeson, R. B., and Sanders-Bush, E. (1999). RNA editing of the human serotonin 5-hydroxytryptamine 2C receptor silences constitutive activity. *J. Biol. Chem.* **274,** 9472–9478.

Price, R. D., Weiner, D. M., Chang, M. S., and Sanders-Bush, E. (2001). RNA editing of the human serotonin 5-HT2C receptor alters receptor-mediated activation of G13 protein. *J. Biol. Chem.* **276,** 44663–44668.

Weiss, S., Pin, J. P., Sebben, M., Kemp, D. E., Sladeczek, F., Gabrion, J., and Bockaert, J. (1986). Synaptogenesis of cultured striatal neurons in serum-free medium: A morphological and biochemical study. *Proc. Natl. Acad. Sci. USA* **83,** 2238–2242.

Werry, T. D., Gregory, K. J., Sexton, P. M., and Christopoulos, A. (2005). Characterization of serotonin 5-HT2C receptor signaling to extracellular signal-regulated kinases 1 and 2. *J. Neurochem.* **93,** 1603–1615.

Werry, T. D., Loiacono, R., Sexton, P. M., and Christopoulos, A. (2008). RNA editing of the serotonin 5HT2C receptor and its effects on cell signalling, pharmacology and brain function. *Pharmacol. Ther.* **119,** 7–23.

CHAPTER FIVE

Differential Inverse Agonism at the Human Muscarinic M_3 Receptor

Paola Casarosa,* Tobias Kiechle,* *and* Remko A. Bakker†

Contents

Abstract

Human muscarinic M_3 receptors (hM_3Rs) induce smooth muscle contraction and mucus gland secretion in response to parasympathetic stimulation. As a consequence of hM_3R function, muscarinic antagonists have wide therapeutic use to treat overactive bladder, abdominal pain (irritable bowel syndrome), and chronic obstructive pulmonary disease (COPD).

In this chapter, we describe the set up and results obtained with different *in vitro* assays to monitor hM_3R activation (agonist-dependent and constitutive) and evaluate functional potencies of different anticholinergics in CHO cells. Given the G_q coupling of hM_3R, assays measuring the second messengers inositol phosphates (InsP) and an AP-1-driven reporter luciferase were developed. In our hands, the reporter gene assay shows advantages: firstly, thanks to the longer incubation times, it allows reaching of pseudo-equilibrium also for ligands with slower receptor dissociation kinetics (e.g., tiotropium). Secondly, the AP-1-driven luciferase detects significant constitutive activity of the hM_3R, which allows characterizing the different anticholinergics for their inverse agonist properties. Given the potential for inverse agonists to cause changes in receptor expression, monitoring hM_3R upregulation is another important

* Respiratory Diseases Research, Boehringer Ingelheim Pharma GmbH & Co. KG, Birkendorferstrasse 65, Biberach an der Riss, Germany

† CardioMetabolic Diseases Research, Boehringer Ingelheim Pharma GmbH & Co. KG, Birkendorferstrasse 65, Biberach an der Riss, Germany

Methods in Enzymology, Volume 485

ISSN 0076-6879, DOI: 10.1016/S0076-6879(10)85005-0

pharmacological parameter. Here, we describe how to measure the effect of chronic exposure to anticholinergics on the expression levels of hM_3R, with particular attention to ensure full antagonist removal from receptor pool before hM_3R quantification.

Taken together, our results indicate that anticholinergics exhibit differential pharmacological behaviors, which are dependent on the pathway investigated, and therefore provide evidence that the molecular mechanism of inverse agonism is likely to be more complex than the stabilization of a single inactive receptor conformation.

1. Introduction to Muscarinic Receptors

The muscarinic receptor (MR) family, which consists in five receptor subtypes (M_1–M_5Rs), belongs to class A rhodopsin-like GPCRs (Wess *et al.*, 2007). The physiological ligand for MRs is acetylcholine (ACh), which exerts the neurotransmitter functions of parasympathetic system. The even-numbered M_2 and M_4 receptors preferentially couple to $G\alpha_{i/o}$ (Fig. 5.1A) and thus inhibit adenylyl cyclase activity as well as prolong potassium channel, nonselective cation channel, and transient receptor potential (TRP) channels opening. Both subtypes are distributed in the central nervous system regulating ACh release from nerve endings as well as in the periphery where they are responsible for heart-rate control and are involved in smooth muscle activity by counteracting the β_2 adrenoceptor-mediated relaxant pathway.

The uneven-numbered M_1, M_3, and M_5 receptors couple instead to the $G\alpha_{q/11}$ pathway (Fig. 5.1B), resulting in activation of phospholipase C (PLC) which in turn hydrolyses phosphoinositides to generate the second messengers inositol 1,4,5-trisphosphate (IP_3) and 1,2-diacylglycerol (DAG) and thus regulates several cellular signaling pathways and functions. M_1R, the most abundant MR in the central nervous system, is implicated in learning and memory processes. In the periphery, it affects the release of catecholamines from adrenal glands and is involved in muscarinic agonist-induced salivary flow. M_5R, the only muscarinic subtype expressed in dopamine containing neurons of the substantia nigra pars compacta and ventral tegmental area, is suggested to facilitate striatal dopamine release. Furthermore, it is expressed in various cerebral and peripheral blood vessels, being responsible for acetylcholine-mediated vasodilator responses in cerebral arteries but not in the periphery.

M_3 receptors are distributed throughout the CNS, where they are involved in modulation of neurotransmitter release, temperature homeostasis, and food intake. In the periphery, they induce smooth muscle contraction and mucus gland secretion in response to parasympathetic stimulation (see paragraph beneath).

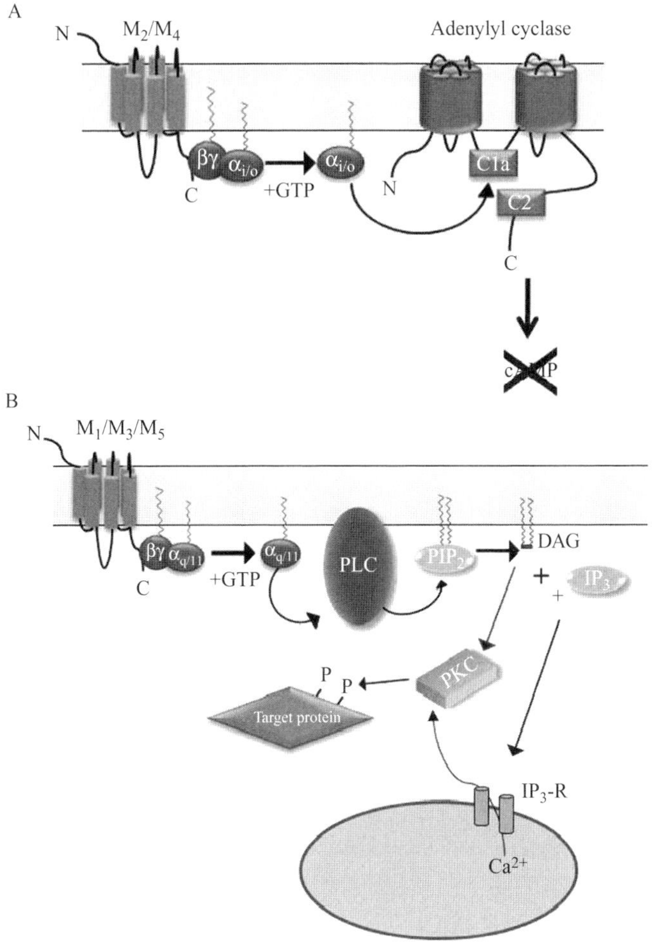

Figure 5.1 Subclassification of muscarinic receptors based on differential G-protein-coupling properties. The M_2 and M_4 muscarinic acetylcholine receptors selectively activate $G_{i/0}$-type G-proteins (A), whereas the M_1, M_3, and M_5 receptor subtypes preferentially couple to G-proteins of the $G_{q/11}$ family (B). In this simplified scheme, only some of the key downstream effector proteins are indicated. For detailed explanation on the different pathways, see the text.

All the above-mentioned effects are based on parasympathetic release of neuronal ACh acting on MRs expressed postsynaptically. However, expression of the cholinergic locus has been reported for several nonneuronal tissues, such as epithelial and endothelial cells, smooth muscle cells, lymphocytes, macrophages, mast cells, eosinophils and neutrophils, implicating also para- and autocrine functions of ACh (Kummer *et al.*, 2008).

2. Role of M_3R in Regulating Smooth Muscle Function and Clinical Use of Anticholinergics

Smooth muscle expresses several MR subtypes, the most important being the M_3 subtype from a functional point of view. At the gastrointestinal level, for example, M_3Rs are localized on smooth muscle of the myenteric plexus and on longitudinal and circular muscle, where they cause contraction in response to ACh released by parasympathetic efferent nerves (Peretto *et al.*, 2009).

Similarly, stimulation of the vagus nerve causes bronchoconstriction and increase in mucus production due to activation of the M_3R on airway smooth muscle and goblet cells, respectively (Barnes, 2004). Also bladder contractility is mainly controlled by the cholinergic input from the parasympathetic nervous system with a predominant role for the M_3 subtype (Peretto *et al.*, 2009).

From a molecular point of view (cf. Fig. 5.1B), hM_3R-induced activation of PLC generates IP_3, which binds a receptor on the endoplasmatic reticulum (ER) to elicit influx of Ca^{2+} from ER stores into the cytosol. Ca^{2+} in smooth muscle cells binds the protein calmodulin to activate the myosin regulatory light chain kinase (MLCK). Phosphorylation of myosin light chain 20 by MLCK facilitates actin/myosin interaction and activation of myosin ATPase, cross-bridge cycling, and muscle contraction (Johnson and Druey, 2002). The activation of PKC by DAG is an additional result of G_q stimulation. Unphosphorylated calponin binds to and prevents activation of the myosin ATPase. PKC-dependent calponin phosphorylation prevents calponin from inhibiting myosin ATPase, suggesting another mechanism by which activation of $G\alpha_q$ might result in smooth muscle contraction.

As a consequence of hM_3R function on smooth muscle, muscarinic antagonists have wide therapeutic use to treat overactive bladder, abdominal pain (irritable bowel syndrome), and chronic obstructive pulmonary disease (COPD).

COPD is characterized by progressive airflow limitation due to persistent inflammatory processes in the airways; the main causal factor in the pathogenesis of COPD is sustained inhalation of cigarette smoke. In COPD, airflow limitation is mainly a result of hyperplasia of mucosal glands, hypertrophy, and in particular constriction of the bronchial smooth muscle in the small airways. An increased cholinergic tone mediates different pathophysiological features of COPD, such as bronchoconstriction and mucus hypersecretion, mostly through activation of the human muscarinic M_3 receptor subtype (Barnes, 2004). Thus, anticholinergic bronchodilators have a particular value in the treatment of COPD because they block the effects of an increased vagal cholinergic tone. Historically, smoke of dried leaves from *Datura stramonium* or *Atropa belladonna* containing antimuscarinics to treat bronchoconstriction can be traced back to antiquity (Gandevia, 1975). However, these compounds easily cross the blood–brain barrier and cause potent hallucinogenic effects. To avoid these severe side effects, several quaternary antimuscarinics were developed by the Pharmaceutical Industry for clinical use. Quaternary compounds are poorly absorbed from mucosal surfaces and do not penetrate the blood–brain barrier to any relevant extent. These compounds are used for the treatment of abdominal pain (e.g., *N*-butyl scopolamine) or COPD (e.g., ipratropium). Tiotropium, introduced in the recent years for the treatment of COPD, is the first example of a long-acting muscarinic antagonist (LAMA): following a single inhaled dose of tiotropium, clinically relevant improvement in lung function lasts for more than 24 h, allowing once-daily dosing (Vincken *et al.*, 2002). The mechanism behind its long duration of action relates to the long kinetics of dissociation from its target, the human M_3 muscarinic receptor (Casarosa *et al.*, 2009; Disse *et al.*, 1999). Long duration of action (preferably 24 h) is an important feature of drugs intended to treat chronic diseases, enabling both prolonged efficacy and a simple, once-daily dosage regime that improves patient compliance.

3. *In Vitro* Assays to Monitor hM_3R Activation and Constitutive Activity

3.1. Inositol phosphate assay

The hM_3R couples to $G\alpha_q$ proteins, resulting in activation of PLC, which can be monitored by measuring changes in the intracellular concentrations of InsPs (Fig. 5.2). Using the measurement of second messengers (here InsPs) to analyze receptor activity offers several advantages over other measurements: firstly, the assay duration of approximately 1 h allows for most compounds to reach pseudo-equilibrium, which means a correct determination of pA_2 values for antagonists (advantage vs., e.g., calcium, a nonequilibrium assay); secondly, the generation of second messengers is directly under the control of G-protein activation, and therefore we do not assist to strong signal amplification (as, e.g., in the case of reporter genes, which usually imply high levels of spare receptors: this can result in overestimation of agonist potency and missing of noncompetitive behavior). Additionally, with the measurement of second messengers, the chances of interference between different pathways are reduced to the minimum, allowing for "clean" analysis of one pathway versus the others (great advantage for GPCRs that show promiscuous coupling, e.g., the human M_2R, see Casarosa *et al.*, 2009).

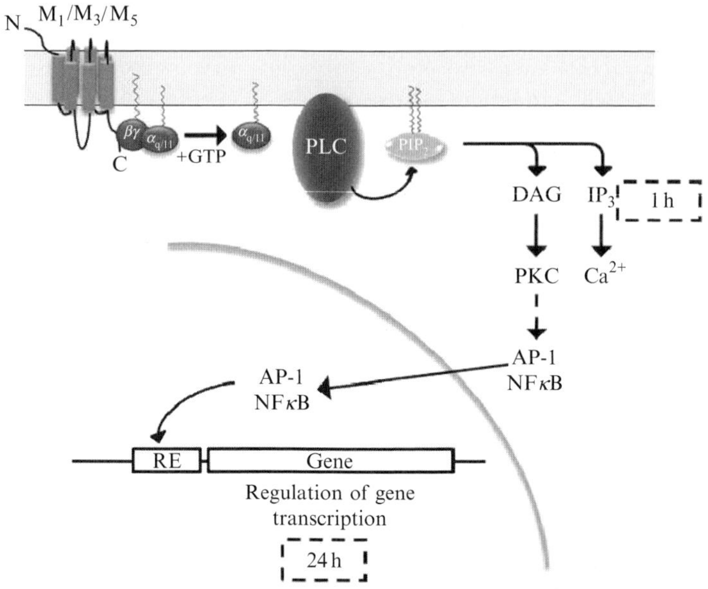

Figure 5.2 Different assays for detection of hM_3R activity. Following phospholipase C (PLC) activation, the phospholipid PIP_2 is hydrolyzed in diacylglycerol (which activates protein kinase C, PKC) and IP_3, which induces calcium release from intracellular stores. To monitor hM_3R activity, different assays are available: measurement of the intracellular second messenger inositol phosphate, or reporter genes under the control of enhancer elements that are responsive to PKC phosphorylation (e.g., NF-κ B and AP-1). The main difference between these two methodologies consists in the different incubation times, that is, 1 versus 24 h.

Historically, InsP assays are performed by first allowing the cells to incorporate tritiated inositol in their cell membranes, then, following receptor activation (from few minutes up to a couple of hours), the produced InsPs are recovered through chromatography (Casarosa *et al.*, 2001). Major limitations of such methodology relate to the presence of radioactive waste, the time-consuming separation phase, and the difficulty to miniaturize the assay format. In recent years, a novel assay has been introduced based on homogeneous time-resolved fluorescence (HTRF technology, CISbio International, France). This is an immunoassay based on the competition between the endogenous InsP and an InsP-D2 dye, which is added to each sample, for binding to an anti-InsP-mAb labeled with Europium (Eu) cryptate. In the absence of endogenous InsPs, the IP-D2 dye generates a complex with the Eu antibody, resulting in a FRET signal (665 nm). Increasing concentrations of endogenous InsPs will progressively decrease the signal. Advantages of such a system include the fact that the assay is homogeneous, prone to high throughput and little time-consuming.

On the downside, costs are significantly higher and sensitivity is possibly lower (also due to the microplate format) when compared to the classical radioactive assay.

In our laboratories, due to the need to test high numbers of compounds and strict guidelines for limiting radioactive waste, we have opted for the FRET assay format. A CHO cell line heterologously expressing human M_3Rs (B_{max} = 2.97 ± 0.03 pmol/mg; Casarosa *et al.*, 2009) was used and assay optimization consisted in testing different cell amounts, either adherent or in suspension, and changing stimulation times. Finally, optimal conditions were met with the following protocol: cells were plated in white 384 well plates (20,000 cells/well), and 1 day after the assay was performed. Cells were stimulated for 1 h at 37 °C. Under these conditions, the signal-to-noise ratio, the read-out window, the stability and reproducibility of the assay were excellent (see, e.g., Fig. 5.3A).

To validate the assay, pA_2 values and Schild plots were determined for several muscarinic antagonists. In practical terms, to construct a Schild plot, the dose-response curve for an agonist is determined in the presence of various concentrations of a competitive antagonist (Fig. 5.3A). Then the shifts of agonist EC_{50} induced by different concentrations of antagonist are plotted (Fig. 5.3B). From this type of experiment, two important values are derived: first, the pA_2 value, which is a measure of affinity of the antagonist for its receptor (i.e., the equilibrium dissociation constant); second, the slope of the regression fit, which by definition needs to be 1, in case the antagonist is competitive. pA_2 values were determined for each antagonist against the agonist carbachol (Table 5.1). Most compounds tested showed a competitive and surmountable antagonistic behavior (Fig. 5.3), and induced a shift of the agonist dose-response curve which could be perfectly fitted with the Gaddum-Schild equation, with a slope not significantly different from 1 (Casarosa *et al.*, 2009). A different profile was seen for compounds with slower kinetics of hM_3 receptor dissociation, the so-called LAMAs (see chapter 2). For these drugs (e.g., tiotropium), the assay incubation time of 1 h is not sufficient to reach equilibrium binding, which results in deviation of the Schild slope from 1 (see Fig. 5.6B). For this reason, we developed reporter gene assays to monitor antagonism at the hM_3R, as described in detail in the next paragraph.

Incubating the CHO-hM_3 cells with antagonists in the absence of any added agonist did not affect the basal signal in the InsP assay, that is, no constitutive activity could be detected with this method. However, in previous studies published by us, the InsP assay proved a valid tool to detect the constitutive activity of other GPCRs, like human histamine H_1R (Bakker *et al.*, 2000), as well as viral encoded receptors like US28 and ORF74 (Casarosa *et al.*, 2001). It is possible that the human M_3R possesses a lower propensity for constitutive activity in comparison to these other GPCRs; more likely explanations relate to the change in methodology from

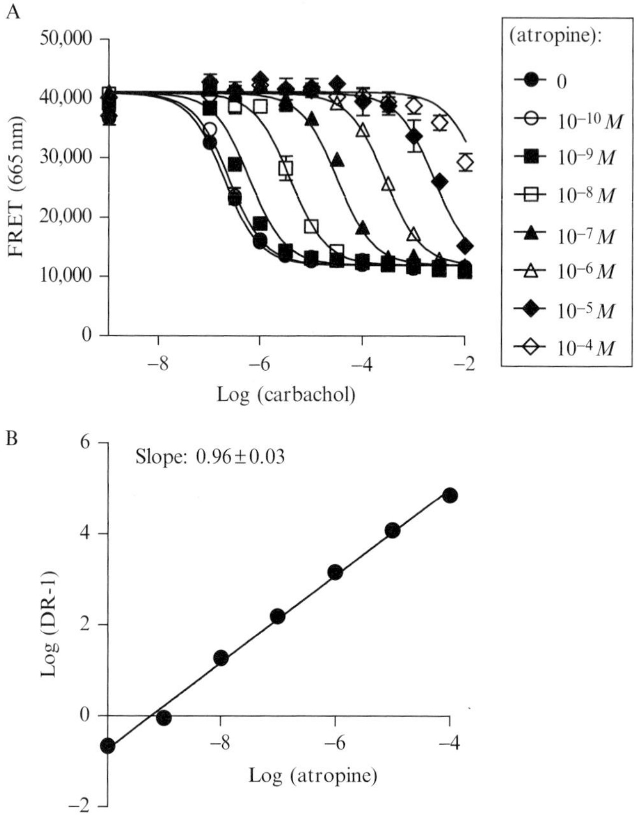

Figure 5.3 Validation of the InsP assay as a tool to detect hM_3R activity and functional antagonism in CHO-hM_3 cells. (A) Atropine induces a concentration-dependent rightward shift (concentrations tested: from 10^{-10} *M* to 10^{-4} *M*) in the agonist curve (here shown: carbachol), measured as inositol phosphates accumulation (FRET signal at 665 nm) in CHO-hM_3 cells. (B) Schild regression of data shown in (A) was calculated.

the radioactive assay (usually performed in 24-well plates) to the proximity assay (performed in the 384-well plate): obviously, the less cells are tested the smaller the chance to detect constitutive activity. To this end, a previous publication from Jakubik and colleagues showed a small but statistically significant inhibition of basal InsP production (measured with the radioactive method) in the presence of the antagonist NMS in cells stably expressing the hM_3R (Jakubik *et al.*, 1995).

In general, however, the measurement of InsP production is not the ideal type of assay to detect constitutive activity, since it is limited in its receptor stimulation times to maximally 1–2 h, as all the read-outs using intracellular second messengers (e.g., cAMP). The reason for this is that in

Table 5.1 Comparison of functional properties of the anticholinergics at the hM_3R in different assays

	InsP pA_2	Reporter-gene pA_2	Reporter-gene pIC_{50}
Atropine	9.21 ± 0.05	9.01 ± 0.08	9.14 ± 0.10
NMS	9.58 ± 0.09	9.83 ± 0.04	9.35 ± 0.05[a]
Pirenzepine	6.67 ± 0.09	7.09 ± 0.06	6.80 ± 0.10
Ipratropium	9.21 ± 0.07	9.25 ± 0.03	9.10 ± 0.07
Tiotropium	10.68 ± 0.10	10.52 ± 0.09	9.61 ± 0.06[a]
4-DAMP	N.D.	9.23 ± 0.10	9.28 ± 0.06

N.D.: not determined. pA_2 values of the different antagonists were determined in CHO-hM_3 cells with the inositol phosphate accumulation (InsP) and the AP-1-driven luciferase reporter gene assays. pA_2 values were experimentally obtained, according to Schild analysis, as shown in Figs. 5.3 and 5.6. Values shown are the average of at least three independent experiments ± SEM, with each point determined in triplicate. The third column of the table reports the inverse agonistic potencies (pIC_{50} ± SEM) of the different anticholinergics, determined as the concentration that blocks 50% of the hM_3R constitutive activity, as measured in the AP-1-driven luciferase assay in the absence of any added agonist. Adapted from Casarosa *et al.* (2010).

[a] Significant differences between pA_2 and pIC_{50} values obtained with the reporter gene assay ($P < 0.05$; Student's *t* test).

the cell there is a continuous basal turn-over of InsPs, which in the presence of lithium chloride (inhibitor of the last step of degradation from inositol-monophosphate to inositol) results in a progressive increase of the baseline, until the detection system is saturated. The time limitation is obviously influencing the chances to detect constitutive activity, since the longer the assay proceeds, the higher the window built by basal signaling. Reporter gene assays, on the other hand, represent ideal tools for detecting constitutive activity. In fact, the reporter gene methodology monitors the upregulation in the expression of a reporter protein (e.g., luciferase) through activation of mRNA transcription by a given transcription factor, a process that requires longer time (see, e.g., agonistic response at the hM_3R after 2, 8, and 18 h, Fig. 5.4). Importantly, such system does not saturate as easily as the InsP assay, as long as enough luciferine (the substrate that is transformed by luciferase, with generation of light) is present in the final read-out, to maintain linearity between enzyme concentration and light emission. The selection of an appropriate reporter gene to analyze hM_3R pharmacology is discussed in the next paragraph.

3.2. Reporter gene assay

We initially intended to set up a reporter gene for the hM_3R as an alternative to the InsP assay, to allow for longer incubation times in the presence of anticholinergics before read-out. This aspect is of particular importance

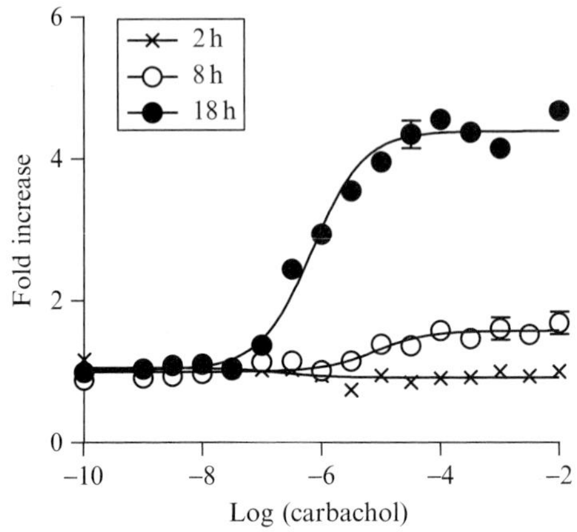

Figure 5.4 Effect of different stimulation times on the hM_3R-mediated induction of AP-1 luciferase. CHO-hM_3 cells were transiently transfected with the AP-1 luciferase reporter gene. Thirty hours after transfection, cells were stimulated with a range of carbachol concentrations, and incubation was allowed to proceed for 2, 8, or 18 h, before cell lysis and luciferase quantification. Data are shown as fold increase in luciferase activity over basal.

when considering ligands with slower kinetics of dissociation from the hM_3 receptor, the so-called long-acting muscarinic antagonists (LAMAs), for example, tiotropium (dissociation $t_{1/2}$ 27 h; Casarosa *et al.*, 2009). LAMAs are unable to reach pseudo-equilibrium within the time frame of the InsP assay, resulting in deviation of the Schild plot from unity (see Fig. 5.6B).

The cell line of choice consisted in the same CHO-K1 stably expressing the human M_3R which was used for the InsP assay, allowing for direct comparison of the results. It has been previously reported that $G\alpha_q$ signaling is capable to activate nuclear factor κB (NF-κB) (Casarosa *et al.*, 2001). On the basis of our previous successful experiences, we chose commercially available constructs from Stratagene, containing the DNA sequence for firefly luciferase which is controlled by 5× NF-κB enhancer elements. As a generic approach to transient transfection, adherent CHO-hM_3 cells were transfected with the reporter vector of interest using the transfection reagent FuGENE 6 (Roche Diagnostics, Mannheim, Germany). One day after transfection, cells were added to 384 well plates in the presence of carbachol and different antagonists and further incubated overnight, to ensure sufficient time for luciferase production. As a control, cells were transfected with an EGFP vector and fluorescence was monitored after 48 h: efficiency of transfection reached approximately 60% of cells according to this protocol. Surprisingly, the NF-κB-driven reporter gene was not activated by stimulating the hM_3 receptor with the agonist carbachol in CHO cells (Fig. 5.5B), although the receptor is expressed at high levels in this cell

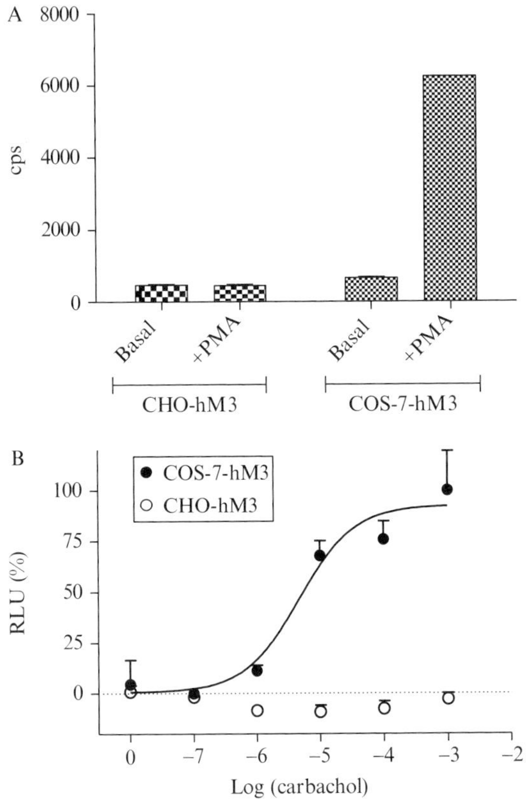

Figure 5.5 NF-κB reporter gene function in CHO and COS-7 cells expressing the hM_3R. Thirty hours after transfection with the reporter vector, cells were stimulated with phorbol myristate acetate (PMA, 100 n*M*) (A), or a range of carbachol concentrations (B). Incubation was allowed to proceed overnight, before cell lysis and luciferase quantification.

line and signals properly in, for example, the InsP assay (see chapter 3.1). As a positive control, phorbol 12-myristate-13-acetate (PMA), which activates directly the protein kinase C (PKC) (and in this way circumvents receptor activation), was tested: as shown in Fig. 5.5A, also PMA could not induce NF-κB-driven transcription in the CHO-hM_3 cell line. In order to check the integrity of the reporter vector, it was tested in a different cell line which had previously given positive results (Casarosa *et al.*, 2001), namely COS-7 cells. In these cells, transiently transfected with pcDNA3.1/hM_3R and the NF-κB luciferase construct, both PMA (i.e., direct PKC activation) and carbachol (i.e., hM_3R-mediated activation of PKC) were able to induce the NF-κB-luciferase reporter gene (Fig. 5.5A and B). Taken together, our results suggest that the PKC pathway leading to NF-κB activation is somehow impaired in CHO cells. To this end, our search in the scientific literature for a reported case of G_q-dependent activation of NF-κB in CHO cells revealed no entries, further supporting our hypothesis.

As an alternative, a second construct was tested, namely the AP-1-driven luciferase reporter gene. The activator protein 1 (AP-1) transcription factor is a dimeric complex of the *jun* and *fos* proteins (Vesely *et al.*, 2009). AP-1 activity is induced by different stimuli, like growth factors, cytokines, and oncoproteins; a well-recognized path of activation involves PKC, which promotes the synthesis, phosphorylation, and activation of members of the *jun* and *fos* families of transcription factors. Upon phosphorylation, the AP-1 complex translocates to the nucleus where it binds to the TPA Responsive element (TRE) and induces transcription of a variety of genes involved in multiple cellular functions such as proliferation, survival, differentiation, and transformation.

This time, carbachol stimulation of CHO-hM_3 cells transiently transfected with the AP1-luciferase construct showed a significant increase in luciferase signal, once incubation was allowed to proceed overnight (Fig. 5.4). To ensure that activation of the reporter gene was solely dependent on hM_3R-mediated signaling, empty CHO-K1 cells (mock control) were transfected in parallel with pAP1-luciferase and stimulated with muscarine: as expected, no effect was observed in mock cells.

To further validate this newly established assay, the functional antagonism of different anticholinergics was measured and compared to the inositol phosphate assay (Fig. 5.6). The pA_2 values obtained for each muscarinic antagonist against carbachol in the InsP assay and the AP-1 reporter gene are in close agreement with each other (Table 5.1; Casarosa *et al.*, 2010), as expected for competitive (i.e., orthosteric, surmountable, and reversible) antagonists inhibiting a common pathway (i.e., G_q coupling). Additionally, the slope of Schild regression obtained with the reporter gene assay was not significantly different from unity over a wide range of antagonists' concentrations also for LAMAs with slower kinetics of receptor dissociation (see, e.g., tiotropium in Fig. 5.6). This represents an advantage to previous results obtained with the InsP assay, and indicates that, in contrast to the InsP accumulation, the reporter gene assay is ideal since its longer incubation time allows also LAMAs to reach pseudo-equilibrium (Casarosa *et al.*, 2010).

Differently from the InsP assay, in the AP1-luciferase assay, the hM_3R showed basal activity, even in the absence of any added agonist, suggesting the detection of constitutive signaling (Fig. 5.7). However, acetylcholine, the physiological agonist for MRs, is known to be released not only by nerve ends as a neurotransmitter, but also by a variety of other cell types (Kummer *et al.*, 2008). To rule out the presence of acetylcholine released in an autocrine fashion by CHO cells, some experiments were performed in

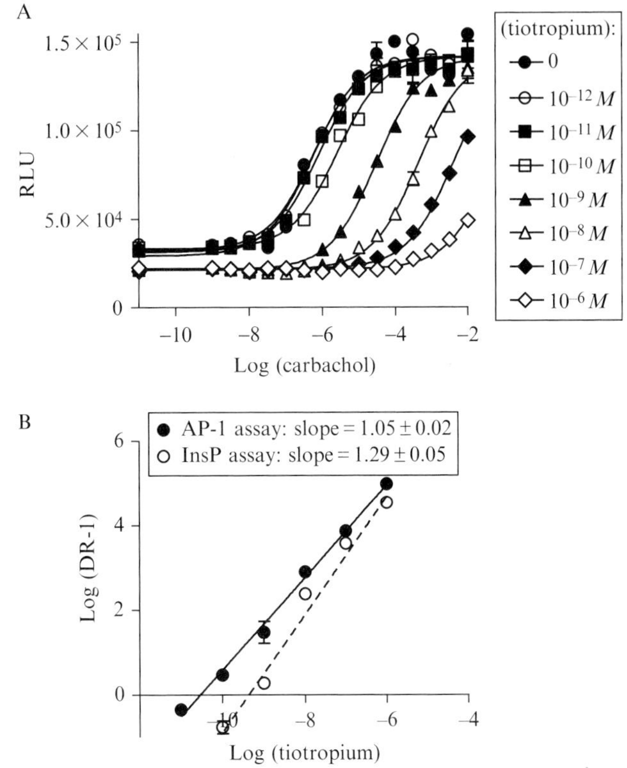

Figure 5.6 The AP-1 luciferase reporter gene allows for correct determination of pA_2 values of muscarinic antagonists in CHO-hM_3 cells. (A) Tiotropium induces a concentration-dependent rightward shift in the carbachol curve, measured as AP-1-driven induction of luciferase activity in CHO-hM_3 cells. (B) Schild regression of the data obtained with tiotropium in the AP-1 reporter gene (filled circles), compared to the data obtained with the inositol phosphate assay (empty circles) are shown.

the presence of acetylcholinesterase from electrophorus electricus (electric eel) Type V-S (Sigma, St. Louis, MO). The addition to the cell culture media of as much as 10 units/ml, which are sufficient to hydrolyze 10 μmol of acetylcholine per minute, did not attenuate the basal hM_3R-mediated signaling nor the observed negative intrinsic activity displayed by muscarinic antagonists in this assay. As an additional control, empty CHO cells transfected with AP-1 luciferase were incubated with NMS: the muscarinic antagonist was unable to affect the basal luciferase activity in the absence of the hM_3 receptor (data not shown). Compared to the agonist-induced response, approximately 30% of the total hM_3 response corresponds to basal, agonist-independent signaling (Fig. 5.7).

These findings imply that hM_3Rs show constitutive activity in the same CHO cell line used for the InsP assay. Possible reasons behind the

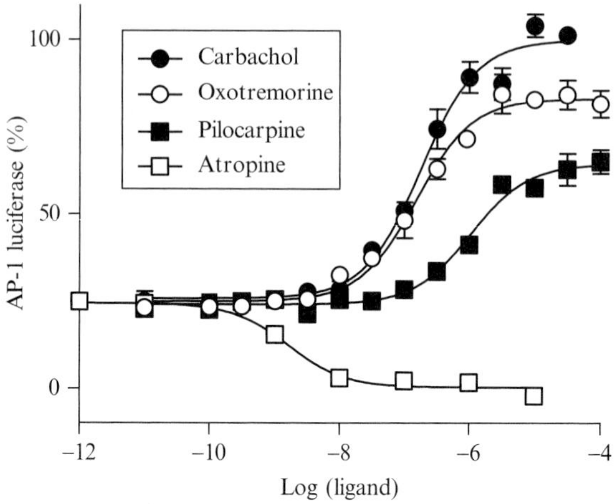

Figure 5.7 The AP-1 reporter gene allows for detection of constitutitve activity of the hM_3R and inverse agonism of the anticholinergics. (A) CHO-hM_3 cells transiently transfected with the AP-1 luciferase reporter gene were stimulated with the indicated concentrations of agonists with different intrinsic activities (i.e., the full agonist carbachol and the partial agonists oxotremorine and pilocarpine), as well as the inverse agonist atropine for 20 h.

differential ability to detect constitutive activity between these two assays have been discussed at the end of previous section.

The ability of the tested muscarinic antagonists to block constitutive activity, or, in other words, the potency of the compounds as inverse agonists was further assessed. Increasing concentrations of muscarinic antagonists were incubated with AP1-luciferase transfected CHO-hM_3 cells in absence of any added agonist (Casarosa *et al.*, 2010). The activity of the reporter decreased in a concentration-dependent manner, which could be fitted by nonlinear regression. From this type of experiments, IC_{50} values for all tested compounds were calculated. All compounds tested behaved as full inverse agonists exerting an intrinsic activity of -1. Inverse agonist's potencies are shown in Table 5.1.

When comparing their potencies in inhibiting the agonist-induced response (pA_2 values against carbachol) and the constitutive signaling of the hM_3 receptor (pIC_{50}) in the same luciferase assay, the tested anticholinergics revealed different behaviors: whereas most were equipotent as antagonists and inverse agonists, others (i.e., NMS and tiotropium) showed a significantly higher propensity to inhibit the agonist response than the constitutive activity of the receptor (3.16- and 8.3-fold difference for NMS and tiotropium, respectively; Casarosa *et al.*, 2010). Taken together, these results suggest that the constitutively active $M_3R^\star$ and the agonist stabilized $M_3AR^\star$, although activating the same pathway, represent distinct conformational states of the receptor, which are differently recognized by some but not all antagonists.

These findings are in agreement with biophysical studies obtained using β_2-adrenoceptors fluorescently labeled at Cys265, a region that is sensitive to receptor conformational changes (Ghanouni *et al.*, 2001). Fluorescence lifetime spectroscopy showed that different β-agonists produced different arrays of receptor conformations, consistent with ligand-selective active states. Instead of a single receptor active state for G-protein activation, there is an ensemble of microconformations (Kenakin, 2003) that are all capable of producing the same pharmacological effect (in this case, G_q-protein activation) but having different overall tertiary conformations.

4. Constitutive Activity and Receptor Upregulation Studies

Cellular populations of GPCRs are not static but can be regulated by several factors: among these, a well-known phenomenon is the capacity of agonists to induce internalization and eventually degradation of the GPCR, resulting in an overall reduction in receptor expression levels, a process called downregulation. Given the opposite pharmacological function of agonists and inverse agonists, based on the selective stabilization of active and inactive receptor conformations, it could be anticipated that equivalent treatment with inverse agonists would result in their upregulation. Indeed, extensive literature exists on such effects being produced by incubation of cells with inverse agonists (Milligan and Bond, 1997). The rationale behind assumes that inverse agonists inhibit the spontaneous formation of receptor active states and consequent phosphorylation and internalization, a process which, coupled with normal receptor neosynthesis and expression, would lead to an increased surface density of receptor (Milligan and Bond, 1997). However, most studies analyzing the effects of inverse agonists on receptor upregulation were performed with constitutively active (CAM) receptor mutants. For example, two studies were published with CAMs of the hM_3R: in the first one (Dowling *et al.*, 2006), a single asparagine-to-tyrosine point mutation at residue 514 (hM_3-$N^{514}Y$) resulted in a marked increase (approximately 300%) in agonist-independent signaling compared with the response observed for the wild-type (WT) receptor; this CAM could be upregulated by incubation with several muscarinic antagonists. In the second study (Zeng *et al.*, 2003), the rat M_3R was modified by swapping the third intracellular loop with the corresponding region of a constitutively active mutant human β_2 adrenergic receptor and attaching *Renilla reniformis* luciferase to its C-terminus; the chimeric fusion receptor displayed constitutive activity and a binding profile comparable with that of the WT receptor for agonists, antagonists, and inverse agonists. However, these mutations are unlikely to represent an accurate model of the agonist-occupied active R* state. Indeed, the creation of a CAM can diminish stabilizing constraints within the receptor, leading to an inherently unstable receptor that is more susceptible to destabilization and/or proteolytic degradation (Milligan and Bond, 1997). Consequently, the expression level of the mutant is often increased by any ligand, either agonist or antagonist, regardless of its efficacy, as shown, for example, with the chimeric M_3R (Zeng *et al.*, 2003). In our laboratory instead, we tested the ability of the different anticholinergics to upregulate the WT hM_3 receptor, ruling out potential artifacts.

For these studies, we used the same CHO-hM_3 cell line as for the functional assays. Given the relatively high B_{max} of this cell line, changes in receptor expression can be easily monitored. As a standard approach, we incubated CHO-hM_3 cells with different concentrations of muscarinic antagonists for 30 h, and then measured cell-surface expression of the hM_3R by radioligand binding. In our laboratory, we used tritiated tiotropium ([^{3}H]-tiotropium; specific activity 65 Ci/mmol) to monitor changes in hM_3 receptor expression. The advantages in using this radioligand consist in its very low unspecific binding and high affinity for the receptor; however, [^{3}H]-tiotropium is not commercially available; as an alternative, [^{3}H]-NMS can be used with good results. The major obstacle that we had to solve to set up this kind of assay was to develop a proper washing procedure which would fully remove the muscarinic antagonists bound to the M_3R and allow [^{3}H]-tiotropium occupation of the entire cell-surface receptor population. Given the slow dissociation kinetics of some of the muscarinic antagonists, the so-called LAMAs, care was taken to develop a washing procedure which would fully remove these compounds bound to the M_3R and allow [^{3}H]-tiotropium occupation of the entire cell-surface receptor population. For this control experiment, cells were incubated for 1 h at 4 °C (to prevent any change in receptor expression) with tiotropium (Fig. 5.8B), then several buffers set at different pH were used to extensively wash the cells, either at 4 or 37 °C. As can be seen in Fig. 5.8B, complete ligand removal could be obtained by washing at 37 °C with buffers at low pH. Briefly, acidic washes with Hams' F12 (with pH set at 2) were allowed to proceed for 40 min, with medium being exchanged every few minutes.

The final protocol consisted in approximately 70–80% confluent cells in 24-well plates, which were incubated with increasing concentrations of antagonists/inverse agonists for 30 h. At the end of this period, cells were subjected to an acidic wash to remove surface-bound ligands, as described above. Afterwards, cells were rinsed twice with ice-cold binding buffer (HBSS, 10 m*M* HEPES, and 0.05% BSA) to set the pH back to neutral, then receptor expression on the cell surface was monitored by binding of [^{3}H]-tiotropium at saturating concentrations (approximately 2 n*M*).

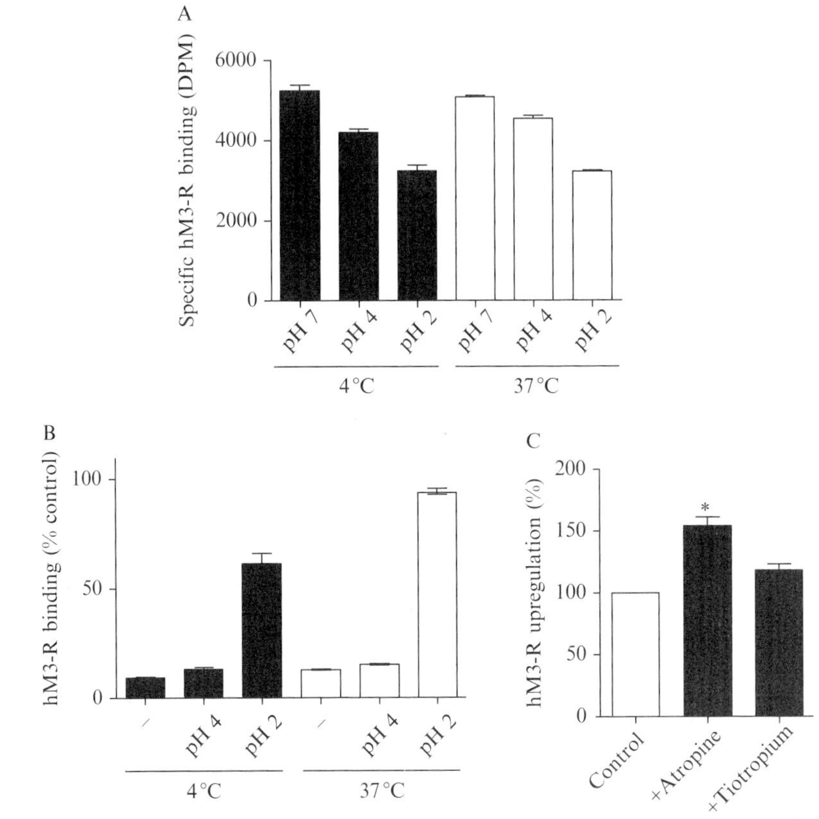

Figure 5.8 Effects of chronic treatment with anticholinergics on hM_3R upregulation. (A, B) Effect of pH and temperature of the wash buffer on antagonist removal. In this control experiment, CHO-hM_3 cells were preincubated 1 h with tiotropium 100 n*M* (B), then extensively washed for 40 min with buffers set at different pH (pH 7, 4, and 2) and temperature (4 °C, shown as filled bars vs. 37 °C, empty bars). Afterwards, total hM_3R quantification at the cell surface was monitored in a radioligand binding assay using [^{3}H]-tiotropium (2 n*M*). Even though a washing procedure at pH 2 reduces total M_3R population at the cell surface (A), this protocol allowed full tiotropium removal (B). In (B) data are presented as percentage of total hM_3R expression, defined for each treatment group as the amount of binding obtained with cells which were not pre-treated with tiotropium but underwent the same washing procedures. (C) CHO-hM_3 cells were incubated in the presence of atropine or tiotropium (100 n*M*) for 30 h. Following an extensive washing procedure with buffer set at pH 2, which ensures fully removal of ligands (as shown in inset B), [^{3}H]-tiotropium was added to establish a B_{max} value. The mean ± SEM for at least three experiments performed in triplicate is shown.

All tested antagonists caused concentration-dependent increases in receptor expression levels (Table 5.2; Casarosa *et al.*, 2010), with EC_{50} values in good agreement with their potency as inverse agonists (cf. Tables 5.1 and 5.2).

Table 5.2 Differential ability of anticholinergics to cause hM_3R upregulation

	% hM_3 upregulation	pEC_{50}
Atropine	54 ± 7%	9.10 ± 0.08
NMS	19 ± 4%[a]	9.03 ± 0.10
Pirenzepine	39 ± 2%	6.14 ± 0.06
Ipratropium	41 ± 2%	9.02 ± 0.07
Tiotropium	18 ± 5%[a]	10.27 ± 0.10
4-DAMP	18 ± 2%[a]	9.79 ± 0.08

The maximal hM_3R upregulation caused by the different anticholinergics and their respective half maximal effective concentration, following 30 h incubation with CHO-hM_3 cells, is reported. The average of at least three independent experiments performed in triplicate is reported. The data concerning receptor upregulation were analyzed by using one-way ANOVA followed by Dunnet's multiple-comparison test. Statistical significance is denoted compared with the atropine group ([a]: $p < 0.05$). Adapted from Casarosa *et al.* (2010).

However, the changes in receptor expression differed significantly in the presence of the different ligands: whereas most antagonists (i.e., atropine, see Fig. 5.8C) caused a significant upregulation above 40% of basal hM_3R expression levels, a second group of antagonists (e.g., tiotropium, Fig. 5.8C), induced less than 20% upregulation. Differences in upregulation do not relate to differences in ligands' negative intrinsic activity, as all tested anticholinergics behaved as full inverse agonists. However, a correlation between induction of upregulation and differential propensity to behave as inverse agonists seems plausible, as most compounds which are causing a less pronounced M_3R upregulation, namely tiotropium and NMS, show a significant difference in their potencies as neutral antagonists (pA_2 values) and as inverse agonists (Table 5.1). The different ability to induce receptor upregulation supports a model where the anticholinergics stabilize distinct M_3R conformations, all equally effective in blocking G-protein signaling (though possibly through different mechanisms), but differently able to affect receptor internalization and its steady-state expression at the cell membrane. This is in agreement with a model where internalization and G-protein activation are mediated by distinct receptor conformations, as suggested by the ability of antagonists or inverse agonists to elicit internalization (Azzi *et al.*, 2003).

5. Physiological Relevance of hM_3R Constitutive Activity

The physiological role of hM_3R constitutive signaling is currently unknown. There is, however, evidence that an increased cholinergic tone may be an important feature of chronic pulmonary diseases such as COPD

(Barnes, 2004), which is the rationale for the use of muscarinic antagonists as bronchodilators. This increased cholinergic tone might, at least in part, result from an increase in the constitutive signaling of hM_3 receptors, as suggested by recent studies indicating an upregulation of M_3R in COPD patients, compared to healthy individuals (Profita *et al.*, 2005, 2009). Additionally, increasing the expression levels of effector proteins also influences the basal signaling of GPCRs: for example, artificial overexpression of $G\alpha_q$ proteins greatly enhances the constitutive activity of hM_3R in a cellular model (Burstein *et al.*, 1995). To this end, it has been shown that proinflammatory stimuli known to play a role in the pathogenesis of COPD, such as TNF-α and cigarette smoke, increase the expression of $G\alpha_q$ and Rho proteins (Chiba *et al.*, 2005; Hotta *et al.*, 1999), which are known signaling partners for the hM_3R. Similarly, a recent publication (Fernandez-Rodriguez *et al.*, 2010) has highlighted an increased muscarinic M_3R activity in airway smooth muscle isolated from a mouse model of allergic asthma, likely related to receptor upregulation. Additionally, emerging evidence indicates that the hM_3 receptor plays key roles in regulating cellular proliferation and cancer progression. hM_3R overexpression has been reported for cancer cells from brain (astrocytoma), breast, colon, lung, and prostate (Shah *et al.*, 2009), and implicated in the promotion of cell proliferation, apoptosis, migration, and other features critical for cancer survival and spread. The increase in receptor expression, together with the fact that chemically blocking acetylcholine synthesis and release reduces but does not fully abrogate proliferation (e.g., Cheng *et al.*, 2008), suggests that constitutive signaling of the human M_3 receptor might play a role in these pro-oncogenic effects.

If, indeed, constitutive activity was operative in these pathologies, inverse agonists would be the pharmacological tool of choice in the retardation of disease progression. An important appendix to the use of inverse agonists in the clinical practice relates to their potential to induce receptor upregulation and tolerance upon chronic treatment. Indeed, studies have indicated the upregulation of MRs as a consequence of chronic atropine administration in animals (Chevalier *et al.*, 1991; Wall *et al.*, 1992); in particular, chronic exposure to atropine resulted in an upregulation of M_3R that was associated with enhanced airway smooth muscle contraction in rabbits (Witt-Enderby *et al.*, 1995), suggesting that the phenomenon might be of relevance with chronic treatment in the clinical use. Taken together, negative efficacy is a molecular property of anticholinergics which might be desirable in clinical use for COPD (and possibly cancer therapy), but attention should be paid to their potential for receptor upregulation, leading to tolerance. Setting up assays which independently monitor these different pharmacological parameters becomes very useful for identifying the right molecule, since, as we have shown, it is possible to dissociate inverse agonistic properties from potential to induce upregulation (e.g., the anticholinergics NMS and tiotropium). In agreement with these results, recent data from the UPLIFT (Understanding Potential Long-term Impacts on Function with Tiotropium) study involving almost 6000 COPD patients indicated no loss of tiotropium bronchodilatory effects throughout the trial, which lasted 4 years, ruling out tolerance to this drug in the clinical setting (Tashkin *et al.*, 2008).

REFERENCES

Azzi, M., Charest, P. G., Angers, S., Rousseau, G., Kohout, T., Bouvier, M., and Pineyro, G. (2003). Beta-arrestin-mediated activation of MAPK by inverse agonists reveals distinct active conformations for G protein-coupled receptors. *Proc. Natl. Acad. Sci. USA* **100,** 11406–11411.

Bakker, R. A., Wieland, K., Timmerman, H., and Leurs, R. (2000). Constitutive activity of the histamine H(1) receptor reveals inverse agonism of histamine H(1) receptor antagonists. *Eur. J. Pharmacol.* **387,** R5–R7.

Barnes, P. J. (2004). The role of anticholinergics in chronic obstructive pulmonary disease. *Am. J. Med.* **117**(Suppl. 12A), 24S–32S.

Burstein, E. S., Spalding, T. A., Brauner-Osborne, H., and Brann, M. R. (1995). Constitutive activation of muscarinic receptors by the G-protein Gq. *FEBS Lett.* **363,** 261–263.

Casarosa, P., Bakker, R. A., Verzijl, D., Navis, M., Timmerman, H., Leurs, R., and Smit, M. J. (2001). Constitutive signaling of the human cytomegalovirus-encoded chemokine receptor US28. *J. Biol. Chem.* **276,** 1133–1137.

Casarosa, P., Bouyssou, T., Germeyer, S., Schnapp, A., Gantner, F., and Pieper, M. (2009). Preclinical evaluation of long-acting muscarinic antagonists: Comparison of tiotropium and investigational drugs. *J. Pharmacol. Exp. Ther.* **330,** 660–668.

Casarosa, P., Kiechle, T., Sieger, P., Pieper, M., and Gantner, F. (2010). The constitutive activity of the human muscarinic M3 receptor unmasks differences in the pharmacology of anticholinergics. *J. Pharmacol. Exp. Ther.* **333,** 201–209.

Cheng, K., Samimi, R., Xie, G., Shant, J., Drachenberg, C., Wade, M., Davis, R. J., Nomikos, G., and Raufman, J. P. (2008). Acetylcholine release by human colon cancer cells mediates autocrine stimulation of cell proliferation. *Am. J. Physiol. Gastrointest. Liver Physiol.* **295,** G591–G597.

Chevalier, B., Mansier, P., Teiger, E., Callen-el, A. F., and Swynghedauw, B. (1991). Alterations in beta adrenergic and muscarinic receptors in aged rat heart. Effects of chronic administration of propranolol and atropine. *Mech. Ageing Dev.* **60,** 215–224.

Chiba, Y., Murata, M., Ushikubo, H., Yoshikawa, Y., Saitoh, A., Sakai, H., Kamei, J., and Misawa, M. (2005). Effect of cigarette smoke exposure in vivo on bronchial smooth muscle contractility in vitro in rats. *Am. J. Respir Cell Mol. Biol.* **33,** 574–581.

Disse, B., Speck, G. A., Rominger, K. L., Witek, T. J., Jr., and Hammer, R. (1999). Tiotropium (Spiriva): Mechanistical considerations and clinical profile in obstructive lung disease. *Life Sci.* **64,** 457–464.

Dowling, M. R., Willets, J. M., Budd, D. C., Charlton, S. J., Nahorski, S. R., and Challiss, R. A. (2006). A single point mutation (N514Y) in the human M3 muscarinic acetylcholine receptor reveals differences in the properties of antagonists: Evidence for differential inverse agonism. *J. Pharmacol. Exp. Ther.* **317,** 1134–1142.

Fernandez-Rodriguez, S., Broadley, K. J., Ford, W. R., and Kidd, E. J. (2010). Increased muscarinic receptor activity of airway smooth muscle isolated from a mouse model of allergic asthma. *Pulm. Pharmacol. Ther.* **23,** 300–307.

Gandevia, B. (1975). Historical review of the use of parasympatholytic agents in the treatment of respiratory disorders. *Postgrad. Med. J.* **51,** 13–20.

Ghanouni, P., Gryczynski, Z., Steenhuis, J. J., Lee, T. W., Farrens, D. L., Lakowicz, J. R., and Kobilka, B. K. (2001). Functionally different agonists induce distinct conformations in the G protein coupling domain of the beta 2 adrenergic receptor. *J. Biol. Chem.* **276,** 24433–24436.

Hotta, K., Emala, C. W., and Hirshman, C. A. (1999). TNF-alpha upregulates Gialpha and Gqalpha protein expression and function in human airway smooth muscle cells. *Am. J. Physiol* **276,** L405–L411.

Jakubik, J., Bacakova, L., El-Fakahany, E. E., and Tucek, S. (1995). Constitutive activity of the M1–M4 subtypes of muscarinic receptors in transfected CHO cells and of muscarinic receptors in the heart cells revealed by negative antagonists. *FEBS Lett.* **377,** 275–279.

Johnson, E. N., and Druey, K. M. (2002). Heterotrimeric G protein signaling: Role in asthma and allergic inflammation. *J. Allergy Clin. Immunol.* **109,** 592–602.

Kenakin, T. (2003). Ligand-selective receptor conformations revisited: The promise and the problem. *Trends Pharmacol. Sci.* **24,** 346–354.

Kummer, W., Lips, K. S., and Pfeil, U. (2008). The epithelial cholinergic system of the airways. *Histochem. Cell Biol.* **130,** 219–234.

Milligan, G., and Bond, R. A. (1997). Inverse agonism and the regulation of receptor number. *Trends Pharmacol. Sci.* **18,** 468–474.

Peretto, I., Petrillo, P., and Imbimbo, B. P. (2009). Medicinal chemistry and therapeutic potential of muscarinic M3 antagonists. *Med. Res. Rev.* **29,** 867–902.

Profita, M., Giorgi, R. D., Sala, A., Bonanno, A., Riccobono, L., Mirabella, F., Gjomarkaj, M., Bonsignore, G., Bousquet, J., and Vignola, A. M. (2005). Muscarinic receptors, leukotriene B4 production and neutrophilic inflammation in COPD patients. *Allergy* **60,** 1361–1369.

Profita, M., Bonanno, A., Siena, L., Bruno, A., Ferraro, M., Montalbano, A. M., Albano, G. D., Riccobono, L., Casarosa, P., Pieper, M. P., and Gjomarkaj, M. (2009). Smoke, choline acetyltransferase, muscarinic receptors, and fibroblast proliferation in chronic obstructive pulmonary disease. *J. Pharmacol. Exp. Ther.* **329,** 753–763.

Shah, N., Khurana, S., Cheng, K., and Raufman, J. P. (2009). Muscarinic receptors and ligands in cancer. *Am. J. Physiol. Cell Physiol.* **296,** C221–C232.

Tashkin, D. P., Celli, B., Senn, S., Burkhart, D., Kesten, S., Menjoge, S., and Decramer, M. (2008). A 4-year trial of tiotropium in chronic obstructive pulmonary disease. *N. Engl. J. Med.* **359,** 1543–1554.

Vesely, P. W., Staber, P. B., Hoefler, G., and Kenner, L. (2009). Translational regulation mechanisms of AP-1 proteins. *Mutat. Res.* **682,** 7–12.

Vincken, W., Van Noord, J. A., Greefhorst, A. P., Bantje, T. A., Kesten, S., Korducki, L., and Cornelissen, P. J. (2002). Improved health outcomes in patients with COPD during 1 yr's treatment with tiotropium. *Eur. Respir. J.* **19,** 209–216.

Wall, S. J., Yasuda, R. P., Li, M., Ciesla, W., and Wolfe, B. B. (1992). Differential regulation of subtypes m1-m5 of muscarinic receptors in forebrain by chronic atropine administration. *J. Pharmacol. Exp. Ther.* **262,** 584–588.

Wess, J., Eglen, R. M., and Gautam, D. (2007). Muscarinic acetylcholine receptors: Mutant mice provide new insights for drug development. *Nat. Rev. Drug Discov.* **6,** 721–733.

Witt-Enderby, P. A., Yamamura, H. I., Halonen, M., Lai, J., Palmer, J. D., and Bloom, J. W. (1995). Regulation of airway muscarinic cholinergic receptor subtypes by chronic anticholinergic treatment. *Mol. Pharmacol.* **47,** 485–490.

Zeng, F. Y., McLean, A. J., Milligan, G., Lerner, M., Chalmers, D. T., and Behan, D. P. (2003). Ligand specific up-regulation of a *Renilla reniformis* luciferase-tagged, structurally unstable muscarinic M3 chimeric G protein-coupled receptor. *Mol. Pharmacol.* **64,** 1474–1484.

CHAPTER SIX

Ghrelin Receptor: High Constitutive Activity and Methods for Developing Inverse Agonists

Sylvia Els, Annette G. Beck-Sickinger, *and* Constance Chollet

Contents

Abstract

The ghrelin receptor is a G protein-coupled receptor (GPCR) mainly distributed in the brain, and also expressed in peripheral tissues. Remarkably, the ghrelin receptor possesses a naturally high constitutive activity representing 50% of its maximal activity. Its endogenous ligand ghrelin is the only known orexigenic gastrointestinal peptide and plays a central role in the regulation of appetite, food intake, and energy homeostasis.

Reducing the constitutive activity of the ghrelin receptor by inverse agonists is the strategy adopted by our group to develop anti-obesity drugs. Therefore, short peptides were synthesized and showed high inverse agonist potency toward the ghrelin receptor.

This review describes the methods used to synthesize the peptides and to evaluate their biological activity. Peptide synthesis was performed on solid

Institute of Biochemistry, Leipzig University, Brüderstrasse, Leipzig, Germany

Methods in Enzymology, Volume 485
ISSN 0076-6879, DOI: 10.1016/S0076-6879(10)85006-2

phase using a Fmoc/tBu-strategy. Peptide potency was measured with a signal transduction assay, the inositol trisphosphate turnover assay, adapted to a receptor expressing constitutive activity.

1. Introduction

The ghrelin receptor (formerly the growth hormone secretagogue receptor, GHSR) was identified in 1996 as the receptor of GHS (growth hormone secretagogues) (Howard *et al.*, 1996; Smith *et al.*, 1999a,b). The receptor is a typical G protein-coupled receptor (GPCR) with seven transmembrane helices, an extracellular N-terminus and an intracellular C-terminus. It is coupled to a G protein of the type $G_{\alpha q/11}$ and belongs to the rhodopsin family of GPCR. The ghrelin receptor is well conserved in vertebrate species and presents typical conserved sites such as cysteines in the first two extracellular loops (Smith *et al.*, 2001). It is mainly expressed in the pituitary gland, the hypothalamus, and the hippocampus but also in other tissues such as stomach, liver, and heart (Kojima and Kangawa, 2005; Van Der Lely *et al.*, 2004).

The endogenous ligand ghrelin (from "ghre," the word-root in Proto-Indo-European languages for "grow") was isolated in 1999. It is a 28-amino acid peptide (Fig. 6.1), octanoylated on Ser^3, and mainly produced in the oxyntic mucosa of the stomach. The N-terminal pentapeptide was identified as the minimal sequence necessary for activation of the ghrelin receptor (Bednarek *et al.*, 2000). The n-octanoylation is carried out posttranslationally by ghrelin-*O*-acyltransferase (GOAT; Gutierrez *et al.*, 2008; Yang *et al.*, 2008) and is essential for binding the ghrelin receptor and for ghrelin circulation in blood (Bednarek *et al.*, 2000; De Vriese and Delporte, 2007).

Physiologically, ghrelin is the only known orexigenic peptide produced in the gastrointestinal tract. It stimulates food intake, and upregulates energy homeostasis through the stimulation of NPY/AGRP-containing neurons

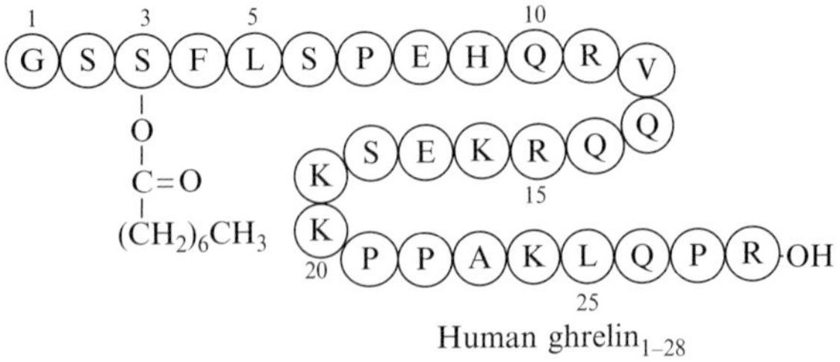

Figure 6.1 Structure of ghrelin. Human ghrelin is a 28-amino acid peptide, octanoylated on Ser^3.

in the arcuate nucleus of hypothalamus (Cummings, 2006; Nakazato *et al.*, 2001).

Ghrelin receptor agonists were widely developed before the discovery of ghrelin and its receptor (Nargund *et al.*, 1998). These molecules were called GHS as they were able to initiate the release of growth hormone, independently from the growth hormone releasing hormone (GHRH) pathway. The major role of ghrelin in the regulation of energy homeostasis orientates the research in the development of ghrelin receptor antagonists that may downregulate energy homeostasis and thus, decrease food intake and body weight (Chollet *et al.*, 2009).

2. Constitutive Activity and Development of Ghrelin Receptor Inverse Agonists

GPCRs can be stimulated by different ligands like partial or full agonists, inverse agonists, and neutral antagonists. According to IUPHAR receptor nomenclature and drug classification (Neubig *et al.*, 2003), an agonist binds its target receptor and generates a biological response. Conventionally, an agonist increases the receptor activity while an inverse agonist decreases it. A full agonist induces a maximal response by switching the receptor into its active conformation. In case of constitutive activity, a fraction of receptor is in its active form without induction of any ligand. It occurs through a spontaneous conformational change from inactive to active state of the receptor. Inverse agonists shift the equilibrium in favor to the inactive conformation, and thereby, reduce the biological response of the receptor. An antagonist reduces the action of an agonist and can be competitive (same binding site than the agonist), noncompetitive (different binding site), or neutral (same affinity for both active and inactive state of the receptor).

Constitutive activity was first discovered for ionotropic receptors, such as GABAa receptors (Ehlert *et al.*, 1983). Since then, basal activity was observed for several GPCRs (Leurs *et al.*, 1998).

The constitutive signaling of the ghrelin receptor is remarkably high and represents up to 50% of its maximal activity *in vitro* (Holst and Schwartz, 2004; Holst *et al.*, 2003). Physiologically, this high constitutive activity is clearly related to energy homeostasis. Indeed, a naturally occurring mutation selectively suppressing the constitutive activity is correlated with the development of short nature and obesity (Holst *et al.*, 2006; Pantel *et al.*, 2006). In addition, the high constitutive activity of the receptor may be the reason for appetite and food intake between meals (Holst and Schwartz, 2004). Therefore, development of inverse agonists acting on the ghrelin receptor appeared to be a very appropriate approach for developing anti-obesity drugs (Fig. 6.2).

The modified substance P (MSP, Fig. 6.3), initially developed as a substance P antagonist for the neurokinin 1 receptor, was identified as a weak antagonist, but a full inverse agonist of the ghrelin receptor (Holst *et al.*, 2003). Although MSP cannot be considered for pharmaceutical applications due to its nonspecific binding to a large amount of receptors, it led to extensive SAR studies for further development of ghrelin receptor inverse agonists.

Systematic truncation of MSP revealed that the C-terminal heptapeptide fQwFwLL-NH_2 was the minimal sequence maintaining high binding affinity and inverse agonist properties (Holst *et al.*, 2006). The study demonstrated that: D-Trp^7, D-Trp^9, and Leu^{10} were necessary for the inverse agonist activity; replacement of Phe^4 reduced the activity; Gln^6 and Leu^{11} were

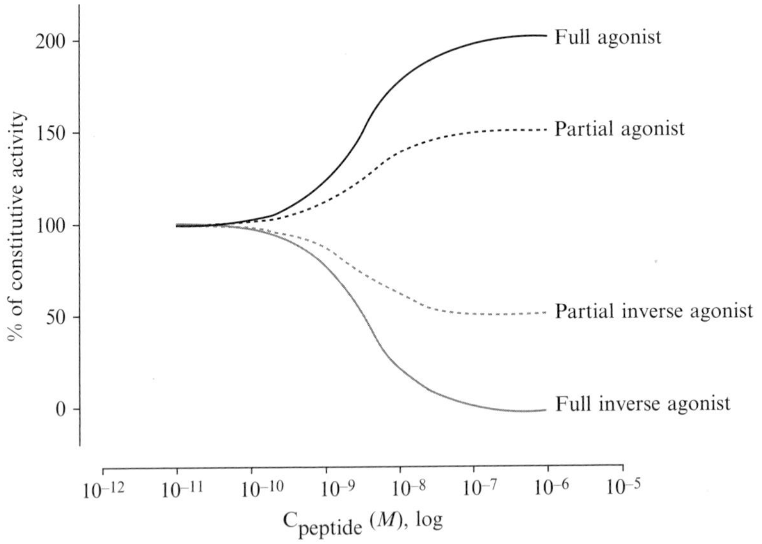

Figure 6.2 Theoretical dose-response curves for agonists and inverse agonists. The graph shows theoretical dose-response curves obtained after stimulation of a receptor with different kinds of ligands. The activity is plotted against the logarithm of concentration, therefore sigmoidal curves are obtained. The point of inflexion is calculated and gives the EC_{50} value which is the concentration of ligand corresponding to the half-maximal receptor activity. In this example, all ligands have the same EC_{50} (10^{-9} *M*), that is, the same activity at the receptor. The maximal receptor activity corresponds to the ligand potency. A full agonist increases the receptor activity to its maximum (Here, 200% of constitutive activity (CA)). With a partial agonist, the receptor activity reaches an intermediate maximum lower than the maximal activity (150% of CA). Inversely, a full inverse agonist reduces the receptor to zero (0% of CA), and a partial inverse agonist decreases partially the receptor activity (50% of CA).

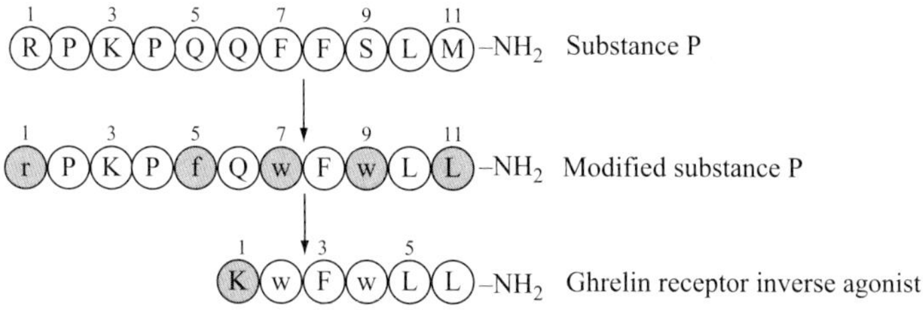

Figure 6.3 Development of short peptides as ghrelin receptor inverse agonists. From the substance P, the modified substance P was identified as a ghrelin receptor inverse agonist. Then, structure–activity relationship studies led to the short hexapeptide, KwFwLL-NH_2, currently the most potent ghrelin receptor inverse agonist.

not essential for binding and function. D-Phe^5 was not essential and could be replaced with L-Ala, L-Phe, or the tetrapeptide D-Arg-Pro-Lys-Pro without important loss of affinity and inverse agonist activity. Thus, the presence of a positive charge next to the core peptide wFwLL-NH_2 was suspected essential for inverse agonist activity. This hypothesis was confirmed by the synthesis of KwFwLL-NH_2, which showed high inverse agonist activity (EC_{50} = 36 ± 8 n*M* in inositol trisphosphate turnover assay). Replacement of Lys^1 with Arg^1 or His^1 had no effect on potency and affinity, whereas the peptide AwFwLL-NH_2 showed high agonist activity (Holst *et al.*, 2007).

3. Synthesis of Ghrelin Receptor Inverse Agonists: Solid-Phase Peptide Synthesis (SPPS)

In this section, we provide information about the synthesis of ghrelin receptor inverse agonists. All peptides are synthesized automatically or manually on solid phase.

3.1. Principle

Amide bonds are formed by condensation of a carboxylic acid with an amine. SPPS was developed by Merrifield in 1963 and consists in the stepwise synthesis of a peptide, by assembling the amino acids on a solid support from the C-terminus to the N-terminus (Merrifield, 1963). The choice of the resin depends on the desired C-terminus. Wang and Rink amide resins can be used to introduce, respectively, a C-terminal carboxylic acid or a carboxamide, released after the final cleavage from the resin. Introduction of orthogonal protecting groups on trifunctional amino acids is required to avoid side reactions during the elongation. For this purpose, the Fmoc/tBu-strategy,

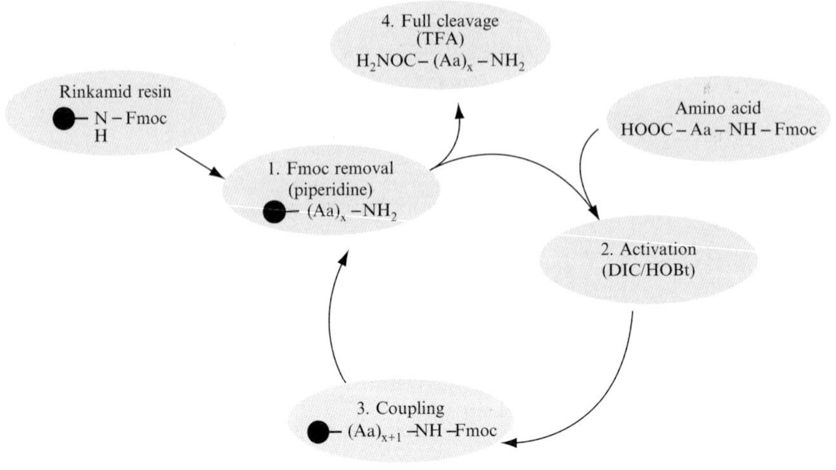

Figure 6.4 Ghrelin receptor inverse agonist synthesis on solid phase. The synthesis cycle starts with the deprotection of the rinkamid resin (1). Then, a Fmoc-protected amino acid is activated at its free carboxyl group with DIC/HOBt (2) and coupled to the free N-terminus of the resin (3). The cycle can be repeated until the completion of synthesis. The peptide is finally cleaved from the resin (4).

based on acidic/basic cleavage of the different protecting groups, is recommended. The N-terminus of each amino acid is protected by a Fmoc group (Fmoc-Aa-OH), while side chains containing functional groups are protected with a *tert*-butyl (tBu) group (Fmoc-Aa(tBu)-OH). After each coupling step, Fmoc can be easily removed using 20% piperidine in DMF, to release the N-terminal amine and continue the peptide elongation. At the end of the synthesis, the *tert*-butyl groups are removed simultaneously to the resin cleavage with 90% TFA. For efficient amide bond formation, the carboxylic group needs to be activated by coupling reagents. In our laboratory, we use a combination of DIC and HOBt, to prevent racemization (Fig. 6.4).

3.2. Methods for solid-phase peptide synthesis

Syringes with filters are employed as reactors to simplify the removal of the reaction mixture and the washing steps. For the synthesis of ghrelin receptor inverse agonists, only the Rink amide resin is used, as a C-terminal carboxamide is required.

3.2.1. Automated solid-phase synthesis

Coupling of standard L-amino acids can be carried out with a synthesis robot with *in situ* activation by DIC/HOBt. In our laboratory, reactors usually contain 15 μmol of Rink amide resin and are placed in the synthesis robot.

A 10-fold excess of each Fmoc-protected amino acid is dissolved in a solution of 0.5 *M* HOBt in DMF. To dissolve Fmoc-L-Phe-OH, NMP is preferred to DMF. Solutions of 1.65 *M* DIC in DMF and 40% piperidine in DMF (v/v) are also prepared. All solutions are placed in the robot and will be used for coupling or deprotection steps, depending on the peptide sequence. At the beginning of the cycle, the resin is let to swell in 800 μl DMF for 15 min. Then, the synthesis starts with the removal of the Fmoc protecting group from the Rink amide resin, followed by the coupling of the amino acids. Each amino acid is coupled twice for 40 min, followed by a washing step and the removal of the Fmoc group (see Table 6.1). After coupling of the last amino acid, the Fmoc protecting group is cleaved, leaving the peptide with a free N-terminus.

3.2.2. Manual solid-phase synthesis

Special amino acids such as D-amino acids or unnatural amino acids should be coupled manually, following the same strategy.

3.2.2.1. Loading of the resin with the first amino acid The resin is let to swell in 500 μl of DMF for 15 min. The Fmoc group is first removed from the Rink amide resin. The resin is treated twice with 500 μl of a fresh solution of 30% piperidine in DMF (v/v) and shaken 20 min at room temperature. The resin is thoroughly washed with DMF.

5eq. of Fmoc-Aa-OH and 5 eq. of HOBt are dissolved in DMF to a concentration of 0.5 *M*. The solution is added to the resin with 5 eq. of

Table 6.1 Synthesis cycle of automated solid-phase peptide synthesis

Reaction step	Reagents	Volume	Reaction time
Fmoc removal	– 40% piperidine in DMF (v/v)	– 400 μl	– 2 min
	– 20% piperidine in DMF (v/v)	– 400 μl	– 10 min
Washing	DMF	600 μl	4 x 1 min
Coupling of Fmoc-Aa-OH	– Fmoc-Aa (10 eq.) + 0.5 *M* HOBt in DMF	– 300 μl	40 min
	– 1.65 *M* DIC in DMF	– 100 μl	
Washing	DMF	800 μl	1 min 30 s
Coupling of Fmoc-Aa-OH	– Fmoc-Aa (10 eq.) + 0.5 *M* HOBt in DMF	– 300 μl	40 min
	– 1.65 *M* DIC in DMF	– 100 μl	
Washing	DMF	800 μl	2 x 1 min 30 s

DIC, and the reaction mixture is shaken overnight at room temperature. Then, the resin is washed five times, successively with DMF, DCM, MeOH, and Et_2O, and dried *in vacuo*.

To measure the loading rate of the resin, a small amount of resin (approximately 3 mg) is transferred in a tube. Exactly 500 μl of a fresh solution of piperidine, 30% (v/v) is added to the resin, and the suspension is shacken 30 min at room temperature. 250 μl of the supernatant is then diluted in 1.5 ml of 30% piperidine in DMF (v/v). The absorbance (A) of the solution is measured at 300 nm and gives the loading rate of the resin (L_{300} in mmol/g) according to the Beer–Lambert law:

$$A = \varepsilon \cdot l \cdot c = \frac{\varepsilon \cdot l \cdot n}{V}$$

$$L_{300} = \frac{n}{m} = \frac{2 \cdot A \cdot V}{\varepsilon \cdot l \cdot m}$$

A = Absorbance; V = Volume (ml); $\varepsilon = 7.8$ = molar extinction coefficient ($M^{-1}\ cm^{-1}$); l = pathlength (cm), m = mass (mg)

3.2.2.2. Manual coupling of amino acids The resin is let to swell in 500 μl of DMF for 15 min. 5 eq. of Fmoc-Aa-OH and 5 eq. of HOBt are dissolved in DMF to a concentration of 0.5 *M*. The solution is added to the resin with 5 eq. of DIC, and the reaction mixture is shaken for at least 3 h at room temperature. Then, the resin is washed five times, successively with DMF, DCM, MeOH, and Et_2O, and dried *in vacuo*. A Kaisertest is carried out to determine the completion of the coupling.

NOTE: For more sophisticated couplings, reaction time can be extended or more reactive coupling reagents like HATU, PyBOP, or COMU can be used. In some cases, addition of a base as DIPEA is required to deprotonate acid and amine groups and increase the yield.

3.2.2.3. Cleavage of the Fmoc protecting group The resin is let to swell in 500 μl of DMF for 15 min and is treated twice with 500 μl of a fresh solution of 20% piperidine in DMF (v/v). The reaction mixture is shaken for 20 min at room temperature. If the cleavage is directly followed by another coupling step, the resin is thoroughly washed with DMF. If the Fmoc removal is the last step of the synthesis, the resin is washed five times, respectively, with DMF, DCM, MeOH, and Et_2O, and dried *in vacuo*.

3.2.3. Cleavage from the resin

The use of a scavenger mixture is required to protect electron-rich amino acids from alkylation or oxidation by highly reactive carbocations or radicals generated during the cleavage. The composition of the scavenger mixture essentially depends on the amino acids contained in the peptide sequence. In our laboratory, we use a standard mixture of thioanisol/thiocresol (1:1, v/v) and a mixture of thioanisol/ethanedithiol (7:3, v/v) for peptides containing Trp, Met, and Cys. The peptides can be precipitated in dry and ice-cooled *n*-hexane or diethyl ether, depending on their hydrophobicity.

3.2.3.1. Sample cleavage of peptides Few beads of dry resin are transferred in a reaction tube. 10 μl of scavenger and 90 μl of TFA are added and the reaction mixture is shaken at room temperature for 3 h. Subsequently, 1 ml of ice-cold, dry Et_2O/hexane (1:3, v/v) is added and the peptide is let to precipitate for 20 min at − 20 °C. After centrifugation (2 min, 4000 rpm, 4 °C), the supernatant is discarded and the pellet is washed five times with 1 ml of Et_2O/hexane (1:3, v/v) and centrifuged. The pellet is then dried on air and dissolved in 300 μl of ACN/H_2O to be analyzed by mass spectrometry and analytical HPLC.

3.2.3.2. Total cleavage of peptides 100 μl of scavenger and 900 μl of TFA are added to the dry resin. The mixture is shaken at room temperature for 3 h and then filtered to remove the resin. 9 ml of ice-cold, dry Et_2O/hexane (1:3, v/v) is added to the solution and the peptide is let to precipitate 1 h at − 20 °C. After centrifugation (2 min, 4000 rpm, 4 °C), the supernatant is discarded and the pellet is washed five times with 9 ml of Et_2O/hexane (1:3, v/v) and centrifuged. The pellet is dried *in vacuo* and is dissolved in ACN/H_2O to be analyzed by mass spectrometry and analytical HPLC.

3.3. Analytical methods

3.3.1. Kaisertest

The completion of a coupling can be checked using the Kaisertest. Few beads of resin are transferred into a reaction tube. One drop of solution I, II, and III (Table 6.2)) are added. A negative control without resin is treated equally. The reaction mixture is incubated at 95 °C for 5 min. The detection of free amino groups results in a color change to blue caused by the reaction of the free N-terminal amine with ninhydrin. If the solution stays yellow, the coupling is complete.

Table 6.2 Composition of Kaisertest solutions

Reaction solution	Ingredients
Solution I	1 g ninhydrin in 20 ml ethanol
Solution II	80 g phenol in 20 ml ethanol
Solution III	0.4 ml 1 m*M* aqueous KCN in 20 ml piperidine

3.3.2. Analytical RP-HPLC

The purity of the synthesized peptides is measured by analytical RP-HPLC with a C18 column and an elution gradient composed by (A): H_2O + 0.1% TFA and (B): ACN + 0.08% TFA. The peptides are detected with a diode array detector at 220 nm.

3.3.3. MALDI-TOF mass spectrometry

The identity of the peptides is determined with mass spectrometry, through their mass-to-charge ratio. Using the excitation method MALDI, the peptide is embedded in a solid, electron-rich matrix and ionized by a laser beam. In the analyzer, the time of flight (TOF), proportional to the mass-to-charge ratio, is determined.

3.4. Peptide purification

The peptides are purified by preparative RP-HPLC with a C18 column. Up to 20 mg of peptide can be dissolved in 1 ml of ACN/H_2O and injected to the column. The peptide is then eluted with a gradient of (A) H_2O + 0.1% TFA and (B) ACN + 0.08% TFA at a flow rate of 10 ml/min. Detection is carried out at 220 nm for peptide bonds and 280 nm for aromatic groups. Each fraction of peptide is then concentrated, identified with MALDI-TOF MS, and the purity is determined by analytical RP-HPLC. The pure fractions are collected together and lyophilized.

3.5. Materials

Chemicals are purchased from Sigma-Aldrich, Merck, and Iris Biotech. The chemicals and devices specific for SPPS are only mentioned.

3.5.1. Chemicals

- Solvents for washing steps and coupling:
- Dimethyl formamide (DMF),
- *N*-Methylpyrrolidone (NMP),
- Dichloromethane (DCM),
- Methanol (MeOH),
- Diethyl ether (Et_2O).

- Reagents for coupling steps:
  - 4-(2′,4′-dimethoxyphenyl-Fmoc-aminomethyl)-phenoxy resin (Rink amide resin),
  - Fmoc protected amino acids,
  - Hydroxybenzotriazole (HOBt),
  - Diisopropylcarbodiimide (DIC).
- Reagent for Fmoc cleavage:
  - Piperidine.

3.5.2. Devices

- Peptide synthesizer: MultiSyroTech Syro II and MultiSyroTech Syro XP
- Vacuum centrifuge: Jouan RC1022
- Analytical RP-HPLC: Merck-Hitachi System with a column Grace Vydac, Vydac RP18
- Preparative RP-HPLC: Shimazu System with a column Jupiter 10u Proteo 90 Å, phenomenex (250 × 21.2 mm; 7.78 μm; 90 Å)
- MALDI-TOF mass spectrometer: Bruker Daltonics Ultraflex III
- Thermomixer: Eppendorf Thermomixer compact, Eppendorf Thermomixer 5436

4. Functional Assays for Ghrelin Receptor Inverse Agonists

The potency of a ligand is defined as its ability to activate or inhibit a receptor and can be evaluated by functional assays. Several functional assays are available for GPCRs. Most of them consist in measuring a component of the signal transduction cascade, generated by the GPCR in response to a ligand.

Briefly, a ligand induces a conformational change of the receptor, leading to the dissociation of the G protein subunits. Depending on the nature of the G subunit, different pathway can be activated or inhibited. The ghrelin receptor is a $G_{\alpha q/11}$-coupled receptor. Thus, it directly activates the phospholipase C (PLC) pathway. Once activated, the PLC cleaves the phosphatidylinositol bisphosphate into inositol trisphosphate (IP_3) and diacyl glycerol (DAG). IP_3 and DAG will induce the releases of second messengers such as Ca^{2+} and the activation of protein kinases. Ca^{2+} indirectly activates the adenylyl cyclase pathway, triggering the release of cAMP (Kojima and Kangawa, 2005; Mizuno and Itoh, 2009).

Ghrelin receptor possesses a high constitutive activity that can only be visualized in certain assays. To measure inverse agonist potencies, a functional assay showing the constitutive activity is mandatory. In literature, several functional assays are described for the ghrelin receptor and will be summarized. In our laboratory, the measurement of the inositol trisphosphate turnover is the assay of choice and will be detailed.

4.1. Overview of functional assays used for the ghrelin receptor

The overview of the functional assays used for the ghrelin receptor is not exhaustive. It includes the main assays used to test ghrelin receptor ligands and/or to elucidate the intracellular signaling systems related to the receptor. For each assay, the basis of a method is given as example to illustrate the principle of the assay. The functional assays are summarized in Table 6.3, with a specific mention for the detection of constitutive activity.

4.1.1. Intracellular calcium mobilization assay

Calcium mobilization assay is currently the most popular functional assays to evaluate the potency of ghrelin receptor agonists and antagonists. It consists in the measurement of intracellular calcium with a calcium-sensitive fluorescent dye. Nevertheless, constitutive activity is not visible with Ca^{2+} assay. Thus, calcium measurement cannot be employed for measuring inverse agonist activity. An example of the assay is given by Moulin *et al.* (2007) to evaluate the potency of triazole ligands. CHO cells were transiently transfected with the ghrelin receptor DNA and seeded. The next day, the cells were washed, loaded with the fluorescent calcium indicator Fluo-4AM, and incubated 1 h in the dark. After washing, the plates were placed in a scanning fluorometer with a plate containing the ligands solution. The compounds were automatically added to the cells after 15 s and the fluorescence output is measured during 60 s.

Table 6.3 Summary of the functional assays for ghrelin receptor

Assay	Detection of constitutive activity	Example for method
Calcium mobilization	−	Moulin *et al.* (2007)
Inositol trisphosphate turnover	+	Holst *et al.* (2006)
CRE-luciferase reporter assay	+	Holst *et al.* (2003)
SRE-luciferase assay	+	Holst *et al.* (2004)
cAMP accumulation	+	Carreira *et al.* (2004)

4.1.2. Inositol trisphosphate turnover assay

The release of IP_3 is directly regulated by G_q activation. In our laboratories, the inositol trisphosphate turnover assay (IP_3 assay) is the functional assay of choice for measuring agonist and inverse agonist potencies. Indeed, the high constitutive signaling of ghrelin receptor was discovered in IP_3 assay. In addition, the MSP was first identified as a ghrelin receptor inverse agonist (Holst *et al.*, 2003).The method for IP_3 assay will be detailed in Section 4.2.

4.1.3. CRE-luciferase reporter assay

Release of the cAMP responsive element (CRE) is induced by the elevation of intracellular cAMP and Ca^{2+} in the G_q pathway (Chen *et al.*, 1995). Activation of the CRE pathway can be monitored by a reporter assay using CRE-driven luciferase activity. Holst *et al.* tested the ability of the ghrelin receptor to activate the CRE pathway (Holst *et al.*, 2003). HEK-293 cells were transiently transfected with the receptor DNA and a mixture of pFA2-CRE binding protein and pFR-Luc reporter plasmid. The next day, cells were stimulated with different ligand for 5 h. Cells were washed with PBS, luciferase assay reagent was added, and related luminescence was measured. They showed that this assay could be used for the ghrelin receptor and detects the constitutive activity.

4.1.4. SRE-luciferase reporter assay

The serum responsive element (SRE) pathway is mainly stimulated by $G_{\alpha 13}$ protein-coupled receptors. The SRE reporter assay follows the same principle than the CRE reporter assay. Holst *et al.* showed that ghrelin receptor stimulates SRE-mediated transcriptional activity (Holst *et al.*, 2004). Therefore, HEK-293 cells were transfected with a SRE-Luc reporter plasmid and the receptor DNA. The next day, cells were stimulated with different ligands for 5 h and then washed with PBS. Luciferase assay reagent was added and related luminescence was measured. Compared with the IP_3 assay and the CRE reporter assay, the constitutive activity of the receptor was decreased and a 10-fold loss of potency was detected for the agonists tested.

4.1.5. cAMP accumulation assay

Measurement of cAMP release is the assay of choice for $G_{\alpha 13}$ protein-coupled receptors. G_s directly stimulates cAMP release through activation of adenylyl cyclase. cAMP concentration can be measured with fluorescence, luminescence, or radioactive-based assays. cAMP accumulation assay has been performed on the ghrelin receptor, although cross-talk between ghrelin receptor and G_s pathway is not elucidated (Carreira *et al.*, 2004; Hermansson *et al.*, 2007).

An example of radioactive-based assay is given in Carreira *et al.* (2004). HEK-293 cells, stably expressing the human ghrelin receptor 1a (HEK-GHSR-1a), were suspended in DMEM and incubated for 30 min with 1.0 m*M* 3-isobutyl-1-methylxanthine to prevent cAMP degradation. Cells were stimulated with the tested ligand for 30 min and then lysed with a buffer solution containing EDTA. The lysate was used to measure cAMP concentration by a competitive protein-binding assay using [8-^{3}H] cAMP.

4.2. Inositol trisphosphate turnover assay

In our laboratory, the assay of choice is the inositol trisphosphate turnover assay. It can detect the potency of both agonists and inverse agonists. For inverse agonists, the decrease of the constitutive activity is measured.

The IP_3 assay consists in the measurement of radioactive [^{3}H]-myo-inositol, released by cells, transiently transfected with the ghrelin receptor, after stimulation with bioactive compounds. For this purpose, cells need to be transfected with a vector containing the ghrelin receptor, then labeled with [^{3}H]-myo-inositol, and finally stimulated with the peptides. In this section, we describe this method in detail.

4.2.1. Cell culture and cell seeding

Several cell lines can be used for functional assays. In the literature, COS-7 cells, HEK-293 cells, and CHO cells were mostly transfected with the ghrelin receptor. In our laboratory, we developed a transfection method with COS-7 cells. All cell experiments are performed under sterile conditions. Media and solutions are warmed to 37 °C prior to use. Incubation of COS-7 cells should always be carried out at 37 °C and 5% CO_2 under humidified atmosphere.

4.2.1.1. Method COS-7 cell culture is performed in 75 cm^{2}-cell culture flasks, with DMEM, 10% FCS (v/v), and 1% antibiotics (v/v) (penicillin+streptomycin). When cells reached confluency, they can be seeded out. Therefore, cells are washed with 10 ml PBS and incubated with 2 ml trypsin/EDTA for 1 min. Subsequently, cells are loosened mechanically from the ground and resuspended in 10 ml of COS-7 medium (DMEM, 10% FCS (v/v)).

COS-7 cells are seeded in a 24-well plate with 80,000–100,000 cells per well on the first day of the assay. The cell concentration is determined with a counting chamber. COS-7 medium is added to the cell suspension to obtain the required dilution. Seeding is carried out by adding successively 500 μl of COS-7 medium and 500 μl of the final cell suspension to each well.

4.2.1.2. Materials

- COS-7: African Green Monkey (Cercopithecus aethiops), SV40-transfected kidney fibroblast cell line,
- DMEM, high glucose (4.5 g/l) with L-glutamine,
- COS-7 medium: DMEM, high glucose (4.5 g/l) with L-glutamine + 10% FCS (v/v),
- DMEM, high glucose (4.5 g/l) with L-glutamine + 10% FCS (v/v) + 1% penicillin/streptomycin (v/v),
- PBS: phosphate buffered saline, without Ca^{2+} and Mg^{2+},
- Trypsin/EDTA: Trypsin: 0.5 mg/ml and EDTA: 0.22 mg/ml in PBS.

4.2.2. Transfection with the human ghrelin receptor

When the confluency reaches about 80% (generally day 2), cells are transfected with a pVitro vector containing the ghrelin receptor (GHS-R1a) and a green-fluorescent protein (EYFP). The transfection is achieved through lipofection with a transfecting reagent, Metafectene™. EYFP allows the use of fluorescence microscopy to control the success of the transfection.

4.2.2.1. Method In the 24-well plate, the medium is removed and each well is filled with 300 μl of COS-7 medium. Solutions of Metafectene™ (0.9 μl in 50 μl of DMEM per well) and of DNA (0.3 μg in 50 μl of DMEM per well) are prepared, mixed together, and incubated for 20 min at room temperature. 100 μl of this final solution is then added to each well. As a negative control, three wells are not transfected and 100 μl DMEM is added to these wells. The transfection takes place overnight for about 14 h. On the next day, the transfection medium is removed and replaced with 300 μl of fresh COS-7 medium.

4.2.2.2. Materials

- DMEM, high glucose (4.5 g/l) with L-glutamine,
- COS-7 medium,
- Metafectene™,
- The vector system GHS-R_EYFP_pVITRO2_mcs (containing hygromycin resistance),
- Fluorescence microscope: Zeiss Axiovert 25 with YFP Fs46 filter.

4.2.3. Radioactive labeling

One day after transfection, cells are labeled with [^{3}H]-myo-inositol and incubated. At the end of the incubation, LiCl is added to inhibit the inositol monophosphatase and prevent the degradation of inositol monophosphate to inositol and phosphate.

4.2.3.1. Method The medium is removed and the wells are filled with 150 μl of fresh COS-7 medium. A solution of [^{3}H]-myo-inositol in COS-7 medium is prepared at a concentration of 4 μCi/ml and 150 μl is added to each well. Cells are incubated at 37 °C, 5% CO_2, and 95% humidity for 18 h. The medium is then aspirated and 500 μl of DMEM + LiCl (10 m*M*) is added to each well.

4.2.3.2. Materials

- [2-^{3}H(N)]-myo-inositol (NEN® Radiochemicals),
- COS-7 medium,
- DMEM, high glucose (4.5 g/l) with L-glutamine + LiCl (10 m*M*).

4.2.4. Receptor stimulation

The stimulation is carried out directly at the end of the labeling, with six to nine different concentrations, in duplicate, for each compound tested. As controls, three wells are not stimulated and filled with 300 μl of medium (DMEM/LiCl + 1% BSA). These wells will represent the constitutive activity. The three wells not transfected represent the radioactive background.

4.2.4.1. Method The peptides are dissolved in DMEM/LiCl + 1% BSA and diluted to obtain the desired concentration. The medium (DMEM+10 m*M* LiCl) is aspirated from the 24-well plate and 300 μl of the peptide solution is added. For the stimulation, the cells are incubated at 37 °C and 5% CO_2 in a humidified atmosphere for 2 h. Then, the medium is aspirated and the cells lysed. For this, 150 μl of a 0.1 *M* NaOH solution is added to each well. After 5 min of incubation, the solution is neutralized with 50 μl of a 0.2 *M* formic acid solution. Then, 1 ml of IP dilution buffer is added to each well. Cell debris is aspirated by pipetting (50 μl for each well). The radioactive mixture is separated by column exchange chromatography.

4.2.4.2. Materials

- DMEM, high glucose (4.5 g/l) with L-glutamine + 10 m*M* LiCl + 1% BSA (m/v),
- 0.1 *M* NaOH solution,
- 0.2 *M* formic acid solution,
- IP dilution buffer: 5.0 m*M* Na-borate + 0.5 m*M* Na-EDTA.

4.2.5. Ion exchange chromatography

Ion exchange chromatography is performed on the radioactive mixture to separate inositol species from the nonspecific radioactive species generated during the labeling. Columns containing a positively charged resin are loaded with the radioactive mixture. Inositol species bind to the resin with their negatively charged phosphate groups and can be eluted after washing steps.

4.2.5.1. Method The columns are regenerated successively with:

- 10 ml of regeneration buffer,
- 10 ml of H_2O,
- 10 ml of regeneration buffer,
- 2 × 10 ml of H_2O.

The samples are then loaded and the columns are washed successively with:

- 2 × 5 ml of H_2O,
- 2 × 5 ml of glycerolphosphate elution buffer,
- 2 × 5 ml of H_2O.

Radioactive inositol phosphate species are eluted with 2.5 ml of $IP/IP_2/IP_3$ elution buffer in 20 ml scintillation vials containing 10 ml of scintillation mix. Vials are strongly mixed and radioactivity is measured with a betacounter.

4.2.5.2. Materials

- Bio-Rad AG 1-X8 anion-exchange resin.
- Regeneration buffer: 3.0 *M* ammoniumformate + 0.25 *M* formic acid.
- Glycerolphosphate elution buffer: 5 m*M* Na-borate + 60 m*M* Na-formate.
- $IP/IP_2/IP_3$ elution buffer: 0.2 *M* ammoniumformate + 0.1 *M* formic acid.
- Betacounter: TRI-CARB 2900TR and TRI-CARB 2910TR (Hewlett-Packard).

4.2.6. Analysis

Data are analyzed with Prism 3.0 software (GraphPad). Therefore, dpm values are assigned to the corresponding peptide concentrations. Nonlinear regression was used to obtain sigmoidal curves. The turning point of the curve represents the EC_{50} value. For better comparability, dpm values can be normalized to the constitutive activity.

REFERENCES

Bednarek, M. A., Feighner, S. D., Pong, S. S., McKee, K. K., Hreniuk, D. L., Silva, M. V., Warren, V. A., Howard, A. D., Van Der Ploeg, L. H., and Heck, J. V. (2000). Structure–function studies on the new growth hormone-releasing peptide, ghrelin: Minimal sequence of ghrelin necessary for activation of growth hormone secretagogue receptor 1a. *J. Med. Chem.* **43,** 4370–4376.

Carreira, M. C., Camina, J. P., Smith, R. G., and Casanueva, F. F. (2004). Agonist-specific coupling of growth hormone secretagogue receptor type 1a to different intracellular signaling systems. Role of adenosine. *Neuroendocrinology* **79,** 13–25.

Chen, W., Shields, T. S., Stork, P. J., and Cone, R. D. (1995). A colorimetric assay for measuring activation of Gs- and Gq-coupled signaling pathways. *Anal. Biochem.* **226,** 349–354.

Chollet, C., Meyer, K., and Beck-Sickinger, A. G. (2009). Ghrelin—A novel generation of anti-obesity drug: Design, pharmacomodulation and biological activity of ghrelin analogues. *J. Pept. Sci.* **15,** 711–730.

Cummings, D. E. (2006). Ghrelin and the short- and long-term regulation of appetite and body weight. *Physiol. Behav.* **89,** 71–84.

De Vriese, C., and Delporte, C. (2007). Influence of ghrelin on food intake and energy homeostasis. *Curr. Opin. Clin. Nutr. Metab. Care* **10,** 615–619.

Ehlert, F. J., Roeske, W. R., Gee, K. W., and Yamamura, H. I. (1983). An allosteric model for benzodiazepine receptor function. *Biochem. Pharmacol.* **32,** 2375–2383.

Gutierrez, J. A., Solenberg, P. J., Perkins, D. R., Willency, J. A., Knierman, M. D., Jin, Z., Witcher, D. R., Luo, S., Onyia, J. E., and Hale, J. E. (2008). Ghrelin octanoylation mediated by an orphan lipid transferase. *Proc. Natl. Acad. Sci. USA* **105,** 6320–6325.

Hermansson, N. O., Morgan, D. G., Drmota, T., and Larsson, N. (2007). Adenosine is not a direct GHSR agonist—artificial cross-talk between GHSR and adenosine receptor pathways. *Acta Physiol. (Oxf.)* **190,** 77–86.

Holst, B., and Schwartz, T. W. (2004). Constitutive ghrelin receptor activity as a signaling set-point in appetite regulation. *Trends Pharmacol. Sci.* **25,** 113–117.

Holst, B., Cygankiewicz, A., Jensen, T. H., Ankersen, M., and Schwartz, T. W. (2003). High constitutive signaling of the ghrelin receptor—identification of a potent inverse agonist. *Mol. Endocrinol.* **17,** 2201–2210.

Holst, B., Holliday, N. D., Bach, A., Elling, C. E., Cox, H. M., and Schwartz, T. W. (2004). Common structural basis for constitutive activity of the ghrelin receptor family. *J. Biol. Chem.* **279,** 53806–53817.

Holst, B., Lang, M., Brandt, E., Bach, A., Howard, A., Frimurer, T. M., Beck-Sickinger, A., and Schwartz, T. W. (2006). Ghrelin receptor inverse agonists: Identification of an active peptide core and its interaction epitopes on the receptor. *Mol. Pharmacol.* **70,** 936–946.

Holst, B., Mokrosinski, J., Lang, M., Brandt, E., Nygaard, R., Frimurer, T. M., Beck-Sickinger, A. G., and Schwartz, T. W. (2007). Identification of an efficacy switch region in the ghrelin receptor responsible for interchange between agonism and inverse agonism. *J. Biol. Chem.* **282,** 15799–15811.

Howard, A. D., Feighner, S. D., Cully, D. F., Arena, J. P., Liberator, P. A., Rosenblum, C. I., Hamelin, M., Hreniuk, D. L., Palyha, O. C., Anderson, J., Paress, P. S., Diaz, C., *et al.* (1996). A receptor in pituitary and hypothalamus that functions in growth hormone release. *Science* **273,** 974–977.

Kojima, M., and Kangawa, K. (2005). Ghrelin: Structure and function. *Physiol. Rev.* **85,** 495–522.

Leurs, R., Smit, M. J., Alewijnse, A. E., and Timmerman, H. (1998). Agonist-independent regulation of constitutively active G-protein-coupled receptors. *Trends Biochem. Sci.* **23,** 418–422.

Merrifield, R. B. (1963). Solid phase peptide synthesis. I. The synthesis of a tetrapeptide. *J. Am. Chem. Soc.* **85,** 2149–2154.

Mizuno, N., and Itoh, H. (2009). Functions and regulatory mechanisms of Gq-signaling pathways. *Neurosignals* **17,** 42–54.

Moulin, A., Demange, L., Berge, G., Gagne, D., Ryan, J., Mousseaux, D., Heitz, A., Perrissoud, D., Locatelli, V., Torsello, A., Galleyrand, J. C., Fehrentz, J. A., *et al.* (2007). Toward potent ghrelin receptor ligands based on trisubstituted 1, 2, 4-triazole structure. 2. Synthesis and pharmacological in vitro and in vivo evaluations. *J. Med. Chem.* **50,** 5790–5806.

Nakazato, M., Murakami, N., Date, Y., Kojima, M., Matsuo, H., Kangawa, K., and Matsukura, S. (2001). A role for ghrelin in the central regulation of feeding. *Nature* **409,** 194–198.

Nargund, R. P., Patchett, A. A., Bach, M. A., Murphy, M. G., and Smith, R. G. (1998). Peptidomimetic growth hormone secretagogues. Design considerations and therapeutic potential. *J. Med. Chem.* **41,** 3103–3127.

Neubig, R. R., Spedding, M., Kenakin, T., and Christopoulos, A. (2003). International Union of Pharmacology Committee on Receptor Nomenclature and Drug Classification. XXXVIII. Update on terms and symbols in quantitative pharmacology. *Pharmacol. Rev.* **55,** 597–606.

Pantel, J., Legendre, M., Cabrol, S., Hilal, L., Hajaji, Y., Morisset, S., Nivot, S., Vie-Luton, M. P., Grouselle, D., de Kerdanet, M., Kadiri, A., Epelbaum, J., *et al.* (2006). Loss of constitutive activity of the growth hormone secretagogue receptor in familial short stature. *J. Clin. Invest.* **116,** 760–768.

Smith, R. G., Feighner, S., Prendergast, K., Guan, X., and Howard, A. (1999a). A new orphan receptor involved in pulsatile growth hormone release. *Trends Endocrinol. Metab.* **10,** 128–135.

Smith, R. G., Palyha, O. C., Feighner, S. D., Tan, C. P., McKee, K. K., Hreniuk, D. L., Yang, L., Morriello, G., Nargund, R., Patchett, A. A., and Howard, A. D. (1999b). Growth hormone releasing substances: Types and their receptors. *Horm. Res.* **51**(Suppl. 3), 1–8.

Smith, R. G., Leonard, R., Bailey, A. R., Palyha, O., Feighner, S., Tan, C., McKee, K. K., Pong, S. S., Griffin, P., and Howard, A. (2001). Growth hormone secretagogue receptor family members and ligands. *Endocrine* **14,** 9–14.

van der Lely, A. J., Tschop, M., Heiman, M. L., and Ghigo, E. (2004). Biological, physiological, pathophysiological, and pharmacological aspects of ghrelin. *Endocr. Rev.* **25,** 426–457.

Yang, J., Brown, M. S., Liang, G., Grishin, N. V., and Goldstein, J. L. (2008). Identification of the acyltransferase that octanoylates ghrelin, an appetite-stimulating peptide hormone. *Cell* **132,** 387–396.

CHAPTER SEVEN

Constitutive Activity and Inverse Agonism at the α_{1a} and α_{1b} Adrenergic Receptor Subtypes

Susanna Cotecchia[*,†]

Contents

Abstract

The α_{1b}-adrenergic receptor (AR) was, after rhodopsin, the first G protein-coupled receptor (GPCR) in which point mutations were shown to trigger constitutive (agonist-independent) activity. Constitutively activating mutations have been found in other AR subtypes as well as in several GPCRs. This chapter briefly summarizes the main findings on constitutively active mutants of the α_{1a}- and α_{1b}-AR subtypes and the methods used to predict activating mutations, to measure constitutive activity of Gq-coupled receptors and to investigate inverse agonism. In addition, it highlights the implications of studies on constitutively active AR mutants on elucidating the molecular mechanisms of receptor activation and drug action.

* Department of General and Environmental Physiology, University of Bari, Italy
† Department of Pharmacology and Toxicology, University of Lausanne, Switzerland

Methods in Enzymology, Volume 485
ISSN 0076-6879, DOI: 10.1016/S0076-6879(10)85007-4

Abbreviations

AR	adrenergic receptor
CAM	constitutively active mutant
GPCR	G protein-coupled receptor
IP	inositol phosphate
MD	molecular dynamics

1. Introduction

Within the large family of G protein-coupled receptors (GPCRs), the adrenergic receptors (ARs) include nine gene products: three β (β_1, β_2, β_3), three α_2 (α_{2A}, α_{2B}, α_{2C}), and three α_1 (α_{1a}, α_{1b}, α_{1d}) receptor subtypes. The α_1-AR subtypes (Schwinn *et al.*, 1995) mediate several important central and peripheral effects of epinephrine and norepinephrine including the control of blood pressure, heart growth, glucose metabolism, and various behavioral responses.

Mutational analysis performed by various investigators contributed to identify some of the structural determinants of the α_1-AR subtypes involved in each of the three main "classical" functional properties of GPCRs: (1) ligand binding (Cavalli *et al.*, 1996; Hwa and Perez, 1996); (2) coupling to G protein-effector systems (Cotecchia *et al.*, 1992; Greasley *et al.*, 2001); (3) desensitization (Diviani *et al.*, 1996, 1997; Lattion *et al.*, 1994; Vazquez-Prado *et al.*, 2000).

Activation of the α_1-AR subtypes causes polyphosphoinositide hydrolysis catalyzed by phospholipase C *via* pertussis toxin-insensitive G proteins of the $G_{q/11}$ family in almost all tissues where this effect has been examined. Polyphosphoinositide hydrolysis results in the increase of intracellular inositol phosphate (IP) production. Several lines of evidence demonstrated that the third intracellular (i3) loop contains the main structural determinants involved in α_{1b}-AR coupling to G proteins of the $G_{q/11}$ family. In particular, the results from site-directed mutagenesis in conjunction with the predictions of molecular modeling suggested that specific residues, that is, R^{254} and K^{258} in the i3 loop as well as L^{151} in the second intracellular loop, are directly involved in receptor-G protein interaction and/or receptor-mediated activation of the G protein (Fanelli *et al.*, 1999; Greasley *et al.*, 2001). Specific residues involved in receptor-G protein coupling have not been identified yet in the α_{1a}- and α_{1d}-AR subtypes.

In the context of a study in which a number of residues in the i3 loop of the α_{1b}-AR were mutated, it was fortuitously discovered that a conservative substitution (A293L) in the cytosolic extension of helix 6 (Fig. 7.1) resulted in agonist-independent receptor activation of polyphosphoinositide hydrolysis (Cotecchia *et al.*, 1990). The α_{1b}-AR was, after rhodopsin, the first GPCR in which point mutations were shown to trigger constitutive (agonist-independent) receptor activation. To further assess the role of A293, this amino acid was systematically mutated by substituting each of the other 19 amino acids (Kjelsberg *et al.*, 1992). Remarkably, all possible amino acid substitutions of A293$^{(6.34)}$ (the superscript refers to numbering from Ballesteros and Weinstein, 1995) in the α_{1b}-AR induced variable levels of constitutive activity which was the highest for the A293E mutant (Fig. 7.2).

Similar mutations were performed in the β_2 (Samama *et al.*, 1993) and α_{2A}AR (Ren *et al.*, 1993) leading to increased or decreased agonist-independent adenylyl cyclase activity, respectively.

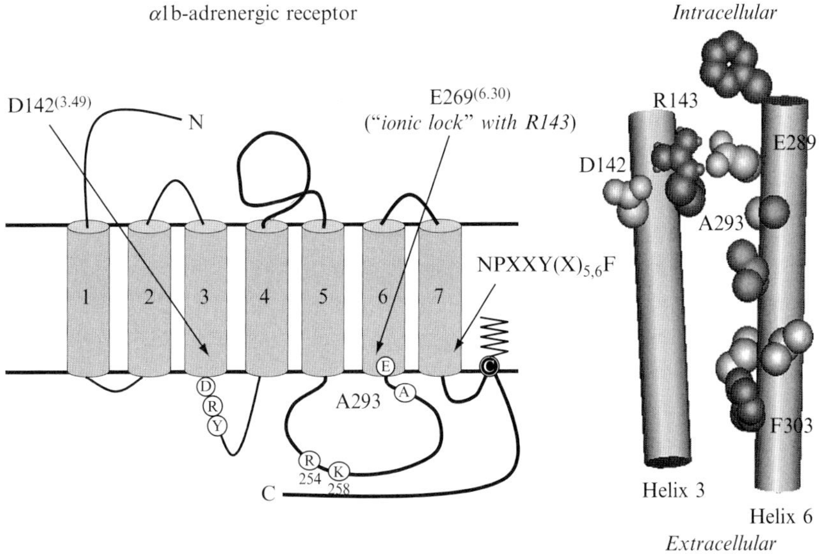

Figure 7.1 The α_{1b}-adrenergic receptor (AR). (Left) Topographical model of the receptor displaying key amino acids involved in receptor function mentioned in the text. (Right) Relative position of helices 3 and 6 in the homology model of the wild-type α_{1b}-AR. Comparative modeling and molecular dynamic simulations were performed as described in Greasley *et al.* (2002). The view displays the amino acids of helices 3 and 6 involved in receptor activation. Van der Waals spheres, whose radius has been reduced by 40%, depict each side chain. The color indicates the effect of mutations: white/no effect, green/constitutively activating, red/impairing receptor signaling, violet/either impairing or constitutively activating depending upon the substituent amino acid. (See Color Insert.)

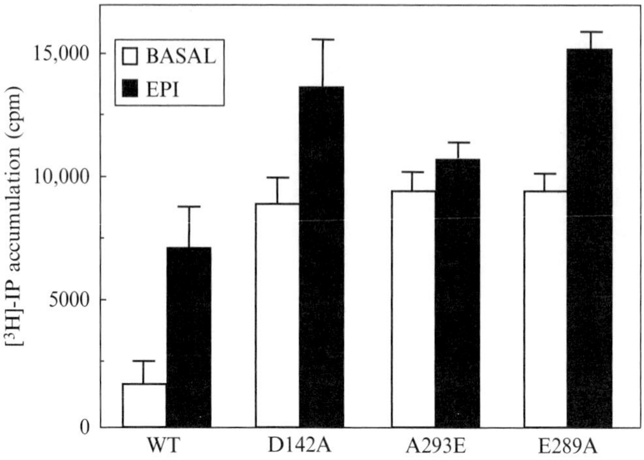

Figure 7.2 Constitutively active mutants of the α_{1b}-AR. COS-7 cells (0.3×10^6) were seeded in 6-well plates and transfected with different amounts of DNA encoding the receptors (0.2 μg/million cells for wild type, A293E, and E289A, 2 μg/million cells for D142A) to obtain comparable levels of expression. The receptor densities measured on membrane preparations were ≈200 fmol/million of cells. Cells were harvested 48 h after transfection. Transfected cells were labeled for 15–18 h with myo-[^{3}H]inositol at 5 μCi/ml in inositol-free DMEM supplemented with 1% fetal bovine serum. Cells were preincubated for 10 min in PBS containing 20 m*M* LiCl and then treated with different ligands. Total inositol phosphates (IPs) were extracted and separated as previously described (Cotecchia *et al.*, 1992). IP accumulation was measured in cells expressing the wild-type α_{1b}-AR and its mutants D142A, A293E, and E289A in the absence (BASAL) or in the presence of 10^{-4} *M* epinephrine (EPI) for 45 min.

The discovery of constitutively active mutants (CAMs) in the AR family had several implications (Costa and Cotecchia, 2005). First, it highlighted constitutive activity as a potential intrinsic feature of GPCRs. Second, it encouraged the search for spontaneous activating mutations involved in the pathogenesis of diseases. Third, it lead to the identification of inverse agonism, that is, the capacity of ligands, previously characterized as AR antagonists, to inhibit the constitutive activity of the receptors.

In this chapter, we will review some of the findings obtained in the past several years on constitutive activity and inverse agonism at the α_{1a}- and α_{1b}-AR subtypes. A different chapter of this volume is devoted to the α_{1d}-AR subtype. We will focus our attention on three methodological issues: (i) combination of computational modeling and site-directed mutagenesis of the receptor to identify constitutively activating mutations; (ii) measuring constitutive activity of receptor-mediated Gq activation; (iii) measuring inverse agonism.

2. Combination of Computational Modeling and Site-Directed Mutagenesis of the Receptor to Identify Constitutively Activating Mutations

2.1. The predictive ability of molecular modeling: The E/DRY motif

A detailed analysis of the pharmacological properties of a β_2-AR CAM was instrumental to propose the "allosteric ternary complex model" (Samama *et al.*, 1993). This extended version of the ternary complex model explicitly recognized the allosteric transition between at least two interconvertible allosteric "states," **R** (inactive or ground state) and **R★** (active). In the absence of the agonist, **R** predominates, whereas agonists trigger the equilibrium toward **R★** thus favoring its stabilization. This analysis led to the suggestion that constitutively activating mutations mimic, at least to some extent, the conformational change triggered by agonist binding to a wild-type GPCR.

It therefore made the hypothesis that, in the absence of agonist, structural "constraints" keep the wild-type receptor in its ground or inactive state (**R**), whereas agonist binding or activating mutations release such constraints triggering the conversion of the receptor into its active state (**R★**), which couples to G proteins. Once this hypothesis was made, we combined 3-D model building of the α_{1b}-AR structure with molecular dynamics (MD) simulations of the receptor to compare the structural/dynamic features of the wild-type structure (ground or inactive state) with that of CAMs carrying mutations of A293 (Scheer *et al.*, 1996).

The first α_{1b}-AR model was built using an iterative *ab initio* procedure consisting of upgrading and complicating the model to incorporate the increasing number of experimental data on rhodopsin and the homologous GPCRs, as previously described (Fanelli *et al.*, 1998). The *ab initio* model of the α_{1b}-AR was extremely useful for developing a computational approach which allowed to define the structural/dynamic features differentiating the inactive from the active receptor states. These studies highlighted that the cationic residue $R143^{(3.50)}$ of the highly conserved E/DRY motif at the cytosolic end of helix 3 (Fig. 7.1) undergoes the most conspicuous changes. Whereas in the inactive state $R143^{(3.50)}$ is involved in persistent hydrogen bonding interactions with residues forming a highly conserved "polar pocket" (Scheer *et al.*, 1996), the active states are characterized by a progressive shift of $R143^{(3.50)}$ out of this pocket. The model predicted that protonation of the aspartate ($D142^{(3.49)}$) of the E/DRY motif would induce a conformational change of the receptor leading to the shift of $R143^{(3.50)}$ out of the "polar pocket," a hall mark of the receptor active states. Site-directed mutagenesis of $D142^{(3.49)}$ confirmed this prediction since mutations of the aspartate generated a family of CAMs (Fig. 7.2; Scheer *et al.*, 1996, 1997).

The interpretative and predictive abilities of the computational approach improved as soon as the first crystal structure of rhodopsin became available, as extensively described elsewhere (Fanelli and De Benedetti, 2005). Despite the structural differences between the *ab initio* and comparative models, they both suggested that in the α_{1b}-AR the highly conserved E/DRY motif is an important switch of receptor activation (Scheer *et al.*, 2000). Increased constitutive activity was also found after mutating the acidic residue of the E/DRY motif in other receptors including rhodopsin (Cohen *et al.*, 1993) and the β_2-AR (Rasmussen *et al.*, 1999). Altogether, these findings strongly suggest that mutations of the highly conserved E/DRY sequence might represent a general strategy to increase constitutive activity of GPCRs belonging to the rhodopsin-like class of receptors.

2.2. Microdomains involved in receptor activation

In the homology model of the α_{1b}-AR (Greasley *et al.*, 2002), the arginine of the E/DRY motif displays a double salt bridge with both the adjacent D142$^{(3.49)}$ and a glutamate (E289$^{(6.30)}$) on the cytosolic end of helix 6 (Fig. 7.1). This interaction pattern of R143$^{(3.50)}$, inherited from the rhodopsin structure, constitutes an important feature of the ground or inactive state of the receptor (Hofmann *et al.*, 2009; Palczewski *et al.*, 2000). Similar to the effect induced by mutating the aspartate of the E/DRY motif, mutations of E289$^{(6.30)}$ on helix 6 markedly increased the constitutive activity of the α_{1b}-AR (Fig. 7.2; Greasley *et al.*, 2002). Also mutations of F$^{6.44}$, belonging to an aromatic cluster on helix 6 modulating the helix 3/helix 6 packing, increased the constitutive activity of the α_{1b}-AR.

These findings provided a strong basis in favor of a theoretical model of receptor activation in which the E/DRY microdomain is part of a network linking helices 3 and 6—the so-called helix 3/helix 6 "ionic lock" (Hofmann *et al.*, 2009). Whereas this "ionic lock" is a fundamental structural constraint keeping the receptor inactive, the release or weakening of the interactions involving R143$^{(3.50)}$ results in receptor activation. It is now well established that this putative model of receptor activation is shared by other GPCRs belonging to the rhodopsin-like class, as recently reviewed (Gether, 2000).

Recent results on rhodopsin structures indicate that the NPXXY(X)$_{5,6}$F on helices 7 and 8 (Fig. 7.1) is another microdomain which, in concert with the E/DRY motif, controls the structural changes underlying photoreceptor activation (Hofmann *et al.*, 2009). In particular, the NPXXY(X)$_{5,6}$F motif provides two constraints: one, involving N$^{(7.49)}$, is represented by the H-bonding network linking helices 1, 2, and 7, and the other, involving Y$^{(7.53)}$ and F$^{(7.60)}$, establishes a link between helices 7 and 8.

These interactions, which might be conserved in different GPCRs, could be interesting targets of site-directed mutagenesis studies to generate CAMs.

2.3. Constitutively activating mutations of the α_{1a}- and α_{1b}-AR subtypes

Combining molecular modeling and site-directed mutagenesis, our group has found a number of activating mutations in the human α_{1a} and in the hamster α_{1b}-AR subtypes. CAMs of the α_{1d}AR subtype have not been reported so far. In our studies, constitutive activity was mainly assessed measuring IP accumulation in whole cells, as described in Section 3.

Most of the activating mutations are predicted to perturb the helix 3/helix 6 packing of the receptor (Figs. 7.1 and 7.2). These include the following findings:

(i) All possible amino acid substitutions of A293$^{(6.34)}$ in the cytosolic extension of helix 6 in the α_{1b}-AR induced variable levels of constitutive activity (Kjelsberg *et al.*, 1992); the homologous mutation of A271$^{(6.34)}$ in the α_{1a}-AR into glutamate or lysine also increases the constitutive activity of the receptor (Rossier *et al.*, 1999).
(ii) All possible natural amino acid substitutions of D142$^{(3.49)}$ of the E/DRY motif in the α_{1b}-AR resulted in variable levels of agonist-independent activity (Scheer *et al.*, 1997); CAMs displayed increased affinity for the full agonist epinephrine which was, at least to some extent, correlated with their degree of constitutive activity; mutating the homologous D123$^{(3.49)}$ into isoleucine in the α_{1a}-AR activated the receptor (Rossier *et al.*, 1999).
(iii) In the α_{1b}-AR mutations of E289$^{(6.30)}$, at the cytosolic end of helix, into A, D, F, K, and R resulted in a marked increase in the constitutive activity of the receptor as well as in its affinity for epinephrine (Greasley *et al.*, 2002); these mutations are predicted to release the "ionic lock" linking helices 3 and 6.
(iv) Replacement of the highly conserved F303$^{(6.44)}$ in helix 6 of the α_{1b}-AR with leucine also resulted in increased constitutive activity (Greasley *et al.*, 2002).

Interesting findings were obtained by mutating the arginine of the E/DRY sequence of the α_{1b}-AR into K, H, D, E, A, I, and N (Scheer *et al.*, 2000). The charge-conserving mutation of R143$^{(3.50)}$ into lysine and histidine conferred, respectively, high and low degree of constitutive activity to the receptor. In contrast, all the other replacements of R143$^{(3.50)}$ were not constitutively active and were dramatically impaired in their ability to mediate agonist-induced IP response.

Few other activating mutations were found by other groups in the extracellular half of the seven-helix bundle: mutations of C128$^{(3.35)}$ on helix 3 (Porter and Perez, 1999) and of A204$^{(5.39)}$ on helix 5 of the α_{1b}-AR (Hwa *et al.*, 1996), mutations of M292 in the rat α_{1a}-AR (Hwa *et al.*, 1996).

3. Measuring Constitutive Activity of Receptor-Mediated Gq Activation

3.1. Constitutive activity of α_1-AR CAMs

In our studies, constitutive activity of the α_1-AR CAMs was mainly assessed measuring agonist-independent accumulation of IPs in whole cells (COS-7 or HEK293) expressing the recombinant receptors. Experiments measuring constitutive activity in whole cells have the main advantage of amplifying the receptor output signal in experiments measuring the accumulation of second messengers under conditions in which their degradation is inhibited. However, their results should be interpreted with caution for several reasons including the presence of endogenous ligands activating the receptor or receptor regulatory events (e.g., desensitization or upregulation).

A crucial problem in experiments measuring constitutive activity in whole cells concerns the relationship between the receptor activity and the expression levels of different receptors. As predicted by the "allosteric ternary complex model" (Samama *et al.*, 1993), the constitutive activity measured in cells depends on the amount of expressed receptor in its active form. Therefore, to compare the constitutive activity of wild-type and mutated receptors, these should be expressed at the same level, which is often difficult to achieve in transfected cells. Two main approaches have been used to overcome this problem.

In most studies on CAMs of the α_{1b}AR, different amounts of receptor DNA have been transfected to obtain similar expression levels (Scheer *et al.*, 1996), as shown in Fig. 7.3. In these studies, the constitutive activity could be directly compared among different receptors in the same experiment. In other studies, the constitutive activity of different receptors was normalized to their receptor number after assessing the existence of a linear relationship between constitutive activity and receptor number for each receptor (Rossier *et al.*, 1999). Both methods rely on the careful measurement of receptor expression. In fact, if the expression level of a receptor is underestimated, its constitutive activity might be overestimated. This could happen for receptor mutants expressed at levels too low to obtain reliable binding data or displaying changes in the affinity for the ligand. In the vast majority of studies on CAMs, receptor expression has been measured using ligand binding. It might be useful in some cases to monitor the expression of the protein by Western blotting to confirm the ligand-binding data.

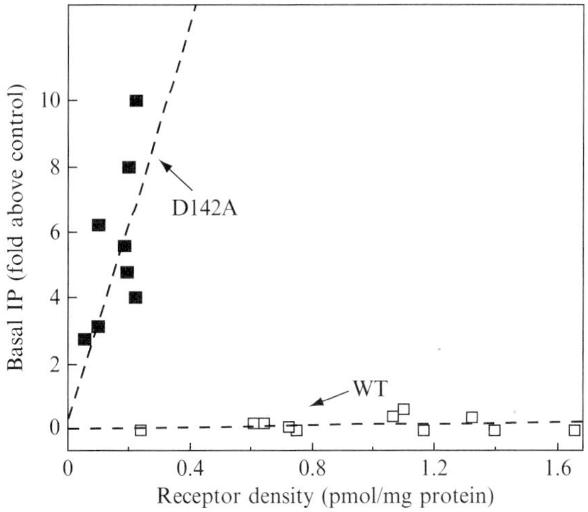

Figure 7.3 Relationship between constitutive activity and receptor expression. Cell transfection and total inositol phosphate (IP) determination was as described in Fig. 7.2. Membrane preparations derived from cells expressing the different receptors and ligand-binding experiments using [^{125}I]HEAT were performed as described elsewhere (Cotecchia *et al.*, 1992). COS-7 cells were transfected with increasing amounts of DNA which ranged from 0.2 to 2.0 μg/million of cells for the wild type and from 1.0 to 6.0 μ/million of cells for D142A, each performed in triplicate. Basal IPs are expressed as percentage above control which indicates cells not expressing the receptors. The results are from several independent experiments.

To overcome these problems, it would be ideal to measure constitutive activity in membrane assays or directly monitoring receptor-mediated G protein activation.

For GPCRs coupled to G_i, receptor-mediated binding of [^{35}S]GTPγS to $G\alpha_i$ in membrane preparations has been successfully used to measure constitutive activity. However, for GPCRs linked to the G_s or $G_{q/11}$ signaling pathway receptor-mediated binding of [^{35}S]GTPγS gives smaller signals probably because of the lower abundance of the G proteins in the cells.

To detect constitutive activity of the α_{1b}-AR and its CAMs measuring guanine nucleotide exchange, the receptors have been fused to Gα11 subunit and [^{35}S]GTPγS binding was measured in membranes from cells in which these constructs were overexpressed (Carrillo *et al.*, 2002). To increase the signal, the radiolabeled Gα11 was immunoprecipitated with antibodies specific for its C-tail. The results from this study showed that agonist-independent [^{35}S]GTPγS binding was significantly greater at the CAMs compared to the wild-type α_{1b}-AR. Another approach could be to coexpress receptors with Gαq or 11 and immunoprecipitate the [^{35}S] GTPγS bound. These approaches are certainly useful to demonstrate constitutive activity in a cell-free system, but they are more time consuming and less sensitive compared to assays in intact cells.

CAMs of the AR display also a number of interesting features that could be used to define the profile of constitutively active receptors: increased binding affinity for agonists (Samama *et al.*, 1993; Scheer *et al.*, 1997), greater efficacy of partial agonists (Samama *et al.*, 1993), increased basal phosphorylation (Pei *et al.*, 1994; Cotecchia & Mhaouty-Kodja, 1999), enhanced endocytosis (Mhaouty-Kodja *et al.*, 1999). However, these properties differ among receptor mutants depending also on the localization of the mutations and cannot be generalized to all CAMs.

3.2. Constitutive activity of wild-type α_1-ARs

The "allosteric ternary complex model" predicts that the constitutive activity measured in cells depends on the amount of expressed receptor in its active state **R*** (Samama *et al.*, 1992). This prediction is clearly supported by findings demonstrating that increasing density of GPCRs results in the progressive elevation of basal (agonist-independent) receptor-mediated production of second messengers. This implies also that native GPCRs might display spontaneous activity in physiological systems which has been demonstrated only for few receptors. There are documented differences in the extent of constitutive activity among even highly related GPCRs and this might depend on differences in their intrinsic activation properties, on the cellular environment in which signaling is measured as well as on experimental factors.

For the α_1-AR subtypes, we reported constitutive activity of the wild-type α_{1a}- and α_{1b}-AR when the receptors were overexpressed in COS-7 cells (Rossier *et al.*, 1999). Constitutive or basal activity of the wild-type receptor was assessed comparing the IP accumulation in transfected cells with that of mock transfected cells. The spontaneous activity of the α_{1b} was greater than that of the α_{1a}-AR expressed at similar levels (2–3 pmol/mg of protein) (Fig. 7.4). Constitutive activity of the α_{1a}- or α_{1b}-AR in physiological systems has not been investigated, whereas studies on the α_{1d}-AR are summarized in another chapter of this volume. This does not exclude that native α_1-AR subtypes might have some constitutive activity which could fulfill specific roles in signaling. Such issues should be further explored and the elucidation of their physiological implications might represent an important area of investigation.

4. Inverse Agonism at the α_1-ARs

Negative efficacy, that is, the capacity of an antagonist binding to its receptor to repress its spontaneous activity, has been a concept implicit in receptor theory from the start (Costa and Cotecchia, 2005). However, it

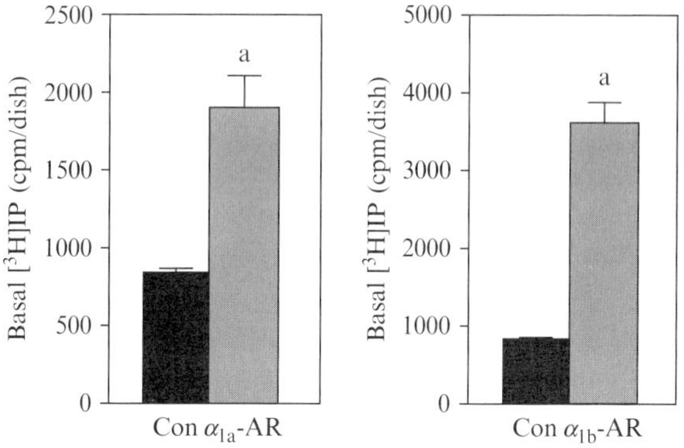

Figure 7.4 Constitutive activity of the wild-type α_{1a}- and α_{1b}-AR. Total inositol phosphate (IP) accumulation was measured, as described in Fig. 7.2, in control COS-7 cells (Con) not expressing the receptors or in cells expressing the wild-type α_{1a}- or α_{1b}-AR. Basal IPs were measured after 100 min incubation in the absence of ligands. Receptor expression ranged 2–3 pmol/mg of proteins. Statistical significance was analyzed by unpaired Student's t test a, $P < 0.05$ Bas of cells expressing the receptor versus control cells. The figure is from Rossier *et al.* (1999).

remained an undeveloped idea for many years largely because of the difficulty to find adequate experimental systems to test it. The availability of CAMs and of cell systems overexpressing wild-type GPCRs represents a useful tool to identify drugs with negative efficacy which is commonly measured as their ability of inhibiting the agonist-independent receptor effect. Ligands displaying negative efficacy have been indicated with the term of inverse agonists or negative antagonists without any specific reference to the mechanistic basis of their effect which might differ among ligands.

To identify inverse agonists at the α_1-AR subtypes, we tested 24 alpha-antagonists differing in their chemical structures for their capacity to decrease the basal activity of CAMs of the α_{1a}- and α_{1b}-AR subtypes (Fig. 7.5; Rossier *et al.*, 1999). The receptors were expressed in COS-7 cells and constitutive activity was assessed measuring agonist-independent accumulation of IPs. Drugs, previously characterized for their binding affinity, were tested at saturating concentrations.

As shown in Fig. 7.5, the vast majority of alpha-antagonists displayed inverse agonism. However, the various alpha-antagonists differed in their negative efficacy (varying from about 20% to 90% of inhibition of basal receptor response) and some of these differences depended on the α_1-AR subtype. In fact, a large number of structurally different alpha-antagonists including all the tested quinazolines were inverse agonists at

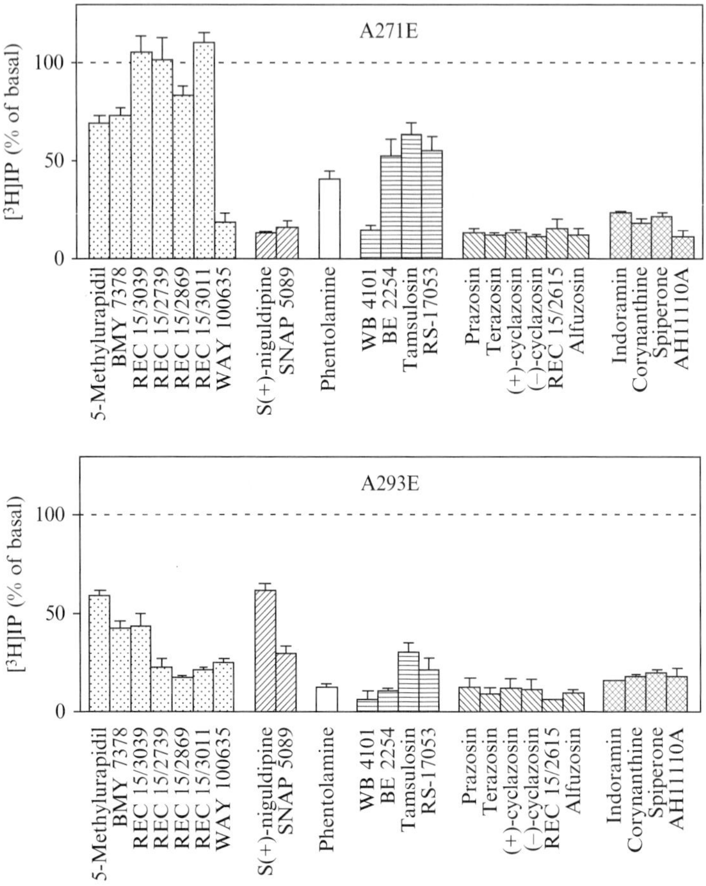

Figure 7.5 Inverse agonists at constitutively active α_1-AR mutants. Total inositol phosphate (IP) accumulation was measured, as described in Fig. 7.2, in COS-7 cells expressing the α_{1a}-AR CAM, A271E, or α_{1b}-AR CAM, A293E, in the absence (basal) or presence of different ligands at a concentration of 10^{-5} *M* (10^{-4} *M* for REC 15/3039) for 45 min. The ligands are grouped according to their structural similarities (from left to right the groups include *N*-arylpiperazines; 1,4-dihydropyridines; imidazolines; benzodioxanes and phenylalkylamines; quinazolines; various structures). The results are expressed as % of basal which indicates the basal levels of IP measured in untreated cells. Results are the mean ± S.E. of three to six independent experiments. The figure is from Rossier *et al.* (1999).

both the α_{1a}- and α_{1b}-AR subtypes. In contrast, several *N*-arylpiperazines displayed different properties at the two α_1-AR subtypes being inverse agonists with profound negative efficacy at the α_{1b}-AR, but not at the α_{1a}. To assess whether the effect of the alpha-blockers on CAMs reflected their activity of wild-type receptors, few ligands were tested also for their capacity to inhibit the basal activity of the wild-type α_{1a}- and α_{1b}-AR. These findings indicated that the behavior of the drugs was similar at CAM or wild-type receptor suggesting that the use of CAMs is a good tool to screen for inverse agonists.

An important problem in experiments measuring inverse agonism in whole cells is to confirm that the effect of the ligand is mediated by the receptor and not by other unknown mechanisms on signaling. For this purpose, we further investigated the effect on the α_1-AR CAMs of two ligands, prazosin and 5-methylurapidil, the first being almost a full inverse agonist and the second having only modest negative efficacy. The inhibition of basal IP accumulation by prazosin was competitively inhibited by increasing concentrations of 5-methylurapidil thus confirming that the inhibitory effect of the inverse agonist prazosin was mediated by the receptor (Rossier *et al.*, 1999).

Studies on the structure–activity relationship of alpha-blockers differing in their negative efficacy in combination with molecular modeling of the receptors might help delineating receptor domains crucially involved in the inhibition of receptor isomerization and activation.

The therapeutic benefit of inverse agonists in diseases related to spontaneous activating mutations of GPCRs is quite obvious. However, a question which remains to be answered is whether therapeutic differences and benefits exist in the clinical use of drugs having negative efficacy versus those that behave as neutral blockers at native receptors. With respect to this question, it is important to highlight that drugs with different degrees of negative efficacy might differ in their ability to induce upregulation of GPCRs upon chronic treatment (Lee *et al.*, 1997). Answering to many open questions about the therapeutic implications of inverse agonism will remain an interesting and important area of investigation in the next years.

5. Conclusions

The studies on CAMs of the α_1AR subtypes and of other GPCRs had an important impact on our understanding of the molecular mechanisms underlying GPCR activation and inverse agonism. Structural information at high resolution on other GPCRs than rhodopsin will be necessary to improve our understanding of GPCR activation and drug action at a molecular level. In addition, it will be important to understand the implications of drugs with negative efficacy *in vivo*, evaluate their benefits and improve future therapeutic strategies.

ACKNOWLEDGMENTS

The work in S. Cotecchia's laboratory was supported by the Fonds National Suisse de la Recherche Scientifique (grant no. 31000A0-10073).

REFERENCES

Ballesteros, J. A., and Weinstein, H. (1995). Integrated methods for the construction of three dimensional models and computational probing of structure-function relations in G-protein coupled receptors. *Methods Neurosci.* **125,** 366–428.

Carrillo, J. J., Stevens, P. A., and Milligan, G. (2002). Measurement of agonist-dependent and independent signal initiation of alpha(1b)-adrenoceptor mutants detected by direct analysis of guanine nucleotide exchange on the G protien-galpha(11). *J. Pharmacol. Exp. Ther.* **302,** 1080–1088.

Cavalli, A., Fanelli, F., Taddei, C., De Benedetti, P. G., and Cotecchia, S. (1996). Amino acids of the alpha1B-adrenergic receptor involved in agonist binding: Differences in docking catecholamines to receptor subtypes. *FEBS Lett.* **399,** 9–13.

Cohen, G. B., Yang, T., Robinson, P. R., and Oprian, D. D. (1993). Constitutive activation of opsin: Influence of charge at position 134 and size at position 296. *Biochemistry* **32,** 6111–6115.

Costa, T., and Cotecchia, S. (2005). Historical review: Negative efficacy and the constitutive activity of G-protein coupled receptors. *Trends Pharmacol. Sci.* **26,** 618–624.

Cotecchia, S., and Mhaouty-Kodja, S. (1999). Regulatory mechanisms of alpha1b-adrenergic receptor function. *Biochem. Soc. Trans.* **27,** 154–157.

Cotecchia, S., Exum, S., Caron, M. G., and Lefkowitz, R. J. (1990). Regions of alpha1-adrenergic receptor involved in coupling to phosphatidylinositol hydrolysis and enhanced sensitivity of biological function. *Proc. Natl. Acad. Sci. USA* **87,** 2896–2900.

Cotecchia, S., Ostrowski, J., Kjelsberg, M. A., Caron, M. G., and Lefkowitz, R. J. (1992). Discrete amino acid sequences of the α1-adrenergic receptor determine the selectivity of coupling to phosphatidylinositol hydrolysis. *J. Biol. Chem.* **267,** 1633–1639.

Diviani, D., Lattion, A. L., Larbi, N., Kunapuli, P., Pronin, A., Benovic, J. L., and Cotecchia, S. (1996). Effect of different G protein-coupled receptor kinase- and protein kinase C-mediated desensitization of the alpha1B-adrenergic receptor. *J. Biol. Chem.* **271,** 5049–5058.

Diviani, D., Lattion, A. L., and Cotecchia, S. (1997). Characterization of the phosphorylation sites involved in G protein-coupled receptor kinase- and protein kinase C-mediated desensitization of the alpha1B-adrenergic receptor. *J. Biol. Chem.* **272,** 28712–28719.

Fanelli, F., and De Benedetti, P. G. (2005). Computational modeling approaches to structure–function analysis of G protein-coupled receptors. *Chem. Rev.* **105,** 3297–3351.

Fanelli, F., Menziani, C., Scheer, A., Cotecchia, S., and De Benedetti, P. G. (1998). Ab initio modeling and molecular dynamics simulation of the alpha1b-adrenergic receptor activation. *Methods* **14,** 302–317.

Fanelli, F., Menziani, C., Scheer, A., Cotecchia, S., and De Benedetti, P. G. (1999). Theoretical study on the electrostatically driven step of receptor–G protein recognition. *Proteins Struct. Funct. Genet.* **37,** 145–156.

Gether, U. (2000). Uncovering molecular mechanisms involved in activation of G protein-coupled receptors. *Endocr. Rev.* **21,** 90–113.

Greasley, P. J., Fanelli, F., Scheer, A., Abuin, L., Nenniger-Tosato, M., De Benedetti, P. G., and Cotecchia, S. (2001). Mutational and computational analysis of the alpha1b-adrenergic receptor: Involvement of basic and hydrophobic residues in receptor activation and G protein coupling. *J. Biol. Chem.* **276,** 46485–46494.

Greasley, P., Fanelli, F., Rossier, O., Abuin, L., and Cotecchia, S. (2002). Mutagenesis and modelling of the alpha1b-adrenergic receptor highlights the role of the helix 3/helix 6 interface in receptor activation. *Mol. Pharmacol.* **61,** 1025–1032.

Hofmann, K. P., Scheerer, P., Hildebrand, P. W., Choe, H.-W., Park, J. H., Heck, M., and Ernst, O. P. (2009). A G protein-coupled receptor at work: The rhodopsin model. *Trends Biochem. Sci.* **34,** 540–552.

Hwa, J., and Perez, D. M. (1996). Identification of critical determinants of alpha1-adrenergic receptor subtype selective agonist binding. *J. Biol. Chem.* **271,** 6322–6327.

Hwa, J., Graham, R. M., and Perez, D. M. (1996). Chimeras of alpha1-adrenergic receptor subtypes identify critical residues that modulate active state isomerization. *J. Biol. Chem.* **271,** 7956–7964.

Kjelsberg, M. A., Cotecchia, S., Ostrowski, J., Caron, M. G., and Lefkowitz, R. J. (1992). Constitutive activation of the alpha1B-adrenergic receptor by all amino acid substitution at a single site. *J. Biol. Chem.* **267,** 1430–1433.

Lattion, A. L., Diviani, D., and Cotecchia, S. (1994). Truncation of the receptor impairs agonist-induced phosphorylation and desensitization of the alpha1B-adrenergic receptor. *J. Biol. Chem.* **269,** 22887–22893.

Lee, T. W., Cotecchia, S., and Milligan, G. (1997). Up-regulation of the levels of expression and function of a constitutively active mutant of the hamster alpha1B-adrenoceptor by ligands that act as inverse agonists. *Biochem. J.* **325,** 733–739.

Mhaouty-Kodja, S., Barak, L. S., Scheer, A., Abuin, L., Diviani, D., Caron, M. G., and Cotecchia, S. (1999). Constitutively active alpha1b-adrenergic receptor mutants display different phosphorylation and internalization properties. *Mol. Phamacol.* **55,** 339–347.

Palczewski, C., Kumasaka, T., Hori, T., Behnke, C. A., Motoshima, H., Fox, B. A., Le Trong, I., Teller, D. C., Okada, T., Stenkamp, R. E., Yamamoto, M., and Miyano, M. (2000). Crystal structure of rhodopsin: A G protein-coupled receptor. *Science* **289,** 739–745.

Pei, G., Samama, P., Lohse, M., Wang, M., Codina, J., and Lefkowitz, R. J. (1994). A constitutively active mutant b2-adrenergic receptor is constitutively desensitized and phosphorylated. *Proc. Natl. Acad. Sci. USA* **91,** 2699–2702.

Porter, J. E., and Perez, D. M. (1999). Characteristics for a salt-bridge switch mutation of the alpha(1b) adrenergic receptor. Altered pharmacology and rescue of constitutive activity. *J. Biol. Chem.* **274,** 34535–34538.

Rasmussen, S. G. F., Jensen, A. D., Liapakis, G., Ghanouni, P., Javitch, J. A., and Gether, U. (1999). Mutation of highly conserved aspartic acid in the beta2 adrenergic receptor: Constitutive activation, structural instability and conformational rearrangement of transmembrane segment 6. *Mol. Pharmacol.* **56,** 175–184.

Ren, Q., Kurose, H., Lefkowitz, R. J., and Cotecchia, S. (1993). Constitutively active mutants of the a_{2A}-adrenergic receptor. *J. Biol. Chem.* **268,** 16483–16487.

Rossier, O., Abuin, L., Fanelli, F., Leonardi, A., and Cotecchia, S. (1999). Inverse agonism and neutral antagonism at alpha(1a)- and alpha(1b)-adrenergic receptor subtypes. *Mol. Pharmacol.* **56,** 858–866.

Samama, P., Cotecchia, S., Costa, T., and Lefkowitz, R. J. (1993). A mutation-induced activated state of the $beta_2$-adrenergic receptor: Extending the ternary complex model. *J. Biol. Chem.* **268,** 4625–4636.

Scheer, A., Fanelli, F., Costa, T. De., Benedetti, P. G., and Cotecchia, S. (1996). Constitutively active mutants of the alpha1B-adrenergic receptor: Role of highly conserved polar amino acids in receptor activation. *EMBO J.* **15,** 3566–3578.

Scheer, A., Fanelli, F., Costa, T., De Benedetti, P. G., and Cotecchia, S. (1997). The activation process of the alpha1B-adrenergic receptor: Potential role of protonation and hydrophobicity of a highly conserved aspartate. *Proc. Natl. Acad. Sci. USA* **94,** 808–813.

Scheer, A., Costa, T., Fanelli, F., De Benedetti, P. G., Mhaouty-Kodja, S., Abuin, L., Nenniger-Tosato, M., and Cotecchia, S. (2000). Mutational analysis of the highly conserved arginine within the Glu/Asp-Arg-Tyr motif of the alpha(1b)-adrenergic receptor: Effects on receptor isomerization and activation. *Mol. Pharmacol.* **57,** 219–231.

Schwinn, D. A., Johnston, G. I., Page, S. O., Mosley, M. J., Wilson, K. H., Worman, N. P., Campbell, S., Fidock, M. D., Furness, L. M., Parry-Smith, D. J., Beate, P., and Bailey, D. S. (1995). Cloning and pharmacological characterization of human alpha-1 adrenergic receptors: Sequence corrections and direct comparison with other species homologues. *J. Pharmacol. Exp. Ther.* **272,** 134–142.

Vazquez-Prado, J., Medina, L. C., Romero-Avila, M. T., Gonzalez-Espinosa, C., and Garcia-Sainz, J. A. (2000). Norepinephrine- and phorbol ester-induced phosphorylation of alpha(1a)-adrenergic receptors. Functional aspects. *J. Biol. Chem.* **275,** 6553–6559.

CHAPTER EIGHT

Measurement of Inverse Agonism of the Cannabinoid Receptors

Tung M. Fong

Contents

Abstract

The cannabinoid receptors are G protein-coupled receptors that are activated by endocannabinoids or exogenous agonists such as tetrahydrocannabinol. Upon agonist binding, cannabinoid receptors will activate Gi which in turn inhibits adenylyl cyclase. Recently, inverse agonists for the cannabinoid receptors have been identified, demonstrating constitutive activity of the cannabinoid receptors. Several methods have been used to measure inverse agonist activity of ligands for the cannabinoid receptors, including Gi-cAMP second messenger assay, GTPγS binding assay, and electrophysiological assays. Each assay has its advantages and limitations, and the Gi-cAMP second messenger assay appears to provide the best overall measurement of inverse agonism in a cellular environment.

1. Introduction

Cannabinoid receptors are G protein-coupled receptors, including two well-characterized subtypes. The CB1 receptor (CB1R) is expressed mainly in central and peripheral neurons, although lower level of expression

Department of Pharmacology, Forest Research Institute, Harborside Financial Center, Jersey City, New Jersey, USA

Methods in Enzymology, Volume 485
ISSN 0076-6879, DOI: 10.1016/S0076-6879(10)85008-6

is also detected in isolated adipocytes and hepatocytes (Howlett *et al.*, 2002; Mackie, 2005). The CB2 receptor (CB2R) is expressed mainly in immune cells. Endogenous agonists of cannabinoid receptors include anandamide and 2-arachidonyl glycerol, and exogenous agonists are exemplified by the natural product tetrahydrocannabinol and synthetic agonist CP55940. Both CB1R and CB2R are Gi coupled. Gi is a class of inhibitory heterotrimeric GTP binding proteins which inhibit the plasma membrane adenylyl cyclase. Hence, activated CB1R or CB2R will activate Gi, which in turn inhibits adenylyl cyclase, preventing the increase of intracellular cAMP.

In 1994, SR141716 was discovered as a synthetic CB1R ligand. At the time of its discovery, it was called CB1R antagonist because it antagonizes the action of CB1R agonists (Rinaldi-Carmona *et al.*, 1994). As our understanding of G protein-coupled receptors evolved and additional detailed assays conducted, it became clear that SR141716 is a CB1R inverse agonist, because it can produce a cellular effect in the absence of agonists and the effect is in the opposite direction of that produced by agonists (Bouaboula *et al.*, 1997; Fong and Heymsfield, 2009; Landsman *et al.*, 1997; MacLennan *et al.*, 1998; Pan *et al.*, 1998; Pertwee, 2005). This chapter will describe various assays that can be used to measure the inverse agonism of compounds for the cannabinoid receptors, and outline the limitation of each assay.

2. Gi-cAMP Assay

Since the cannabinoid receptors are Gi-coupled receptors, the best way to measure agonist and inverse agonist activity is to perform the Gi-cAMP assay. In this cell-based assay, cells expressing CB1R or CB2R are exposed to forskolin, which will activate the adenylyl cyclase and thus increase intracellular cAMP concentration (Fong *et al.*, 2007). If a compound is an agonist, it will inhibit the forskolin-stimulated cAMP response. If a compound is an inverse agonist, it will further increase the forskolin-stimulated cAMP response. If a compound is a neutral antagonist, it alone will not affect the forskolin-stimulated cAMP response.

2.1. Assay protocol

The Gi-cAMP assay is conducted with cells expressing the CB1R or CB2R (either endogenously or heterologously expressed). Before testing unknown compounds for their inverse agonist or agonist activity, it will be necessary to establish what concentration of forskolin is optimal for the assay associated with a specific cell line. If the forskolin concentration is too high, it usually will not be easy to detect inverse agonist activity due to the maximal activity of adenylyl cyclase. If the concentration of forskolin is too low, it will be impossible to detect significant agonist activity because the cyclase is not sufficiently activated to allow the inhibition by Gi. Hence, a best compromise is to select a forskolin concentration that would allow the detection of both agonist activity and inverse agonist activity.

To select an optimal forskolin concentration, three forskolin dose–response curves are constructed, one in the absence of agonist or inverse agonist, one in the presence of agonist CP55940 (typically 1 n*M*), and one in the presence of inverse agonist AM251 (typically 100 n*M*). Cells are incubated with various concentrations of forskolin (usually in the range of 1–100 μ*M*) in the presence of 200 μ*M* 3-isobutyl-1-methylxanthine in the assay buffer (Earle's balanced salt solution supplemented with 5 m*M* $MgCl_2$, 10 m*M* HEPES, pH 7.3, and 1 mg/ml BSA) at room temperature for 30 min with or without a fixed concentration of CP55940 or AM251. Cells were lysed by boiling, and intracellular cAMP level was determined using the cAMP Scintillation Proximity Assay kit (GE Healthcare, Little Chalfont, Buckinghamshire, UK) by scintillation counting (Fong *et al.*, 2007). The amount of cells in each well may need to be adjusted so that the cpm value per well is near the center of the cAMP standard curve. The cAMP level at various forskolin concentration is plotted against the forskolin concentration, and the optimal forskolin concentration would be selected to allow a maximal window of both CP55940-mediated inhibition and AM251-mediated stimulation of forskolin-stimulated cAMP responses.

To measure the inverse agonist activity or agonist activity of an unknown compound, cells are incubated with various concentrations of a test compound in the presence of forskolin (typically 10 μ*M*, determined in advance for each batch of forskolin and each specific cell line), 200 μ*M* 3-isobutyl-1-methylxanthine in the assay buffer (see above) at room temperature for 30 min. In these assays, positive control dose–response curves of CP55940 and AM251 are always conducted to ensure the appropriateness of the assay condition (Fig. 8.1).

2.2. Data analysis

Nonlinear curve fitting is performed to determine the EC50 value and the maximal inverse agonism efficacy for the test compound. The maximal CP55940-mediated inhibition of forskolin-stimulated cAMP increase is defined as 100% agonist efficacy, and the intrinsic activity of all other test compounds is relative to the efficacy of CP55940. For inverse agonist, the ratio of maximal enhancement of forskolin-stimulated cAMP mediated by the test compound to the maximal CP55940-mediated inhibition of forskolin-stimulated cAMP is the maximal inverse agonism. A negative sign is added to denote inverse agonism when describing efficacy (Fong *et al.*, 2007).

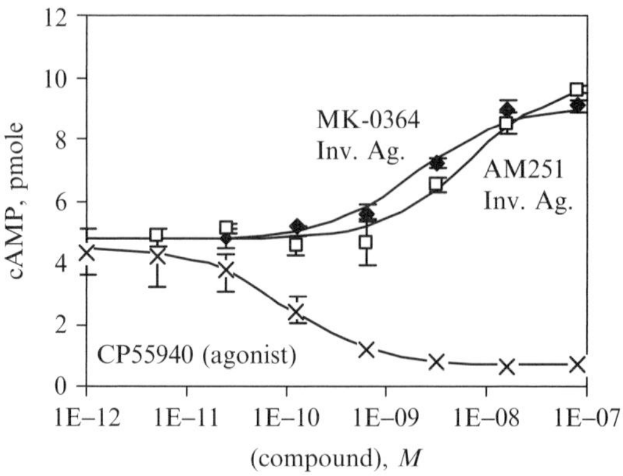

Figure 8.1 Inverse agonist activity of taranabant/MK-0364 and AM251. The data were derived from cell-based assay measuring intracellular cAMP level utilizing recombinant expression system. The figure is reproduced, with permission, from Fong *et al.* (2007).

3. GTPγS Binding Assay

An alternative assay for inverse agonist activity at Gi-coupled receptors is the GTPγS binding assay. This assay uses membrane preparation instead of intact cells, which can be advantageous from the perspective of material consistency. However, membrane preparation does not mimic the intact cellular environment and does not have the associated transmembrane potential and membrane protein organization. The cell-free condition may explain that the EC50 value determined from GTPγS binding assay is usually higher than that from the Gi-cAMP assay. In addition, GTPγS can bind to many other GTP binding proteins, which may further reduce the sensitivity of the assay.

The GTPγS binding assay is based on the reaction that when the receptor is activated by agonist, GDP is released from Gi, allowing GTP to bind. Using $[^{35}S]$GTPγS as the nonhydrolyzable analog of GTP, an increase in $[^{35}S]$GTPγS binding can be observed upon agonist binding. Conversely, a decrease of GTPγS binding can be observed upon inverse agonist binding (Landsman *et al.*, 1997; MacLennan *et al.*, 1998).

To perform the typical GTPγS binding assay, membranes are incubated with $[^{35}S]$GTPγS (typically 0.1 or 1 n*M*) in 10 m*M* HEPES buffer supplemented with 100 m*M* NaCl, 32 m*M* $MgCl_2$, 0.32 m*M* GDP, and various concentrations of the test compound (Landsman *et al.*, 1997; MacLennan *et al.*, 1998). The concentrations of $[^{35}S]$GTPγS and GDP need to be optimized for each membrane preparation (Harrison and

Traynor, 2003). The bound radioactivity is then separated from free radioactivity by membrane filtration. Data are analyzed by nonlinear curve fitting.

4. Electrophysiological Assays

There are several types of electrophysiological assays that can demonstrate the inverse agonism of cannabinoid ligands. One can use cells which express CB1R and calcium channel either endogenously or heterologously (Pan *et al.*, 1998). In this system, CB1R agonist inhibits the calcium current while inverse agonist increases the calcium current in the absence of agonists. Another system to measure inverse agonism activity is the brain slice preparation, where CB1R agonist inhibits neurotransmission while inverse agonist enhances neurotransmission (Auclair *et al.*, 2000; Fong *et al.*, 2009; Melis *et al.*, 2004; Wallmichrath and Szabo, 2002). The effect on neurotransmission is the same whether the synapse is excitatory or inhibitory.

As discussed in Section 2, *in vitro* cell-based assay carries the limitation of an artificial system. However, inverse agonism cannot be demonstrated by *in vivo* studies because of the requirement of the absence of agonist if one were to observe inverse agonist activity. The brain slice assay thus represents an *ex vivo* system where the native neuronal environment is mostly preserved. Although the lack of endocannabinoid release cannot be ruled out, specific slice recording configuration can be used to minimize the likelihood of endocannabinoid release. For instance, if one uses whole cell patch recording, voltage clamping the cell at polarization can minimize endocannabinoid release (Fong *et al.*, 2009). In general, brain slice electrophysiological studies have corroborated the *in vitro* cell-based assays and cell-free assays for the existence of constitutive activity in CB1R as demonstrated by the inverse agonist property of rimonabant (SR141716), taranabant (MK-0364), and AM251.

5. Summary

Of the three types of assays reviewed here, the Gi-cAMP assay has the advantage of cell-based assay and measuring second messenger directly. The limitation is the need to use forskolin to activate adenylyl cyclase, which obviously is not the same as activating a Gs-coupled receptor in the native environment. The GTPγS binding assay has the advantage of a membrane preparation with a better day-to-day reproducibility than a cell culture assay. The limitation of the GTPγS assay is the lack of cellular structure which can be important for a transmembrane signaling event mediated by membrane receptors. If using recombinant systems in either Gi-cAMP assay or GTPγS assay, the expression level of the receptor needs to approximate that in the native tissues in order to minimize artifacts. The brain slice electrophysiological assay preserves most of the native environment, although it is a low throughput assay. For routine assay of newly synthesized compounds, the Gi-cAMP assay offers the best solution when considering both biological rationale and technical operation. When there is a need to understand the mechanistic basis of a specific compound, brain slice electrophysiological assay will add significant value and provide an independent confirmation of those data derived from recombinant systems.

REFERENCES

Auclair, N., Otani, S., Soubrie, P., and Crepel, F. (2000). Cannabinoids modulate synaptic strength and plasticity at glutamatergic synapses of rat prefrontal cortex pyramidal neurons. *J. Neurophysiol.* **83,** 3287–3293.

Bouaboula, M., Perrachon, S., Milligan, L., Canat, X., Rinaldi-Carmona, M., Portier, M., Barth, F., Calandra, B., Pecceu, F., Lupker, J., Maffrand, J. P., Le Fur, G., *et al.* (1997). A selective inverse agonist for central cannabinoid receptor inhibits mitogen-activated protein kinase activation stimulated by insulin or insulin-like growth factor 1. Evidence for a new model of receptor/ligand interactions. *J. Biol. Chem.* **272,** 22330–22339.

Fong, T. M., and Heymsfield, S. B. (2009). Cannabinoid-1 receptor inverse agonists: Current understanding of mechanism of action and unanswered questions. *Int. J. Obes.* **33,** 947–955.

Fong, T. M., Guan, X.-M., Marsh, D. J., Shen, C. P., Stribling, D. S., Rosko, K. M., Lao, J., Yu, H., Feng, Y., Xiao, J. C., Van der Ploeg, L. H. T., Goulet, M. T., *et al.* (2007). Anti-obesity efficacy of a novel cannabinoid-1 receptor inverse agonist MK-0364 in rodents. *J. Pharmacol. Exp. Ther.* **321,** 1013–1022.

Fong, T. M., Shearman, L. P., Stribling, D. S., Shu, J., Lao, J., Huang, C. R. R., Xiao, J. C., Shen, C. P., Tyszkiewicz, J., Strack, A. M., DeMaula, C., Hubert, M. F., *et al.* (2009). Pharmacological efficacy and safety profile of taranabant in preclinical species. *Drug Dev. Res.* **70,** 349–362.

Harrison, C., and Traynor, J. R. (2003). The [35S]GTP-gamma-S binding assay: Approaches and applications in pharmacology. *Life Sci.* **74,** 489–508.

Howlett, A. C., Barth, F., Bonner, T. I., Cabral, G., Casellas, P., Devane, W. A., Felder, C. C., Herkenham, M., Mackie, K., Martin, B. R., Mechoulam, R., and Pertwee, R. G. (2002). International Union of Pharmacology. XXVII. Classification of cannabinoid receptors. *Pharmacol. Rev.* **54,** 161–202.

Landsman, R. S., Burkey, T. H., Consroe, P., Roeske, W. R., and Yamamura, H. I. (1997). SR141716 is an inverse agonist at the human cannabinoid CB1 receptor. *Eur. J. Pharmacol.* **334,** R1–R2.

Mackie, K. (2005). Distribution of cannabinoid receptors in the central and peripheral nervous system. *Handb. Exp. Pharmacol.* **168,** 299–325.

MacLennan, S. J., Reynen, P. H., Kwan, J., and Bonhaus, D. W. (1998). Evidence for inverse agonism of SR141716A at human recombinant cannabinoid CB1 and CB2 receptors. *Br. J. Pharmacol.* **124,** 619–622.

Melis, M., Pistis, M., Perra, S., Muntoni, A. L., Pillolla, G., and Gessa, G. L. (2004). Endocannabinoids mediate presynaptic inhibition of glutamatergic transmission in rat ventral tegmental area dopamine neurons through activation of CB1 receptors. *J. Neurosci.* **24,** 53–62.

Pan, X., Ikeda, S. R., and Lewis, D. L. (1998). SR 141716A acts as an inverse agonist to increase neuronal voltage-dependent Ca2+ currents by reversal of tonic CB1 cannabinoid receptor activity. *Mol. Pharmacol.* **54,** 1064–1072.

Pertwee, R. G. (2005). Inverse agonism and neutral antagonism at cannabinoid CB1 receptors. *Life Sci.* **76,** 1307–1324.

Rinaldi-Carmona, M., Barth, F., Heaulme, M., Shire, D., Calandra, B., Congy, C., Martinez, S., Maruani, J., Neliat, G., Caput, D., Ferrara, P., Soubrie, P., *et al.* (1994). SR141716A, a potent and selective antagonist of the brain cannabinoid receptor. *FEBS Lett.* **350,** 240–244.

Wallmichrath, I., and Szabo, B. (2002). Analysis of the effect of cannabinoids on GABAergic neurotransmission in the substantia nigra pars reticulata. *Naunyn. Schmiedebergs Arch. Pharmacol.* **365,** 326–334.

CHAPTER NINE

Constitutively Active Thyrotropin and Thyrotropin-Releasing Hormone Receptors and Their Inverse Agonists

Susanne Neumann, Bruce M. Raaka, *and* Marvin C. Gershengorn

Contents

Abstract

Receptors for thyrotropin-releasing hormone (TRH) and thyrotropin (thyroid-stimulating hormone—TSH) are important regulators of the function of the TSH-producing cells of the anterior pituitary gland and the thyroid gland, respectively, and thereby play a central role in thyroid hormone homeostasis. Although the roles of TRH- and TSH-stimulated signaling in these endocrine glands are well understood, these receptors are expressed in other sites and their roles in these extraglandular tissues are less well known. Moreover, one of the two subtypes of TRH receptors (TRH-R2) and the single TSH receptor (TSHR) exhibit constitutive signaling activity and the roles of constitutive signaling by these receptors are poorly understood. One approach to studying constitutive

Clinical Endocrinology Branch, National Institute of Diabetes and Digestive and Kidney Diseases, National Institutes of Health, Bethesda, Maryland, USA

Methods in Enzymology, Volume 485
ISSN 0076-6879, DOI: 10.1016/S0076-6879(10)85009-8

signaling is to use inverse agonists. In this chapter, we will describe the experimental procedures used to measure constitutive signaling by TRH-R2 and TSHR and the effects of their specific inverse agonists.

1. Introduction

The hypothalamic–pituitary–thyroid axis plays the major role in regulating thyroid hormone homeostasis in the body. Thyrotropin-releasing hormone (TRH), which is a tripeptide, is secreted by the hypothalamus into a portal system that directs TRH to the pituitary gland where it can bind to its seven transmembrane-spanning receptors (7TMRs or G protein-coupled receptors—GPCRs). In most mammalian species, there are two subtypes of TRH receptors, subtype 1 (TRH-R1) that is found in the anterior pituitary gland and in extrapituitary tissues and subtype 2 (TRH-R2) that is found in extrapituitary tissues but not in the pituitary (see below); in humans, however, there is only a single TRH receptor that is homologous structurally to TRH-R1. Activation of TRH-R1 in the anterior pituitary leads to stimulation of the synthesis and secretion into the systemic circulation of thyrotropin (thyroid-stimulating hormone—TSH), which is an $\sim$30 kDa heterodimeric glycoprotein, from where it binds to TSH receptors (TSHRs), which are 7TMRs, on thyroid hormone-producing cells (thyrocytes) of the thyroid gland. Activation of TSHRs on thyrocytes leads to stimulation of the synthesis and secretion into the general circulation of thyroid hormones. Thyroid hormones, in turn, regulate the biology of virtually every cell of the body. In a classical endocrine negative feedback loop, thyroid hormones bind to nuclear receptors in the hypothalamus and pituitary to inhibit synthesis of both TRH and TSH, respectively.

In addition to the above-noted difference in expression of TRH-R1 and TRH-R2 in the pituitary gland, TRH-R1 and TRH-R2 are expressed in different sites within the rat brain and in different cells outside of the central nervous system (Sun *et al.*, 2003). Although a number of effects of administration of TRH and its analogs to intact animals and to cells in culture have been observed, we think it is fair to state that the physiological roles of TRH receptors in extrapituitary tissues are still not understood. TRH-R1 and TRH-R2 exhibit very similar binding profiles for numerous TRH analogs and, therefore, most peptidic TRH analogs cannot distinguish between TRH-R1 and TRH-R2. However, some unnatural peptidic analogs (Monga *et al.*, 2008) and a nonpeptidic, small molecule antagonist (Engel *et al.*, 2008) that bind preferentially to TRH-R1 or TRH-R2 have been described and these ligands may help distinguish the unique physiological roles of each receptor. For this chapter, the most important distinction between TRH-R1 and TRH-R2 is that TRH-R2 is constitutively active whereas TRH-R1 is not. While homology modeling and related receptor mutagenesis studies have suggested a structural basis for its increased basal activity (Deflorian *et al.*, 2008), the functional role of this constitutive signaling by TRH-R2 is not understood. Inverse agonists inhibit constitutive activity. Midazolam, a short-acting drug in the benzodiazepine class, is an inverse agonist for TRH-R2 that can be used in studies *in vitro* (Lu *et al.*, 2004) but its depressant effects prevent it from being used in animal studies.

In addition to its expression on the surface of thyrocytes, TSHR is also found in extrathyroidal sites including bone, and orbital and peripheral adipose tissue (Neumann *et al.*, 2009) but its physiologic or pathologic roles in these extrathyroidal tissues is not well understood. Moreover, as for TRH-R2, the role of constitutive signaling by TSHR has not been elucidated. TSH analogs, antibodies that bind to TSHR and, more recently, small molecule TSHR ligands have been used to study TSHR signaling (Neumann *et al.*, 2009). Of note, several TSHR-binding antibodies have been shown to be inverse agonists (Chen *et al.*, 2007; Moriyama *et al.*, 2003; Sanders *et al.*, 2008) and we have recently described a small molecule TSHR inverse agonist (Neumann *et al.*, 2010). However, inverse agonists have not yet been used to delineate the role of TSHR constitutive signaling in animal studies.

2. TRH-R2 and Its Inverse Agonist Midazolam

With many 7TMRs, in particular, those coupled via G_s to adenylyl cyclase (see TSHR below), it is possible to measure constitutive signaling by measuring the production of the proximal second messenger molecules. In our experience, it has been more difficult to measure constitutive signaling by 7TMRs that couple via $G_{q/11}$ to phospholipase C. This is primarily due to the rapid turnover of second messenger inositol-1,4,5-trisphosphate. We were able to quantify constitutive signaling in a 7TMR expressed by the Kaposi's sarcoma virus (human herpesvirus 8) by measuring the more stable inositol monophosphate, a metabolic breakdown product of inositol-1,4,5-trisphosphate (Arvanitakis *et al.*, 1997). However, we could not measure constitutive TRH-R2 signaling by measuring inositol monophosphate production using cells prelabeled with radioactive inositol or an ELISA that measures unlabeled inositol monophosphate even though the degradation of inositol monophosphate was inhibited by LiCl. We, therefore, measured constitutive signaling of TRH-R2 using a reporter gene assay, which takes advantage of the amplification that occurs during signal transduction to gene transcription.

2.1. Use of protein kinase C-activated reporter genes in cells transfected to express TRH-R2

Over the years, we have used a number of transfection cocktails to coexpress TRH-R2 at controlled levels with several protein kinase C (PKC) reporter genes in a number of different adherent mammalian cell lines. We have had success with several of these combinations. Of note, although we have not made an exhaustive comparison between mouse and rat TRH-R2s, we have not found substantive differences in their signaling characteristics.

2.1.1. Required materials

a. *Cells used*: HEK 293 EM cells (Robbins and Horlick, 1998) were obtained from Dr R. A. Horlick because they are very adherent to tissue culture wells. We have also used HEK 293 and COS-1 cells that can be obtained from the American Type Culture Collection (ATCC).
b. Plasmids expressing mouse TRH-R1, TRH-R2, and mutants of the wild-type receptors that we constructed. DNA of high quality (endotoxin-free plasmid DNA, use of Plasmid Maxi kits) is recommended.
c. Many useful reporter plasmids for 7TMR signaling through PKC or other kinases are available within the PathDetect® *In Vivo* Signal Transduction Pathway *cis*- or *trans*-Reporting System from Agilent (Santa Clara, CA). For constitutive signaling by TRH-R2, we have used the AP-1-Luciferase *cis*-Reporting System (Cat # 219073).
d. *Transfection cocktail*: Many available transfection reagents will provide satisfactory results. We have used FuGene® 6 Transfection Reagent from Roche (Mannheim, Germany).
e. *Growth medium*: Dulbecco's Modified Eagle's Medium (DMEM) containing 10% fetal bovine serum.
f. *Ligand-binding assays*: [^{3}H][*methyl*-His]TRH ([^{3}H]MeTRH) is available from Perkin Elmer (Waltham, MA) and [^{3}H]TRH is available from American Radiolabeled Chemicals (St. Louis, MO).
g. Hank's Balanced Salt Solution (HBSS) containing 10 m*M* HEPES, pH 7.4
h. Trypsin (0.05%)–EDTA (0.53 m*M*) solution
i. Phosphate buffered saline (PBS), pH 7.4
j. Disposables
   1. Tissue culture—175-cm^2 flasks, 24-well plates
   2. Test tubes—50, 15, 1.5 ml (Eppendorf)
   3. Pipettes

2.1.2. Measurement of constitutive signaling activity of TRH-R2

Most investigators (see other chapters) measure constitutive signaling by relating the agonist-independent signaling parameter to a measurement of the number of receptors when it is feasible to measure both in the same cell or cell population. Receptor number is usually measured by ligand or antireceptor antibody binding. Some investigators will make these measurements at a single level of receptor expression. This is sometimes unavoidable, for example, if one is measuring constitutive signaling of receptors expressed endogenously. However, most measurements of constitutive signaling are made using receptors expressed exogenously and we believe that showing a direct correlation between the level of constitutive signaling and the number of receptors is the most definitive way of estimating the level of constitutive signaling activity for any receptor. Moreover, if ligand or antibody binding cannot be used, it is possible to vary the level of receptor expression plasmid input to show constitutive activity although this will not permit definitive quantitative comparisons amongst different receptors.

Constitutive activity of TRH-R2 was demonstrated definitively as a direct correlation between receptor number and luciferase reporter activity in cells transiently transfected with increasing amounts of receptor expression vector (Fig. 9.1). This finding allowed studies of TRH-R2 signaling to be performed in cells stably expressing TRH-R2. Because there is no cell

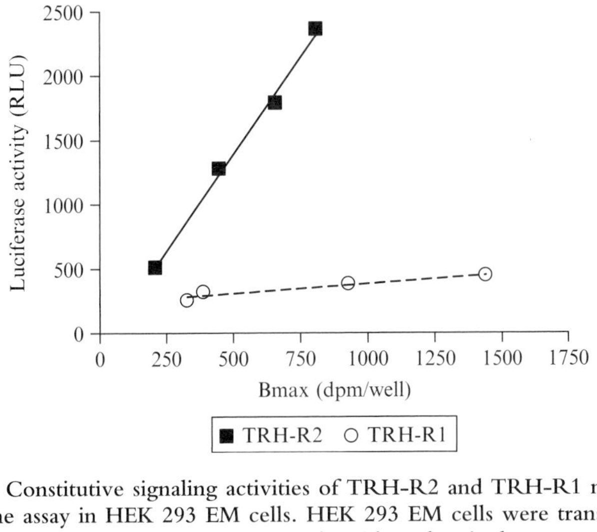

Figure 9.1 Constitutive signaling activities of TRH-R2 and TRH-R1 measured by reporter gene assay in HEK 293 EM cells. HEK 293 EM cells were transfected with increasing amounts of receptor expression plasmids and a single amount of reporter gene plasmid. After 24 h, luciferase activities (relative light units, RLU) and maximum specific ligand binding (receptor density, B_{max}) were measured. The figure is adapted from Wang and Gershengorn (1999).

line that expresses TRH-R2 endogenously, we have used HEK 293 EM cells stably transfected with mouse TRH-R2 for these studies.

One day prior to transfection

Harvest HEK 293 EM cells from a 175-cm^2 flask with 10 ml of Trypsin–EDTA, count cells, and seed 80,000 cells per well of a 24-well plate in 0.5 ml DMEM with 10% fetal bovine serum.

On the day of transfection

1. Prepare Fugene® 6 suspension: Add Fugene® 6 directly to DMEM without serum to a final volume of 100 μl/well and mix by vortexing for 1–3 s. The amount of Fugene® 6 required is calculated so that the ratio of Fugene® 6 to DNA (added in step 3) will be 3:1 (e.g., 3 μl Fugene® 6 per 1 μg DNA).
2. Incubate the suspension for 5 min at RT.
3. Prepare transfection cocktail: Add receptor plasmid DNA (0.01–0.2 μg) and 0.2 μg AP-1-Luciferase DNA to ~100 μl of Fugene® 6 suspension to achieve the desired 3:1 ratio. Mix thoroughly by vortexing for 1–3 s and incubate for 15–60 min at RT. If dose-dependent constitutive activity is measured (as in Fig. 9.1), add empty vector plasmid DNA to the wells with less than 0.2 μg receptor plasmid DNA to achieve a final total DNA concentration of 0.4 μg in all wells.
4. Add the transfection cocktail (100 μl/well) dropwise to the wells containing 0.5 ml DMEM with 10% fetal bovine serum, dispersing it evenly throughout the well.
5. Incubate in 5% CO_2 at 37 °C.

After 48–72 h

- Measure luciferase activity
  1. Wash wells with 1 ml PBS three times.
  2. Add 0.5 ml lysis buffer (25 m*M* Gly-Gly, pH 7.8, 15 m*M* $MgSO_4$, 4 m*M* EGTA, 1% Triton X-100) and incubate for 15 min at RT.
  3. Scrape cells, place in 1.5-ml tube and spin at maximum speed in Eppendorf centrifuge for 5 min in the cold.
  4. Decant supernatant into a fresh tube.
  5. Immediately prior to measuring luciferase activity add 0.1 ml sample to 0.5 ml assay reagent (25 m*M* Gly-Gly, pH 7.8, 15 m*M* $MgSO_4$, 4 m*M* EGTA, 15 m*M* KH_2PO_4, 2 m*M* DTT, 2 m*M* ATP, 0.4 m*M* luciferin).
  6. Measure the light emitted from the reaction in a luminometer using an integration time of 10–30 s.

- Measure receptor density
  1. Wash wells with 0.5 ml HBSS three times.
  2. Add 0.25 ml HBSS containing 0.1–10 n*M* [^{3}H]MeTRH or 1–10 n*M* [^{3}H]TRH without or with 1000-fold excess unlabeled MeTRH or TRH. For some mutant receptors with lower binding affinities, it may be necessary to increase the concentration of [^{3}H]MeTRH or [^{3}H] TRH.
  3. Incubate for 1 h at 37 °C.
  4. Wash cell monolayers three times with 0.5 ml ice-cold HBSS.
  5. Solubilize the cells with 0.5 ml 0.4 N NaOH.
  6. Count 0.4 ml in a beta scintillation counter.
  7. Specific binding is calculated as the difference between total binding (^{3}H-labeled ligand alone) minus nonspecific binding (^{3}H-labeled ligand plus unlabeled ligand). Maximum binding (total receptor binding) is calculated from nonlinear regression analysis; we use the PRISM program (GraphPad Software, Inc., San Diego, CA). The number of TRH receptors is calculated using the specific radioactivity of the ^{3}H-labeled ligand assuming a stoichiometry of ligand:receptor of 1:1.

Measurements of constitutive signaling activities in cells expressing increasing levels of TRH-R2 and TRH-R1 are illustrated in Fig. 9.1 (Wang and Gershengorn, 1999).

Midazolam is a benzodiazepine drug that is an inverse agonist at TRH receptors. In a parallel series of experiments to that illustrated in Fig. 9.1, we showed that midazolam inhibited constitutive signaling by TRH-R2 by 42% at 10 μ*M* and by 74% at 50 μ*M* (Wang and Gershengorn, 1999).

3. TSHR and Its Inverse Agonist NCGC00161856

In contrast to 7TMRs that couple via $G_{q/11}$ to phospholipase C where it is difficult to measure constitutive signaling by measuring proximal messenger molecules, it is relatively easy to measure the production of the second messenger cAMP for 7TMRs coupled via G_s to adenylyl cyclase. These studies are generally performed in the presence of the phosphodiesterase inhibitor isobutylmethylxanthine (IBMX), and the accumulated cAMP can be measured readily by a number of antibody-based assays. We routinely have used IBMX and measured cAMP by immunoassay. In addition, we occasionally have used reporter genes (see Section 2.1.1), in particular the PathDetect® AP-1 *cis*- and CREB *trans*-Reporting Systems.

To monitor TSHR constitutive signaling in a more physiologically relevant cell type, we also use primary cultures of human thyroid cells (thyrocytes) that express TSHRs endogenously. In thyrocytes, we measure cAMP production along with expression levels of mRNA transcripts for TSH-regulated genes by reverse transcription followed by quantitative real time polymerase chain reaction (qRT-PCR).

3.1. Use of immunoassays to measure cAMP in cells endogenously expressing TSHRs and in cells stably or transiently transfected to express TSHR

3.1.1. Required materials

a. *Cells used*: HEK 293 EM cells (Robbins and Horlick, 1998) were obtained from Dr R. A. Horlick because they are very adherent to tissue culture wells. As we described the method for transient transfection above (Section 2.1.1) and TSHR has been shown to exhibit constitutive signaling by showing a direct correlation between receptor number and agonist-independent cAMP production (Vassart *et al.*, 1991), we will describe studies using HEK 293 EM cells stably expressing TSHR (HEK-TSHR cells). The same protocol can be used in cells transiently expressing human TSHR and mutants of TSHR and in primary cultures of human thyrocytes.

   Human thyrocytes: Normal thyroid tissue was collected from patients undergoing total thyroidectomy for thyroid cancer at the National Institutes of Health Clinical Center. Patients provided informed consent on an Institutional Review Board-approved protocol and materials were received anonymously via approval of research activity through the Office of Human Subjects Research. The tissue was kept in HBSS on ice during transport. Monodispersed cells were obtained by mincing the tissue into small pieces, washing three times with ice-cold HBSS, and digesting with HBSS containing 3 mg/ml Collagenase Type IV (GIBCO, Carlsbad, CA) for 30 min or longer with constant shaking in a water bath at 37 °C. After centrifugation for 5 min at 160× *g*, the supernatant was removed, the cells were washed with DMEM once and resuspended in 10 ml DMEM with 10% fetal bovine serum, plated in 10-cm tissue culture dishes and incubated at 37 °C in a humidified 5% CO_2 incubator. After 24 h, the cells were rinsed with DMEM two times, fresh growth medium was added and the primary cultures of adherent thyrocytes formed a confluent monolayer within 5–7 days.
b. Measurement of cAMP: cAMP-Screen Direct® System from Applied Biosystems (Bedford, MA).
c. Growth medium: DMEM containing 10% fetal bovine serum.
d. Ligand-binding assays: ^{125}I-TSH (bovine) from BRAHMS Aktiengesellschaft (Hennigsdorf, Germany).
e. Antibody binding: Mouse antihuman TSHR antibody 2C11 was obtained from Serotec Ltd (Oxford, UK) and Alexa Fluor®

488-conjugated $F(ab')_2$ rabbit antimouse IgG antibody was from Molecular Probes (Invitrogen, Carlsbad, CA).

f. HBSS containing 10 mM HEPES, pH 7.4
g. Trypsin (0.05%)–EDTA (0.53 mM) solution
h. EGTA (1 mM)/EDTA (1 mM) buffer
i. PBS without Ca^{2+} or Mg^{2+}
j. Fluorescence activated cell sorter (FACS) buffer (PBS with 0.1% bovine serum albumin and 0.1% sodium azide)
k. 1% paraformaldehyde (PFA)
l. Disposables
   1. Tissue culture—175-cm^2 flasks, 24-well plates
   2. Test tubes—50, 15, 5, 1.5 ml (Eppendorf)
   3. Pipettes

3.1.2. Measurement of constitutive signaling activity of TSHR by cAMP production

One day prior to beginning the experiment:

Harvest HEK-TSHR cells from a 175-cm^2 flask with 10 ml of Trypsin–EDTA, count cells and seed 220,000 cells per well of a 24-well plate in 0.5 ml DMEM with 10% fetal bovine serum. For thyrocytes, seed 80,000 cells per well.

On the day of the experiment:

1. Aspirate the growth medium and wash the cell monolayer three times with 0.5 ml HBSS at RT.
2. Incubate cells in 0.25 ml HBSS in 5% CO_2 incubator at 37 °C for 30 min.
3. Aspirate the buffer and replace it with HBSS containing 1 mM IBMX and test substance(s) and incubate for desired times.
4. At the desired times, either add 0.25 ml lysis buffer (from the cAMP-Screen Direct® Kit) directly to the wells to measure total cAMP or remove the media from the cells by aspiration and add 0.25 ml lysis buffer to the wells to measure intracellular cAMP.
5. Measure cAMP using the chemiluminescent immunoassay system (Applied Biosystems cAMP-Screen Direct® System) according to the manufacturer's instructions.

Measuring receptor density by ligand (^{125}I-TSH) binding to HEK-TSHR cells

1. Wash wells with 0.5 ml HBSS three times.

2. Add 0.25 ml HBSS containing 0.2 g BSA and 2.5 g milk powder per 100 ml and ~60,000 cpm ^{125}I-TSH without or with 1000-fold excess unlabeled bovine TSH. For some mutant receptors with lower affinities, it may be necessary to increase the concentration of ^{125}I-TSH.
3. Incubate for 2 h at 4 °C.
4. Wash cell monolayers with 0.5 ml ice-cold HBSS three times.
5. Solubilize the cells with 0.5 ml 0.4 N NaOH.
6. Count 0.4 ml in a gamma counter.
7. Specific binding is calculated as the difference between total binding (^{125}I-TSH alone) minus nonspecific binding (^{125}I-TSH plus unlabeled TSH). As the apparent affinity of TSHR for ^{125}I-TSH is relatively low, the binding assay is usually performed with a fixed concentration of ^{125}I-TSH and various concentrations of unlabeled TSH. Maximum binding (total receptor binding) is calculated from nonlinear regression analysis; we use the PRISM program (GraphPad Software, Inc., San Diego, CA). The number of TSHRs is calculated using the various specific radioactivities (decreased by adding unlabeled TSH) of the ^{125}I-TSH at each dose assuming a stoichiometry of ligand:receptor of 1:1.

Estimating relative TSHR expression using antihuman TSHR antibody

1. Wash cell monolayer with 0.5 ml PBS.
2. Add 0.5 ml EGTA/EDTA buffer and incubate for 5 min at RT.
3. Detach cells and spin at 1500 rpm for 2 min.
4. Discard supernatant and add 0.1 ml 5% donkey serum in FACS buffer and incubate for 10 min at 4 °C.
5. Wash with 2 ml FACS buffer and spin at 1500 rpm for 2 min.
6. Discard supernatant and add 0.1 ml anti-TSHR antibody 2C11 (final concentration 10 μg/ml) or 0.1 ml mouse IgG (final concentration 10 μg/ml).
7. Incubate for 1 h at 4 °C.
8. Wash twice with 2 ml FACS buffer.
9. Add 0.1 ml secondary antibody (Alexa Fluor® 488-conjugated $F(ab')_2$ anti-IgG).
10. Incubate in the dark for 30 min at 4 °C.
11. Wash twice with 2 ml FACS buffer at 4 °C.
12. Add 0.15–0.2 ml 1% PFA (depending on cell pellet size).
13. Measure fluorescence in a FACS analyzer. The TSHR-expressing cells are those whose fluorescence levels are greater than cells labeled with an "isotype control" mouse IgG.

3.2. Use of qRT-PCR to measure mRNAs for thyroglobulin (TG), thyroperoxidase (TPO), TSHR, sodium-iodide symporter (NIS), and Type 2 deiodinase (DIO2) in primary cultures of human thyrocytes endogenously expressing TSHRs

3.2.1. Required materials

a. Human thyrocytes: see Section 3.1.1b
b. Measurement of mRNA levels—primers and probes for TG, TPO, TSHR, NIS, DIO2, and GAPDH were obtained from Applied Biosystems (Foster City, CA).
c. Growth medium—DMEM containing 10% fetal bovine serum.
d. Trypsin (0.05%)–EDTA (0.53 m*M*) solution
e. Disposables
   1. Tissue culture—175-cm^2 flasks, 24-well plates, 12-well plates
   2. Test tubes—50, 15, 5, 1.5 ml (Eppendorf)
   3. Pipettes

3.2.2. Measurement of constitutive signaling activity by effects on thyroid-selective genes

Three days prior to mRNA isolation

Harvest thyrocytes with Trypsin–EDTA and seed 60,000 cells per well of a 24-well plate or 100,000 cells per well of a 12-well plate in DMEM with 10% fetal bovine serum.

On the first day of the experiment (24 h after seeding of the cells)

Change medium to DMEM with 2% fetal bovine serum containing 1 m*M* IBMX without or with test factors (e.g., inverse agonist).

On the third day of the experiment (48 h after addition of test factors)

1. Aspirate buffer and extract total RNA using the RNeasy® Micro Kit from Qiagen (Valencia, CA).
2. Prepare cDNA using a High Capacity cDNA Archive Kit from Applied Biosystems (Foster City, CA).
3. qRT-PCR is performed in 0.025 ml reactions using cDNA prepared from ~100 ng of RNA and Universal Master Mix (Applied Biosystems, Foster City, CA). The mRNA level of each sample is normalized to GAPDH to correct for differences in cDNA input.

NCGC00161856 is an inverse agonist of TSHR (Neumann *et al.*, 2010). Figure 9.2 illustrates the effects of NCGC00161856 on cAMP production

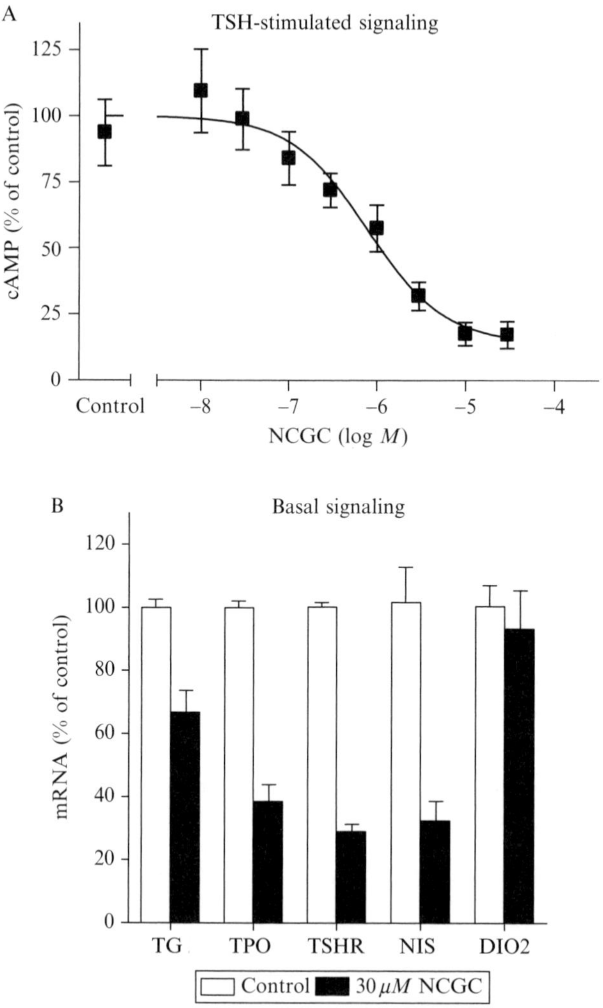

Figure 9.2 Effect of inverse agonist NCGC00161856 (NCGC) on constitutive signaling by TSHR measured as cAMP production and mRNA levels for TG, TPO, TSHR, NIS, and DIO2 in primary cultures of human thyrocytes. (A) Thyrocytes were incubated in HBSS with 1 m*M* IBMX and without (control) or with increasing concentrations of NCGC for 2 h. The cAMP production measured without 1 m*M* IBMX was subtracted from all samples. (B) Thyrocytes were incubated in DMEM containing 2% fetal bovine serum and 1 m*M* IBMX without or with 30 μ*M* NCGC for 48 h. This figure is from Neumann *et al.* (2010).

(over 2 h) and on mRNA levels for TG, TPO, TSHR, NIS, and DIO2 after 48 h exposure in primary cultures of human thyrocytes. The inverse agonist reduces basal cAMP levels in a dose-dependent fashion. Interestingly, the transcript analyses suggest that basal signaling by TSHR plays an important role in maintaining expression levels of several key thyroid-specific genes. By reducing basal signaling, the inverse agonist NCGC00161856 decreases expression of these key genes.

ACKNOWLEDGMENTS

This research was supported by the Intramural Research Program of the National Institute of Diabetes and Digestive and Kidney Diseases, National Institutes of Health.

REFERENCES

Arvanitakis, L., Geras-Raaka, E., Varma, A., Gershengorn, M. C., and Cesarman, E. (1997). Human herpesvirus KSHV encodes a constitutively active G-protein-coupled receptor linked to cell proliferation. *Nature* **385,** 347–350.

Chen, C. R., McLachlan, S. M., and Rapoport, B. (2007). Suppression of thyrotropin receptor constitutive activity by a monoclonal antibody with inverse agonist activity. *Endocrinology* **148,** 2375–2382.

Deflorian, F., Engel, S., Colson, A.-O., Raaka, B. M., Gershengorn, M. C., and Costanzi, S. (2008). Understanding the structural and functional differences between mouse thyrotropin-releasing hormone receptors 1 and 2. *Proteins* **71,** 783–794.

Engel, S., Skoumbourdis, A. P., Childress, J., Neumann, S., Deschamps, J. R., Thomas, C. J., Colson, A. O., Costanzi, S., and Gershengorn, M. C. (2008). A virtual screen for diverse ligands: Discovery of selective G protein-coupled receptor antagonists. *J. Am. Chem. Soc.* **130,** 5115–5123.

Lu, X., Huang, W., Worthington, S., Drabik, P., Osman, R., and Gershengorn, M. C. (2004). A model of inverse agonist action at thyrotropin-releasing hormone receptor type 1: Role of a conserved tryptophan in helix 6. *Mol. Pharmacol.* **66,** 1192–1200.

Monga, V., Meena, C. L., Kaur, N., and Jain, R. (2008). Chemistry and biology of thyrotropin-releasing hormone (TRH) and its analogs. *Curr. Med. Chem.* **15,** 2718–2733.

Moriyama, K., Okuda, J., Saijo, M., Hattori, Y., Kanamoto, N., Hataya, Y., Matsuda, F., Mori, T., Nakao, K., and Akamizu, T. (2003). Recombinant monoclonal thyrotropin-stimulation blocking antibody (TSBAb) established from peripheral lymphocytes of a hypothyroid patient with primary myxedema. *J. Endocrinol. Invest.* **26,** 1076–1080.

Neumann, S., Raaka, B. M., and Gershengorn, M. C. (2009). Human TSH receptor ligands as pharmacological probes with potential clinical application. *Expert Rev. Endocrinol. Metab.* **4,** 669–679.

Neumann, S., Huang, W., Eliseeva, E., Titus, S., Thomas, C. J., and Gershengorn, M. C. (2010). A small molecule inverse agonist for the human TSH receptor. *Endocrinology* **151,** 3454.

Robbins, A. K., and Horlick, R. A. (1998). Macrophage scavenger receptor confers an adherent phenotype to cells in culture. *Biotechniques* **25,** 240–244.

Sanders, J., Evans, M., Betterle, C., Sanders, P., Bhardwaja, A., Young, S., Roberts, E., Wilmot, J., Richards, T., Kiddie, A., Small, K., Platt, H., *et al.* (2008). A human monoclonal autoantibody to the thyrotropin receptor with thyroid-stimulating blocking activity. *Thyroid* **18,** 735–746.

Sun, Y., Lu, X., and Gershengorn, M. C. (2003). Thyrotropin-releasing hormone receptors—Similarities and differences. *J. Mol. Endocrinol.* **30,** 87–97.

Vassart, G., Parmentier, M., Libert, F., and Dumont, J. (1991). Molecular genetics of the thyrotropin receptor. *Trends Endocrinol. Metab.* **2,** 151–156.

Wang, W., and Gershengorn, M. C. (1999). Rat TRH receptor type 2 exhibits higher basal signaling activity than TRH receptor type 1. *Endocrinology* **140,** 4916–4919.

CHAPTER TEN

Inverse Agonists and Antagonists of Retinoid Receptors

William Bourguet,* Angel R. de Lera,† *and* Hinrich Gronemeyer‡

Contents

* INSERM U554 and CNRS UMR5048, Centre de Biochimie Structurale, Universités Montpellier 1 & 2, Montpellier, France
† Departamento de Química Orgánica, Facultad de Química, Universidade de Vigo, Vigo, Spain
‡ Department of Cancer Biology, Institut de Génétique et de Biologie Moléculaire et Cellulaire (IGBMC), C. U. de Strasbourg, France

Methods in Enzymology, Volume 485
ISSN 0076-6879, DOI: 10.1016/S0076-6879(10)85010-4

Abstract

Nuclear receptors (NRs) are ligand-inducible transcription factors that regulate a plethora of cell biological phenomena, thus orchestrating complex events like development, organ homeostasis, immune function, and reproduction. Due to their regulatory potential, NRs are major drug targets for a variety of diseases, including cancer and metabolic diseases, and had a major societal impact following the development of contraceptives and abortifacients. Not surprisingly in view of this medical and societal importance, a large amount of diverse NR ligands have been generated and the corresponding structural and functional analyses have provided a deep insight into the molecular basis of ligand action. What we have learnt is that ligands regulate, via allosteric conformational changes, the ability of NRs to interact with different sets of coregulators which in turn recruit enzymatically active complexes, the workhorses of the ligand-induced epigenetic and transcription-regulatory events. Thus, ligands essentially direct the communication of a given NR with its intracellular environment at the chromatin and extragenomic level to modulate gene programs directly at the chromatin level or via less well-understood extranuclear actions. Here we will review our current structural and mechanistic insight into the functionalities of subsets of retinoid and rexinoid ligands that act generically as antagonists but follow different mechanistic principles, resulting in "classical" or neutral antagonism, or inverse agonism. In addition, we describe the chemical features and guidelines for the synthesis of retinoids/rexinoids that exert specific functions and we provide protocols for a number of experimental approaches that are useful for studies of the agonistic and antagonistic features of NR ligands.

1. Introduction

Mechanistic studies of retinoic acid receptors (RARs, NR1B1-3), and other members of nuclear receptor (NR) superfamily, have demonstrated that ligand binding modulates the ability of the receptor to "communicate" with its intracellular environment by establishing temporally defined ligand, cell, signaling context, and NR-specific receptor–protein and receptor–DNA/chromatin interactions. This communication repertoire is at the basis of the powerful signaling potential of NRs, which affects complex processes ranging from key phenomena in early development and organogenesis to adult body/organ function. Mechanistically, the binding of a ligand allosterically alters receptor surfaces required for interaction with coactivator (CoA) and corepressor (CoR) complexes that are central to the control of target gene transcription. RAR agonists facilitate the exchange between CoR and CoA complexes by destabilizing CoR and stabilizing CoA–RAR interfaces. The subsequent recruitment of epigenetically active and/or chromatin modifying complexes, such as histone acetyltransferases and methyltransferases, leads to chromatin alterations and posttranslational modifications of important transcription regulators that specify activation of target gene expression by the basal transcriptional machinery. In the absence of agonists, RARs can actively repress gene transcription as nonliganded RAR establishes CoR complexes comprising repressive factors, which are believed to promote chromatin compaction (Gronemeyer *et al.*, 2004; Lonard and O'Malley, 2007; Rosenfeld *et al.*, 2006). Indeed, the physiological impact of CoRs in NR signaling has been recently revealed by gene deletion studies (Astapova *et al.*, 2008; Nofsinger *et al.*, 2008).

This active transcriptional silencing, mediated by RARα in complex with its retinoid X receptor (RXR) heterodimeric partner, is a direct consequence of association with CoRs such as SMRT/TRAC (Chen and Evans, 1995; Jepsen *et al.*, 2007; Sande and Privalsky, 1996) or NCoR/RIP13 (Horlein *et al.*, 1995; Seol *et al.*, 1996; Zamir *et al.*, 1996). These cofactors exhibit modular structures facilitating the assembly of high molecular weight complexes and serve also as CoRs for other NRs (Zamir *et al.*, 1996) and transcription factors (see, e.g., Evert *et al.*, 2006; Melnick *et al.*, 2002). Among others, histone deacetylases are factors present in these complexes (Alland *et al.*, 1997; Heinzel *et al.*, 1997; Nagy *et al.*, 1997). The formation of a CoR complex with associated histone deacetylase activity at the promoters of NR target genes is believed to cause transcriptional silencing through histone deacetylation and chromatin condensation at receptor-targeted loci. Binding of an agonist to RAR induces a conformational switch of the ligand-binding domain (LBD) leading to the release of CoRs and concomitant association of CoA complexes. These CoAs, such as the p160 proteins SRC1, TIF2/GRIP1/SRC2, and RAC3/ACTR/SRC3 (Anzick *et al.*, 1997; Chen *et al.*, 1997; Hong *et al.*, 1997; Li *et al.*, 1997; Onate *et al.*, 1995; Takeshita *et al.*, 1997; Voegel *et al.*, 1996; Yao *et al.*, 1996), serve as recruitment platforms for the "work horses," namely epigenetic enzymes that modify chromatin histones but also other transcription modulators (Bannister and Kouzarides, 1996; Chen *et al.*, 1999; Collingwood *et al.*, 1999; Glass *et al.*, 1997; Gu *et al.*, 1997; Hanstein *et al.*, 1996; Kamei *et al.*, 1996; Korzus *et al.*, 1998; McKenna *et al.*, 1999; Ogryzko *et al.*, 1996; Torchia *et al.*, 1996; Voegel *et al.*, 1998; Westin *et al.*, 2000; Yao *et al.*, 1996).

The allosteric modulation of NR surfaces is at the basis of ligand function and has the promise of allowing for the synthesis of function-specific ligands beyond mere agonists and antagonists. Indeed the concept of selective nuclear receptor modulators (SNuRMs) with tissue-selective activities originating from the divergent expression levels of CoAs, CoRs, and the factors involved in the corresponding complexes has been described (Gronemeyer *et al.*, 2004) and awaits validation, while the biological/pharmaceutical potential of the different types of antagonists, such as inverse agonists, still has to be elucidated. Clearly, the analyses of NR ligand actions have provided an extraordinary deep insight into the possibility to modulate the functions of key regulators by ligand design through allosteric alteration.

2. Functional Classification of Retinoid Receptor Ligands

NR ligands can be classified by their action on the recruitment or dissociation of coregulator complexes, that is, those established by CoAs, such a p160s/SRCs, and CoRs, such as NCoR/SMRT. Table 10.1 gives a (simplified) view on the effects of various classes of ligands. Note that only effects on the receptor LBD, which contains the activation function 2 (AF2), are considered; very little is known about the possible effects of ligand binding on the N-terminal activation function AF1, while its existence and synergy with AF2 has been confirmed (Nagpal *et al.*, 1993). Note also that the so-called "partial agonist/antagonist" is a weaker agonist than the natural or reference agonist; it acts as an antagonist if its affinity allows for an efficient competition and reduces the agonistic activity of the natural/reference agonist to its own activation level.

Table 10.1 Retinoid modulation of coregulator binding[a]

Ligand	Coactivator	Corepressor	Example
RAR agonist	Recruitment	Dissociation	ATRA[b] [**1**], TTNPB[b] [**2**], Am580[c] [**3**]
Inverse pan-RAR agonist	Dissociation	Stabilization	BMS204,493[b] [**4a**]; AGN194,310 [**5**]
Neutral RAR antagonist	Dissociation	Dissociation	BMS195,614[c] [**6**]

[a] Only ligands for which CoA and CoR bindings have been confirmed are shown; other types of ligands exist, in particular, pure AF2 antagonists (which allow for AF1 activity, such as OH-tamoxifen and RU486 in the cases of estrogen receptor-α and progesterone receptor, respectively), partial agonist/antagonists, etc.

[b] pan-RAR agonist.

[c] RARα-selective.

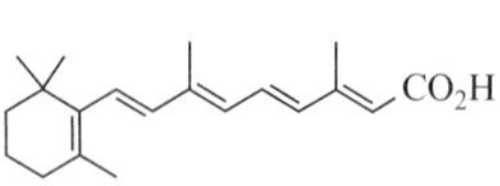

1, all-*trans*-retinoic acid (ATRA)

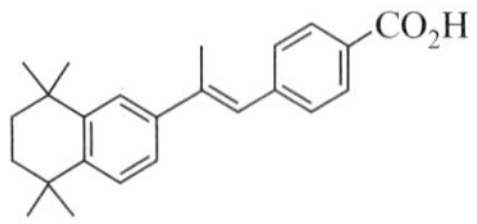

2, TTNPB

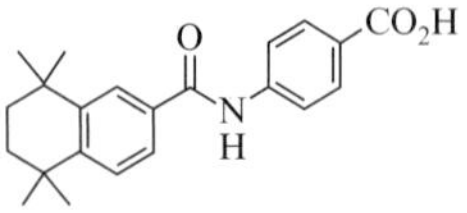

3, Am580

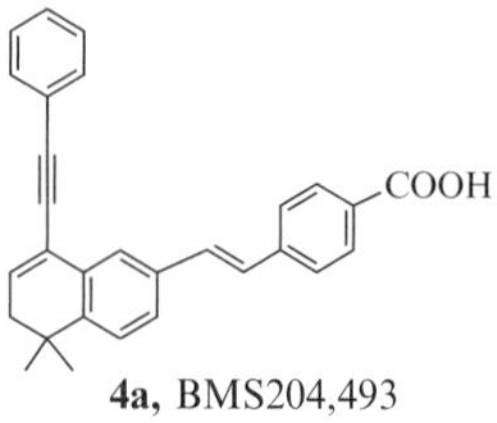

4a, BMS204,493

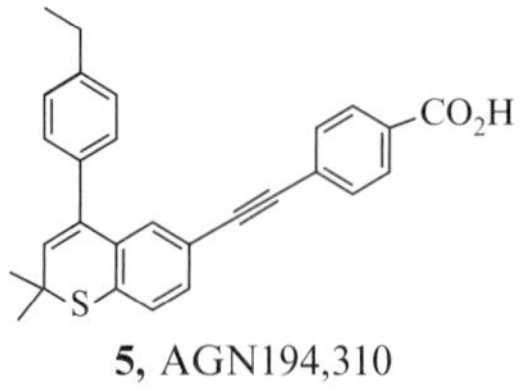

5, AGN194,310

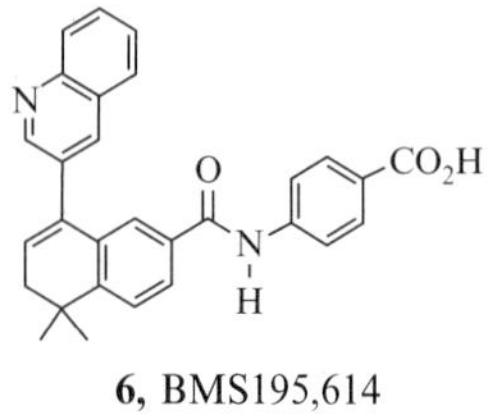

6, BMS195,614

3. Structural Basis of Retinoid Receptor Action

3.1. Common structural paradigms govern ligand-dependent NR functions

Our understanding of how ligand binding leads to the activation of NRs has been greatly advanced by structural studies of NR LBDs and their interactions with CoA- and CoR-derived peptides (Bourguet *et al.*, 2000a; Huang *et al.*, 2010). Indeed, in the case of RARs, agonistic ligands establish a direct contact to H12 residues and maintain the helix in a position which, together with other secondary structural elements, generates a surface with increased affinity for the LxxLL (x stands for any amino acid) motifs (also called NR boxes) that are located in the NR-interacting surface of CoA proteins. Correlative analyses of biochemical data and protein sequences provided evidence that CoA and CoR recruitments share similar molecular features. Evaluating CoR binding to mutants in the CoA binding site of the thyroid hormone receptor α (TRα, NR1A1) Hu and Lazar (1999) demonstrated that mutations that impair activation and CoA recruitment also decrease repression and CoR binding, thus indicating that NCoR binds to an NR surface topologically related to that involved in CoA interaction with the important difference that the AF2 H12 is not involved in the interaction with CoRs (Hu and Lazar, 1999).

Accordingly, examination of the two C-terminal NR-interaction domains ID1 and ID2 in SMRT and NCoR (Cohen *et al.*, 2001; Seol *et al.*, 1996; Zamir *et al.*, 1996) revealed sequences (CoRNR1 and CoRNR2 by analogy to NR boxes of CoAs; note that the numbering of CoRNR1 to CoRNR2 is from N- to C-terminus and that the ID1/2 and CoRNR1/2 nomenclature is used synonymously) similar but not identical to the LxxLL motif of CoAs, which were predicted to adopt a longer amphipathic helical conformation (Nagy *et al.*, 1999; Perissi *et al.*, 1999). CoRNR boxes are not equivalent as, for example, RAR interacts strongly with CoRNR1 but very weakly with CoRNR2, and further biochemical and structural studies revealed that both CoRNRs fold as extended helical motifs with residues flanking the core I/LxxV/II sequence determining NR specificity (Hu *et al.*, 2001; Perissi *et al.*, 1999; Xu *et al.*, 2002). It was thus proposed that discrimination by NR H12 helix of the lengths of CoA and CoR interaction helices occupying partially overlapping binding sites may constitute the molecular basis of ligand-mediated coregulator exchange on NRs. Moreover, the recent discovery of an unforeseen specific interface between RAR and NCoR revealed that a secondary-structure transition affecting helix H11 plays a master role in CoR association and release (le Maire *et al.*, 2010; see below for more details).

3.2. Agonist and antagonist action from a structural perspective

It is now well established that ligand binding has a profound impact on AF2 H12 positioning and dynamics both of which determine receptor activity (Nagy and Schwabe, 2004). The crystal structures of retinoid receptor LBDs (from RARs and RXRs) have been determined in complex with a multitude of natural or synthetic ligands and in the presence of various coregulator fragments. The structures reveal a conserved core of 12 α-helices (H1–H12) and a short two-stranded antiparallel β-sheet (S1 and S2), arranged into a three-layered sandwich fold (Fig. 10.1). This arrangement generates a mostly hydrophobic ligand-binding pocket (LBP) in the lower half of the domain. In all agonist-bound LBD structures, the ligand-binding cavity is sealed by helix H12 (Fig. 10.1A). This conformation is specifically induced by the binding of retinoic acids (ATRA; 9-*cis* retinoic acid) or synthetic agonists and is referred to as the "holo" or "active" conformation because it favors the recruitment of transcription CoAs. Helices H3, H4, and H12 define a hydrophobic binding groove for the short LxxLL (x stands for any amino acid) helical motifs found within CoAs (Fig. 10.1A). It is noteworthy that this holo-LBD conformation is displayed by all agonist-bound NRs crystallized to date and thus corresponds to the canonical active form of receptors. Only a very few NR LBDs have been crystallized in their unliganded "apo" form. In contrast to the agonist-bound conformation, helix H12 in apo-receptors either adopts

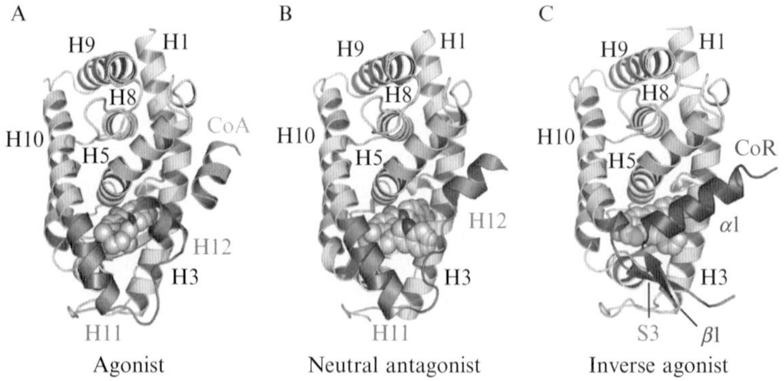

Figure 10.1 Crystal structures of RARα LBD in various functional states. (A) The active form of the receptor is induced by agonist binding and allows coactivator (CoA, green) binding. (B) In contrast, binding of neutral antagonists provokes a displacement of H12 that prevents CoA recruitment and maintains the receptor in an inactive state. (C) The repressive state can be obtained in the presence of an inverse agonist that stabilizes the interaction with corepressors (CoR, violet). The C-terminal portion of RARα LBD that is subjected to important conformational changes is highlighted in red. (See Color Insert.)

different positions or exists as a dynamic ensemble of conformations resulting in both cases in the formation of an incomplete binding surface for the LxxLL motifs and a much lower affinity for CoAs. Hence, agonists act as activators of the AF2 by locking helix H12 in the active conformation.

In contrast, some ligands bind to retinoid receptors with high affinity but fail to stabilize their active form and are classified as partial agonists or antagonists depending on their ability to prevent CoA recruitment. Insight into the structural basis of antagonism was provided by the resolution of the structure of RARα LBD in complex with the synthetic antagonist BMS614 **[6]** (Bourguet *et al.*, 2000b). The structure shows that BMS614 **[6]** binds into the same cavity as agonists (Fig. 10.1B). However, the particular feature of antagonists, including BMS614 **[6]**, is the presence of a side chain, pointing toward helix H12, which is too long to be contained within the buried ligand-binding site. As a consequence, the bulky extension protruding between H3 and H11 prevents helix H12 from adopting the active conformation. Thus, antagonists block receptor's AF2 by (i) avoiding retinoic acid binding and (ii) disrupting the interaction surface with CoAs. Interestingly, the antagonistic capacity of such compounds strongly depends on the size, the position, and the chemical nature of the side chain so that in addition to full antagonists, it is possible to synthesize partial (weak) agonists that incompletely disrupt the receptor holo-conformation and in turn the association with CoAs. Recent studies have revealed that the mixed agonist/antagonist activity of such ligands relies on the lowering of the interaction strength between holo-H12 and the LBD surface which renders the AF2 helix more dynamics as compared to the agonist-bound situation. However, in contrast with antagonists, partial agonists exert moderate steric constraints on helix H12 so that the presence of CoAs helps stabilization of AF2 H12 in the active position (Nahoum *et al.*, 2007; Pérez-Santín *et al.*, 2009). This unique feature allows partial agonists to "sense" intracellular coregulator levels and act as cell-selective modulators with agonist or antagonist properties depending on the cellular context.

3.3. Inverse agonists and neutral antagonists

Compounds with antagonistic activity can be divided into two subclasses on the ground of the mechanism by which they counteract the action of agonists. BMS614 **[6]** belongs to the group of "classical" or "neutral" antagonists that prevent the recruitment of CoAs (see above). The second group contains so-called inverse agonists, such as compound BMS493 **[4a]**, that are less effective in preventing CoA binding but in contrast have the potential to strengthen the interaction with CoRs (Germain *et al.*, 2009). The molecular basis for the differential action on coregulator interaction displayed by the two types of antagonists has been recently revealed by the resolution of the crystal structure of the BMS493-bound RARα LBD in complex with a peptide containing the CoRNR1 sequence of NCoR (Fig. 10.1C; le Maire *et al.*, 2010). An obvious difference with the agonist- and neutral antagonist-bound RAR structures is that helix H12 appears highly mobile and has no defined position in the crystal structure. Moreover, unlike the agonist Am580 [**3**] (Fig. 10.1A) and antagonist BMS614 [**6**] (Fig. 10.1B) complexes in which residues C-terminal of helix H10 adopt a helical structure (H11), these residues (Arg394 to Lys399) assume an extended β-strand (S3) conformation in the CoR complex allowing for the formation of an antiparallel β-sheet with N-terminal residues (β1) of CoRNR1 (Fig. 10.1C). The remaining of CoRNR1 folds as a four-turn α-helix (α1) which docks into the coregulator groove of RAR through the conserved LxxxIxx(I/V)Ixxx(Y/F) motif reminiscent of the short LxxLL sequence found in CoAs. Together with the observation that a similar S3/β1 β-sheet interface has not been observed in previously reported NR/inverse agonist/CoR structures, mutagenesis data strongly support the notion that this mode of interaction with CoRs accounts for the basal repressive activity of apo-RAR and some other NRs including Rev-Erb (le Maire *et al.*, 2010). The reported data also demonstrate that the secondary-structure switch from β-strand S3 to α-helix H11 plays a central role in the agonist-induced CoR dissociation (le Maire *et al.*, 2010).

Although very similar in their chemical structures, BMS614 [**6**] and BMS493 [**4a**] display slightly different binding modes in RAR LBD which most likely account for their differential actions on coregulator recruitment (Fig. 10.2A). While the A rings of both ligands superpose well, a steric clash between the carbonyl group of BMS614 [**6**] and Ile273 (Fig. 10.2A and B) forces rings B, C, and the antagonistic extensions to adopt different positions in BMS614 [**6**] and BMS493 [**4a**]. In the neutral antagonist complex, interactions between BMS614 [**6**] and Val395 and Leu398 stabilize the helical conformation of H11 (Fig. 10.2B). With a chemical structure similar to that of BMS614 [**6**] (with the exception of the antagonistic extension) Am580 [**3**] adopts a comparable binding mode and also displays the stabilizing interactions with H11 residues (Fig. 10.2C). In the inverse agonist complex, these interactions are lost but novel van der Waals contacts of Ile396 and Leu398 with the aromatic side chain of BMS493 [**4a**] stabilize the β-strand S3 conformation thus favoring CoR binding (Fig. 10.2D). Due to their antagonistic extensions, both BMS614 [**6**] and BMS493 [**4a**] prevent the positioning of helix H12 in the active orientation and thus antagonize CoA binding. However, it is noteworthy that the bulky extension of BMS614 [**6**] protrudes markedly more from the LBP toward helix H12 than that of BMS493 [**4a**] (Fig. 10.2A). This difference is very likely to contribute to the higher efficiency of BMS614 [**6**] in preventing CoA binding through a more effective displacement of helix H12 from its active position. Hence, in contrast with agonists that disrupt the RAR/CoR interface via S3 destabilization and formation of helix H11 and induction of the active conformation, BMS493 [**4a**] acts as an inverse agonist by stabilizing residues Arg394 to Lys399 in the β-strand S3 conformation, thus reinforcing CoR interaction.

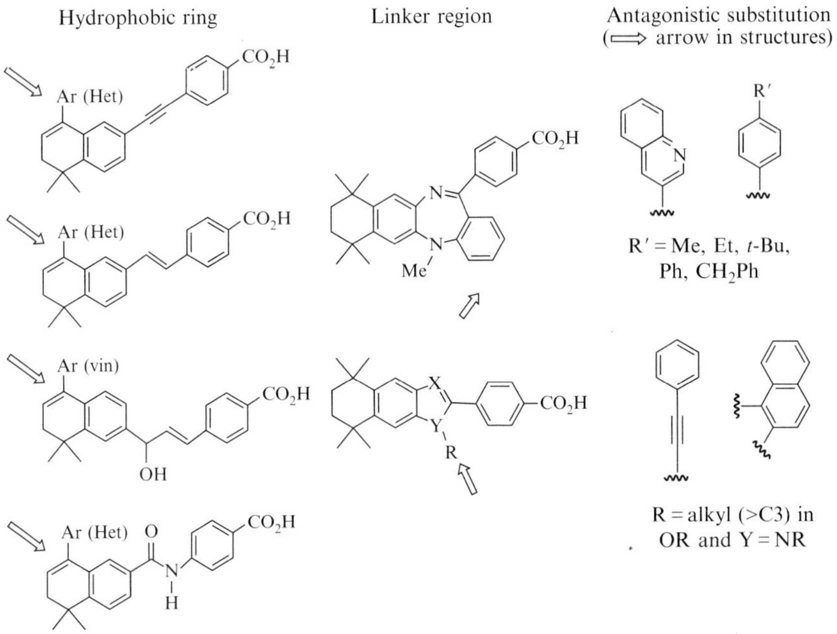

Figure 10.3 Antagonist position (arrow) and antagonist substitution in retinoid scaffolds.

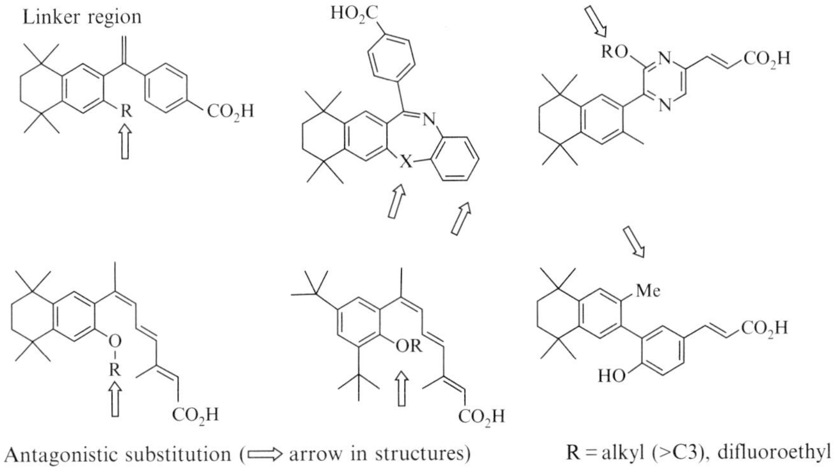

Figure 10.4 Antagonist position (arrow) and antagonist substitution in rexinoid scaffolds.

In addition, using these privileged scaffolds, the design of antagonists and inverse agonists, which are structurally very similar, focuses on the currently accepted ligand-induced H12-repositioning mechanism of antagonistic action in NR–ligand interaction, which was revealed by the crystal structure of the mouse BMS195,614-RARα/oleic acid-RXRα F318A LBD heterodimer (Bourguet *et al.*, 2000b). Accordingly, retinoid/rexinoid antagonists must incorporate extended and/or bulky substituent in selected positions of the parent agonist structures that are oriented toward helix H12 and span this distance (Bourguet *et al.*, 2000b; Hashimoto and Miyachi, 2005). The arrow in Fig. 10.3 (retinoids) and Fig. 10.4 (rexinoids) illustrates the antagonistic region in the reported skeletons (de Lera *et al.*, 2007).

The dissimilar LBPs and orientations of the rexinoids and retinoid ligands inside the LBPs determines also a different position and size of the antagonist/inverse agonists-enforcing substituents in these skeletons. Rexinoids are modified with alkyl (size greater than C_3) groups at defined positions of the linker region, whereas retinoids most often display bulkier aryl (heteroaryl) substituents at the carbo/heterocyclic hydrophobic rings (de Lera *et al.*, 2007).

4.2. Design of retinoid agonists and antagonists

4.2.1. Selective RAR inverse agonists/antagonists

Previous studies on the transcriptional activities of stilbene arotinoids (TTNPB [**2**] and analogues, Fig. 10.5) led to the discovery of ligands with interesting profiles, including agonists, antagonists, inverse agonists, and

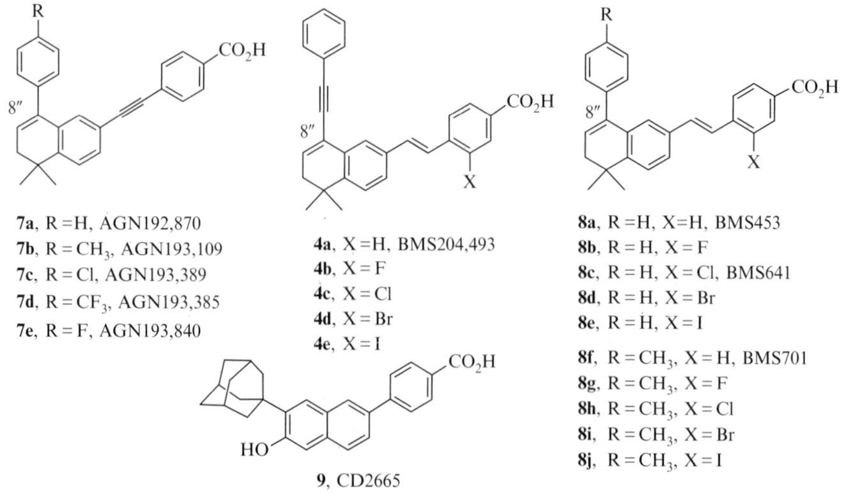

Figure 10.5 Series of diarylacetylene, stilbene arotinoids, and adamantyl arotinoids discussed in the text.

mixed agonists/antagonist, that is, ligands (such as BMS453, **8a**) that display agonistic properties with one RAR isotype and antagonistic properties with a paralogue (Chen *et al.*, 1995; Gehin *et al.*, 1999; Germain *et al.*, 2004). Moreover, some of the congeners, namely the RARβ-agonist BMS641 [**8c**] (Germain *et al.*, 2004) and the RARα-antagonist BMS614 [**6**], display subtype selectivity. General principles for the acquisition of subtype selectivity in stilbene arotinoids through ligand design, that are useful guidelines for other scaffolds, have been discussed previously (de Lera *et al.*, 2007).

Transactivation studies on the entire series of stilbene arotinoids with a C3-halogen and bulky substituents at C8″ position [**4a–e**, **8a–j**] revealed the synergistic effect of these groups in subtype selectivity and activity (Álvarez *et al.*, 2009). Series **8** exhibited in general inverse agonistic profiles with RARα/β subtype selectivity and RARγ antagonism. Halogens in series **8f–j** convert the RARβ/γ weak antagonist **8f** into potent RARα/β inverse agonists [**8g–i**] and RARγ antagonist. Large halogen atoms alter the RARα inverse agonistic activity of parent **4a**. Whereas **4d** (Br) functions as RARβ-selective agonist with moderate RARγ-partial agonist/antagonist activity, the bulkiest **4e** (I) becomes a moderate antagonist of the three subtypes (Álvarez *et al.*, 2009). The formation of "halogen bonds" (Auffinger *et al.*, 2004) by charge-transfer complexes (C–X···O–Y) from a heteroatom lone pair (Lewis base) to the halogen (Lewis acid) might explain some of the effect of halogen atoms in ligand–receptor interactions (Memic and Spaller, 2008).

The first inverse agonists reported in the NR field were based on the related diarylacetylene arotinoid structure which had been modified with *para*-substituted C8-phenyl groups **7** (R = CH_3, CF_3, Cl; Fig. 10.5; Klein *et al.*, 1996). The analogues with smaller groups (R = H, F; Fig. 10.5) instead functioned as neutral antagonists. Recent in-depth functional characterization revealed the subtype selectivity of some members of the series: AGN109 [**7b**] is an inverse agonist of RARγ (and a weak agonist of RARα/β), and AGN870 [**7a**] is a weak pan-inverse agonist of RARα/β/γ with selectivity for RARγ (Germain *et al.*, 2009).

4.2.2. Selective RXR inverse agonists/antagonists

The hypothesis that, within a ligand scaffold, the agonist-to-antagonist transition can be achieved by systematic modification of the length or bulk of a chosen substituent has been demonstrated with a family of analogues of the potent and selective RXR agonist CD3254 [**10a**] (Fig. 10.6; Nahoum *et al.*, 2007). The proposed modification at position C2, guided first by Molecular Modeling studies and later confirmed by the crystal structure of the LBD of the RXRβ-CD3254 [**10a**] complex (Pérez-Santín *et al.*, 2009), was considered to disrupt, after a certain size, the hydrophobic interactions that stabilize the agonist conformation of H12.

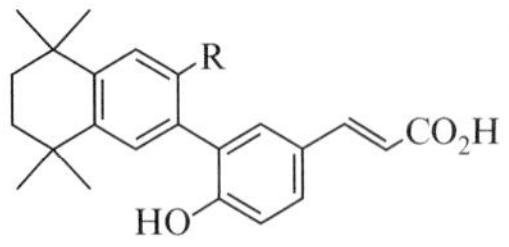

10a, R = CH_3, CD3254
10b, R = OCH_3
10c, R = OCH_2CH_3
10d, R = $O(CH_2)_2CH_3$
10e, R = $O(CH_2)_3CH_3$
10f, R = $O(CH_2)_4CH_3$, UVI3003
10g, R = $O(CH_2)_5CH_3$

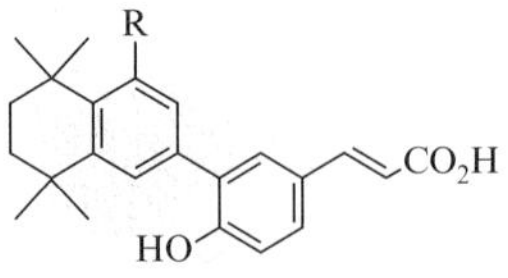

11a, R = OCH_3
11b, R = OCH_2CH_3
11c, R = $O(CH_2)_2CH_3$
11d, R = $O(CH_2)_3CH_3$
11e, R = $O(CH_2)_4CH_3$
11f, R = $O(CH_2)_5CH_3$

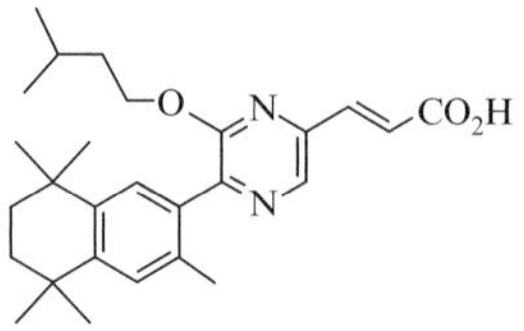

12

Figure 10.6 RXR agonist CD3254 [10a] and the series of alkyl aryl ether homologues (10b–g) that exhibit a gradual switch of their activity from agonists to partial agonists to antagonists (in particular UVI3003, 10f), as well as positional isomers (11a–f). Structure of the dual RAR/RXR inverse agonist 12.

In fact, the progressive lengthening of the alkyl ether chain induced a transition from agonist (**10b** and **10c** are weaker than parent **10a**) to full antagonist (**10f, 10g**) *via* mixed agonist/antagonists (**10d, 10e**). The activities were confirmed by direct *in vivo* photon counting of Gal4-RXR LBD chimera-based luciferase reporter cells (Fig. 10.7), transient transactivation experiments, coregulator recruitment assays (Section 5.3), and fluorescence anisotropy studies (Section 5.4). Instead of a direct displacement of H12, the effect of the alkoxy chain at C2 of **10** on the position of H12 was indirect, and in fact was mediated by the interaction with L436 at H11, which by steric clash with L455 (mediated in the shorter homologues **10b–d** through a water molecule) disrupts the normal mobility of H12, as crystal structure studies of **10b–d** confirmed (Nahoum *et al.*, 2007).

In contrast, the C1′-alkoxy positional isomers **11a–f** behave as less-potent RXR agonists to antagonists, as judged from transactivation assays and fluorescence anisotropy experiments, thus showing the relevance of the directionality of the substituents in the LBP as determinant of the antagonist/inverse agonist profile (Pérez-Santín *et al.*, 2009).

4.2.3. Dual RAR/RXR inverse agonists

A related family of trisubstituted pyrazine-based arotinoids exhibited selective RAR and RXR agonist/antagonist profiles (García *et al.*, 2009). The pyrazine acrylic acid **12** with an isoamyl ether chain showed the unexpected property of being a dual inverse agonist of all RAR subtypes and of RXR (García *et al.*, 2009). These pyrazine arotinoids are the first nonrigid RAR inverse agonists with structural modifications at the central heterocyclic core. The alkyl substituent must be oriented in the LBP similarly than the aryl (**7**) and phenylethynyl (**4**) substituents of the previously described RAR inverse

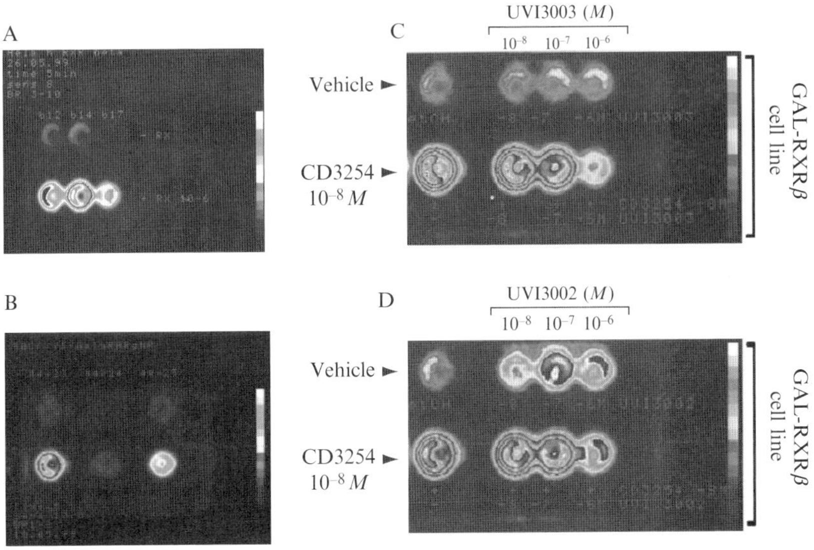

Figure 10.7 Illustration of ligand activities using different types of reporter cell lines. (A) Inducibility of three different HeLa-based Gal-RXRβ reporter cells (b12, b14, and b17); RX, 1 μ*M* 9-*cis* retinoic acid. (B) SQCC/Y1-based RARγ reporter cell line clones 44-23, -24, -25. "−," without ligand; "+" after incubation with 1 μ*M* all-*trans* retinoic acid. (C) UVI3003 [**10f**] is an RXR antagonist. Gal-RXRβ reporter cells were exposed to vehicle (top panel) or the rexinoid agonist CD3254 [**10a**] (10 n*M*; bottom panel) in the absence (top left) or increasing concentrations (10 n*M* to 1 μ*M*) of UVI3003 [**10f**], as indicated. Note the decrease of CD3254-induced luciferase expression in the presence of UVI3003, revealing that this ligand is an antagonist. (D) Analysis as in (C) but with UVI3002 [**10d**]. UVI3002 activates luciferase expression on its own but also decreases CD3254-induced luciferase expression. Thus, UVI3002 [**10d**] acts as a partial agonist/antagonist.

agonists. The RXR inverse agonist profile of **12** indicates an allosteric effect on RXR H12 that allow for increased CoR interaction, since in contrast to RAR, H12 of RXR masks CoR access (Zhang *et al.*, 1999).

4.3. Synthetic challenges and strategies

Modern synthetic methodologies, based on reactions that use substoichiometric quantities of transition metals (de Meijere and Diederich, 2004), have also been adapted to the preparation of retinoids and rexinoids (Domínguez *et al.*, 2003). The synthetic routes conceived choosing these synthetic methods have in general fewer steps than more traditional routes, and proceed with higher yields and efficiencies. As an example, the synthesis of pyrazine arotinoid **12**, the dual RAR/RXR inverse agonist, is shown in Scheme 10.1. It uses a 2-amino-3,5-dibromopyrazine [**13**] as a central

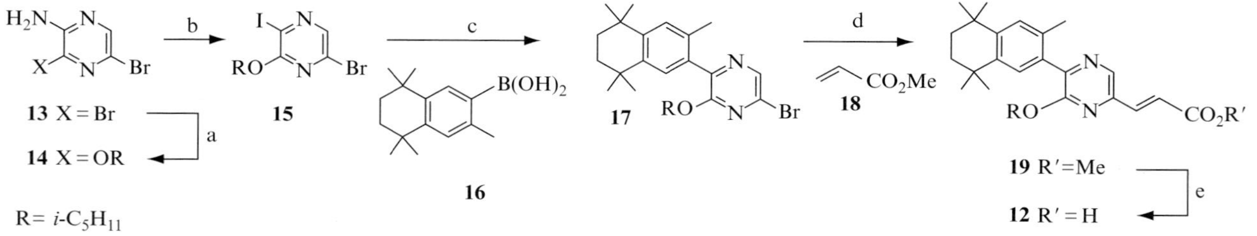

Scheme 10.1 Synthesis of pyrazine arotinoid 12 using key steps that employ substoichiometric quantities of transition metals. (A) isoamyl alcohol, NaH, THF, microwaves (200 W), 15 min, 70%. (B) Aqueous HI, $NaNO_2$, CH_3CN, H_2O, 60 °C, 15 h, 45%. (C) $Pd(PPh_3)_4$, aqueous Na_2CO_3, MeOH, benzene, 70 °C, 16 h, 51%. (D) $Pd(OAc)_2$, *n*-Bu_4NCl, $NaHCO_3$, 4 Å MS, DMF, 70 °C, 17 h, 87%. (E) $LiOH{\cdot}H_2O$, THF/H_2O (1:1), 25 °C, 2 h, 99%.

linchpin connector to which the hydrophobic and polar (cinnamic acid) units will be attached using two palladium-catalyzed cross-coupling reactions. The sodium salt of isoamyl alcohol in anhydrous THF converted **13** into the corresponding alkyl ether **14** under microwave irradiation. A second halogen (I) was incorporated in the structure of **14** through formation of the intermediate diazonium salt. A halogen-selective Pd-catalyzed Suzuki reaction of **15** with aryl boronic acid **16** provided **17**, whereas the subsequent Heck reaction with methyl acrylate **18**, afforded ester **19** (Scheme 10.1), from which cinnamic acid **12** was obtained by saponification.

5. Protocols for the Study of Ligand Function

5.1. Reporter cell lines

5.1.1. Establishment of reporter cell lines

Time lines for the establishment of the cell lines:

- If several cell lines can be used the transfectability is established first
- Transfect the cells by the calcium phosphate method and start the selection 48 h later
- Selection during approximately 3 weeks
- Induce the cells with the appropriate ligand during 6 h or overnight
- Check which clones are luminescent and transfer them into 96-well plates
- Expand and amplify the clones (2 weeks)
- Retest the clones with and without ligand to assess inducibility
- *Amplify the clones again in order to*:
  - Retest for inducibility
  - Freeze them
  - Purify by limiting dilution to get pure clones
- Amplify the pure clones and test them for inducibility
- Keep the "best" inducible clones and freeze them

The entire procedure requires 2–3 months, provided all steps are without problems.

5.1.2. Examples of reporter cell types

Stably transfected "reporter cells" provide a ligand "readout" in the context of a given living cell (Fig. 10.7). The use of Gal4 DNA-binding domain (DBD)-NR LBD chimeras has the advantage that there will be no interference by endogenous NRs on the cognate 17 mer Gal4 response element-driven luciferase reporter gene. Different cell types can be used, thereby providing a cell-specific context to the assay.

5.1.3. Reporter cell assay to define agonist and antagonist activity in living cells

The following protocol is an example given for the characterization of candidate ligands as pan-RAR or RARα-selective agonists/antagonists.

5.1.4. Required materials (*store at −20 °C)

- MicroLumat LB96P luminometer (Berthold)
- Opaque white Optiplate-96-well plates (Perkin Elmer)
- HeLa reporter cells, stably transfected with (17 mer)$_5$-βGlobin-Luc-Neo and Gal4-mRARα receptor chimera expression vectors. Cells have to be maintained in DMEM containing 5% charcoal-stripped fetal calf serum (FCS) and supplemented with the corresponding antibiotics to maintain plasmids.
- *Lysis buffer*★: 25 m*M* Tris phosphate (pH 7.8), 2 m*M* EDTA, 1 m*M* DTT, 10% glycerol, and 1% Triton X-100.
- *Luciferin buffer*★: 20 m*M* Tris phosphate (pH 7.8), 1.07 m*M* $MgCl_2$, 2.67 m*M* $MgSO_4$, 0.1 m*M* EDTA, 33.3 m*M* DTT.
- ATP (0.53 m*M*)★, Luciferin (0.47 m*M*)★, Coenzyme A (0.27 m*M*)★

5.1.5. Determination of RAR/RXR agonist/antagonist activity

The ligand-binding assay is performed with reporter cells grown in DMEM without phenol red containing 5% charcoal-treated FCS.

- To determine the RARα, RARβ, RARγ, and RXRβ agonistic/antagonistic potential of the ligands, seed equal amount (13,000 cells/well) of the corresponding cell line in a 96-well plate.
- Allow cells to attach to the bottom and approximately 4–5 h later add the ligands in respective wells with respective concentrations. To analyze the antagonist activity of the compound, incubate the cells with different concentrations of the test ligands along with the specific agonist keeping respective controls.
- Incubate the cells at 37 °C in 5% CO_2 for 16 h (overnight).
- After overnight incubation, remove the medium containing ligands, wash the cells with PBS, and lyse them with lysis buffer (50 μl) for 15 min.
- Transfer equal aliquots (25 μl) of the cell lysates in an opaque white Optiplate-96-well plate and determine luminescence in RLU (relative luminescence units) on a MicroLumat LB96P luminometer ("Berthold") after automatic injection of 50 μl of luciferin buffer including 0.53 m*M* ATP, 0.47 m*M* luciferin, and 0.27 m*M* coenzyme A.

- The receptor activation potential of each ligand can be presented as fold induction measured as ratio of RLU of the compound over the RLU of the vehicle control.

5.2. Dual reporter cells

For several NRs, no natural or synthetic ligand has been identified (Laudet and Gronemeyer, 2001). To facilitate the screening for orphan receptor ligands, as well as characterize ligand and molecular features of classical NRs, we have engineered cellular reporting systems which comprise two NR LBD fused to the GAL4 DNA DBD (Fig. 10.8A). This setup has two advantages, as it (i) allows for the isolation of ligand-responsive cell clones

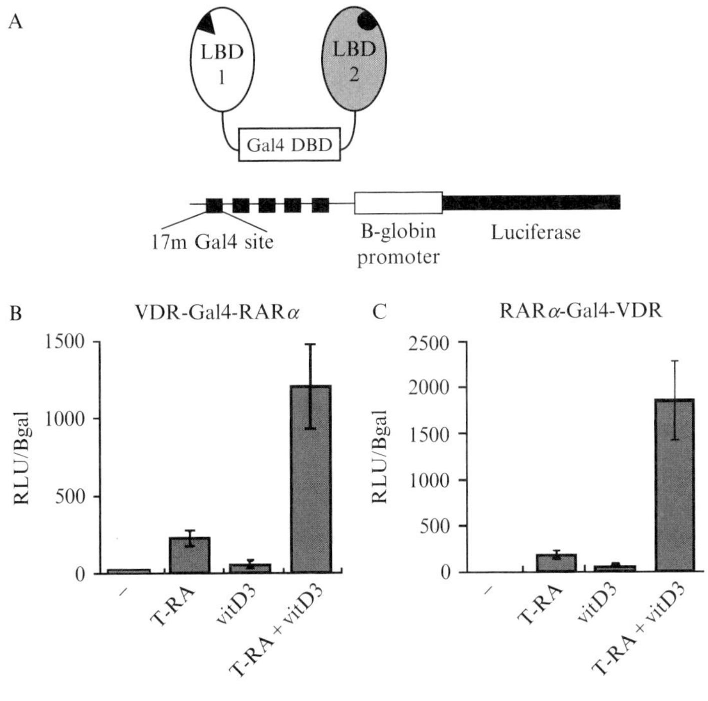

Figure 10.8 In the double ligand-binding chimera, each ligand activates its own LBD and both ligands synergize for transcriptional activation. (A) Description of the reporter system used. Two LBDs are fused in frame with the DNA-binding domain of the yeast transactivator Gal4. The reporter construct contains the firefly luciferase coding sequence under the control of a basic β-globin promoter and five repeats of the 17 bp Gal4 binding element. (B) Transient transfections were performed in HeLa cells with either RARα-Gal4-VDR or VDR-Gal4-RARα construct and activation of the reporter gene after treatment with the various ligands was quantified by measuring luciferase activity and normalized it against β-gal activity.

and (ii) exploits the synergy between the transcription activation functions in the two LBDs, thus facilitating the identification of weak agonists. This synergy can reach several magnitudes, as is apparent from a comparison of the single ligand- and double ligand-induced transient transactivation of VDR-Gal4-RARα and its inverse chimera (Fig. 10.8B).

This ligand synergy is also observed in stably transfected HeLa cells harboring one of the two NR LBD chimeras (Fig. 10.9). Note that, possibly due to the chromatin environment of the site of integration of the double LBD chimeric receptors, the ultimate single ligand response and synergy on coexposure to two ligands may vary among clones. For screening purposes, two strategies may be therefore employed. Either the temporal and freezing/thawing stability of single reporter cell subclones has been established or

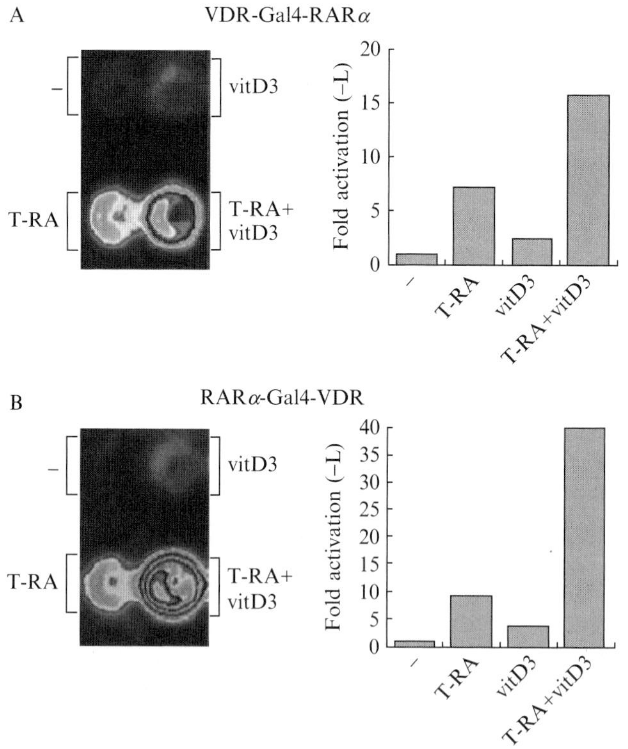

Figure 10.9 Synergy between the two LBDs occurs in stably transfected HeLa cells. (A) False color representation of luciferase-induced activity in HeLa cells stably transfected with the VDR-Gal4-RARα construct (left panel) and quantitation of the data obtained by photon counting (right panel). Data of a representative experiment (out of at least five independent experiments) are shown as fold inductions over basal activity observed without ligand. (B) same as (A), except that the inverse RARα-Gal4-VDR was stably expressed.

retroviral transduction of the reporting system and the use of cell populations should be considered.

The double LBD chimeric systems allow for a rapid definition of the agonistic, antagonistic, and inverse agonistic activity of ligands (Fig. 10.10). Indeed, while in the case of RARα-Gal4-VDR the pure agonist ATRA [**1**] may lead to a near 40-fold increase over the vitamin D3-induced signal, the inverse agonist BMS493 [**4a**] (Germain *et al.*, 2002) decreases this activity and partial agonists like CD2665 [**9**] lead to a moderate increase. Interestingly, the neutral antagonist BMS614 [**6**] induces a weak activity over the vitamin D3-induced level, indicating that the CoRs recruited to the RARα apo-LBD within the chimera attenuate the agonist-induced activity emanating from the VDR holo-LBD in the context of the double LBD chimera.

The signal readout from reporter cells may be further increased up to two orders of magnitude by chromatin derepression. Indeed, the agonist-induced activity in RARα-Gal4-VDR reporter cells was strongly enhanced by coexposure to histone deacetylase inhibitors (Fig. 10.11), such as TSA or VPA.

Taken together, reporter cells are a versatile tool for the definition of agonist, antagonist, and inverse agonist activities, especially when combined with mechanistic assays to monitor the recruitment/dissociation of CoA and CoR complexes and the associated (epigenetic) enzyme activities.

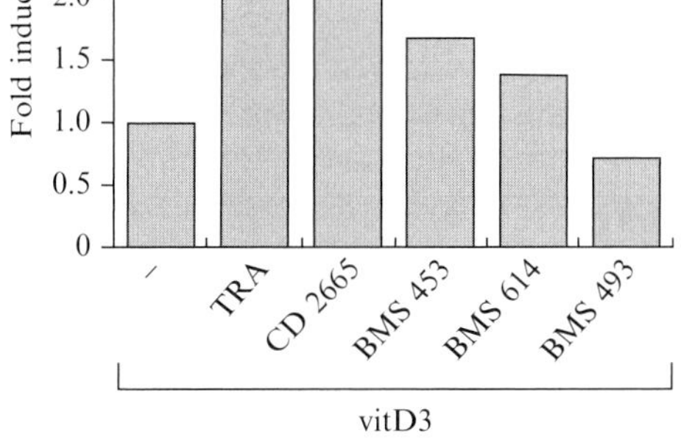

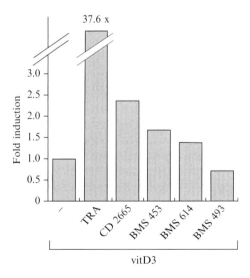

Figure 10.10 Different retinoids have been tested on the RARa-Gal4-VDR stable cell line in addition to vitD3. Luciferase activities were measured and fold inductions represent stimulation over vitD3 alone (set to 1). All ligands were used at 1 μ*M* except CD2665 [**9**] used at 2 μ*M*. Bars represent mean of duplicate samples of one representative experiment with similar results obtained in at least five independent experiments.

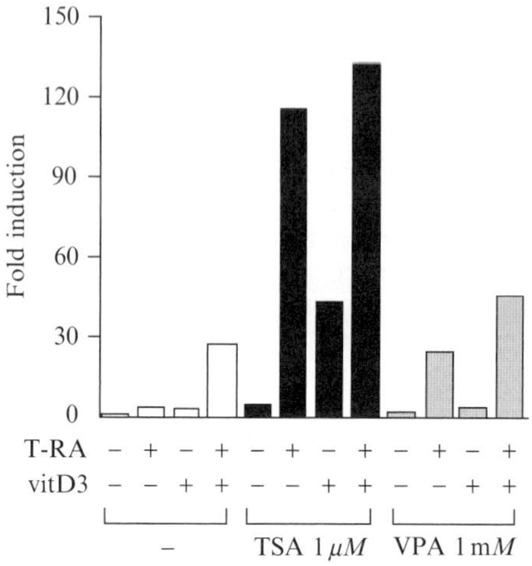

Figure 10.11 Histone deacetylase inhibitors enhance the readout signal. Trichostatin (TSA) and valproic acid (VPA) were tested on the RARα-Gal4-VDR either alone or in addition with the different ligands as indicated. Luciferase activities were measured and fold inductions represent activation over basal level without ligand. Bars represent mean of duplicate samples of one representative experiment with similar results obtained in at least five independent experiments.

Particularly the double LBD chimeras have great promise for the screening and validation of ligands for orphan receptors in the context of living cells. The option to establish reporter systems in different cell types allows for the comparison of the cell-specific differences of a particular ligand. This is of significance when functionally dissociated ligands or SNuRMs (Gronemeyer *et al.*, 2004) are to be identified and characterized.

5.3. Two hybrid analysis by transient cotransfection

The first sphere of NR communications, at least as far as the LBD is concerned, involves CoAs and CoRs. These factors serve as platforms for the assembly of complexes that exert epigenetic activities. The recruitment and dissociation of these complexes in absence of, or in response to, ligand binding can be determined by two-hybrid assays and provided a mechanistic rationale for the classification of a ligand as agonist, inverse agonist, neutral or full antagonist, or inverse agonist (Fig. 10.12).

5.3.1. Transient transfection of HeLa cells

1.2×10^5 cells are seeded per P24 well 2 h before the transfection with JetPei reagent.

Per well add

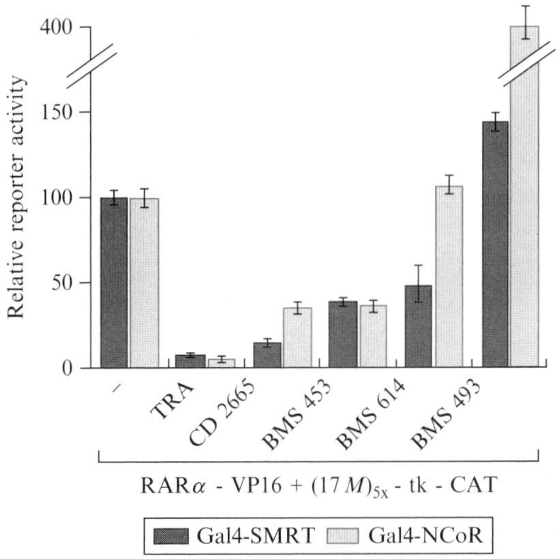

Figure 10.12 Double hybrid experiment revealing the ability of ligands to modulate the SMRT and NCoR corepressor binding capacity of the RARα ligand-binding domain. While all-*trans* retinoic acid (ATRA, **1**) dissociates the corepressors, BMS493 [**4a**] enforces corepressor binding, identifying this ligand as an inverse agonist. Note that BMS614 [**6**] has no effect on NCoR binding.

Plasmid mix diluted in 50 μl of 150 mM NaCl buffer

- 50 ng Gal4 chimera [e.g., Gal4-NCoR]
- 40 ng VP16 chimera [e.g., VP16-hRARα(DEF)]
- 200 ng (17 mer)5×-βglob-luc
- 50 ng CMV-ßGal
- BSK plasmid QSP 1 μg

Vortex gently and spin down briefly

- *JetPei dilution*: 2 μl in 50 μl 150 m*M* NaCl buffer (N/P = 5)

Vortex gently and spin down briefly

- Add the 50-μl JetPei solution to the 50-μl DNA solution all at once (*do not mix the solution in the reverse order*)
- Vortex-mix the solution immediately and spin down briefly
- Incubate for 15–20 min at room temperature
- Add the 100-μl JetPei/DNA mixture drop-wise to the medium in each well and homogenize by gently swirling the plate

- Incubate the cells at 37 °C in a 5% CO_2 incubator for 24 h
- Remove culture medium from the wells and replace with fresh medium containing the ligand at 1 μM
- Incubate the cells at 37 °C in a 5% CO_2 incubator again for 24 h.

Transfect 3 wells per assay to have triplicate measurements

5.3.2. Cell lysis

The cell lysis is performed with the 5× passive lysis buffer from Promega

- Prepare a sufficient quantity of 1× working concentration by adding 1 volume of 5× passive lysis buffer to 4 volumes of distilled water
- Remove the growth medium containing the ligand from the cells and gently apply a sufficient volume of phosphate saline buffer (PBS) to wash the surface of the well. Swirl briefly. Completely remove the PBS from the well
- Dispense into each well 100 µl of 1× passive lysis buffer
- Place the culture plate on an orbital shaker with gentle rocking at room temperature for 20 min
- Transfer the lysate to a 96-well plate.

5.3.3. Luciferase assay

Preparation of luciferase assay reagent

- Prepare 50 µl per sample + 3 ml of luciferase assay reagent just before use.
- For 1 ml mix
    - 500 µl 2× luciferase buffer
    - 47 µl luciferine 10^{-2} M
    - 53 µl ATP 10^{-2} M
    - 27 µl Coenzyme A (lithium salt) 10^{-2} M
    - 430 µl distilled water

      Note: This buffer is light sensitive.
- 2× luciferase buffer (40 mM Tris–phosphate, 2.14 mM $MgCl_2$, 5.4 $MgSO_4$, 0.2 mM EDTA, 66.6 mM DTT, pH 7.8) stable at −20 °C.

5.3.4. Luciferase measurement using the Berthold Technologie luminometer

- Distribute 10 µl of lysate per well of a 96-well white plate.
- Wash the injector and prime with the luciferase assay reagent.
- Program the luminometer to perform 2-s premeasurement delay followed by a 10-s measurement period with 50 µl luciferase assay reagent.
- Perform the measurement and record the luciferase activity measurement.

5.3.5. Measurement of βgal activity

Preparation of βgal assay buffer, per sample:

- 150 μl βgal buffer (Na_2HPO_4 ($12H_2O$) 60 m*M*, NaH_2PO_4 40 m*M*, KCl 10 m*M*, $MgCl_2$ ($6H_2O$) 1 m*M*, β-mercaptoethanol 50 m*M*)
- 30 μl ONPG (4 mg/ml)
- Distribute 10 μl lysate in a 96-well plate
- Add 180 μl of βgal assay buffer
- Incubate the plate at 37°C and measure the time until the yellow color has developed
- Stop the reaction by adding 75 μl of 1 *M* Na_2CO_3; avoid bubbles
- Read the absorbance immediately at 420 nm in a plate reader.

βgal unit = 100 × total volume × absorbance/(assay volume × time (in h))

5.3.6. Normalization

The luciferase values are normalized for transfection efficiency with the βgal values. [*Remark*: Transient transfection aficionados make sure that the expressed factors do not modulate the expression of the internal control, which can occur under conditions that affect the βgal promoter; one such condition can be AP1 repression; this remark is true for all transient transactivation experiments.]

5.4. Fluorescence anisotropy

The effect of ligands on the recruitment of coregulators can be easily and rapidly monitored by measuring the affinities (K_d) of fluorescein-labeled peptides derived from CoAs and CoRs for RAR and RXR LBDs in various ligation states. The following protocol is an example given for the determination of the binding affinities of a peptide containing the NR box2 of the CoA SRC-1 (SRC-1 NR2) for RARα LBD in its unliganded form and in the presence of the agonist Am580 (see also Fig. 10.13).

5.4.1. Required materials

- Safire microplate reader (TECAN)
- Corning NBS 384 well low volume microplates (Corning Incorporated)
- *Buffer A*: 20 m*M* Tris–HCl (pH 7.5), 150 m*M* NaCl, 1 m*M* EDTA, 5 m*M* DTT, 10% glycerol
- Fluorescein-labeled SRC-1 NR2 (686-RHKILHRLLQEGS-698) at 2 μ*M* in buffer A
- Ligands (Am580) at 10^{-2} *M* in EtOH or DMSO
- Purified RARα LBD at 5–10 mg/ml

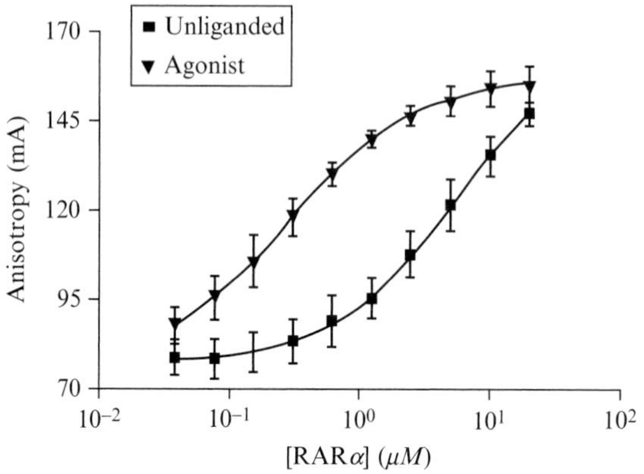

Figure 10.13 Titrations of fluorescein-labeled SRC-1 NR2 by RARα LBD in the absence of ligand (unliganded) and in the presence of the agonist Am580 [3]. The K_d values derived from the curves are 4.74 ± 1.20 μ*M* and 0.39 ± 0.15 μ*M* for the apo and the agonist-bound receptors, respectively.

5.4.2. Determination of K_ds

1. Prepare a 1-ml solution (Solution A) containing 4 n*M* of fluorescent peptide and 80 μ*M* of ligand in buffer A
2. Dispense 30 μl in 10 microplate wells
3. Prepare a 60-μl solution containing 40 μ*M* of RARα LBD in buffer A and dispense 30 μl in the first well
4. Add 30 μl of solution A into the remaining 30 μl of the solution prepared in Step 3, mix gently and dispense 30 μl in the second well
5. Repeat Step 4 until the last (10th) well
6. Dispense 60 μl of buffer A in the 11th well as a reference measurement
7. Determine anisotropy values on the Safire^2 microplate reader with the excitation wavelength set at 470 nm and emission measured at 530 nm
8. Repeat two times from Step 1 so that reported data are the average of at least three experiments
9. Fit binding data and determine K_d using a sigmoidal dose–response model in GRAPHPAD PRISM (Graphpad Software, San Diego)

5.5. Electrospray ionization mass spectrometry

Supramolecular mass spectrometry is a powerful tool to rapidly and unambiguously determine if a particular coregulator-derived peptide can interact with RAR and RXR LBDs and to directly visualize the influence of various ligands on the stability of the corresponding ternary complexes.

The following protocol is an example given for the use of electrospray ionization mass spectrometry (ESI-MS) under nondenaturating conditions to monitor the interaction between the RXR/RAR heterodimer and the motif CoRNR1 of NCoR (see example in Fig. 10.14).

5.5.1. Required materials

- Electrospray time-of-flight mass spectrometer (LCT, Waters)
- *Buffer A*: 50 m*M* ammonium acetate pH 6.5
- Ligands (Am580 [**3**] and BMS493 [**4a**]) at 10^{-2} *M* in EtOH or DMSO
- Purified RXR/RAR LBD heterodimer at 5–10 mg/ml
- CoRNR1 peptide (2045-THRLITLADHICQIITQDFARNQV-2068) at 20 m*M* in DMSO

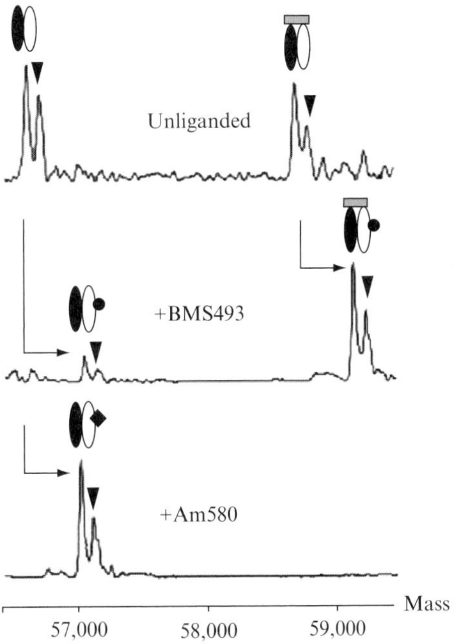

Figure 10.14 Positive ESI mass spectra of RXR LBD/RAR LBD/CoRNR1 complexes in the absence of ligand or in the presence of twofold molar excess of the inverse agonist BMS493 [4a] or the agonist Am580 [3]. Peaks labeled with a black triangle correspond to species with an additional N-terminal methionine. As expected, the two ligands display opposite effects on CoRNR1 recruitment. Whereas the inverse agonist BMS493 [4a] strongly stabilizes the interaction between the heterodimer and CoRNR1, the Am580 [3] induces a quantitative release of the peptide.

5.5.2. ESI-MS assays

- Calibrate the instrument using the multiply charged ions produced by an injection of horse heart myoglobin diluted to 2 pmol/μl in a water/acetonitrile mixture (1:1, v/v) acidified with 1% (v/v) formic acid
- Mix the purified RXR/RAR LBD with a twofold molar excess of CoRNR1 peptide and a twofold molar excess of ligands and keep on ice for 1 h
- Prior to ESI-MS analysis, samples have to be desalted on Centricon PM30 microconcentrators (Amicon, Millipore) in buffer A
- Verify purity and homogeneity of the samples in denaturing conditions by diluting the complex solution to 5 pmol/μl in a water/acetonitrile mixture (1:1, v/v) acidified with 1% (v/v) formic acid
- Record spectra in the positive ion mode on the mass range 500–2500 *m/z*
- Verify that the measured molecular masses are in agreement with those calculated from the amino acid sequences
- Dilute samples to 10 pmol/μl in buffer A and continuously infuse it into the ESI ion source at a flow rate of 6 μl/min through a Harvard syringe pump
- In order to prevent dissociation of the complexes in the gas phase, the accelerating voltage (Vc) must be set to 50 V in order to preserve ternary complex formation and good mass accuracy
- Acquire ESI-MS data in the positive ion mode on the mass range 1000–5000 *m/z*
- Measure the relative abundance of the different species present on ESI mass spectra from their respective peak intensities, assuming that the relative intensities displayed by the different species reflect the actual distribution of these species in solution

5.6. Limited proteolysis

Limited proteolytic-digestion assays can be used to study protein conformations, and then to investigate possible ligand-induced conformational changes in RARα. Protease digestion studies can not only reveal ligand-induced conformational changes but also distinguish between distinct structural alteration induced by agonists and antagonists.

Limited proteolysis has, for example, been used previously to assess conformational changes induced by ligand binding to RARα (Germain *et al.*, 2004). [^{35}S]methionine-labeled RARα translated *in vitro* was incubated with retinoids or ethanol alone and then digested with trypsin (endopeptidase) or carboxypeptidase Y (exopeptidase) and analyzed by sodium dodecyl sulfate-polyacrylamide gel electrophoresis (SDS-PAGE). Binding of an RARα agonist to *in vitro* translated RARα followed by trypsin digestion results in protection of 30-kDa fragment of the receptor and is indicative for ligand binding. Carboxypeptidase Y is a vacuolar serine carboxypeptidase from *Saccaromyces cerevisiae*, which catalyzes a stepwise removal of C-terminal amino acids from proteins. This exopeptidase can be used to reveal altered positioning by protease sensitivity of the C-terminal H12 of RARα in the presence of a particular ligand (Germain *et al.*, 2004).

5.6.1. *In vitro* transcription–translation

Prepare [^{35}S]-labeled human RARα by *in vitro* transcription–translation using the TNT rabbit reticulocyte lysate system (Promega) programmed with pSG5-hRARα according to the instructions provided by the manufacturer.

5.6.2. Limited proteolysis with trypsin

- Incubate on ice for 1 h 2 μl of TNT reaction products containing the labeled receptor with either 10 μ*M* retinoids in ethanol or an equivalent amount of ethanol
- Digest receptor proteins at 25 °C for 10 min with 100 μg/ml trypsin (*avoid bubbles*)
- Stop reactions by adding one volume of 2× sample buffer (125 m*M* Tris–HCl, pH 6.8, 4% (w/v) sodium dodecyl sulfate, 1.4 *M* β-mercaptoethanol, 25% (v/v) glycerol, 0.1% (w/v) bromophenol blue) and by boiling.

5.6.3. Limited proteolysis with carboxypeptidase Y

- Incubate on ice for 1 h 1 μl of TNT reaction products and 7 μl of 50 m*M* Tris–HCl (pH 6.8) and either 10 μ*M* of ligand or an equivalent amount of ethanol
- Digest receptor proteins at 25 °C for either 10, 30, or 60 min with 2 μl carboxypeptidase Y at 1 mg/ml (*avoid bubbles*)
- Stop reactions by adding one volume of 2× sample buffer (125 m*M* Tris–HCl, pH 6.8, 4% (w/v) sodium dodecyl sulfate, 1.4 *M* β-mercaptoethanol, 25% (v/v) glycerol, 0.1% (w/v) bromophenol blue) and by boiling.

5.6.4. Sodium dodecyl sulfate-polyacrylamide gel electrophoresis

- Electrophorese on 10–15% denaturing polyacrylamide gels
- Dry the gels
- Expose gels to X-ray film overnight

6. Chemical Syntheses

6.1. Synthesis of the RAR inverse agonist BMS493 4a and the RAR/RXR inverse agonist 12

The synthesis of **12** has been detailed in a recent publication (García *et al.*, 2009). The synthesis of BMS493 [**4a**] follows the same methodology described recently for the preparation of the halogenated derivatives. Briefly, the condensation of the anion of phosphonate **21** and aldehyde **20** was followed by the saponification of ester **22** to produce BMS493 [**4a**] (Álvarez *et al.*, 2009).

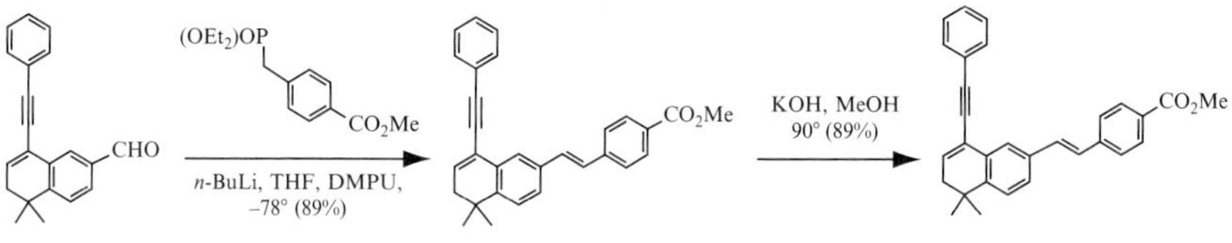

6.2. Synthesis of 4-[(1*E*)-2-(5,5-dimethyl-8-(2-phenylethynyl)-5,6-dihydronaphthalen-2-yl)-vin-1-yl] benzoic acid BMS493 [4a]

To a cooled (0 °C) solution of methyl 4-[(diethylphosphonyl)methyl]benzoate **21** (0.85 g, 2.97 mmol) in THF (4 ml) was added DMPU (1.06 ml, 6.12 mmol) and *n*-BuLi (2.8 ml, 1.05 *M* in hexane, 2.97 mmol) and the mixture was stirred for 30 min. The mixture was cooled down to −78 °C and a solution of 5,5-dimethyl-8-(2-phenylethynyl)-5,6-dihydronaphthalen-2-al **20** (0.5 g, 1.75 mmol) in THF (8 ml) was added. After 5 h, a saturated aqueous solution of NH_4Cl was added and the mixture was extracted with *t*-BuOMe (3×). The combined organic layers were washed with H_2O (3×) and brine (3×), dried (Na_2SO_4), and the solvent was evaporated. The residue was purified by chromatography (silica gel, 95:5 hexane/AcOEt) to afford 0.65 g (89%) of a white solid identified as methyl 4-[(1*E*)-2-(5,5-dimethyl-8-(2-phenylethynyl)-5,6-dihydronaphthalen-2-yl)-vin-1-yl]benzoate **22**. **M.p.**: 165 °C (Et2O/hexane). ^{1}**H NMR** (400.16 MHz, $CDCl_3$): δ 7.94 (d, *J* = 8.0 Hz, 2H, $H_2 + H_6$), 7.78 (d, *J* = 1.6 Hz, 1H, $H_{1''}$), 7.52–7.49 (m, 3H, H3, H_5, $H_{4''}$), 7.39 (dd, *J* = 8.0, 1.7 Hz, 1H, $H_{3''}$), 7.33–7.25 (m, 5H, HPh), 7.19 (d, *J* = 16.3 Hz, 1H, $H_{2'}$), 7.06 (d, *J* = 16.3 Hz, 1H, $H_{1'}$), 6.46 (t, *J* = 4.8 Hz, 1H, $H_{7''}$), 3.85 (s, 3H, CO_2CH_3), 2.32 (d, *J* = 4.9 Hz, 2H, $H_{6''}$), 1.25 (s, 6H, $C_{5''}$-$(CH_3)_2$)

ppm. ^{13}**C NMR** (100.62 MHz, $CDCl_3$): δ 166.9 (s), 144.1 (s), 142.0 (s), 135.1 (s), 135.0 (d), 134.8 (s), 131.6 (d), 131.2 (d, 2×), 130.0 (d, 2×), 128.8 (s), 128.4 (d, 2×), 128.2 (d), 127.2 (d, 2×), 126.4 (d), 126.3 (d), 124.3 (d), 124.2 (d), 123.5 (s), 121.2 (s), 90.6 (s), 87.3 (s), 52.0 (q), 39.2 (t), 33.5 (s), 28.5 (q, 2×) ppm. **MS** (EI^+): *m/z* (%) 418 (M^+, 100), 404 (19), 403 (56), 328 (8), 326 (8), 186 (12). **HRMS** (EI^+): calcd. for $C_{30}H_{26}O_2$,418.1933; found, 418.1946. **IR** (NaCl): υ 3025–2810 (m), 1708 (s) cm^{-1}. **UV** (MeOH): λ_{max} 330, 297, 240 nm.

To a solution of methyl 4-[(1*E*)-2-(5,5-dimethyl-8-(2-phenylethynyl)-5,6-dihydronaphthalen-2-yl)-vin-1-yl)benzoate **22** (0.5 g, 1.20 mmol) in MeOH (15 ml) was added a 2 *M* aqueous KOH solution (19.2 ml) in MeOH (77 ml) and the mixture was heated to 90 °C for 1 h. The reaction was cooled down to ambient temperature, CH_2Cl_2 and brine were added, and the layers were separated. The aqueous layer was treated with a 10% aqueous HCl solution until acidic pH and the mixture was extracted with CH_2Cl_2 (3×). The combined organic layers were dried (Na_2SO_4) and the solvent was removed. The residue was purified by chromatography (silica-gel, CH_2Cl_2/MeOH, 95:5) to afford 0.43 g (89%) of a white solid identified as 4-((1*E*)-2-(5,5-dimethyl-8-(2-phenylethynyl)-5,6-dihydronaphthalen-2-yl)-vin-1-yl)benzoic acid **BMS493 [4a]. M.p.**: 111 −C (acetone/hexane). **Elemental analysis**: calcd. for $C_{29}H_{24}O_2$: C, 82.44; H, 6.20. Found: C, 82.86; H, 5.78. ^{1}**H NMR** (400.16 MHz, $(CD_3)_2CO$)): δ 8.04 (d, *J* = 8.2 Hz, 2H, H_3, and H_5), 7.96 (s, 1H, $H_{1''}$), 7.75 (d, *J* = 7.9 Hz, 2H, H_2, and H_6), 7.64 (d, *J* = 7.3 Hz, 2H, $H_{3''}$ and $H_{4''}$), 7.51 (d, *J* = 16.3 Hz, 1H, $H_{2'}$), 7.5–7.4 (m, 5H, ArH), 7.34 (d, *J* = 16.3 Hz, 1H, $H_{1'}$), 6.58 (t, *J* = 4.7 Hz, 1H, $H_{7''}$), 2.44 (d, *J* = 4.8 Hz, $2H_{6''}$), 1.33 (s, 6H, $C_{5''}$-$(CH_3)_2$) ppm. ^{13}**C NMR** (100.62 MHz, $(CD_3)_2CO$)): δ 168.4 (s), 147.9 (s), 143.9 (s), 137.0 (d), 133.4 (d, 2×), 132.9 (d), 131.9 (d, 2×), 131.1 (s), 130.6 (d, 2×), 130.5 (s), 130.3 (d), 129.0 (d), 128.5 (s), 128.4 (d), 128.3 (d, 2×), 126.3 (d), 126.1 (d), 125.2 (s), 122.9 (s), 92.4 (s), 88.9 (s), 40.6 (t), 35.1 (s), 29.7 (q, 2×) ppm. **MS** (EI^+): *m/z* (%) 404 (M^+, 84), 391 (14), 390 (33), 389 (100), 264 (28), 328 (25), 327 (20), 326 (21), 315 (20), 267 (31), 254 (21), 253 (38), 252 (49), 239 (42), 228 (25), 215 (23), 78 (11). **HRMS** (EI^+): calcd. for $C_{29}H_{24}O_2$, 404.1776; found, 404.1766. **UV** (MeOH): λ_{max} 294 nm.

ACKNOWLEDGMENTS

We thank Pierre Germain for the trypsin and carboxypeptidase Y protocols (Section 5.6), Claudine Gaudon and Pierre Germain for the dual reporter cell studies (Section 5.2), Cathy Erb for the two-hybrid analysis protocol (Section 5.3), Albane le Maire for the fluorescence anisotropy protocol (Section 5.4), Sarah Sanglier for the ESI-MS protocol (Section 5.5), Harshal Khanwalkar for the reporter cell protocol (Section 5.1.3), and Susana Álvarez for the optimized synthesis of BMS493. We are grateful to Koichi Shudo, Bristol-Myers Squibb, and Galderma for providing synthetic retinoids. Work from our laboratories which is mentioned here was supported by funds from the Association for International Cancer Research (HG), the Ligue National Contre le Cancer (H. G.; laboratoire labelisé), the French National Research Agency (W. B.; ANR-07-PCVI-0001-01), the MICINN (SAF-2007-63880-FEDER, A. d. L.), the Xunta de Galicia (INBIOMED, A. d. L.), and the European Community contracts QLK3-CT2002-02029 "Anticancer Retinoids" (A. d. L., H. G.), LSHM-CT-2005-018652 "Crescendo" (H. G.) and LSHC-CT-2005-518417 "Epitron" (A. d. L., H. G.).

REFERENCES

Alland, L., Muhle, R., Hou, H., Potes, J., Chin, L., SchreiberAgus, N., *et al.* (1997). Role for N-CoR and histone deacetylase in Sin3-mediated transcriptional repression. *Nature* **387,** 49–55.

Álvarez, S., Khanwalkar, H., Álvarez, R., Erb, C., MartÌnez, C., Rodríguez-Barrios, F., *et al.* (2009). C3 halogen and C8″ substituents on stilbene arotinoids modulate retinoic acid receptor subtype function. *ChemMedChem* **4,** 1630–1640.

Anzick, S. L., Kononen, J., Walker, R. L., Azorsa, D. O., Tanner, M. M., Guan, X. Y., *et al.* (1997). AIB1, a steroid receptor coactivator amplified in breast and ovarian cancer. *Science* **277,** 965–968.

Astapova, I., Lee, L. J., Morales, C., Tauber, S., Bilban, M., and Hollenberg, A. N. (2008). The nuclear corepressor, NCoR, regulates thyroid hormone action in vivo. *Proc. Natl. Acad. Sci. USA* **105,** 19544–19549.

Auffinger, P., Hays, F. A., Westhof, E., and Ho, P. S. (2004). Halogen bonds in biological molecules. *Proc. Natl. Acad. Sci. USA* **101,** 16789–16794.

Bannister, A. J., and Kouzarides, T. (1996). The CBP co-activator is a histone acetyltransferase. *Nature* **384,** 641–643.

Bourguet, W., Germain, P., and Gronemeyer, H. (2000a). Nuclear receptor ligand-binding domains: Three-dimensional structures, molecular interactions and pharmacological implications. *Trends Pharmacol. Sci.* **21,** 381–388.

Bourguet, W., Vivat, V., Wurtz, J. M., Chambon, P., Gronemeyer, H., and Moras, D. (2000b). Crystal structure of a heterodimeric complex of RAR and RXR ligand-binding domains. *Mol. Cell* **5,** 289–298.

Chen, J. D., and Evans, R. M. (1995). A transcriptional co-repressor that interacts with nuclear hormone receptors. *Nature* **377,** 454–457.

Chen, J. Y., Penco, S., Ostrowski, J., Balaguer, P., Pons, M., Starrett, J. E., *et al.* (1995). RAR-specific agonist/antagonists which dissociate transactivation and AP1 transrepression inhibit anchorage-independent cell proliferation. *EMBO J.* **14,** 1187–1197.

Chen, H., Lin, R. J., Schiltz, R. L., Chakravarti, D., Nash, A., Nagy, L., *et al.* (1997). Nuclear receptor coactivator ACTR is a novel histone acetyltransferase and forms a multimeric activation complex with P/CAF and CBP/p300. *Cell* **90,** 569–580.

Chen, D., Ma, H., Hong, H., Koh, S. S., Huang, S. M., Schurter, B. T., *et al.* (1999). Regulation of transcription by a protein methyltransferase. *Science* **284,** 2174–2177.

Cohen, R. N., Brzostek, S., Kim, B., Chorev, M., Wondisford, F. E., and Hollenberg, A. N. (2001). The specificity of interactions between nuclear hormone receptors and corepressors is mediated by distinct amino acid sequences within the interacting domains. *Mol. Endocrinol.* **15,** 1049–1061.

Collingwood, T. N., Urnov, F. D., and Wolffe, A. P. (1999). Nuclear receptors: Coactivators, corepressors and chromatin remodeling in the control of transcription. *J. Mol. Endocrinol.* **23,** 255–275.

de Lera, A. R., Bourguet, W., Altucci, L., and Gronemeyer, H. (2007). Design of selective nuclear receptor modulators: RAR and RXR as a case study. *Nat. Rev. Drug Discov.* **6,** 811–820.

de Meijere, A., and Diederich, F. (2004). Metal-Catalyzed Cross-Coupling Reactions. 2nd edn. Weinheim, Wiley-VCH.

Domínguez, B., Alvarez, R., and de Lera, A. R. (2003). Recent advances in the synthesis of retinoids. *Org. Prep. Proc. Int.* **35,** 239–306.

Evert, B. O., Araujo, J., Vieira-Saecker, A. M., de Vos, R. A., Harendza, S., Klockgether, T., *et al.* (2006). Ataxin-3 represses transcription via chromatin binding, interaction with histone deacetylase 3, and histone deacetylation. *J. Neurosci.* **26,** 11474–11486.

García, J., Khanwalkar, H., Pereira, R., Erb, C., Voegel, J. J., Collette, P., *et al.* (2009). Pyrazine arotinoids with inverse agonist activities on the retinoid and rexinoid receptors. *Chembiochem* **10,** 1252–1259.

Gehin, M., Vivat, V., Wurtz, J. M., Losson, R., Chambon, P., Moras, D., *et al.* (1999). Structural basis for engineering of retinoic acid receptor isotype-selective agonists and antagonists. *Chem. Biol.* **6,** 519–529.

Germain, P., Iyer, J., Zechel, C., and Gronemeyer, H. (2002). Co-regulator recruitment and the mechanism of retinoic acid receptor synergy. *Nature* **415,** 187–192.

Germain, P., Kammerer, S., Pérez, E., Peluso-Iltis, C., Tortolani, D., Zusi, F. C., *et al.* (2004). Rational design of RAR-selective ligands revealed by RARbeta crystal structure. *EMBO Rep.* **5,** 877–882.

Germain, P., Gaudon, C., Pogenberg, V., Sanglier, S., Van Dorsselaer, A., Royer, C. A., *et al.* (2009). Differential action on coregulator interaction defines inverse retinoid agonists and neutral antagonists. *Chem. Biol.* **16,** 479–489.

Glass, C. K., Rose, D. W., and Rosenfeld, M. G. (1997). Nuclear receptor coactivators. *Curr. Opin. Cell Biol.* **9,** 222–232.

Gronemeyer, H., Gustafsson, J. A., and Laudet, V. (2004). Principles for modulation of the nuclear receptor superfamily. *Nat. Rev. Drug Discov.* **3,** 950–964.

Gu, W., Shi, X. L., and Roeder, R. G. (1997). Synergistic activation of transcription by CBP and p53. *Nature* **387,** 819–823.

Hanstein, B., Eckner, R., Di Renzo, J., Halachmi, S., Liu, H., Searcy, B., *et al.* (1996). p300 is a component of an estrogen receptor coactivator complex. *Proc. Natl. Acad. Sci. USA* **93,** 11540–11545.

Hashimoto, Y., and Miyachi, H. (2005). Nuclear receptor antagonists designed based on the helix-folding inhibition hypothesis. *Bioorg. Med. Chem.* **13,** 5080–5093.

Heinzel, T., Lavinsky, R. M., Mullen, T. M., Soderstrom, M., Laherty, C. D., Torchia, J., *et al.* (1997). A complex containing N-CoR, mSin3 and histone deacetylase mediates transcriptional repression. *Nature* **387,** 43–48.

Hong, H., Kohli, K., Garabedian, M. J., and Stallcup, M. R. (1997). GRIP1, a transcriptional coactivator for the AF-2 transactivation domain of steroid, thyroid, retinoid, and vitamin D receptors. *Mol. Cell. Biol.* **17,** 2735–2744.

Horlein, A. J., Naar, A. M., Heinzel, T., Torchia, J., Gloss, B., Kurokawa, R., *et al.* (1995). Ligand-independent repression by the thyroid hormone receptor mediated by a nuclear receptor co-repressor. *Nature* **377,** 397–404.

Hu, X., and Lazar, M. A. (1999). The CoRNR motif controls the recruitment of corepressors by nuclear hormone receptors. *Nature* **402,** 93–96.

Hu, X., Li, Y., and Lazar, M. A. (2001). Determinants of CoRNR-dependent repression complex assembly on nuclear hormone receptors. *Mol. Cell. Biol.* **21,** 1747–1758.

Huang, P., Chandra, V., and Rastinejad, F. (2010). Structural overview of the nuclear receptor superfamily: Insights into physiology and therapeutics. *Annu. Rev. Physiol.* **72,** 247–272.

Jepsen, K., Solum, D., Zhou, T., McEvilly, R. J., Kim, H. J., Glass, C. K., *et al.* (2007). SMRT-mediated repression of an H3K27 demethylase in progression from neural stem cell to neuron. *Nature* **450,** 415–419.

Kamei, Y., Xu, L., Heinzel, T., Torchia, J., Kurokawa, R., Gloss, B., *et al.* (1996). A CBP integrator complex mediates transcriptional activation and AP-1 inhibition by nuclear receptors. *Cell* **85,** 403–414.

Klein, E. S., Pino, M. E., Johnson, A. T., Davies, P. J. A., Nagpal, S., Thacher, S. M., *et al.* (1996). Identification and functional separation of retinoic acid receptor neutral antagonists and inverse agonists. *J. Biol. Chem.* **271,** 22692–22696.

Korzus, E., Torchia, J., Rose, D. W., Xu, L., Kurokawa, R., McInerney, E. M., *et al.* (1998). Transcription factor-specific requirements for coactivators and their acetyltransferase functions. *Science* **279,** 703–707.

Laudet, V., and Gronemeyer, H. (2001). The Nuclear Receptor Factsbook. Academic Press, San Diego, 462 pp.

le Maire, A., Teyssier, C., Erb, C., Grimaldi, M., Alvarez, S., de Lera, A. R., *et al.* (2010). A unique secondary structure switch controls constitutive gene repression by retinoic acid receptor. *Nat. Struct. Mol. Biol.* **17,** 801–807.

Li, H., Gomes, P. J., and Chen, J. D. (1997). RAC3, a steroid/nuclear receptor-associated coactivator that is related to SRC-1 and TIF2. *Proc. Natl. Acad. Sci. USA* **94,** 8479–8484.

Loeliger, P., Bollag, W., and Mayer, H. (1980). Arotinoids: A new class of highly active retinoids. *Eur. J. Med. Chem.* **15,** 9–15.

Lonard, D. M., and O'Malley, B. W. (2007). Nuclear receptor coregulators: Judges, juries, and executioners of cellular regulation. *Mol. Cell* **27,** 691–700.

McKenna, N. J., Lanz, R. B., and O'Malley, B. W. (1999). Nuclear receptor coregulators: Cellular and molecular biology. *Endocr. Rev.* **20,** 321–344.

Melnick, A., Carlile, G., Ahmad, K. F., Kiang, C. L., Corcoran, C., Bardwell, V., *et al.* (2002). Critical residues within the BTB domain of PLZF and Bcl-6 modulate interaction with corepressors. *Mol. Cell Biol.* **22,** 1804–1818.

Memic, A., and Spaller, M. R. (2008). How do halogen substituents contribute to protein-binding interactions? A thermodynamic study of peptide ligands with diverse aryl halides. *ChemBioChem* **9,** 2793–2795.

Nagpal, S., Friant, S., Nakshatri, H., and Chambon, P. (1993). RARs and RXRs: Evidence for two autonomous transactivation functions (AF-1 and AF-2) and heterodimerization in vivo. *EMBO J.* **12,** 2349–2360.

Nagy, L., and Schwabe, J. W. (2004). Mechanism of the nuclear receptor molecular switch. *Trends Biochem. Sci.* **29,** 317–324.

Nagy, L., Kao, H. Y., Chakravarti, D., Lin, R. J., Hassig, C. A., Ayer, D. E., *et al.* (1997). Nuclear receptor repression mediated by a complex containing SMRT, mSin3A, and histone deacetylase. *Cell* **89,** 373–380.

Nagy, L., Kao, H. Y., Love, J. D., Li, C., Banayo, E., Gooch, J. T., *et al.* (1999). Mechanism of corepressor binding and release from nuclear hormone receptors. *Genes Dev.* **13,** 3209–3216.

Nahoum, V., Pérez, E., Germain, P., Rodríguez-Barrios, F., Manzo, F., Kammerer, S., *et al.* (2007). Modulators of the structural dynamics of the retinoid X receptor to reveal receptor function. *Proc. Natl. Acad. Sci. USA* **104,** 17323–17328.

Nofsinger, R. R., Li, P., Hong, S. H., Jonker, J. W., Barish, G. D., Ying, H., *et al.* (2008). SMRT repression of nuclear receptors controls the adipogenic set point and metabolic homeostasis. *Proc. Natl. Acad. Sci. USA* **105,** 20021–20026.

Ogryzko, V. V., Schiltz, R. L., Russanova, V., Howard, B. H., and Nakatani, Y. (1996). The transcriptional coactivators p300 and CBP are histone acetyltransferases. *Cell* **87,** 953–959.

Onate, S. A., Tsai, S. Y., Tsai, M. J., and O'Malley, B. W. (1995). Sequence and characterization of a coactivator for the steroid hormone receptor superfamily. *Science* **270,** 1354–1357.

Pérez-Santín, E., Germain, P., Quillard, F., Khanwalkar, H., Rodríguez-Barrios, F., Gronemeyer, H., *et al.* (2009). Modulating retinoid X receptor with a series of (E)-3-[4-hydroxy-3-(3-alkoxy-5, 5, 8, 8-tetramethyl-5, 6, 7, 8-tetrahydronaphthalen-2-yl) phenyl]acrylic acids and their 4-alkoxy isomers. *J. Med. Chem.* **52,** 3150–3158.

Perissi, V., Staszewski, L. M., McInerney, E. M., Kurokawa, R., Krones, A., Rose, D. W., *et al.* (1999). Molecular determinants of nuclear receptor–corepressor interaction. *Genes Dev.* **13,** 3198–3208.

Rosenfeld, M. G., Lunyak, V. V., and Glass, C. K. (2006). Sensors and signals: A coactivator/corepressor/epigenetic code for integrating signal-dependent programs of transcriptional response. *Genes Dev.* **20,** 1405–1428.

Sande, S., and Privalsky, M. L. (1996). Identification of TRACs (T3 receptor-associating cofactors), a family of cofactors that associate with, and modulate the activity of, nuclear hormone receptors. *Mol. Endocrinol.* **10,** 813–825.

Seol, W., Mahon, M. J., Lee, Y. K., and Moore, D. D. (1996). Two receptor interacting domains in the nuclear hormone receptor corepressor RIP13/N-CoR. *Mol. Endocrinol.* **10,** 1646–1655.

Takeshita, A., Cardona, G. R., Koibuchi, N., Suen, C. S., and Chin, W. W. (1997). TRAM-1, A novel 160-kDa thyroid hormone receptor activator molecule, exhibits distinct properties from steroid receptor coactivator-1. *J. Biol. Chem.* **272,** 27629–27634.

Torchia, J., Rose, D. W., Inostroza, J., Kamei, Y., Westin, S., Glass, C. K., *et al.* (1996). The transcriptional co-activator p/CIP binds CBP and mediates nuclear–receptor function. *Nature* **387,** 677–684.

Voegel, J. J., Heine, M. J., Zechel, C., Chambon, P., and Gronemeyer, H. (1996). TIF2, a 160 kDa transcriptional mediator for the ligand-dependent activation function AF-2 of nuclear receptors. *EMBO J.* **15,** 3667–3675.

Voegel, J. J., Heine, M. J., Tini, M., Vivat, V., Chambon, P., and Gronemeyer, H. (1998). The coactivator TIF2 contains three nuclear receptor-binding motifs and mediates transactivation through CBP binding-dependent and -independent pathways. *EMBO J.* **17,** 507–519.

Westin, S., Rosenfeld, M. G., and Glass, C. K. (2000). Nuclear receptor coactivators. *Adv. Pharmacol.* **47,** 89–112.

Xu, H. E., Stanley, T. B., Montana, V. G., Lambert, M. H., Shearer, B. G., Cobb, J. E., *et al.* (2002). Structural basis for antagonist-mediated recruitment of nuclear co-repressors by PPARalpha. *Nature* **415,** 813–817.

Yao, T. P., Ku, G., Zhou, N., Scully, R., and Livingston, D. M. (1996). The nuclear hormone receptor coactivator SRC-1 is a specific target of p300. *Proc. Natl. Acad. Sci. USA* **93,** 10626–10631.

Zamir, I., Harding, H. P., Atkins, G. B., Horlein, A., Glass, C. K., Rosenfeld, M. G., *et al.* (1996). A nuclear hormone receptor corepressor mediates transcriptional silencing by receptors with distinct repression domains. *Mol. Cell. Biol.* **16,** 5458–5465.

Zhang, J., Hu, X., and Lazar, M. A. (1999). A novel role for helix 12 of retinoid X receptor in regulating repression. *Mol. Cell. Biol.* **19,** 6448–6457.

CHAPTER ELEVEN

γ-Aminobutyric Acid Type A ($GABA_A$) Receptor Subtype Inverse Agonists as Therapeutic Agents in Cognition

Guerrini Gabriella *and* Ciciani Giovanna

Contents

Abstract

The gabaergic system has been identified as a relevant regulator of cognitive and emotional processing. In fact, the discovery that negative allosteric regulators (or inverse agonists) at $GABA_A$ (γ-aminobutyric acid) α5 subtype receptors improve learning and memory tasks, has further validated this concept. The localization of these extrasynaptic subtype receptors, mainly in the hippocampus, has suggested that they play a key role in the three stages of memory: acquisition, consolidation, and retrieval.

The "α5 inverse agonist" binds to an allosteric site at $GABA_A$ receptor, provoking a reduction of chlorine current, but to elicit this effect, the necessary condition is the binding of agonist neurotransmitter (gamma-amino butyric

Dipartimento di Scienze Farmaceutiche, Laboratorio di Progettazione, Sintesi e Studio di Eterocicli Biologicamente attivi (HeteroBioLab) Università degli Studi di Firenze, Via U. Schiff, 6, 50019 Polo Scientifico, Sesto Fiorentino, Firenze, Italy

Methods in Enzymology, Volume 485
ISSN 0076-6879, DOI: 10.1016/S0076-6879(10)85011-6

acid) at its orthosteric site. In this case, the $GABA_A$ receptor is not a "constitutively active receptor" and, however, the presence of spontaneous opening channels for native $GABA_A$ receptors is rare.

Here, we present various classes of nonselective and α5 selective $GABA_A$ receptor ligands, and the *in vitro* and *in vivo* tests to elucidate their affinity and activity.

The study of the $GABA_A$ α5 inverse agonists is one of the important tools, although not the only one, for the development of clinical strategies for treatment of Alzheimer disease and mild cognitive impairment.

1. Introduction

Deficits in the functional expression of the GABA (γ-aminobutyric acid) system are critical in anxiety disorders, epilepsy, depression, schizophrenia, and cognitive impairment. GABA is the main inhibitory neurotransmitter in the brain and exerts its action through activation of GABA receptors (GABA-Rs). Fast inhibitory action is mediated via GABA type A ionotropic receptors ($GABA_A$-Rs), while slow and prolonged action is mediated via metabotropic GABA type B receptors ($GABA_B$-Rs).

1.1. $GABA_A$ receptors

$GABA_A$-Rs are members of the Cys-loop ligand-gated ion channel (LGIC) superfamily, including nicotinic acetylcholine receptors, inhibitory glycine receptors, ionotropic $5HT_3$ (serotonin) receptors, and zinc-activated receptors (ZAC). $GABA_A$-R forms a pentameric complex of subunits that are arranged around a central chloride channel. There are 19 subunits (α1-6, β1-3, γ1-3, δ, ε, τ, π, ρ) combined in $GABA_A$-Rs taking into account the Nomenclature Committee of IUPHAR (NC IUPHAR) recommendations that consider receptors containing the rho (ρ) subunits as subtypes of $GABA_A$-R instead of the use of the term "GABAc-R" (Olsen and Sieghart, 2009).

All members belonging to the LGIC superfamily are homologous not only in membrane topology but also at the functional domain. Thus, the polypeptidic subunits are structurally related and consist of a large extracellular region in which there is the characteristic "Cys-loop" (formed by a covalent bond between two conserved cysteines) and there are the ligand binding sites (Clayton *et al.*, 2007).

The transmembrane domain consists of four membrane spanning segments (TM1–4) of which TM2 forms the ion channel. A leucine residue in TM2 and conserved through all subunit isoforms, has been proposed to play an important role in the channel gate. The interaction with intravenous and volatile anesthetics in an intrahelical hydrophobic pocket, involves aminoacids within TM2 and TM3. Between TM3 and TM4, an intracellular domain is present that contains portals through which ions access the cell cytoplasm and phosphorylation sites necessary for receptor regulation (Fig. 11.1).

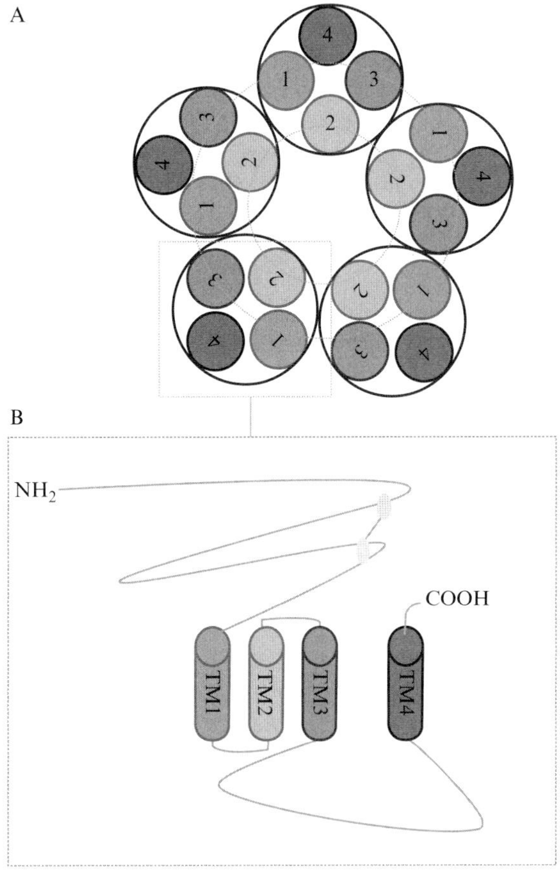

Figure 11.1 Schematic representation of the structure of Cys-loop receptor ion channel. Structural topology of the GABA$_A$ receptor, belonging to the ligand-gated ion channel family. (A) Cross-section of the ion channel: the five subunits to form the receptor channel; the inner ring consists of the five TM2 domains that lined the pore. The outer ring is formed by the alternating TM1 and TM3 domains. In the groove between these segments lies the TM4 domain. It does not contact with the TM2 domain. (B) Individual subunit (shown from the side) is formed by four hydrophobic transmembrane domains α-helices (TM1–TM4). The large N-terminal β-folded extracellular domain, of approximately 200 amino acids, contains the intrachain disulphide bond (the so-called Cys-loop, yellow circles) and neurotransmitter and modulator binding sites. (For interpretation of the references to color in this figure legend, the reader is referred to the web version of this chapter.)

$GABA_A$-R presents complex heterogeneity due to the theoretically possible combinations of the known 19 subunits into pentameric structure. Olsen and Sieghart (2008) suggest a list of 26 combinations of $GABA_A$-Rs divided into the groups: "identified," "existence with high probability," and "tentative." Criteria for identifying which subtypes to insert are well described in a review by the same authors. Unequivocally, it is possible to assert that the most frequent colocalization involves α, β, γ, and δ subunits. Only 9–11 different subtypes are identified in the brain, but the list could be updated continuously as more information becomes available.

The majority of $GABA_A$-R_S are composed of two α-, two β-, and one γ-subunit (sometimes substituted by δ subunit) in absolute arrangement γ-α-β-α-β (clockwise) when viewed top-down from the extracellular domain (Fig. 11.2).

Two-orthosteric sites for GABA are present in the $GABA_A$-R at the interface of α- and β-subunits and other "allosteric" sites for barbiturates, ethanol, steroids, picrotoxin, general anesthetics, benzodiazepines (BZs) and β-carbolines. The ability of BZ ligands to modulate GABA-mediated channel activity depends on the presence of the γ-subunit in the pentamer, whereas their selective modulator effects are related to the type of the α-subunit, from the moment that the BZ site lies between α/γ subunits.

The amino acids that contribute to ligand binding sites belong to the so-called "loops A–F" (Ernst *et al.*, 2003). GABA binding site residues

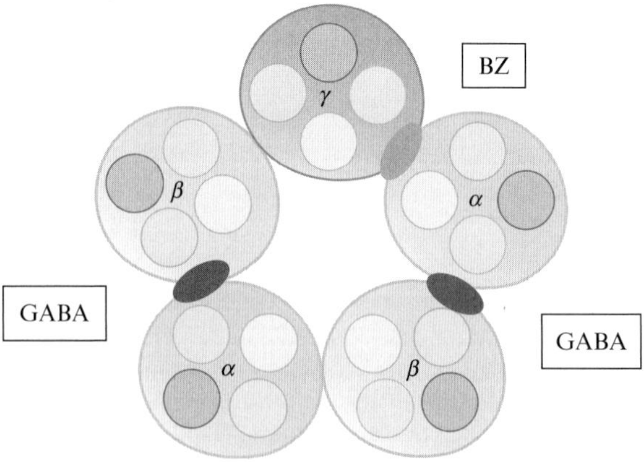

Figure 11.2 αβγ containing $GABA_A$ receptors. The subunit arrangement is γ-α-β-α-β (clockwise when viewed from extracellular space). In the figure the heteromeric receptor, composed by α(1–6), β(1–3) and γ(1–3) subunits, is depicted; γ could be substituted by a δ subunit, at the extrasynaptic receptor, and colocalized with α(4, 6). In the homomeric receptor, five homologous ρ subunits are present. At the interface of α and γ subunits is located the BZ binding site whereas at the interfaces of α and β subunits are located the GABA binding sites.

belong to loops B and C for the principal subunit (β^+) and to loops D and E for the complementary subunits (α^-; Boileau *et al.*, 2002). In the BZ-site loops A, B, and C of the principal subunit (α^+) and loops D and E of the complementary subunit (γ^-) are involved in the binding (Hanson *et al.*, 2008).

1.2. GABA$_A$ receptor expression and localization in the brain

The GABA$_A$-Rs are expressed throughout the brain and localized either synaptically, mediating the "phasic inhibition," or extrasynaptically, mediating the "tonic inhibition" (Goetz *et al.*, 2007, 2009; Jacob *et al.*, 2008). GABA$_A$-Rs, composed of α1, α2, and α3 subunits together with β and γ, are thought to be primarily synaptically localized, whereas the α5βγ receptors are predominantly extrasynaptic. Both these types of GABA$_A$-Rs are BZ sensitive. By contrast, receptors composed of α4 or α6 subunits, together with β and δ subunits, are not sensitive to BZ modulation and are localized at extrasynaptic sites. The lack of sensitivity of these latter receptors to BZ is due to the presence of arginine residue instead of histidine residue at position 101 of the α-subunit: this residue is a key amino acid for binding diazepam (the classical BZ full agonist).

It has been possible to determine the distributions of the "identified" GABA$_A$-Rs in the brain:

- The α1β2γ2 combination is the most abundant type (60%) and is located in the cortex, hippocampal interneurons, thalamus, and cerebellum and it is very plausible that it mediates the sedative effect of diazepam (a full agonist at GABA$_A$ subtype receptors; Mohler, 2006).
- α2βγ2 and α3βγ2 combinations are moderately abundant (10–20%) and are found in the hippocampal formation, hypothalamus, amygdala and are responsible for the anxiolytic-like and/or myorelaxant effect of diazepam (Mohler, 2006).
- α4β2/3δ and α4βγ2 with α6β2/3δ and α6βγ2 combinations (<5%) represent a minor population of receptors in the striatum and cerebellum (Uusi-Oukari and Korpi, 2010). These receptor subtypes are called BZ-insensitive because are not activated by diazepam.
- The α5βγ2 combination is expressed in the hippocampus (pyramidal cells), amygdala, and hypothalamus (5%) and is thought to be responsible for cognition and memory (Martin *et al.*, 2009).

1.3. GABA$_A$ receptor functionality

GABA$_A$-Rs convert chemical messages into electrical signals: the binding of two small neurotransmitter molecules of GABA, induces a conformational change in the multisubunit receptor that opens the central channel and chlorine ions move into the cells, causing a strong inhibitory hyperpolarization.

Although little is known about local movements within the GABA binding site and about the mechanism underlying allosteric coupling of BZs and other allosteric ligands, several studies based on molecular cloning have played an important role in the understanding of receptor activation.

A recent review (Miller and Smart, 2010) details the processes, structural variations and events that are present in the activation of Cys-loop receptors, to which the $GABA_A$-R belongs, and how neurotransmission evolves. By making a scheme of the model that describes how a typical ligand (e.g., GABA) evokes response, we can identify three states that the receptor can assume, as the result of the conformational changes induced by binding. The closed (resting, R) state is the main one in the absence of GABA (agonist), the open (active, A) is when exposure to GABA occurs and ion-conducting is present, and finally the desensitized (D) is due to prolonged GABA exposure. In this latter state, the receptor cannot be activated by subsequent agonist application. The transitions between the various states (R, A, D) have different "energy barriers" and each state can be stabilized by ligand binding (Hogg *et al.*, 2005).

Certain types of receptors have been found to be constitutively active (CA or spontaneously active without ligand) and it is likely that, in this case, the open state is energetically relatively more stable (Findlay *et al.*, 2000; Hogg *et al.*, 2005).

The existence of constitutive activity for G protein-coupled receptors (GPCRs) was first described by Costa (Costa and Herz, 1989) and is now firmly rooted in receptor pharmacology (Milligan, 2003).

It is worth noting that the first naturally occurring constitutively active mutations in GPCRs, were reported in Rhodopsin (Keen *et al.*, 1991) and since then, numerous constitutively active mutations that cause human diseases have been reported in several additional receptors (Parma *et al.*, 1993; Shenker *et al.*, 1993). More recently, the loss of constitutive activity was postulated to cause diseases (Vaisse *et al.*, 2000), suggesting the critical importance of the CA receptors in normal physiological conditions.

Speaking of the LGICs, there are only a few examples of CA receptors even if the preliminary evidence was reported in the early 1980s (Sakmann *et al.*, 1983). The spontaneous opening was shown by the nACh receptor channel (Jackson, 1984), by the new ZAC receptors (Davies *et al.*, 2003) and glycine receptors (Maksay and Bíró, 2002).

The presence of spontaneous opening channels for native $GABA_A$-Rs is rare (Chang and Weiss, 1999); there is only evidence for homomeric β subunits in which constitutive activity appears to represent the major form of their functionality (Miko *et al.*, 2004).

Since the functionality of the channel is related to the TM2 (that lines the channel) of each subunit, studies based on recombinant receptors or point mutation, performed to clarify the mechanism of "gating," indicated spontaneously active receptors. In general, it is plausible to think that reduced hydrophobicity at the TM2 level could generate channel opening in the absence of GABA. In fact, several papers report that, for example, the hydrophilic substitution of the conserved TM2 leucine with serine in heteromeric α1β2γ2 subtypes could generate a reduced energy barrier for channel opening (Chang and Weiss, 1999; Dalziel *et al.*, 2000; Pan *et al.*, 1997). On the other hand, it has also been shown that cloned α1β3ε receptors (Neelands *et al.*, 1999) and homomeric ρ receptors (Pan *et al.*, 1997) can exhibit spontaneous activity although their pharmacological characterization is not yet defined. However, the physiological significance of the constitutive $GABA_A$-R activity is unclear given the difficulty of identifying *in vivo*, this interesting behavior of the receptor.

The study of the possible patho-physio-pharmacological roles of CA and the importance of inverse agonists (antagonists with negative efficacy) acting on these receptors could have relevance and implications for drug discovery, especially for treating CNS-related disorders. In fact, studies performed during the past decade have provided evidence that, inverse agonists are expected to have better therapeutic values compared with neutral antagonists for diseases caused by CA mutations (Bond and IJzerman, 2006; Parra and Bond, 2007).

2. Inverse Agonism: Definition

The concept of inverse agonism was for the first time used for ligands acting as modulators at the "BZ site" (now better named $GABA_A$ receptor subtypes) that showed *in vivo* pharmacological profiles opposite to those of classical BZ-agonists (Wood *et al.*, 1984).

On the other hand, in the1980s evidence of constitutive activity of δ-opioid receptors (Koski *et al.*, 1982) and β_2-adrenoceptors (Cerione *et al.*, 1984) induced researchers to focus on the concept of "inverse agonist" also for the GPCRs.

Since the terms "inverse agonist" or "negative antagonist" are used in general to describe a ligand with negative efficacy, a reclassification was proposed (Costa and Cotecchia, 2005).

This schematization is based on the "topography" of the interaction (orthosteric/synoptic or allosteric/ectopic site) and on the effect that the inverse agonist has on the receptor. Thus, three types of inverse agonist can be individuated: ligands that bind to orthosteric or allosteric sites and affect the receptor signaling, reducing the constitutive receptor activity (types 1 and 2) on their own. Ligands of type 3, which bind to allosteric site, need agonist-induced receptor activity to manifest their negative effect; these types of ligands have little or no influence on channel gating *per se*. The best example of this third type is represented by ligands that bind to $GABA_A$-R subtypes, previously identified with the term "ligands at BZ site on $GABA_A$ receptors."

Thus, the best proposed name for this latter type of "inverse agonist" is "negative allosteric regulator." Although the "inverse agonism effect" is evocated in a different manner (types 1, 2, 3), the final effect on receptor signaling is the reduction of both constitutive receptor activity and agonist effects in GPCRs or only reduction of agonist effect in, for example, $GABA_A$-Rs in which a reduction of chlorine current is recorded.

3. Negative Allosteric Regulators of $GABA_A$-R in Cognitive Impairment

Here we report the recent developments in "negative allosteric regulators" of $GABA_A$-Rs, useful in the treatment of cognitive impairment associated with depression and anxiety, psychosis, stroke, traumatic brain injury, attention deficit disorder or age-related cognitive decline related to vascular dementia and neurodegenerative disorders. To date, in medicinal chemistry, compounds with these features are called "$GABA_A$ $\alpha5$ subtype inverse agonists" (Atack, 2010).

3.1. Non selective $GABA_A$ inverse agonists

It is interesting to note that one of the first patents for treatment of the senile dementia in Alzheimer disease was registered in 1987 and the authors proposed the use of a BZ inverse agonist in a drug combination (Rodríguez-Puertas and Barreda-Gómez 2006). In 1988, the potential beneficial role of β-carboline (ZK 93426) inverse agonist as tool for cognition improvement was noted (Duka *et al.*, 1988). Since then, the design and development of several compounds belonging to different chemical classes that bind to the "BZ site" having non selective inverse agonist effects (DMCM, FG4172, S-8510, RY023, RY080, RY024) have been shown to enhance cognitive performance in animal models. Their therapeutic potential is limited by severe side effects (anxiogenic, convulsant, and proconvulsant) (Fig. 11.3).

On the other hand, the chemical optimization of these classes gives rise to development of highly selective substances.

The knowledge generated by the use of genetically modified mice has permitted scientists to identify the $GABA_A$ subtypes responsible for each pharmacological effect evoked by diazepam, when binding to the $GABA_A$ receptor: anticonvulsant, anxiolytic-like, sedation, myorelaxant, and impaired memory. Thus, it has been evidenced that the $\alpha1$ subtype mediates

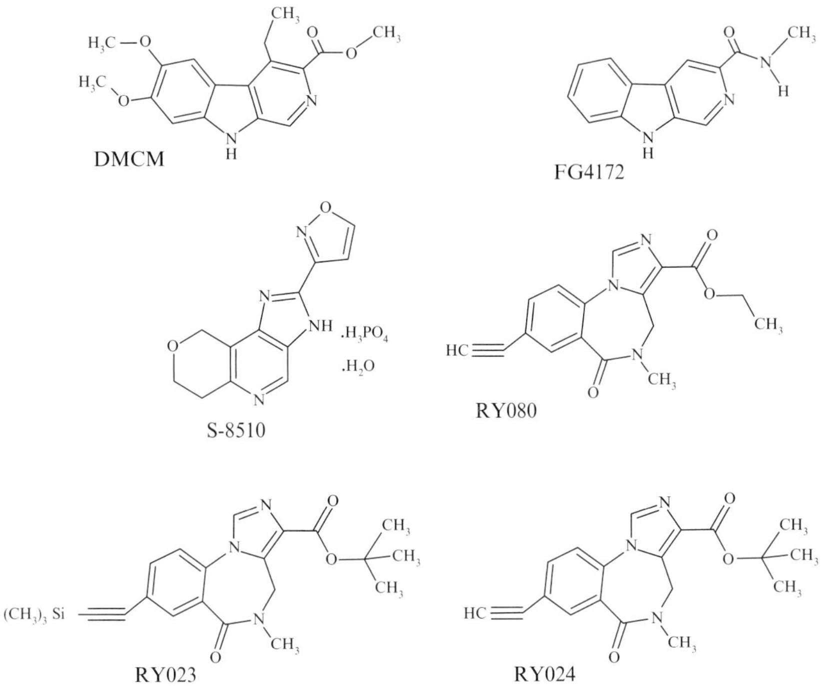

Figure 11.3 Non selective GABA$_A$ inverse agonist.

the sedative action of diazepam, whereas α2 the anxiolytic-like action and/or myorelaxant effect. The α5 subunit may be associated with cognition and memory. In fact, in mice lacking or genetically with reduced α5 subunit, hippocampus dependent-memory and spatial learning were enhanced (Collinson *et al.*, 2002; Crestani *et al.*, 2002).

3.2. Selective GABA$_A$ α5 inverse agonists

Two different approaches were used in medicinal chemistry to obtain compounds that are more GABA$_A$ α5 inverse agonist selective: selective affinity and selective efficacy. In the first case, the ligand is endowed with affinity for only a single GABA$_A$ subtype (α5) and in the second the ligand has affinity for all GABA$_A$ subtypes but functional selectivity, that is, the inverse agonism at the α5 subtype, and low efficacy at the other GABA$_A$ receptor subtypes.

Few compounds are reported to exhibit binding selectivity for GABA$_A$ α5 subtype of the many different structural classes of GABA$_A$ receptor ligands. The first identified selective ligand was L-655708 (also known as

FG8094; Quirk *et al.*, 1996) a compound with an imidazobenzodiazepine core that displays 50–100-fold α5 selectivity over the other subtypes. This compound is also the only commercially available radioligand ([^{3}H]-L-655708, from the Tocris Company), used as a selective probe for the $GABA_A$ α5 subtype.

Following an approach based on chemical similarity, the Merck Sharp & Dohme Company, in the early 2000s identified a series of benzothiophene derivatives that displayed a notable degree of binding selectivity for the α5 subtype, 10–13-fold over the other subtypes (Chambers *et al.*, 2003). Again, as result of Merck research, compounds containing the triazolopyridazine (Sternfeld *et al.*, 2004), the triazolophthalazine (Chambers *et al.*, 2002), and pyrazolotriazine (Chambers *et al.*, 2004) core showed either high affinity or high inverse efficacy at the α5 subtype. Several bicyclic (Buettelmann *et al.*, 2007, 2009) and tricyclic compounds were synthesized (Street *et al.*, 2004) in the evolution or lead optimization of the triazolophthalazine ligands, and among them, emerge the ligands MRK-016, α5IA-II, and α5IA (Atack *et al.*, 2009; Dawson *et al.*, 2006). These compounds showed functionally selective profile *in vitro* and in animal models of cognitive function. The compound α5IA is being tested in preclinical and clinical pharmacology studies (Atack, 2010).

Another research group that works on ligands having a pyrazolobenzotriazine structure (Guerrini *et al.*, 2006, 2009), which is strictly correlated to α5IA-II, has obtained very important results about memory enhancing ligands. The ligands, evaluated in *in vivo* pharmacological test showed selective activity in murine memory tasks.

In the same years, Ro-4938581 and Ro-4882224 emerged from research of Hoffmann-La Roche Ltd. Company, that evidenced the importance of the imidazotriazolobenzodiazepine scaffold in the search for molecules having an α5 subtype selective inverse agonism. These ligands have either selective affinity (the affinity at α5 is 20–30-fold over other subtypes) or significantly modulate chlorine-flux, showing functional inverse agonism (Knust *et al.*, 2009; Fig. 11.4).

4. Methods for Evaluating the Affinity and Efficacy at $GABA_A$ Receptor Subtypes

4.1. *In vitro* methods

The affinity for the BZ binding site of recombinant human or rat α1, α2, α3, and α5 subtype receptors has been evaluated in radioligand binding studies. The measurement of the displacement of [^{3}H]Ro 15-1788 binding, by the test compounds, and evaluating the IC_{50}, the K_i was calculated assuming respective K_D values of [^{3}H]Ro 15-1788 binding at all subtype

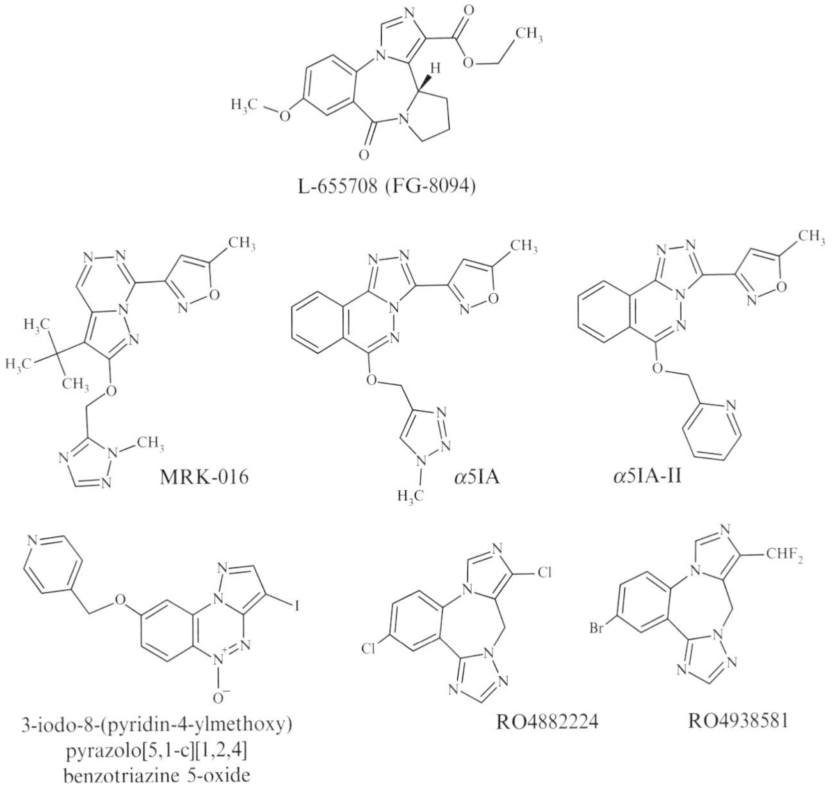

Figure 11.4 Selective GABA$_A$ α5 subtype inverse agonists.

receptors. Nonspecific binding was defined by inclusion of 10 μ*M* of flunitrazepam for all receptor subtypes.

The direct measurement of *in vitro* efficacy at GABA$_A$ receptor subtypes can be evaluated by $^{36}Cl^-$ uptake assay or by electrophysiology experiments in recombinant subtype receptors or in *Xenopus* oocytes expressing the appropriate combination of subtype receptors. In both cases, the chlorine current was measured in response to GABA alone or with that obtained in the presence of drugs. Agonists show an increase of chlorine influx while inverse agonists show a decrease. Compounds that do not substantially affect GABA-evoked Cl^- influx can be considered antagonists.

The mouse hippocampal slice model is an electrophysiological model useful for the study of cognitive processes that involve long-term changes in synaptic efficacy, such as LTP (long-term potentiation; Bliss and Collingridge, 1993). The enhancement of LTP indicates an α5 selective inverse agonism (Dawson *et al.*, 2006).

4.2. Behavioral models

Memory tasks are a behavioral procedure to which an animal is repeatedly (usually twice) exposed. The first exposure is the learning or acquisition session, whereas the effects of training are then assessed after an appropriate delay in retention session(s). Depending on administration-time (before or just after the learning session, or before the retention session) it is possible to modulate acquisition, consolidation, and retrieval stages of memory.

Usually, the animals learn to avoid a painful or otherwise disagreeable stimulus in the tests by performing (active avoidance) or refraining from performing (inhibitory or passive avoidance) a specific response (http://nbc.jhu.edu/BehTaskNBC.aspx).

5. Conclusion

Because cognitive processes are modulated by different neurotransmitters, several attempts have been made to improve human acquisition, or to reduce cognitive deficits in humans by using multitargeted approaches. Modulation of the gabaergic system, by the use of $GABA_A$ $\alpha5$ selective inverse agonists, may be one of the important tools for the development of clinical treatment strategies in Alzheimer disease or mild cognitive impairment.

REFERENCES

Atack, J. R. (2010). Preclinical and clinical pharmacology of the GABAA receptor a5 subtype-selective inverse agonist a5IA. *Pharmacol. Ther.* **125,** 11–26.

Atack, J. R., *et al.* (2009). In vitro and in vivo properties of 3-tert-butyl-7-(5-methylisoxazol-3-yl)-2-(1-methyl-1H–1, 2, 4-triazol-5-ylmethoxy)-pyrazolo[1, 5-d][1, 2, 4]triazine (MRK-016), a GABAA receptor a5 subtype selective inverse agonist. *J. Pharmacol. Exp. Ther.* **331,** 470–484.

Bliss, T. V., and Collingridge, G. L. (1993). *Nature* **361,** 31.

Boileau, A. J., *et al.* (2002). GABAA receptor b2Tyr97 and Leu99 line the GABA-binding site. *J. Biol. Chem.* **277,** 2931–2937.

Bond, R. A., and IJzerman, A. P. (2006). Recent developments in constitutive receptor activity and inverse agonism, and their potential for GPCR drug discovery. *Trends Pharmacol. Sci.* **27,** 92–96.

Buettelmann, B., *et al.* (2007). Aryl-Isoxazol-4-yl-Imidazo[1,5a]pyridine derivatives. PCT Int. Appl. WO2007/074089 A1, database: CAPLUS.

Buettelmann, B., *et al.* (2009). Isoxazolo-pyridazine derivatives. U.S. Pat. Appl. Publ. US 2009/0143385 A1, database: CAPLUS.

Cerione, R. A., *et al.* (1984). Mammalian.beta.2-adrenergic receptor: Reconstitution of functional interactions between pure receptor and pure stimulatory nucleotide binding protein of the adenylate cyclase system. *Biochemistry* **23,** 4519–4525.

Chambers, M. S., *et al.* (2002). Nitrogen substituted 1,2,4-triazolo[3,4-a]phtalazine derivatives for enhancing cognition. PCT Int. Appl. WO 02/42305 A1, database: CAPLUS.

Chambers, M. S., *et al.* (2003). Identification of a novel, selective GABAA alpha5 receptor inverse agonist which enhances cognition. *J. Med. Chem.* **46,** 2227–2240.

Chambers, M. S., *et al.* (2004). An orally bioavailable, functionally selective inverse agonist at the benzodiazepine site GABA$_A$ a5 receptors with cognition enhancing properties. *J. Med. Chem.* **47,** 5829–5832.

Chang, Y., and Weiss, D. S. (1999). Allosteric activation mechanism of the [alpha]1[beta]2 [gamma]2 [gamma]-aminobutyric acid type a receptor revealed by mutation of the conserved M2 leucine. *Biophys. J.* **77,** 2542–2551.

Clayton, T., *et al.* (2007). An updated unified pharmacophore model of the benzodiazepine binding site on -aminobutyric acida receptors: Correlation with comparative models. *Curr. Med. Chem.* **14,** 2755–2775.

Collinson, N., *et al.* (2002). Enhanced learning and memory and altered GABAergic synaptic transmission in mice lacking the alpha 5 subunit of the GABAA receptor 20026436. *J. Neurosci.* **22,** 5572–5580.

Costa, T., and Cotecchia, S. (2005). Historical review: Negative efficacy and the constitutive activity of G-protein-coupled receptors. *Trends Pharmacol. Sci.* **26,** 618–624.

Costa, T., and Herz, A. (1989). Antagonists with negative intrinsic activity at delta opioid receptors coupled to GTP-binding proteins. *Proc. Natl. Acad. Sci. USA* **86,** 7321–7325.

Crestani, F., *et al.* (2002). Trace fear conditioning involves hippocampal a5 GABAA receptors. *Proc. Natl. Acad. Sci. USA* **99,** 8980–8985.

Dalziel, J. E., *et al.* (2000). Mutating the highly conserved second membrane-spanning region 9′ leucine residue in the a1 or b1 subunit produces subunit-specific changes in the function of human a1b1 γ-aminobutyric acid$_a$ receptors. *Mol. Pharmacol.* **57,** 875–882.

Davies, P. A., *et al.* (2003). A novel class of ligand-gated ion channel is activated by Zn2+. *J. Biol. Chem.* **278,** 712–717.

Dawson, G. R., *et al.* (2006). An inverse agonist selective for alpha5 subunit-containing GABAA receptor enhances cognition. *JPET* **316,** 1335–1345.

Duka, T., *et al.* (1988). Beta-carbolines as tools in memory research:human data with the beta-carboline ZK 93426. *Psychopharmacol. Ser.* **6,** 246–260.

Ernst, M., *et al.* (2003). Comparative modeling of GABAA receptors: Limits, insights, future developments. *Neuroscience* **119,** 933–943.

Findlay, G. S., *et al.* (2000). Allosteric modulation in spontaneously active mutant [gamma]-aminobutyric acidA receptors in frogs. *Neurosci. Lett.* **293,** 155–158.

Goetz, T., *et al.* (2007). GABA$_A$ receptors: Structure and function in the basal ganglia. *Prog. Brain Res.* **160,** 21–41.

Goetz, T., *et al.* (2009). GABA$_A$ receptors: Molecular biology, cell biology, and pharmacology. *In* "Encyclopedia of Neuroscience," pp. 463–470, Larry R. Squire Editor, Elsevier Academic Press, Amsterdam.

Guerrini, G., *et al.* (2006). Benzodiazepine receptor ligands. 8: Synthesis and pharmacological evaluation of new pyrazolo[5, 1-c] [1, 2, 4]benzotriazine 5-oxide 3- and 8-disubstituted: High affinity ligands endowed with inverse-agonist pharmacological efficacy. *Bioorg. Med. Chem.* **14,** 758–775.

Guerrini, G., *et al.* (2009). Synthesis, in vivo evaluation and molecular modeling studies of new pyrazolo[5, 1-c][1, 2, 4]benzotriazine 5-oxides derivatives. Identification of a bi-functional hydrogen bond area related to the inverse agonism. *J. Med. Chem.* **52,** 4668–4682.

Hanson, S. M., *et al.* (2008). Structural requirements for eszopiclone and zolpidem binding to the g-aminobutyric acid type-A (GABAA) receptor are different. *J. Med. Chem.* **51,** 7243–725210.1021/jm800889m.

Hogg, R. C., *et al.* (2005). Allosteric modulation of ligand-gated ion channels. *Biochem. Pharmacol.* **70,** 1267–1276.

Jackson, M. B. (1984). Spontaneous openings of the acetylcholine receptor channel. *Proc. Natl. Acad. Sci. USA* **81,** 3901–3904.

Jacob, T. C., *et al.* (2008). GABAA receptor trafficking and its role in the dynamic modulation of neuronal inhibition. *Nat. Rev. Neurosci.* **9,** 331–343.

Keen, T. J., *et al.* (1991). Autosomal dominant retinitis pigmentosa: Four new mutations in rhodopsin, one of them in the retinal attachment site. *Genomics* **11,** 199–205.

Knust, H., *et al.* (2009). The discovery and unique pharmacological profile of RO4938581 and RO4882224 as potent and selective GABAA [alpha]5 inverse agonists for the treatment of cognitive dysfunction. *Bioorg. Med. Chem. Lett.* **19,** 5940–5944.

Koski, G., *et al.* (1982). Modulation of sodium-sensitive GTPase by partial opiate agonists. An explanation for the dual requirement for Na+ and GTP in inhibitory regulation of adenylate cyclase. *J. Biol. Chem.* **257,** 14035–14040.

Maksay, G., and Bíró, T. (2002). Dual cooperative allosteric modulation of binding to ionotropic glycine receptors. *Neuropharmacology* **43,** 1087–1098.

Martin, L. J., *et al.* (2009). The physiological properties and therapeutic potential of a5-$GABA_A$ receptors. *Biochem. Soc. Trans.* **037,** 1334–1337.

Miko, A., *et al.* (2004). A TM2 residue in the b1 subunit determines spontaneous opening of homomeric and heteromeric g-aminobutyric acid-gated Ion channels. *J. Biol. Chem.* **279,** 22833–22840.

Miller, P. S., and Smart, T. G. (2010). Binding, activation and modulation of Cys-loop receptors. *Trends Pharmacol. Sci.* **31,** 161–174.

Milligan, G. (2003). Constitutive activity and inverse agonists of G protein-coupled receptors: A current perspective. *Mol. Pharmacol.* **64,** 1271–1276.

Mohler, H. (2006). $GABA_A$ receptor diversity and pharmacology. *Cell Tissue Res.* **326,** 505–516.

Neelands, T. R., *et al.* (1999). Spontaneous and g-aminobutyric acid (GABA)-activated GABAA receptor channels formed by e subunit-containing isoforms. *Mol. Pharmacol.* **55,** 168–178.

Olsen, R. W., and Sieghart, W. (2008). International union of pharmacology. LXX. Subtypes of {gamma}-aminobutyric acida receptors: Classification on the basis of subunit composition, pharmacology and function. *Pharmacol. Rev.* **60,** 243–26010.1124/pr.108.00505.

Olsen, R. W., and Sieghart, W. (2009). GABAA receptors: Subtypes provide diversity of function and pharmacology. *Neuropharmacology* **56,** 141–148.

Pan, Z.-H., *et al.* (1997). Agonist-induced closure of constitutively open g-aminobutyric acid channels with mutated M2domains. *Proc. Natl. Acad. Sci. USA* **94,** 6490–6495.

Parma, J., *et al.* (1993). Somatic mutations in the thyrotropin receptor gene cause hyperfunctioning thyroid adenomas. *Nature* **365,** 649–651.

Parra, S., and Bond, R. A. (2007). Inverse agonism: From curiosity to accepted dogma, but is it clinically relevant? *Curr. Opin. Pharmacol.* **7,** 146–150.

Quirk, K., *et al.* (1996). [3H]L-655, 708, a novel ligand selective for the benzodiazepine site of GABAA receptors which contain the a5 subunit. *Neuropharmacology* **35,** 1331–1335.

Rodríguez-Puertas, R., and Barreda-Gómez, G. (2006). Development of new drugs that act through membrane receptors and involve an action of inverse agonism. *Recent Pat. CNS Drug Discov.* **1,** 207–217.

Sakmann, B., *et al.* (1983). Acetylcholine activation of single muscarinic K+ channels in isolated pacemaker cells of the mammalian heart. *Nature* **303,** 250–253.

Shenker, A., *et al.* (1993). A constitutively activating mutation of the luteinizing hormone receptor in familial male precocious puberty. *Nature* **365,** 652–654.

Sternfeld, F., *et al.* (2004). Selective orally active g-aminobutyric acid a-5 receptor inverse agonists as cognition enhancer. *J. Med. Chem.* **47,** 2176–2179.
Street, L. J., *et al.* (2004). Synthesis and biological evaluation of 3-heterocyclyl-7, 8, 9, 10-tetrahydro-(7, 10-ethano)-1, 2, 4-triazolo[3, 4-a]phtalazines and analogues as subtype-selective inverse agonists for GABAA a5 benzodiazepine binding site. *J. Med. Chem.* **47,** 3642–3657.
Uusi-Oukari, M., and Korpi, E. R. (2010). Regulation of GABAA receptor subunit expression by pharmacological agents. *Pharmacol. Rev.* **62,** 97–135.
Vaisse, C., *et al.* (2000). Melanocortin 4 receptor mutations are a frequent and heterogeneous cause of morbid obesity. *J. Clin. Investig.* **106,** 253–262.
Wood, P. L., *et al.* (1984). In vitro characterization of benzodiazepine receptor agonists, antagonists, inverse agonists and agonist/antagonists. *J. Pharmacol. Exp. Ther.* **231,** 572–576.

CHAPTER TWELVE

Assays for Inverse Agonists in the Visual System

Masahiro Kono

Contents

Abstract

Visual pigment proteins belong to the superfamily of G protein-coupled receptors and are the light-sensitive molecules in rod and cone photoreceptor cells. The protein moiety is known as opsin and the ligand in the dark is 11-*cis* retinal, which serves as both the photon detector and an inverse agonist. While much is known about properties of the rod pigment rhodopsin, much less is understood about cone visual pigments. Being able to identify ligands that effect opsins give an insight into structure–activity relationships. The action of some ligands indicates that there are differences between not only rod and cone opsins but also among the different classes of cone opsins. Furthermore, inverse agonists of cone opsins may have potential therapeutic uses under conditions when the native 11-*cis* retinal ligand is absent. A method for determining the effects of ligands on rod and cone opsin activity is described.

Department of Ophthalmology, Medical University of South Carolina, Charleston, South Carolina, USA

Methods in Enzymology, Volume 485
ISSN 0076-6879, DOI: 10.1016/S0076-6879(10)85012-8

1. Introduction

Visual pigments are the light-sensitive components in rod and cone photoreceptor cells and are comprised of two parts: opsin, the G protein-coupled receptor (GPCR), and the 11-*cis* aldehyde form of vitamin A (11-*cis* retinal), the chromophore. Visual pigments differ from other GPCRs in that the ligand is covalently bound to the receptor via a Schiff base linkage to a strictly conserved lysine residue in the seventh transmembrane helix (Bownds, 1967; Morton and Pitt, 1955; Wang *et al.*, 1980). The opsins themselves are constitutively active (Cohen *et al.*, 1993; Isayama *et al.*, 2006; Kono, 2006; Melia *et al.*, 1997; Surya *et al.*, 1995). The native 11-*cis* retinal ligand is an inverse agonist, which limits spontaneous activation of the visual signaling cascade in the dark. The very fast response of visual pigments as GPCRs is that light isomerizes the 11–12 bond from *cis* to *trans* within 200 fs (Schoenlein *et al.*, 1991). This is followed by conformational changes of the protein enabling it to activate its G protein transducin. Thus, light converts the bound ligand from an inverse agonist to an agonist. The all-*trans* ligand then dissociates from the opsin and is transported to the retinal pigment epithelium where it is converted back to 11-*cis* retinal and shuttled back to the photoreceptor to regenerate new pigment.

Rod photoreceptor cells contain the visual pigment rhodopsin and are used for dim light vision; cone photoreceptor cells contain cone visual pigments and are used for bright light and color vision. Although cone pigments are structurally and functionally highly homologous to rhodopsin, there are differences. Because the chromophore of both rod and cone opsins is 11-*cis* retinal, the specific interactions with the receptor result in tuning the absorption properties, resulting in sensitivity across the visible spectrum. Furthermore, cone pigments regenerate, activate, and inactivate faster than rhodopsin. The effects of retinal analogs on rod opsin's activity have been studied (Bartl *et al.*, 2005; Buczylko *et al.*, 1996; Han *et al.*, 1997); again, less is known about structure–activity relationships with cone opsins and ligands. One reason for this is the lack of methods and sources for pure cone opsins, whereas rod opsins in good purity are easily isolated. Another reason is the perceived instability of the protein (Ramon *et al.*, 2009). Cone pigments have been shown to lose its chromophore or, in the presence of analogs, exchange chromophores in the dark (Crescitelli, 1984; Kefalov *et al.*, 2005; Matsumoto *et al.*, 1975), unlike the rod opsin where the pigment (rhodopsin) is quite stable even to hydroxylamine. However, the cone opsin proteins themselves are quite stable in a membrane environment, as pigments can be generated and regenerated repeatedly when fresh chromophore is supplied.

Furthermore, the importance of 11-*cis* retinal *in vivo* appears to extend beyond visual signaling for cones and less so with rods. When there is a limited supply of 11-*cis* retinal, as is the case with some forms of a childhood-onset blinding diseases Leber Congenital Amaurosis (LCA), cones appear to degenerate rapidly, while the rods survive for quite a bit longer, based on photoreceptor cell layer thicknesses in patients with LCA (Jacobson *et al.*, 2007, 2009). Histological studies of mouse models of LCA clearly show rapid loss of cone cells preceded by opsins that are no longer localized solely in the outer segments of cones but rather distributed throughout (Rohrer *et al.*, 2005; Znoiko *et al.*, 2005). Supplementing mouse models of LCA with 11-*cis* retinal improved cone cell survival and cone opsin localization presumably by their abilities to form pigment (Rohrer *et al.*, 2005). Thus, cone opsin inverse agonists have the potential as therapeutic agents in preserving cone photoreceptor cells in LCA. A practical problem with using the native ligand as treatment is that light abolishes any benefits of 11-*cis* retinal in mouse models (manuscript in preparation). Thus, our recent interests in identifying compounds that can inactivate cone opsins have extended beyond structure–function characterization of cone opsins. We have begun to characterize the effects of analogs of retinal on the abilities of cone opsins to activate transducin. A partial list of retinoids and analogs that are inverse agonists to long wavelength-sensitive cone opsins is shown in Fig. 12.1. We have found that 11-*cis* retinol, however, is an agonist to rod opsin (Ala-Laurila *et al.*, 2009;

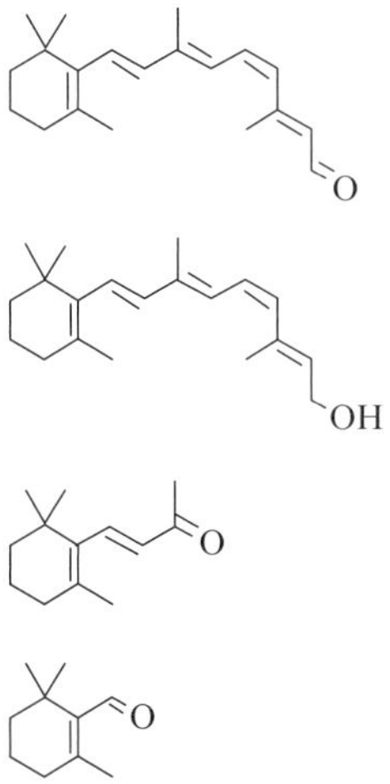

11-*cis* Retinal

11-*cis* Retinol

β-Ionone

Cyclocitral

Figure 12.1 Compounds that have been found to act as inverse agonists to human red cone opsins. 11-*cis* retinal is the native ligand in the eye and was obtained from the National Eye Institute and Dr Rosalie Crouch of the Medical University of South Carolina; 11-*cis* retinol was obtained from Dr Crouch; β-ionone and cyclocitral were purchased from Sigma-Aldrich (St. Louis, MO). Even though these compounds were found to be inverse agonists to the red cone opsin from our *in vitro* assays, 11-*cis* retinol and β-ionone are agonists to rod opsin (Isayama *et al.*, 2006; Kono, 2006; Kono *et al.*, 2008), and β-ionone is also an agonist to a short wavelength-sensitive class of cone opsins (Isayama *et al.*, 2006).

Kono *et al.*, 2008). Also, β-ionone is an agonist to rod opsin and a class of short wavelength-sensitive cone opsins (Buczylko *et al.*, 1996; Isayama *et al.*, 2006). Thus, not all opsins respond to ligands in the same manner, and generalizations about which ones act as inverse agonists or agonists should be made with caution in the absence of data.

1.1. Assay considerations

There are a few techniques that can be and have been used to study ligand effects on rod and cone opsins and cells with strengths and limitations of each one. Perhaps the most sensitive is single cell electrophysiology where single photoreceptor cells are monitored for light- and/or ligand-dependent changes to ion permeability (Estevez *et al.*, 2006; Jin *et al.*, 1993; Jones *et al.*, 1989; Kefalov *et al.*, 1999, 2001). Advantages include the high sensitivity, fast time resolution, and physiological environment of the opsins. However, potential complications could arise from the presence of the native ligand, which needs to be washed out; the test ligand affecting ion channel properties directly (Dean *et al.*, 2002; McCabe *et al.*, 2004); the presence of multiple types of opsins within the same photoreceptor cell (Applebury *et al.*, 2000; Makino and Dodd, 1996); and modification of ligand by the cell (e.g., 11-*cis* retinol is oxidized to 11-*cis* retinal in cone cells and regenerate cone visual pigments; Ala-Laurila *et al.*, 2009; Jones *et al.*, 1989).

In vitro studies with heterologously expressed opsins have their own set of advantages. There is only the single opsin that is expressed; there is no need to wash out native chromophore because it was never present; mutants can be easily constructed and expressed. The ability of testing mutant opsins demonstrated that some rhodopsin mutants associated with retinitis pigmentosa are constitutively active which could be deactivated with either 11-*cis* retinal or 11-*cis* retinyl Schiff base (Cohen *et al.*, 1992; Robinson *et al.*, 1992; Zhukovsky *et al.*, 1991). Limitations include the fact that the opsins are not in their native *in vivo* environment and assays may not be as sensitive as electrophysiological measurements (Melia *et al.*, 1997).

We use a radioactive filter binding assay to assess the abilities of the opsins to activate transducin in the absence and presence of ligand (Fig. 12.2). Opsins are expressed in COS-1 cells, and membranes isolated. Binding of the native ligand is easy to demonstrate as the receptor becomes able to activate G proteins in a light-dependent manner. To determine inverse agonist behavior, we need to demonstrate that activation with ligand is lower than without ligand. The opsins are maintained in a lipid environment as the purified apoprotein is not stable. The constitutive activity of the apoprotein is difficult to distinguish at physiological pH values. Figure 12.3 illustrates the pH dependence of transducin activation by the human red cone opsin and pigment in the dark and light. This follows a similar pattern

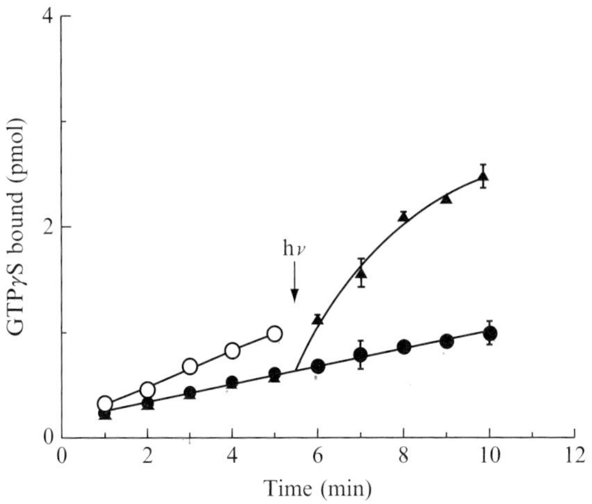

Figure 12.2 Transducin activation by human red cone opsin at pH 6.4 as a function of time. Activation by opsin (open circles), opsin with 11-*cis* retinol (filled circles), and opsin with 11-*cis* retinal (triangles) were determined in the dark for the first 5.5 min, at which time the samples were illuminated with a 6 s pulse of light from a 300 W slide projector passed through a 530 nm long pass filter. Both 11-*cis* retinoids act as inverse agonists in the dark, but only the sample with 11-*cis* retinal forms a pigment that absorbs visible light and results in photoactivation of the pigment. The figure is adapted from Ala-Laurila *et al.* (2009). © The American Society for Biochemistry and Molecular Biology.

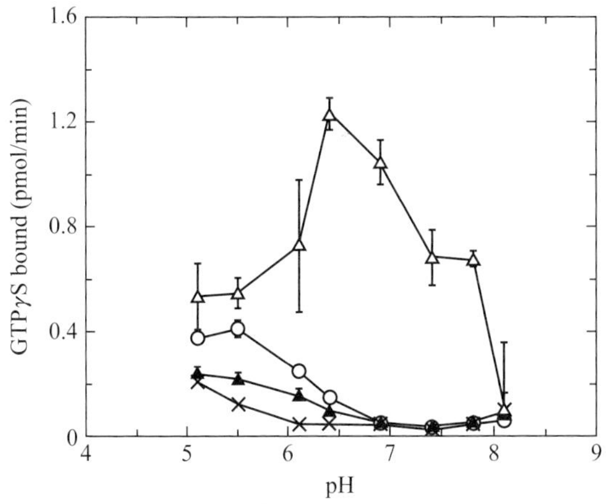

Figure 12.3 pH dependence of transducin activation by human red cone opsin in the absence and presence of 11-*cis* retinal and light. Human cone opsins were incubated with ethanol for opsin alone (open circles), and 11-*cis* retinal in the dark (filled triangles), 11-*cis* retinal followed by photobleaching (open triangles). Membranes from mock transfected COS-1 cells were used to measure activity of transducin in the absence of opsin and ligand (X). The following buffers were used to set the pH: MES for pH 5.1–6.4 and HEPES for pH 6.8–8.1.

as with the bovine rod opsin pH dependence of transducin activation (Cohen *et al.*, 1992). Thus, we assay at a slightly acidic pH of 6.4 where we can differentiate a change in activity between opsin and opsin with an inverse agonist. In order to carry out these assays, we use bovine rod transducin as our G protein for both rod and cone opsins because of the availability of bovine retinas and relative ease of purifying large quantities in a cost-effective manner. Both preparations are briefly described here, followed by the radioactive filter binding assay.

2. Opsin Preparation

For a quick preparation of opsins, we isolate the membrane fractions from transfected COS-1 cells. The methods are straightforward and relatively inexpensive. The opsin genes generally have the last codons for the last eight amino acid residues of bovine rhodopsin, the 1D4 epitope (Molday and MacKenzie, 1983), to help with quantification on slot or western blots, although it is not absolutely necessary. Typically, 2 μg of plasmid is transfected for each 10-cm plate of confluent COS-1 cells using the DEAE-dextran method described previously (Oprian, 1993), although other transfection methods work without problems. After 3 days, the cells are harvested and up to 10 plates are pooled into 15 mL conical tubes at a volume of 1 mL per 10-cm plate, then pelleted and stored at −80 °C until needed. Cells are resuspended in a hypotonic buffer of 10 m*M* Tris–HCl (pH 7.4) buffer with 30 μ*M* phenylmethanesulfonyl fluoride (PMSF) and passed through a 25 G needle attached to a 10–20 mL syringe four times. The homogenate is then layered on top of a 20 mL 37.7% (w/v) sucrose cushion in a 25 × 89 mm polyallomer tube (Beckman-Coulter, Palo Alto, CA) and spun in a Beckman SW32 rotor at 15,000 rpm for 30 min (Kono, 2006; Kono *et al.*, 2005; Robinson, 2000). The membrane fraction at the interface is collected by piercing the polyallomer tube with an 18 G needle attached to a 5 mL syringe and sucrose diluted with 10 m*M* Tris–HCl (pH 7.4) and recentrifuged. The pellet is resuspended in a buffer containing 150 m*M* NaCl, 1 m*M* $MgCl_2$, 1 m*M* $CaCl_2$, 0.1 m*M* EDTA, 10 m*M* Tris–HCl (pH 7.4) at a volume of 25 μL per plate of COS-1 cells used and stored at −80 °C in 25 μL aliquots. The amount of opsin in the membrane preparations are determined by slot blot analysis (Kono *et al.*, 2005) using known amounts of bovine rhodopsin as reference and probed with the rhodopsin 1D4 antibody (available through a number of vendors such as catalog number MA1-722 from Affinity BioReagents/Thermo Fisher Scientific, Rockford, IL).

3. Transducin Preparation

Transducin is purified from 100 bovine retinas (W. L. Lawson, Lincoln, NE). The procedure is essentially a rod outer segment (ROS) preparation using a discontinuous sucrose gradient (Wessling-Resnick and Johnson, 1987) conducted in the light such that transducin is bound to the bleached rhodopsin and will pellet with the membranes. Transducin is released into solution by incubating the ROS with 40 μ*M* GTP on ice for 30 min. After centrifugation, the supernatant containing transducin is subjected to further purification by DEAE-cellulose anion exchange chromotography (Baehr *et al.*, 1982). Transducin fractions are pooled and dialyzed against and stored at −20 °C in 50% glycerol, 10 m*M* Tris–HCl (pH 7.4), 2 m*M* $MgCl_2$, and 1 m*M* DTT at a concentration of 50 μ*M* (Yu *et al.*, 1995).

4. Transducin Activation Assay

4.1. Stock solutions

1. Reaction/wash buffer (10×): 100 m*M* 2-(*N*-morpholino)ethanesufonic acid (MES), pH 6.5 (final pH will be 6.4 in reaction), 1 *M* NaCl, 50 m*M* $MgCl_2$.
2. DTT (50 m*M*) in milli-Q water.
3. Bovine rod transducin (50 μ*M*, see above).
4. Opsin in membranes (see above). The opsin concentration is typically 10–50 n*M*.
5. Test ligand dissolved in ethanol. As a starting point, we often make stock concentrations of 2 m*M* or 200 μ*M*.
6. 150 μ*M* GTPγS. A mixture of cold GTPγS with $GTP\gamma S^{35}$ at ~0.25 mCi/mL is prepared as follows: to 100 μL of a 150 μ*M* solution of GTPγS from ~3 m*M* stock solution is added 2 μL (25 μCi) GTPγS-35. (GTPγS-35: catalog number NEG030H250UC, PerkinElmer Life and Analytical Sciences, Waltham, MA).

4.2. Measurement

The ability of the specific opsin to activate bovine rod transducin is determined using a radioactive filter binding assay with membrane preparations of opsin expressed in COS-1 cells essentially as described previously with a few modifications (Kono, 2006; Robinson, 2000). We use a Millipore 1225 sampling vacuum manifold (Millipore, Billerica, MA) with 25 mm diameter

Millipore mixed cellulose ester membranes (HAWP 02500, Millipore, Billerica, MA). The filter membranes are first moistened in water and then placed in the vacuum manifold.

The reaction is set up in 1.5 mL microcentrifuge tube without retinoid and GTPγS as follows:

Volume (μL)	
70	Deionized water
10	10× buffer (100 m*M* Tris buffer, pH 7.5, 1 *M* NaCl, 50 m*M* $MgCl_2$)
2	50 m*M* DTT
5	Transducin (50 μ*M* stock)
10	Opsin/visual pigment (typically, 10–50 n*M* stock concentration) in membranes

If the test ligand is a full-length or light-sensitive retinoid, then at this point the room lights are turned off and dim red light (photographic dark room filters such as Kodak number 2 or GBX2) conditions are used. One microliter of ligand solution (or ethanol for as a control) is added and mixed with a micropipettor. 1 min later, 2 μL of 150 μ*M* GTPγS solution is added and mixed, and a timer is started. At 1 min, a 10 μL aliquot of the reaction mixture is transferred to the vacuum manifold. Filters are immediately rinsed three times with 4 mL rinse buffer (10 m*M* Tris, pH 6.4, 100 m*M* NaCl, 5 m*M* $MgCl_2$) with a repeating pipettor, which leaves proteins including transducin and bound GTPγS on the filter and unbound GTPγS to flow through. This process is repeated at typically 1 min intervals. After 3–6 time points, the filter membranes are placed into scintillation vials and filled with 10 mL Amersham BCS scintillation cocktail (catalog number: NBCS104, GE Healthcare, Piscataway, NJ). An additional vial is filled with 10 mL scintillant, into which 10 μL of the reaction mixture is pipetted. These counts will be used to convert counts per minute (cpm) to mol GTPγS. The vials are shaken for at least 1 h and often overnight for convenience and measured in a scintillation counter (usually 1–5 min counts).

The cpm from the 10 μL reaction mixture pipetted directly into the scintillation fluid is used to convert cpm to pmol GTPγS taken up as a result of G protein activation. If aliquots from more than one reaction were measured, they can be averaged. Ten microliters of the reaction mixture contains 30 pmol of GTPγS. Thus, 30 pmol divided by these counts is used to determine the pmol GTPγS bound to transducin. The number of moles GTPγS bound is plotted against time, and the activity is the slope from this plot. If the concentration of opsin has been quantified, then the specific activity can be reported.

Because GTPγS is a nonhydrolyzable analog of GTP, the data represents an accumulation of transducin activated with time. This assay can, of course, be used to assay agonists as well. If light-dependent activity is to be determined, then the total reaction mixture is scaled to 150 μL total volume, and dark points first taken (i.e., see Fig. 12.2). A brief pulse of bright light is then given at a convenient time such as at 5.5 min after initiation of the reaction and measurement continued. A change in activity before and after light will indicate light-dependent activity. If assaying light activity of cone pigments, then one should take care to consider that the lifetime of the active intermediate of wild-type cone pigments is on the order of tens of seconds and a long photobleach will result in a large fraction of decayed product (Das *et al.*, 2004; Kono, 2006). Furthermore, if starting with cone opsins in membranes and adding an excess of chromophore such as 11-*cis* retinal according to the protocol described in this chapter, then it is best to keep the room lights off except during the photobleach as cone pigments can be regenerated after the rapid decay of the active species; thus with continued light, the activity would result from a steady state of light activated cone pigments which would give an overestimate of activity.

Other opsin preparations such as immunoaffinity purification after solubilizing in a detergent such as CHAPS with lipid can also be used, and with dilution or removal of CHAPS, purified opsins will be in vesicles (Rim and Oprian, 1995; Rim *et al.*, 1997). If pigments are formed from incubation of COS-1 cell suspensions with 11-*cis* retinal, then they can similarly be immunopurified. We use the 1D4 antibody conjugated to sepharose routinely to isolate and purify pigments (Das *et al.*, 2004; Kono, 2006). If detergent-solubilized and purified pigment is to be used, then care must be taken with detergent type and concentration for the transducin activation assay. For example, we try to keep the final concentration of dodecyl maltoside to be about 0.01%; otherwise, activity decreases precipitously.

ACKNOWLEDGMENTS

This work was supported in part by NIH grant R01-EY019515 and an unrestricted grant to the Department of Ophthalmology at MUSC from the Research to Prevent Blindness. 11-*cis* retinal and 11-*cis* retinol were gifts from Rosalie Crouch of the Medical University of South Carolina.

REFERENCES

Ala-Laurila, P., Cornwall, M. C., Crouch, R. K., and Kono, M. (2009). The action of 11-*cis*-retinol on cone opsins and intact cone photoreceptors. *J. Biol. Chem.* **284,** 16492–16500.

Applebury, M. L., Antoch, M. P., Baxter, L. C., Chun, L. L. Y., Falk, J. D., Farhangfar, F., Kage, K., Krzystolik, M. G., Lyass, L. A., and Robbins, J. T. (2000). The murine cone photoreceptor: A single cone type expresses both S and M opsins with retinal spatial patterning. *Neuron* **27,** 513–523.

Baehr, W., Morita, E. A., Swanson, R. J., and Applebury, M. L. (1982). Characterization of bovine rod outer segment G-protein. *J. Biol. Chem.* **257,** 6452–6460.

Bartl, F. J., Fritze, O., Ritter, E., Herrmann, R., Kuksa, V., Palczewski, K., Hofmann, K. P., and Ernst, O. P. (2005). Partial agonism in a G protein-coupled receptor. Role of the retinal ring structure in rhodopsin activation. *J. Biol. Chem.* **280,** 34259–34267.

Bownds, D. (1967). Site of attachment of retinal in rhodopsin. *Nature* **216,** 1178–1181.

Buczylko, J., Saari, J. C., Crouch, R. K., and Palczewski, K. (1996). Mechanisms of opsin activation. *J. Biol. Chem.* **271,** 20621–20630.

Cohen, G. B., Oprian, D. D., and Robinson, P. R. (1992). Mechanism of activation and inactivation of opsin: Role of Glu^{113} and Lys^{296}. *Biochemistry* **31,** 12592–12601.

Cohen, G. B., Yang, T., Robinson, P. R., and Oprian, D. D. (1993). Constitutive activation of opsin: Influence of charge at position 134 and size at position 296. *Biochemistry* **32,** 6111–6115.

Crescitelli, F. (1984). The gecko visual pigment: The dark exchange of chromopohore. *Vision Res.* **24,** 1551–1553.

Das, J., Crouch, R. K., Ma, J.-X., Oprian, D. D., and Kono, M. (2004). Role of the 9-methyl group of retinal in cone visual pigments. *Biochemistry* **43,** 5532–5538.

Dean, D. M., Nguitragool, W., Miri, A., McCabe, S. L., and Zimmerman, A. L. (2002). All-*trans*-retinal shuts down rod cyclic nucleotide-gated ion channels: A novel role for photoreceptor retinoids in the response to bright light? *Proc. Natl. Acad. Sci. USA* **99,** 8372–8377.

Estevez, M. E., Ala-Laurila, P., Crouch, R. K., and Cornwall, M. C. (2006). Turning cones off: The role of the 9-methyl group of retinal in red cones. *J. Gen. Physiol.* **128,** 671–685.

Han, M., Groesbeek, M., Sakmar, T. P., and Smith, S. O. (1997). The C9 methyl group of retinal interacts with glycine-121 in rhodopsin. *Proc. Natl. Acad. Sci. USA* **94,** 13442–13447.

Isayama, T., Chen, Y., Kono, M., DeGrip, W. J., Ma, J.-X., Crouch, R. K., and Makino, C. L. (2006). Differences in the pharmacological activation of visual opsins. *Vis. Neurosci.* **23,** 899–908.

Jacobson, S. G., Aleman, T. S., Cideciyan, A. V., Heon, E., Golczak, M., Beltran, W. A., Sumaroka, A., Schwartz, S. B., Roman, A. J., Windsor, E. A. M., Wilson, J. M., Aguirre, G. D., *et al.* (2007). Human cone photoreceptor dependence on RPE65 isomerase. *Proc. Natl. Acad. Sci. USA* **104,** 15123–15128.

Jacobson, S. G., Aleman, T. S., Cideciyan, A. V., Roman, A. J., Sumaroka, A., Windsor, E. A. M., Schwartz, S. B., Heon, E., and Stone, E. M. (2009). Defining the residual vision in Leber congenital amaurosis caused by *RPE65* mutations. *Invest. Ophthalmol. Vis. Sci.* **50,** 2368–2375.

Jin, J., Crouch, R. K., Corson, D. W., Katz, B. M., MacNichol, E. F., and Cornwall, M. C. (1993). Noncovalent occupancy of the retinal-binding pocket of opsin diminishes bleaching adaptation of retinal cones. *Neuron* **11,** 513–522.

Jones, G. J., Crouch, R. K., Wiggert, B., Cornwall, M. C., and Chader, G. J. (1989). Retinoid requirements for recovery of sensitivity after visual-pigment bleaching in isolated photoreceptors. *Proc. Natl. Acad. Sci. USA* **86,** 9606–9610.

Kefalov, V. J., Cornwall, M. C., and Crouch, R. K. (1999). Occupancy of the chromophore binding site of opsin activates visual transduction in rod photoreceptors. *J. Gen. Physiol.* **113,** 491–503.

Kefalov, V. J., Crouch, R. K., and Cornwall, M. C. (2001). Role of noncovalent binding of 11-*cis*-retinal to opsin in dark adaptation of rod and cone photoreceptors. *Neuron* **29,** 749–755.

Kefalov, V. J., Estevez, M. E., Kono, M., Goletz, P. W., Crouch, R. K., Cornwall, M. C., and Yau, K.-W. (2005). Breaking the covalent bond—A pigment property that contributes to desensitization in cones. *Neuron* **46,** 879–890.

Kono, M. (2006). Constitutive activity of a UV cone opsin. *FEBS Lett.* **580,** 229–232.

Kono, M., Crouch, R. K., and Oprian, D. D. (2005). A dark and constitutively active mutant of the tiger salamander UV pigment. *Biochemistry* **44,** 799–804.

Kono, M., Goletz, P. W., and Crouch, R. K. (2008). 11-*cis* and all-*trans* retinols can activate rod opsin: Rational design of the visual cycle. *Biochemistry* **47,** 7567–7571.

Makino, C. L., and Dodd, R. L. (1996). Multiple visual pigments in a photoreceptor of the salamander retina. *J. Gen. Physiol.* **108,** 27–34.

Matsumoto, H., Tokunaga, F., and Yoshizawa, T. (1975). Accessibility of the iodopsin chromophore. *Biochim. Biophys. Acta* **404,** 300–308.

McCabe, S. L., Pelosi, D. M., Tetreault, M., Miri, A., Nguitragool, W., Kovithvathanaphong, P., Mahajan, R., and Zimmerman, A. L. (2004). All-trans-retinal is a closed-state inhibitor of rod cyclic nucleotide-gated ion channels. *J. Gen. Physiol.* **123,** 521–531.

Melia, T. J., Jr., Cowan, C. W., Angleson, J. K., and Wensel, T. G. (1997). A comparison of the efficiency of G protein activation by ligand-free and light-activated forms of rhodopsin. *Biophys. J.* **73,** 3182–3191.

Molday, R. S., and MacKenzie, D. (1983). Monoclonal antibodies to rhodopsin: Characterization, cross-reactivity, and application as structural probes. *Biochemistry* **22,** 653–660.

Morton, R. A., and Pitt, G. A. J. (1955). Studies on rhodopsin. 9. pH and the hydrolysis of indicator yellow. *Biochem. J.* **59,** 128–134.

Oprian, D. D. (1993). Expression of opsin genes in COS cells. *Methods Neurosci.* **15,** 301–306.

Ramon, E., Mao, X., and Ridge, K. D. (2009). Studies on the stability of the human cone visual pigments. *Photochem. Photobiol.* **85,** 509–516.

Rim, J., and Oprian, D. D. (1995). Constitutive activation of opsin: Interaction of mutants with rhodopsin kinase and arrestin. *Biochemistry* **34,** 11938–11945.

Rim, J., Faurobert, E., Hurley, J. B., and Oprian, D. D. (1997). In vitro assay for transphosphorylation of rhodopsin by rhodopsin kinase. *Biochemistry* **36,** 7064–7070.

Robinson, P. R. (2000). Assays for the detection of constitutively active opsins. *Methods Enzymol.* **315,** 207–218.

Robinson, P. R., Cohen, G. B., Zhukovsky, E. A., and Oprian, D. D. (1992). Constitutively active mutants of rhodopsin. *Neuron* **9,** 719–725.

Rohrer, B., Lohr, H. R., Humphries, P., Redmond, T. M., Seeliger, M. W., and Crouch, R. K. (2005). Cone opsin mislocalization in $Rpe65^{-/-}$ mice: A defect that can be corrected by 11-*cis* retinal. *Invest. Ophthalmol. Vis. Sci.* **46,** 3876–3882.

Schoenlein, R. W., Peteanu, L. A., Mathies, R. A., and Shank, C. V. (1991). The first step in vision: Femtosecond isomerization of rhodopsin. *Science* **254,** 412–415.

Surya, A., Foster, K. W., and Knox, B. E. (1995). Transducin activation by the bovine opsin apoprotein. *J. Biol. Chem.* **270,** 5024–5031.

Wang, J. K., McDowell, J. H., and Hargrave, P. A. (1980). Site of attachement of 11-*cis*-retinal in bovine rhodopsin. *Biochemistry* **19,** 5111–5117.

Wessling-Resnick, M., and Johnson, G. L. (1987). Allosteric behavior in transducin activation mediated by rhodopsin. *J. Biol. Chem.* **262,** 3697–3705.

Yu, H., Kono, M., McKee, T. D., and Oprian, D. D. (1995). A general method for mapping tertiary contacts between amino acid residues in membrane-embedded proteins. *Biochemistry* **34,** 14963–14969.

Zhukovsky, E. A., Robinson, P. R., and Oprian, D. D. (1991). Transducin activation by rhodopsin without a covalent bond to the 11-*cis*-retinal chromophore. *Science* **251,** 558–560.

Znoiko, S. L., Rohrer, B., Lu, K., Lohr, H. R., Crouch, R. K., and Ma, J.-x. (2005). Downregulation of cone-specific gene expression and degeneration of cone photoreceptors in the *rpe65*$^{-/-}$ mouse at early ages. *Invest. Ophthalmol. Vis. Sci.* **46,** 1473–1479.

CHAPTER THIRTEEN

Receptor-Driven Identification of Novel Human A3 Adenosine Receptor Antagonists as Potential Therapeutic Agents

Silvia Paoletta,* Stephanie Federico,† Giampiero Spalluto,† *and* Stefano Moro*

Contents

Abstract

The field of therapeutic application of the A_3 adenosine receptor (A_3AR) antagonists represents a rapidly growing and intense area of research in the adenosine field. Even if there are currently no A_3AR antagonists in clinical phases, in light of the plethora of biological effects attributed to A_3ARs, substantial efforts in medicinal chemistry have been directed toward developing antagonists for the A_3AR subtype. In this review, we summarize the more recent and promising evidences of the possible A_3AR application as drug candidates, and the role of the receptor-driven design in their *in silico* characterization.

* Molecular Modeling Section (MMS), Dipartimento di Scienze Farmaceutiche, Università di Padova, via Marzolo, Padova, Italy
† Dipartimento di Scienze Farmaceutiche, Università degli Studi di Trieste, Piazzale Europa, Trieste, Italy

Methods in Enzymology, Volume 485
ISSN 0076-6879, DOI: 10.1016/S0076-6879(10)85013-X

1. Introduction

The A_3 adenosine receptor (A_3AR) is the last member of the adenosine family of G protein-coupled receptors (GPCR) to have been cloned (Jacobson and Gao, 2006). Considering receptor distribution, the highest levels of human A_3AR mRNA have been found in lung and liver (Borea *et al.*, 2009). However, A_3ARs have been detected in various tissues including testis, lung, kidney, placenta, heart, brain, spleen, liver, uterus, bladder, jejunum, aorta, proximal colon, and eyes (Borea *et al.*, 2009; Jacobson and Gao, 2006).

The A_3AR has been mapped on human chromosome 1 and consists of 318 amino acid residues (Borea *et al.*, 2009). Differently to other adenosine receptors, the C-terminal region presents multiple serine and threonine residues, which may serve as potential sites of phosphorylation that are important for receptor desensitization upon agonist application. In particular, the amino acid residues in the C-terminus responsible for rapid desensitization were Thr(307), Thr(318), and Thr(319) (Borea *et al.*, 2009).

The first second-messenger systems found to be associated with A_3AR activation were adenylyl cyclase (AC) activity, which is inhibited, and phospholipase C (PLC), which is stimulated, through G_i and G_q protein coupling, respectively. In addition, other intracellular pathways have been described as being linked with A_3AR activation as summarized in Fig. 13.1 (Borea *et al.*, 2009; Jacobson and Gao, 2006).

2. Newer Potential Therapeutic Role of A_3 Adenosine Receptors

Unfortunately, there are currently no A_3AR antagonists in clinical phases. However, in light of the plethora of biological effects attributed to A_3ARs, substantial efforts in medicinal chemistry have been directed toward developing antagonists for the A_3AR subtype (Jacobson and Gao, 2006; Moro *et al.*, 2006a,b). In fact, a number of molecules are in biological testing as therapeutic agents for asthma and chronic obstructive pulmonary disease (COPD), glaucoma, stroke, cardiac hypoxia, and cerebral ischemia (Borea *et al.*, 2009; Jacobson and Gao, 2006).

The role of A_3AR in inflammatory diseases is currently controversial, and both anti- and proinflammatory effects have been attributed to its activation. One of the first therapeutic applications that was hypothesized for A_3AR antagonists was the treatment of asthma. In fact, it was reported that in rodents, A_3AR activation was responsible for mast cell degranulation (Borea *et al.*, 2009; Jacobson and Gao, 2006).

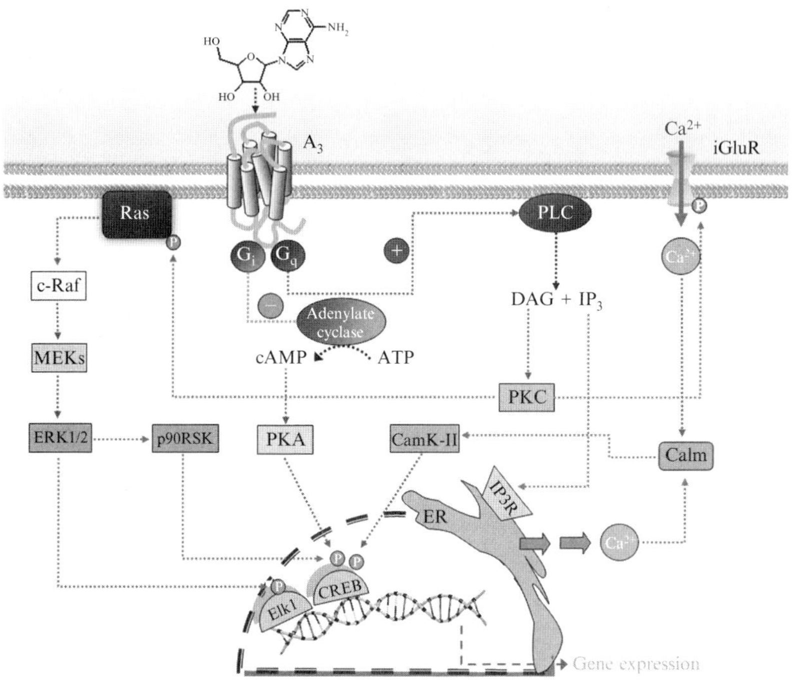

Figure 13.1 Signal transduction pathways associated with the activation of the human A_3 adenosine receptor. Abbreviations: ATP, adenosine triphosphate; Calm, calmodulin; CamK-II, Ca2+/calmodulin-dependent protein kinase; cAMP, cyclic adenosine monophosphate; c-Raf, RAF proto-oncogene serine/threonine-protein kinase; CREB, cAMP response element-binding; DAG, diacylglycerol; ERK, extracellular signal-regulated kinase; G_i, G_i family of G proteins; G_q, G_q family of G proteins; iGluR, ionotropic glutamate receptors; IP3, inositol (1,4,5)-trisphosphate; MEK, mitogen-activated protein kinase/extracellular signal-regulated kinase kinase; P, phosphate moiety; PKA, protein kinase A; PKC, protein kinase C; PLC, phospholipase C.

Activation of A_3AR leads to the regulation of chloride channels in nonpigmented ciliary epithelial cells, suggesting that A_3AR agonists would increase aqueous humor secretion and thereby intraocular pressure *in vivo*, whilst A_3AR antagonists may represent a specific approach for treating ocular hypertension (Borea *et al.*, 2009; Jacobson and Gao, 2006). Another important topic in the area of A_3AR-targeted therapy is the protective role of this adenosine receptor subtype in cardiac ischemia. To date, several studies have pointed to the evidence that the A_3AR is a key player in adenosine-induced cardioprotection during and following ischemia–reperfusion. Finally, A_3AR have been demonstrated to be over-expressed in some tumor cell lines, thus suggesting this receptor as a potential target in cancer therapy (Borea *et al.*, 2009; Jacobson and Gao, 2006).

3. A_3 Adenosine Receptor Antagonists

In the past years, many efforts have been made for searching potent and selective human A_3 adenosine receptor (hA_3AR) antagonists (Jacobson and Gao, 2006; Jacobson *et al.*, 2009; Moro *et al.*, 2006a,b).

Historically, xanthines, such as caffeine and theofilline that are considered the natural antagonists for adenosine receptors, show in general very low affinity for the A_3AR subtype (high micromolar range). Nevertheless, very recent SAR studies on these compounds indicated that a cyclization between the 7- and 8- or 3- and 4-positions led to A_3AR antagonists, such as 2-phenylimidazopurin-5-one PSB-10, **1**. Other classes of extended xanthine structures have been reported as A_3AR antagonists such as triazolopurines (**2**) which proved to be potent and selective hA_3AR antagonists (Fig. 13.2).

A large amount of hA_3AR antagonists possess polyheterocyclic nucleus, which can be classified into six families of derivatives: (i) flavonoids; (ii) 1,4-dihydropyridines and pyridines; (iii) triazoloquinazolines; (iv) isoquinolines and quinazolines; (v) pyrazolo-triazolo-pyrimidines (PTPs), and (vi) various. In Figs. 13.2 and 13.3 are summarized the most representative members of these family of compounds, which have extensively reviewed (Jacobson and Gao, 2006; Jacobson *et al.*, 2009; Moro *et al.*, 2006a,b).

Optimization of flavonoid nucleus, through a classical structure–activity-relationship study led to MRS1067 (**3**), which proved to be the most potent (hA_3 561 n*M*) and selective compound of this series at hA_3AR subtype.

A very similar approach was utilized for studying dihydropyridines which are typically antagonists of the L-type calcium channel, but a combination of substitutions on the 1,4-dihydropyridines skeleton led to MRS1334 (**4**) which proved to be the most potent derivative of this series. Simultaneously, the same authors studied the affinity of the pyridines, derived from the oxidation of the corresponding 1,4-dihydropyridines, obtaining MRS1523 (**5**), which showed quite good affinity at the hA_3AR (18 n*M*) but could be considered the first derivative which possessed discrete affinity for the rat A_3AR. Acylation at the 5-position of the triazoloquinazoline derivative CGS 15943 (9-chloro-2-(2-furanyl)-[1,2,4]triazolo[1,5-*c*]quinazoline-5-amine), a classic unselective adenosine receptor antagonist, led to the discovery of MRS1220 (**6**) a highly potent (0.65 n*M*) and quite selective hA_3AR antagonist. In a screening program of compounds, quinazoline derivatives, such as VUF5574 (**7**), showed good affinity at hA_3AR while resulted to be ineffective at A_1 and A_{2A} receptor subtypes. Further library screenings led to novel heterocyclic A_3 antagonists which showed quite good affinity, such as L-249313 (**8**) and L-268605 (**9**).

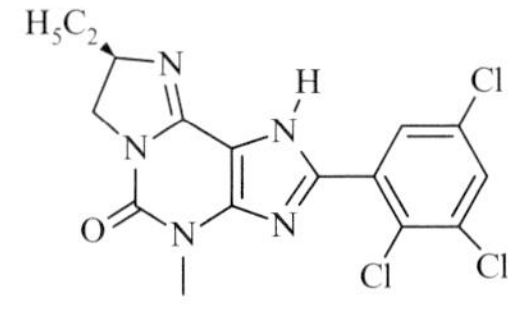

PSB-10, **1**

$hA_1 = 1700\,nM$
$hA_{2A} = 2700\,nM$
$hA_3 = 0.43\,nM$

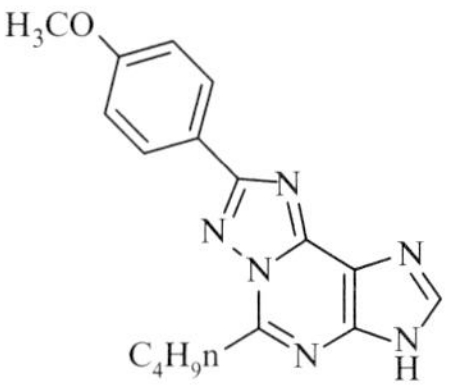

2

$hA_1 = 398\,nM$
$hA_{2A} = 822\,nM$
$hA_{2B} = 1030\,nM$
$hA_3 = 0.18\,nM$

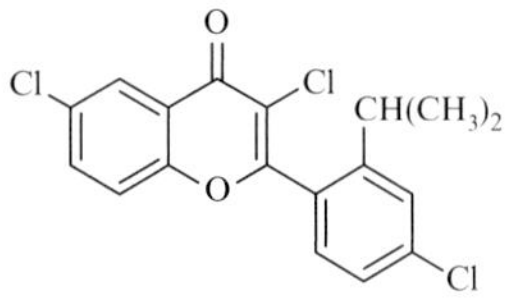

MRS 1067, **3**

$rA_1 = 36\%$ (% displ. at $10\,\mu M$)
$rA_{2A} = 19\%$ (% displ. at $10\,\mu M$)
$hA_3 = 561\,nM$

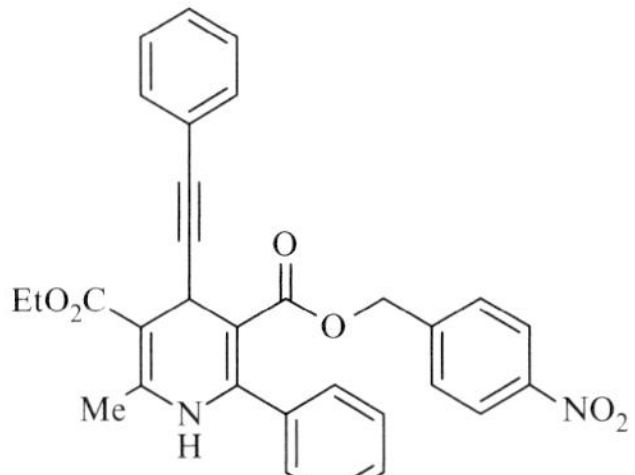

MRS 1334, **4**

$rA_1 = >100\,\mu M$
$rA_{2A} = >100\,\mu M$
$hA_3 = 2.69\,nM$

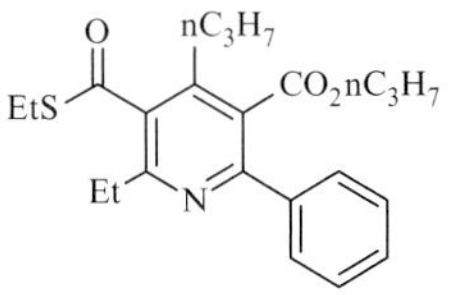

MRS 1523, **5**

$rA_1 = 15{,}600\,nM$
$rA_{2A} = 2050\,nM$
$rA_3 = 113\,nM$
$hA_3 = 18.9\,nM$

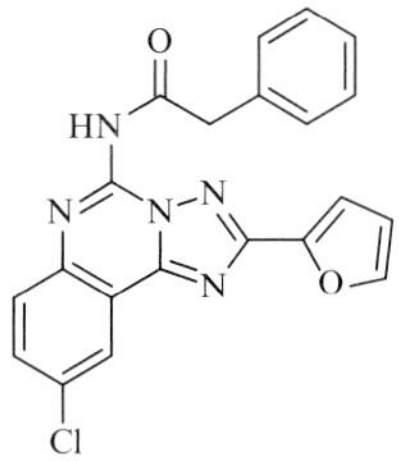

MRS 1220, **6**

$rA_1 = 52.7\,nM$
$rA_{2A} = 10.3\,nM$
$hA_3 = 0.65\,nM$

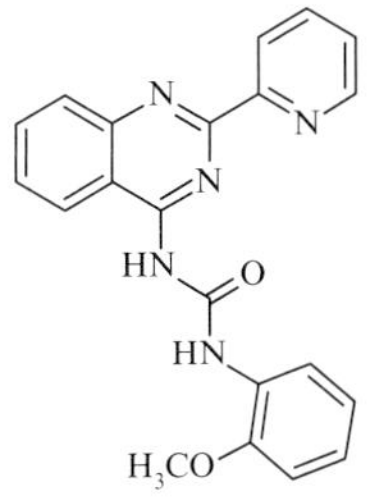

VUF 5574, **7**

$rA_1 = 3870\,nM$
$rA_{2A} = 22\%$ (10–5 *M*)
$hA_3 = 4\,nM$

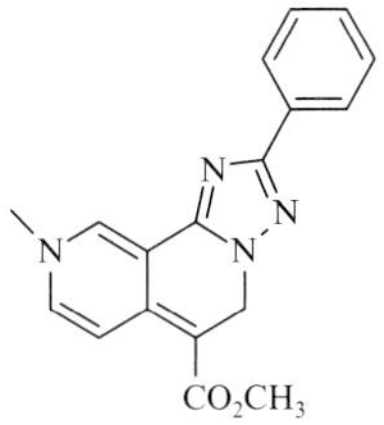

L-249313, **8**

$hA_1 = 4000\,nM$
$hA_{2A} = 19{,}000\,nM$
$hA_3 = 13\,nM$

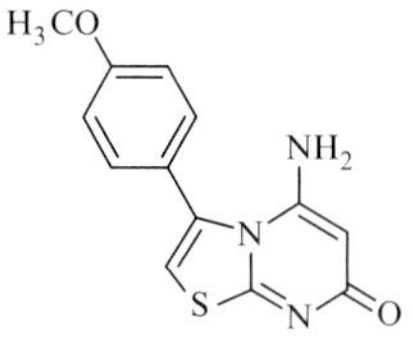

L-268605, **9**

$hA_1 = >10{,}000\,nM$
$hA_{2A} = >10{,}000\,nM$
$hA_3 = 18\,nM$

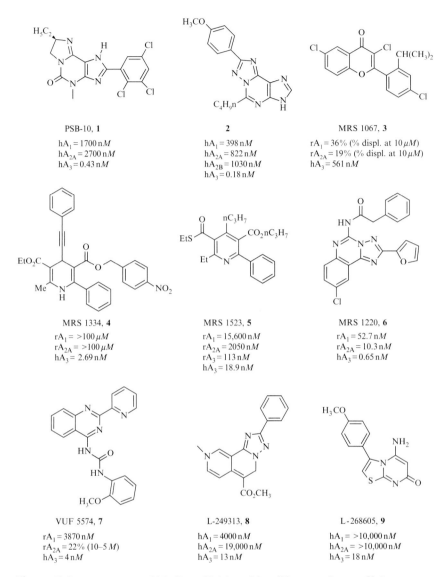

Figure 13.2 Structures and binding affinities of the all known classes of hA_3 receptor antagonists.

The best results in terms of potency and selectivity at the hA_3AR were obtained with the PTPs derivatives. In particular, introduction at the N5 position of a phenyl carbamoyl moiety and a methyl group at the N8 pyrazole nitrogen led to compound **10** which can be considered one of the most potent and selective hA_3AR antagonists ever reported.

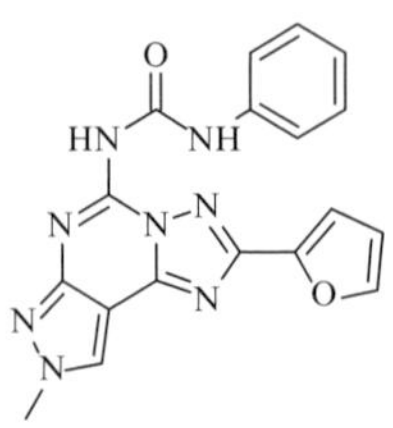

10

$hA_1 = 594\,nM$
$hA_{2A} = 381\,nM$
$hA_{2B} = 222\,nM$
$hA_3 = 0.16\,nM$

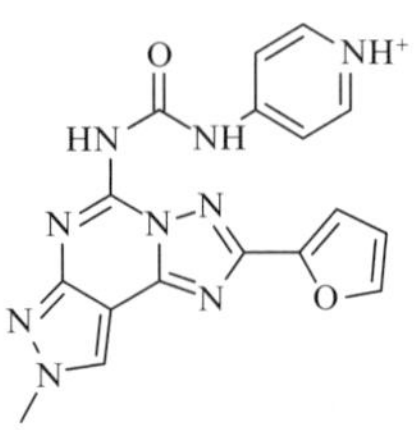

11

$hA_1 = 350\,nM$
$hA_{2A} = 100\,nM$
$hA_{2B} = 250\,nM$
$hA_3 = 0.01\,nM$

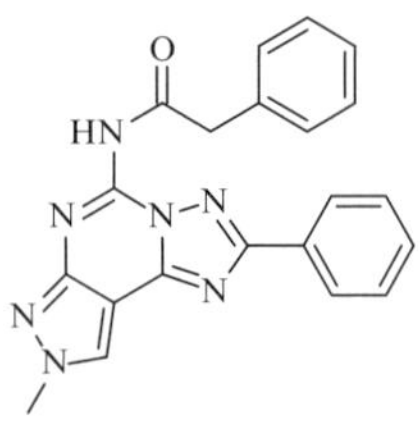

12

$hA_1 = 562\,nM$
$hA_{2A} = 778\,nM$
$hA_{2B} = >10{,}000$ (functional)
$hA_3 = 0.108\,nM$

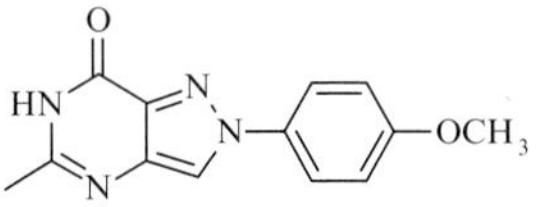

13

$hA_1 = 5\%$ (% displ. at $1\,\mu M$)
$hA_{2A} = 1\%$ (% displ. at $1\,\mu M$)
$hA_3 = 1.2\,nM$

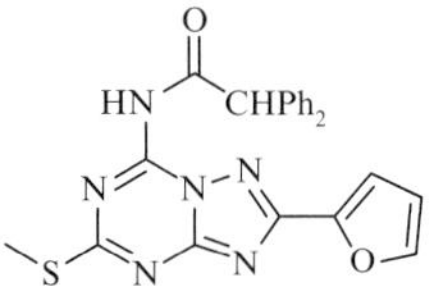

14

$rA_1 = 557\,nM$
$rA_{2A} = >10{,}000\,nM$
$hA_{2B} = >100{,}000\,nM$ (functional)
$hA_3 = 170\,nM$

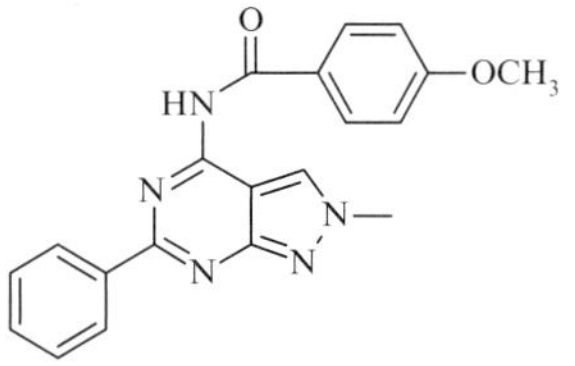

15

$hA_1 = 1037\,nM$
$hA_{2A} = 3179\,nM$
$hA_3 = 0.18\,nM$

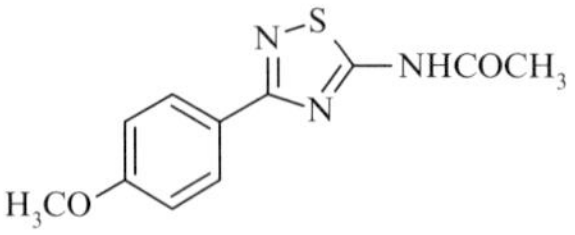

16

$hA_1 = 24\%$ (% displ. at $10\,\mu M$)
$hA_{2A} = 28\%$ (% displ. at $10\,\mu M$)
$hA_3 = 0.79$

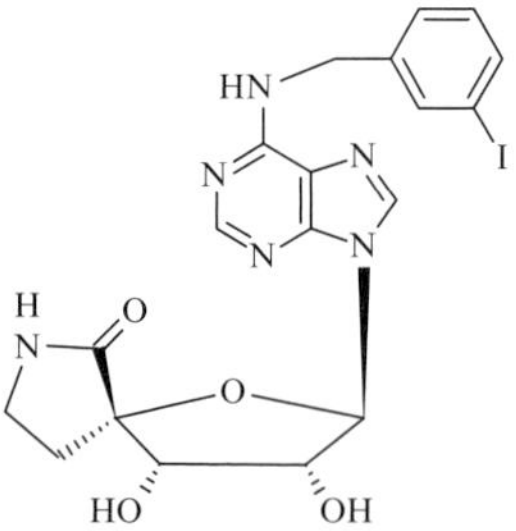

MRS 1292, **17**

$hA_3 = 29.3\,nM$
$rA_3 = 51\,nM$

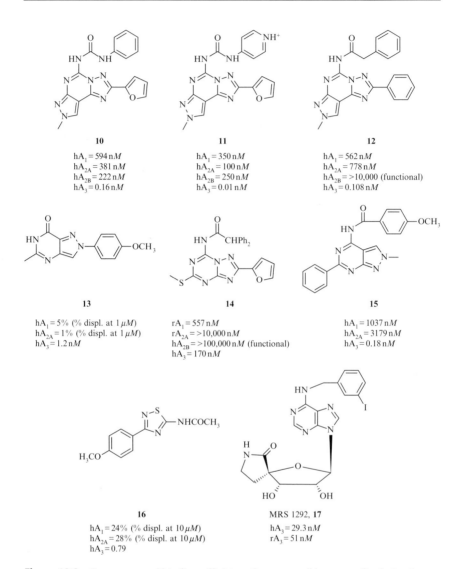

Figure 13.3 Structures and binding affinities of some novel heterocyclic derivatives as hA_3 receptor antagonists.

Interestingly, the bioisosteric substitution of the phenyl ring with a pyridinium salt led to a completely water soluble (15 m*M*) derivative **11** and with an increased affinity and selectivity for the hA_3AR (Jacobson and Gao, 2006; Jacobson *et al.*, 2009; Moro *et al.*, 2006a,b). An extended study on this class of compounds revealed that substitution of furyl ring at C2 position (a metabolically unstable group) with phenyl ring permit to retain potency at the hA_3AR with an increased selectivity versus the other receptors (compound **12**; Cheong *et al.*, 2010). Other derivatives structurally related to this family have been reported, in particular, the triazolo-quinoxalines. In this class, several compounds have been synthesized as antagonists for the different adenosine receptor subtypes (Jacobson and Gao, 2006; Jacobson *et al.*, 2009; Moro *et al.*, 2006a,b).

Structural simplified A_3AR antagonists have been reported, and in particular, bicyclic scaffolds were investigated. The molecular simplification of the pyrazolo-[3,4-*c*]quinolin-4-one (PQ) structure yields to a novel class of 2-arylpyrazolo[4,3-*d*]pyrimidin-7-one derivatives. These studies permitted to identify in compound **13**, a potent and highly selective hA_3AR antagonist (Lenzi *et al.*, 2009).

Interestingly, the [1,2,4]triazolo[1,5-*a*][1,3,5]triazine moiety was intensively studied toward A_{2A} adenosine receptor ($A_{2A}AR$) subtype but the introduction of a diphenylacetyl moiety at N7 position and a methylthio group at the 5-position (derivative **14**) gave an improvement of hA_3 affinity. Further modifications on this type of chemical scaffold can tune the spectrum of selectivity from A_{2A} to A_3AR subtype (Pastorin *et al.*, 2010). Furthermore, recent investigation on pyrazolo-pyrimidine scaffold led to compound **15** that shows a good affinity and selectivity toward A_3AR (Taliani *et al.*, 2010). Also other bicyclic compounds, such as the thiadiazole (derivative **16**) seem to be promising agent considering the very simple synthetic preparation and their low hydrophobic propensities (Moro *et al.*, 2006a,b).

Finally, it should be underlined that all the reported compounds showed significant potency and selectivity at the hA_3AR in human model, limiting studies *in vivo*. This aspect has been partially avoided working on the adenosine core. In fact, adenosine derivatives are considered for the presence of ribose moiety receptor agonists, but the introduction of extended substituents at 8 position or the ribose ring constrains (MRS1292, **17**) led to quite potent A_3AR antagonist also in a rat model (Gao *et al.*, 2002a,b).

4. Receptor-Based Antagonist Design

The investigation of human A_3AR and its ligands is rapidly growing with an increasing impact in the drug discovery process. The development of antagonists, in particular, for the A_3AR has been directed mainly by traditional medicinal chemistry, but the influence of structure-based approaches is increasing. The evolution of the field of computer-aided design of GPCR ligands, including A_3 agonists and antagonists, has depended on the availability of suitable molecular receptor templates (Martinelli and Tuccinardi, 2008; Moro *et al.*, 2006a,b).

Based on the assumption that GPCRs share similar overall topology, homology models of hA_3AR have been proposed by different groups (Martinelli and Tuccinardi, 2008; Morizzo *et al.*, 2009; Moro *et al.*, 2005). In fact, all family A GPCRs have in common a central core domain consisting of seven transmembrane helices (TM1–TM7) that are connected by three intracellular (IL1, IL2, and IL3) and three extracellular (EL1, EL2, and EL3) loops. Two cysteine residues (one in TM3 and one in EL2), which are conserved in most GPCRs belonging to class A, form a disulfide link. Aside from sequence variation, GPCRs differ in the length and function of their N-terminal extracellular domain, their C-terminal intracellular domain and their intra- and extracellular loops. Each of these domains provides very specific properties to these receptor proteins.

However, despite the enormous biomedical relevance of GPCRs, high resolution structural information on their active and inactive states is still lacking. In fact, being GPCRs integral membrane proteins, they are not easy to crystallize and hence to characterize through X-ray diffraction.

Rhodopsin had represented for many years the only 3D structural information available for GPCRs and rhodopsin-based homology modeling had been the most widely used approach to obtain 3D models of GPCRs, including A_3AR. The first highly resolved structure of rhodopsin was published by Palczewski *et al.* (2000). This 2.8 Å resolution structure of bovine rhodopsin (PDB code: 1F88) showed all major structural features of GPCRs as predicted from years of biochemical, biophysical, and bioinformatics studies.

Subsequently, crystallographic structures of other GPCRs have been reported such as the human β2-adrenergic receptor in 2007 (Cherezov *et al.*, 2007; Rosenbaum *et al.*, 2007) and the turkey β1-adrenergic receptor in 2008 (Warne *et al.*, 2008).

Finally, in 2008 the crystal structure of the human A_{2A} adenosine receptor ($hA_{2A}AR$) in complex with a high-affinity subtype-selective antagonist, ZM241385, has been determined (PDB code: 3EML; Jaakola *et al.*, 2008). Interestingly, to crystallize, the 2.60 Å resolution structure was applied the T4L fusion strategy, where most of the third cytoplasmic loop was replaced with lysozyme and the C-term tail was truncated from Ala317 to Ser412. This crystal structure presents three features different from previously reported GPCR structures. First of all, the EL2 is considerably different from β1/β2-adrenergic receptors and bovine/squid rhodopsins; in fact, it lacks any clearly secondary structural element and possesses three disulfide linkages, one with TM3 (Cys77–Cys166), that is conserved among most of the members of family A GPCRs, and two with EL1 (Cys71–Cys159 and Cys74–Cys146) that are unique to the $A_{2A}AR$. Moreover, the crystallographic structure of $hA_{2A}AR$ shows a disulfide bond between Cys259 and Cys262 in the EL3 of the receptor. The presence of all these bridges contributes to the formation of a disulfide bond network that creates a rigid, open structure exposing the ligand-binding cavity to solvent, possibly allowing free access for small molecule ligands (Jaakola *et al.*, 2008).

Second, ZM241385 is perpendicular to the membrane plane, colinear with TM7 and it interacts with both EL2 and EL3. The ligand position is significantly different from the position of retinal and amine ligands of adrenergic receptors. Finally, even if the helical arrangement is similar to other GPCRs, the binding pocket of the $A_{2A}AR$ is shifted closer to TM6 and TM7 and less interactions are allowed with TM3 and TM5 (Jaakola *et al.*, 2008). This means that even though GPCRs share a common topology, ligands may bind in a different fashion and interact with different positions of the receptor.

Following these structural considerations, it is clear that the recently published crystallographic structure of $hA_{2A}AR$ provide a new useful template to perform homology modeling of other family A GPCRs, and in particular of adenosine receptors. The first important step in the homology modeling protocol is related to the sequence alignment that is guided by the most conserved residues in each TM helix (at least one highly conserved residue). This peculiar residue is used as reference for the Ballesteros and Weinstein nomenclature system: every amino acid of TM regions is identified by a number that refers to the transmembrane segment of the GPCR, followed by a number that refers to the position relative to reference residue that has arbitrarily the number 50 (Asn1.50, Asp2.50, Arg3.50, Trp4.50, Pro5.50, Pro6.50, and Pro7.50 in TM1–7, respectively; Morizzo *et al.*, 2009).

The $A_{2A}AR$ can be considered the best template for homology modeling of hA_3AR according to the percentages of identity of the aligned sequences ($hA_3AR/hA_{2A}AR$ identity percentage $\approx$40%), but there are some important differences between these two adenosine receptor subtypes that have to be considered while building homology models (Morizzo *et al.*, 2009). In particular, the loops constitute the most variable region. Among them, EL2 is of great interest, while building homology models of GPCRs used for drug design, because of its possible role in the ligand recognition process.

The presence of three disulfide links on EL2 is a peculiarity of human $A_{2A}AR$. This is an important point that has to be considered when $A_{2A}AR$ is used as template for homology modeling of other adenosine receptors used then for drug design. In fact, the conformation of the binding pocket and the extracellular surface properties are influenced by the second extracellular loop, that is considerably different from that of other family A GPCRs because of the presence of three disulfide linkages (Morizzo *et al.*, 2009).

In contrast, the sequences of hA_3AR present only one cysteine residue in the EL2 (Cys166) and this residue forms the disulfide bridge common to family A GPCRs with the respective cysteine of TM3 (Cys83). Then, no other disulfide bonds involving extracellular loops can be formed in the hA_3AR structure (Morizzo *et al.*, 2009).

4.1. Mutagenesis analysis

The high resolution 3D structure of $hA_{2A}AR$ elucidated the characteristic of this receptor subtype and clarified the role of the amino acids involved in antagonist binding. Next to the structural information provided by the crystallographic structure, mutagenesis studies help to identify the residues that can be involved in the ligand recognition. It is interesting to compare the mutagenesis information available with the 3D information. Site-directed mutagenesis studies highlighted several residues important for antagonist binding at the $A_{2A}AR$.

Mutagenesis studies have involved the TM3 domain and specifically the mutation of Gln89 (3.37) with histidine or arginine. Experimentally, the mutant clearly affects the antagonist binding (Jiang *et al.*, 1996). At the same time, the investigation of EL2 revealed the important role of Glu169 and Glu151 whose mutations with alanine determine the loss of agonist and antagonist binding (Kim *et al.*, 1996).

Mutagenesis analysis of $A_{2A}AR$ anticipated the key role of TM5, TM6, and TM7 for agonist and antagonist interactions as successively confirmed by the $A_{2A}AR$ crystallographic structure. The substitution of Phe182 (5.43) with alanine determine the loss of agonist and antagonist binding (Kim *et al.*, 1995). In contrast, the mutation of Ser277 (7.42) determines only marginal changes in ligand binding (Kim, *et al.*,1995; Jiang, *et al.*,1996).

The important role of TM6 is supported by the fact that the mutations of His250 (6.52), Asn253 (6.55), and Phe257 (6.59) with alanine determine the loss of agonist and antagonist binding (Kim *et al.*, 1995). In the $A_{2A}AR$ crystal structure, TM7 clearly plays a key role, according to the previously reported mutagenesis data; in fact, the mutations of Ile274 (7.39), His278 (7.43), and Ser281 (7.46) with alanine determine the loss of agonist and antagonist binding (Kim *et al.*, 1995).

Therefore, many residues shown to be important for ligand binding in previously published mutagenesis studies, such as Glu169 (EL2), His250 (6.52), Asn253 (6.55), and Ile274 (7.39), were also shown to have important direct contacts with the bound ligand in the $A_{2A}AR$ crystal structure.

Recently, some residues shown to be important for the binding of ZM241385 in the crystal structure but for which no mutagenesis data have been previously reported, namely Phe168 (5.29), Met177 (5.38), and Leu249 (6.51), were mutated (Jaakola *et al.*, 2010). The results of these mutagenesis studies confirm the critical role of Phe168 (5.29) and Leu249 (6.51) interactions with antagonists such as ZM241385 as observed in the crystal structure. In fact, the mutations of Phe168 (5.29) and Leu249 (6.51) with alanine lead to loss of antagonist binding while the mutation of Met177 (5.38) with alanine reduce the binding affinity of ZM241385 (Jaakola *et al.*, 2010).

Moreover, considering a multiple sequence alignment of adenosine receptor subtypes, it was noted that the lower transmembrane part of the ZM241385-binding cavity is highly conserved among adenosine subtypes, and the extracellular domains and upper part of the ZM241385 binding site are somewhat less conserved (Jaakola *et al.*, 2010). This observation suggests that the interactions that determine subtype selectivity reside in the more divergent "upper" region of the binding cavity.

Considering the human A_3AR, site-directed mutagenesis studies show an important role of TM3, TM5, TM6, and TM7 for interactions of both agonists and antagonists. The mutations of His95 (3.37), Met177 (5.38), Val178 (5.39), Ser271 (7.42), His272 (7.43), and Asn274 (7.45) with alanine determine a decrease of affinity of both agonists and antagonists in human A_3AR subtype (Gao *et al.*, 2003). The substitution of Asn250 (6.55) with alanine causes the loss of binding of agonists and antagonists (Gao *et al.*, 2002a,b).

Antagonist binding is affected also by the substitutions of Thr94 (3.36) (Gao *et al.*, 2003), Lys152 (EL2) (Gao *et al.*, 2002a,b, 2003), and Phe182 (5.43) with alanine (Gao *et al.*, 2003). Finally, the mutation of Tpr243 (6.48), residue conserved among all four subtypes of adenosine receptors and a variety of other GPCRs, with alanine or phenylalanine determines a substantial decrease of antagonists affinity (Gao *et al.*, 2002a,b, 2003).

Therefore, as shown in Fig. 13.4, the comparison of available mutagenesis data for amino acids affecting antagonist binding on hA_{2A} and hA_3 adenosine receptor subtypes shows that the binding pocket of hA_3 receptor can be slightly different from the one of hA_{2A} subtype.

More precisely, some residues whose site-directed mutagenesis affects antagonist binding at the hA_{2A}AR, namely Phe168 (EL2), Leu249 (6.51), and Ile268 (7.39), are conserved at the corresponding positions in the hA_3 subtype. Therefore, although no mutagenesis data are available for these residues at the hA_3AR, they could be crucial for ligand binding at this subtype too.

Nevertheless, it can be noted that Glu169 (EL2), His250 (6.52), and Phe257 (6.59), whose mutation with alanine also affects antagonist binding at hA_{2A}AR, are not present in the corresponding position of the hA_3 receptor, where these residues are replaced by Val169 (EL2), Ser247 (6.52), and Tyr254 (6.59), respectively. These particular amino acids could play a key role in determining the selectivity profile of adenosine antagonists.

In particular, the difference in position 169 of EL2 was supposed to influence not only the binding mode but also the entrance of ligands to the TM region of the receptors (Lenzi *et al.*, 2009).

In addition, the electrostatic potentials of amino acids present at the binding pocket entrance from the extracellular site are very different in the two receptors. In the hA_3AR model, the binding pocket gate is surrounded essentially by side chains of hydrophobic residues including Phe168 (EL2),

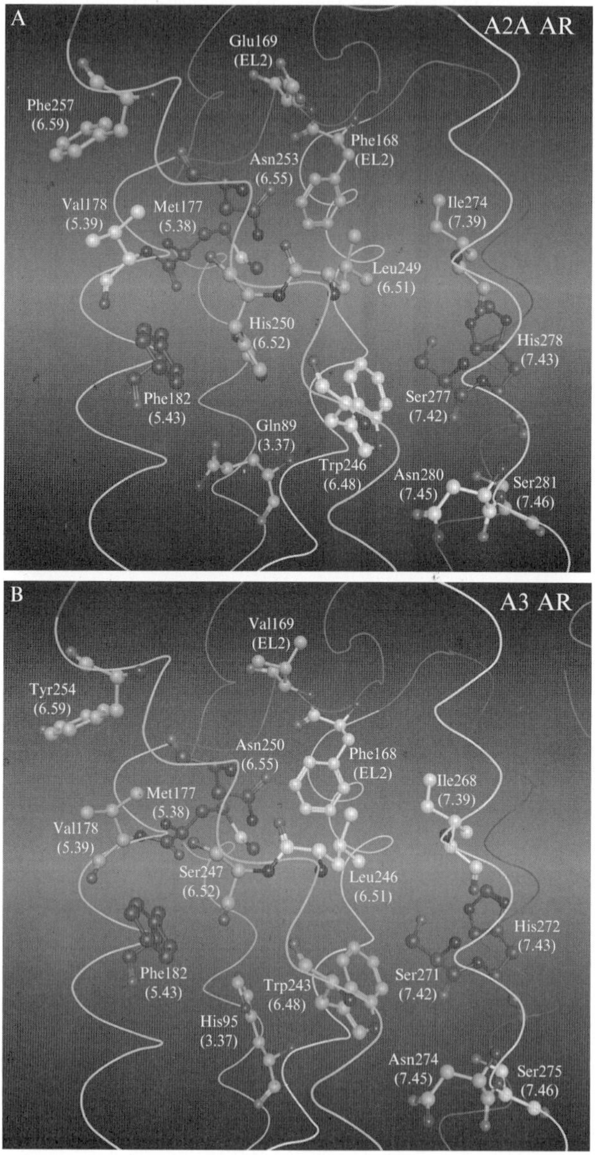

Figure 13.4 Comparison of available mutagenesis data for amino acids affecting antagonists binding on (A) hA_{2A} and (B) hA_3 adenosine receptor subtypes. Amino acids color legend: *light green*: mutagenesis data available for one subtype; *dark green*: mutagenesis data available for both subtypes; *light pink*: conserved residues whose mutagenesis data are available only for the other subtype; *dark pink*: not conserved residues whose mutagenesis data are available only for the other subtype. (See Color Insert.)

Val169 (EL2), Ile253 (6.58), Val259 (EL3), and Leu264 (7.35). In the $hA_{2A}AR$ crystal structure, there are ionic residues, such as Glu169 (EL2) and His264 (EL3), among the amino acids delimiting the binding site access. It was hypothesized that the presence of this charged gate may affect both the ligand orientation, while approaching the binding pocket, and its accommodation into the final TM binding cleft (Lenzi *et al.*, 2009).

4.2. Human A_3 adenosine receptor models

Homology models of the A_3AR have been helpful in providing structural hypothesis for the design of new ligands. Many hA_3AR models, built using different templates, have been published describing the hypothetical interactions with known A_3AR ligands with different chemical scaffolds. In particular, the recent publication of the $hA_{2A}AR$ provides important structural information for the adenosine receptor family.

Based on the assumption that GPCRs share similar TM boundaries and overall topology, a homology model of the hA_3AR was constructed, as previously reported (Lenzi *et al.*, 2009; Martinelli and Tuccinardi, 2008; Morizzo *et al.*, 2009), using as template the recently published crystal structure of hA_{2A} receptor.

First, the amino acid sequences of TM helices of the hA_3AR were aligned with those of the template, guided by the highly conserved amino acid residues, including the DRY motif (Asp3.49, Arg3.50, and Tyr3.51) and three proline residues (Pro4.60, Pro6.50, and Pro7.50) in the TM segments of GPCRs. The same boundaries were applied for the TM helices of hA_3AR as they were identified from the 3D structure for the corresponding sequences of the template, the coordinates of which were used to construct the seven TM helices for hA_3AR. Then, the loop domains were constructed on the basis of the structure of compatible fragments found in the Protein Data Bank.

The structure-based drug design approach using hA_3AR models is mainly affected by differences in EL2, because residues of this loop can directly interact with ligands in the binding pocket; therefore special caution has to be given to this extracellular loop. A driving force to the peculiar fold of the EL2 loop might be the presence of a disulfide bridge between cysteines in TM3 and EL2. Since this covalent link is conserved in both hA_{2A} and hA_3 receptors, the EL2 loop was modeled using a constrained geometry around the EL2-TM3 disulfide bridge. In particular, Cys166 (EL2) and Cys77 (3.25) of the hA_{2A} receptor were constrained, respectively, with Cys166 (EL2) and Cys83 (3.25) of the hA_3 receptor (Lenzi *et al.*, 2009).

The helical arrangement of the model built starting from the $hA_{2A}AR$ template is similar to hA_3AR models built using different templates, but the helices are slightly shifted. The main difference among different models is in the loop region.

Comparing the model built starting from the $A_{2A}AR$ template with the previous models of A_3AR, it can be noted that the binding pocket is closer to TM6 and TM7 and open to the extracellular side. Therefore, the volume of the binding site is difficult to be measured and was estimated to be 1930 $Å^3$ (Morizzo *et al.*, 2009).

Even if the percentage of identity of the hA_3AR is higher with respect to the $A_{2A}AR$ than with the previously reported crystallographic structures, the conformation of the EL2 and consequently of the binding pocket of the hA_3AR might be different from the $A_{2A}AR$. The peculiarity of the $A_{2A}AR$ is the presence of three disulfide bridges on EL2, which are not conserved among ARs. Also, mutagenesis data support the hypothesis of different roles of TM helices in different AR subtypes.

Therefore, the use of the $A_{2A}AR$ as template for the homology modeling of other AR subtypes, and in particular of hA_3AR, is a powerful technique but for some aspects is still imprecise. Especially, more efforts are necessary to elucidate the correct topological organization of the EL2 and its role in the recognition of both agonists and antagonists.

4.3. Docking studies of A_3 adenosine receptor antagonists

In order to rationalize the structure–affinity relationships and the selectivity profiles of new hA_3AR antagonists, a receptor-driven molecular modeling investigation, based on the previously proposed model of hA_3 receptor derived from the crystallographic structure of human A_{2A} receptor, was performed (Lenzi *et al.*, 2009).

In the process of selecting a reliable docking protocol to be employed in the docking studies of new hA_3AR antagonists, the ability of different docking softwares in reproducing the crystallographic pose of ZM241385 inside the binding cavity of hA_{2A} receptor was evaluated (Lenzi *et al.*, 2009; Pastorin *et al.*, 2010).

Among the four different docking suites, with different docking algorithms and scoring functions, used to calibrate the docking protocol, the GOLD program was finally chosen because it showed the best performance with regard to the calculated rmsd values relative to the crystallographic pose of ZM241385 (Lenzi *et al.*, 2009; Pastorin *et al.*, 2010).

Consequently, based on the selected docking protocol, docking simulations were performed to identify the hypothetical binding mode of the new series of adenosine receptor antagonists inside both the binding site of the hA_3AR model and the binding site of the $hA_{2A}AR$ crystal structure.

Then, to analyze in a more quantitative way the possible ligand–receptor recognition mechanism, the individual electrostatic (ΔE_{int}^{el}) and hydrophobic (ΔE_{int}^{hyd}) contributions to the interaction energy (ΔE_{int}) of each receptor residue have been calculated.

Hypothetical binding modes, obtained after docking simulations, of selected hA_3AR antagonists are shown in Fig. 13.5.

Among the several diverse structures that have demonstrated affinity at the hA_3AR, there are compounds bearing a PTP nucleus. The exploitation of substituents, mainly at positions N8 and N5 of such structure, has given rise to highly potent and moderately selective hA_3 antagonists. Recently, the bioisosteric replacement of the furan ring at the C2 position with a phenyl ring has led to the identification of new series of 2-(*para*-substituted)phenyl-pyrazolo-triazolo-pyrimidine derivatives as hA_3 antagonists with good affinity and remarkably improved selectivity profile toward the other adenosine receptor subtypes in comparison to the 2-furyl PTP derivatives (Cheong *et al.*, 2010). The derivative of this class with

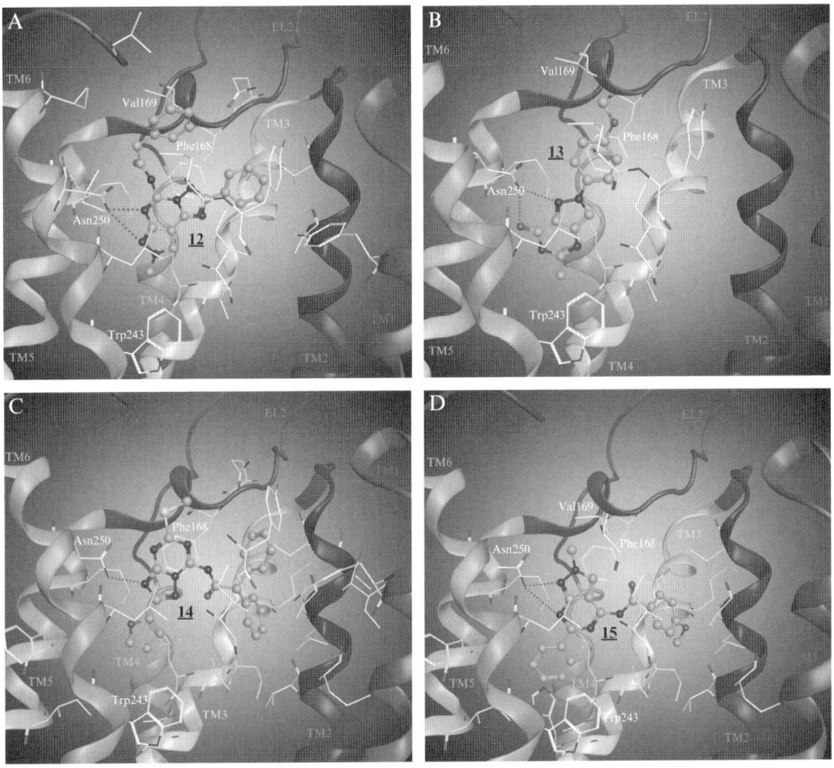

Figure 13.5 Hypothetical binding modes, obtained after docking simulations inside the hA_3AR binding site, of (A) compound **12**; (B) compound **13**; (C) compound **14**; (D) compound **15**. Poses are viewed from the membrane side facing TM6, TM7, and TM1. The view of TM7 is voluntarily omitted. Side chains of some amino acids important for ligand recognition and H-bonding interactions are highlighted. Hydrogen atoms are not displayed. (See Color Insert.)

highest affinity for the hA_3 receptor subtypes is compound **12** (K_i $hA_3AR = 0.108$ n*M*).

As shown in Fig. 13.5A, for compound **12** recognition at hA_3 receptor occurs in the upper region of the TM bundle, and the PTP scaffold is surrounded by TMs 3, 5, 6, and 7 with the 2-aryl ring oriented toward TM2 and the substituent at N8 located deep into the binding cavity. At the hA_3 receptor, compounds **12** forms two stabilizing H-bonding interactions with Asn250 (6.55), a π–π stacking interaction between the triazole ring and Phe168 (EL2) and hydrophobic interactions with several residues of the binding site including Ala69 (2.61), Val72 (2.64), Thr87 (3.29), Leu90 (3.32), Leu91 (3.33), Phe168 (EL2), Val169 (EL2), Trp243 (6.48), Leu246 (6.51), Ile249 (6.54), Ile253 (6.58), Val259 (EL3), Leu264 (7.35), Tyr265 (7.36), and Ile268 (7.39).

Other human A_3AR antagonists possessing a tricyclic scaffold are the pyrazolo[3,4-*c*]quinolin-4-one (PQ) derivatives. The PQ derivatives show high affinities for the hA_3 receptor and also high hA_3 versus hA_{2A} selectivity. A molecular simplification of the PQ structure yielded the 2-arylpyrazolo [4,3-*d*]pyrimidin-7-one derivatives, such as compound **13** (Lenzi *et al.*, 2009). This compound displays hA_3AR affinity in the low nanomolar range (K_i $hA_3AR = 1.2$ n*M*) and is totally inactive at the other three adenosine receptor subtypes. From the docking simulation analysis, it resulted that for compound **13** ligand recognition occurs in the upper region of the TM region of the hA_3AR (Fig. 13.5B). The pyrazolo[4,3-*d*] pyrimidin-7-one scaffold is surrounded by TMs 3, 5, 6, 7 with the 2-phenyl ring pointing toward EL2 and the substituent at the 5-position oriented toward the intracellular environment. Compound **13** is anchored, inside the binding cleft, by two stabilizing hydrogen-bonding interactions with the amide moiety of Asn250 (6.55) side chain and it forms hydrophobic interactions with many residues of the binding cleft including Leu90 (3.32), Leu91 (3.33), Phe168 (EL2), Val169 (EL2), Met177 (5.38), Trp243 (6.48), Leu246 (6.51), Leu264 (7.35), Tyr265 (7.36), and Ile268 (7.39). In particular, the 2-phenyl ring forms an aromatic π–π stacking interaction with Phe168 (EL2), while the pyrazolo[4,3-*d*]pyrimidin-7-one core interacts with the highly conserved Trp243 (6.48), an important residue in receptor activation and in antagonist binding.

The 5,7-disubstituted-[1,2,4]triazolo[1,5-*a*][1,3,5]triazine derivatives, which were designed as simplified analogs of previously reported hA_{2A} and hA_3 antagonists, constitute a new class of promising AR antagonists. To this class of derivatives belongs compound **14** (K_i $hA_3AR = 170$ n*M*). Critical interactions seen in the hypothetical binding pose of this compound inside the TM cavity of hA_3AR (Fig. 13.5C) are one hydrogen-bonding interaction with the side chain of Asn250 (6.55) and a π–π stacking interaction with Phe168 (EL2). In addition, the furan ring is located deep in the binding cavity and interacts with Trp243 (6.48) while the diphenylacetyl group is directed toward a hydrophobic pocket delimited by Val65 (2.57), Ala69 (2.61), Val72 (2.64), Leu90 (3.32), Tyr265 (7.36), and Ile268 (7.39).

Finally, Fig. 13.5D displays the docking pose at the hA_3AR obtained for a compound bearing a pyrazolo-pyrimidine scaffold (compound **15**, K_i $hA_3AR = 0.18$ n*M*; Taliani *et al.*, 2010). This binding mode shows some interactions already seen for the other compounds analyzed, such as two H-bonds with Asn250 (6.55) and a π–π stacking interaction with Phe168 (EL2). Moreover, some hydrophobic interactions contribute to stabilize the ligand-binding pose; in particular, the 6-phenyl ring of the ligand interacts with the highly conserved Trp243 (6.48), while the methoxyphenyl group interacts with several hydrophobic residues, including Val65 (2.57), Leu68 (2.60), Ala69 (2.61), Val72 (2.64), Leu90 (3.32) and Ile268 (7.39).

By analyzing the calculated individual electrostatic contribution to the interaction energy of each receptor residue for all the considered binding poses (collected in Fig. 13.6), it is evident that the Asn250 (6.55) strongly stabilized the ligand–hA_3AR complex (negative electrostatic interaction energy) due to the hydrogen-bonding interactions. This asparagine residue 6.55, conserved among all AR subtypes, was already found to be important for ligand binding at both the hA_3 and hA_{2A} ARs. Therefore, from the electrostatic point of view, one of the most critical residues affecting the affinity at hA_3AR seems to be Asn250 (6.55).

Docking simulations at the $hA_{2A}AR$ binding site were also performed for these series of hA_3AR antagonists. Comparing the binding poses at the $hA_{2A}AR$ with those obtained at the hA_3AR, the selectivity profile of these antagonists was explained. In particular, the mutation of the valine at the position 169 (EL2) with a glutamate was hypothesized to be critical for the hA_3 versus hA_{2A} selectivity profile of some antagonists (Lenzi *et al.*, 2009; Cheong *et al.*, 2010).

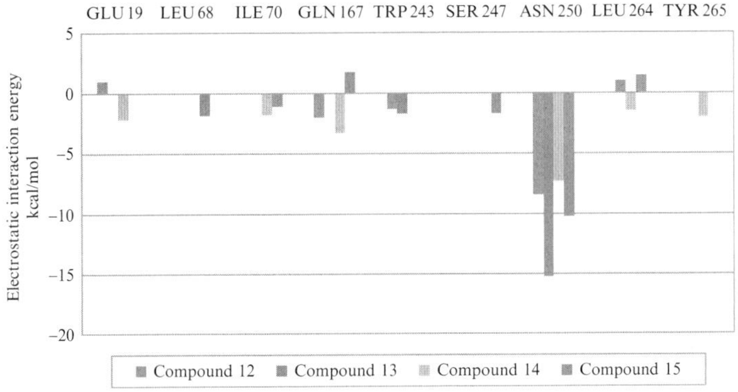

Figure 13.6 Electrostatic interaction energy (in kcal/mol) between compounds 12, 13, 14, and 15 and particular hA_3AR residues involved in ligand recognition. (See Color Insert.)

In conclusion, the analysis of the docking results permitted to identify some structural features of hA_3AR antagonists important for the binding at this receptor subtype (see Fig. 13.7).

In fact, the planar, bicyclic, or tricyclic, scaffold of these ligands seems to play a key role in the binding process due to the interaction with Phe168 (EL2). Moreover, the presence of one or more H-bonds acceptor groups allow these ligands to interact with the amide moiety of Asn250 (6.55) side chain through stabilizing H-bonding interactions. Three additional hydrophobic groups in the ligand structures contribute to the stability of the ligand–receptor complex. In particular, one hydrophobic/aromatic group can interact with the highly conserved Trp243 (6.48); the second one can interact with several hydrophobic residues at the entrance of the binding site, including Phe168 (EL2), Val169 (EL2), Ile249 (6.54), Ile253 (6.58), Val259 (EL3), and Leu264 (7.35); and the third one can occupy a hydrophobic side cleft delimited by Val65 (2.57), Leu68 (2.60), Ala69 (2.61), Val72 (2.64), Leu90 (3.32), Tyr265 (7.36), and Ile268 (7.39). Therefore, all these considerations can guide the rational design of new class hA_3AR antagonists.

Concluding, GPCRs represent a fabulous assembly of nanomachines. The insightful understanding of their structures and functions is one of the most challenging objectives of all biosciences. However, several important

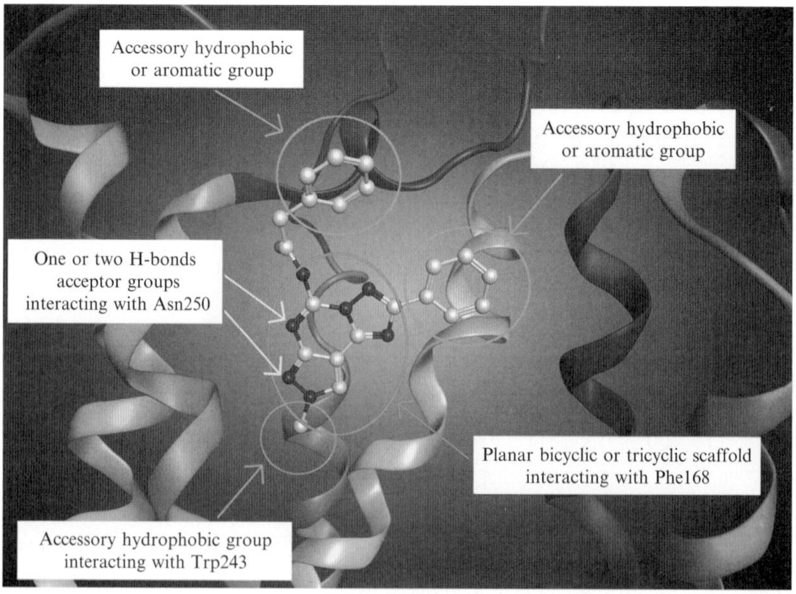

Figure 13.7 Structural features of hA_3 adenosine receptor antagonists important for the interaction with the hA_3AR binding site.

questions will require additional structural information at higher resolution. From a drug discovery point of view, one of the most intriguing is "what is the precise structural basis of ligand specificity in a particular receptor, and how can the basic seven-helical structure be tuned to bind such a large and chemically diverse spectrum of ligands?" We do believe that the successful answer to these questions will ultimately result from a synergistic interaction between theory and experiment, and we consider our finding in adenosine receptor subtypes a nice example of the applicability of this integrated approach.

ACKNOWLEDGMENTS

This work was funded by MIUR Cofin PRIN 2008. S. M. is also very grateful to Chemical Computing Group for the scientific and technical partnership.

REFERENCES

Borea, P. A., Gessi, S., Bar-Yehuda, S., and Fishman, P. (2009). A3 adenosine receptor: pharmacology and role in disease. *Handb. Exp. Pharmacol.* **193,** 297–327.

Cheong, S. L., Dolzhenko, A., Kachler, S., Paoletta, S., Federico, S., Cacciari, B., Dolzhenko, A., Klotz, K. N., Moro, S., Spalluto, G., and Pastorin, G. (2010). The significance of 2-furyl ring substitution with a 2-(para-substituted) aryl group in a new series of pyrazolo-triazolo-pyrimidines as potent and highly selective hA(3) adenosine receptors antagonists: new insights into structure-affinity relationship and receptor-antagonist recognition. *J. Med. Chem.* **53,** 3361–3375.

Cherezov, V., Rosenbaum, D. M., Hanson, M. A., Rasmussen, S. G., Thian, F. S., Kobilka, T. S., Choi, H. J., Kuhn, P., Weis, W. I., Kobilka, B. K., and Stevens, R. C. (2007). High-resolution crystal structure of an engineered human beta2-adrenergic G protein-coupled receptor. *Science* **318,** 1258–1265.

Gao, Z. G., Kim, S. K., Biadatti, T., Chen, W., Lee, K., Barak, D., Kim, S. G., Johnson, C. R., and Jacobson, K. A. (2002a). Structural determinants of A(3) adenosine receptor activation: nucleoside ligands at the agonist/antagonist boundary. *J. Med. Chem.* **45,** 4471–4484.

Gao, Z. G., Chen, A., Barak, D., Kim, S. K., Muller, C. E., and Jacobson, K. A. (2002b). Structural determinants of A(3) adenosine receptor activation: nucleoside ligands at the agonist/antagonist boundary. *J. Biol. Chem.* **277,** 19056–19063.

Gao, Z. G., Kim, S. K., Gross, A. S., Chen, A., Blaustein, J. B., and Jacobson, K. A. (2003). Structural determinants of A(3) adenosine receptor activation: nucleoside ligands at the agonist/antagonist boundary. *Mol. Pharmacol.* **63,** 1021–1031.

Jaakola, V. P., Griffith, M. T., Hanson, M. A., Cherezov, V., Chien, E. Y., Lane, J. R., Ijzerman, A. P., and Stevens, R. C. (2008). *Science* **322,** 1211–1217.

Jaakola, V. P., Lane, J. R., Lin, J. Y., Katritch, V., Ijzerman, A. P., and Stevens, R. C. (2010). *J. Biol. Chem.* **285,** 13032–13044.

Jacobson, K. A., and Gao, Z. G. (2006). Adenosine receptors as therapeutic targets. *Nat. Rev. Drug Discov.* **5,** 247–264.

Jacobson, K. A., Klutz, A. M., Tosh, D. K., Ivanov, A. A., Preti, D., and Baraldi, P. G. (2009). Medicinal chemistry of the A3 adenosine receptor: agonists, antagonists, and receptor engineering. *Handb. Exp. Pharmacol.* **193,** 123–159.

Jiang, Q., Van Rhee, A. M., Kim, J., Yehle, S., Wess, J., and Jacobson, K. A. (1996). *Mol. Pharmacol.* **50,** 512–521.

Kim, J., Wess, J., van Rhee, A. M., Schoneberg, T., and Jacobson, K. A. (1995). *J. Biol. Chem.* **270,** 13987–13997.

Kim, J., Jiang, Q., Glashofer, M., Yehle, S., Wess, J., and Jacobson, K. A. (1996). *Mol. Pharmacol.* **49,** 683–691.

Lenzi, O., Colotta, V., Catarzi, D., Varano, F., Poli, D., Filacchioni, G., Varani, K., Vincenzi, F., Borea, P. A., Paoletta, S., Morizzo, E., and Moro, S. (2009). 2-Phenylpyrazolo[4,3-d]pyrimidin-7-one as a new scaffold to obtain potent and selective human A3 adenosine receptor antagonists: new insights into the receptor-antagonist recognition. *J. Med. Chem.* **52,** 7640–7652.

Martinelli, A., and Tuccinardi, T. (2008). Molecular modeling of adenosine receptors: new results and trends. *Med. Res. Rev.* **28,** 247–277.

Morizzo, E., Federico, S., Spalluto, G., and Moro, S. (2009). Human A3 adenosine receptor as versatile G protein-coupled receptor example to validate the receptor homology modeling technology. *Curr. Pharm. Des.* **15,** 4069–4084.

Moro, S., Spalluto, G., and Jacobson, K. A. (2005). Techniques: Recent developments in computer-aided engineering of GPCR ligands using the human adenosine A3 receptor as an example. *Trends Pharmacol. Sci.* **26,** 44–51.

Moro, S., Gao, Z. G., Jacobson, K. A., and Spalluto, G. (2006a). Progress in the pursuit of therapeutic adenosine receptor antagonists. *Med. Res. Rev.* **26,** 131–159.

Moro, S., Deflorian, F., Bacilieri, M., and Spalluto, G. (2006b). Novel strategies for the design of new potent and selective human A3 receptor antagonists: an update. *Curr. Med. Chem.* **13,** 639–645.

Palczewski, K., Kumasaka, T., Hori, T., Behnke, C. A., Motoshima, H., Fox, B. A., Le Trong, I., Teller, D. C., Okada, T., Stenkamp, R. E., Yamamoto, M., and Miyano, M. (2000). Crystal structure of rhodopsin: A G protein-coupled receptor. *Science* **289,** 739–745.

Pastorin, G., Federico, S., Paoletta, S., Corradino, M., Cateni, F., Cacciari, B., Klotz, K. N., Gao, Z. G., Jacobson, K. A., Spalluto, G., and Moro, S. (2010). Synthesis and pharmacological characterization of a new series of 5,7-disubstituted-[1,2,4]triazolo[1,5-a][1,3,5] triazine derivatives as adenosine receptor antagonists: A preliminary inspection of ligand-receptor recognition process. *Bioorg. Med. Chem.* **18,** 2524–2536.

Rosenbaum, D. M., Cherezov, V., Hanson, M. A., Rasmussen, S. G., Thian, F. S., Kobilka, T. S., Choi, H. J., Yao, X. J., Weis, W. I., Stevens, R. C., and Kobilka, B. K. (2007). GPCR engineering yields high-resolution structural insights into beta2-adrenergic receptor function. *Science* **318,** 1266–1273.

Taliani, S., La Motta, C., Mugnaini, L., Simorini, F., Salerno, S., Marini, A. M., Da Settimo, F., Cosconati, S., Cosimelli, B., Greco, G., Limongelli, V., Marinelli, L., *et al.* (2010). Novel N2-substituted pyrazolo[3,4-d]pyrimidine adenosine A3 receptor antagonists: inhibition of A3-mediated human glioblastoma cell proliferation. *J. Med. Chem.* **53,** 3954–3963.

Warne, T., Serrano-Vega, M. J., Baker, J. G., Moukhametzianov, R., Edwards, P. C., Henderson, R., Leslie, A. G., Tate, C. G., and Schertler, G. F. (2008). Structure of a beta1-adrenergic G-protein-coupled receptor. *Nature* **454,** 486–491.

CHAPTER FOURTEEN

Inverse Agonist Activity of Steroidogenic Factor SF-1

Fabrice Piu* *and* Andria L. Del Tredici†

Contents

Abstract

Steroidogenic factor 1 (SF-1) is a key regulator of endocrine function, especially steroidogenesis and reproduction. Unlike most nuclear receptors, SF-1 is constitutively activated and still remains an orphan receptor. To study its function, it is imperative to have reliable assays that can assess potential pharmacological modulators. Here we describe in detail three different cell-based assays that evaluate distinct aspects of SF-1 function: a cellular proliferation assay R-SAT® that monitors events far downstream of the receptor/ligand interaction, a

* Otonomy Inc., San Diego, California, USA
† Trianode, Inc., San Diego, California, USA

Methods in Enzymology, Volume 485
ISSN 0076-6879, DOI: 10.1016/S0076-6879(10)85014-1

transcriptional assay that focuses on the gene-modulating properties of SF-1, and an assay in adrenocortical cultures that constitutes a surrogate measure of SF-1 function in native tissues.

1. Introduction

Receptors are sometimes capable of producing biological responses in the absence of a bound ligand. This pharmacological property is known as constitutive activity. Unlike an antagonist, which inhibits the activation of receptors by an agonist ligand, an inverse agonist is a bioactive substance that blocks the constitutive activity displayed by a receptor.

Steroidogenic factor 1 (SF-1, AD4BP, NR5A1) belongs to the family of nuclear hormone receptors and has emerged in recent years as a key regulator of endocrine function. It is expressed primarily in steroidogenic tissues such as testes and ovaries (Ikeda *et al.*, 1994; Morohashi *et al.*, 1992). SF-1 modulates the transcriptional activity of many genes involved in steroidogenesis and reproduction, especially steroidogenic acute regulatory protein (StAR) and cytochrome P450 steroid hydroxylases (reviewed in Hoivik *et al.*, 2010; Parker and Schimmer, 1997).

Unlike most members of the nuclear receptor family, SF-1 is an orphan receptor that is constitutively activated. In various cell systems, heterologously expressed SF-1 modulates the transcription of target genes in the absence of a ligand (Ito *et al.*, 1998). Several additional lines of investigation have confirmed the constitutively activated nature of SF-1. For instance, non-ligand modes of regulation have been reported, including tissue-specific repressors and phosphorylation of the hinge region (Desclozeaux *et al.*, 2002; Ito *et al.*, 1998). Furthermore, the crystal structure of the ligand-binding domain revealed that SF-1, when bound to a coactivator-derived peptide in the absence of a ligand, adopts a transcriptionally active conformation (Krylova *et al.*, 2005; Li *et al.*, 2005; Madauss *et al.*, 2004; Wang *et al.*, 2005). Interestingly enough, more recent investigations have uncovered a large binding pocket that is filled by a variety of phospholipids, especially phosphatidyl inositol bis- and triphosphates and C12–C16 fatty acids including sphingosine (Krylova *et al.*, 2005; Ortlund *et al.*, 2005; Urs *et al.*, 2007; Wang *et al.*, 2005). Finally, it has been demonstrated that synthetic SF-1 inverse agonists lead to the pharmacological modulation of steroidogenic enzymes (Del Tredici *et al.*, 2008).

Thus, there is a critical need for reliable means of investigating the pharmacological properties of SF-1 and potential ligands, especially their constitutive and inverse agonist activities. Here, the detailed procedural methods for several cell-based assays are presented.

2. SF-1 Inverse Agonism in the R-SAT® Assay of Cellular Proliferation

The Receptor Selection and Amplification Technology (R-SAT®) assay is a highly coupled, functional assay that measures receptor-dependent cellular proliferation (reviewed in Burstein *et al.*, 2006). The technology has been validated for many receptor families, including G-protein coupled receptors, receptor tyrosine kinases, cytokine receptors, and nuclear receptors (Piu *et al.*, 2002). Compared to other assays, the R-SAT® assay has a long 6-day incubation time, and the output measured, proliferation, is far downstream of the initial ligand-receptor binding event. Thus, R-SAT® is especially useful for the detection of constitutive activity (Burstein *et al.*, 1995, 1997a). Indeed, the R-SAT® assay has been used to identify and pharmacologically characterize inverse agonists of the serotonin 5-HT2A (Vanover *et al.*, 2006) and cannabinoid CB-1 receptors (Pettersson *et al.*, 2009), in addition to the SF-1 nuclear receptor (Del Tredici *et al.*, 2008). The technique is, furthermore, ideally suited for screening of large groups of molecules to identify pharmacological modulators of SF-1 receptor activity.

This functional cell-based assay allows one to monitor receptor-mediated cellular proliferation in a ligand-dependent or -independent manner. Its principle resides in the genetic selection and the amplification of the target receptor. This process is achieved by partial cellular transformation through the overcome of contact inhibition and the loss of growth factor dependency. Monitoring is achieved by transfecting the cells with a reporter gene vector whose expression is under a constitutively active promoter.

The R-SAT® assay can be summarized by its day-to-day procedure:

Day 1: Cell plating and growth
Day 2: Transfection
Day 3: Addition of ligands
Days 3–6: Incubation
Day 6: Measurement of reporter activity

2.1. Required materials

Cells: NIH/3T3 cells are available from American Tissue Type Collection (ATCC).

Growth medium: DMEM (Dulbecco's modified eagle medium) supplemented with 10% bovine calf serum and 1% penicillin/streptomycin/glutamine (Invitrogen).

Drug medium: DMEM supplemented with 30% Ultraculture (Hyclone), 0.5% bovine calf serum, and 1% penicillin/streptomycin/glutamine.

Plasmid constructs: Highly purified DNA is recommended.

Transfection reagent: We recommend the dendrimer transfection reagent, Polyfect (Qiagen) which has been optimized for use on NIH/3T3 cells. However, other transfection reagents have been shown to work in the R-SAT® assay, including Superfect (Qiagen), calcium phosphate, and Lipofectamine (Invitrogen) (Burstein *et al.*, 1998; Piu *et al.*, 2002; Weiner *et al.*, 2001). Because the assay relies on signal amplification via cellular proliferation, there is less of a need for high transfection rates compared to other cell-based assays.

Cell lysis and detection reagent: 3.5 m*M* *o*-nitrophenyl-D-galacto pyranoside (ONPG) in phosphate buffered saline with 5% Nonidet P-40 detergent.

Other reagents: Trypsin-EDTA solution (Invitrogen), Hanks' buffered salt solution (HBSS) (Invitrogen).

Disposables: 96-well plates (Corning).

2.2. Cell growth and transfection

Day 1: Cell plating.

Culture cells in 96-well plates. The cells should reach 60–70% confluence on the day of transfection. The R-SAT® assay can also be performed in a high-throughput manner using frozen cells. This procedure is described in Section 2.5.

Note: NIH/3T3 cells have been cloned from Swiss 3T3 mouse embryos to acquire highly contact-inhibited properties as well as a large susceptibility to virus-induced focus formation (Jainchill *et al.*, 1969). According to ATCC, NIH/3T3 cells must be split at 80% confluence. In our experience, splitting cells at greater than 80% confluence resulted in loss of contact inhibition, and thus a high background signal in the R-SAT® assay. Cells with earlier passage numbers tend to provide better signal-to-noise values. Moreover, active monitoring of cell growth is required for a strong signal to noise in the assay. Finally, R-SAT® requires an intact cellular monolayer, and as such, care must be taken not to disturb the cellular monolayer during media changes.

Day 2: Transfection.

On the day of transfection, prewarm growth medium to 37 °C. Bring DMEM medium and plasmid DNAs to room temperature. Use Polyfect reagent according to the manufacturer's protocol recommendations. Briefly, mix the transfection reagent and DNAs in a tube with DMEM, and incubate at room temperature for 10–15 min. Use volumes and Polyfect/DNA ratios as listed in the product brochure ("Fast-forward protocol for transfection of NIH/3T3 cells in 96-well plates using Polyfect

Transfection Reagent"; available at www.qiagen.com). Do not add serum containing-medium during this incubation, because serum can inhibit formation of the transfection reagent:DNA complex. Then add preheated growth medium containing serum to quench the reaction. Remove medium from cells, and replace with transfection mix (50 μl/well). Return cells to incubator (37 °C, 5% CO_2) for overnight incubation.

DNA reagents: DNAs required for transfection include plasmids encoding the human SF-1 receptor, coactivators, and the reporter gene. A vector with a strong constitutive promoter and multiple cloning sites should be used. We recommend the pCI Expression Vector (Promega), which contains the human cytomegalovirus (CMV) major immediate-early gene enhancer/promoter region, or the pSI Expression Vector (Promega), which contains the simian virus 40 (SV40) enhancer and early promoter region. The SF-1 R-SAT® assay minimally requires SF-1 receptor and beta-galactosidase expression vectors. For the SF-1 receptor DNA, we have demonstrated that increased levels of receptor DNA correlated with increased signal, that is increased constitutive activity (Fig. 14.1A). Furthermore, the maximum signal was achieved with 40 ng/well, indicating that there is a limit for receptor level dependent activation. We also uncovered that the intrinsic activity of SF-1 in the R-SAT® assay, as shown in other assays (Borud *et al.*, 2003; Ito *et al.*, 1998; Li *et al.*, 2005),

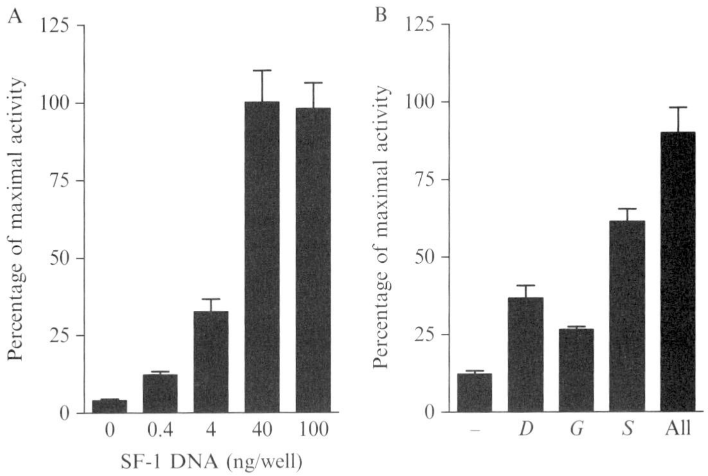

Figure 14.1 Conditions in the cellular proliferation R-SAT® assay. (A) Constitutive activity of SF-1 is dependent upon the amounts of SF-1 transfected in NIH/3T3 cells. (B) Coactivators increase the constitutive activity displayed by SF-1. D, DRIP205; G, GRIP1; S, SRC-1; All, all three coactivators. Activity is reported as the percentage of maximal activity. 100% is defined as the activity observed in the 40 ng/well SF-1 DNA condition.

could be modulated by the addition of DNA encoding coactivators. As shown in Fig. 14.1B, when low amounts of SF1 DNA were transfected (0.4 ng/well), transfection of DNA encoding the coactivators SRC-1, DRIP205, and GRIP1 (0.4 ng for each coactivator DNA per well) resulted in increased observed activity (Del Tredici *et al.*, 2008). When DNA encoding all three coactivators was included together, an additive effect was observed.

2.3. Addition of ligands and measurement of R-SAT® response

Day 3: Ligand addition.

Prepare 96-well plates with drug medium containing serial dilutions of ligand, such as AC-45594 (4-(heptyloxy)phenol). Ligands are typically stored in DMSO, and should be diluted with drug medium to well below 1% DMSO concentration. Sixteen hours after transfection, remove the medium from cells and replace with drug medium containing ligand. Return cells to incubator (37 °C, 5% CO_2).

Note: Ultraculture (Hyclone) is a defined serum-free medium consisting of a DMEM:F12 base, supplemented with bovine insulin, bovine transferrin, and a purified mixture of bovine serum proteins. We find that this is critical to promoting growth while maintaining low background signal in the assay. 2% Cyto-SF3, a defined supplement (Kemp Biotechnologies), has also given good results.

Day 6: Measurement of proliferative response.

Remove the medium from cells. Add 150 μl of the lysis/detection reagent to each well and incubate plates at room temperature. The detection reagent includes 3.5 m*M* ONPG, a colorimetric substrate for β-galactosidase, and 5% NP-40 detergent, which quenches the reaction by lysing the cells (Burstein *et al.*, 1997b). A standard microplate reader should be utilized to detect absorbance at 420 nm. Incubation time can vary between reagent lot numbers and cell batches. The first time the assay is performed, readings should be taken at intervals (15 min to 24 h) to determine the optimal incubation time.

2.4. Curve fitting calculations

Fit data from R-SAT® assays to the equation:

$$r = A + B\,(x/(x + c))$$

where A is the minimum response; B, maximum response minus minimum response; c, EC_{50}; r, response; and x, concentration of ligand. We recommend the use of curve-fitting software such as Excel Fit (Microsoft) or Graph Pad Prism (Graph Pad Software).

2.5. Large-scale R-SAT® assays

R-SAT® can easily be configured for larger scale cellular preparations. Transfections are performed in large cell batches, and cells harvested and frozen in single-use vials which can be plated onto 96-well (half well) plates containing ligand. The use of frozen transfected cell batches ensures assay reproducibility by minimizing the variability observed in transfection efficiency and cell passage.

For dishes (from 60 to 150 mm), roller bottles (up to 850 cm^2), or cell factories (from 1- to 40-stack), transfection conditions can be appropriately scaled, with concentrations and volumes adjusted for surface area of the culture vessel. The day following transfection, rinse cells with HBSS, trypsinize, and freeze cells in vials as per usual cell protocols in growth medium containing 10% DMSO. After initially freezing cells at -70 °C, frozen transfected cells can be stored for several months in liquid nitrogen or at -135 °C without any loss in cell integrity.

To use frozen transfected cells, thaw rapidly in a 37 °C water bath and add to previously warmed drug medium. Then centrifuge at low speed (1000 rpm for 5 min) to pellet the cells. Resuspend the cells in drug medium and add directly to ligands plated in 96-well plates. Then return to the incubator (37 °C, 5% CO_2) and proceed with detection on day 6 as described above.

3. SF-1 Inverse Agonism in Luciferase Transcriptional Assay

The primary function of SF-1 as a transcription factor can be directly assessed in a reporter-gene assay. Such an assay measures the ability of a protein (nuclear receptor, transcription factor), in a ligand-dependent or -independent manner, to bind to a DNA sequence and modulate the transcription of a reporter gene (typically LacZ, β-galactosidase, or luciferase).

We found that luciferase reporters constituted a more sensitive and reliable approach than colorimetric readouts. The ability to use either synthetic (SF-1 response element, SFRE) or natural promoters (StAR) was confirmed, and was found to be a suitable method to measure the constitutive activity of SF-1, and to assess pharmacological modulators of SF-1. In the absence of any ligands, transient transfection of increasing amounts of SF-1 translates into higher transcriptional activity through a synthetic SFRE. In conditions where the constitutive activity of SF-1 was maximal, addition of the AC-45594 ligand led to a dose-dependent inhibition of SFRE or StAR promoter mediated transcription (Fig. 14.2).

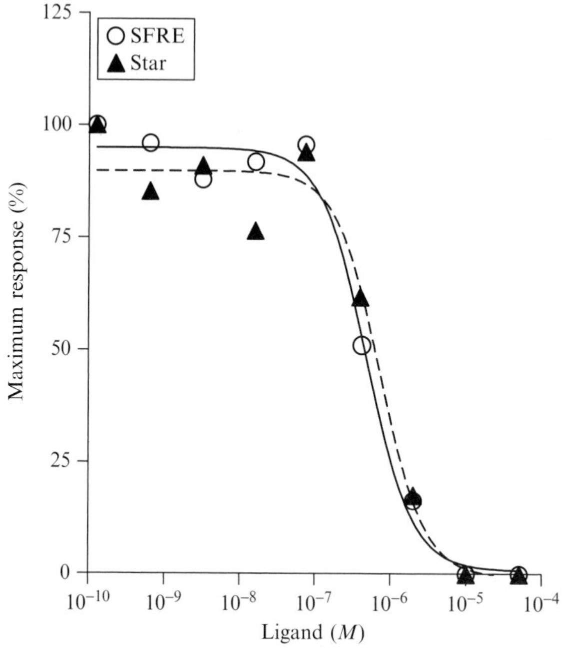

Figure 14.2 Inverse agonist properties of AC-45594 in a luciferase transcriptional assay. Conditions where SF-1 constitutive activity was optimal were used. NIH/3T3 cells were transfected with high amounts of SF-1 and the coactivators DRIP205, GRIP1, and SRC-1, along with a luciferase reporter gene coupled to either SFRE (open circles, solid line) or the StAR promoter (closed triangles, dashed line). Cells were exposed to various doses of AC-45594.

SF-1 binds to DNA sequences characterized by the consensus half-site PyCAAGGTCA (Wilson *et al.*, 1993). Among its target genes is the steroidogenic acute regulatory protein StAR, a master regulator of steroidogenesis that contains several SF-1 binding sites in its promoter (Sugawara *et al.*, 1996).

3.1. Required materials

Cells: Human embryonic kidney HEK-293T available from American Type Culture Collection.

Transfection reagent: Polyfect (Qiagen).

Assay reagent: Luciferase Assay kit SteadyGlo (Promega).

Growth medium: DMEM supplemented with 10% fetal bovine serum and 1% penicillin/streptomycin/glutamine (Invitrogen).

Chemicals: 4-(hexyloxy)phenol, 4-(heptyloxy)phenol (aka AC-45594) and 4-(octyloxy)phenol were synthesized in-house at ACADIA Pharmaceuticals.

Plasmid constructs: The SFRE luciferase vector was generated by modification of the ER2 luciferase reporter vector (Panomics). The ER2 motif, containing three estrogen response element (ERE) sequences, was removed via restriction enzyme digestion. It was replaced by three SFRE sequences generated by the hybridization and subsequent insertion of the following primers: 5′-CTAGCTCAAGGTCACAGTCAAGGTCACAGTCAAGGTCAA and 3′-GAGTTCCAGTGTCAGTTCCAG TGTCAGTTCCAGTTC-TAG (SFRE underlined). The StAR promoter-luciferase construct contained a large portion of the proximal and distal promoter (−1300, +39) and has been described elsewhere. Expression vectors for the nuclear receptor SF-1, the coactivators GRIP1, SRC-1, and DRIP205 were from commonly used eukaryotic expression systems.

Detection apparatus: Luminescence plate reader (Lumistar Galaxy, BMG Labtechnologies)

3.2. Cell growth and transfection

Maintain HEK-293T cells in T150 cm^2 cell culture flasks, using DMEM supplemented with 10% fetal bovine serum. Passage cells as needed by standard trypsinization procedures, making sure that the cultures do not reach overconfluence. We recommend maintaining cells in culture up to passage 25, at which point they should be discarded and a newer passage used (typically passage 5). This procedure minimizes derivation of the HEK-293T cultures and ensures consistent results.

Prior to the day of transfection, harvest HEK-293T cells and plate overnight (37 °C, 5% CO_2) in 96-well plates at 20,000 cells/well in 100 μl of DMEM containing 10% charcoal-stripped fetal bovine serum. The following day, remove media and replace with fresh DMEM containing 10% charcoal-stripped fetal bovine serum (50 μl/well). Perform transient transfections using Polyfect, according to manufacturer's instructions.

The transfection mix should include, per well, 60 ng SF-1 encoding plasmid, 20 ng reporter gene (SFRE or StAR promoter-luciferase), 0.4 ng each of vectors encoding GRIP1, SRC-1, and DRIP205 in the presence of 0.5 μl of Polyfect reagent. These DNAs dissolved in TE buffer (10 m*M* Tris, EDTA 1 m*M*) should be added to DMEM containing 10% charcoal-stripped fetal bovine serum. Then, add Polyfect reagent, vortex briefly, and incubate at room temperature for 10 min. Finally, distribute transfection mix (50 μl/well) onto the 96-well plates containing HEK-293T cells and incubate at 37 °C, 5% CO_2 overnight.

3.3. Ligand addition and reporter activity measurement

Sixteen hours post transfection, replace the media with serum-free DMEM with or without drugs, added at different concentrations (100 μl). We recommend a no-drug control (serum-free DMEM) for each compound tested. Return cells to the incubator for an additional 36 h at 37 °C, 5% CO_2.

Determine luciferase activity using the commercially available kit SteadyGlo. Briefly, remove media and replace with 100 μl/well of equilibrated, room temperature lysis buffer. Following a 5 min incubation period, add 100 μl/well of equilibrated, room temperature assay reagent and incubate for an additional 5 min. Finally, within 30 min after assay reagent addition, measure luminescence activity using a luminescence plate reader.

4. SF-1 Inverse Agonism in Adrenocortical Cultures

Human adrenocortical cultures represent a physiological cell-based model for testing SF-1 inverse agonism. Unlike previously described cell-based assays, human adrenocortical cells, specifically NCI-H295R cells, express many types of steroid hormones, including glucocorticoids, mineralocorticoids, estrogens, and androgens (Gazdar *et al.*, 1990). In addition, these cells endogenously express SF-1 so that transfection of SF-1 receptor encoding DNA is not necessary (Val *et al.*, 2003). Moreover, in these cells, expression of SF-1 target genes can also be induced by cAMP in a SF-1-dependent manner (Brand *et al.*, 2000; Val *et al.*, 2003).

In this section, we describe the use of NCI-H295R cells to assay for inverse agonism of the SF-1 receptor. This method is distinct from a similar assay utilizing NCI-H295R cells, in which SF-1 overexpression leads to cell proliferation, and thus, can be considered a model for adrenal tumorigenesis (Doghman *et al.*, 2007, 2009). In contrast, we have used nontransfected NCI-H295R cells to demonstrate that a class of alkyloxyphenol compounds can inhibit cAMP induced expression of SF-1 target genes (and SF-1 itself) at both the mRNA and protein levels (Del Tredici *et al.*, 2008). Our study describes inhibition of SF-1-dependent expression of two genes, CYP11A1 and StAR. CYP11A1 (CYP450scc) catalyzes the side chain cleavage of cholesterol, the first step in the steroid biosynthetic pathway, while StAR constitutes the rate-limiting factor in the steroid biosynthetic pathway. More than likely the technique can be applied to other SF-1 target genes.

4.1. Required materials

Growth medium: DMEM/F12 supplemented with 10% fetal bovine serum, 1/1000th ITS mix (insulin, transferrin, selenium) (Invitrogen), and 1% penicillin/streptomycin/glutamine.

Assay medium: DMEM/F12 supplemented with 10% charcoal-stripped fetal bovine serum, 1/1000th ITS mix, and 1% penicillin/streptomycin/glutamine.

Assay quantification: RNAqueous-4PCR kit (Ambion).

Antibodies: StAR (ABR Affinity Reagents), GAPDH (Santa Cruz Biotechnology), SF-1 (Santa Cruz Biotechnology), CYP11A1 (Santa Cruz Biotechnology).

Additional reagents: Laemmli buffer, BCA protein assay kit (Pierce Biotechnology), $(Bu)_2cAMP$ that is $(Bu)_2cAMPN^6$,2′-O-dibutyryladenosine 3′,5′-cyclic monophosphate sodium salt (Sigma).

Disposables: Kontes pestles (Fisher Scientific).

4.2. Cell treatment

Day 1: Plate 1.7×10^6 cells per well of a 6-well dish in growth media and incubate overnight at 37 °C, 5% CO_2.

Day 2: Replace growth media with assay media. All subsequent additions are made in assay media. Incubate cells for 1 h. Treat cells with compound at desired concentration, or DMSO as negative control. Incubate cells for 48 h at 37 °C, 5% CO_2.

Note: The observed reduction of SF-1 induced activity requires pretreatment with compound prior to addition of cAMP.

Day 4: Remove media. Add assay media with compound (or DMSO) and $(Bu)_2cAMP$ (300 μM).

Day 5: After an incubation of 24 h, harvest RNA and/or protein extracts as described below (Sections 4.3 and 4.4, respectively).

4.3. cAMP-induced mRNA expression of SF-1 target genes

On Day 5 (see Section 4.2), isolate total RNA using RNAqueous-4PCR kit (Ambion) according to manufacturer's instructions.

Note: Care should be taken to prevent contamination with RNAses. We recommend the use of RNAse Away (Ambion) sprayed on bench area, racks, and on gloves prior to working with the RNA. In addition, all tubes, pipette tips, and other disposables should be restricted for use in RNA experiments only.

Convert 2 μg of extracted total RNA to cDNA using SuperScript III Reverse Transcriptase and oligo dT (Invitrogen), according to manufacturer's instructions. Perform PCR reactions using Platinum Taq (Invitrogen) using manufacturer's instructions. We have optimized the conditions so that the StAR reaction should be performed in the presence of 5% DMSO, the GAPDH and CYP11A1 reactions in the presence of 2.5% DMSO and the SF-1 reaction in the presence of 3 mM $MgSO_4$. The following cycling conditions are recommended: StAR 95 °C (50 s), 58 °C (30 s), 68 °C (50 s) for 30 cycles; GAPDH 95 °C (50 s), 58 °C (30 s), 68 °C (50 s) for 20 cycles; SF-1 95 °C (50 s), 56 °C (30 s), 68 °C (50 s) for 40 cycles; CYP11A1 95 °C (50 s), 58 °C (30 s), 68 °C (50 s) for 25 cycles.

Primer sequences and expected size of amplified band are listed below:

StAR

Sense 5′-CCAGATGTGGGCAAGGTG-3′
Antisense 5′-CAGCGCACGCTCACAAAG-3′
Size of amplified band: 227 bp

GAPDH

Sense 5′-CGAGCCACATCGCTCAGACAC-3′
Antisense 5′-GCTAAGCAGTTGGTGGTGCAGG-3′
Size of amplified band: 495 bp

CYP11A1

Sense 5′-CAAGACCTGGAAGGACCATGTG-3′
Antisense 5′-GATATCTCTGCAGGGTCACGGAG-3′
Size of amplified band: 413 bp

SF-1

Sense 5′-GACAAGGTGTCCGGCTACCAC-3′
Antisense 5′-GTCTCCAGCTTGAAGCCATTGG-3′
Size of amplified band: 314 bp

After PCR, visualize products on a 1.5% agarose gel and quantitate bands. We used Scion Image to measure pixels (Scion Corporation). Normalize band intensities to GADPH, which is the loading control.

Note: The amount of cDNA and the number of cycles have been optimized so that band intensities are in the linear range.

4.4. cAMP-induced protein expression of SF-1 target genes

On Day 5 (see Section 4.2), warm Laemmli buffer by placing in 37 °C water bath. To harvest cells, rinse cells in PBS with protease inhibitors, and then add prewarmed Laemmli buffer (300 μl) directly onto cells. Scrape cells into

1.5 ml Eppendorf tubes. Rinse dish with an additional 50 μl Laemmli buffer. Use clean plastic pestle to homogenize harvested cells. Sonicate briefly. Then spin at 15,000 rpm and harvest the supernatant. Determine the total protein concentration in the supernatant using the BCA protein assay kit (Pierce Biotechnology). Load 20–25 μg of protein per lane on standard SDS-PAGE gel. Perform Western blotting using standard conditions. Quantify expression levels by scanning blots and measuring pixels in each band (using Scion Image). Normalize band intensities, using GADPH as a loading control.

ACKNOWLEDGMENTS

This work was performed at and supported by ACADIA Pharmaceuticals.

REFERENCES

Borud, B., Mellgren, G., Lund, J., and Bakke, M. (2003). Cloning and characterization of a novel zinc finger protein that modulates the transcriptional activity of nuclear receptors. *Mol. Endocrinol.* **17,** 2303–2319.

Brand, C., Nury, D., Chambaz, E. M., Feige, J. J., and Bailly, S. (2000). Transcriptional regulation of the gene encoding the StAR protein in the human adrenocortical cell line, H295R by cAMP and TGFbeta1. *Endocr. Res.* **26,** 1045–1053.

Burstein, E. S., Spalding, T. A., Brauner-Osborne, H., and Brann, M. R. (1995). Constitutive activation of muscarinic receptors by the G-protein Gq. *FEBS Lett.* **363,** 261–263.

Burstein, E. S., Spalding, T. A., and Brann, M. R. (1997a). Pharmacology of muscarinic receptor subtypes constitutively activated by G proteins. *Mol. Pharmacol.* **51,** 312–319.

Burstein, E. S., Spalding, T. A., and Brann, M. R. (1997b). Use of random-saturation mutagenesis to study receptor-G protein coupling. *Methods Mol. Biol.* **83,** 143–157.

Burstein, E. S., Hesterberg, D. J., Gutkind, J. S., Brann, M. R., Currier, E. A., and Messier, T. L. (1998). The ras-related GTPase rac1 regulates a proliferative pathway selectively utilized by G-protein coupled receptors. *Oncogene* **17,** 1617–1623.

Burstein, E. S., Piu, F., Ma, J. N., Weissman, J. T., Currier, E. A., Nash, N. R., Weiner, D. M., Spalding, T. A., Schiffer, H. H., Del Tredici, A. L., and Brann, M. R. (2006). Integrative functional assays, chemical genomics and high throughput screening: Harnessing signal transduction pathways to a common HTS readout. *Curr. Pharm. Des.* **12,** 1717–1729.

Del Tredici, A. L., Andersen, C. B., Currier, E. A., Ohrmund, S. R., Fairbain, L. C., Lund, B. W., Nash, N., Olsson, R., and Piu, F. (2008). Identification of the first synthetic steroidogenic factor 1 inverse agonists: Pharmacological modulation of steroidogenic enzymes. *Mol. Pharmacol.* **73,** 900–908.

Desclozeaux, M., Krylova, I. N., Horn, F., Fletterick, R. J., and Ingraham, H. A. (2002). Phosphorylation and intramolecular stabilization of the ligand binding domain in the nuclear receptor steroidogenic factor 1. *Mol. Cell. Biol.* **22,** 7193–7203.

Doghman, M., Karpova, T., Rodrigues, G. A., Arhatte, M., De Moura, J., Cavalli, L. R., Virolle, V., Barbry, P., Zambetti, G. P., Figueiredo, B. C., Heckert, L. L., and Lalli, E. (2007). Increased steroidogenic factor-1 dosage triggers adrenocortical cell proliferation and cancer. *Mol. Endocrinol.* **21,** 2968–2987.

Doghman, M., Cazareth, J., Douguet, D., Madoux, F., Hodder, P., and Lalli, E. (2009). Inhibition of adrenocortical carcinoma cell proliferation by steroidogenic factor-1 inverse agonists. *J. Clin. Endocrinol. Metab.* **94,** 2178–2183.

Gazdar, A. F., Oie, H. K., Shackleton, C. H., Chen, T. R., Triche, T. J., Myers, C. E., Chrousos, G. P., Brennan, M. F., Stein, C. A., and La Rocca, R. V. (1990). Establishment and characterization of a human adrenocortical carcinoma cell line that expresses multiple pathways of steroid biosynthesis. *Cancer Res.* **50,** 5488–5496.

Hoivik, E. A., Lewis, A. E., Aumo, L., and Bakke, M. (2010). Molecular aspects of steroidogenic factor 1 (SF-1). *Mol. Cell. Endocrinol.* **315,** 27–39.

Ikeda, Y., Shen, W. H., Ingraham, H. A., and Parker, K. L. (1994). Developmental expression of mouse steroidogenic factor-1, an essential regulator of the steroid hydroxylases. *Mol. Endocrinol.* **8,** 654–662.

Ito, M., Yu, R. N., and Jameson, J. L. (1998). Steroidogenic factor-1 contains a carboxy-terminal transcriptional activation domain that interacts with steroid receptor coactivator-1. *Mol. Endocrinol.* **12,** 290–301.

Jainchill, J. L., Aaronson, S. A., and Todaro, G. J. (1969). Murine sarcoma and leukemia viruses: Assay using clonal lines of contact-inhibited mouse cells. *J. Virol.* **4,** 549–553.

Krylova, I. N., Sablin, E. P., Moore, J., Xu, R. X., Waitt, G. M., MacKay, J. A., Juzumiene, D., Bynum, J. M., Madauss, K., Montana, V., Lebedeva, L., Suzawa, M., *et al.* (2005). Structural analyses reveal phosphatidyl inositols as ligands for the NR5 orphan receptors SF-1 and LRH-1. *Cell* **120,** 343–355.

Li, Y., Choi, M., Cavey, G., Daugherty, J., Suino, K., Kovach, A., Bingham, N. C., Kliewer, S. A., and Xu, H. E. (2005). Crystallographic identification and functional characterization of phospholipids as ligands for the orphan nuclear receptor steroidogenic factor-1. *Mol. Cell* **17,** 491–502.

Madauss, K., Juzumiene, D., Waitt, G., Williams, J., and Williams, S. (2004). Generation and characterization of human steroidogenic factor 1 LBD crystals with and without bound cofactor peptide. *Endocr. Res.* **30,** 775–785.

Morohashi, K., Honda, S., Inomata, Y., Handa, H., and Omura, T. (1992). A common trans-acting factor, Ad4-binding protein, to the promoters of steroidogenic P-450s. *J. Biol. Chem.* **267,** 17913–17919.

Ortlund, E. A., Lee, Y., Solomon, I. H., Hager, J. M., Safi, R., Choi, Y., Guan, Z., Tripathy, A., Raetz, C. R., McDonnell, D. P., Moore, D. D., and Redinbo, M. R. (2005). Modulation of human nuclear receptor LRH-1 activity by phospholipids and SHP. *Nat. Struct. Mol. Biol.* **12,** 357–363.

Parker, K. L., and Schimmer, B. P. (1997). Steroidogenic factor 1: A key determinant of endocrine development and function. *Endocr. Rev.* **18,** 361–377.

Pettersson, H., Bulow, A., Ek, F., Jensen, J., Ottesen, L. K., Fejzic, A., Ma, J. N., Del Tredici, A. L., Currier, E. A., Gardell, L. R., Tabatabaei, A., Craig, D., *et al.* (2009). Synthesis and evaluation of dibenzothiazepines: A novel class of selective cannabinoid-1 receptor inverse agonists. *J. Med. Chem.* **52,** 1975–1982.

Piu, F., Magnani, M., and Ader, M. E. (2002). Dissection of the cytoplasmic domains of cytokine receptors involved in STAT and Ras dependent proliferation. *Oncogene* **21,** 3579–3591.

Sugawara, T., Holt, J. A., Kiriakidou, M., and Strauss, J. F. 3rd. (1996). Steroidogenic factor 1-dependent promoter activity of the human steroidogenic acute regulatory protein (StAR) gene. *Biochemistry* **35,** 9052–9059.

Urs, A. N., Dammer, E., Kelly, S., Wang, E., Merrill, A. H. Jr., and Sewer, M. B. (2007). Steroidogenic factor-1 is a sphingolipid binding protein. *Mol. Cell. Endocrinol.* **265–266,** 174–178.

Val, P., Lefrancois-Martinez, A. M., Veyssiere, G., and Martinez, A. (2003). SF-1 a key player in the development and differentiation of steroidogenic tissues. *Nucl. Recept.* **1,** 8.

Vanover, K. E., Weiner, D. M., Makhay, M., Veinbergs, I., Gardell, L. R., Lameh, J., Del Tredici, A. L., Piu, F., Schiffer, H. H., Ott, T. R., Burstein, E. S., Uldam, A. K., *et al.* (2006). Pharmacological and behavioral profile of N-(4-fluorophenylmethyl)-N-(1-methylpiperidin-4-yl)-N'-(4-(2-methylpropylo xy)phenylmethyl) carbamide (2R, 3R)-dihydroxybutanedioate (2:1) (ACP-103), a novel 5-hydroxytryptamine(2A) receptor inverse agonist. *J. Pharmacol. Exp. Ther.* **317,** 910–918.

Wang, W., Zhang, C., Marimuthu, A., Krupka, H. I., Tabrizizad, M., Shelloe, R., Mehra, U., Eng, K., Nguyen, H., Settachatgul, C., Powell, B., Milburn, M. V., *et al.* (2005). The crystal structures of human steroidogenic factor-1 and liver receptor homologue-1. *Proc. Natl. Acad. Sci. USA* **102,** 7505–7510.

Weiner, D. M., Burstein, E. S., Nash, N., Croston, G. E., Currier, E. A., Vanover, K. E., Harvey, S. C., Donohue, E., Hansen, H. C., Andersson, C. M., Spalding, T. A., Gibson, D. F., *et al.* (2001). 5-hydroxytryptamine2A receptor inverse agonists as antipsychotics. *J. Pharmacol. Exp. Ther.* **299,** 268–276.

Wilson, T. E., Fahrner, T. J., and Milbrandt, J. (1993). The orphan receptors NGFI-B and steroidogenic factor 1 establish monomer binding as a third paradigm of nuclear receptor-DNA interaction. *Mol. Cell. Biol.* **13,** 5794–5804.

CHAPTER FIFTEEN

Methods to Measure G-Protein-Coupled Receptor Activity for the Identification of Inverse Agonists

Gabriel Barreda-Gómez, M. Teresa Giralt, *and* Rafael Rodríguez-Puertas

Contents

Abstract

Before the concept of constitutive or intrinsic activity of the biological systems, which was formulated about thirty years ago, it was thought that agonist compounds were the only drugs capable of activating physiological responses,

Department of Pharmacology, Faculty of Medicine, University of the Basque Country, Vizcaya, Spain

Methods in Enzymology, Volume 485
ISSN 0076-6879, DOI: 10.1016/S0076-6879(10)85015-3

while antagonists were the ones capable of blocking them. However, this basic classification of drugs in pharmacology started to change only at the end of the eighties, when bioactive ligands, with negative efficacy, were developed. The G-protein-coupled receptors (GPCR) were promptly selected as one of the most useful types of pharmacological targets to study this inverse efficacy. This family of receptors is responsible for the signaling and control of many physiological processes, from the peripheral nervous system to the central. Therefore, the GPCR have become the most studied family of receptors in drug discovery. It has been estimated that around a third of the drugs actually used act via the GPCR, nevertheless there are still many orphan GPCR encoded by the human genome.

During the last decade, reports and patents have described new methods to detect GPCR inverse agonist compounds. The detection of the G-protein constitutive activity and the quantification of the positive or negative efficacies induced by agonists or inverse agonists respectively has been studied by analyzing the binding of the nonhydrolyzable GTP analog, [^{35}S]GTPγS. The present chapter describes an optimized method to detect GPCR inverse agonist ligands such as cannabinoid compounds, in both membrane homogenates and tissue sections (autoradiography).

1. Introduction

For many years, agonists were considered to be the only drugs capable of inducing pharmacological responses, but during the last thirty years the concept that some bioactive ligands are able to induce negative efficacy has progressively been accepted. According to this theory, the binding of an inverse agonist to its receptor can inhibit the intrinsic or constitutive (basal) activity of that receptor. These compounds would induce an opposite (inverse) effect in contrast to the "positive" agonists. Accordingly, full inverse agonists could completely block the basal activity. In this context, the term antagonist would be applied to the compounds that block the effects of agonists, and these have been redefined as "neutral antagonists," or simply referred to as "antagonists," which blocked both the effect of agonists and inverse agonists (Fig. 15.1). However, the above-described model is a simplification, and it is indeed difficult to experimentally characterize a compound as a pure antagonist or inverse agonist, and the method and assay conditions that are used have to be carefully evaluated.

Previously, the methods used to characterize effects in drug discovery were designed to detect agonism, and the basal activity measured using pharmacological techniques, as the recording of twitches of isolated organs, was sometimes attributed to experimental artifacts. Therefore, and in addition to the obligatory change of mind of the pharmacologists during the last

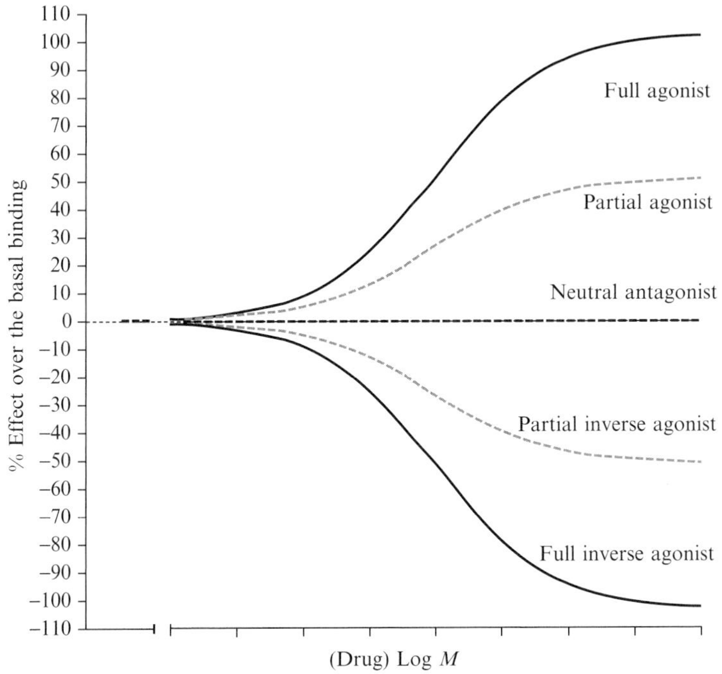

Figure 15.1 The discovery of the constitutive activity of some transduction systems such as the complex GPCR-G-protein, in the absence of ligand, increased the possibilities of pharmacological action from the classically limited agonists and antagonists to inverse agonists. The negative efficacy induced by an inverse agonist reduces the constitutive activity of the receptor and produces a concentration-dependent inhibition curve of the basal activity. As happens with an agonist, a full inverse agonist elicits a complete inhibition, while a partial inverse agonist inhibits basal activity in a percentage that will depend on the kind of compound and system used.

few decades, it has also been necessary to develop new methods and/or modify experimental conditions in order to identify inverse agonism effects.

The described effects of some drugs have been reevaluated, and numerous compounds, which were originally classified as neutral antagonists, are now referred to as inverse agonists. Like agonists, inverse agonist compounds could also induce regulatory responses to modify efficacy.

Although the inverse agonist effect is also present in other kinds of receptors such as ion channels, this chapter describes one of the most useful methods to quantify the inverse agonism effect mediated by GPCR, the measurement of G-protein activity. Some examples applied specifically to cannabinoid receptors are shown. The CB1 subtype of cannabinoid receptor is one of the most abundant GPCR that is evenly distributed throughout the brain and shows a high efficiency in its coupling to Gi/o proteins (Howlett *et al.*, 2004). Curiously, the antagonist compounds developed for CB1 receptors also induce a high negative efficacy below their constitutive activity.

1.1. G-protein-coupled receptors

GPCR genes represent up to 1% of the total genome of mammals and it is estimated that around 1500 genes for GPCR are expressed in the human genome. In fact, up to 5% of the total cell proteins correspond to GPCR. This high number and great diversity of receptors has caused the superfamily of GPCR to become one of the principal drugs targets used in pharmacology, especially in the central nervous system (CNS). Indeed, almost 40% of the most commonly used drugs act through GPCR and the profits that they generate for the pharmaceutical industry amount to about thirty billion dollars annually. However, marketed drugs target less than 50 types of GPCR; therefore, it is predictable that drug discovery in this field will increase over during the following years (Eglen *et al.*, 2007). In this context, the study of the mechanism of action for inverse agonist compounds has led to the search for pharmacological tools, which are able to decrease a pathological high basal or constitutive activity, or even induce a prolonged hypersensitivity, opposite to the down regulation that is produced by chronic agonist treatments.

Although the relative weight or contribution of the basal activity mediated by cannabinoid receptors to the total GPCR–mediated constitutive activity measured in the CNS has not been determined, the potency of the cannabinoid "inverse agonists," such as the molecule rimonabant (SR141716A), to inhibit the basal activity of GPCR is very high in *in vitro* assays. To date, two cannabinoid receptors (CB) have been described: CB1 and CB2. The CB1 subtype is mainly located in brain and nervous tissue, while the CB2 receptor is found primarily in immune cells (macrophages, monocytes, leukocytes, etc.) and in peripheral tissues (spleen, bone marrow, pancreas, etc.). According to this anatomical distribution, the cannabinoid receptors participate in different neurophysiological functions such as the control of neuropathic pain, feeding, analgesia, intestinal motility, memory and cognition. 2-Arachidonoylglycerol and anandamide are the endogenous ligands for cannabinoid receptors and several selective compounds have been isolated from natural sources, the most studied being the Δ9-tetrahydrocannabinol (THC). Many cannabinoid synthetic compounds have been developed: agonists such as WIN55,212-2 and CP55,940 and inverse agonist ligands such as SR141716A, AM281 and AM251 (Pertwee, 2005). The last compound could be considered as a partial inverse agonist, while SR141716A and AM281 are full inverse agonists, since they are able to induce the maximal inhibition of the basal G-protein activity recorded *in vitro*.

1.2. Techniques used for the detection of inverse agonist compounds

Diverse neurochemical methodologies previously used to study receptor-mediated effects and neurotransmitter release have been adapted for the study of the inhibition of constitutive activity. Some of these methods consist of quantifying intracellular second messenger levels such as cyclic adenosine monophosphate (cAMP) or inositol phosphate derivatives (IP). Other pharmacological techniques have also been adopted to study inverse agonist compounds as isolated organ preparations, that is the measurement of the inhibition of basal cardiac contraction rate (Engelhardt *et al.*, 2001). More recently, methods used in cellular and molecular biology have been developed, using fluorescence resonance energy transfer (FRET) analysis for the study of receptor internalization in living cells (Stanasila *et al.*, 2003) or the molecular conformation of the receptor triggered by inverse agonists (Vilardaga *et al.*, 2005). However, one of the most widely used techniques to detect inverse agonist drugs for GPCR is the measurement of receptor coupling to G-proteins by the quantification of the binding of the radioligand guanosine 5′-(γ-[^{35}S]thio)triphosphate ([^{35}S]GTPγS) both in membrane homogenates and in tissue sections, the latter method also being known as [^{35}S]GTPγS autoradiography. This nonhydrolyzable GTP analog is able to bind to the α subunit of the G-protein and stops the GPCR activation cycle. The model of GPCR cycle starts when a ligand activates the GPCR, inducing the binding of the receptor to the heterotrimeric G-protein. This fact promotes the interchange of GDP for GTP in the α subunit and the dissociation of the trimer in the αsubunit and the $\beta\gamma$ dimmer, which are then able to interact with intracellular or membrane effectors (enzymes, ion channels, etc.). Then, the α-subunit, due to its intrinsic GTPase activity, hydrolyzes GTP into GDP, restoring its initial inactive conformation as well as its affinity for the $\beta\gamma$ complex, and the heterotrimer becomes ready for a new activation cycle. Nevertheless, if the nonhydrolyzable GTP analog [^{35}S]GTPγS is used in an *in vitro* assay instead of GTP, the α subunit cannot hydrolyze it and the GPCR cycle stops at that point. This cycle also occurs spontaneously, without the intervention of any ligand or drug, and the intrinsic activity is known as constitutive or basal activity of the G-proteins (Fig. 15.2).

We describe the procedures to detect inverse agonism using the [^{35}S]GTPγS nonhydrolyzable GTP analog on isolated cell membranes and in tissue sections, and focus on the inverse agonism mediated by cannabinoid drugs, but the protocols described here could be used to study the negative activity of different compounds from other types of GPCR.

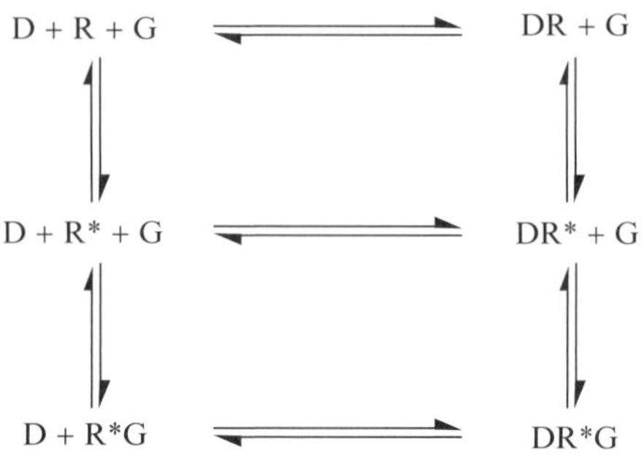

Figure 15.2 Extension of the ternary complex model of receptor activation, which differentiates distinct receptor forms or active states: active (R*) and inactive (R). In the presence of the agonist the receptor is prone to be converted into an active conformation and the equilibrium shifts towards DR* because the agonists have a higher affinity for R*. The inverse agonists stabilize the inactive conformation (R) of receptors. The R state is uncoupled from the G-protein (G), whereas the R* state is coupled and can activate the G-protein (G). The active conformation (R*) of GPCR, has a high affinity for G. According to the ternary complex model of receptor activation, the R*G complex is in equilibrium with R* + G. Inverse agonists, with a higher affinity for R, shift this equilibrium from R*G towards R + G, decreasing the basal activity. Neutral antagonists have the same affinity for R and R* and are able to compete with both agonists and inverse agonists.

2. [^{35}S]GTPγS Binding Assay in Membrane Homogenates

2.1. Chemicals

[^{35}S]GTPγS (1250 Ci/mmol) was purchased from DuPont NEN (at present from Perkin-Elmer), and WIN55,212-2 and SR141716A were supplied by Tocris. All other chemicals were obtained from standard sources, mainly from Sigma and Merck, and were of the highest purity available for *in vitro* assays.

2.2. Equipment

The main apparatus is a Harvester instrument for tubes, which allows liquid suspensions (membrane homogenates in buffer solution) to pass through glass-fiber filters. We used an Innotech apparatus that is able to filter 24, 48, or 96 (ELISA plates) simultaneously. A bench centrifuge (we recommend one that fixes Eppendorf vials and reaches up to 14,000 rpm) and an ultracentrifuge are necessary for the isolation of the membranes. A liquid scintillation counter and a computer with which to obtain the data are also necessary. Finally, common small equipment and consumables of a normal biochemistry laboratory are required, such as micropipettes, homogenizers, and a temperature-controlled thermostated bath for the incubations.

2.3. Tissue samples

Male rat tissue has been chosen to exemplify these methods, although obviously, the procedure followed will be basically the same for any other kind of fresh tissue source.

Sprague Dawley rats, weighing 250–275 g, were purchased from Harlan (Barcelona, Spain). Animals were treated in accordance with institutional guidelines and Directive 86/609/EEC.

After being anesthetized with chloral hydrate (400 mg/kg, i.p.), they were decapitated and their brains were rapidly removed. The cortex area was dissected from both hemibrains and hermetically packed (into a plastic vial) and either used immediately to isolate the cell membranes, or kept frozen at −80 °C until use. It is important to maintain the tissue at 4 °C during the dissection procedure, since the receptor proteins can suffer denaturation and lose their native conformation, which maintains the binding pocket that recognizes the ligands. In addition, both the coupling of the receptor to the G-protein and the alpha subunit site for GTP and GDP, can also suffer conformational changes if we are unaware of the need to avoid temperature shocks and to maintain the temperature at −80 °C. Obviously, tissue fixation with formaldehyde or any other fixative commonly used in immunochemistry must be avoided for similar reasons.

2.4. Membrane preparation

The cortex samples were thawed, when frozen, and homogenized using a Teflon-glass grinder (10 up-and-down strokes at 1500 rpm) in 30 volumes of homogenization buffer (1 m*M* EGTA, 3 m*M* $MgCl_2$, and 50 m*M* Tris–HCl, pH 7.4) supplemented with 0.25 m*M* sucrose, always keeping the temperature at 4 °C. The crude homogenate was centrifuged for 5 min at 3,000 rpm, and the obtained supernatant was centrifuged for 10 min at 18,000 rpm. The resultant pellet was washed in 10 volumes of homogenization buffer and centrifuged in similar conditions. When the membranes are not going to be used immediately for the radioligand binding experiment, it is preferable to do this last centrifugation in Eppendorf tubes in order to obtain a pellet and to remove the supernatants. Then, aliquots of dried membrane preparation can be stored at −80 °C until assay in the same hermetically closed Eppendorf vials. Protein content is measured according to the method of Bradford (1976) using BSA as standard, since the data are calculated as a fraction of the amount of proteins in the samples, usually milligrams of proteins.

2.5. [^{35}S]GTPγS binding assay

Membrane aliquots that had been frozen were thawed and resuspended in 10 volumes of buffer containing 1 m*M* EGTA, 3 m*M* $MgCl_2$, 100 m*M* NaCl, 0.5% BSA, and 50 m*M* Tris–HCl, at pH 7.4, at 4 °C (resuspension buffer), and homogenized using an Ultra-Turrax disperser homogenizer. Incubation is started by the addition of the membrane suspension (a concentration of 0.1 mg/ml of membrane proteins is recommended) to the incubation buffer, which is the same as that described above, but in the presence of 0.5 n*M* [^{35}S]GTPγS, 0.5 m*M* GDP for the "basal" conditions (tubes), and the appropriate concentrations of drugs in order to have a wide enough range of different concentrations with which to build a concentration–response curve. Around five drug concentrations ranging from 0.1 n*M* to 1 m*M* are recommended. Nonspecific binding is defined as the remaining [^{35}S]GTPγS binding in the presence of 10 μ*M* unlabelled GTPγS. Additional assay tubes must be prepared to determine this nonspecific binding. All the tubes containing the different experimental conditions should be duplicated or triplicated whenever possible. The incubation step was performed at 30 °C for 120 min with shaking. The assayed compound that is shown in Fig. 15.3 is the CB1 receptor antagonist/inverse agonist, SR141716A, which was diluted in pure DMSO for every concentration from 1 n*M* to 1 m*M*, and basal activity was measured in the same conditions with the same amount of DMSO.

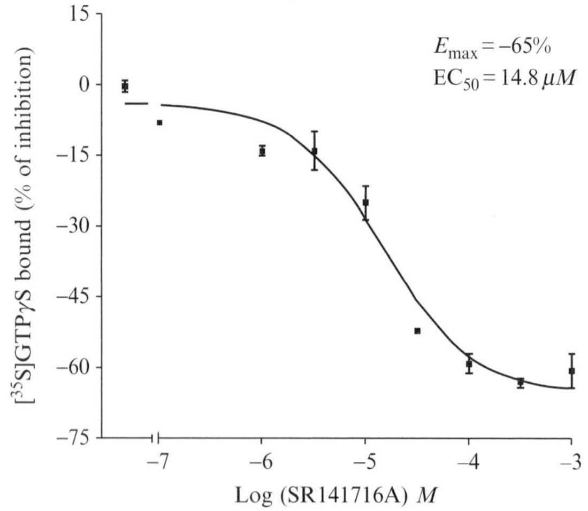

Figure 15.3 SR141716A concentration–response curve expressed in percentages of inhibition of the [^{35}S]GTPγS basal binding. Nonspecific binding was determined in the presence of 10 μ*M* GTPγS. Data were mean ± S.E.M. of triplicates from five rats.

Note: the cannabinoid compounds are very lipophilic and they usually bind unspecifically to plastic material. Therefore, we advise borosilicate tubes previously treated with "sigmacote" (Sigma) to be used. For the same reason it is necessary to use DMSO as a solvent and the same DMSO concentration should be added to the "basal" and "nonspecific" tubes. Incubations were terminated by adding 3 ml of the same ice-cold buffer followed by rapid filtration through Whatman GF/C filters, also presoaked in this buffer, using the Harvester automatic system previously described. This apparatus is connected to a vacuum pump and a safety bottle in which to deposit the filtered radioactive residues. The filters were rinsed twice with the ice-cold resuspension buffer and they retained the membranes with the bound radioligand.

The filters were transferred to vials containing 5 ml of OptiPhase HiSafe II cocktail and radioactivity was determined by liquid scintillation spectrometry (Packard 2200CA). The nonspecific binding was subtracted from the total bound radioactivity to determine [^{35}S]GTPγS specific binding.

2.6. Quantification and statistical analysis

The counts per minute (cpm) values were calculated and transformed to nCi values using the radioligand specifications (concentration and specific activity) and also finally expressed in nCi/mg proteins, as the protein concentration had previously been calculated. The percentage of stimulation or inhibition on the basal [^{35}S]GTPγS specific binding was calculated. Note that this basal binding can directly be interpreted as the intrinsic or constitutive activity contributed by the whole population of GPCR present in the membrane samples. Furthermore, to display the pharmacological parameters of the agonist-induced stimulation, concentration–response curves were plotted and analyzed by nonlinear regression analysis. The maximal effect (E_{max}) and the concentration of the agonist that induces the half-maximal effect (EC_{50}) were calculated using Prism Graphad software. On the contrary, the inverse agonist, SR141716A, induced a concentration-dependant decrease under basal [^{35}S]GTPγS specific binding. Consequently, the maximal inhibitory effect and the concentration of the drug necessary to reach 50% of inhibition (IC_{50}), was able to be calculated (Fig. 15.3).

2.7. [^{35}S]GTPγS binding in the presence of the inverse agonist compound

The experimental conditions described above allow the study of agonists, neutral antagonists (no effect or inhibition of the effect elicited by other compounds), and also of inverse agonist compounds, as previously mentioned. In this context, the concentration of the cannabinoid inverse agonist ligand, SR141716A, which was able to inhibit 50% of the maximal effect (IC_{50}), was 14.8 μ*M* and the maximal inhibitory effect (E_{max}) was 65% (Fig. 15.3). These data are similar to those published by Sim-Selley and coworkers using the [^{35}S]GTPγS technique with this cannabinoid compound (Sim-Selley *et al.*, 2001). The slight differences observed between the results might be due to the different concentration of GDP employed in the experiments. As was stated before, we recommend using a high concentration of GDP to detect inverse agonism, because the inverse agonistic properties of the molecules are enhanced, although the constitutive activity is reduced. This controversial effect may be due to the fact that at a high concentration of GDP, the equilibrium of the ternary complex (drug-receptor-G-protein) is displaced to the inactive state of GPCR, facilitating the action of the inverse agonist compounds (Fig. 15.2). The inverse agonist would be characterized by a higher affinity for the "inactive state" of the receptor that could be translated as the uncoupled receptor from the G-protein. The real molecular processes, which take place in the presence of an inverse agonist compound for GPCR, are still not well understood and the ternary complex model should be taken as such, just a model.

3. [^{35}S]GTPγS Autoradiography in Brain Sections

3.1. Tissue sections

Rat brains were obtained as previously described (Section 2.3). The freezing of the samples was rapid at −80 °C and the samples were immediately enveloped in aluminum foil and plastic film (Parafilm) to avoid dehydration. It is advisable to wash away any blood with saline solution (0.9% NaCl) before the freezing procedure. In the present example, sagittal sections of 20 μm were cut in a cryostat (Microm) for the autoradiographic assays and then mounted on gelatine-coated slides. It is important to allow the tissue to go from −80 °C to −20 °C before sectioning, otherwise it is not possible to obtain completely flat sections. The tissue sections were stored at −20 °C until the experiments were carried out, although, in order to maintain the functionality of the receptor-G-protein complex as well as possible, it is very important that the storage period is no longer than three months.

3.2. [^{35}S]GTPγS autoradiography assay

The tissue sections were air-dried for up to 60 min and then incubated in a buffer containing 0.2 m*M* EGTA, 3 m*M* $MgCl_2$, 100 m*M* NaCl, 0.1% BSA, 2 m*M* GDP, 1 m*M* DTT, and 50 m*M* Tris–HCl, at pH 7.4 for 30 min at room temperature. The purpose of this first incubation is to modify the conformational state of the receptor and the G-protein to favor the uncoupling and, in addition, to wash out the endogenous ligands present in the tissue slices. In the case of the cannabinoid receptor CB1 assay, the preincubation washes the endocannabinoids, such as the anandamide and the 2-arachidonoylglycerol (2-AG). After this preincubation, the slides were changed to other coupling jars with the same buffer, but supplemented with 0.04 n*M* [^{35}S]GTPγS in the absence (basal binding) or in the presence of the CB1 antagonist/inverse agonist molecule SR141716A for 120 min at 30 °C in different conditions. Nonspecific binding was determined in the presence of 10 μ*M* GTPγS in a consecutive slice. The basal conditions are duplicated, which means that two consecutive slides are incubated in two different coupling jars in identical conditions. The cannabinoid compound was dissolved in pure DMSO to obtain its final concentration and the same amount of DMSO was added to each assay condition including the basal and nonspecific conditions. The same precautions as in the previous protocol in relation to the affinity of the cannabinoid compound for plastic tubes should be taken in the autoradiographic assays. Following the incubation, the tissue sections were washed twice in a buffer 50 m*M* Tris–HCl for 15 min, dipped in distilled water and dried under a cold airflow (a small fan placed in a freezer at 4 °C was used). The sections were then exposed to β radiation-sensitive films (Kodak Biomax MR) with a set of [^{14}C]-microscales in a dark room. In order to compare them and to avoid any possible variations between different films, it is important to place all the consecutive slides on the same film.

3.3. Quantification and statistical analysis

Exposure time mainly depends on the density of binding sites and the specific activity of the radioligand. The [^{35}S]GTPγS habitually marketed by NEN has 1250 Ci/mmol for the day 1 of labeling, but [^{35}S] isotope decay must be taken into consideration. An exposure time of around 48 h is usually enough to get a good contrast of the images. Films were developed following the recommended instructions for Kodak MR films, and scanned afterwards. Note that a special scanner for transparencies must be used that allows simultaneous scanning of both sides of the film and recognizes a real transparent image in the computer. The images are quantified by transforming optical densities into nCi/g tissue equivalent (nCi/g t.e.) using the calibration mode in the software (NIH-IMAGE, Bethesda, MA, USA). The slide background and nonspecific densities were subtracted. The percentages of inhibition were calculated from the basal and SR141716A-inhibited [^{35}S]GTPγS binding densities. Data were expressed as mean values ±S.E.M.

Statistical analyses were carried out using the Student's *t*-test to calculate the statistical significance of cannabinoid ligand-mediated inhibition (*p* values).

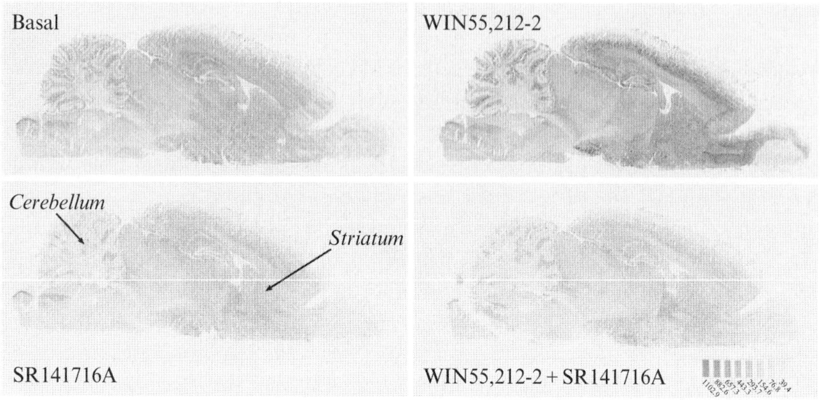

Figure 15.4 Representative autoradiograms of rat brain sagittal sections, corresponding to the [^{35}S]GTPγS binding in the absence (basal) or in the presence of WIN55,212-2 (10 μ*M*) or SR141716A (100 μ*M*) and in the presence of both compounds (WIN55,212-2 and SR141716A). [^{14}C] standards are shown at the right bottom corner (nCi/g t.e.).

3.4. [^{35}S]GTPγS binding autoradiograms in the presence of cannabinoid drugs

Figure 15.4 shows the results obtained after testing agonist, antagonist, and inverse agonist ligands in tissue sections. On the one hand, the cannabinoid agonist WIN55,212-2 stimulated the [^{35}S]GTPγS binding in 61% and 55% above the basal binding, in the *striatum* and *cerebellum* respectively. On the other hand, the inverse agonist compound, SR141716A induced 48% and 78% reduction of [^{35}S]GTPγS basal binding in the same brain areas. However, the presence of WIN55,212-2 was not able to block the inverse agonist effect induced by SR141716A in both areas. These results are in agreement with those published using the autoradiographic experimental conditions described above (Breivogel *et al.*, 1997; Sim-Selley *et al.*, 2001). Therefore, the [^{35}S]GTPγS autoradiography technique can be considered a useful tool to detect the action of inverse agonist drugs in discrete brain areas.

ACKNOWLEDGMENTS

This study was supported by the Spanish Ministry of Education and Science (SAF2007-60211), Basque Government IT 440-10 grant to the Neurochemistry and Neuroscience research group and UPV/EHU Research Groups (GIU07/50). G. Barreda-Gómez is supported by a grant of the UPV/EHU Researchers Specialization Program.

REFERENCES

Bradford, M. M. (1976). A rapid and sensitive method for the quantitation of microgram quantities of protein utilizing the principle of protein-dye binding. *Anal. Biochem.* **72,** 248–254.

Breivogel, C. S., Sim, L. J., and Childers, S. R. (1997). Regional differences in cannabinoid receptor/G-protein coupling in rat brain. *J. Pharmacol. Exp. Ther.* **282,** 1632–1642.

Eglen, R. M., Bosse, R., and Reisine, T. (2007). Emerging concepts of guanine nucleotide-binding protein-coupled receptor (GPCR) function and implications for high throughput screening. *Assay Drug Dev. Technol.* **5,** 425–451.

Engelhardt, S., Grimmer, Y., Fan, G. H., and Lohse, M. J. (2001). Constitutive activity of the human beta(1)-adrenergic receptor in beta(1)-receptor transgenic mice. *Mol. Pharmacol.* **60,** 712–717.

Howlett, A. C., Breivogel, C. S., Childers, S. R., Deadwyler, S. A., Hampson, R. E., and Porrino, L. J. (2004). Cannabinoid physiology and pharmacology: 30 years of progress. *Neuropharmacology* **47,** 345–358.

Pertwee, R. G. (2005). Inverse agonism and neutral antagonism at cannabinoid CB1 receptors. *Life Sci.* **76,** 1307–1324.

Sim-Selley, L. J., Brunk, L. K., and Selley, D. E. (2001). Inhibitory effects of SR141716A on G-protein activation in rat brain. *Eur. J. Pharmacol.* **414,** 135–143.

Stanasila, L., Perez, J. B., Vogel, H., and Cotecchia, S. (2003). Oligomerization of the alpha 1a- and alpha 1b-adrenergic receptor subtypes. Potential implications in receptor internalization. *J. Biol. Chem.* **278,** 40239–40251.

Vilardaga, J. P., Steinmeyer, R., Harms, G. S., and Lohse, M. J. (2005). Molecular basis of inverse agonism in a G protein-coupled receptor. *Nat. Chem. Biol.* **1,** 25–28.

SECTION TWO

NOVEL STRATEGIES AND TECHNIQUES FOR CONSTITUTIVE ACTIVITY AND INVERSE AGONISM

CHAPTER SIXTEEN

Use of Pharmacoperones to Reveal GPCR Structural Changes Associated with Constitutive Activation and Trafficking

Jo Ann Janovick* *and* P. Michael Conn*,†

Contents

Abstract

The gonadotropin-releasing hormone (GnRH) receptor (GnRHR), because of its small size among G-protein-coupled receptors (GPCRs), is amenable to facile preparation of mutants. This receptor is used in our laboratory as a structural model for this super-family of protein receptors and has helped us understand the requirements for proper trafficking. We have demonstrated that

* Divisions of Reproductive Sciences and Neuroscience (ONPRC), Beaverton, Oregon, USA
† Departments of Pharmacology and Physiology, Cell and Developmental Biology, and Obstetrics and Gynecology (OHSU), Beaverton, Oregon, USA

Methods in Enzymology, Volume 485

ISSN 0076-6879, DOI: 10.1016/S0076-6879(10)85016-5

pharmacoperones ("pharmacological chaperones"), small target-specific drugs that diffuse into cells, are capable of rescuing misfolded and misrouted GnRHR mutants and restoring them to function. By rescuing these proteins, these drugs enable the plasma membrane expression of such mutants in living cells and allow examination of mutants that would otherwise be retained in the endoplasmic reticulum and would not be available for ligand binding and signal transduction. As an example of the efficacy of this method, we have shown that mutant $E^{90}K$, which breaks a salt bridge (E^{90}–K^{121}) normally found in the GnRHR, results in constitutive activity when rescued by pharmacoperones. A second method of rescue, involving a mutation that increases the expression of GnRHRs, is shown to have a similar effect. Normally, in the absence of rescue by either of these methods, this mutant, associated with human hypogonadotropic hypogonadism, is misrouted and this constitutive activity has gone unrecognized. This observation [Janovick, J. A., and Conn, P. M. (2010). Salt bridge integrates GPCR activation with protein trafficking. *Proc. Natl. Acad. Sci. USA* **107**, 4454–4458.] showed that the cell normally recognizes this protein as defective and prevents its routing to the plasma membrane.

1. Introduction

Hundreds of GnRHR mutants have been reported but none has been shown to have constitutive activity (CA). This is surprising, since many G-protein-coupled receptors (GPCRs) have mutants (Bond and Ijzerman, 2006) or WT receptors (Seifert and Wenzel-Seifert, 2002) with CA. We considered that one reason for this might be that constitutively active mutants might be recognized as misfolded and retained in the ER.

We have characterized the features of the GnRHR needed for correct trafficking to the plasma membrane and shown that single point mutants are often retained in the ER and unable to traffic to the plasma membrane (Conn and Ulloa-Aguirre, 2010).

A highly conserved (Janovick *et al.*, 2009) structural feature of the GnRHR is a salt-bridge between TM2 and TM3 (residues E^{90} and K^{121} in the human sequence). GnRH agonists, but not peptide antagonists, bind receptor residues D^{98} and K^{121} (Flanagan *et al.*, 2000; Zhou *et al.*, 1995) and alter the relation between TM2 and TM3; this same relation is important for trafficking of the hGnRHR to the plasma membrane (Blomenrohr *et al.*, 2001; Janovick *et al.*, 2009).

Another way to perturb this TM2–TM3 relation, other than ligand binding, is to alter the charge of one of its constituents; mutant $E^{90}K$ would result in charge repulsion between K^{90} and K^{121}. This mutant, which occurs in some cases of human hypogonadotropic hypogonadism (Janovick *et al.*, 2002), is typically recognized by the QCS as a misfolded protein (Maya-Nunez *et al.*, 2002), retained in the ER (Brothers *et al.*, 2004) and does not traffic to the plasma membrane. The patient, lacking functional GnRHR at the plasma membrane, has a reproductive failure.

$E^{90}K$ can be rescued by pharmacoperones, drugs that diffuse into cells and provide a folding template (Figs. 16.1 and 16.2). This template enables (otherwise) misfolded mutants to fold or refold correctly (Janovick *et al.*, 2007), pass a quality control system (constituted of chemically heterogeneous chaperone proteins of the ER that either promote correct folding or retain misfolded proteins) (Conn and Janovick, 2009a) traffic to the plasma membrane, bind agonist, and produce a signal (Fig. 16.1). Pharmacoperones of the hGnRHR rescue mutant $E^{90}K$ by creating a surrogate bridge between D^{98} and K^{121} that also stabilizes the relation between TM2 and TM3 (Janovick *et al.*, 2009). We took advantage of pharmacoperone rescue of mutants and a genetic modification that enhances trafficking to the plasma membrane, in order to examine the relation between this bridge, receptor trafficking, and activation.

2. Methods for Measuring Receptors and Receptor Activity

2.1. Cell transfection

Cos-7 cells were cultured in growth medium (Dulbecco's Minimal Essential Medium, containing 10% fetal calf serum, 20 μg/ml gentamicin) at 37 °C in a 5% CO_2 humidified atmosphere. For transfection of WT or mutant receptors into cells, 5×10^4 cells were plated in 0.25 ml growth medium in 48-well Costar cell culture plates. Twenty-four hours after plating, the cells were washed with 0.5 ml OPTI-MEM then transfected with WT or mutant receptor DNA with pcDNA3.1 (empty vector) to keep the total DNA constant (100 ng/well). Lipofectamine was used according to the manufacturer's instructions. Five hours after transfection, 0.125 ml DMEM with 20% FCS and 20 μg/ml gentamicin was added. Twenty-three hours after transfection the medium was replaced with 0.25 ml fresh growth medium. Where indicated, pharmacoperones (indicated concentration) in 1% DMSO ("vehicle") were added for 4 h in respective media to the cells, and then removed 18 h before agonist treatment (Leanos-Miranda *et al.*, 2003). In this study, we used trypan blue exclusion to show cell viability after drug exposure.

2.2. Inositol phosphate assays

Twenty-seven hours after transfection, cells were washed twice with 0.50 ml DMEM/0.1% BSA/20 μg/ml gentamicin, "preloaded" for 18 h with 0.25 ml of 4 μCi/ml myo-[2-^{3}H(N)]-inositol in inositol free

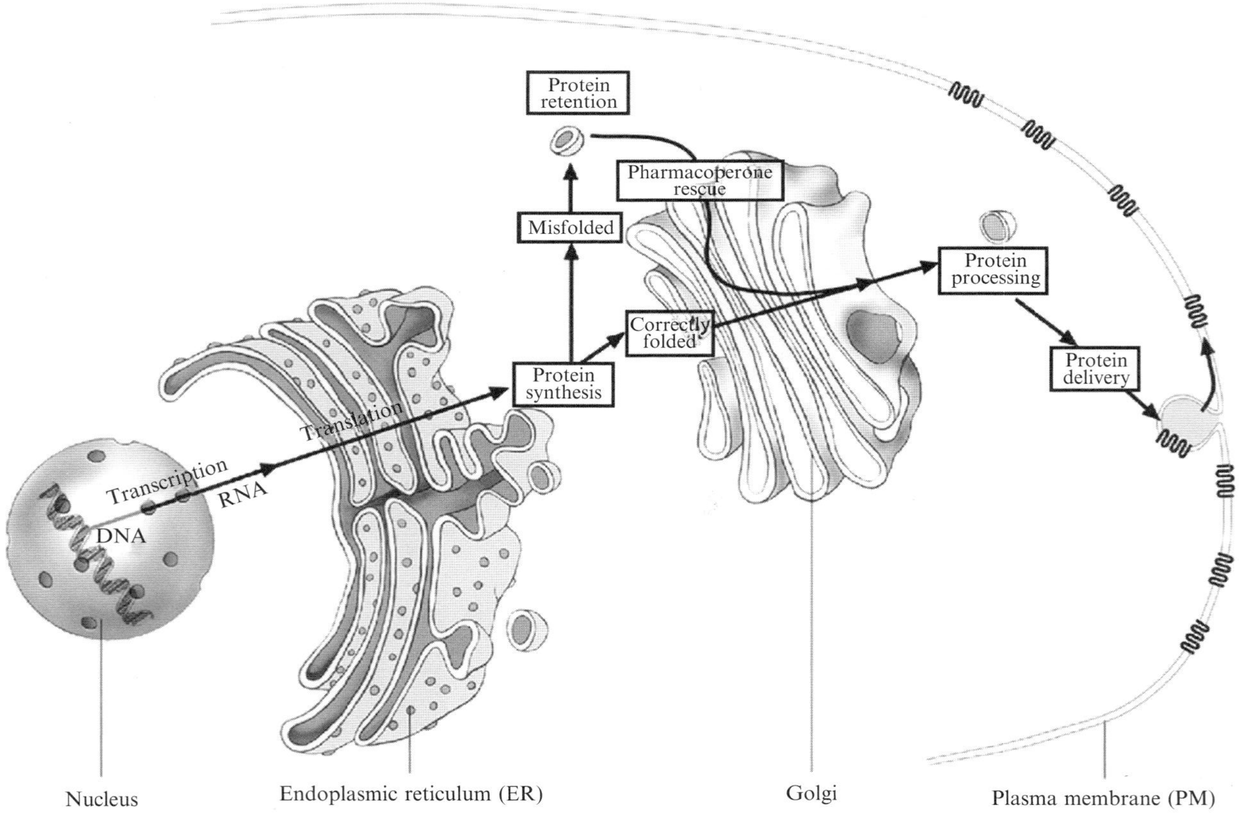
Protein
retention
Misfolded
Pharmacoperone
rescue
Protein
processing
Correctly
folded
Protein
delivery
Protein
synthesis
Translation
Transcription
RNA
DNA
Nucleus
Endoplasmic reticulum (ER)
Golgi
Plasma membrane (PM)

DMEM, and then washed twice with 0.30 ml DMEM (inositol free) containing 5 m*M* LiCl and treated for 2 h with 0.25 ml of a saturating concentration of Buserelin (10^{-7} *M*) in the same medium. When constitutive activity was assessed, Buserelin was omitted from the assessment period. Total inositol phosphate (IP) was then determined (Huckle and Conn, 1987). This assay

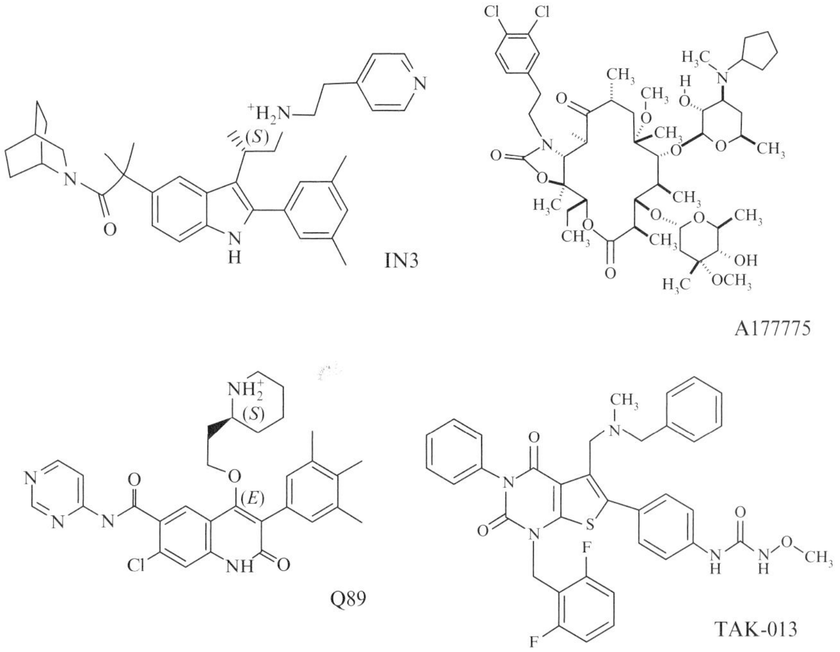

Figure 16.2 Structures of four known GnRHR pharmacoperones from different chemical classes. [Reprinted, with permission, from Conn and Janovick, 2009a.]

Figure 16.1 Cellular sites associated with protein synthesis. Proteins, including GPCRs, are synthesized in the ER and assessed for overall quality. Folding is facilitated by interaction of the nascent polypeptide with molecular chaperones. Misfolded and misassembled products are retained in the ER and exposed to resident chaperones to attempt folding. Eventually, misfolded proteins are dislocated into the cytoplasm for proteosomal degradation after dissociation of the molecular chaperones. Alternatively, defective proteins may be exported to and retained by the Golgi apparatus, retro-translocated to the ER where correct folding is again attempted, or diverted to lysosomes for degradation. Mature products are then exported to their final destination (the PM). Pharmacoperones can frequently rescue misfolded proteins by correcting folding and allowing them to escape retention by the QCS and route to the plasma membrane where they are able to bind ligand and couple to the effector system. Reprinted, with permission, from Conn PM, Ulloa-Aguirre A, Ito J, Janovick JA. G protein-coupled receptor trafficking in health and disease: lessons learned to prepare for therapeutic mutant rescue *in vivo. Pharmacol Rev.* 59(3):225–250, 2007.

has been validated as a sensitive measure of PME for functional receptors when expressed at low amounts of DNA (< 100 ng/well) and stimulated by excess agonist (Castro-Fernandez *et al.*, 2005; Conn *et al.*, 2002; Cook and Eidne, 1997; Janovick *et al.*, 2003a,b; Knollman *et al.*, 2005; Leanos-Miranda *et al.*, 2002, 2003; Ulloa-Aguirre *et al.*, 2003, 2004).

2.3. Cellular assays

Cells were cultured and plated in growth medium as described above, except 10^5 cells in 0.5 ml growth medium were added to 24-well Costar cell culture plates (cell transfection and medium volumes were doubled accordingly). Twenty-three hours after transfection, the medium was replaced with 0.5 ml fresh growth medium with or without pharmacoperone (1 μg/ml In3). Twenty-seven hours after transfection, cells were washed twice with 0.5 ml DMEM containing 0.1% BSA and 20 μg/ml gentamicin, then 0.5 ml of DMEM was added. After 18 h, cells were washed twice with 0.5 ml DMEM/0.1% BSA/10 mM HEPES, then 2×10^6 cpm/ml of [^{125}I]-Buserelin, prepared in our laboratory (specific activity, 700–800 μCi/μg), was added to the cells in 0.5 ml of the same medium and allowed to incubate at room temperature for 90 min, consonant with maximum binding (Brothers *et al.*, 2002). New receptor synthesis during this period is negligible at room temperature. After 90 min, the media was removed and radioactivity was measured (Brothers *et al.*, 2003). To determine nonspecific binding, the same concentrations of radioligand were added to similarly transfected cells in the presence of 10 μg/ml unlabeled GnRH.

2.4. Statistics

Data ($n \geq 3$) were analyzed with one-way ANOVA and then Holm-Sidak test; paired with Student's *t*-test (SigmaStat 3.1, Jandel Scientific Software, Chicago, IL). SEMs are shown.

2.5. Sources of materials

pcDNA3.1 (Invitrogen, San Diego, CA), GnRH analog, D-*tert*-butyl-Ser^6-des-Gly^{10}-Pro^9-ethylamide-GnRH (Buserelin, Hoechst-Roussel Pharmaceuticals, Somerville, NJ), myo-[2-^{3}H(N)]-inositol (PerkinElmer, Waltham, MA; NET-114A), DMEM, OPTI-MEM, lipofectamine, phosphate buffered saline (GIBCO, Invitrogen), competent cells (Promega, Madison, WI), and Endofree maxi-prep kits (Qiagen, Valencia, CA) were obtained as indicated. *Mutant receptors:* WT and mutant GnRHR cDNAs for transfection were prepared as reported (Janovick *et al.*, 2002); the purity and identity of plasmid DNAs were verified by dye terminator cycle sequencing (Applied Biosystems, Foster City, CA). GnRH analogs were obtained as indicated: DPhe^{2}-DAla^{6}-GnRH; DLeu^{2}-DAla^{6} GnRH; Des-His^{2}-GnRH; DPhe^{2}, DPhe^{6}-GnRH (Wyeth-Ayerst Laboratories, Andover, MA); "Nal-Arg," [Ac-DNal^{1}-DCpa^{2}-DPal^{3}, Arg^{5}-D-Arg^{6}-D-Ala^{10}]-GnRH; "Nal-Glu," [Ac-DNal^{1}, DCpa^{2}, D3Pal^{3}, Arg^{5}, DGlu^{6}, DAla^{10}]-GnRH; acyline, [Ac-d-2Nal^{1}, D4Cpa^{1,2}-D3Pal^{3}, Ser^{4}Aph(Ac), d-4Aph(Ac)^{6}-Leu^{7}-ILys^{8}-Pro^{9}-D-Ala^{10}-NH$_{2}$]-GnRH; azaline B, [Ac-DNal^{1}-DCpa^{2}-DPal^{3}-Aph^{5}(atz)-DAph^{6} (atz)-ILys^{8}-DAla^{10}]-GnRH; Asp^{2}-GnRH, Glu^{2}-GnRH, Gly^{2}-GnRH (Dr Jean Rivier, Salk Institute, La Jolla, CA); "FE486," [Ac-D2Nal^{1}-D4Cpa^{2}-D3Pal^{3}-Ser^{4}-4Aph(l-hydroorotyl)^{3}-D4Aph (carbamoyl)^{6}-Leu^{7}-ILys^{8}-Pro^{9}-DAla^{10}]-GnRH (Ferring Research Institute, Southampton, UK); Antide, [Ac-(DNal)-(DpClPhe)-(DPal)-Ser-Lys(nicotinoyl)-[D-Lys(nicotinoyl)]-Leu-Lys(isopropyl)-Pro-[DAla]-NH2 (Serono Laboratories, Aubonne, Switzerland), and DpGlu^{1}-DPhe^{2}-DTrp^{3}-DLys^{6}-GnRH (John Stewart, U Colo, Denver). Irrelevant molecules were obtained as follows: GHRP6 (His^{1}-DTrp^{2}-Ala^{3}-Trp^{4}-DPhe^{5}-Lys^{6}-NH$_{2}$), human galanin, salmon calcitonin, and thyrotropin-releasing hormone (Phoenix Pharmaceuticals, Inc., Belmont, CA); vapreotide was obtained from Debiopharm (Lausanne, Switzerland). Pharmacoperones In3, Q89, Q103 (Merck and Company), A177775 (Abbott Laboratories), and TAK-013 (Takeda Pharmaceuticals) were obtained as indicated (Conn and Janovick, 2009a; Janovick *et al.*, 2009). The drugs are, respectively, from the following chemical classes: indoles, quinolones (both Q89 and Q103), erythromycin macrolides, and (*N*-1-(2,6-difluorobenzyl)-2,4-dioxo-3-phenyl-1,2,3,4-tetrahydrothieno [2,3-*d*]pyrimidin-6-yl]phenyl}-*N*′-methoxyurea). Chemical structures are shown in Fig. 16.2 and the mechanism of action on the hGnRHR (Janovick *et al.*, 2009) has been reported.

3. Assessment of Results

3.1. E^{90}K is constitutively active

Mutations from patients with hypogonadotropic hypogonadism are widely distributed over the 328 residues of the hGnRHR structure (Fig. 16.3A). Most are misfolded and misrouted molecules (Brothers *et al.*, 2004) that are rescuable with pharmacoperone drugs. The dashed lines show the range of basal activity (no agonist) of cells transfected with each of the mutants, but which have *not* been rescued. These dashed lines show that activity does not exceed the vector-only control. When rescued by pharmacoperone In3, only a single mutant, E^{90}K, shows constitutive activity, producing IP in the absence of agonist (Fig. 16.3B). Figure 16.3C shows pharmacoperone-rescued mutants subsequently incubated with the stable GnRHR agonist,

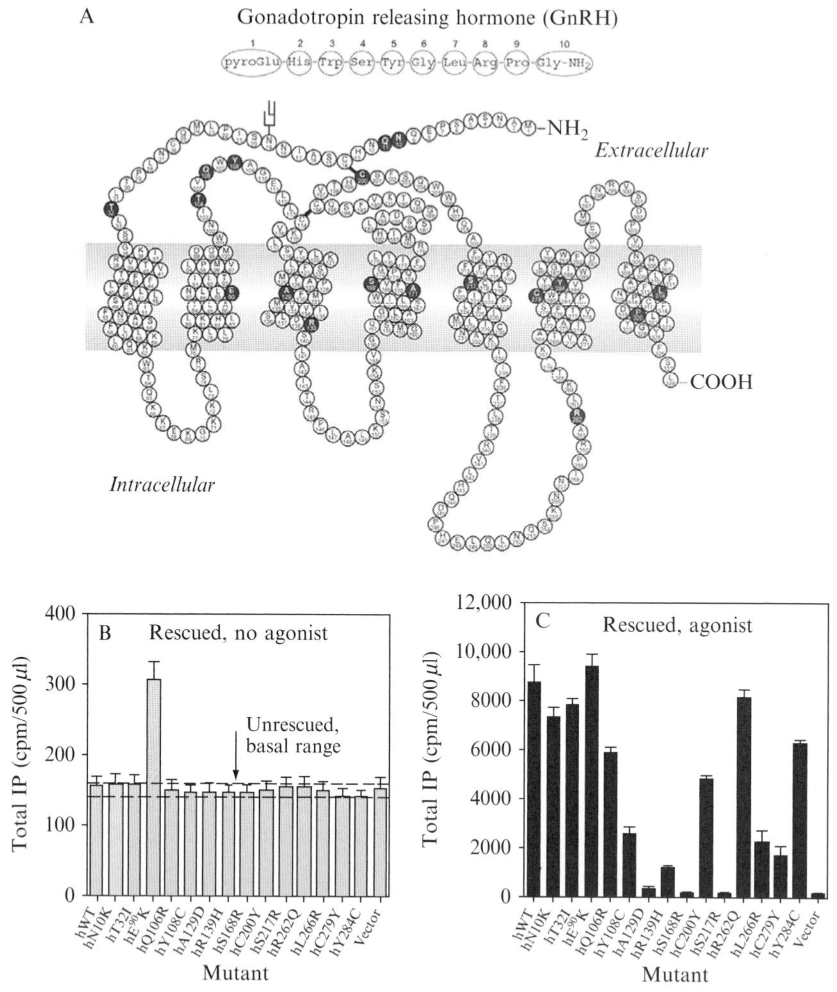

Figure 16.3 (A) Structures of the GnRHR, showing its ligands and sites of mutations (dark circles) that are associated with human disease. Assessment of constitutive activity in IP production of 14 naturally occurring mutations of the hGnRHR following pharmacoperone rescue. COS-7 cells were transfected with 100 ng WT or mutant cDNA as described in Janovick and Conn, 2010. Mutants were (B) incubated in media alone (shown as two dashed lines showing upper and lower level of IP production) or rescued with pharmacoperone (In3) as described in Methods; In3 was then washed out and IP production was measured in response to media alone (no agonist added) or (C) rescued with pharmacoperone (In3), then In3 was washed out and agonist added. In all figures, SEMs are shown for least three independent experiments performed in replicates of 4–6. [Reprinted, with permission, from Janovick and Conn, 2010.]

10^{-7} *M* Buserelin. This confirms that rescue occurs for most mutants and that they are able to couple to IP, but most exhibit no constitutive activity. $S^{168}R$ and $S^{217}R$ cannot be rescued for thermodynamic reasons (Janovick *et al.*, 2006) and $A^{129}D$ is minimally rescuable for unknown reasons.

3.2. Rescue of $E^{90}K$ by two methods: Demonstration by binding and IP analysis

Figure 16.4A (radioligand binding) and B (IP production) shows cells expressing WT human GnRHR (hGnRHR), $E^{90}K$, or sequences from which K^{191} was deleted then either rescued by pharmacoperone In3 or unrescued. Like pharmacoperone rescue, deletion of residue K^{191} is known to rescue misrouted GnRHR mutants, including $E^{90}K$ (Maya-Nunez *et al.*, 2002). Like pharmacoperone rescue, deletion of K^{191} also reveals constitutive activation by $E^{90}K$. When $E^{90}K$–desK^{191} was subjected to pharmacoperone rescue, the constitutive activity (IP) was increased more dramatically than for $E^{90}K$–desK^{191} alone, since routing to the plasma membrane is increased by both deletion of the K^{191} and pharmacoperone. Radioligand binding confirms that pharmacoperones and deletion of residue K^{191} increase the number of mutant receptors at the plasma membrane (Fig. 16.4A). Because In3 was identified in a screen relying on rat GnRH (which does not contain K^{191}), it is difficult to wash out of cells expressing hWT-desK^{191}; accordingly the specific binding after In3 rescue may not quantitatively reflect the number of receptors at the PM. The increased fold-stimulation (and responsiveness) of $E^{90}K$–desK^{191} to agonist, compared with constitutive activity shown by the mutant, suggests that the K^{90}–K^{121} repulsion precludes attaining the optimal activation structure, but that this is corrected by the presence of the agonist (Fig. 16.4B).

3.3. Mutation causes a change in ligand specificity

Because of the close relation of the agonist binding site and the salt bridge, we compared the ligand specificity of the CA mutant to that of the WT receptor. Figure 16.5 shows the effects of GnRH peptide analogs (groups A, B, D) and selected irrelevant compounds (group C) on signaling by the WT hGnRHR (Fig. 16.5A) and by the constitutively active mutant, $E^{90}K$–desK^{191} (Fig. 16.5B). High-affinity antagonists of the WT (Fig. 16.5A, group D) inhibit responsiveness to the agonist, Buserelin; they are also inverse agonists of the mutant (i.e., they inhibit constitutive activity). Irrelevant peptides that have no agonistic or antagonistic activity on the WT also have no activity on the rescued mutant (Fig. 16.5B, group C). Replacement of the His^2 in the natural ligand by Gly^2, Asp^2, or Glu^2 results in little or no agonism with the WT receptor (i.e., in medium without Buserelin) (Fig. 16.5A, group A (Grant *et al.*, 1973)), *but strong agonism with*

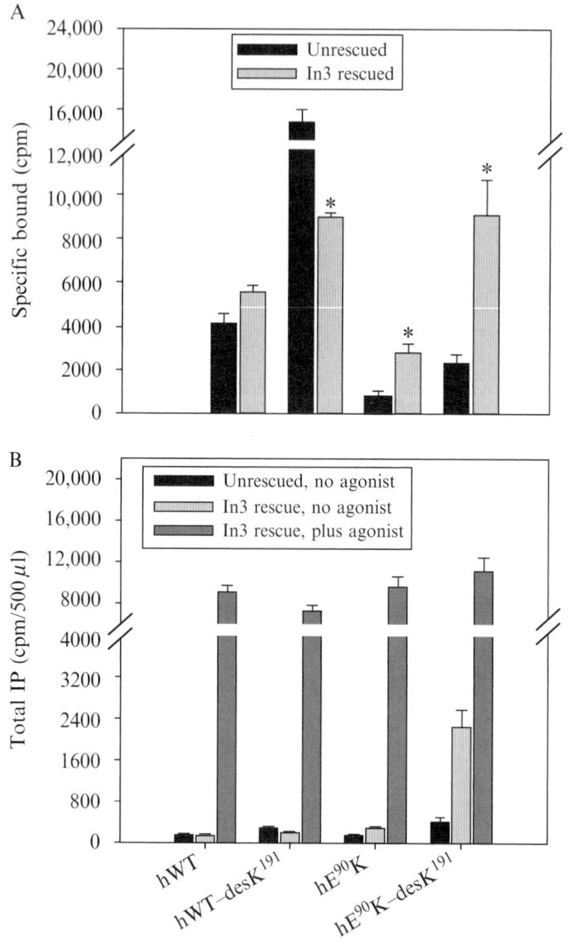

Figure 16.4 Effect on constitutive activity of deleting Lys^{191} on E^{90}K mutant. Cells were transfected with 25 ng WT or mutant (each with or without K^{191}) cDNA and rescued with or without pharmacoperone (In3), as described in Janovick and Conn, 2010 for binding studies. The In3 was washed out and (B) specific binding was determined using 2×10^6 cpm/ml [^{125}I-Buserelin] for 90 min at room temperature. The tracer was removed, cells were washed twice and radioactivity was measured ("*" means that $p < 0.05$ for the comparison of each mutant that is rescued or not). (C) Cells were transfected with 100 ng WT or mutant (hE^{90}K or hE^{90}K–desK^{191}) cDNA and rescued with or without pharmacoperone (In3), as described in Janovick and Conn, 2010 and total IP production was measured in response to medium alone or the addition of agonist (10^{-7} *M* Buserelin). The values for hE^{90}K and hE^{90}K–desK^{191}, difficult to see in the figure, are 156 ± 19 and 440 ± 50 (DMSO) and 292 ± 33 and 2246 ± 300 (In3), respectively. [Reprinted, with permission, from Janovick and Conn, 2010.]

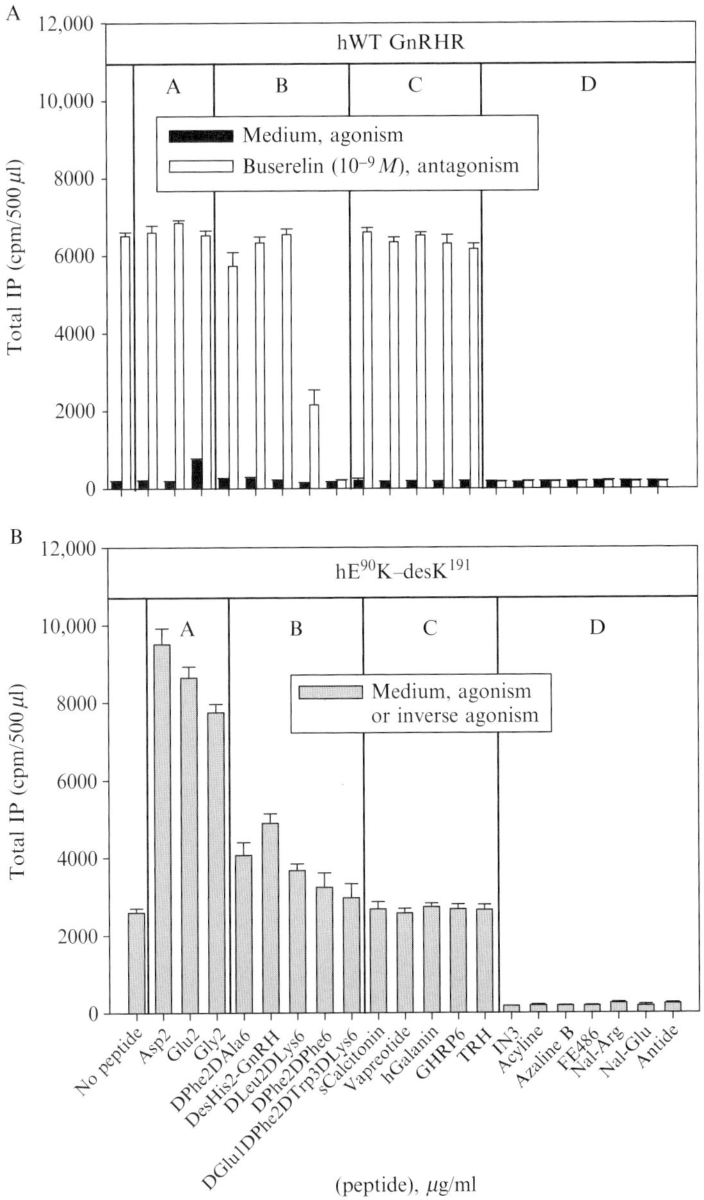

Figure 16.5 Assessment of GnRH peptides and irrelevant compounds on WT and $hE^{90}K$–$desK^{191}$ mutant IP production. (A) Cells were transfected with 100 ng human WT GnRHR cDNA and left unrescued. Various GnRH peptides or irrelevant compounds were added in medium alone (1 μg/ml) or with 10^{-9} *M* Buserelin to show agonistic or antagonistic activity on total IP production. For comparison, a maximal concentration (10^{-9} *M*) of the GnRH agonist, Buserelin, produces a response of

the mutant (Fig. 16.5B, group A). Deletion of His^2 or its replacement by D-amino acid residues, along with D-amino acids at position 6, actions that stabilize the peptide against degradation, produces weak antagonists or no activity on the WT, *but weak agonists of the mutant* (group B). This suggests that the mutant possesses an altered ligand-binding site compared to WT, that it has reduced specificity required for receptor activation by ligands, and that both agonistic and antagonistic peptides which bind near the active site of the mutant can produce the configuration associated with activated receptor.

3.4. Different chemical interactions of pharmacoperones are needed for mutant rescue and development of CA

In order to determine if the same chemical interactions of pharmacoperones with $hE^{90}K$ were involved in both mutant rescue and development of CA, we assessed (a) whether these agents had a similar efficacy order for the development of CA and for mutant rescue and (b) whether pharmacoperones from different chemical classes acted as inverse agonists (inhibiting constitutive activity). In the protocols described thus far, once the mutant is rescued, In3 was removed so that the unoccupied receptor could be assessed for constitutive activity. Figure 16.6A shows, using this approach, that the five different pharmacoperones tested (from four different chemical classes, Fig. 16.3) rescue mutant $E^{90}K$ and lead to some degree of constitutive activity, although this is quite modest for A177775. The concentrations used were selected as the optimum for rescue from prior studies (Janovick *et al.*, 2003a, 2009). In Fig. 16.6B all cells were rescued by In3, then In3 was washed out and each of the five pharmacoperones was added to assess constitutive activation; this shows that the pharmacoperones also behave as inverse agonists, agents that inhibit constitutive activity (compare to dashed line, showing IP production by mutant $E^{90}K$ that was not rescued).

Because the pharmacoperones used in this study all form a bridge between receptor residues D^{98} and K^{121} (Conn and Janovick, 2009a,b; Janovick *et al.*, 2009), other interactions notwithstanding, they stabilize

6517 ± 93 cpm. (B) Cells were transfected with 100 ng $hE^{90}K$–$desK^{191}$ cDNA, then rescued with pharmacoperone (In3). The In3 was then washed out. After 18 h of preloading the cells with ^{3}H-inositol, the cells were incubated with 1 μg/ml of GnRH analogs or irrelevant compounds to assess agonistic or antagonistic activity on total IP production. Group A (Asp^2, Glu^2, Gly^2); Group B ($DPhe^2DAla^6$, $DGlu^1DPhe^2DTrp^3DLys^6$, $DesHis^2$-GnRH, $DLeu^2DLys^6$, $DPhe^2DPhe^6$); Group C (sCalcitonin, Vapreotide, hGalanin, $GHRP^6$, TRH); Group D (In3, Acyline, Azaline B, FE486, Nal-Arg, Nal-Glu, Antide). For comparison, a maximal concentration (10^{-7} *M*) of the GnRH agonist, Buserelin, produces a response of 11,144 ± 1098 cpm. [Reprinted, with permission, from Janovick and Conn, 2010.]

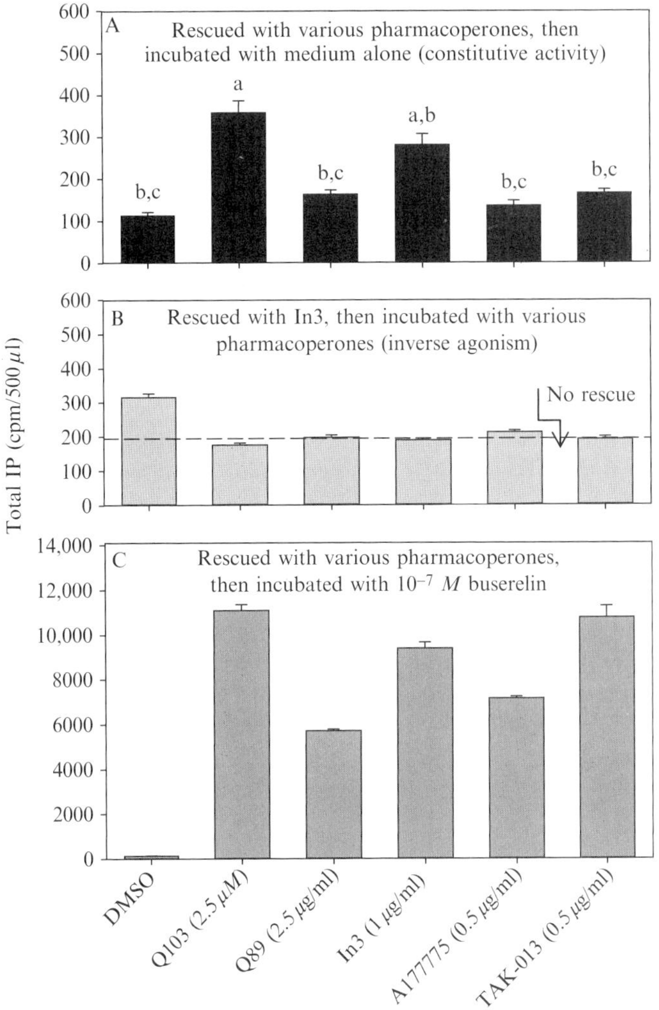

Figure 16.6 Comparison of constitutive activity of $hE^{90}K$ with four classes of pharmacoperones on rescue. All cells were transfected with 100 ng cDNA of $hE^{90}K$ mutant and rescue was attempted with each indicated pharmacoperone. The cells were then incubated with (A) media alone to assess constitutive activity. ("a" means $p < 0.05$ compared with DMSO only; "b" means $p < 0.05$ compared with Q103; "c" means $p < 0.05$ compared with In3) (B) Rescued with In3 and then incubated with various pharmacoperones to assess their actions as inverse agonists. (C) Stimulated with 10^{-7} *M* of the GnRH agonist, Buserelin to assess the ability to couple. (All values are significant compared with DMSO only). [Reprinted, with permission, from Janovick and Conn, 2010.]

the TM2–TM3 relation, an event that allows correct trafficking but precludes constitutive activity, due to occupancy. When the pharmacoperones are washed out and the metabolically stable GnRH agonist, Buserelin, is present following rescue of $E^{90}K$ by each of the pharmacoperones, effector coupling is observed. It is interesting to note that In3 is more effective than TAK-013 in producing constitutive activity (Fig. 16.6A), while it is less effective in rescuing Buserelin-stimulated activity (Fig. 16.6C). This suggests that the binding or steric interactions required for production of constitutive activity and for mutant rescue may be subtly different between different classes of pharmacoperones.

4. Conclusions

The approach described in this work allows examination of mutants of a GPCR that would otherwise be retained in the ER and degraded. In this approach, pharmacoperones are used to refold the misfolded mutant and cause it to traffic to the plasma membrane where it can bind ligand and couple to effector. In the case of hGnRHR $E^{90}K$, this mutant shows constitutive activity and altered ligand specificity. In addition to pharmacoperones, a second method based on a mutation that increases plasma membrane routing (deletion of amino acid K^{191}) is used to confirm that $E^{90}K$ has constitutive activity. The constitutive activity of $E^{90}K$, along with its altered ligand specificity, would have gone unnoticed without this approach, suggesting the value for other proteins.

ACKNOWLEDGMENTS

This work was supported by NIH grants: HD-19899, RR-00163, and HD-18185. We thank Jo Ann Binkerd for formatting the manuscript; this is an extended version of a prior publication (Janovick and Conn, 2010).

REFERENCES

Blomenrohr, M., *et al.* (2001). Proper receptor signalling in a mutant catfish gonadotropin-releasing hormone receptor lacking the highly conserved Asp(90) residue. *FEBS Lett.* **501,** 131–134.

Bond, R. A., and Ijzerman, A. P. (2006). Recent developments in constitutive receptor activity and inverse agonism, and their potential for GPCR drug discovery. *Trends Pharmacol. Sci.* **27,** 92–96.

Brothers, S. P., *et al.* (2002). Conserved mammalian gonadotropin-releasing hormone receptor carboxyl terminal amino acids regulate ligand binding, effector coupling and internalization. *Mol. Cell. Endocrinol.* **190,** 19–27.

Brothers, S. P., *et al.* (2003). Unexpected effects of epitope and chimeric tags on gonadotropin-releasing hormone receptors: Implications for understanding the molecular etiology of hypogonadotropic hypogonadism. *J. Clin. Endocrinol. Metab.* **88,** 6107–6112.

Brothers, S. P., *et al.* (2004). Human loss-of-function gonadotropin-releasing hormone receptor mutants retain wild-type receptors in the endoplasmic reticulum: Molecular basis of the dominant-negative effect. *Mol. Endocrinol.* **18,** 1787–1797.

Castro-Fernandez, C., *et al.* (2005). Beyond the signal sequence: Protein routing in health and disease. *Endocr. Rev.* **26,** 479–503.

Conn, P. M., and Janovick, J. A. (2009a). Drug development and the cellular quality control system. *Trends Pharmacol. Sci.* **30,** 228–233.

Conn, P. M., and Janovick, J. A. (2009b). Trafficking and quality control of the gonadotropin releasing hormone receptor in health and disease. *Mol. Cell. Endocrinol.* **299,** 137–145.

Conn, P. M., and Ulloa-Aguirre, A. (2010). Trafficking of G-protein-coupled receptors to the plasma membrane: Insights for pharmacoperone drugs. *Trends Endocrinol. Metab.* **23** (3), 190–197.

Conn, P. M., *et al.* (2002). Protein origami: Therapeutic rescue of misfolded gene products. *Mol. Interv.* **2,** 308–316.

Cook, J. V., and Eidne, K. A. (1997). An intramolecular disulfide bond between conserved extracellular cysteines in the gonadotropin-releasing hormone receptor is essential for binding and activation. *Endocrinology* **138,** 2800–2806.

Flanagan, C. A., *et al.* (2000). Multiple interactions of the Asp(2.61(98)) side chain of the gonadotropin-releasing hormone receptor contribute differentially to ligand interaction. *Biochemistry* **39,** 8133–8141.

Grant, G., *et al.* (1973). Control of anterior pituitary hormone secretion. *In* "Recent Studies of Hypothalamic Function International Symposium," (K. Lederis and K. E. Cooper, eds.), pp. 180–195. Karger, Calgary.

Huckle, W. R., and Conn, P. M. (1987). Use of lithium ion in measurement of stimulated pituitary inositol phospholipid turnover. *Methods Enzymol.* **141,** 149–155.

Janovick, J. A., and Conn, P. M. (2010). Salt bridge integrates GPCR activation with protein trafficking. *Proc. Natl. Acad. Sci. USA* **107,** 4454–4458.

Janovick, J. A., *et al.* (2002). Rescue of hypogonadotropic hypogonadism-causing and manufactured GnRH receptor mutants by a specific protein-folding template: Misrouted proteins as a novel disease etiology and therapeutic target. *J. Clin. Endocrinol. Metab.* **87,** 3255–3262.

Janovick, J. A., *et al.* (2003a). Structure-activity relations of successful pharmacologic chaperones for rescue of naturally occurring and manufactured mutants of the gonadotropin-releasing hormone receptor. *J. Pharmacol. Exp. Ther.* **305,** 608–614.

Janovick, J. A., *et al.* (2003b). Evolved regulation of gonadotropin-releasing hormone receptor cell surface expression. *Endocrine* **22,** 317–327.

Janovick, J. A., *et al.* (2006). Regulation of G protein-coupled receptor trafficking by inefficient plasma membrane expression: Molecular basis of an evolved strategy. *J. Biol. Chem.* **281,** 8417–8425.

Janovick, J. A., *et al.* (2007). Refolding of misfolded mutant GPCR: Post-translational pharmacoperone action in vitro. *Mol. Cell. Endocrinol.* **272,** 77–85.

Janovick, J. A., *et al.* (2009). Molecular mechanism of action of pharmacoperone rescue of misrouted GPCR mutants: The GnRH receptor. *Mol. Endocrinol.* **23,** 157–168.

Knollman, P. E., *et al.* (2005). Parallel regulation of membrane trafficking and dominant-negative effects by misrouted gonadotropin-releasing hormone receptor mutants. *J. Biol. Chem.* **280,** 24506–24514.

Leanos-Miranda, A., *et al.* (2002). Receptor-misrouting: An unexpectedly prevalent and rescuable etiology in gonadotropin-releasing hormone receptor-mediated hypogonadotropic hypogonadism. *J. Clin. Endocrinol. Metab.* **87,** 4825–4828.

Leanos-Miranda, A., *et al.* (2003). Dominant-negative action of disease-causing gonadotropin-releasing hormone receptor (GnRHR) mutants: A trait that potentially coevolved with decreased plasma membrane expression of GnRHR in humans. *J. Clin. Endocrinol. Metab.* **88,** 3360–3367.

Maya-Nunez, G., *et al.* (2002). Molecular basis of hypogonadotropic hypogonadism: Restoration of mutant (E(90)K) GnRH receptor function by a deletion at a distant site. *J. Clin. Endocrinol. Metab.* **87,** 2144–2149.

Seifert, R., and Wenzel-Seifert, K. (2002). Constitutive activity of G-protein-coupled receptors: Cause of disease and common property of wild-type receptors. *Naunyn Schmiedebergs Arch. Pharmacol.* **366,** 381–416.

Ulloa-Aguirre, A., *et al.* (2003). Misrouted cell surface receptors as a novel disease aetiology and potential therapeutic target: The case of hypogonadotropic hypogonadism due to gonadotropin-releasing hormone resistance. *Expert Opin. Ther. Targets* **7,** 175–185.

Ulloa-Aguirre, A., *et al.* (2004). Pharmacologic rescue of conformationally-defective proteins: Implications for the treatment of human disease. *Traffic* **5,** 821–837.

Zhou, W., *et al.* (1995). A locus of the gonadotropin-releasing hormone receptor that differentiates agonist and antagonist binding sites. *J. Biol. Chem.* **270,** 18853–18857.

CHAPTER SEVENTEEN

Application of Large-Scale Transient Transfection to Cell-Based Functional Assays for Ion Channels and GPCRs

Jun Chen,[*,1] Sujatha Gopalakrishnan,[†,1] Marc R. Lake,[†] Bruce R. Bianchi,[*] John Locklear,[†] *and* Regina M. Reilly[*]

Contents

Abstract

Despite increasing use of cell-based assays in biomedical research and drug discovery, one challenge is the adequate supply of high-quality cells expressing the target of interest. To this end, stable cell lines expressing the target are often established, maintained, and expanded in large-scale cell culture. These steps require significant investment of time and resources. Moreover, variability

[*] Neuroscience Research, Global Pharmaceutical Research and Development, Abbott Laboratories, Abbott Park, Illinois, USA
[†] Advanced Technology, Global Pharmaceutical Research and Development, Abbott Laboratories, Abbott Park, Illinois, USA
[1] Corresponding authors: Jun Chen jun.x.chen@abbott.com and Sujatha Gopalakrishnan sujatha.m.gopalakrishnan@abbott.com

Methods in Enzymology, Volume 485
ISSN 0076-6879, DOI: 10.1016/S0076-6879(10)85017-7

occurs regularly in cell yield, viability, expression, and target activities. In particular, stable expression of many targets, such as ion channels, causes toxicity, cell line degeneration, and loss of functional activity. To circumvent these problems, we utilize large-scale transient transfection (LSTT) to generate a large quantity of cells, which are cryopreserved and readily available for use in cell-based functional assays. Here we describe the application of LSTT cells to ion channel and G protein-coupled receptor (GPCR) assays in a drug discovery setting. This approach can also be applied to many other assay formats and target classes.

1. Introduction

One of the major challenges in drug discovery is identification of lead compounds with optimal chemical and pharmacological properties. This is often achieved by primary high-throughput screening (HTS) followed by secondary screening to enable lead optimization. At the HTS stage, libraries of 100,000 to several million compounds are screened against the target of interest. The major considerations are assay robustness, throughput, reagent supply, and cost. During lead optimization, HTS hits and derivative analogs (often in the thousands) are interrogated in greater detail (e.g., IC_{50} or EC_{50} determination), in different assay formats, against various orthologs (e.g., human, rat, and mouse) and against related family members. The data quality is the major consideration. Therefore, the prosecution of a drug discovery program requires a large number of *in vitro* assays, especially cell-based functional assays (Moore and Rees, 2001; Johnston, 2002). These assays involve expression of the gene of interest, most often in mammalian cells; and target activities are measured by their cellular functions or impacts on cellular pathways. Compared to cell-free assays using purified proteins, cell-based assays measure ligand–target interactions in a physiologically relevant environment. They are also capable of distinguishing mechanisms of action (e.g., agonism, antagonism, allosteric modulation). In addition, cell-based assays can accommodate poorly defined targets with no known ligands, such as orphan receptors (Moore and Rees, 2001).

The versatility and efficiency of cell-based functional assays are often hindered by the demand for high-quality live cells. In drug discovery, a large number of cells are required to support the screening campaign and lead optimization, which are ongoing processes lasting from months to years. Typically, stable cell lines expressing the gene of interest are used. However, establishing/characterizing/selecting stable clones with desired expression levels and activities, maintaining, and scaling up stable clones for HTS can be extremely laborious, time consuming, and expensive. Intrinsic to living cells, significant variability occurs in growth rate, yield, viability, and target activities. Moreover, the stable expression of certain targets can lead to cellular toxicity, which manifests as cell damage and loss of target activity. To overcome toxicities associated with constitutive expression of some genes, inducible expression system has been attempted (Xia *et al.*, 2003). However, it still requires generation of stable cell lines. The induction of gene expression on a large scale is often tedious, expensive, and inconsistent. Moreover, "leaky" expression of some proteins, even at low levels, can lead to cell damage when cells are cultured for extended periods of time. Baculovirus expression systems (e.g., BacMam) can be used for expression in mammalian cells, though extra steps and optimizations are required (Kost *et al.*, 2010). Over the past several years, we have used transient transfection as an efficient, versatile, and reliable method for many drug discovery programs.

2. Large-Scale Transient Transfection

Transient transfection in mammalian cells was traditionally conducted on a small scale (e.g., 10^7 cells) using adherent cells; therefore its application was generally limited to characterization of the gene product, assay development, and pilot screens. Recently, transfection on a large scale (e.g., $>10^{10}$ cells) has become available, due to simplified handling steps, reduced reagent consumption, and most importantly, development of host cells amenable to suspension growth. Several cell lines, including HEK293-EBNA (constitutively expressing the Epstein-Barr virus nuclear antigen), HEK293-F, and CHO-F (Invitrogen), have been adapted to growth in suspension, therefore a large number of cells can be prepared and transfected in a short period of time. In our hands, the FreeStyle™ 293 Expression System (Invitrogen) is particularly straightforward and reliable. It has been extremely useful in advancing our ion channel and GPCR (G protein-coupled receptor) efforts. A protocol using HEK293-F cells is described below (Chen *et al.*, 2007).

The gene of interest is cloned in a vector that contains a promoter (e.g., CMV in pCDNA3, Invitrogen) to drive a high level of mammalian expression. HEK293-F cells are grown in suspension in flasks (cell volume 30 mL to 1 L) or in a Wave Bioreactor (GE Healthcare, Piscataway, NJ) (2–10 L). To support high-density growth of suspension culture and transfection, cells are grown in FreeStyle™ 293 media, an optimized, serum-free formulation. 293fectin™ (Invitrogen) is used as transfection reagent. Cells are grown or diluted to a density of 1.0×10^6 cells/mL, and then transfected with the plasmid DNA of choice by using 1 mg/L cells and 1.33 mL 293fectin/mg DNA mixed in OPTI-MEM medium (5% final expression volume) for 25 min. The mixture is then added to the cells and incubated at 37 °C, 125 rpm, and 8% CO_2. Subsequently cells (usually at a density of $2.5–3 \times 10^6$ cells/mL) are harvested after 48 h by centrifugation ($1000\times g$, 5 min).

Many ion channels and GPCRs are heteromultimers composed of different subunits. To mimic their function and pharmacology in native tissues, it is necessary to express all subunits at the appropriate levels and stoichiometry. This can be extremely difficult to replicate in stable cell lines; but relatively straightforward through transient transfection. To do so we perform pilot transfections using mixes of several genes in ratios (e.g., from 10:1 to 1:10); choose the optimal ratio based on protein expression, functional activities, and pharmacology; then prepare a large-scale transfection. This approach has been successfully applied to NR2B and NR1-containing NMDA receptors. Previous efforts failed to establish stable cell lines with appropriate stoichiometry (NR2B/NR1) and pharmacology properties. However, by varying the ratio of two plasmids (each containing NR1 or NR2B) in transient transfection, we found that a 1:6 ratio produced comparable amounts of proteins, appropriate activities, and pharmacology. Subsequently assays were developed and HTS proceeded smoothly. The same approach has been extended to other heteromultimeric proteins.

3. Cryopreservation of Cells

LSTT cells can be used fresh; but most often, they are cryopreserved, stored at −80 °C up to several years, and revived on demand. This approach greatly reduces cell culture work and also circumvents batch-to-batch variation. Controlled rate freezing is critical to retaining cell viability and target activities. A traditional freezing procedure utilizes Cryo 1 °C "Mr. Frosty" Freezing Container (Nalgene Labware, Rochester, NY). The polycarbonate container is filled with isopropyl alcohol, and vials (18 at a time) are submerged in an internal rack in the isopropyl alcohol. The container is then placed in a −80 °C freezer with samples cooled at roughly 1 °C/min. This simple device usually works well on a small scale, but not consistently when freezing hundreds of aliquots. To overcome this issue, we employ a liquid nitrogen driven controlled rate freezer (CryoMed, Thermo Scientific) with an integral sample probe that measures the temperature and cools the sample vial at the programmed rate. This instrument has proved to be reliable and the superior choice for controlled rate freezing for large samples.

For LSTT cells, we supplement the 293 Freestyle medium with 10% DMSO and 10% fetal bovine serum and freeze cells in the CryoMed freezer at a rate of 1 °C/min to −40 °C, and 10 °C/min to −80 °C. Cell densities can range from 10–40 million cells/mL and volumes can vary from 2 mL (standard cryovials) up to 10 mL (15 mL polypropylene centrifuge tubes). Cells can then be stored at −80 °C or in a liquid nitrogen tank. When needed, vials are removed from −80 °C and quickly thawed in a 37 °C water bath. Cells are aseptically transferred into conical tubes containing DPBS (10 mL/vial). After centrifugation at 1000×g for 5 min, supernatant is removed by aspiration and cells are resuspended in 293 medium. Usually the viability of cells is ~ 90% as assessed by a ViCell cell counter (Beckman Coulter, Brea, CA). Routinely, we obtain 1.5–2.0 × 10^{10} cells from a 6 L transfection, which is sufficient to prepare 1950–2600 384-well plates (2 × 10^4 cells/well).

4. Application to Ion Channel Assays

Ion channels play important physiological functions in neuronal signaling, neurotransmitter release, sensory function, muscle contraction, gene transcription, and ionic homeostasis; and genetic mutations in many ion channel genes cause human diseases (channelopathies) (Hubner and Jentsch, 2002). Ion channels have been well recognized as important therapeutic targets for treating a number of human diseases (Kaczorowski *et al.*, 2008). So far, approximately 180 FDA-approved drugs exert their effects through interaction with ion channels, with several drugs reaching multibillion dollar sales (e.g., amlodipine for hypertension and gabapentin for epilepsy/pain). There are more than 400 ion channel genes in the human genome, with ~300 pore-forming subunits (Treherne, 2006). The number of functional ion channels is further increased by alternative splicing, heteromeric assembly, and association with auxiliary subunits. Due to the large number of ion channels, their structural homology and diverse physiological functions, subtype-specific targeting is often required. Therefore, it is necessary to test compounds against many ion channels, often in cellular assays, to ensure target selectivity.

There are many cell-based functional assays available for ion channels (Zheng *et al.*, 2004). Most of these assays are based on the conductivity to inorganic ion species (e.g., Ca^{2+}, Na^+, K^+, Cl^-), but they differ in measurement parameters, information content, throughput and cost, and are suited for different targets and different stages of drug discovery. For example, the fluorescence-based Ca^{2+} influx assay and membrane potential assays are the predominant formats for HTS, due to high throughput and low costs; but they are indirect measurements of channel function and prone to artifacts. Patch clamp electrophysiology is the gold standard for studying channel function, but it is extremely labor intensive and notoriously low throughput. Regardless of their differences, these assays universally require consistent supply of high-quality cells expressing the channel of interest, which can be provided through transient transfection.

The utility of transient transfection is here illustrated by using TRPA1 as an example. TRPA1 is involved in sensation, pain, neurogenic inflammation and represents an attractive drug target. It is permeable to Ca^{2+}, Na^{+}, and K^{+}; activation of TRPA1 simultaneously mediates increase in intracellular Ca^{2+} and membrane depolarization. Therefore, a number of assay formats are suitable to monitor TRPA1 function. As the first step in the TRPA1 drug discovery program, we established HEK293 stable cell lines expressing TRPA1. However, after single colonies were picked, the cells grew very slowly and appeared unhealthy. This was consistent with a report by another group that the constitutive expression of TRPA1 was toxic (Story *et al.*, 2003). We also noted a loss of TRPA1 expression and activity during cell passage. As shown in Fig. 17.1A, a stable cell line exhibited robust expression at passage 5; but not 28 weeks later at passage 33. In contrast, robust TRPA1 expression was detected in transiently transfected cells, either fresh, or cryopreserved at −80 °C for 35 weeks (Fig. 17.1B). We have used cryopreserved LSTT cells to develop a variety of cell-based assays, including Ca^{2+} influx, membrane potential, fluorescence energy transfer-based membrane potential (FRET-MP), Yo-Pro uptake, ratiometric Ca^{2+} imaging, and patch clamp electrophysiology. Described below are background information and assay protocols.

4.1. Ca^{2+} influx assay

This assay uses Ca^{2+} indicators (e.g., Fluo-4) and a fluorometric imaging plate reader, or FLIPR (MDS Analytic Technology, Sunnyvale, CA) to measure intracellular Ca^{2+}. When excited at a wavelength of 480 nm, Ca^{2+} dyes give an emission at 525 nm, with intensities corresponding to the free

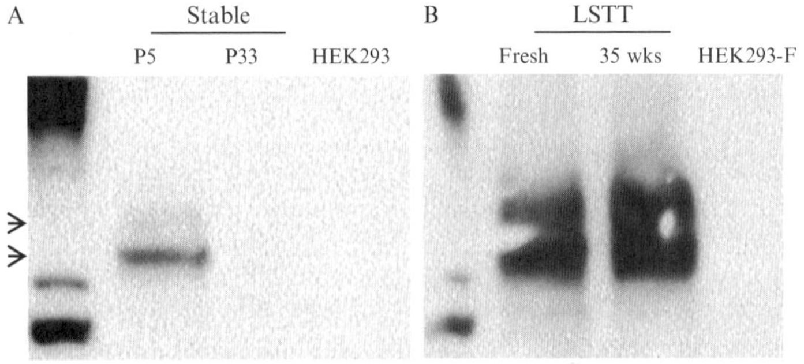

Figure 17.1 TRPA1 protein expression in a stable cell line and LSTT cells. (A) In the stable cell line, TRPA1 expression (127 and 150 Da as indicated by arrows) is present at passage 5 (P5) but not at passage 33 (P33). Protein extracts were immunoprecipitated with one human TRPA1 polyclonal antibody and detected by another. Protein markers are shown in the left lane. (B) In LSTT cells, robust protein expression is present in fresh cells or cells frozen for 35 weeks at −80 °C). (Adapted from Chen *et al.*, 2007 with permission.)

intracellular Ca^{2+} concentration. The CCD camera equipped in the FLIPR instrument measures fluorescence signals from a microplate (96-, 384-, or 1536-well) so a large data set can be generated. In LSTT cells expressing TRPA1, the agonist AITC (allyl isothiocyanate) activates the channel and elicits Ca^{2+} influx in a dose-dependent manner, as represented by increases in fluorescent signals (Fig. 17.2). To monitor the effect of long time storage at −80 °C, a single batch of LSTT cells, either fresh, or cryopreserved for 1, 4, and 42 weeks were tested. Under similar conditions (e.g., cell density,

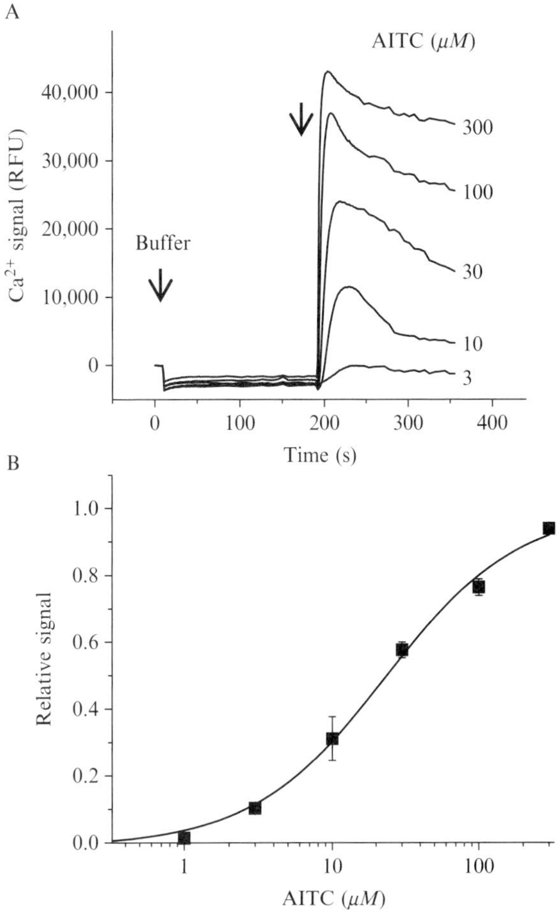

Figure 17.2 Ca^{2+} influx assay using cryopreserved LSTT cells expressing TRPA1. (A) Representative traces of fluorescence signals (RFU, relative fluorescence units) in responses to varying concentrations of AITC. Buffer and AITC additions are indicated by arrows. (B) Concentration–effect relationship of AITC. (Adapted from Chen *et al.*, 2007 with permission.)

Table 17.1 Comparison of TRPA1 activities in fresh and cryopreserved LSTT cells in the Ca^{2+} influx assays ($n = 6$–8)

	Fresh	Frozen 1w	Frozen 4w	Frozen 42w
Signals	30,917 ± 1829	27,761 ± 1631	25,579 ± 1061	25,184 ± 1329
Signal/background	15	13.3	11.8	12.0
AITC EC_{50} (μM)	21.3 ± 0.4	23.0 ± 3.8	15.1 ± 3.1	16.9 ± 2.4
RR IC_{50} (μM)	13.8 ± 0.9	14.0 ± 2.1	17.2 ± 3.5	14.8 ± 2.5

Samples include: fresh LSTT cells (Fresh), same batch of cells cryopreserved for 1 week (Frozen 1w), 4 weeks (Frozen 4w), and 42 weeks (Frozen 42w). Variability in cell densities and dye loading were compensated by adjusting the basal fluorescent signal to 15,000. 30 μM AITC evoked-signals are reported. Signal to background ratios were determined by normalizing signals elicited by 30 μM AITC against signals induced by buffer addition. IC_{50} for block of AITC-induced Ca^{2+} influx by RR was determined using 30 μM AITC. (Adapted from Chen *et al.*, 2007 with permission.)

incubation time, dye loading, and basal signals), robust signals and signal to background ratios were retained for all samples (Table 17.1). In addition, similar EC_{50} for AITC and IC_{50} for ruthenium red (RR, a nonselective antagonist) were maintained. Therefore cryopreserved cells maintained robust signals and appropriate pharmacology.

Following assay optimization, we embarked on a HTS campaign to identify TRPA1 modulators by screening >700,000 compounds (Chen *et al.*, 2007) (Fig. 17.2A). From a 6 L transfection, 1.1×10^{10} cells were generated, harvested, and cryopreserved, sufficient for plating ~1400 384-well microplates. Compounds were screened as 10 compound mixtures against 30 μM AITC responses. The scatter was minimal and hits were readily identifiable (Fig. 17.2B). Hits were retested and their potencies determined. Selected hits were further confirmed by whole cell patch clamp electrophysiology. These efforts identified antagonists and agonists suitable for target validation. Throughout the whole process (assay development, screening, hit confirmation, and potency determination), only a single batch of LSTT cells was used; robust and consistent signals were maintained. Similar assays were also established for other TRP channels such as TRPM8 and TRPV1-V4 (Bianchi *et al.*, 2007). Described below is protocol for TRPA1 Ca^{2+} influx assay in 384-well format.

Frozen aliquots of LSTT cells are quickly thawed, washed, and resuspended in 293 medium. Cells are seeded into black-walled clear-bottom Biocoat™ poly-D-lysine 384-well plates (BD Biosciences, Bedford, MA) at a density of 2×10^4 cells/well, and incubated overnight at 37 °C under a humidified 5% CO_2 atmosphere. Ca^{2+} influx is measured using a FLIPR calcium assay kit (R8033; MDS Analytic Technology). The Ca^{2+} indicator dye is dissolved in HBSS/HEPES, Hanks' balanced salt solution (Invitrogen) supplemented with 20 mM HEPES buffer (Invitrogen). Before starting the assay, the medium is removed by aspiration, and cells are loaded with 30 μL Ca^{2+} dye for 2–3 h at room temperature. Known agonists, antagonists, and test compounds are prepared in HBSS/HEPES as 4× concentration. Two additions are made with a delivery rate of 20 μL/s. To determine agonist activities, buffer or a known antagonist (15 μL) is added at the 10 sec time point; compounds (15 μL) are added at the 3 min time point followed by another 3 min reading. To determine antagonist activities, compounds are added first, AITC or a known agonist is added 3 min later to activate TRPA1 channel. Changes in fluorescence are measured over time with an excitation wavelength of 485 nm and emission at 525 nm. Max–min signals before the second addition to the end of experiments are used for data analysis.

4.2. Membrane potential assay

The membrane depolarization associated with TRPA1 activation can be detected by using a membrane potential-sensitive dye combined with a fluorescence quencher. At hyperpolarized potentials, the negatively charged oxonol dye is localized at outleaf of the lipid bilayer; upon excitation, the emission is absorbed by the quencher near the membrane. At depolarized potentials (i.e., opening of TRP channels), the oxonol dye moves to the inner layer of bilayer and a strong fluorescent emission occurs. The fluorescent signals can be detected by the CCD camera in the FLIPR instrument. In the TRPA1 membrane potential assay, we used the same cells and nearly identical conditions to the Ca^{2+} influx assay, except for using a membrane potential assay kit (R8034, MDS Analytic Technology) and a different optical setting (excitation at 488 nm and emission at 550 nm). The ACTOne membrane potential dye (Codex BioSolutions, Montgomery Village, MD) is an excellent alternative, since it is reasonably priced and performs equally well in terms of signals and pharmacology determination.

4.3. FRET-MP assay

The fluorescence resonance energy-transfer based membrane potential assay (FRET-MP) utilizes the energy transfer between the fluorescence donor coumarin-phospholipid (CC2-DMPE) and acceptor oxonol (e.g., Dis-BAC_2) to measure membrane potential. Coumarin binds only to the cell membrane exterior, while oxonol is negatively charged and translocates according to membrane potential. At the resting state, the two probes are colocalized at the exterior membrane; an excitation of coumarin results in energy transfer to oxonol and an emission at 580 nm. At the depolarized state, oxonol translocates to the cell interior and coumarin emits at 460 nm. Therefore, the ratio between emissions at 460 and 580 nm represents membrane depolarization. Compared to the fluorescence membrane potential assay, FRET-MP requires more dye loading/washing steps and the data analysis is considerably more tedious. However, FRET-MP is a more accurate measurement of membrane potential and also is less prone to artifacts or well-to-well variation. Described below is a FRET-MP protocol for the TRPA1 using LSTT cells, FLIPR-Tetra and Voltage Sensor Probe optics (MDS Analytic Technology).

LSTT cells are plated in 384-well plates similar to the Ca^{2+} influx assay. Cells are first loaded with 15 μL, 10 μ*M* of CC2-DMPE (Invitorgen) for 40 min in the dark, washed with the HBSS/HEPES buffer, then loaded with 15 μL, 10 μ*M* of $DiSBAC_2$ (Invitrogen) for a minimum of 40 min. In experiments to identify antagonists, compounds are added first; followed by known agonists. In experiments to identify agonists, buffer or known antagonists were added first, followed by compounds. Addition of AITC or other agonists activates the channel and increases the F460/F580 ratio, indicating membrane depolarization. For data analysis, the ratio is further normalized against baseline. Routinely a ~40% increase in normalized ratio (i.e., from 1 to 1.4) is observed following agonist stimulation, and the data are very tight.

4.4. Yo-Pro uptake assay

The function of ion channels is attributed to their ability to pass certain ion species (e.g., Ca^{2+}, K^+, Na^+) across the plasma membrane. It has been assumed that ion selectivity is an invariant signature of each channel. However, this notion has been revised recently for a number of channels including P2X7, TRPV1, and TRPA (Chung *et al.*, 2008; Chen *et al.*, 2009). One key finding is the increased permeability to Yo-Pro (Invitrogen). Yo-Pro is a 376 Da cation dye that is impermeable to the plasma membrane under normal circumstance. However, when the pore of TRPA1 is dilated, Yo-Pro enters into the cell, binds nucleic acids and emits fluorescence signals. Described below is a protocol for studying TRPA1-mediated Yo-Pro uptake by using LSTT cells (Chen *et al.*, 2009).

Fluorescent signals are measured using the FLIPR instrument. Yo-Pro (1 m*M* in dimethyl sulfoxide, Invitrogen) is diluted to a final concentration of 2 μ*M* and added to LSTT cells in microplates. A two-addition protocol is used for evaluating agonist activity (i.e., Yo-Pro uptake) and antagonist activities (i.e., inhibition of agonist-evoked Yo-Pro uptake): 10 s baseline readout, addition of assay buffer or antagonist, 3 min readout, addition of agonists, and readout for 60 min. Max–min signals before the second addition to the end of the experiment are obtained. Under these conditions, TRPA1 agonists such as AITC evoke antagonist-sensitive Yo-Pro uptake, indicating pore-dilation. This result is corroborated by an independent experiment measuring $NMDG^+$ permeability (Chen *et al.*, 2009). Through Yo-Pro uptake assays using LSTT cells, we have demonstrated that pore dilation occurs in TRPA1, TRPV1, TRPV3, TRPV4, but not in the TRPM8 channel.

4.5. Ratiometric Ca^{2+} imaging

This assay uses Fura 2-AM (Fura 2-acetoxymethyl ester, Invitrogen) to calculate intracellular Ca^{2+} within a single cell based on the relative emission between two excitation wavelengths (i.e., 340 and 380 nm), which automatically cancels out differences in dye loading and cell thickness. It also allows dynamic measurement, since compounds can be applied/washed off many times over a period, for example, 1 h. This technique is widely used to assess Ca^{2+} dynamics in acutely dissociated cells (e.g., dorsal root ganglion neurons), and can also be applied to LSTT cells.

Briefly, LSTT cells expressing TRPA1 are seeded in poly-D-lysine coated coverslips and incubated with 5 μ*M* Fura 2-AM for 30 min at 37 °C. Intracellular Ca^{2+} concentration is measured on an imaging system consisting of the X-Cite™ 120 Fluorescence Illumination System (EXFO) and a Digital Camera (C4742-95, Hamamatsu Photonics, Hamamatsu, Japan) connected to an Olympus 1X71 microscope. Through the control of Slide Book 4.2 software, fluorescence is measured during excitation at 340 and 380 nm, and the ratio of the fluorescence at both excitation wavelengths (F340/F380) is monitored. HBSS/HEPES is used as assay buffer. Agonists and antagonists are applied through a ValveLink system (AutoMate Scientific, Berkeley, CA). Figure 17.3A shows Ca^{2+} responses in five individual cells. Application of a noncovalent agonist evoked robust Ca^{2+} responses, which is blocked by preapplication of an antagonist.

4.6. Ion channel electrophysiology

Transient transfection of channel cDNA into mammalian cells is often the first step to characterize channel function using electrophysiological techniques such as whole cell patch clamp. Often GFP (green fluorescence protein) is cotransfected and channel function is studied in fluorescenct cells several days after transfection. While this procedure is routinely used, we have gone a step further by generating a large batch of frozen LSTT cells and thawing them on demand for electrophysiological recordings. Briefly, 30 mL of HEK 293-F cells are transfected with a plasmid encoding an ion channel (e.g., TRPA1) and a plasmid encoding GFP (total 0.5–1 μg/mL). Forty-eight hours after transfection, cells are harvested, frozen, and stored at -80 °C at 1×10^6 cells/mL. Routinely >60 vials of cells can be generated from a 30 mL transfection. 2–24 h prior to recordings, cells are revived and plated on poly-D-lysine-coated coverslips. Using LSTT cells, lasting, high-resistance seals (between patch pipette and cell membrane) and break-in are routinely achieved to enable whole cell recordings. As shown in Fig. 17.4B, AITC induces robust inward currents in a TRPA1/GFP-transfected cell (-60 mV), which are blocked by an antagonist in a dose-dependent

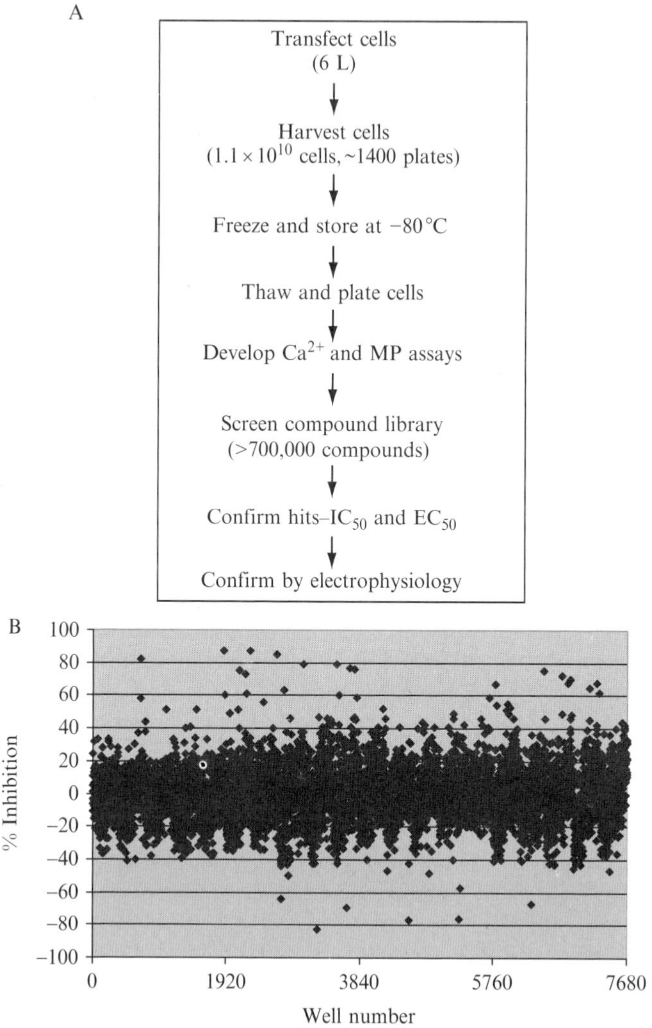

Figure 17.3 Application of LSTT cells for TRPA1 assay development and HTS campaign. (A) An overview of HTS campaign. (B) Scatterplot of a collection of 38,400 compounds located in twenty 384-well plates containing 10 compounds/well and tested twice. Hits above the scatter (40%) were picked for retesting and IC_{50} determination.

manner. LSTT cells can also be applied to the recently developed automated electrophysiology platforms. The transfection rate in LSTT cells is relatively low (60–70% positive vs. 90% in high-quality stable cell lines), but this should not be a concern for the population patch clamp platforms, which record many cells in a single well simultaneously (Finkel *et al.*, 2006).

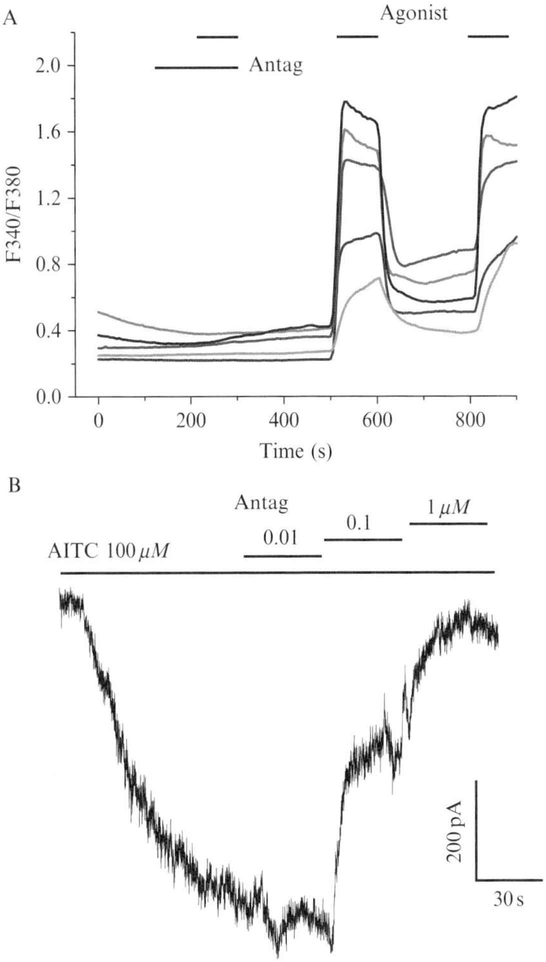

Figure 17.4 Ratiometric Ca^{2+} imaging and patch-clamp recordings using LSTT cells expressing TRPA1. (A) In ratiometric Ca^{2+} imaging, applications of a noncovalent TRPA1 agonist increase intracellular Ca^{2+}, as represented by increase in F340/F380 ratio; while preapplication of an antagonist blocks the agonist response. Traces for five cells are shown. (B) In whole cell patch clamp recording, AITC evokes inward currents in a TRPA1/GFP transfected cells. The currents are blocked by an antagonist in a concentration-dependent manner.

5. Application to GPCR Assays

Approximately 30% current drugs in the market target GPCRs, therefore they are among the most important classes of drug targets. The design and implementation of robust, reliable, and cost-effective functional screens for both known and orphan GPCRs to identify novel drug candidates is a major focus of the pharmaceutical industry. When a ligand binds to the GPCR, it causes a conformational change in the receptor and activation of heterotrimeric G protein, initiating signal transduction and ultimately cellular responses. G proteins are divided into four subfamilies (Gs, Gi/o, Gq/11, and G12/13) based on the structural and functional similarity of their α subunits. Members of the Gαs subfamily primarily act to stimulate adenylyl cyclase to produce the secondary messenger cAMP, while the Gαi/o subfamily inhibits adenylyl cyclase. Increases in cellular cAMP lead to activation of protein kinase A (PKA) and the phosphorylation of specific cellular substrates. The Gαq/11 subfamily members act to stimulate phospholipase Cβ, to form IP3 and diacylglycerol, which ultimately increases intracellular Ca^{2+} and activates PKC. The main effector system activated by the Gα12/13 subfamily is the guanine nucleotide exchange factor RhoGEF, which in turn activates the small G protein RhoA. In addition, GPCR signaling can be regulated by G protein independent pathways which cause desensitization, receptor complex formation with β-arrestin, and receptor internalization.

Based on the coupling of GPCRs to G proteins, there are different functional assay formats available to identify novel ligands. Transient transfection has been proved to be useful for many assay formats. The receptors coupled to Gαs are easily amenable to HTS because the cAMP assays are straightforward, automation friendly, and easier to execute in HTS. Ca^{2+} assay formats are routinely used for Gq coupled receptors since the Ca^{2+} flux can be easily measured using Ca^{2+} dyes or aequorine-based technology. As a member of the Gq family, Gα16 acts through phospholipase Cβ to trigger the release of intracellular calcium. Therefore, Gα16 is typically used as a promiscuous protein to convert the response of various G proteins to Ca^{2+} signaling. This is typically an advantage when screening modulators of Gαi protein coupled receptors due to the difficulty in monitoring negative modulation of cAMP. Often, the optimal coexpression of a GPCR with accessory proteins (e.g., G proteins, chimeric G proteins, or reporters) is essential to develop robust assays for HTS. This can be very challenging for stable cell lines, but achievable through transient cotransfection of a GPCR and accessory proteins, or introducing one protein into a cell line already expressing the other protein.

Here we illustrate the assay development and HTS campaign for a GPCR target coupled to the adenylate cyclase pathway using LSTT cells. Briefly, HEK293-F cells grown in suspension were transiently transfected with the GCPR of interest and Gαs in various ratios, and grown for 24 h in 293 media. The transfected cells are harvested, resuspended in freezing medium, aliquoted, and frozen in 2 mL vials. Cells are stored in -80 °C until use. Intracellular cAMP levels are measured using the LANCE cAMP-384 kit (Perkin Elmer, Waltham, MA). The kit contains an Alexa Fluor 647-labeled anti-cAMP antibody that binds biotin-cAMP and Europium-labeled streptavidin and produces time resolved fluorescence energy transfer (TR-FRET) signals at 665 nm when excited at 340 nm intracellular cAMP reduces the formation of Alexa/biotin-cAMP/europium complex and subsequently reduces light emissions at 665 n*M*.

Prior to performing assays, frozen cell aliquots are quickly thawed in a 37 °C water bath, diluted in 293 media and spun down to remove any residual DMSO. Cells are then resuspended in a stimulation buffer (DPBS/ 0.1% BSA) at 2×10^6 cells/mL along with IBMX, a phosphodiesterase inhibitor at 200 μ*M*, and Alexa Fluor 647 anti-cAMP antibody at 1:125 dilution. Using a Multidrop dispenser (Thermoscientific, Waltham, MA), 10 μL of cell suspension is added to the assay plates containing 10 μL of compound and incubated for 1 h. Using a multidrop, 20 μL of detection mix is added to the plate and allowed to incubate for 2.5 h before reading on EnVision (Perkin Elmer).

GPCR and Gαs are cotransfected in HEK293-F cells for the cAMP assay. In some cases, overexpression of Gαs is necessary for a robust signal, but it may also increase basal cAMP and mask the agonist response. Therefore, expression levels of the receptor to G protein should be optimized. For our target, a GPCR/Gαs ratio of 8 is optimal. As shown in Fig. 17.5A, the reference agonist increases cAMP in a concentration-dependent manner, with an expected EC_{50} (80 n*M*) and a signal to background of 2.7-fold. This agonist does not elicit any response in untransfected HEK293-F cells. The assay was robust and reproducible ($Z' = 0.7$). Using LSTT cells, we screened 450,000 compounds as 10-mixtures of 10 compounds/well at 1 μ*M*, followed by retesting at 10 μ*M* as single compounds. The screen and retest had excellent correlation and reproducibility (Fig. 17.4B). This screen identified 1021 hits, with a hit rate (0.24%) typical of a GPCR screen. To ensure the receptor specificity, hits were tested against mock transfected HEK293-F cells. Sixteen compounds were positive in mock transfected cells and were removed from further characterization. The extremely low false positive rate (0.004%) indicates assay specificity using LSTT cells.

In addition to Gs-coupled receptors, we have also been able to transfect Gα16 into cell lines and promiscuously couple the targets to Ca^{2+} signaling (data not shown). Because the transfection and banking of cells are relatively straightforward, we are able to transfect, evaluate, and select cells generated under a variety of transfection conditions. Once the optimal condition is chosen, large-scale preparation of sufficient number of cells for HTS and lead optimization is performed. Under liquid nitrogen storage, the cells remain viable and perform well throughout the multiyear process.

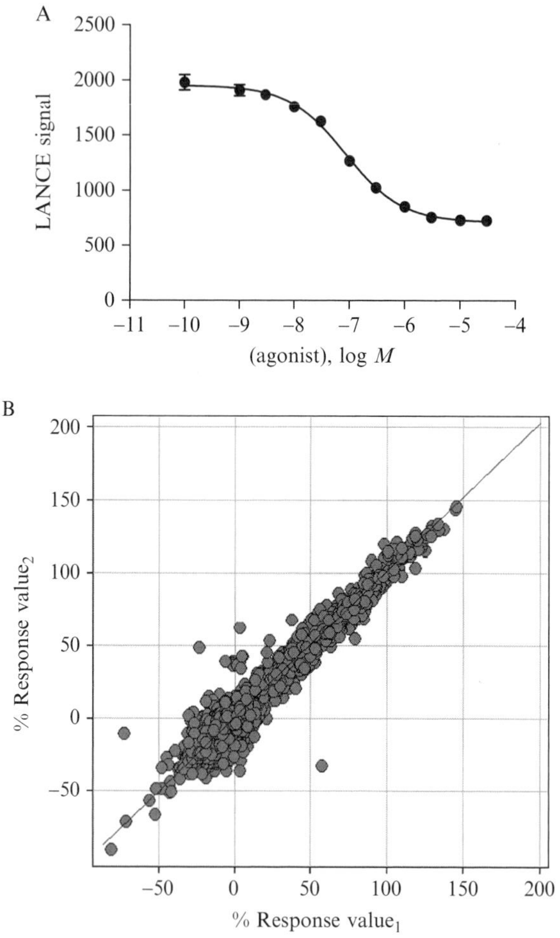

Figure 17.5 A cAMP assay and HTS campaign for a GPCR target. (A) In LSTT cells expressing GPCR and Gαs, a reference agonist exhibits concentration-dependent effects on cAMP levels. (B) Correlation between retest duplicates with R^2 value of 0.97.

6. Conclusion

Large-scale transient transfection has proved to be an extremely valuable technique for cell-based functional assays. Compared with the use of stable cell lines, this method offers the following advantages: (1) It eliminates the need to establish, select, and maintain stable cell lines, thereby enables assay development within days instead of months. (2) It circumvents problems associated with stable cell lines such as toxicity and loss of activity. (3) It is especially useful for heteromeric proteins which require expressing several subunits with certain stoichiometry. (4) A large number of cells can be prepared to support many assay formats, from pilot through lead optimization, thereby significantly reducing labor and cost. Besides the aforementioned assays for ion channels and GPCRs, this method can be applied to other cell types, assay formats and target classes.

ACKNOWLEDGMENT

The research relevant to this chapter was supported by Abbott Laboratories.

REFERENCES

Bianchi, B. R., Moreland, R. B., Faltynek, C. R., and Chen, J. (2007). Application of large-scale transiently transfected cells to functional assays of ion channels: Different targets and assay formats. *Assay Drug Dev. Technol.* **5,** 417–424.

Chen, J., Lake, M. R., Sabet, R. S., Niforatos, W., Pratt, S. D., Cassar, S. C., Xu, J., Gopalakrishnan, S., Pereda-Lopez, A., Gopalakrishnan, M., Holzman, T. F., Moreland, R. B., *et al.* (2007). Utility of large-scale transiently transfected cells for cell-based high-throughput screens to identify transient receptor potential channel A1 (TRPA1) antagonists. *J. Biomol. Screen.* **12,** 61–69.

Chen, J., Kim, D., Bianchi, B. R., Cavanaugh, E. J., Faltynek, C. R., Kym, P. R., and Reilly, R. M. (2009). Pore dilation occurs in TRPA1 but not in TRPM8 channels. *Mol. Pain* **5,** 3.

Chung, M. K., Guler, A. D., and Caterina, M. J. (2008). TRPV1 shows dynamic ionic selectivity during agonist stimulation. *Nat. Neurosci.* **11,** 555–564.

Finkel, A., Wittel, A., Yang, N., Handran, S., Hughes, J., and Costantin, J. (2006). Population patch clamp improves data consistency and success rates in the measurement of ionic currents. *J. Biomol. Screen.* **11,** 488–496.

Hubner, C. A., and Jentsch, T. J. (2002). Ion channel diseases. *Hum. Mol. Genet.* **11,** 2435–2445.

Johnston, P. (2002). Cellular assays in HTS. *Methods Mol. Biol.* **190,** 107–116.

Kaczorowski, G. J., McManus, O. B., Priest, B. T., and Garcia, M. L. (2008). Ion channels as drug targets: The next GPCRs. *J. Gen. Physiol.* **131,** 399–405.

Kost, T. A., Condreay, J. P., and Ames, R. S. (2010). Baculovirus gene delivery: A flexible assay development tool. *Curr. Gene Ther.* **10,** 168–173.

Moore, K., and Rees, S. (2001). Cell-based versus isolated target screening: How lucky do you feel? *J. Biomol. Screen.* **6,** 69–74.

Story, G. M., Peier, A. M., Reeve, A. J., Eid, S. R., Mosbacher, J., Hricik, T. R., Earley, T. J., Hergarden, A. C., Andersson, D. A., Hwang, S. W., McIntyre, P., Jegla, T., *et al.* (2003). ANKTM1, a TRP-like channel expressed in nociceptive neurons, is activated by cold temperatures. *Cell* **112,** 819–829.

Treherne, J. M. (2006). Exploiting high-throughput ion channel screening technologies in integrated drug discovery. *Curr. Pharm. Des.* **12,** 397–406.

Xia, M., Imredy, J. P., Santarelli, V. P., Liang, H. A., Condra, C. L., Bennett, P., Koblan, K. S., and Connolly, T. M. (2003). Generation and characterization of a cell line with inducible expression of Ca(v)3.2 (T-type) channels. *Assay Drug Dev. Technol.* **1,** 637–645.

Zheng, W., Spencer, R. H., and Kiss, L. (2004). High throughput assay technologies for ion channel drug discovery. *Assay Drug Dev. Technol.* **2,** 543–552.

CHAPTER EIGHTEEN

Quantification of RNA Editing of the Serotonin 2C Receptor (5-HT$_{2C}$R) *Ex Vivo*

Maria Fe Lanfranco,[*,†] Noelle C. Anastasio,[*,†] Patricia K. Seitz,[*,†] *and* Kathryn A. Cunningham[*,†]

Contents

Abstract

The 5-HT$_{2C}$R receptor (5-HT$_{2C}$R) exerts tonic and phasic inhibitory influence over brain circuitry, and dysregulation of this influence contributes to the neurochemical underpinnings in the etiology of a variety of neuropsychiatric disorders including addiction, depression, and schizophrenia. A strategically important regulator of the 5-HT$_{2C}$R function and protein diversity is mRNA editing, a type of posttranscriptional modification that alters codon identity and thus the translation of distinct, though closely related, isoforms of 5-HT$_{2C}$R from a single, original transcript. The 5-HT$_{2C}$R mRNA can be edited at five closely spaced sites, altering the identity of up to three amino acids in the predicted second intracellular loop of the receptor to modulate receptor: G-protein coupling and constitutive activity. Methods to study changes in mRNA

[*] Center for Addiction Research, University of Texas Medical Branch, Galveston, Texas, USA
[†] Department of Pharmacology and Toxicology, University of Texas Medical Branch, Galveston, Texas, USA

Methods in Enzymology, Volume 485
ISSN 0076-6879, DOI: 10.1016/S0076-6879(10)85018-9

editing based upon direct DNA sequencing are both time and labor intensive. To streamline the acquisition of mRNA editing data and improve quantification, we have adapted real-time reverse transcription polymerase chain reaction (qRT-PCR) to detect and quantify mRNA editing in 5-HT_{2C}R transcripts by utilizing TaqMan® probes modified with a minor groove binder (MGB). The method is very sensitive, detecting as little as 10^{-18} g (1 attogram) of standard cDNA template and can discriminate closely related 5-HT_{2C}R mRNA edited isoforms. This technique expands the breadth of available quantification methods for mRNA editing and is particularly useful for the *ex vivo* analyses of mRNA editing of the 5-HT_{2C}R by allowing the rapid collection of data on large numbers of tissue samples. In addition, the general technique can be adapted easily to investigate edited mRNA from other genes, thus facilitating the development of a broader knowledge base of the physiological role of mRNA editing.

1. RNA Editing of the 5-HT_{2C}R

The serotonin 2C receptor (5-HT_{2C}R) is a member of the superfamily of G-protein coupled receptors (GPCR) that is well characterized to regulate cognition, feeding, satiety, and mood (Kaye *et al.*, 2005; Serretti *et al.*, 2004). Dysregulation of this system contributes to the neuromolecular mechanisms underlying the etiology of a variety of neuropsychiatric disorders including addiction, depression, and schizophrenia (Bubar and Cunningham, 2008; Giorgetti and Tecott, 2004). Expression of the 5-HT_{2C}R is widely distributed in the mammalian brain and its function and expression are dynamically regulated by mRNA editing, a posttranscriptional modification (Niswender *et al.*, 1998; Sanders-Bush *et al.*, 2003) that results in an insertion, deletion (Cruz-Reyes *et al.*, 1998; Igo *et al.*, 2002), and/or modifications (Samuel, 2003) of single or small number of ribonucleotides in pre-mRNA. As a consequence, mRNA editing alters the information content and coding properties of mRNA molecules (Bass, 2002; Hoopengardner *et al.*, 2003; Maydanovych and Beal, 2006). Of importance, basal patterns of 5-HT_{2C}R isoform expression in brain circuitry contribute to vulnerability to and/or expression of various psychiatric disorders (Dracheva *et al.*, 2003, 2008a,b; Gurevich *et al.*, 2002b; Iwamoto *et al.*, 2005b; Schmauss, 2003). Furthermore, it is conceivable that abnormal 5-HT_{2C}R function based upon altered patterns of 5-HT_{2C}R isoform expression may be normalized by therapeutic treatment with psychiatric medications (Niswender *et al.*, 2001; Sanders-Bush *et al.*, 2003).

The 5-HT_{2C}R is the only GPCR currently known to undergo mRNA editing (Burns *et al.*, 1997; Gurevich *et al.*, 2002a), although several ionotropic neurotransmitter receptors [e.g., α-amino-3-hydroxyl-5-

methyl-4-isoxazole-propionate (AMPA) receptors] also undergo this posttranscriptional process. The 5-HT$_{2C}$R is unique in that it contains five closely spaced adenosine (A) editing sites (designated A, B, E, C, and D; Burns *et al.*, 1997; Gurevich *et al.*, 2002b; Niswender *et al.*, 1998) within a 13-base region of the 5-HT$_{2C}$R pre-mRNA (Fig. 18.1). These sites are substrates for base modification via hydrolytic deamination by *a*denosine *d*eaminases that *a*ct on *R*NA (ADARs) to yield inosines. In the course of translation, the ribosome recognizes the resultant inosine as a guanosine (G), resulting in an A → G substitution. Thirty-two mRNA isoforms are possible and can lead to a predicted 24 potential resultant protein isoforms; however, fewer isoforms are routinely detected and their pattern of expression is influenced by variables such as species and brain region (Burns *et al.*, 1997; Fitzgerald *et al.*, 1999; Morabito *et al.*, 2010; Niswender *et al.*, 1998; Sanders-Bush *et al.*, 2003).

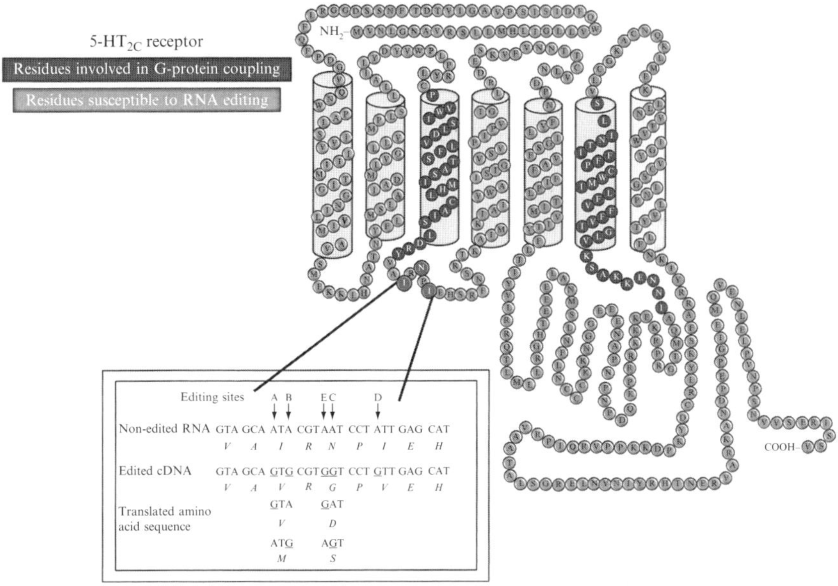

Figure 18.1 Schematic representation of the 5-HT$_{2C}$R (adapted from Julius *et al.*, 1988, 1989). Predicted amino acid residues linked to G-protein coupling are denoted in blue (Roth *et al.*, 1998; Wess, 1998) and amino acid residues at which RNA editing occurs are denoted in red (Niswender *et al.*, 1998). RNA, cDNA nucleotide, and predicted amino acid sequences for the nonedited and fully edited transcripts are shown with the editing sites indicated by arrows [↓] (Berg *et al.*, 2008b; Herrick-Davis *et al.*, 2005; Sanders-Bush *et al.*, 2003). (See Color Insert.)

2. Functional Properties of 5-HT$_{2C}$R Edited Isoforms

Editing of 5-HT$_{2C}$R pre-mRNA results in a single nucleic acid substitution at each of the five adenosine editing sites, substitutions that lead to the formation of proteins that differ by up to three amino acids (Fig. 18.1, residues highlighted in red), thereby modulating receptor: G-protein coupling, constitutive activity and receptor trafficking (Berg *et al.*, 2001; Herrick-Davis *et al.*, 1999; Marion *et al.*, 2004; Niswender *et al.*, 1999). The mRNA editing region is located within the second intracellular (i2) loop of the receptor, two amino acid residues downstream of the highly conserved E/DRY motif, which has been strongly linked to G-protein coupling (Fig. 18.1, residues highlighted in blue; Berg *et al.*, 2001, 2008a; Burns *et al.*, 1997; Marion *et al.*, 2004; Wess, 1998). The impact of RNA editing on 5-HT$_{2C}$R function is evidenced by the lowered potency of 5-HT$_{2C}$R agonists to activate intracellular signaling (Burns *et al.*, 1997; Niswender *et al.*, 1999; Wang *et al.*, 2000). For example, the human partially-edited isoform 5-HT$_{2C\text{-}VSV}$R exhibited a fivefold rightward shift in potency for 5-HT-stimulated accumulation of [^{3}H]-inositol monophosphates (^{3}H-IP) compared to the non-edited isoform 5-HT$_{2C\text{-}INI}$R (Niswender *et al.*, 1999). Furthermore, in those same studies, the fully-edited human 5-HT$_{2C\text{-}VGV}$R isoform presented an even more substantial rightward shift in the dose–response curve for 5-HT, with an EC_{50} value of 59 n*M* versus 2.3 n*M* for the 5-HT$_{2C\text{-}INI}$R (Niswender *et al.*, 1999). Stimulation of the partially-edited 5-HT$_{2C\text{-}VSV}$R and fully-edited 5-HT$_{2C\text{-}VGV}$R isoforms with the 5-HT$_{2C}$R partial agonist (±)-1-(4-iodo-2,5-dimethoxyphenyl)-2-aminopropane (DOI) produced 9- and 43-fold less potent signal activation, respectively, compared to the 5-HT$_{2C\text{-}INI}$R (Niswender *et al.*, 1999). These *in vitro* results strongly suggest that the efficiency of 5-HT$_{2C}$R agonists to activate phospholipase C (PLC) is dependent on the degree of editing of the 5-HT$_{2C}$R (Fig. 18.2).

5-HT$_{2C}$R mRNA editing also controls the degree of constitutive activity or ligand-independent stimulation and the subsequent basal levels of function (Fig. 18.2; Berg *et al.*, 2001; Herrick-Davis *et al.*, 1999; Niswender *et al.*, 1999). For example, in the absence of 5-HT$_{2C}$R ligands, a fivefold higher level of ^{3}H-IP accumulation was seen in COS-7 cells transfected with the non-edited 5-HT$_{2C\text{-}INI}$R compared to those transfected with the fully-edited 5-HT$_{2C\text{-}VGV}$R isoform (Herrick-Davis *et al.*, 1999). More recently, the I156V substitution, a naturally occurring partially-edited 5-HT$_{2C}$R isoform, has been shown to alter the functional selectivity of both ligand-dependent and ligand-independent 5-HT$_{2C}$R

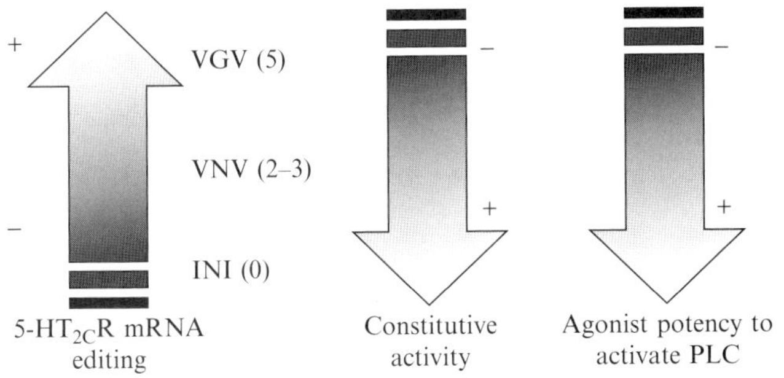

Figure 18.2 Schematic representation of the functional consequences of RNA editing of the 5-HT$_{2C}$R. As the number of edited sites increases (5-HT$_{2C\text{-}INI}$R → 5-HT$_{2C\text{-}VGV}$R), constitutive activity, G-protein coupling, and the ability of agonists to activate PLC decreases.

signaling (Berg *et al.*, 2008a). This single substitution is sufficient to produce profound changes in agonist selectivity, favoring signaling via PLC over phospholipase A2. In keeping with this concept, the partially-edited 5-HT$_{2C\text{-}VSV}$R and the fully-edited 5-HT$_{2C\text{-}VGV}$R retained the ability to couple to the MAP kinase pathway, but exhibited a loss of efficacy compared to the non-edited 5-HT$_{2C\text{-}INI}$R isoform (Werry *et al.*, 2008b). Thus, the non-edited 5-HT$_{2C\text{-}INI}$R isoform has a high affinity for G-protein coupling and a high level of constitutive activity, while increased editing of 5-HT$_{2C}$R mRNA serves to reduce the efficiency of the interaction between the receptor and its G-protein, resulting in lower constitutive activity (Fig. 18.2; Burns *et al.*, 1997; Marion *et al.*, 2004). Accordingly, mRNA editing has a profound influence over the fate of intracellular signaling following 5-HT$_{2C}$R activation and may play a key role in providing functional selectivity to the signal transduction machinery of the cell (Werry *et al.*, 2008a).

3. Current Methods for Quantification of 5-HT$_{2C}$R Editing Events *Ex Vivo*

Quantitative measurement of final protein products of edited 5-HT$_{2C}$R transcripts is most desirable; however, protein capture reagents (e.g., antibodies) capable of discriminating among 5-HT$_{2C}$R proteins with the minor sequence differences due to RNA editing are not currently available. Hence, there is an obvious utility in accurately detecting and quantifying changes in the levels of mature mRNA isoforms in *ex vivo* brain samples (see Table 18.1 for a summary of the advantages and limitations of current methods).

Table 18.1 Summary of available methods to detect RNA editing events *ex vivo*

	Direct sequencing	Pyrosequencing	qRT-PCR	Capillary electrophoresis	Deep sequencing
Accuracy	++++	++++	+++	+++	++++
Sensitivity	+	+	+++	+++	++++
RNA population	+	+	+++	+++	++++
Isoform profile	++	++	+	+++	++++
Equipment accessibility	+++	+++	+++	++	+

Accuracy, ability to unequivocally detect individual isoforms; sensitivity, ability to detect low abundance populations; RNA population, # molecules sampled relative to entire mRNA population; isoform profile, ease of measuring entire isoform profile; equipment accessibility, availability of necessary equipment.
+, Low; +++, medium; ++++, high.

We adapted qRT-PCR as a method of detecting and quantifying changes in RNA editing events of the 5-HT_{2C}R by utilizing TaqMan® probes modified with a minor groove binder (MGB; De La Vega *et al.*, 2005; Lanfranco *et al.*, 2009) to facilitate analyses of 5-HT_{2C}R edited isoform expression in brain *ex vivo* (detailed method described below). We have verified that this technique can sensitively measure probe target isoforms and can discriminate closely related isoforms by testing four probes (ABECD, ABD, AD, and non-edited) against standard cDNA templates. When necessary, the addition of unlabeled competing oligonucleotides ("competitors") that differ from the complementary target sequence for each respective TaqMan® MGB probe at a single (editing site) nucleotide can be added to increase specificity even further (Lanfranco *et al.*, 2009). This concept is based on Taqman® single nucleotide polymorphism (SNP) genotyping assays which utilize two separate probes (one for each SNP and each labeled with a different fluorophore) to distinguish the single nucleotide difference (De La Vega *et al.*, 2005).

The two most commonly used methods to quantify edited 5-HT_{2C}R transcripts *ex vivo* are direct DNA sequencing (Burns *et al.*, 1997; Gurevich *et al.*, 2002b) and pyrosequencing (Iwamoto *et al.*, 2005a; Sodhi *et al.*, 2005). Direct DNA sequencing and pyrosequencing methods provide unequivocal results, but are labor intensive and involve multiple steps that can increase the possibility for error and introduce sampling bias. Importantly, the number of sequences identified is often small compared to the actual number of 5-HT_{2C}R mRNA molecules in the population resulting in statistical uncertainty (Table 18.1; Sodhi *et al.*, 2005). Our new TaqMan® MGB method addresses several drawbacks for the study of 5-HT_{2C}R mRNA editing *ex vivo* compared to traditional direct sequencing (Burns *et al.*, 1997; Gurevich *et al.*, 2002b) and/or pyrosequencing techniques (Iwamoto *et al.*, 2005a; Sodhi *et al.*, 2005). For example, measurements are performed directly on cDNA reverse transcribed from isolated RNA, thus increasing the speed of data collection and eliminating several sources of error. Furthermore, the measurement of the entire population of mRNA transcripts in a sample reduces the errors generated due to sampling bias, resulting in more precise measurements and greater statistical discrimination among experimental groups (Table 18.1; Sodhi *et al.*, 2005; Wong and Medrano, 2005; Yao *et al.*, 2006). Also, this method allows the investigator to focus on only the isoforms of interest based upon the experimental question in hand, bypassing the need to characterize the entire isoform profile for each experimental subject. In addition, the qRT-PCR method enables investigators to study the impact of treatment-elicited changes in less abundant isoforms because of the increased precision of the assay. Following an initial survey of isoform profiles by DNA sequencing, assays can be designed to specifically target one (or a few) isoform(s) of interest and rapidly assess changes in expression levels following treatment or to correlate with disease state or behavioral activity.

Potential limitations with qRT-PCR are that it is relatively low-throughput and standardization of the assay can be potentially time consuming. Recently introduced RNA deep sequencing technology represents a rapid, high-throughput method and affords more comprehensive and quantitative data for quantifying 5-HT_{2C}R RNA editing patterns with substantially increased accuracy and sensitivity for detection of RNA editing events (Table 18.1; Abbas *et al.*, 2010; Bentley *et al.*, 2008; Morabito *et al.*, 2010). *H*igh-*T*hroughput *M*ultiplexed *T*ranscript *A*nalysis (HTMTA) is a novel deep sequencing method (Morabito *et al.*, 2010) which was adapted from the massively parallel, short-read sequencing technology available on the Illumina/Solexa platform (Abbas *et al.*, 2010; Bentley *et al.*, 2008). Although these techniques are more cost-effective on a "per transcript" basis, they are extremely expensive on a "per experiment" basis and require access to a large core facility with the appropriate equipment. Capillary electrophoresis (CE) is an additional promising technique which incorporates reverse transcription and nested PCR (in which the second round of PCR incorporates different fluorescent labels onto the forward and reverse cDNA strands) to separate the single stranded products (Poyau *et al.*, 2007) to characterize the entire 5-HT_{2C}R isoform profile, but also requires access to the appropriate machinery (Table 18.1). In spite of these various limitations, new sequencing tools (qRT-PCR, deep sequencing, CE) offer a considerable advance in the quantification and identification of isoform-specific 5-HT_{2C} mRNA editing events *ex vivo* and provide the field with powerful approaches to identify previously undetected variations in mRNA editing patterns in neuropsychiatric diseases and further characterization of the functional role of a constitutively active 5-HT_{2C}R in the manifestation of these diseases.

4. Quantification of 5-HT$_{2C}$R Editing Events *Ex Vivo* with qRT-PCR

The steps involved in setting up and characterizing the qRT-PCR assay are (1) choose 5-HT$_{2C}$R edited isoforms to be measured; (2) create probes and optimize assay conditions for chosen isoforms; (3) isolate RNA from experimental samples; (4) perform reverse transcription; and (5) perform TaqMan® qRT-PCR assay with MGB modified probes.

4.1. Choose 5-HT$_{2C}$R isoform(s) to be analyzed

There are frequently *a priori* reasons (e.g., preliminary data, literature support) that dictate which edited isoforms should be quantified with qRT-PCR in a given *ex vivo* experiment. For example, the profile of edited 5-HT$_{2C}$R isoforms is significantly modified in a rat model of depression (Iwamoto *et al.*, 2005b) and in addiction-related phenotypes (Dracheva *et al.*, 2009). In humans, changes in 5-HT$_{2C}$R mRNA editing profiles have been observed in the prefrontal cortex of depressed suicide victims (Dracheva *et al.*, 2008a,b; Gurevich *et al.*, 2002b; Schmauss, 2003) and schizophrenics (Dracheva *et al.*, 2003). However, for instances in which evidence-based information is not available it may be necessary to characterize the entire 5-HT$_{2C}$R isoform profile using more comprehensive methods, such as direct sequencing (Burns *et al.*, 1997; Gurevich *et al.*, 2002b), pyrosequencing (Iwamoto *et al.*, 2005a; Sodhi *et al.*, 2005) or deep sequencing (Abbas *et al.*, 2010; Bentley *et al.*, 2008; Morabito *et al.*, 2010).

4.2. Create probes and optimize assay for each isoform

We adapted the qRT-PCR assay by utilizing TaqMan® MGB probes (Kutyavin *et al.*, 2000) to improve specificity and yield greater discrimination between targeted sequences (De La Vega *et al.*, 2005). In addition to usual primers, the Taqman® method requires a third oligonucleotide known as a probe, which hybridizes to a region between the two primers. This probe contains both a reporter fluorescent dye attached to the 5′ end, and a quencher dye attached to the 3′ end (Hembruff *et al.*, 2005). While the probe is intact, the proximity of the quencher greatly reduces the fluorescence emitted by the reporter dye. During amplification, the $5' \rightarrow 3'$ exonuclease activity of the Taq DNA polymerase cleaves nucleotides from the probe. When the reporter dye diffuses away from the quencher, fluorescence increases proportionately to the amount of accumulated PCR product. This dependence on polymerization ensures that cleavage of the probe occurs only if the correct target sequence is being amplified. Furthermore, with fluorogenic probes, there is little chance of nonspecific amplification due to mispriming or primer–dimer artifacts.

We outline here the general steps to optimize and perform qRT-PCR assays for analyzing any 5-HT$_{2C}$R mRNA isoform. We have previously optimized assays for the non-edited, AD, ABD, and ABECD (fully-edited) isoforms (Lanfranco *et al.*, 2009). The probes designed for those isoforms (Lanfranco *et al.*, 2009) are longer than customary MGB probes because of the tight restraints on the sequences to be measured, but all worked well to discriminate the tested isoforms. Modifications of the sequences listed in Table 18.2 at the appropriate sites should provide initial sequences for probes for other isoforms.

Required materials: The materials listed below are available from Applied Biosystems, Foster City, CA; other vendors may have suitable equivalents.

- 6-Carboxyfluorescein (FAM)-labeled TaqMan® and TaqMan® MGB probes for RNA isoforms of 5-HT$_{2C}$R (custom synthesized)
- Primers flanking the 5-HT$_{2C}$R edited region (accession # M21410; bases 1014–1192):
  - SN = 5′-CCT GTC TCT GCT TGC AAT TCT-3′
  - ASN = 5′-GCG AAT TGA ACC GGC TAT G-3′
- Primers and probe for a housekeeping gene, such as cyclophilin (accession # M_19533; bases 224–293); primer sequences:
  - SN = 5′-TGT GCC AGG GTG GTG ACT T-3′
  - ASN = 5′-TCA AAT TTC TCT CCG TAG ATG GAC TT-3′
- cDNA standard template for each isoform to be measured (see below)
- Reagents for:
  - RNA isolation: RiboPure Kit; includes TRI-Reagent®

Table 18.2 TaqMan® MGB probes

Probe name	RNA isoform detected (amino acid)	Sequence
ABECD	ABECD (VGV)	[FAM] TAGCAGTGCGTGGTCCTG TTGA [MGB/NFQ]
ABD	ABD (VNV)	[FAM] TAGCAGTGCGTAATCCTGT TGA [MGB/NFQ]
AD	AD (VNV)	[FAM] TGTAGCAGTACGTAATCCTG TTGA [MGB/NFQ]
Nonedited	Nonedited (INI)	[FAM] TAGCAATACGTAATCCTA TTG [MGB/NFQ]

  - Reverse transcription: TaqMan® Reverse Transcription Kit
  - qRT-PCR: TaqMan® Genotyping PCR Master Mix
- *Disposables*: MicroAmp® Optical 96-Well Reaction Plates, MicroAmp® Optical Adhesive Film, DNase–RNase free microfuge tubes (1.5 and 2.0 ml), pipette tips with filters, MicroAmp® Optical 8-Tube Strip (0.2 ml), and MicroAmp® 8-Cap Strip
- *Equipment*: PCR thermo cycler (for the reverse transcription reaction); Real-time PCR thermo cycler (e.g., 7500 Fast Real-Time PCR System; Applied Biosystems); appropriate data collection and analysis software (e.g., 7500 Fast System Detection Software (SDS) version 1.3.1)

4.2.1. Prepare standard templates

A cDNA library containing each independent 5-HT_{2C}R mRNA isoform can be used to produce the standard templates needed to assess the sensitivity and the specificity of the TaqMan® MGB probes. The standard templates can be produced using conventional PCR with Taq DNA polymerase (Fisher Scientific, Houston, TX) and the primers described above. Following PCR, all amplimers can be characterized by agarose gel electrophoresis for the correct size (180 bp), then excised, gel purified, and sequenced to confirm the identities of the 5-HT_{2C}R isoforms. Estimates of the DNA concentrations of standard template stock solutions can be determined by absorbance at 260 nm and by comparison to quantitative DNA standards in polyacrylamide gel electrophoresis. Standard templates can then be aliquoted and frozen at −20 °C for future use.

4.2.2. Standard assay conditions

We tested a number of different assay parameters to determine standard conditions for the qRT-PCR assay (Lanfranco *et al.*, 2009). The program suggested by the manufacturer (Applied Biosystems) proved optimal: activation step (95 °C, 10 min) followed by 40 cycles of denaturation (95 °C, 15 s), annealing, and elongation (60 °C, 1 min). Specific qRT-PCR components for each type of test are described below.

Tips: To ensure replicate reproducibility, prepare probes, primer mix, experimental samples, and standard templates no earlier than 1 day before the experiment. Probes are photosensitive; therefore, they should remain protected from light until immediately prior to use. To reduce pipetting errors, we recommend making a master mix (composed of all reagents except sample) so that only a single pipetting step into each well of the MicroAmp® Optical 96-Well Reaction Plates is required before sample addition. To minimize cross-contamination of samples, arrange pipettes, tips boxes, plates, and solution so that your hands never pass over the top of the

MicroAmp® Optical 96-Well Reaction Plates when setting up the qRT-PCR. Keep a separate master sheet with the loading information, as any marks on the plates can alter the signal. Always make sure that there are no bubbles on the bottom of the wells. When running multiple plates on the same day, prepare all the plates in the morning and cover the plates with foil (to protect the probe from the light) and place the plates at 4 °C until needed.

4.2.3. Determine sensitivity of each probe

Test probe sensitivity by performing qRT-PCR assays for each probe against serial dilutions of the matching standard template stock solution (e.g., 10^6- to 10^{12}-fold) using the standard assay conditions and the reaction mix is described below. Dilute target templates until the Ct (or crossing-threshold) value (see below) of the most dilute sample is not significantly different from that of no template controls (NTCs). The Ct value is defined as the PCR cycle at which the sample fluorescence crosses a threshold set above baseline fluctuations (background noise) and within the logarithmic portion of the amplification plot (Livak and Schmittgen, 2001; Schefe *et al.*, 2006; Wong and Medrano, 2005).

Reaction components: 10 μL TaqMan® Genotyping Master Mix, 125 n*M* sense and antisense primers, 100 n*M* Taqman® MGB probe, and 6 μL of standard template in a final reaction volume of 20 μL.

- Calculate the efficiency (*E*) of each assay with the following equation (Schefe *et al.*, 2006; Wong and Medrano, 2005):

$$\mathbf{E} = 10^{(-1/\text{slope})} - 1$$

The slope of the standard template dilution curve can be calculated by plotting Ct value measurements as a function of the log DNA concentration and should vary logarithmically over at least six orders of magnitude of dilution (Lanfranco *et al.*, 2009).

Tips. Test each sample in triplicate; if a replicate differs from the other two by >0.5 cycles, then this replicate can be considered an outlier and be removed from the data analysis. If the objective of an experiment is to compile quantitative 5-HT$_{2C}$R mRNA isoform profiles, adjust measurements of individual 5-HT$_{2C}$R isoforms to account for differences in efficiencies of the TaqMan® 5-HT$_{2C}$R MGB probes. The algorithms to correct for probe-to-probe and sample-to-sample differences in efficiency vary among the available online software packages (e.g., LinRegPCR, GenEx, and REST; http://www.efficiency.gene-quantification.info; University of Amsterdam, Amsterdam, The Netherlands).To our knowledge, there is currently no consensus for the choice of one correction method over the others (Schmittgen and Livak, 2008). However, when the experimental goal is to determine changes in a given isoform(s) in response to treatment or disease, there is no need to correct for efficiency differences. Such experiments are the ideal application of the proposed method.

4.2.4. Determine specificity of each probe

Our published data indicate that all of the TaqMan® MGB probes tested unequivocally discriminated templates that differed by two or more nucleotides even without addition of competitors (Lanfranco *et al.*, 2009). However, any new probes must be tested for ability to distinguish single nucleotide mismatches.

Perform a series of qRT-PCR assays using each probe with all templates that differ by a single nucleotide from the perfectly matched template. From the sensitivity information collected above, use a concentration of standard template with a Ct value between 20 and 30 cycles (since expected Ct values for *ex vivo* samples usually fall in this range). Perform separate qRT-PCRs for matched and all singly mismatched templates as described above (Section 4.2.3, reaction components). Compare the Ct values for mismatched templates to those for perfectly matched templates (Ct Difference) using the following equations:

$$\text{Ct Difference} = (\text{Ct of mismatched template}) - (\text{Ct of matched template})$$

Then, calculate the percentage of cross-hybridization using the following equation:

$$\text{Cross-hybridization} = \left[1/\left(2^{\text{Ct Difference}}\right)\right] \times 100\%$$

Tips: A Ct Difference of five cycles corresponds to a cross-hybridization of 3.125%, which provides a sufficient level of discrimination between closely related isoforms. We adopted this criterion as our minimum acceptable value for probe specificity; however, the user can adopt a lower value of cross-hybridization to make the test more stringent. If the Ct Difference falls below the adopted criterion, addition of competitors (see below) is not necessary to discriminate that single mismatch. If the Ct Difference is above the adopted criterion, then consider the addition of unlabeled competitor oligonucleotide(s) to the reaction mix to ensure discrimination between the desired target and particular mismatch(es).

4.2.5. TaqMan® MGB assay in the presence of competitors

Unlabeled competing probes (competitors) are oligonucleotides that differ from the complementary target sequence for each respective TaqMan® MGB 5-HT_{2C}R probe at a single (editing site) nucleotide. [We use the same nomenclature regarding editing sites for competitor probes as for the TaqMan® 5-HT_{2C}R MGB probes.] The addition of unlabeled probes complementary to the singly mismatched templates (competitors) competes with the labeled probe to increase the Ct value of the mismatched template, thus reducing cross-hybridization, if it occurs (Lanfranco *et al.*, 2009).

Test fixed concentrations of the TaqMan® MGB probe and its matched template with various concentrations of each singly mismatched competitor (ranging from 0.0125 to 1 μ*M* final concentration). As an example, use the D standard template as the mismatched template for the TaqMan® MGB nonedited probe. Once the necessity or concentration of competitors is determined, make a competitor cocktail master mix to decrease pipetting errors.

Reaction components: Ten microliters of TaqMan® Genotyping Master Mix, 125 n*M* sense and antisense primers, 100 n*M* TaqMan® MGB probe, 6 μL of standard template (at the concentration described above), and 2 μL 0.1–10 μ*M* stock solution of singly mismatched competitor(s) in a final reaction volume of 20 μL.

4.3. Preparation of RNA from *ex vivo* samples

The first step in any qRT-PCR assay is the isolation of high-quality RNA followed by reverse transcription of the RNA template into cDNA, then its exponential amplification and detection by PCR. There are a number of commercially available kits—utilizing either alcohol precipitation or filter adsorption—that yield suitable RNA. To ensure that the final RNA product is free of DNA contamination, it is necessary that the protocol includes either an initial extraction with a phenol-containing reagent or a final DNase treatment. The method we describe here utilizes the RiboPure Kit (Applied Biosystems) that has worked well for us.

4.3.1. RNA extraction

Homogenize freshly dissected or frozen brain tissue in 20× w/v TRI-Reagent® using a Tissue-Tearor™ or Polytron™ homogenizer. Use a minimum volume of 700 μL (even for small samples) to ensure reasonable separation of aqueous and organic phases (below). Next, vortex the samples for 5 min at room temperature. Centrifuge the sample at 12,000×*g* for 5 min at 4 °C. This step is described as optional, but ensures that the sample will not clog the filter with cell debris. Isolate RNA using the RiboPure Kit according to the manufacturer's directions, with the following considerations: all solutions need to be at room temperature prior to RNA extraction; perform all centrifugation steps at room temperature to increase the yield of RNA. It is useful to run a 1% agar gel (MOPS buffer) at 80 V for 30 min to confirm the integrity of the RNA samples.

Tips: RNA is notorious for being unstable in the face of contaminants. To maintain RNA integrity use gloves, RNase-free reagents, and filtered pipet tips. Rinse the homogenizer with sterile nuclease-free water and wipe with clean laboratory KimWipes after each sample is homogenized. Determine the final RNA concentration by absorbance at 260 nm and purity by the 260/280 ratio.

4.4. Reverse transcription

Perform the reverse transcription reaction on 0.25–0.5 μg of RNA using the TaqMan® Reverse Transcription Kit with random hexamer primers according to the manufacturer's directions. The following conditions ensure optimal cDNA synthesis: annealing step (25 °C, 10 min); elongation step (48 °C, 30 min); enzyme inactivation step (95 °C, 5 min). We add four volumes of DNase-free, RNase-free water to the reverse transcription reaction and use the samples immediately or store at −80 °C for future use. Dilution of the cDNA product (e.g., four- to fivefold) decreases the final concentration of reverse transcription reagents in the qRT-PCR step and reduces error due to pipetting of very small volumes of sample into the reaction plate.

Perform a reverse transcription in the absence of enzyme to test for genomic DNA contamination. There should be no amplification observed with this sample during conventional PCR amplification. If any samples exhibit DNA contamination, treat with DNase (e.g., TURBO DNA-*free*™ Kit, Applied Biosciences) according to manufacturer's directions.

4.5. Application of qRT-PCR assay to assess 5-HT$_{2C}$R RNA isoform expression in mouse brains *ex vivo*

We describe here an experiment to assess 5-HT$_{2C}$R RNA isoform expression *ex vivo* as an example of the utility of the qRT-PCR method. Total whole brain RNA from 129S6 mice was extracted as described above and analyzed with TaqMan® qRT-PCR assay in the presence and absence of competitors for quantification of 5-HT$_{2C}$R isoforms and compared to the 5-HT$_{2C}$R isoform profile from sequencing studies previously published (Du *et al.*, 2006). Our assay indicates that the ABD isoform is the most prevalent; this agrees with results obtained by direct DNA sequencing for 129S6 mice (Fig. 18.3). In our hands, the presence of competitors did not change the detected amounts of the four isoforms compared to control (absence of competitor): ABECD [$X^2_{(\mathrm{DF}=2,\ n=9)} = 7.200$, $p = 0.0036$], ABD [$X^2_{(\mathrm{DF}=2,\ n=9)} = 5.600$, $p = 0.05$], AD [$X^2_{(\mathrm{DF}=2,\ n=9)} = 5.600$, $p = 0.05$], and nonedited probes [$X^2_{(\mathrm{DF}=2,\ n=9)} = 4.622$, $p = 0.1$] for this strain (Lanfranco *et al.*, 2009).

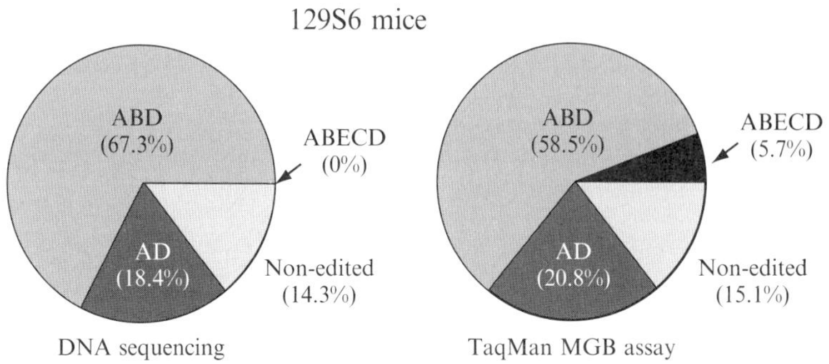

Figure 18.3 129S6 mouse whole brain expression pattern of 5-HT_{2C}R isoforms obtained from direct DNA Sequencing (Du *et al.*, 2006) compared to our TaqMan® qRT-PCR assay (Lanfranco *et al.*, 2009).

Reaction components: Ten microliters of TaqMan® Genotyping Master Mix, 125 n*M* sense and antisense primers, 100 n*M* TaqMan® MGB probe, 6 μL of cDNA ± 2 μL 0.1–1 μ*M* stock solution of competitor cocktail in a final reaction volume of 20 μL.

ACKNOWLEDGMENTS

This research was supported by the National Institute on Drug Abuse grants DA006511, DA000260, DA020087; the Peter F. McManus Charitable Trust (KAC); and the Jeane B. Kempner Postdoctoral Scholar Award (NCA). We also thank Drs William Clarke and Kelly Berg from the University of Texas Health Science Center at San Antonio for kindly providing the plasmid containing the cDNA for the 5-HT_{2C}R non-edited isoform and Dr Ronald Emeson from Vanderbilt University for providing the plasmids containing the cDNA for the A, B, D, AD, ABD, ABCD, and ABECD edited isoforms used in our studies.

REFERENCES

Abbas, A. I., Urban, D. J., Jensen, N. H., Farrell, M. S., Kroeze, W. K., Mieczkowski, P., Wang, Z., and Roth, B. L. (2010). Assessing serotonin receptor mRNA editing frequency by a novel ultra high-throughput sequencing method. *Nucleic Acids Res.* **38**(10), 1–13.

Bass, B. L. (2002). RNA editing by adenosine deaminases that act on RNA. *Annu. Rev. Biochem.* **71,** 817–846.

Bentley, D. R., Balasubramanian, S., Swerdlow, H. P., Smith, G. P., Milton, J., Brown, C. G., Hall, K. P., Evers, D. J., Barnes, C. L., Bignell, H. R., Boutell, J. M., Bryant, J., *et al.* (2008). Accurate whole human genome sequencing using reversible terminator chemistry. *Nature* **456,** 53–59.

Berg, K. A., Cropper, J. D., Niswender, C. M., Sanders-Bush, E., Emeson, R. B., and Clarke, W. P. (2001). RNA-editing of the 5-HT(2C) receptor alters agonist–receptor–effector coupling specificity. *Br. J. Pharmacol.* **134,** 386–392.

Berg, K. A., Dunlop, J., Sanchez, T., Silva, M., and Clarke, W. P. (2008a). A conservative, single-amino acid substitution in the second cytoplasmic domain of the human Serotonin2C receptor alters both ligand-dependent and -independent receptor signaling. *J. Pharmacol. Exp. Ther.* **324,** 1084–1092.

Berg, K. A., Harvey, J. A., Spampinato, U., and Clarke, W. P. (2008b). Physiological and therapeutic relevance of constitutive activity of 5-HT 2A and 5-HT 2C receptors for the treatment of depression. *Prog. Brain Res.* **172,** 287–305.

Bubar, M. J., and Cunningham, K. A. (2008). Prospects for serotonin 5-HT_2R pharmacotherapy in psychostimulant abuse. *Prog. Brain Res.* **172,** 319–346.

Burns, C. M., Chu, H., Rueter, S. M., Hutchinson, L. K., Canton, H., Sanders-Bush, E., and Emeson, R. B. (1997). Regulation of serotonin-2C receptor G-protein coupling by RNA editing. *Nature* **387,** 303–308.

Cruz-Reyes, J., Rusche, L. N., Piller, K. J., and Sollner-Webb, B. (1998). *T. brucei* RNA editing: Adenosine nucleotides inversely affect U-deletion and U-insertion reactions at mRNA cleavage. *Mol. Cell* **1,** 401–409.

De La Vega, F. M., Lazaruk, K. D., Rhodes, M. D., and Wenz, M. H. (2005). Assessment of two flexible and compatible SNP genotyping platforms: TaqMan SNP genotyping assays and the SNPlex genotyping system. *Mutat. Res.* **573,** 111–135.

Dracheva, S., Elhakem, S. L., Marcus, S. M., Siever, L. J., McGurk, S. R., and Haroutunian, V. (2003). RNA editing and alternative splicing of human serotonin 2C receptor in schizophrenia. *J. Neurochem.* **87,** 1402–1412.

Dracheva, S., Chin, B., and Haroutunian, V. (2008a). Altered serotonin 2C receptor RNA splicing in suicide: Association with editing. *Neuroreport* **19,** 379–382.

Dracheva, S., Patel, N., Woo, D. A., Marcus, S. M., Siever, L. J., and Haroutunian, V. (2008b). Increased serotonin 2C receptor mRNA editing: A possible risk factor for suicide. *Mol. Psychiatry* **13,** 1001–1010.

Dracheva, S., Lyddon, R., Barley, K., Marcus, S. M., Hurd, Y. L., and Byne, W. M. (2009). Editing of serotonin 2C receptor mRNA in the prefrontal cortex characterizes high-novelty locomotor response behavioral trait. *Neuropsychopharmacology* **34,** 2237–2251.

Du, Y., Davisson, M. T., Kafadar, K., and Gardiner, K. (2006). A-to-I pre-mRNA editing of the serotonin 2C receptor: Comparisons among inbred mouse strains. *Gene* **382,** 39–46.

Fitzgerald, L. W., Iyer, G., Conklin, D. S., Krause, C. M., Marshall, A., Patterson, J. P., Tran, D. P., Jonak, G. J., and Hartig, P. R. (1999). Messenger RNA editing of the human serotonin 5-HT2C receptor. *Neuropsychopharmacology* **21,** 82S–90S.

Giorgetti, M., and Tecott, L. H. (2004). Contributions of 5-HT(2C) receptors to multiple actions of central serotonin systems. *Eur. J. Pharmacol.* **488,** 1–9.

Gurevich, I., Englander, M. T., Adlersberg, M., Siegal, N. B., and Schmauss, C. (2002a). Modulation of serotonin 2C receptor editing by sustained changes in serotonergic neurotransmission. *J. Neurosci.* **22,** 10529–10532.

Gurevich, I., Tamir, H., Arango, V., Dwork, A. J., Mann, J. J., and Schmauss, C. (2002b). Altered editing of serotonin 2C receptor pre-mRNA in the prefrontal cortex of depressed suicide victims. *Neuron* **34,** 349–356.

Hembruff, S. L., Villeneuve, D. J., and Parissenti, A. M. (2005). The optimization of quantitative reverse transcription PCR for verification of cDNA microarray data. *Anal. Biochem.* **345,** 237–249.

Herrick-Davis, K., Grinde, E., and Niswender, C. M. (1999). Serotonin 5-HT2C receptor RNA editing alters receptor basal activity: Implications for serotonergic signal transduction. *J. Neurochem.* **73,** 1711–1717.

Herrick-Davis, K., Grinde, E., Harrigan, T. J., and Mazurkiewicz, J. E. (2005). Inhibition of serotonin 5-hydroxytryptamine2c receptor function through heterodimerization: Receptor dimers bind two molecules of ligand and one G-protein. *J. Biol. Chem.* **280,** 40144–40151.

Hoopengardner, B., Bhalla, T., Staber, C., and Reenan, R. (2003). Nervous system targets of RNA editing identified by comparative genomics. *Science* **301,** 832–836.

Igo, R. P., Jr., Weston, D. S., Ernst, N. L., Panigrahi, A. K., Salavati, R., and Stuart, K. (2002). Role of uridylate-specific exoribonuclease activity in *Trypanosoma brucei* RNA editing. *Eukaryot. Cell* **1,** 112–118.

Iwamoto, K., Bundo, M., and Kato, T. (2005a). Estimating RNA editing efficiency of five editing sites in the serotonin 2C receptor by pyrosequencing. *RNA* **11,** 1596–1603.

Iwamoto, K., Nakatani, N., Bundo, M., Yoshikawa, T., and Kato, T. (2005b). Altered RNA editing of serotonin 2C receptor in a rat model of depression. *Neurosci. Res.* **53,** 69–76.

Julius, D., MacDermott, A. B., Axel, R., and Jessell, J. M. (1988). Molecular characterization of a functional cDNA encoding the serotonin 1C receptor. *Science* **241,** 558–564.

Julius, D., Livelli, T. J., Jessell, T., and Axel, R. (1989). Ectopic expression of the serotonin 1C receptor and the triggering of malignant transformation. *Science* **244,** 1057–1062.

Kaye, W. H., Frank, G. K., Bailer, U. F., Henry, S. E., Meltzer, C. C., Price, J. C., Mathis, C. A., and Wagner, A. (2005). Serotonin alterations in anorexia and bulimia nervosa: New insights from imaging studies. *Physiol. Behav.* **85,** 73–81.

Kutyavin, I. V., Afonina, I. A., Mills, A., Gorn, V. V., Lukhtanov, E. A., Belousov, E. S., Singer, M. J., Walburger, D. K., Lokhov, S. G., Gall, A. A., Dempcy, R., Reed, M. W., *et al.* (2000). 3′-Minor groove binder-DNA probes increase sequence specificity at PCR extension temperatures. *Nucleic Acids Res.* **28,** 655–661.

Lanfranco, M. F., Seitz, P. K., Morabito, M. V., Emeson, R. B., Sanders-Bush, E., and Cunningham, K. A. (2009). An innovative real-time PCR method to measure changes in RNA editing of the serotonin 2C receptor (5-HT(2C)R) in brain. *J. Neurosci. Methods* **179,** 247–257.

Livak, K. J., and Schmittgen, T. D. (2001). Analysis of relative gene expression data using real-time quantitative PCR and the 2(-delta delta C(T)) method. *Methods* **25,** 402–408.

Marion, S., Weiner, D. M., and Caron, M. G. (2004). RNA editing induces variation in desensitization and trafficking of 5-hydroxytryptamine 2c receptor isoforms. *J. Biol. Chem.* **279,** 2945–2954.

Maydanovych, O., and Beal, P. A. (2006). Breaking the central dogma by RNA editing. *Chem. Rev.* **106,** 3397–3411.

Morabito, M. V., Ulbricht, R. J., O'Neil, R. T., Airey, D. C., Lu, P., Zhang, B., Wang, L., and Emeson, R. B. (2010). High-throughput multiplexed transcript analysis yields enhanced resolution of 5-hydroxytryptamine 2C receptor mRNA editing profiles. *Mol. Pharmacol.* **77,** 895–902.

Niswender, C. M., Sanders-Bush, E., and Emeson, R. B. (1998). Identification and characterization of RNA editing events within the 5-HT2C receptor. *Ann. NY Acad. Sci.* **861,** 38–48.

Niswender, C. M., Copeland, S. C., Herrick-Davis, K., Emeson, R. B., and Sanders-Bush, E. (1999). RNA editing of the human serotonin 5-hydroxytryptamine 2C receptor silences constitutive activity. *J. Biol. Chem.* **274,** 9472–9478.

Niswender, C. M., Herrick-Davis, K., Dilley, G. E., Meltzer, H. Y., Overholser, J. C., Stockmeier, C. A., Emeson, R. B., and Sanders-Bush, E. (2001). RNA editing of the human serotonin 5-HT2C receptor. Alterations in suicide and implications for serotonergic pharmacotherapy. *Neuropsychopharmacology* **24,** 478–491.

Poyau, A., Vincent, L., Berthomme, H., Paul, C., Nicolas, B., Pujol, J. F., and Madjar, J. J. (2007). Identification and relative quantification of adenosine to inosine editing in serotonin 2c receptor mRNA by CE. *Electrophoresis* **28,** 2843–2852.

Roth, B. L., Willins, D. L., Kristiansen, K., and Kroeze, W. K. (1998). 5-Hydroxytryptamine2-family receptors (5-hydroxytryptamine2A, 5-hydroxytryptamine2B, 5-hydroxytryptamine2C): Where structure meets function. *Pharmacol. Ther.* **79,** 231–257.

Samuel, C. E. (2003). RNA editing minireview series. *J. Biol. Chem.* **278,** 1389–1390.

Sanders-Bush, E., Fentress, H., and Hazelwood, L. (2003). Serotonin 5-HT2 receptors: Molecular and genomic diversity. *Mol. Interv.* **3,** 319–330.

Schefe, J. H., Lehmann, K. E., Buschmann, I. R., Unger, T., and Funke-Kaiser, H. (2006). Quantitative real-time RT-PCR data analysis: Current concepts and the novel "gene expression's CT difference" formula. *J. Mol. Med.* **84,** 901–910.

Schmauss, C. (2003). Serotonin 2C receptors: Suicide, serotonin, and runaway RNA editing. *Neuroscientist* **9,** 237–242.

Schmittgen, T. D., and Livak, K. J. (2008). Analyzing real-time PCR data by the comparative C(T) method. *Nat. Protoc.* **3,** 1101–1108.

Serretti, A., Artioli, P., and De, R. D. (2004). The 5-HT2C receptor as a target for mood disorders. *Expert Opin. Ther. Targets* **8,** 15–23.

Sodhi, M. S. K., Airey, D. C., Lambert, W., Burnet, P. W. J., Harrison, P. J., and Sanders-Bush, E. (2005). A rapid new assay to detect RNA editing reveals antipsychotic-induced changes in serotonin-2C transcripts. *Mol. Pharmacol.* **68,** 711–719.

Wang, Q., O'Brien, P. J., Chen, C. X., Cho, D. S., Murray, J. M., and Nishikura, K. (2000). Altered G protein-coupling functions of RNA editing isoform and splicing variant serotonin2C receptors. *J. Neurochem.* **74,** 1290–1300.

Werry, T. D., Loiacono, R., Sexton, P. M., and Christopoulos, A. (2008a). RNA editing of the serotonin 5HT2C receptor and its effects on cell signalling, pharmacology and brain function. *Pharmacol. Ther.* **119,** 7–23.

Werry, T. D., Stewart, G. D., Crouch, M. F., Watts, A., Sexton, P. M., and Christopoulos, A. (2008b). Pharmacology of 5HT(2C) receptor-mediated ERK1/2 phosphorylation: Agonist-specific activation pathways and the impact of RNA editing. *Biochem. Pharmacol.* **76,** 1276–1287.

Wess, J. (1998). Molecular basis of receptor/G-protein-coupling selectivity. *Pharmacol. Ther.* **80,** 231–264.

Wong, M. L., and Medrano, J. F. (2005). Real-time PCR for mRNA quantitation. *Biotechniques* **39,** 75–85.

Yao, Y., Nellaker, C., and Karlsson, H. (2006). Evaluation of minor groove binding probe and Taqman probe PCR assays: Influence of mismatches and template complexity on quantification. *Mol. Cell Probes.* **20**(5), 311–316.

CHAPTER NINETEEN

Strategies for Isolating Constitutively Active and Dominant-Negative Pheromone Receptor Mutants in Yeast

Mercedes Dosil* *and* James B. Konopka†

Contents

* Centro de Investigación del Cáncer and Instituto de Biología Molecular y Celular del Cáncer, CSIC-University of Salamanca, Campus Unamuno, Salamanca, Spain
† Department of Molecular Genetics and Microbiology, State University of New York, Stony Brook, New York, USA

Methods in Enzymology, Volume 485
ISSN 0076-6879, DOI: 10.1016/S0076-6879(10)85019-0

Abstract

Mating pheromone receptors of the yeast *Saccharomyces cerevisiae* are useful models for the study of G protein-coupled receptors. The mating pheromone receptors, Ste2 and Ste3, are not essential for viability so they can be readily targeted for analysis by a variety of genetic approaches. This chapter will describe methods for identification of two kinds of mutants that have been very informative about the mechanisms of receptor signaling: constitutively active mutants and dominant-negative mutants. Interestingly, these distinct types of mutants have revealed complementary information. Constitutive signaling is caused by mutations that are thought to weaken interactions between the seven transmembrane domains (TMDs), whereas the dominant-negative mutants apparently stabilize contacts between TMDs and lock receptors in the off conformation. In support of these conclusions, certain combinations of constitutively active and dominant-negative mutants restore nearly normal signaling properties.

1. Introduction

Saccharomyces cerevisiae mating pheromone receptors stimulate a G protein signal pathway that induces conjugation. The pheromone-induced responses include transcriptional induction, cell division arrest, and morphological changes. The ability of the yeast pheromone receptors to induce such robust and diverse responses has made them an interesting model for the analysis of G protein-coupled receptors (GPCRs). Some of the advantages of studying the pheromone receptors are that yeast cells grow rapidly and that the pheromone receptors can be easily manipulated since they are not essential for growth. However, the main advantage is the ease with which yeast cells can be manipulated by genetic approaches, ranging from classical genetic screens to targeted mutations. This has permitted detailed analysis of GPCR function by a wide range of genetic approaches, including the isolation of constitutive signaling and dominant-negative mutants that will be described in this chapter.

The pheromone pathway is activated when haploid cells of opposite mating type (*MAT***a** and *MAT*α) stimulate each other with secreted mating pheromones to conjugate and form an **a**/α diploid cell (reviewed in Bardwell, 2005; Dohlman and Thorner, 2001). The *MAT***a** and *MAT*α haploid cells differ primarily in the combination of receptor and pheromone they produce (Fig. 19.1). *MAT***a** cells produce a-factor and the receptors for α-factor (*STE2*); *MAT*α cells produce α-factor and the receptors for a-factor (*STE3*). Haploid cells are used for analysis of pheromone signaling because the **a**/α diploid cells do not express the pheromone receptor genes and are insensitive to pheromone. Most studies focus on the α-factor

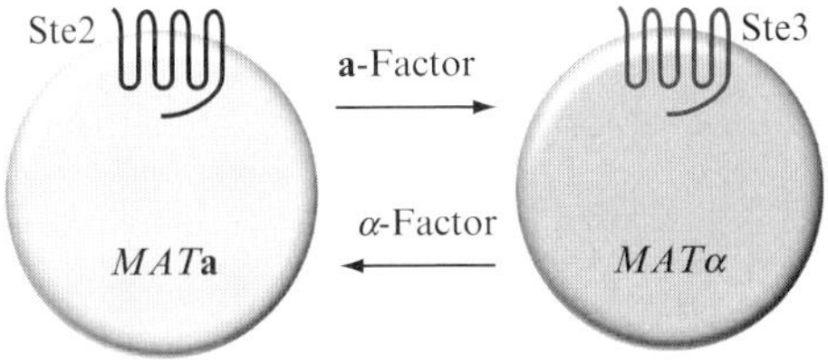

Figure 19.1 Yeast cell identity. The cell types are named for the type of pheromone they produce. *MAT*a cells produce a-factor pheromone and the receptors for α-factor (Ste2). In contrast *MAT*α cells produce α-factor pheromone and the receptors for a-factor Ste3.

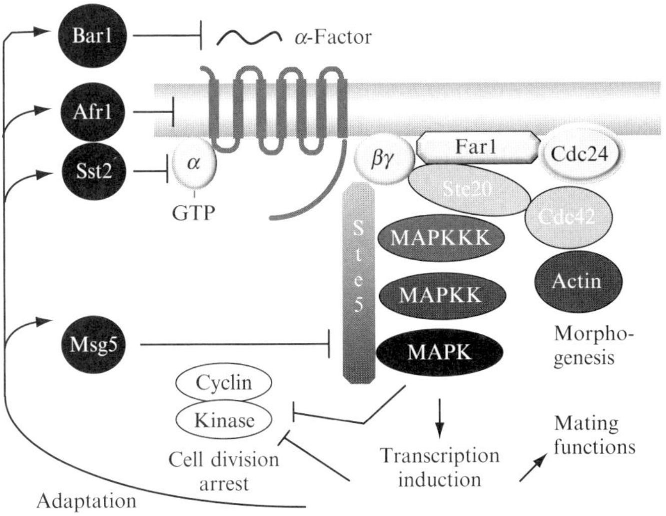

Figure 19.2 Yeast pheromone signal transduction pathway. The major proteins involved in pheromone signaling are indicated.

receptor (Ste2) in *MAT***a** cells since α-factor is a soluble peptide that is commercially available (e.g., Sigma-Aldrich Co. or Bachem, Inc.). The a-factor receptor (Ste3) has been less studied because a-factor is a lipid-modified peptide that is difficult to use in conventional dose-response and ligand binding assays.

The key elements of the pheromone signal pathway are summarized in Fig. 19.2. Ligand binding stimulates receptors to form an activated complex with a G protein that leads to the exchange of GDP for GTP on the α subunit (Gpa1p) and the subsequent release of the βγ complex (Ste4p, Ste18p) (Bardwell 2005; Dohlman and Thorner, 2001). The free βγ subunits then recruit Ste5p to the membrane. Ste5p, a scaffolding protein, recruits the components of a MAP kinase cascade including a MAPKKK (Ste11p), a MAPKK (Ste7p), and a MAPK (Fus3p). This pathway ultimately activates the pheromone-responsive transcription factor Ste12p that induces genes such as *FUS1*.

Genetic strategies have been used to identify a wide range of mutant phenotypes for *STE2* that have contributed to models for ligand binding, signaling across the plasma membrane, G protein activation, receptor desensitization, and ligand-induced endocytosis (Konopka and Thorner, 2004). Two strategies that have been very successful are the identification of constitutively active mutants and dominant-negative mutants. Constitutively active mutants are primarily caused by mutations that are predicted to change the packing of the seven α-helical transmembrane domains (TMDs) (Dube and Konopka, 1998; Konopka *et al.*, 1996; Parrish *et al.*, 2002). For example, the strongest constitutively active mutant identified thus far results from a substitution of Pro-258 for Leu in TMD six, which is expected to alter the structure of this domain in a way that impacts the way that the adjacent third intracellular loop interacts with the G protein (Konopka *et al.*, 1996). In contrast, dominant-negative mutations tend to cluster near the extracellular ends of the TMDs, where they affect ligand binding or receptor activation (Dosil *et al.*, 1998; Leavitt *et al.*, 1999). These inactive receptors are dominant because they can sequester the G protein away from the wild-type receptors (Dosil *et al.*, 2000). It should be possible to extend some of the genetic approaches described in this chapter to the study of GPCRs that can be heterologously expressed in yeast. We have previously described methods for expressing heterologous GPCRs in yeast (Mentesana *et al.*, 2002).

2. Selecting a Yeast Strain and Expression Vector

2.1. Choosing a yeast strain

S. cerevisiae genes are designated by three letters, in italics, followed by a number corresponding to a specific gene (e.g., *STE2*). Dominant alleles are written in capital letters; recessive alleles are written in lower case letters. Mutant alleles are distinguished by the addition of a dash followed by a specific identifying allele number (e.g., *ste2-3*). Deletion of a gene or disruption by insertion of another DNA sequence is indicated either with a Δ symbol (e.g., *ste2*Δ) or a double colon followed by the name of the marker gene used to create the deletion allele (e.g., *ste2::LEU2*). Proteins are usually designated with a capital first letter and not italics (e.g., Ste2 or Ste2p).

Isolation of constitutively active receptor mutants requires a specialized yeast strain to avoid the mating pheromone induced cell division arrest, which would kill off the desired mutant cells. To avoid this, the *FAR1* gene should be mutated to prevent cell division arrest in response to activation of the pathway without affecting induction of transcriptional responses (Chang and Herskowitz, 1990). The endogenous *STE2* should be deleted to facilitate analysis of mutagenized versions introduced on plasmids. The sensitivity of cells for detection of the pheromone signal can also be increased by deleting genes that promote adaptation to α-factor (Fig. 19.2). A major increase in sensitivity can be obtained by deleting the *SST2* gene, which encodes an RGS protein that promotes GTP hydrolysis by the Gα subunit (Bardwell, 2005; Dohlman and Thorner, 2001), but in practice this does not work that well since the *sst2*Δ mutation causes a high basal activity that can interfere with detection of constitutive receptor mutants. For studies in which cells will be treated with α-factor pheromone, it will also be important to delete the *BAR1* gene, which encodes a secreted protease that degrades α-factor in the medium. Finally, it is important to introduce a sensitive reporter gene to detect signaling. Most commonly used reporter genes use the promoter from the *FUS1* gene, which is dramatically induced by pheromone signaling (Trueheart *et al.*, 1987). The *FUS1* promoter is fused to reporter genes such as *lacZ* or *HIS3* to provide convenient assays for detection of receptor activation (Konopka *et al.*, 1996; Trueheart *et al.*, 1987).

Different considerations must be used for the isolation of dominant-negative *STE2* mutants (Dosil *et al.*, 1998). The cells must contain a wild-type *FAR1* gene and a wild-type *STE2* gene so that they can undergo cell division arrest in response to pheromone. As described above, the cells should carry a mutation of *BAR1* to prevent recovery from cell division arrest by degradation of pheromone and they should also contain a *FUS1-lacZ* reporter gene.

2.2. Selecting an expression vector

Mutagenesis strategies in yeast are greatly facilitated by the availability of shuttle plasmids that can be propagated as episomal plasmids in both yeast and *E. coli*. A variety of plasmids are available, so care should be taken to choose an appropriate vector. Yeast episomal plasmids (YEp) replicate to high copy number. (YEp plasmids are derived from the 2 μ circle, an autonomously replicating plasmid, and are thus also called 2 μ plasmids.) YEp plasmids are the first choice for expression of *STE2* in mutagenesis studies because they lead to overproduction of the Ste2 protein. This compensates for the fact that many mutant receptors are not stable at the cell surface and therefore increases the phenotypic effects of many constitutively active and dominant-negative mutants. Overexpression of *STE2* does not result in activation of the pathway, so it does not interfere with genetic screens for constitutively active mutants. Yeast centromeric plasmids (YCp) are present in low copy number (1 or 2 copies per cell) because they contain a centromeric element (CEN) that mediates even partitioning at mitosis. This type of vector can be advantageous for experiments where a consistent level of expression in each cell is desired. An HA (hemagglutinin) epitope tag should be introduced at the C terminal end of the receptor coding region to facilitate detection of the receptor protein.

Most currently used yeast plasmids are based on pBluescript or pUC19 plasmids (Gietz and Sugino, 1988; Sikorski and Hieter, 1989) to facilitate replication in *E. coli*. These shuttle plasmids typically carry one of several different selectable marker genes that can be used to select for plasmid maintenance in yeast by complementing a specific auxotrophy in the host strain. The most commonly used selectable markers are *URA3*, *HIS3*, *LEU2*, and *TRP1*, which are involved in the biosynthesis of uracil, histidine, leucine, and tryptophan, respectively. We have had good success using vectors containing *URA3*, since there are media that can be used to select for or against the presence of *URA3* plasmids in yeast (Sherman, 2002).

2.3. General methods for growth and storage of yeast

A rich medium (YPD) is used for general propagation of yeast strains under nonselective conditions (Sherman, 2002). Plasmids are maintained by growth of cells on chemically defined synthetic medium. The starting point for synthetic media is "Yeast Nitrogen Base," which consists of salts, vitamins, and ammonium sulfate as nitrogen source (Sherman, 2002). The medium is supplemented with amino acids and other appropriate constituents except for the one that will provide the selective pressure. Dextrose is used as an energy source for optimal growth unless the experiment specifically involves the use of other sugars to regulate inducible promoters etc. Laboratories that do not routinely grow yeast can purchase commercially made media. In our experience, media purchased from Qbiogene (Carlsbad, CA) is comparable to our homemade media. Recipes for preparing commonly used media are given below.

2.3.1. Nonselective YPD medium

1. Add 10 g yeast extract, 20 g Bacto-Peptone, and 0.120 g adenine[1] to 900 mL distilled water. Mix with a magnetic stir bar in a 2 L flask. If making solid medium, add 20 g Bacto Agar.
2. Autoclave for 30 min.
3. Add 100 mL of sterile 20% dextrose.

[1] Adenine should be added to the medium for strains that carry the *ade2* mutation to suppress the formation of a toxic red pigment that accumulates due to disruption of the adenine biosynthetic pathway.

2.3.2. Selective media

10× media supplement

1. To 1 L of distilled water, add 0.2 g of histidine, arginine, and methionine, 0.3 g of tyrosine, isoleucine and lysine, 0.4 g each of adenine sulfate, uracil, and tryptophan, 0.5 g phenylalanine, 0.6 g leucine, and 1.0 g each of glutamic acid and aspartic acid, 1.5 g valine, 2.0 g threonine, and 4.0 g serine.
2. Omit any of the above constituent(s) that will be used for selection of plasmids.
3. Aliquot into bottles and autoclave for 30 min.

Selective media (dropout media)

1. Add 6.7 g Bacto Yeast Nitrogen Base (e.g., BD Diagnostics Systems cat. # 239210) to 800 mL distilled water. If making solid agar medium add 20 g bacteriological grade agar.
2. Autoclave 300 mL aliquots for 30 min.
3. Add 37 mL sterile 20% dextrose to cooled aliquots.
4. Add 37 mL of the appropriate 10× media supplemental solution.

2.3.3. Storage of yeast cultures

Yeast can typically be stored at 4 °C for 1–3 months. Most strains will remain viable for even longer periods of time, but there is the potential danger of selecting a mutant strain that has adapted for survival on old plates. Yeast go into a deep resting state so they should be restreaked onto a fresh plate to allow a period of logarithmic growth prior to starting a new experiment. Long-term stocks can be prepared by adjusting a fresh culture of yeast to 15% glycerol and then freezing at −70 °C. Frozen cultures are viable for many years.

3. Transforming Plasmids into Yeast

The most common method for the introduction of plasmid DNA into yeast cells involves treating cells with lithium acetate (Gietz *et al.*, 1995). This method yields up to 1×10^6 transformants/μg of DNA. The steps that are important for a successful transformation include the addition of properly prepared single-stranded carrier DNA, the use of the proper-sized polymer of polyethylene glycol, and the heat shock at 42 °C. There is a very useful internet site that describes recent variations on these protocols: http://www.umanitoba.ca/medicine/biochem/gietz/.

3.1. Preparation of single-stranded carrier DNA

1. Dissolve Salmon Testes DNA (Sigma-D1626) in TE (10 m*M* Tris–HCl, pH 8.0, 1 m*M* EDTA) to 2 mg/mL by mixing on a magnetic stirrer overnight at 4 °C.
2. Aliquot DNA into microcentrifuge tubes and store at −20 °C.
3. Prior to use, boil an aliquot for 5 min and then cool in an ice water bath. Small aliquots should be used so that the sample is not thawed too many times. If kept frozen or on ice it is not necessary to boil before each use, but samples can be reheated to generate single-stranded DNA.

3.2. Transformation protocol

Special reagent solutions needed

LiAc/TE: 10 m*M* Tris, pH 7.5; 1 m*M* EDTA; 100 m*M* lithium acetate, pH 7.5.
PEG/LiAc/TE: 10 m*M* Tris, pH 7.5; 1 m*M* EDTA; 100 m*M* lithium acetate, pH 7.5; 40% PEG 3350 (e.g., Sigma cat.# P-3640; note that size of PEG is critical).

Procedure

1. Inoculate 50 mL YPD to a cell density of 2.5×10^6 cells/mL ($OD_{660} \sim 0.15$) with cells from a fresh overnight culture.
2. Incubate with shaking for ~4 h at 30 °C until cells reach 5 to 10×10^6 cells/mL ($OD_{660} = 0.4$–0.7).
3. Pellet cells by centrifugation at 3000×*g* for 5 min.
4. Wash cell pellet once with sterile water and then once with LiAc/TE solution.
5. Resuspend the final cell pellet in 0.5 mL LiAc/TE (1/100 volume).
6. Add 50 μL of the yeast cell suspension to a tube containing 10 μL (20 μg) carrier DNA and 2–4 μL transforming DNA (up to 5 μg) and vortex. Minimize the volume of added DNA to avoid diluting the other reagents.
7. Add 0.4 mL PEG/LiAc/TE and mix well.
8. Incubate at 30 °C for 30 min.
9. Heat shock by incubating at 42 °C for 15 min.
10. Add 1 mL selective media to dilute the PEG, mix, and centrifuge for 10 s to pellet cells.
11. Resuspend pellet in 0.2 mL selective medium and plate on selective agar plate.
12. Incubate at 30 °C for 2–3 days then check for growth of transformed colonies. Be sure to include a control reaction lacking plasmid DNA to determine if the transformants are genuine and not due to a contamination in the yeast culture.

4. "Gap-Repair" Approach for Targeting Mutagenesis to Genes on Plasmids

The starting point for isolating constitutively active and dominant-negative mutants in *STE2* is to mutagenize a plasmid-borne copy of *STE2* (Dosil *et al.*, 1998; Konopka *et al.*, 1996). This facilitates rapid recovery of the plasmid into *E. coli* so that the mutant phenotype can be confirmed by transforming the plasmid back into yeast. It also facilitates recovery of mutant genes for DNA sequence analysis to identify the specific mutation. To save time mapping the sites of the mutations within the *STE2* gene, we target the mutagenesis to small regions of *STE2* using a "Gap Repair" method (Kunes *et al.*, 1987; Parrish *et al.*, 2002). The basis of this technique is outlined in Fig. 19.3. First, a gap is introduced by digesting a plasmid with two enzymes that remove a small 250–300 bp region of the *STE2* coding region. Specially designed PCR primers are then used to amplify a DNA fragment that spans the gap and contains about 50 bp of homology to the ends of the digested plasmid.

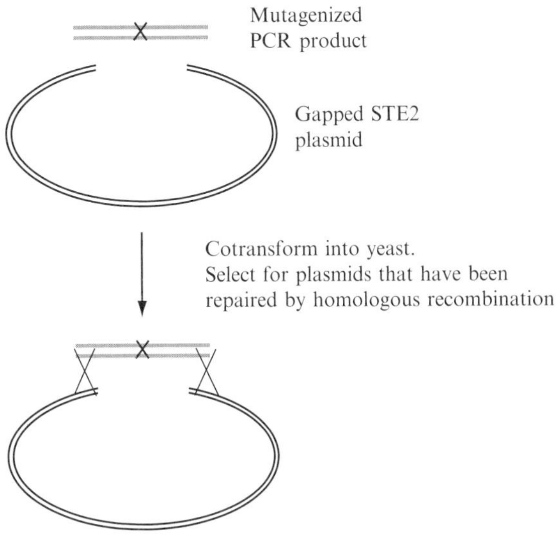

Figure 19.3 Gap-Repair method for mutagenizing *STE2* on plasmids. PCR amplification under suboptimal conditions for the Taq polymerase is used to introduce random mutations within the GPCR gene. The PCR products are then transformed into yeast along with a GPCR plasmid that has been linearized within the same region. Homologous recombination that occurs between the plasmid and the PCR product repairs the gap and introduces mutations into the targeted region. Yeast colonies carrying the mutagenized plasmids can then be replica-plated to appropriate media to assay for constitutive activity or dominant interference of the pheromone pathway.

The PCR is performed under sub-optimal conditions using error-prone Taq polymerase to introduce a wide range of random mutations. (*Note*: Do not mutagenize to such a high level that multiple mutations are introduced or it will take a lot of effort later on to determine which mutations actually contribute to the phenotype.) This fragment is then cotransformed into yeast along with the gapped plasmid whose ends are homologous to those of the PCR DNA. The gap in the plasmid is then repaired by homologous recombination with the ends of the PCR product (Kunes *et al.*, 1987). Other procedures have also been developed to use oligonucleotide-directed mutagenesis to introduce a broad range of mutations in *STE2* (Martin *et al.*, 2002).

Procedure

1. PCR amplify a region of *STE2* using Taq polymerase lacking the proofreading function.
2. Adjust the standard conditions by using 1/5 the normal concentration of one nucleotide. This will cause a higher error rate in the polymerase and introduce mutations.
3. Amplify using a standard protocol, such as 94° 1 min, 55° 1 min, 72° 1 min/kb, 30 cycles, 72 °C 10 min.
4. Digest the plasmid with restriction enzymes that will create a gap within the region of *STE2* that is being mutagenized. Design the primers so that there is at least 50 bp of homology between the PCR product and the ends of the gap on the plasmid.
5. Cotransform the purified PCR fragment and the gapped plasmid (Section 3.2). We typically use a molar ratio of about least 3:1, although good results are also obtained with a 1:1 ratio.
6. Plate transformations on selective medium and then incubate at 30 °C for 2–3 days until colonies are visible.
7. A control transformation of gapped plasmid alone should be performed to ensure that the plasmid was digested efficiently. Linearized plasmids do not propagate in yeast so the gapped plasmid should yield only a few colonies.

5. Isolation of Constitutively Active Mutants

Constitutively active mutants can be readily identified by performing the Gap-Repair mutagenesis strategy with a yeast strain that is *MATa ste2Δ far1Δ FUS1-lacZ*. The yeast strain should also contain a mutation such as *ura3Δ* to select the plasmid. We have had best results by first plating the transformation mixture on selective media to allow for colonies to grow, and then replica plating onto special medium containing X-gal to detect colonies in which the *FUS1-lacZ* reporter gene is constitutively stimulated resulting in higher levels of β-galactosidase. Although the transformation mixture can be plated directly onto the X-gal medium, the transformation efficiency and growth are not optimal on this type of medium so it is more difficult to identify the constitutive mutants. The exact amount of X-gal in the plates should be titrated so that control cells carrying a wild-type *STE2* plasmid display a barely detectable blue color after 3 or 4 days. Constitutively active mutants are readily apparent as darker blue colonies. The initial colonies should then be restreaked to purify them away from the nontransformed cells and retested to confirm that the phenotype is consistent. Plasmids can be recovered from the mutant cells and retransformed into a fresh yeast strain to confirm that the mutation is linked to the plasmid and not to a chromosomal mutation in another gene.

We have also tried using a *FUS1-HIS3* gene to directly select for constitutively active mutant receptors on medium lacking histidine (Mentesana *et al.*, 2002). This approach can be optimized by adding aminotriazole to the plates, which is a competitive inhibitor of the *HIS3*-encoded enzyme and selects for cells with the highest level of pheromone pathway activation. However, we found that this strategy yielded a high degree of false positives that were apparently caused by mutations that somehow activated the *FUS1-HIS3* reporter gene but not the pheromone pathway. These false positives take considerable effort to weed out, so we found it to be more efficient to screen for mutants on X-gal medium. In most experiments, we were able to titrate the efficiency of mutagenesis so that we identified about 1 constitutive mutant per 1000 colonies screened. In cases where multiple mutations were identified within the DNA fragment, we typically found it was more efficient to continue screening until the corresponding single mutant was identified rather than taking the time to recreate the single mutations.

In order to quantify the degree of constitutive activity, the mutant strains are then compared to the wild-type control in liquid assays for β-galactosidase activity. This is a relatively easy assay that gives a reliable comparison of the relative level of constitutive activity that will be described in detail below (see Section 7).

5.1. Identification of constitutively active mutants

1. Replica-plate colonies obtained after Gap-Repair mutagenesis to indicator plates containing X-gal (see Section 4). Incubate at 30 °C for 3–5 days. Constitutive mutants will be identified as blue colonies.
2. Restreak candidates to new X-gal plates to verify phenotype.
3. Recover the plasmid using a previously described procedure (Hoffman and Winston, 1987). Basically, the cell pellet from 1.5 mL of liquid culture is resuspended in 200 μL lysis buffer (2% Triton X-100, 1% SDS, 100 m*M* NaCl, 10 m*M* Tris, pH 8.0, 1 m*M* EDTA) and then mixed

with 200 μL phenol/chloroform. Add 200 μL glass beads and vortex for 2 min. Spin 5 min in a microcentrifuge. One microliter of the supernatant can be used to transform *E. coli*. The efficiency is low because of an inhibitor in the yeast extract.

4. Retransform yeast to confirm that the observed phenotype is plasmid-dependent.

5.2. X-gal plates for β-galactosidase reporter gene detection

To detect activation of the *FUS1-lacZ* (β-galactosidase) reporter gene, add X-gal (5-bromo-4-chloro-3-indoyl β D-galactoside) to the agar medium at a final concentration of 40 μg/mL. The X-gal should be added when the agar is cooled to about 50 °C, just prior to pouring to prevent inactivation. The X-gal plates differ from standard plates in that they are buffered to pH 7.0 for optimal β-galactosidase activity.

Special reagent solutions needed

20 mg/mL stock of X-gal in dimethylformamide.
KPI solution: 1 *M* KH_2PO_4, 0.15 *M* $(NH_4)_2SO_4$, 0.75 *M* KOH, adjusted to pH 7.0.

1. Prepare solid synthetic medium lacking the appropriate amino acid as described above in Section 2.3.2.
2. Add 100 mL of the appropriate 10× Media Supplemental solution (minus the appropriate additive), 100 mL 20% dextrose, 100 mL KPI.
3. Allow medium to cool, and then just prior to pouring plates add 2 mL of a 20 mg/mL stock of X-gal.

5.3. Quantitative assay for *FUS1-lacZ* reporter gene activity

The colorimetric substrate ONPG can be conveniently used to assay β-galactosidase activity (Miller, 1972) if the yeast cells are permeabilized to allow access of the substrate to the enzyme. Since wild type *S. cerevisiae* cells do not produce an endogenous β-galactosidase, reporter genes that make use of the *E. coli* β-galactosidase gene (*lacZ*) provide a sensitive and convenient indicator of gene expression. Sample data for a constitutive mutant is shown in Fig. 19.4. Note that increased constitutive activity is detected when the *STE2-P258L* mutant is carried on a high copy YEp plasmid. Overproduction of this mutant receptor compensates for the fact that it is not stably produced and is present at lower levels at the cell surface (Konopka *et al.*, 1996).

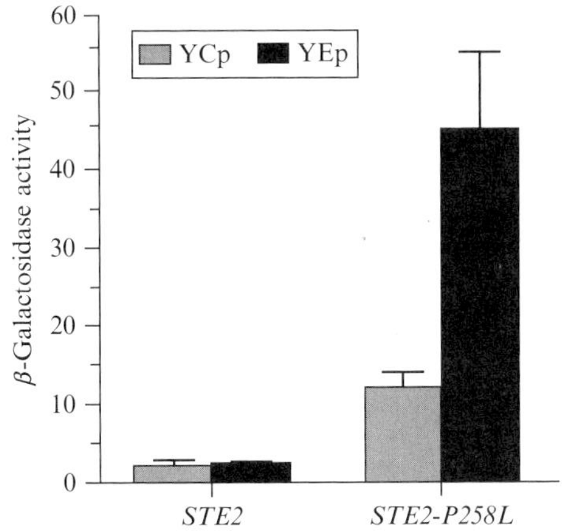

Figure 19.4 Elevated *FUS1-lacZ* reporter gene activity in constitutively signaling mutant strains. β-Galactosidase activity resulting from the *FUS1-lacZ* pheromone-responsive reporter gene is shown for cells carrying the indicated *STE2* gene on either a low copy YCp plasmid or a high copy YEp plasmid. Overproduction of the *STE2-P258L* mutant results in higher β-galactosidase activity because it compensates for the reduced stability of this mutant receptor at the cell surface.

Special reagents

Z buffer (10 m*M* KCl, 1 m*M* $MgSO_4$, 60 m*M* $Na_2HPO_4 \cdot 7H_2O$, 40 m*M* $NaH_2PO_4 \cdot H_2O$, 50 m*M* β-mercaptoethanol). (*Note*: add β-mercaptoethanol just prior to use.)

ONPG (4 mg/mL *o*-nitrophenyl-β-D-galactopyranoside in 0.1 *M* phosphate buffer pH 7.0).

β-galactosidase assay procedure

1. Grow cells overnight at 30 °C so that they remain in log phase (i.e., $OD_{660} < 7$; less than 10^7 cells/mL). To ensure that the basal level remains at a consistently low level, cells should be kept growing for two days before performing the assay.
2. Adjust 2 mL of culture to 2.5×10^6 cells/mL and then add α factor (1×10^{-7} *M*) and incubate for 2 h. A longer incubation can be used for greater sensitivity.
3. Place the cells on ice. For careful measurements add cycloheximide to 10 μg/mL.
4. Measure the OD_{600} of the cells.
5. Add 0.1 mL of the cell culture to a 1.5 mL tube containing 0.7 mL Z buffer with 2-mercaptoethanol added.
6. Add 50 μL chloroform and 50 μL 0.1% SDS. Vortex for 30 s.

7. Add 0.16 mL ONPG and mix by vortexing.
8. Incubate at 37 °C for 1 h or until the reactions turn visibly yellow.
9. Quench the reactions by adding 0.4 mL 1 *M* Na_2CO_3.
10. Spin tubes for 10 min to remove debris and then read OD_{420}.
11. Calculate the β-galactosidase units using the formula:
12. Units = $(1000 \times OD_{420})/(t \times V \times OD_{600})$
13. *t*, time of incubation in minutes and *V*, volume of cell culture added to Z buffer in mL.

6. Isolation of Dominant-Negative Mutants

Identification of dominant-negative mutants begins with the Gap-Repair mutagenesis procedure described above (Section 4). However, in this case the yeast strain used is *MATa STE2 FAR1 FUS1-lacZ* and *ura3Δ*. The colonies carrying the mutagenized plasmids are then replica-plated to agar plates containing different concentrations of α-factor in the medium. We have found it useful to replica plate the colonies to two sets of agar medium plates. One set contains the lowest dose of α-factor that causes cell division arrest and the second set of plates contain a 10-fold higher dose. The low dose helps to identify even weakly dominant mutants, but may also give some false positives. The higher dose helps to more quickly identify the strongest mutants (see Fig. 19.5).

One special caution about this approach is that mutation of genes that are part of the downstream signal pathway can also cause resistance to α-factor induced cell division arrest. To avoid wasting time studying such false-positives, only pick candidates for further analysis where it looks like many cells in the colony are resistant to α-factor, and not just that there is a rare spontaneous mutant in one part of the colony. Furthermore, cells picked for retesting should be taken off of a control plate that lacks α-factor so there has not been any selective pressure to become resistant to α-factor. A second caution is that an unmodified *STE2* gene should be used for this approach. The C-terminal tail of Ste2 is critical for dominant-negative function (Dosil *et al.*, 2000). Thus, tagging the C terminus of Ste2 with an epitope tag or GFP prior to the mutagenesis might limit the range of mutations that are identified.

The dominant-negative *STE2* mutant plasmids that are recovered should be examined in a wild-type *STE2* strain and also in a *ste2Δ* strain. The *STE2* strain will be used to determine the properties of the dominant-negative mutant, such as the strength of the interference with normal receptor function. In contrast, the use of the *ste2Δ* strain will permit analysis of the new *STE2* mutants in the absence of wild-type receptors. Our previous studies identified at least two types of receptor mutants

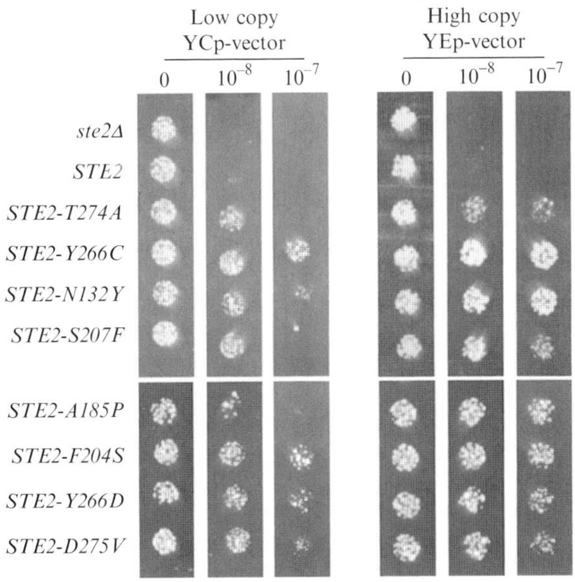

Figure 19.5 Dominant-negative *STE2* mutants are resistant to α-factor-induced cell division arrest. Yeast strain JKY25 carrying a wild-type copy of *STE2* in the genome were transformed with either low copy or high copy plasmids carrying the *STE2* genes indicated on the left. For simplicity, the empty vector control was labeled as *ste2*Δ. Equal amounts of cells were then spotted on medium lacking α-factor, or containing 10^{-8} or 10^{-7} *M* α-factor as indicated above. Growth of cells was photographed after incubation for 2 days at 30 °C. [Reproduced from Dosil *et al.*, 1998 with permission from the American Society for Microbiology (doi:0270-7306/98/$04.00+0)].

(Dosil *et al.*, 1998). One type fails to bind ligand and the other class binds ligand but cannot transduce a signal to the G protein effectively. An advantage of isolating mutants in this manner is that they are very stably produced at the plasma membrane, since they have been selected to compete with the wild-type receptors. In contrast, many previously identified loss of function mutants have been difficult to study, since they are present at low levels at the cell surface.

Special reagents needed

α-Factor stock solution: α-Factor is dissolved at 0.5 m*M* final concentration in a buffer with 10 m*M* HCl, 0.2 m*M* EDTA, and 1 m*M* β-mercaptoethanol.

Synthetic medium plates containing α-factor: These plates are made as described in Section 2 and then the appropriate amount of α-factor is added just before pouring when the agar is cooled to about 50 °C.

Procedure

1. The colonies carrying the mutagenized *STE2* plasmids are created as described in Section 4.
2. The resulting colonies are replica-plated to agar plates containing different concentrations of α-factor in the medium. One plate contains the lowest dose of α-factor that causes cell division arrest (1×10^{-8} *M*) and another plate contains a 10-fold higher dose (10^{-7} *M*). Be sure to include a third plate that lacks α-factor that can be used for future studies. Since these colonies have not been exposed to α-factor there is less chance to select for a rare spontaneous mutation in a chromosomal gene that will cause resistance to α-factor.
3. The plates are incubated at 30 °C and inspected after 48 and 74 h for signs of growth. The low dose helps to identify even weakly dominant mutants, but may also give some false positives. The higher dose helps to more quickly identify the strongest mutants.
4. All mutants should be retested by purifying the plasmids into *E. coli* and retesting following transformation into fresh yeast cells as described above for the constitutive mutants.

7. Further Methods for Analysis of Mutant Receptors

Mutant receptors identified by the above strategies can be further characterized by the strategies presented in this section. Protocols for assaying the ligand binding properties of α-factor receptors have been described previously (Ding *et al.*, 2001; Dosil *et al.*, 2000; Jenness *et al.*, 1983).

7.1. Analysis of receptor protein production by GFP-tagging

A very convenient way to detect mutant receptors is to fuse them to the *Aequorea victoria* green fluorescent protein (GFP). Fusion of GFP to the C terminus of the α-factor receptor results in a receptor protein that displays wild-type functional properties yet can be detected by fluorescence microscopy. Fluorescence is seen as a prominent ring around the cell demonstrating that the protein is stable and properly transported to the plasma membrane. Mutants that do not localize properly or receptors that are down-regulated by endocytosis are generally found in punctate-looking intracellular compartments (Choi and Konopka, 2006; Li *et al.*, 1999). Best results are obtained using a version of GFP that is enhanced for brightness and whose codons are optimized for expression in yeast codons (Longtine *et al.*, 1998).

7.2. Western blot analysis of receptor protein production

Western blot analysis is another very convenient method to detect Ste2 proteins produced in yeast. We use a triple HA tag, which provides a very sensitive level of detection (Dosil *et al.*, 2000). The preparation of yeast cell extracts requires disruption of the cell wall. Mechanical disruption by vortexing cells with glass beads is the simplest and will be described here. Whole cell extracts can be analyzed, but sensitivity can be increased by analyzing a crude membrane fraction. Methods to separate the protein extract by polyacrylamide gel electrophoresis and transfer to nitrocellulose are common practices that are described in detail elsewhere (Sambrook *et al.*, 1989).

Special reagents

Acid treated 450 μm glass beads (e.g., Sigma cat.# G8772).
TE/PP buffer: 100 m*M* NaCl, 50 m*M* Tris–HCl (pH 7.5), 1 m*M* EDTA, 100 μg/ml PMSF, 2 μg/ml pepstatin A.
PAGE sample buffer (8 *M* urea, 3% SDS, 25 m*M* Tris (pH 6.8), 0.01% bromophenol blue, 0.01% xylene cyanol).

Procedure

1. Grow a 50 mL culture of yeast to a density of about 1×10^7 cells/ml ($OD_{660} = 0.5$–1). Either YPD or synthetic medium can be used.
2. Cool centrifuge tubes and buffers on ice.
3. Transfer about 2.5×10^8 cells to a 50 ml conical tube and harvest cell pellet by centrifugation at $1000\times g$ for 5 min.
4. Wash cell pellet with 10 mL sterile water and then pellet cells again by centrifugation for 5 min.
5. Resuspend cell pellet in 1 mL sterile water and transfer to 1.5 mL microfuge tube.
6. Centrifuge at $14{,}000\times g$ for 30 s in a microfuge and then remove the supernatant. (Cells may be stored at −70 °C at this point or placed on ice until needed.)
7. Resuspend 2.5×10^8 cells in 250 μL TE/PP buffer and add 250 μL glass beads.
8. Vortex cells at very high speed for 1 min. Tilt tube so that there is maximum smashing action of the glass beads to break the yeast cell walls. Incubate on ice for 1 min to keep the sample cool and prevent proteolysis. Repeat three times.
9. Transfer cell extract to fresh tube (leaving behind the glass beads).
10. Centrifuge at low speed ($330\times g$) for 5 min at 4 °C to remove unbroken cells.

11. Transfer supernatant to a new tube. Centrifuge at top speed for 15 min at 4 °C in a microfuge to pellet membranes.
12. The crude membrane pellet can be used immediately or stored at −70 °C.
13. Dissolve the crude membrane pellet in 100 μL 2× PAGE loading buffer. Warm tubes at 37 °C for 10 min prior to loading gel. Do not boil GPCR samples or they will aggregate.
14. Centrifuge at top speed in a microfuge for 3 min prior to loading and then perform gel electrophoresis and Western blot according to standard procedures.

7.3. Halo assay for cell division arrest

The halo assay is a useful way to assay α-factor-induced cell division arrest (Dosil *et al.*, 1998). In brief, cells are spread on the surface of an agar plate and then filter disks containing α-factor are placed on the surface. The α-factor diffuses out into the medium and causes a zone of growth inhibition that is proportional to the dose of α-factor in the disk.

Special reagents

α-Factor stock solution: α-Factor is dissolved at 0.5 m*M* final concentration in a buffer with 10 m*M* HCl, 0.2 m*M* EDTA, and 1 m*M* β-mercaptoethanol.

Blank filter discs: 6 mm diameter discs from BD Diagnostic Systems (cat.# 231039).

Procedure

1. Dilute an overnight culture of yeast cells to 10^6 cells/mL.
2. Spread 150 μL of cells on the surface of an agar petri plate. If plates are dry, add extra sterile water so that the cells will spread evenly. (*Note*: some labs mix the yeast cells with relatively cool molten agar and then pour a thin layer across the top an agar plate. This gives nice-looking results, but takes more time to set up.)
3. Allow the plate to dry so that the excess liquid has soaked into the plate. It should take about 20–30 min for the plates to dry.
4. While the plates are drying, prepare appropriate dilutions of α-factor. For a *bar1* mutant strain, prepare a series of twofold dilutions between 10^{-5} and 10^{-6} *M* α-factor. α-Factor sticks to plastic, so it should be diluted into culture medium to help reduce nonspecific sticking.
5. Place four sterile filter discs on a clean sheet of Saran Wrap and then add 10 μL of the appropriate α-factor solution. When the liquid has been

absorbed by the filter, transfer the filter to the agar plate. Arrange the discs so that they are well separated and the halos will be distinct.

6. Incubate until the edge of the halo is well-defined and then measure the diameter of the halo.

ACKNOWLEDGMENTS

We thank our colleagues for contributing suggestions for protocols. J. B. K. gratefully acknowledges research grant support from the American Heart Association and the National Institutes of Health (GM087368).

REFERENCES

Bardwell, L. (2005). A walk-through of the yeast mating pheromone response pathway. *Peptides* **26**(9), 339–350.

Chang, F., and Herskowitz, I. (1990). Identification of a gene necessary for cell cycle arrest by a negative growth factor of yeast: FAR1 is an inhibitor of a G1 cyclin, CLN2. *Cell* **63,** 999–1011.

Choi, Y., and Konopka, J. B. (2006). Accessibility of Cys residues substituted into the cytoplasmic regions of the α-factor receptor identifies the intracellular residues that are available for G protein interaction. *Biochemistry* **45**(51), 15310–15317.

Ding, F. X., Lee, B. K., Hauser, M., Davenport, L., Becker, J. M., and Naider, F. (2001). Probing the binding domain of the *Saccharomyces cerevisiae* α-mating factor receptor with fluorescent ligands. *Biochemistry* **40**(4), 1102–1108.

Dohlman, H. G., and Thorner, J. W. (2001). Regulation of G protein-initiated signal transduction in yeast: Paradigms and principles. *Annu. Rev. Biochem.* **70,** 703–754.

Dosil, M., Giot, L., Davis, C., and Konopka, J. B. (1998). Dominant-negative mutations in the G protein-coupled α-factor receptor map to the extracellular ends of the transmembrane segments. *Mol. Cell. Biol.* **18**(10), 5981–5991.

Dosil, M., Schandel, K., Gupta, E., Jenness, D. D., and Konopka, J. B. (2000). The C-terminus of the *Saccharomyces cerevisiae* α-factor receptor contributes to the formation of preactivation complexes with its cognate G protein. *Mol. Cell. Biol.* **20,** 5321–5329.

Dube, P., and Konopka, J. B. (1998). Identification of a polar region in transmembrane domain 6 that regulates the function of the G protein-coupled α-factor receptor. *Mol. Cell. Biol.* **18,** 7205–7215.

Gietz, R. D., and Sugino, A. (1988). New yeast-*Escherichia coli* shuttle vectors constructed with in vitro mutagenized yeast genes lacking six-base pair restriction sites. *Gene* **74,** 527–534.

Gietz, R. D., Schiestl, R. H., Willems, A. R., and Woods, R. A. (1995). Studies on the transformation of intact yeast cells by the LiAc/SS-DNA/PEG procedure. *Yeast* **11**(4), 355–360.

Hoffman, C. S., and Winston, F. (1987). A ten-minute DNA preparation from yeast efficiently releases autonomous plasmids for transformation of *Escherichia coli*. *Gene* **57,** 267–272.

Jenness, D. D., Burkholder, A. C., and Hartwell, L. H. (1983). Binding of α-factor pheromone to yeast **a** cells: Chemical and genetic evidence for an α-factor receptor. *Cell* **35,** 521–529.

Konopka, J. B., and Thorner, J. (2004). Pheromone receptors (Yeast). *Encyclopedia Biochem.* **3,** 256–261.

Konopka, J. B., Margarit, M., and Dube, P. (1996). Mutation of pro-258 in transmembrane domain 6 constitutively activates the G protein-coupled α-factor receptor. *Proc. Natl. Acad. Sci. USA* **93,** 6764–6769.

Kunes, S., Ma, H., Overbye, K., Fox, M. S., and Botstein, D. (1987). Fine structure recombinational analysis of cloned genes using yeast transformation. *Genetics* **115**(1), 73–81.

Leavitt, L. M., Macaluso, C. R., Kim, K. S., Martin, N. P., and Dumont, M. E. (1999). Dominant negative mutations in the α-factor receptor, a G protein-coupled receptor encoded by the *STE2* gene of the yeast *Saccharomyces cerevisiae*. *Mol. Gen. Genet.* **261**(6), 917–932.

Li, Y., Kane, T., Tipper, C., Spatrick, P., and Jenness, D. D. (1999). Yeast mutants affecting possible quality control of plasma membrane proteins. *Mol. Cell. Biol.* **19**(5), 3588–3599.

Longtine, M. S., McKenzie, A., 3rd, Demarini, D. J., Shah, N. G., Wach, A., Brachat, A., Philippsen, P., and Pringle, J. R. (1998). Additional modules for versatile and economical PCR-based gene deletion and modification in *Saccharomyces cerevisiae*. *Yeast* **14**(10), 953–961.

Martin, N. P., Celic, A., and Dumont, M. E. (2002). Mutagenic mapping of helical structures in the transmembrane segments of the yeast α-factor receptor. *J. Mol. Biol.* **317**(5), 765–788.

Mentesana, P. E., Dosil, M., and Konopka, J. B. (2002). Functional assays for mammalian G-protein-coupled receptors in yeast. *Methods Enzymol.* **344,** 92–111.

Miller, J. H. (1972). *Experiments in Molecular Genetics*. Cold Spring Harbor Laboratory Press, Cold Spring Harbor, pp. 325–355.

Parrish, W., Eilers, M., Ying, W., and Konopka, J. B. (2002). The cytoplasmic end of transmembrane domain 3 regulates the activity of the *Saccharomyces cerevisiae* G-protein-coupled α-factor receptor. *Genetics* **160**(2), 429–443.

Sambrook, J., Fritsch, E. F., and Maniatis, T. (1989). Molecular Cloning: A Laboratory Manual. Cold Spring Harbor Laboratory Press. Cold Spring Harbor.

Sherman, F. (2002). Getting started with yeast. *Methods Enzymol.* **350,** 3–41.

Sikorski, R. S., and Hieter, P. (1989). A system of shuttle vectors and yeast host strains designed for efficient manipulation of DNA in *Saccharomyces cerevisiae*. *Genetics* **122,** 19–27.

Trueheart, J., Boeke, J. D., and Fink, G. R. (1987). Two genes required for cell fusion during yeast conjugation: Evidence for a pheromone-induced surface protein. *Mol. Cell. Biol.* **7,** 2316–2328.

CHAPTER TWENTY

Development of a GPR23 Cell-Based β-Lactamase Reporter Assay

Paul H. Lee* *and* Bonnie J. Hanson†

Contents

Abstract

GPR23 is a G protein-coupled receptor (GPCR) proposed to play a vital role in neurodevelopment processes such as neurogenesis and neuronal migration. To date, no small molecule GPR23 agonists or antagonists have been reported, except for the natural ligand, lysophosphatic acid, and its analogs. Identification of ligands selective for GPR23 would provide valuable tools for studying the pharmacology, physiological function, and pathophysiological implications of this receptor. This report describes how a tetracycline-inducible system was utilized in conjunction with a sensitive β-lactamase reporter gene to develop an assay in which constitutive activity of the receptor could be monitored. This assay was then utilized to screen a 1.1 million compound library to identify the

* Lead Discovery, Amgen, Inc., Thousand Oaks, California, USA
† Cell Systems Division, Discovery Assays and Services, Life Technologies Corp., Madison, Wisconsin, USA

Methods in Enzymology, Volume 485
ISSN 0076-6879, DOI: 10.1016/S0076-6879(10)85020-7

first small molecule inverse agonists for the receptor. We believe that these compounds will be invaluable tools in the further study of GPR23. In addition, we believe that the assay development techniques utilized in this report are broadly applicable to other receptors exhibiting constitutive activity.

1. Introduction

G protein-coupled receptors (GPCRs) constitute a family of membrane proteins characterized by seven transmembrane domains oriented with an extracellular N-terminus and an intracellular C-terminus. They are activated by a diverse array of extracellular substances, including biogenic amines, neuropeptides, hormones, chemokines, odorants, amino acids, free fatty acids, photons, and metabolic intermediates (He *et al.*, 2004, Pierce *et al.*, 2002, Wise *et al.*, 2004). GPCRs are activated when an agonist binds to its recognition site on the receptors. This leads to a conformational change in the receptor and forms an agonist–receptor complex which interacts with heterotrimeric $G_{\alpha\beta\gamma}$ proteins.

The activated GPCR promotes the exchange of guanosine diphosphate (GDP) for guanosine triphosphate (GTP) on the G_α subunit. The GTP-bound G_α subunit then dissociates from the agonist–receptor complex and the $G_{\beta\gamma}$ dimer. Afterward, both the G_α subunit and $G_{\beta\gamma}$ dimer are free to activate specific effector proteins such as adenylate cyclase, phospholipases, phosphodiesterases, and ion channels, leading to the activation of downstream signaling processes. The G_α subunit is inactivated when GTP is hydrolyzed to GDP. The resulting GDP-bound G_α subunit can reassociate with the $G_{\beta\gamma}$ dimer, allowing the heterotrimeric $G_{\alpha\beta\gamma}$ complex to be available for subsequent rounds of receptor activation. Readers can refer to recent reviews for a more detailed discussion of the GPCR signaling mechanism (Williams and Hill, 2009; Xiao *et al.*, 2008).

Historically, ligands for GPCRs were categorized into two main classes: agonists and antagonists. Agonists (and partial agonists) were defined as ligands able to bind to the receptor and promote a receptor conformational change, resulting in an active ligand–receptor complex and G protein activation. Conversely, neutral antagonists were defined as ligands that are able to bind to receptors which cannot cause any modulation of the receptor function. However, in the last two decades, it has become apparent that this simple classification of GPCR ligands is insufficient to describe the receptor molecular pharmacology. There is now evidence supporting the existence of inverse agonists (which bind to the receptor and negatively modulate constitutive receptor functions) (Kenakin, 2004) and allosteric ligands (which are distinct from the natural ligand of the receptor in that they evoke their effects indirectly via a discrete binding pocket to cause a variety of effects, i.e., positive, negative, or modulatory) (Christopolous, 2002).

It is now well known that GPCRs are able to exist in a spontaneously active state that leads to G protein activation in the absence of agonist stimulation (Kenakin, 2004). In experimental or pathological conditions, receptor overexpression may produce significant levels of spontaneously activated receptors that exceed the threshold for detectable constitutive activity. One simple molecular mechanism for inverse agonism is selective affinity for the inactive state of the receptor. It is important to note that inverse agonists behave as simple competitive antagonists in nonconstitutively active receptors. Determining whether a drug produces full or partial inverse agonism is a great deal more complicated than characterizing agonists and antagonists, depending on the dynamic range of the measuring systems in which inverse agonism can be detected and the number of active receptors.

Lysophosphatidic acid (LPA) is a bioactive phospholipid naturally synthesized by many cell types and is involved in multiple physiological processes, including cell proliferation, cell migration, smooth muscle contraction, cell survival, and immune responses. Three GPCRs belonging to the endothelial cell differentiation gene (EDG) family were the first identified LPA receptors (Noguchi *et al.*, 2009). These are LPA_1 (EDG2), LPA_2 (EDG4), and LPA_3 (EDG7) receptors. All three receptors are coupled to the G_i and $G_{q/11}$ signal transduction pathways. GPR23 (or P_2Y_9 receptor) was identified as the fourth LPA receptor (LPA_4); it shares only a 20–24% amino acid identity with LPA_{1-3} (Noguchi *et al.*, 2003). Subsequently, GPR92 was nominated as the fifth LPA receptor (LPA_5), sharing only a 35% amino acid identity with GPR23 (Lee *et al.*, 2006). Both receptors induce cyclic adenosine monophosphate (cAMP) production and intracellular Ca^{2+} mobilization via activation of the G_s and G_q signal transduction pathways. GPR23 closely resembles the P_2Y_5 receptor identified recently in sequence homology, sharing a 56% amino acid identity (Aoki *et al.*, 2008).

GPR23 is expressed at a high level in reproductive, brain, and adipose tissues. LPA has been demonstrated to induce GPR23 internalization, $G_{12/13}$- and Rho-mediated neurite retraction and stress fiber formation, G_q-mediated and pertussis toxin-sensitive Ca^{2+} mobilization, and G_s-mediated cAMP increase (Lee *et al.*, 2007). It is postulated that GPR23 may also play a vital role in neurodevelopment processes such as neurogenesis and neuronal migration (Yanagida *et al.*, 2007). To date, no small molecule GPR23 agonists or antagonists have been reported except the natural ligand, LPA, and its analogs. Identification of ligands selective to GPR23 would provide valuable tools for studying the pharmacology, physiological function, and pathophysiological implication of the receptor.

2. GPCR Cell-Based Assays

For many years, ligand binding assays were among the most popular methods to study the interactions of drugs with GPCRs. However, in the past two decades, multiple novel cellular assay technologies have emerged, making it possible to measure both proximal and distal events that result from GPCR activation; including G protein activation, accumulation or depletion of intracellular second messengers, protein–protein interactions, and gene transcription.

2.1. Choice of GPCR cell-based assays

An ideal GPCR cell-based assay for high-throughput screening should be simple, nonradioactive, robust, homogeneous, contain minimal reagent addition steps, and be amenable to a 384-well or a higher density to facilitate robotic automation. Table 20.1 shows the cell-based assays commonly used for GPCR screens (readers interested in details of individual assays can refer to a recent review article by Xiao *et al.*, 2008). A measurement of events proximal to GPCR activation (β-arrestin binding or receptor internalization) will reduce the incidence of false positives; however, measuring an event further down a signal transduction pathway can provide a higher signal-to-noise ratio as a result of signal amplification (reporter gene or label-free measurement).

Cell-based assays measuring intracellular second messengers, such as Ca^{2+}, inositol phosphate, and cAMP, were developed many years ago to measure the functional activity of GPCRs. These original second messenger

Table 20.1 GPCR cell-based assays

Assays	Assay formats
β-Arrestin binding	EFC, fluorescence imaging, reporter gene (β-lactamase, luciferase)
Receptor internalization	Fluorescence imaging
Ca^{2+} flux	Aequorin, fluorescence Ca^{2+}-sensitive dye, reporter gene (β-lactamase, luciferase)
Inositol phosphate accumulation	EFC, FP, HTRF, SPA
cAMP measurement	Alphascreen®, ECL, EFC, FP, HTRF, SPA, reporter gene (β-lactamase, luciferase)
Label-free measurement	Impedance, refractive index

ECL, electrochemiluminescence; EFC, enzyme fragment complementation; FP, fluorescence polarization; HTRF, homogeneous time-resolved fluorescence; SPA, scintillation proximity assay.

assays were tedious, difficult to perform, and often in radioactive format. Recent advances in fluorescence technologies have enabled these measurements to be performed in a homogeneous assay using fluorescent detection platforms, and have significantly improved the sensitivity and throughput of the assays. For this reason, the majority of GPCR functional screens are currently performed using either Ca^{2+} flux assays (for $G_{\alpha q}$- and $G_{\alpha i}$-coupled GPCRs) or cAMP assays (for $G_{\alpha s}$- and $G_{\alpha i}$-coupled GPCRs).

Reporter gene assays are another way to monitor the activities of intracellular second messengers. These cellular assays are cost-effective, high-throughput, homogeneous, and have been proved to be automation friendly and can be miniaturized to a 1536-well or even a 3456-well format (for β-lactamase) (Kornienko *et al.*, 2004). Reporter gene assays measure the activation of a response element placed upstream of a minimal promoter that regulates the expression of a selected reporter protein, such as β-lactamase or luciferase. With $G_{\alpha s}$- and $G_{\alpha q}$-coupled GPCRs, an increase in the cellular cAMP or Ca^{2+} levels may in turn activate the cAMP reponse element (CRE) or NFAT response element, respectively.

β-Arrestin-based assays constitute another assay platform for GPCRs. Upon GPCR activation, most GPCRs recruit β-arrestin. By differentially tagging the GPCR and the β-arrestin, homogeneous, high-throughput assays have been developed. In one such assay (the Tango™ assay from Life Technologies), the GPCR is tagged with a protease site followed by a non-native transcription factor, while the β-arrestin is tagged with the corresponding protease. Upon β-arrestin recruitment, the transcription factor is proteolytically released from the GPCR, enabling its translocation to the nucleus to activate a reporter gene. The Tango™ assay differs from traditional reporter gene assays in that the released transcription factor leads to immediate activation of the reporter gene, whereas in traditional reporter gene assays, an endogenous signal transduction pathway must be activated prior to the reporter gene activation. In a separate assay (the PathHunter assay from DiscoveRx), the GPCR and the β-arrestin are tagged with complementary fragments of the β-galactosidase enzyme, such that when β-arrestin is recruited to the GPCR, enzyme fragment complementation occurs forming an active β-galactosidase enzyme.

In recent years, a number of innovative label-free cell-based assays have also been developed. These include the impedance-based cellular dielectric spectroscopy (CDS) and the refractive index-based dynamic mass redistribution (DMR). Both technologies measure an integrated cellular response and can be used to monitor $G_{\alpha s}$-, $G_{\alpha i}$-, and $G_{\alpha q}$-coupled GPCR activation without the need for chimeric or promiscuous G proteins, or the loading of fluorescent/luminescent detection reagents (Ciambrone *et al.*, 2004; Fang *et al.*, 2006).

The choice of which cell-based assay to use may be driven in part by the type of ligand being sought. Some receptors display naturally high levels of constitutive activity, and if an agonist for these receptors is being sought, an assay system that is sensitive enough to detect the further increase in signal over the basal level of constitutive activity would be desired. Alternatively, if an inverse agonist is being sought, an assay system in which the constitutive activity can be maximized would be beneficial. In this sense, technologies that offer flexibility and advantages in the development of the cell-based assay may guide the assay choice rather than the readout itself.

For assay development, assay systems which easily allow the selection of rare event clones based upon functional parameters are ideal. One of the most useful techniques for this purpose is fluorescence activated cell sorting (FACS). With FACS, the response profile of a control population can be compared to that of the experimental population. Sorting "gates" can then be applied to isolate just those cells exhibiting the desired phenotype. Single cells exhibiting the desired phenotype can then be deposited into individual wells of a microtiter plate and expanded as individual clones. For example, in the case of constitutively active receptors, this technique could be used to compare the basal fluorescence in a parental cell line to that of a cell line expressing the receptor of interest. Therefore, cells with higher basal fluorescence could be isolated.

For an assay to be most suited for FACS, there are three parameters which should ideally be met. First, the assay should be fluorescent as opposed to chemiluminescent. Chemiluminescent assays, in general, do not produce sufficient photons per second for a significant number to be collected by the flow cytometer in the brief time period during which the cell is in the focal point of the collection lens. Second, the assay should not require cell permeabilization/lysis as clones isolated by this technique need to be viable for expansion. Finally, the fluorescent signal should persist over the course of time required for the cell sorting to avoid having to sort in multiple batches. In these respects, β-lactamase reporter gene is ideally suited for FACS. First, signal amplification occurs with the β-lactamase reporter protein, making this system extremely sensitive and allowing detection of low levels of receptor constitutive activation. Second, it is a live cell assay that does not require cell lysis as many of the second messenger assays do. Finally, the fluorescent β-lactamase substrate and product are retained within cells and the signal is stable for hours, unlike the transient nature of signals stemming from fluorogenic Ca^{2+}-sensitive dyes.

2.2. Use of tetracycline-inducible β-lactamase reporter assays for constitutively active GPCRs

β-lactamase reporter assays have been widely used for HTS of GPCRs (Bercher *et al.*, 2009; Kunapuli *et al.*, 2003; Oosterom *et al.*, 2005). The β-lactamase reporter system makes use of the TEM-1 β-lactamase from *Escherichia coli*. Since there are no mammalian homologs for this enzyme, one does not have to worry about cleavage of the substrate from endogenous β-lactamases in mammalian cells. A ratiometric, fluorescent substrate was developed for β-lactamase (Zlokarnik *et al.*, 1998). This substrate consists of coumarin and fluorescein moieties separated by a β-lactam ring. The molecule was developed as an acetyoxymethyl (AM) ester derivative to lend membrane permeability (Fig. 20.1). As such, the substrate can readily diffuse into cells where intracellular esterases cleave the AM groups, leaving the substrate with a negative charge that helps retain the substrate within the cell. In the absence of β-lactamase, the coumarin and fluorescein remain in close proximity. An excitation of the coumarin leads to efficient Förster resonance energy transfer (FRET), resulting in emission at 520 nm. In the presence of β-lactamase, the β-lactam ring is cleaved, separating the coumarin from the fluorescein resulting in a loss of FRET and an increase in emission at 447 nm. The ratio of 447 to 520 nm fluorescence can then be used as an indicator of the level of β-lactamase present. With this type of ratiometric measurement, the effects of many experimental variations can be reduced or eliminated. These variations may stem from differences in cell number per well, in the substrate loading and retention among cells, or in the excitation intensity or detection efficiency between wells.

For many GPCRs, constitutive activity is observed with receptor overexpression (Chen *et al.*, 2000; Thomsen *et al.*, 2005). If GPCR expression is driven by an inducible expression system, then the constitutive activity can be regulated. An ability to regulate receptor expression can be advantageous for

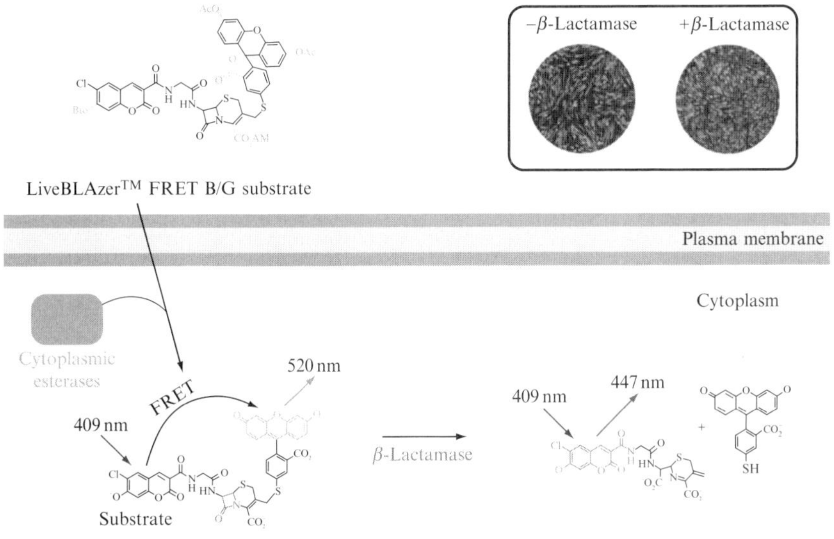

Figure 20.1 Schematic diagram of the β-lactamase system. Copyright Life Technologies Corporation. Used with permission.

several reasons. First, it allows regulation of receptor expression level and therefore constitutive activity to be tailored to a particular experiment. For example, if an inverse agonist screen is being developed, it may be desirable to have very high constitutive activity in order to increase the assay window in the screen. On the other hand, if an agonist screen is being developed, it may be desirable to keep the constitutive activity relatively low in order to allow a larger ligand-activated assay window. Second, the inducible expression can also be used as an internal control within the assay by providing the limits for high constitutive activity (positive control for activation of the receptor) and no/low constitutive activity (negative control for activation of the receptor).

The T-REx^{TM} System (Life Technologies) is a tetracycline (or doxycycline) inducible mammalian expression system in which the gene of interest is repressed in the absence and induced in the presence of tetracycline (Yao *et al.*, 1998). The system makes use of regulatory elements from the *E. coli* Tn-*10*-encoded tetracycline resistance operon (Hillen and Berens, 1994; Hillen *et al.*, 1983). There are three main components to the system which are required to achieve the inducible expression (Fig. 20.2).

- *Inducible Expression Vector Containing TetO$_2$ sites*: Expression of the gene of interest is controlled by a CMV promoter into which two copies of the tet operator 2 (TetO$_2$) sequence have been inserted in tandem. Each TetO$_2$ sequence serves as the binding site for two molecules of the Tet repressor.
- *Tet Repressor (TetR)*: In the absence of tetracycline (or doxycycline), TetR forms a homodimer that binds with very high affinity to each TetO$_2$ sequence in the promoter of the inducible expression vector, repressing transcription of the gene of interest.
- *Tetracycline (or doxycycline)*: Tetracycline (or doxycycline) binds with high affinity to each TetR homodimer causing a conformational change resulting in the inability of the TetR to bind to the TetO$_2$ sites. The removal of the TetR from the TetO$_2$ sites allows transcription of the gene of interest to begin.

By combining the sensitive β-lactamase reporter system with an inducible expression system, the advantages of both systems can be realized as previously shown with the G2A receptor (Bercher *et al.*, 2009) and the GHSR (Hanson *et al.*, 2009). With the G2A receptor, inducible constitutive activity served as a control to show that the $G_{\alpha q}$-coupled orphan G2A receptor activation could be detected with the β-lactamase reporter system, in which β-lactamase expression was controlled by an NFAT response element. As a larger assay window was desired for this agonist screen of the G2A receptor, the expression levels of G2A (and therefore the constitutive activity) were kept low by leaving the receptor in a minimally induced state. This technique allowed novel agonists for the G2A receptor to be identified (Bercher *et al.*, 2009). With the GHSR, it was again shown that the constitutive activity (and the resulting β-lactamase signal) could be

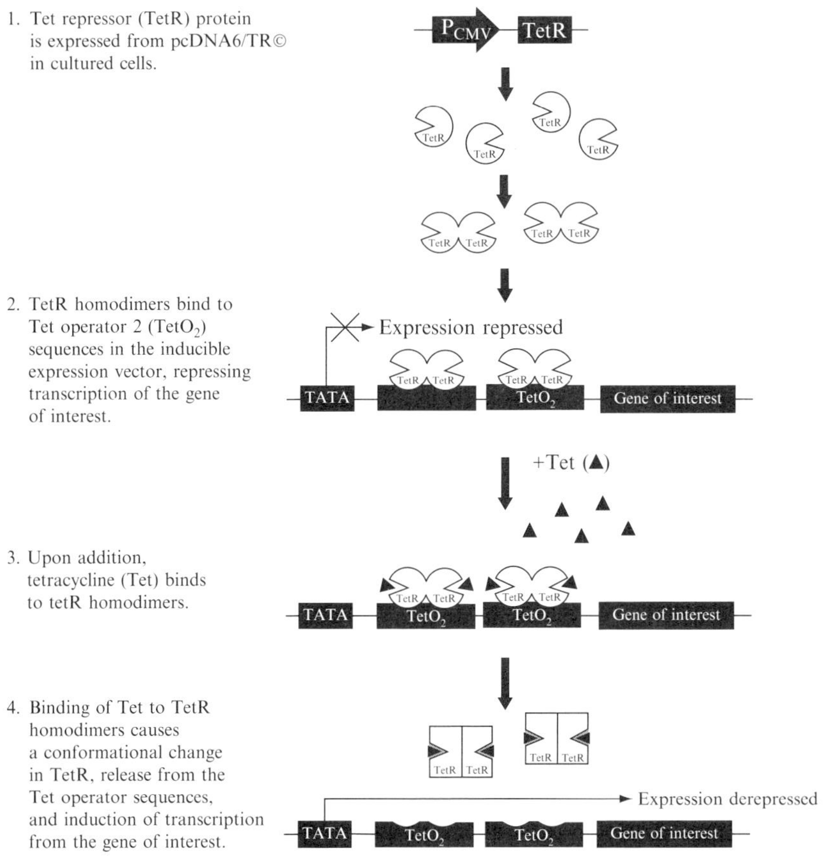
1. Tet repressor (TetR) protein is expressed from pcDNA6/TR© in cultured cells.
P CMV
TetR
2. TetR homodimers bind to Tet operator 2 ($TetO_2$) sequences in the inducible expression vector, repressing transcription of the gene of interest.
Expression repressed
TATA
$TetO_2$
Gene of interest
+Tet (▲)
3. Upon addition, tetracycline (Tet) binds to tetR homodimers.
4. Binding of Tet to TetR homodimers causes a conformational change in TetR, release from the Tet operator sequences, and induction of transcription from the gene of interest.
Expression derepressed

Figure 20.2 Schematic diagram of the T-REx™ system. Copyright Life Technologies Corporation. Used with permission.

regulated using this inducible system. By adding doxycycline to induce high levels of constitutive activity, a large assay window could be obtained to enable screening for inverse agonists.

3. Development of a Cell-Based β-Lactamase Reporter Assay for Constitutively Active GPR23

Live cell β-lactamase reporter assays are well suited for the development of cellular assays for constitutively active GPCRs. With β-lactamase, both the fluorescent substrate and product are retained within the cells, lending single cell resolution to the assay. This is not possible with other reporter genes which lead to a secreted product or which require cell permeabilization for detection. Furthermore, since the substrate for β-lactamase is fluorescent, FACS can also be used to isolate individual cells expressing the desired phenotype (such as high levels of constitutive activity) from a large number of cells. In our study, GPR23 exhibited constitutive activity through coupling to native $G_{\alpha s}$ protein in CHO cells. Therefore we decided to develop a β-lactamase reporter assay for GPR23 using a T-REx™ inducible system.

3.1. Implementation of the T-REx™ (Tet-On) system

A T-REx™-CHO cell line was obtained from Life Technologies. This cell line stably expresses the TetR protein. The GPR23 gene was then cloned into the pcDNA4/TO inducible expression vector. The vector was transfected into the T-REx™-CHO cell background and antibiotic selection was performed with zeocin for two weeks. Individual clones were isolated and selected based upon their ability to cause an increase in cAMP production in response to LPA.

3.2. Introduction of the β-lactamase reporter system and isolation of stable cell clones

An expression vector (p4X-CRE-BlaX) subcloned with the CRE-*bla* reporter gene in which expression of a β-lactamase gene is controlled by CRE (TGACGTCA) was prepared. The T-REx™-GPR23-CHO cell line was transfected with this CRE-*bla* plasmid and selected with geneticin for two weeks. Functionality of the CRE-*bla* reporter was tested with the adenylyl cyclase activator, forskolin (data not shown). Briefly, cells were left unstimulated or were stimulated with 10 μ*M* forskolin at 37 °C in an atmosphere of 5% CO_2 for 5 h. The FRET-based substrate, LiveBLAzer™ FRET-B/G, was loaded at room temperature for 2 h according to manufacturer's directions. The medium only wells (no cell wells) were included for background subtraction of the fluorescence observed. Forskolin directly activates adenylyl cyclases, leading to an increase in cAMP and subsequent phosphorylation/translocation of the CRE binding protein, which binds to the CRE to drive β-lactamase gene expression. The expressed β-lactamase protein in turn cleaves the substrate, disrupting FRET between the coumarin and fluorescein fluorophores in the substrate, which can be detected with a standard fluorescent plate reader with bottom-read capabilities. For data analysis, the average signals obtained at 460 nm (blue) and 520 nm (green) for the no cell wells is subtracted from the signal for all cell-containing wells and expressed as a ratio of blue/green (B/G).

As the fluorescent β-lactamase substrate loads readily into live cells, individual stable cell clones could be isolated by FACS based upon CRE-*bla* functionality. To do this, cells were again left unstimulated or stimulated with 10 μ*M* forskolin prior to loading with the 2 μ*M* LiveBLAzer^{TM} FRET-B/G substrate in a FACS sorting buffer (PBS without Ca^{2+} or Mg^{2+}, 1% glucose, 1 m*M* EDTA, 1 m*M* HEPES, pH 7.4). Prior to sorting, the cells were centrifuged at 100 *g* and resuspended in the FACS sorting buffer to remove any unloaded β-lactamase substrate. A FACS Vantage DiVa cell sorter (Becton Dickinson) equipped with a Krypton laser with violet excitation (407 nm at 60 mW) and fitted with a 100 μm nozzle tip was utilized for sorting. The cell sorter was further configured with an HQ460/50nm (blue) emission filter, an HQ535/40nm (green) emission filter, and a 490 nm dichroic mirror (all available from Omega Optical). After the instrument had been optically aligned and optimized, a dot plot with the level of green fluorescence on the *X*-axis versus blue fluorescence on the *Y*-axis was obtained for the unstimulated and stimulated samples. Clones were isolated by sorting individual cells from the stimulated population into individual wells of three 96-well plates using a sorting gate that was set around the cells containing the highest level of blue fluorescence and lowest level of green fluorescence (i.e., the highest CRE-*bla* activation) (Fig. 20.3).

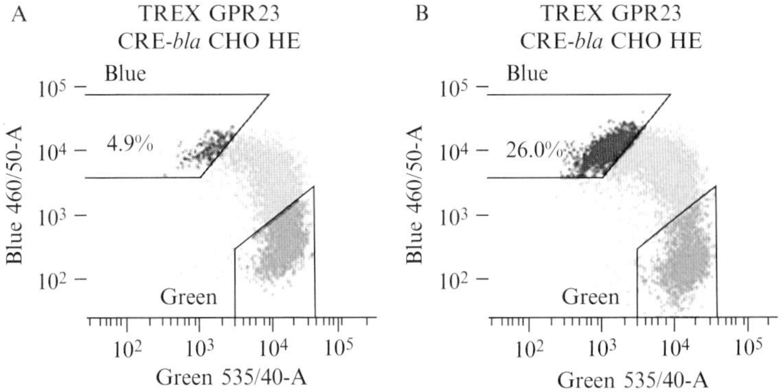

Figure 20.3 Dot plot of T-REx^{TM}-GPR23-CRE-*bla*-CHO cells that have been left unstimulated (A) or been stimulated for 5 h with forskolin (B) to activate adenylate cyclase prior to loading with the LiveBLAzer^{TM} FRET-B/G substrate. Cells in which β-lactamase have been activated have higher intensity of blue fluorescence and less intensity of green fluorescence signals. Individual clones were sorted from the population of cells falling within the blue gate shown in (B). (For interpretation of the references to color in this figure legend, the reader is referred to the Web version of this chapter.)

3.3. Determination of inducible constitutively active GPR23 clones

The individual clones were tested for their level of inducible constitutively active GPR23 by determining the basal level of CRE-*bla* activation observed in the presence (induced) or absence (uninduced) of tetracycline. As serum contains micromolar amounts of LPA (Noguchi *et al.*, 2003), a serum-free assay medium was utilized to reduce any potential serum-activated background signals. In this test, the cells were incubated with or without 1 μg/mL tetracycline in the assay medium (DMEM with 0.1% BSA, 100 ng/mL pertussis toxin) for 16 h. Pertussis toxin was added to block any signals that might be generated by endogenous $G_{\alpha i}$-coupled LPA receptors.

Six clones showing at least threefold higher levels of β-lactamase reporter activity in the presence of tetracycline were selected for further evaluation in a doxycycline (a stable analog of tetracycline) concentration–response study. It was reasoned that as GPR23 expression level increases with higher concentrations of doxycycline, more receptors would become constitutively active and hence activate the CRE-*bla* reporter gene. All six clones showed a doxycycline concentration-dependent β-lactamase response and one clone, H6E2, which demonstrated the largest response ratio (induced/uninduced) was selected as the T-REx™-GPR23-CRE-*bla*-CHO cell line for further assay optimization (Fig. 20.4).

3.4. Optimization of β-lactamase reporter assay

Additional assay parameters were investigated using the T-REx™-GPR23-CRE-*bla*-CHO cells to determine the optimum assay conditions. Generally, the following procedure was followed during assay optimization, which involved plating the cells in a growth medium at 37 °C in an atmosphere of 5% CO_2 for 6 h to allow the cells to attach to the assay plate (BD black-wall clear-bottom poly-D-lysine-coated 384-well plates). The growth medium was then removed and replaced with the assay medium containing doxycycline to induce GPR23 expression for 16 h. After doxycycline induction, the cells were loaded with the LiveBLAzer™ FRET-B/G substrate containing 1 m*M* probenecid to reduce export of the substrate through organic anion transporters.

The parameters investigated included cell density per well in 384-well assay plates, DMSO tolerance, induction time with doxycycline, and substrate loading time. From these experiments, it was determined that the cell line showed a similar assay window (fourfold) at a density of 5000, 10,000, and 20,000 cells per well but the highest density had the least well-to-well variation compared to the other two densities. In an experiment determining the effect of various DMSO concentrations on the assay, the cells were found to tolerate up to 1% DMSO without significant changes in the induced β-lactamase signals. As compounds

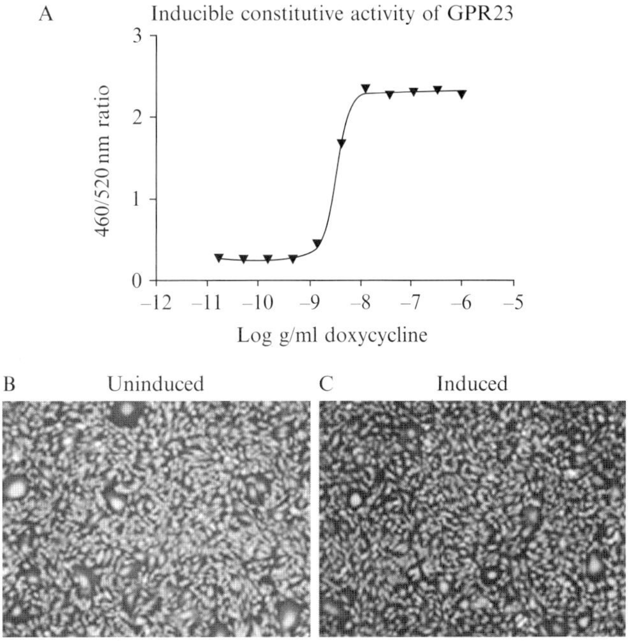

Figure 20.4 Inducible constitutive activity of T-REx™-GPR23-CRE-*bla*-CHO cells. (A) Incubation with various concentrations of doxycycline for 16 h induced an increase in β-lactamase activity (460 nm/520 nm ratio) in a concentration-dependent manner. (B) Uninduced cells loaded with the LiveBLAzer™ FRET-B/G substrate for 2 h appear mostly green as there are only low levels of β-lactamase expression. (C) Doxycycline-induced cells loaded with the substrate appear mostly blue as there are high levels of β-lactamase produced due to the constitutive activity of expressed GPR23, which activates expression of the β-lactamase gene driven by a cAMP response element. (For interpretation of the references to color in this figure legend, the reader is referred to the Web version of this chapter.)

are usually dissolved in DMSO, this is important information as it determines the highest compound concentration that can be tested in the assay. The induction of GPR23 expression with doxycycline at 16, 20, or 24 h were studied and 24-h induction led to a slightly larger assay window than 16 and 20 h. However, 16-h induction was chosen to allow flexibility, and time for experiments that required compounds to be added post-doxycycline stimulation and substrate loading. In addition, the substrate loading times of 1, 1.5, and 2 h were evaluated. All three loading periods were sufficient to generate robust signals ($Z' > 0.5$) but with a general trend that longer times gave larger assay windows. Therefore, the optimized assay conditions involved plating 20,000 cells per well with doxycycline stimulation for 16 h. The cells were loaded with the substrate for 2 h before fluorescence signals were determined. The assay was also evaluated using cryopreserved cells, in which frozen cells were thawed and immediately plated for

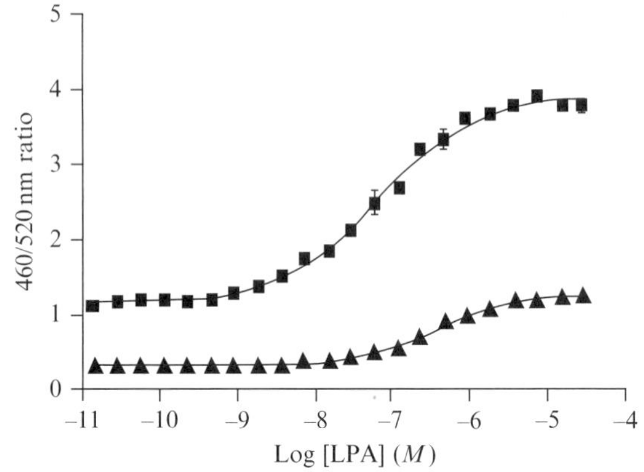

Figure 20.5 Response of T-REx™-GPR23-CRE-*bla*-CHO cells to LPA. The cells were induced with 10 ng/mL doxycycyline for 16 h (squares) or were left uninduced (triangles) prior to stimulation with various concentrations of LPA for 5 h. The cells were then loaded with the LiveBLAzer™ FRET-B/G substrate for 2 h and the resulting 460/520 nm ratios were plotted. Data are mean ± S.D. values of triplicate wells in a representative experiment.

the assay. The results were similar to those using cells maintained in culture. This is a particularly useful technique for HTS as the cells necessary for the screen can be scaled up and frozen in advance into a single batch and thawed on the day of screening.

The LPA-stimulated response in an uninduced versus induced state was studied (Fig. 20.5). Doxycycline induction led to a large increase in basal cAMP level as evidenced by the increased β-lactamase activity at negligible LPA concentrations. LPA further activated β-lactamase activity in a concentration-dependent manner with an EC_{50} of 0.07 μ*M*. In the absence of doxycycline, LPA stimulated a modest level of β-lactamase activity and the response barely reached saturation at 30 μ*M* with an EC_{50} of 0.54 μ*M*. An increased basal β-lactamase activity in these cells upon doxycycline treatment demonstrates the expressed GPR23 are constitutively active and functionally coupled to the endogenous $G_{\alpha s}$ proteins, which activate the cAMP signaling and subsequent β-lactamase gene expression.

4. Identification of GPR23 Inverse Agonists Using a β-Lactamase Reporter Screen

An HTS was performed using doxycycline-induced GPR23 expression and constitutive activity to identify small molecule inverse agonists of the receptor. In the screen, the T-REx™-GPR23-CRE-*bla*-CHO cells were incubated overnight with doxycycline together with individual test compounds. The β-lactamase activity was then determined the next day (Table 20.2). A total of 1.1 million compounds were screened and the results are summarized in Figure 20.6A. The GPR23 β-lactamase inverse agonist screen was robust and showed an average Z' of 0.68. Compounds showing greater than 50% inhibition of the doxycycline-induced β-lactamase activity were defined as primary hits. A cytotoxicity filter was applied using the formula $RFU_{signal} = RFU_{green} + C \times RFU_{blue}$, where C is set at 1 in this assay condition. When the calculated RFU_{signal} is less than 50% of the positive control wells, the compound is considered cytotoxic and

Table 20.2 GPR23 β-lactamase inverse agonist HTS protocol

Step	Parameter	Value	Description
1	Library compounds	10 μL	20 μ*M*, diluted in culture medium
2	Plate cells	40 μL	20,000 cells per well, ±doxycycline
3	Incubation time	16 h	37 °C, 5% CO_2
4	Equilibration time	15 min	Room temperature
5	Reporter reagent	10 μL	β-Lactamase detection
6	Incubation time	3 h	Room temperature
7	Assay readout	450 and 530 nm	Envision, fluorescence mode

Step	Notes
1	Library compounds (1 μL, 1 m*M*) diluted with 50 μL culture medium (phenol red-free DMEM with 0.1% BSA, 25 m*M* HEPES and 1% L-glutamine). 10 μL of diluted compounds was transferred to black-walled clear-bottom 384-well poly-D-lysine coated plates using Vprep.
2	The cells suspended in culture medium containing 0.2 μ*M* doxycycline were plated using bulk dispenser WellMate. The last column was plated with cells without doxycycline.
3	Plates covered with lids and kept as single layer without stacking in the incubator.
4	Equilibrate plates to room temperature before adding reporter reagent.
5	Reporter reagent contains 3 μ*M* CCF-AM substrate and 6 m*M* probenecid.
6	The plates were kept in the dark at room temperature.
7	Data were analyzed and expressed as a ratio of 450/530 nm (B/G) fluorescence and converted to percentage of control (POC) using the formula $[100 \times (\text{sample}_{B/G} - \text{no doxycycline}_{\text{meanB/G}}) / (\text{doxycycline}_{\text{meanB/G}} - \text{no doxycycline}_{\text{meanB/G}})]$.

Reproduced with permission from Wong *et al.* (2010).

removed from the hit list. The screen identified 8709 primary hits. These compounds were retested under the same conditions in triplicate and 993 compounds were confirmed active. The confirmed compounds were counter-screened at the same concentration against a $G_{\alpha s}$-coupled VPAC1-CRE-*bla*-CHO cell line to eliminate nonspecific hits. Subsequently, 109 hits were identified with $EC_{50} \leq 2\ \mu M$ and these compounds represented ten different chemical structural classes (Wong *et al.*, 2010).

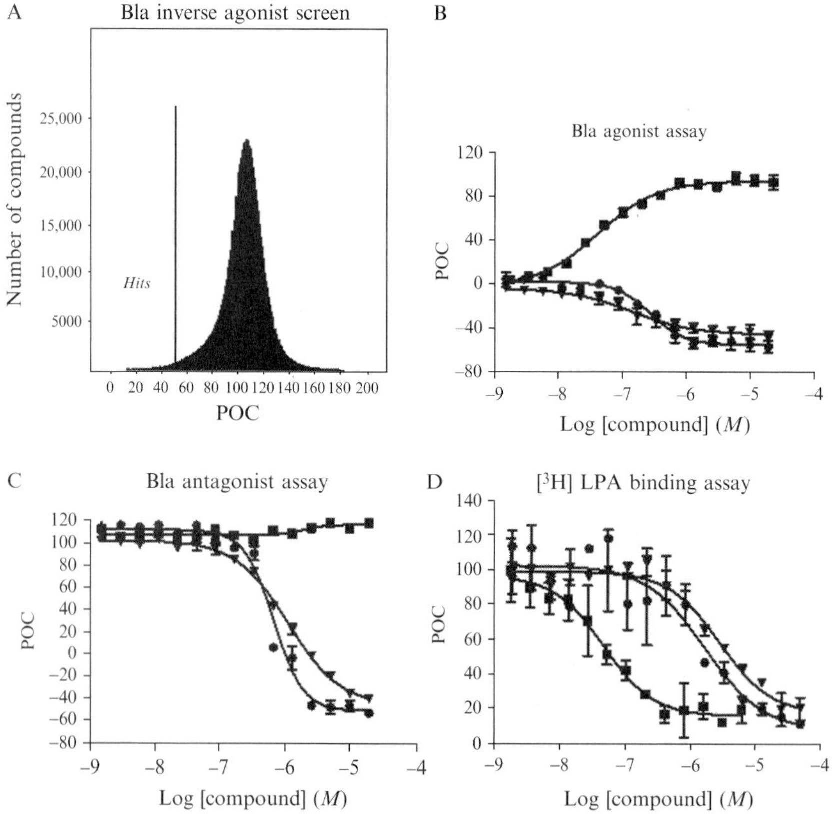

Figure 20.6 GPR23 β-lactamase reporter screen and inverse agonists identified. (A) Frequency distribution profile of screening compounds' inhibitory effects (percentage of control, POC) on doxycycline-induced cells. Data represent results from individual compounds for the entire library of 1.1 million compounds. LPA (squares) and screening hits, compound 1 (circles) and compound 2 (inverted triangles), were tested in cells stimulated overnight with 10 ng/mL doxycycline (B) or in cells stimulated overnight with 10 ng/mL doxycycline and 0.18 μM LPA (C). The compounds were also tested in a GPR23 binding assay using scintillation proximity assay format in the presence of 20 nM [^{3}H]LPA (D). Data are mean ± S.D. values of duplicate wells in a representative experiment. Reproduced with permission from Wong *et al.* (2010).

Figure 20.6 shows the *in vitro* pharmacology of the two most potent GPR23 hits, representing two different chemotypes. Doxycycline induction of the GPR23 expression and subsequent constitutive GPR23-cAMP signaling (in the absence of LPA) provides an assay condition where an agonist would produce a stimulatory response and an inverse agonist would show an inhibitory response. Compounds 1 and 2 demonstrated an inverse agonistic effect with an EC_{50} of 0.18 ± 0.17 and 0.19 ± 0.08 μM ($n = 3$), respectively. However, LPA showed an agonistic activity ($EC_{50} = 0.05 \pm 0.01$ μM, $n = 3$) in this assay condition.

Under the antagonist assay condition where LPA was added at an EC_{80} concentration (0.18 μM) to activate the expressed GPR23, compounds 1 and 2 showed a full inhibition of LPA-induced β-lactamase activity. Their inhibitory effects extended beyond the LPA-induced activity and into negative POC with an IC_{50} of 0.43 ± 0.26 and 1.26 ± 0.17 μM ($n = 3$) for compounds 1 and 2, respectively. The result suggests that the two compounds inhibited both LPA-activated and constitutive GPR23 activities. LPA, at high concentrations, stimulated an additional 10–20% β-lactamase response in this antagonist assay condition (Fig. 20.6C).

The two compounds were further tested in the [^{3}H]LPA binding assay and were found to block [^{3}H]LPA binding to cell membranes prepared from doxycycline-induced cells in a concentration-dependent manner (Fig. 20.6D). The IC_{50} values for compound 1, compound 2, and LPA are 1.59 ± 0.36, 4.93 ± 2.36, and 0.05 ± 0.01 μM ($n = 3$), respectively. The effects of compounds 1 and 2 on the cAMP levels were also tested in doxycycline-induced cells. Both compounds inhibited the elevated cellular cAMP levels upon doxycycline treatment, as well as an increase in cAMP levels under LPA stimulations (Wong *et al.*, 2010). These results demonstrate that compounds 1 and 2 behave as inverse agonists by binding GPR23 to attenuate the constitutive activity of the receptors and they compete with LPA at the same binding domain on the receptor.

5. Concluding Remarks

GPCRs can be spontaneously activated in the absence of ligands. In experimental or pathological conditions, receptors expressed at high levels may produce significant levels of spontaneously activated receptors and exceed the threshold for detectable constitutive activity. One simple molecular mechanism for inverse agonism is selective affinity for the inactive state of the receptor. It is important to note that inverse agonists behave as simple competitive antagonists in ligand-activated receptors (Kenakin, 2004). In tightly coupled functional assays, the effects of even a low concentration of activated receptors would result in measurable levels of constitutive activity.

Generally, all ligands with affinity for receptors would be expected to produce either agonism or inverse agonism. If not, it would require the ligand to recognize the two receptor conformational states (active and inactive) as being identical (Kenakin and Onaran, 2002). A recent survey of 380 previously named antagonists indicates that 322 (85%) are inverse agonists and 58 (15%) are neutral antagonists, suggesting neutral antagonists are the minority category of GPCR ligands (Kenakin, 2004).

There are a wide variety of cell-based assays available for constitutively active GPCRs; however, the assay of choice can influence the results observed. For example, assays that detect effects further down the signal transduction cascade have increased sensitivity due to amplification of the response. Furthermore, the β-lactamase reporter assay is highly sensitive due to an enzymatic amplification effect, in which as few as 100 β-lactamase enzyme molecules can cleave and alter the fluorescence of many more β-lactamase substrate molecules (Zlokarnik *et al.*, 1998). Therefore, the β-lactamase reporter assay provides a sensitive method for measuring GPCR signal transduction pathway activation, particularly in detecting receptor constitutive activity. This report demonstrates a strategy of using an inducible T-Rex™ system for the development of a β-lactamase reporter assay to identify inverse agonists of GPR23. We also believe that the method employed here is applicable to other constitutively active receptors.

ACKNOWLEDGMENTS

The authors thank Justin Wetter, Soo Hang Wong, Rommel Mallari, and Xiaoning Zhao for their technical support of this project.

REFERENCES

Aoki, J., Inoue, A., and Okudaira, S. (2008). Two pathways for lysophosphatidic acid production. *Biochim. Biophys. Acta* **1781,** 513–518.

Bercher, M., Hanson, B., van Staden, C., Wu, K., Ng, G. Y., and Lee, P. H. (2009). Agonists of the orphan human G2A receptor identified from inducible G2A expression and beta-lactamase reporter screen. *Assay Drug Dev. Technol.* **7,** 133–142.

Chen, G., Way, J., Armour, S., Watson, C., Queen, K., Jayawickreme, C. K., Chen, W.-J., and Kenakin, T. (2000). Use of constitutive G protein-coupled receptor activity for drug discovery. *Mol. Pharmcol.* **57,** 125–134.

Christopolous, A. (2002). Allosteric binding sites on cell-surface receptors: Novel targets for drug discovery. *Nat. Rev. Drug Discov.* **1,** 198–210.

Ciambrone, G. J., Liu, V. F., Lin, D. C., McGuinness, R. P., Leung, G. K., and Pitchford, S. (2004). Cellular dielectric spectroscopy: A powerful new approach to label-free cellular analysis. *J. Biomol. Screen.* **9,** 467–480.

Fang, Y., Ferrie, A. M., Fontaine, N. H., Mauro, J., and Balakrishnan, J. (2006). Resonant waveguide grating biosensor for living cell sensing. *Biophys. J.* **91,** 1925–1940.

Hanson, B. J., Wetter, J., Bercher, M. R., Kopp, L., Fuerstenau-Sharp, M., Vedvik, K. L., Zielinski, T., Doucette, C., Whitney, P. J., and Revankar, C. (2009). A homogeneous fluorescent live-cell assay for measuring 7-transmembrane receptor activity and agonist functional selectivity through beta-arrestin recruitment. *J. Biomol. Screen.* **14,** 798–810.

He, W., Maio, F. J., Lin, D. C., Schwandner, R. T., Wang, Z., Gao, J., Chen, J. L., Tian, H., and Ling, L. (2004). Citric acid cycle intermediates as ligands for orphan G-protein-coupled receptors. *Nature* **429,** 188–193.

Hillen, W., and Berens, C. (1994). Mechanisms underlying expression of Tn10 encoded tetracycline resistance. *Annu. Rev. Microbiol.* **48,** 345–369.

Hillen, W., Gatz, C., Altschmied, L., Schollmeier, K., and Meier, I. (1983). Control of expression of Tn10-encoded tetracycline resistance genes: Equilibrium and kinetic investigations of the regulatory reactions. *J. Mol. Biol.* **169,** 707–721.

Kenakin, T. (2004). Efficacy as a vector: The relative prevalence and paucity of inverse agonism. *Mol. Pharmacol.* **65,** 2–11.

Kenakin, T., and Onaran, O. (2002). The ligand paradox between affinity and efficacy: Can you be there and not make a difference? *Trends Pharmacol. Sci.* **23,** 275–280.

Kornienko, O., Lacson, R., Kunapuli, P., Schneeweis, J., Hoffaman, I., Smith, T., Alberts, M., Inglese, J., and Strulovici, B. (2004). Miniaturization of whole live cell-based GPCR assays using microdispensing and detection systems. *J. Biomol. Screen.* **9,** 186–195.

Kunapuli, P., Ransom, R., Murphy, K. L., Pettibone, D., Kerby, J., Grimwood, S., Zuck, P., Hodder, P., Lacson, R., Hoffman, I., Inglese, J., and Strulovici, B. (2003). Development of an intact cell reporter gene beta-lactamase assay for G protein-coupled receptors for high-throughput screening. *Anal. Biochem.* **314,** 16–29.

Lee, C. W., Rivera, R., Gardell, S., Dubin, A. E., and Chun, J. (2006). GPR92 as a new $G_{12/13}$- and G_q-coupled lysophosphatidic acid receptor that increases cAMP, LPA_5. *J. Biol. Chem.* **281,** 23589–23597.

Lee, C. W., Rivera, R., Dubin, A. E., and Chun, J. (2007). LPA_4/GPR23 is a lysophosphatidic acid (LPA) receptor utilizing G_s-, G_q/G_i-mediated calcium signaling and $G_{12/13}$-mediated Rho activation. *J. Biol. Chem.* **282,** 4310–4317.

Noguchi, K., Ishii, S., and Shimizu, T. (2003). Identification of $p2y_9$/GPR23 as a novel G protein-coupled receptor for lysophosphatidic acid, structurally distant from the Edg family. *J. Biol. Chem.* **278,** 25600–25606.

Noguchi, K., Herr, D., Mutoh, T., and Chun, J. (2009). Lysophosphatidic acid (LPA) and its receptors. *Curr. Opin. Pharmacol.* **9,** 15–23.

Oosterom, J., van Doornmalen, E. J., Lobregt, S., Blomenröhr, M., and Zaman, G. J. (2005). High-throughput screebing using beta-lactamase reporter-gene technology for identification of low-molecular-weight antagonists of the human gonadotropin releasing hormone receptor. *Assay Drug Dev. Technol.* **3,** 143–154.

Pierce, K. L., Premont, R. T., and Lefkowitz, R. J. (2002). Seven-transmembrane receptors. *Nat. Rev. Mol. Cell Biol.* **3,** 639–650.

Thomsen, W., Frazer, J., and Unett, D. (2005). Functional assays for screening GPCR targets. *Curr. Opin. Biotechnol.* **16,** 655–665.

Williams, C., and Hill, S. J. (2009). GPCR signaling: Understanding the pathway to successful drug discovery. *Methods Mol. Biol.* **552,** 39–50.

Wise, A., Jupe, S. C., and Rees, S. (2004). The identification of ligands at orphan G protein-coupled receptors. *Annu. Rev. Pharmacol. Toxicol.* **44,** 43–66.

Wong, S. H., Mallari, R., Graham, M., Atangan, L., Lu, S. C., Zhao, X., Gu, W., and Lee, P. H. (2010). Strategy for the identification of GPR23/LPA_4 receptor agonists and inverse agonists. *Assay Drug Dev. Technol.* **8,** 459–470.

Xiao, S. H., Reagan, J. D., Lee, P. H., Fu, A., Schwandner, R., Zhao, X., Knop, J., Beckmann, H., and Young, S. W. (2008). High throughput screening for orphan and liganded GPCRs. *Comb. Chem. High Throughput Screen.* **11,** 195–215.

Yanagida, K., Ishii, S., Hamano, F., Noguchi, K., and Shimizu, T. (2007). $LPA_4/p2y_9$/GPR23 mediates Rho-dependent morphological changes in a rat neuronal cell line. *J. Biol. Chem.* **282,** 5814–5824.

Yao, F., Svensjö, T., Winkler, T., Lu, M., Eriksson, C., and Eriksson, E. (1998). Tetracycline represor, tetR, rather than the tetR-mammalian cell transcription factor fusion derivatives, regulates inducible gene expression in mammalian cells. *Hum. Gene Ther.* **9,** 1939–1950.

Zlokarnik, G., Negulescu, P. A., Knapp, T. E., Mere, L., Burres, N., Feng, L., Whitney, M., Roemer, K., and Tsien, R. Y. (1998). Quantitation of transcription and clonal selection of single living cells with β-lactamase as reporter. *Science* **279,** 84–88.

CHAPTER TWENTY-ONE

Computational Modeling of Constitutively Active Mutants of GPCRs: C5a Receptor

Gregory V. Nikiforovich* *and* Thomas J. Baranski[†,‡]

Contents

Abstract

In the past decade, an increasing number of studies using computational modeling procedures have focused on the structural aspects of constitutive activity in G protein-coupled receptors (GPCRs). This chapter reviews various conceptual approaches in computational modeling of constitutively active mutants (CAMs) including analyzing three-dimensional models of the ground

* MolLife Design LLC, St. Louis, Missouri, USA
† Department of Medicine, Washington University Medical School, St. Louis, Missouri, USA
‡ Department of Developmental Biology, Washington University Medical School, St. Louis, Missouri, USA

Methods in Enzymology, Volume 485
ISSN 0076-6879, DOI: 10.1016/S0076-6879(10)85021-9

states of GPCRs based on structural homology with the known X-ray templates; molecular dynamics simulations starting from the ground states; and modeling of CAMs based on the experimentally suggested templates of the possible activated states. The developed buildup procedure of rotational sampling of the TM regions of GPCRs is highlighted in more detail. Experimental data on CAMs of the complement factor 5a receptor (C5aR) are used to validate the rotational sampling results.

1. Introduction

G protein-coupled receptors (GPCRs) comprise a vast protein family involved in variety of physiological functions. GPCRs are embedded in the cell membrane and include seven-helical transmembrane stretches (TM helices, TMs), the N- and C-terminal fragments and the extra- and intracellular loops connecting the TM helices. Agonist-binding activates GPCRs by triggering receptor interactions with corresponding G proteins inside the cell. Some mutant GPCRs display constitutive activity, that is, ligand-independent activity that produces a second messenger even in the absence of an agonist. Constitutively active mutants (CAMs) are known for many GPCRs (see, e.g., a review Parnot *et al.*, 2002). Since activation of GPCRs is believed to require conformational changes from their inactive ground states to their activated states, it is generally assumed that conformations of CAMs can mimic the activated conformations of GPCRs. Obviously, knowledge of molecular mechanisms of GPCR's transition from the ground conformational states to the activated states would tremendously benefit molecular biophysics, pharmacology, and drug design.

Direct experimental structural data on the activated states of GPCRs are very sparse. Currently, only 17 crystal structures of six GPCRs are available in the PDB (see Mustafi and Palczewski, 2009 for a minireview), and most captured the inactive ground conformational states of GPCRs. The possible exceptions are structures of photoactivated rhodopsin (the PDB entry 2I37; Salom *et al.*, 2006), of the β_2 adrenergic receptor crystallized with the partial inverse agonist carazol (the PDB entry 2RH1; Cherezov *et al.*, 2007), and of opsin (retinal-free rhodopsin) crystallized with the C-terminal fragment of α-transducin (the PDB entry 3DQB; Scheerer *et al.*, 2008). Additionally, some structural data on the light-activated rhodopsin were obtained by spin labeling (Altenbach *et al.*, 2008; Hubbell *et al.*, 2003). Very recently, light activation of rhodopsin was tracked using infrared probes (Ye *et al.*, 2010).

In view of the limited amount of experimental data, computational modeling has become an especially important investigative tool for studies of GPCRs. This chapter highlights the efforts involved in molecular modeling of CAMs during the past decade. It overviews numerous studies and arranges them based on the conceptual approaches used for modeling CAMs. Also, the chapter reviews more closely the buildup procedure of rotational sampling employed for establishing structural features characteristic for constitutive activation of the complement factor 5a receptors (C5aRs).

2. Modeling CAMs Based on Experimental Data for the Ground and Activated States of GPCRs

This section briefly reviews some representative studies explicitly discussing modeling of the TM regions of CAMs including analyzing the models of the ground states of GPCRs and molecular dynamics (MD) simulations starting from the ground states (see the highlights in Table 21.1), as well as employing the specific models of the activated states of CAMs. The positions of mutations mentioned in this section are generally numbered not only by their sequential numbers, but also by the universal nomenclature (in superscript) referring to the TM number and the number of the mutated residue in this TM (Ballesteros and Weinstein, 1992).

2.1. Analyzing 3D models of the ground states

Computational modeling rationalizing the known mutagenic data on CAMs was based on the 3D structures of the ground states of GPCRs (mostly of TM regions) deduced initially from the experimental data on bacteriorhodopsin and rhodopsin obtained by electron microscopy and, further, by the X-ray crystallography. The underlying assumption was that constitutively activating mutations disrupt some specific residue–residue interactions stabilizing the ground states of GPCRs (see Parnot *et al.*, 2002). From this perspective, one does not need to know (or hypothesize) a specific model for the activated state(s); the models of the ground states are built by homology modeling and refined by energy minimization.

A typical example of this approach was presented in an early study where the 3D model of the ground state of the angiotensin receptor type 1 (AT1R) was built based on the experimental structure of bacteriorhodopsin (Groblewski *et al.*, 1997). This modeling suggested that the ground state of AT1R is stabilized by interaction between N111$^{3.35}$ and Y292$^{7.45}$, and the constitutive activity of mutants N111A and N111G was explained by disruption of this interaction. Similarly, a model of mouse delta-opioid receptor (mDOR) structure based on the low-resolution structure of rhodopsin proposed that the constitutive activity of mutants D128$^{3.33}$N/A,

Table 21.1 Highlights of CAM modeling based on analyzing of 3D models of the ground states (upper part) or on short MD simulations (lower part)

Receptor	CAMs	Key stabilizing interaction(s) in ground state	Reference
AT1R	N111$^{3.35}$G	N111$^{3.35}$–Y292$^{7.45}$	Groblewski *et al.* (1997)
m DOR	D128$^{3.33}$A/N Y129$^{3.34}$A/F Y308$^{7.43}$F	D128$^{3.33}$–Y308$^{7.43}$	Befort *et al.* (1999)
hB2R	N113$^{3.35}$A Y115$^{3.37}$A W256$^{6.48}$F/Q	N113$^{3.35}$–W256$^{6.48}$	Marie *et al.* (1999, 2001)
5HT2C	Y368$^{7.53}$X	Y368$^{7.53}$–Y375$^{7.60}$	Prioleau *et al.* (2002)
5HT4R	Basal activity of WT	D100$^{3.32}$–W272$^{6.48}$–F275$^{6.51}$	Joubert *et al.* (2002)
TSHR	M626$^{6.37}$I	M626$^{6.37}$–I515$^{3.46}$ (destabilizing)	Ringkananont *et al.* (2006)
MOR	L275$^{6.30}$E/ T279$^{6.34}$K	L275$^{6.30}$E–R165$^{3.50}$	Huang *et al.* (2002)
α1aAR	A271$^{6.34}$E	D72$^{2.50}$–R124$^{3.50}$	Cotecchia *et al.* (2000)
α1bAR	A293$^{6.34}$E	D91$^{2.50}$–R143$^{3.50}$	Cotecchia *et al.* (2000)
α1bAR	D142$^{3.49}$A E289$^{6.30}$K	D142$^{3.49}$–R143$^{3.50}$–E289$^{6.30}$	Cotecchia *et al.* (2003), Greasley *et al.* (2002)
FSHR	D567$^{6.30}$X T449$^{3.32}$X	E576$^{6.30}$–R162$^{3.50}$ T449$^{3.32}$–H615$^{7.42}$	Montanelli *et al.* (2004)
TSHR	N674$^{7.49}$D	D581$^{6.44}$–N674$^{7.49}$	Govaerts *et al.* (2001)
TSHR	N674$^{7.49}$D	D581$^{6.44}$–N674$^{7.49}$/ T670$^{6.43}$	Urizar *et al.* (2005)
LHR	D556$^{6.44}$H/Q	D556$^{6.44}$–N615$^{7.45}$	Angelova *et al.* (2002)
LHR	L457$^{3.43}$R	D578$^{6.44}$–N615$^{7.45}$	Zhang *et al.* (2005)

Y129$^{3.34}$F/A, and Y308$^{7.43}$F occurred by disrupting of interaction between D128$^{3.33}$ and Y308$^{7.43}$ (Befort *et al.*, 1999). For human B2 bradykinin receptor (hB2R), the constitutive activity of N113$^{3.35}$/A, W256$^{6.48}$F/Q, and Y115$^{3.37}$A was explained by breaking interactions N113$^{3.35}$–W256$^{6.48}$ as well as those involving Y115$^{3.37}$ and Y295$^{7.43}$ in the ground state of hB2R based on the X-ray structure of rhodopsin (Marie *et al.*, 1999, 2001). Disrupting interactions between Y368$^{7.53}$ and Y375$^{7.60}$ was suggested as cause of constitutive activity of the series of mutants Y368$^{7.53}$X in the serotonin 5HT2C receptor (5HT2CR) based on the model of 5HT2CR derived from the X-ray structure of rhodopsin (Prioleau *et al.*, 2002). In another serotonin receptor, 5HT4 (5HT4R), molecular modeling based on the X-ray structure of rhodopsin indicated a network of interactions between D100$^{3.32}$, F275$^{6.51}$, and W272$^{6.48}$ as stabilizing element of the ground state (Joubert *et al.*, 2002). In contrast, other modeling found that a steric conflict between I515$^{3.46}$ and I626$^{6.37}$ destabilizes the ground state of the CAM M626$^{6.37}$I of the thyrotropin receptor (TSHR; Ringkananont *et al.*, 2006). Also, according to molecular modeling, electrostatic repulsion between residues K279$^{6.34}$ and R165$^{3.50}$ disrupts the ground state of the CAM T279$^{6.34}$K of the mu-opioid receptor (MOR; Huang *et al.*, 2002).

The models of the ground states based on the X-ray structure of rhodopsin were employed also in our studies of CAMs in AT1R (Nikiforovich *et al.*, 2005) and the C5a receptor (C5aR; Sen *et al.*, 2008). Molecular modeling of several AT1 receptors with mutations of N111$^{3.35}$ by G, A (CAMs) or W (non-CAM) found cascade of conformational changes, which were characteristic for CAMs, in orientations of the side chains of L112$^{3.36}$, Y113$^{3.37}$, F117$^{3.41}$, I152$^{4.49}$, and M155$^{4.52}$. That allowed the correct prediction of the novel CAMs of AT1R, namely L112$^{3.36}$F and L112$^{3.36}$C, as well as the double mutants N111$^{3.35}$G/L112$^{3.36}$A and N111$^{3.35}$G/F117$^{3.41}$A (Nikiforovich *et al.*, 2005). For C5aR, our modeling analyzed possible orientations of different side chains in two CAMs, I124$^{3.40}$N/L127$^{3.43}$Q (NQ), and F251$^{6.44}$A, suggesting that steric clashes between the side chains of L127$^{3.43}$ and F251$^{6.44}$ prevent certain orientation of F251$^{6.44}$ that is a prerequisite for displaying constitutive activity (Sen *et al.*, 2008). As a result, the novel CAM of C5aR, L127$^{3.43}$A, was predicted and then validated experimentally.

Generally, analyzing the 3D models of the ground states elucidated residue–residue interactions that may be responsible for stabilizing or destabilizing the ground states of GPCRs. Since such interactions were different for different GPCRs, no universal mechanism for inducing constitutive activity in GPCRs was suggested by this simple approach. Presently, the approach remains a widely used modeling tool for GPCR studies, since the homology-based 3D models for the ground states of GPCRs became readily available through a variety of the Internet servers.

2.2. Molecular dynamics simulations starting from the ground states

The models of ground states for various CAMs have been used also as starting points for straightforward MD simulations aimed at finding possible conformational trajectories of transitions from the ground state to the activated states; however, no specific models of the activated states were assumed in these calculations. These early simulations (on scale of 100–1000 ps) were successful mainly in revealing specific interactions that can be disrupted during the short-scale movements of the ground states. These studies were limited since experimental estimations of the transition time are on scale of milliseconds (light activation of rhodopsin; Borhan *et al.*, 2000) or even seconds (activation of β2 adrenoreceptor (β2AR) by diffusible ligand; Ghanouni *et al.*, 2001), and MD simulations on such large scale are unattainable even employing the current state-of-art computational resources (see also Section 5).

Some examples of CAM modeling employing short MD simulations include CAMs of α1a adrenoreceptor (α1aAR), A271$^{6.34}$E, and α1b adrenoreceptor (α1bAR), A293$^{6.34}$E, which suggested that breaking the salt bridge D$^{3.49}$–R$^{3.50}$ in the highly conserved DRY fragment induced constitutive activity in both mutants (Cotecchia *et al.*, 2000). Further studies by the same authors revealed additional involvement of an interaction between R143$^{3.50}$ and E289$^{6.30}$ (elements of the so-called ionic lock, see below) in stabilization of the ground state of α1bAR; disruption of this interaction induces constitutive activity of mutants D142$^{3.49}$A and E289$^{6.30}$K (Cotecchia *et al.*, 2003; Greasley *et al.*, 2002). Weakening of the same type of interaction (between R162$^{3.50}$ and D567$^{6.30}$) was found in several CAMs of the D567$^{6.30}$X series by MD simulations performed for the human follitropin receptor (FSHR; Montanelli *et al.*, 2004). MD simulations showed also that another interaction, D581$^{6.44}$–N674$^{7.49}$, stabilized the ground states of TSHR; the mutant N674$^{7.49}$D, where this interaction was disrupted, showed constitutive activity (Govaerts *et al.*, 2001; Urizar *et al.*, 2005). Residue D$^{6.44}$ may be also involved in stabilization of the ground states by interaction with N$^{7.45}$: this particular interaction was disrupted in MD simulations performed for the series of the CAMs D556$^{6.44}$X of the human lutropin receptor (LHR; Angelova *et al.*, 2002; Zhang *et al.*, 2005).

Compared to the modeling results reviewed in the previous subsection, short MD simulations performed for CAMs pointed out possibility of more general mechanisms of inducing functional activity of GPCRs by disrupting the specific networks of interactions in the ground states (D$^{3.49}$–R$^{3.50}$–E$^{6.30}$ in adrenoreceptors or D$^{6.44}$–N$^{7.49}$/N$^{7.45}$ in hormone receptors). However, due to limitation of short MD trajectories, residue–residue interactions stabilizing the activated states, was, in fact, not explored by these simulations.

2.3. Modeling of CAMs based on the presumed models of the activated states

Several possible models of the activated states of GPCRs were suggested based on collection of experimental structural data and mutagenic data over the course of the past decade. The majority of the structural data related almost exclusively to rhodopsin/opsin system. For example, experimental data on the light-activated structure of rhodopsin obtained in solution by site-directed spin labeling (Hubbell *et al.*, 2003) and double electron–electron resonance technique (Altenbach *et al.*, 2008) showed that the cytoplasmic portion of TM6 undergoes significant movement away from the core TM bundle. That observation was in agreement with modeling results that featured involvement of D/E$^{6.30}$ located in the cytoplasmic part of TM6 in stabilization of the ground states of the CAMs of α1bAR (Cotecchia *et al.*, 2003) or FSHR (Montanelli *et al.*, 2004), and with predicted repulsion of the cytoplasmic parts of TM3 and TM6 in the CAM of the MOR (Huang *et al.*, 2002). However, these biophysical data were somewhat in contradiction with the electron crystallography structure of the MI state of rhodopsin (the transition state preceding the activated MII state), which showed only small structural changes compared to the dark-adapted rhodopsin (Ruprecht *et al.*, 2004). At the same time, the MI structure agrees well with the very recent data of infrared spectroscopy that revealed rotation of TM6 characteristic for this transition state (Ye *et al.*, 2010). The recent determination of the X-ray structure of opsin crystallized with the C-terminal fragment of α-transducin, which stabilizes the photoactivated structure of rhodopsin (Scheerer *et al.*, 2008), produced yet another possible template for the models of the activated states of GPCRs (see also Section 2.3.4).

2.3.1. Models of activated states: TM helix rotations

The site-directed spin-labeling data on the light-activated structure of rhodopsin were initially interpreted as indication of possible rotation of the kinked TM6 helix around the long transmembrane axis (Farrens *et al.*, 1996). Indeed, these data were consistent with rotation of TM6 by *ca.* 120°, as was modeled in our previous study (Nikiforovich and Marshall, 2003). Further modeling based on rotational sampling described in the next section in details suggested that this specific rotation of TM6 may rationalize strong constitutive activity of the rhodopsin mutants G90$^{2.57}$D/M257$^{6.40}$Y and E113$^{3.28}$Q/M257$^{6.40}$Y (Nikiforovich and Marshall, 2006). Earlier, rotation of TM6 was suggested by modeling as an important feature explaining the high basal activity of TSHR, which was eliminated in series of the Y601$^{5.74}$X mutants (Biebermann *et al.*, 1998).

2.3.2. Models of activated states: "Ionic lock," "Pro-kink," and "rotamer toggle switch"

The ionic lock concept originated based on mutations of highly conserved residue E268$^{6.30}$ in β2AR that yielded several strong CAMs (Ballesteros *et al.*, 2001) and was bolstered by the observation that the side chain of R135$^{3.50}$ was involved in the salt bridges with D134$^{3.49}$ and E247$^{6.30}$ in the X-ray structure of dark-adapted rhodopsin. Thus, breaking the ionic lock between R$^{3.50}$ and E/D$^{6.30}$ would lead to GPCR activation by allowing movement of the cytoplasmic part of TM6 away from TM3, which is corroborated by the spin-labeling studies of light-activated rhodopsin. The salt bridge between the side chains of R$^{3.50}$ and E$^{6.30}$ was also disrupted in the recent X-ray structures of genetically engineered mutant of β2AR (Cherezov *et al.*, 2007) and opsin (Scheerer *et al.*, 2008), which may represent the partly or fully activated states of these GPCRs. However, the ionic lock model cannot be universal for GPCRs since only 32% of GPCRs feature residues E or D in position 6.30, while 34% of GPCRs could not form the ionic lock since they have positively charged residues K or R in position 6.30 (Mirzadegan *et al.*, 2003).

The model of the "Pro-kink" was proposed by MD simulations of the isolated TM6 helix of β2AR showing possible straightening of the kinked TM6 helix in the region of the highly conserved $P288^{6.50}$ (Ballesteros *et al.*, 2001). Such straightening may result in movement of the cytoplasmic end of TM6 if the ionic lock $R131^{3.50}$–$E268^{6.30}$ is broken (Ballesteros *et al.*, 2001). Another type of computer simulations performed for the isolated TM6 helix of β2AR have found differences in distributions of rotamers of the side chains in positions $C285^{6.47}$, $W286^{6.48}$, and $F290^{6.52}$ between the wild-type (WT) receptor and the CAM $C285^{6.47}T$ (Shi *et al.*, 2002). Calculations showed also conformational changes in rotamers of $W286^{6.48}$ and $F290^{6.52}$ in $C285^{6.47}T$ that may modulate the Pro-kink (the so-called rotamer toggle switch). Modeling studies based on the Pro-kink/rotamer toggle switch model of the activated states of GPCRs rationalized the high basal activity of the cannabinoid CB1 receptor (CB1R; Singh *et al.*, 2002) and were discussed in more general context of GPCR activation (Visiers *et al.*, 2002). However, the X-ray structure of the opsin cocrystallized with the transducin peptide (Scheerer *et al.*, 2008) did not display specific rotamers of $W^{6.48}$ associated with the activated state by the toggle switch model; also, TM6 in this structure is rather more bended than in the dark-adapted rhodopsin and certainly is not straightened.

2.3.3. Models of activated states deduced from experimental constraints

By applying known experimental constraints to the models of the ground states and by further optimization of the constrained structures, several groups deduced models for the activated states (Choi *et al.*, 2002; Fowler *et al.*, 2004; Pogozheva *et al.*, 2005). Distances between residues obtained by spin labeling (Farrens *et al.*, 1996), or by reasonable estimations of spatial proximity of certain residues suggested by mutagenic data served as constraints. Examples are the model of the activated state of rhodopsin (Choi *et al.*, 2002) and the models for two peptidergic GPCRs, MOR (Fowler *et al.*, 2004) and the melanocortin 4 receptor (MC4R; Pogozheva *et al.*, 2005). In the latter cases, the constraints used to build the models included limitations on docking modes of the corresponding peptide agonists, JOM6 and NDP-MSH, respectively. The MC4R model was employed for rationalizing the constitutive activity of a series of mutants $L250^{6.40}X$ (Proneth *et al.*, 2006). According to the model, TM6 undergoes rotation around the long transmembrane axis, bringing the side chain of R147$^{3.50}$ in close contact with the side chain in position 250$^{6.40}$ (see also below). This MC4R model was based on the ground state of rhodopsin as a template; since β2AR has higher sequence homology to MC4R, a recent study employed the X-ray structure of β2AR (Cherezov *et al.*, 2007) as a template for rationalizing the mutagenic results on various mutants of MC4R, not including the L250$^{6.40}$X series (Tan *et al.*, 2009).

2.3.4. Model of the activated state based on the structure of opsin cocrystallized with the transducin peptide

Recent X-ray studies revealed the structure of opsin in complex with the C-terminal peptide of transducin stabilizing the photoactivated structure of rhodopsin (Scheerer *et al.*, 2008). Using this new prototype structure (the PDB entry 3DQB) as a template for the activated state of MC4R, recent modeling suggested that the cytoplasmic part of TM6 moves away from TM7 and toward TM5 upon transition from the presumed ground state (template provided by the X-ray structure of β2AR; Cherezov *et al.*, 2007) to the presumed activated state of MC4R (Tao *et al.*, 2010). Also, in the activated state, the side chain of D146$^{3.49}$ is involved in a strong salt bridge with the side chain of R165$^{4.42}$, and the R147$^{3.50}$ side chain changes orientation possibly contacting L250$^{6.40}$ (similar to the model above; Proneth *et al.*, 2006). These modeling results explained the constitutive activity of mutants D146$^{3.49}$N and L250$^{6.40}$Q, which are naturally occurring mutations identified in severe childhood obesity.

3. Rotational Sampling of the TM Regions of GPCRs

3.1. Why rotational sampling?

Several considerations support sampling possible configurations of the TM bundles of GPCRs rotated around the long transmembrane axes. Experimental data briefly reviewed in the previous sections clearly indicated that the structural mechanisms of GPCR activation based on rhodopsin or β2AR (breaking the ionic lock, the rotamer toggle switch, the Pro-kink) cannot be universal for GPCRs. On the other hand, conformational transitions occurring in GPCRs upon activation may be sampled by computational modeling without predefining specific models of the activated states.

In principle, this task could be fulfilled by straightforward MD simulations that start from the models of the ground state. However, as was mentioned above, MD simulations (especially those accurately accounting for the complex membrane environment of GPCRs, interactions with water and ion molecules, etc.) currently are unable to follow trajectories on the scale from milliseconds to seconds, even using state-of-art computer resources (Grossfield *et al.*, 2007; Hurst *et al.*, 2010). The task of conformational sampling becomes more realistic if one is willing to sacrifice a detailed description of GPCR environment in order to sample a large number of the possible 3D structures of a GPCR that would include a plausible model of the activated state(s). Our previous studies demonstrated the feasibility of this approach in four separate studies: determining conformational transitions in the TM region of rhodopsin that were consistent with the experimental data on spin labeling (Nikiforovich and Marshall, 2003), deducing the possible activated states of rhodopsin from conformational sampling of its CAMs (Nikiforovich and Marshall, 2006), and successful predictions of novel CAMs of AT1R (Nikiforovich *et al.*, 2005) and C5aR (Sen *et al.*, 2008).

At the same time, many results of site-directed mutagenesis studies for various GPCRs have been interpreted as support for concerted rotations of various TM helices upon constitutive activation. For instance, accessibility of selected cysteine residues to the sulfhydryl reagents in the WT AT1R and in the CAM $N111^{3.35}G$ indicated possible rotation of TM3 (Martin *et al.*, 2004); similar studies supported movements of TM2 (Miura and Karnik, 2002) or TM7 (Boucard *et al.*, 2003). In β2AR, cysteine accessibility data for CAMs were interpreted as evidence for rotation of TM6 (Javitch *et al.*, 1997; Rasmussen *et al.*, 1999). Constitutive activity of the α1BAR mutants with modifications of $D142^{3.49}$ was explained in terms of rotation of TM3 (Scheer *et al.*, 1997). Also, a recent study presented evidence for possible rotation of TM5 in the CAM of AT1R (Domazet *et al.*, 2009). In addition, as it was mentioned above, the independent data on spin labeling (Altenbach *et al.*, 2008; Hubbell *et al.*, 2003) and infrared probe labeling (Ye *et al.*, 2010) were interpreted as indications of rotation of TM6 during light activation of rhodopsin.

Therefore, we explored sampling of possible rotations of TM helices performed employing simple force field and certain limitations as one of the reasonable possibilities to determine the plausible models for the activated states of the TM regions of GPCRs without requiring excessive computational resources. Below, this approach is explained in detail and validated by modeling CAMs of the C5aR.

3.2. General hypothesis

Our main hypothesis postulates that activation of GPCRs is associated mainly with specific configurations of TM helices differing by rotations along their long transmembrane axes. Additionally, we assume that the TM bundles of CAMs should possess certain configurations characterized by relative energies that are lower or at least comparable to those of the ground state configurations. At the same time, these specific configurations (the

CAM states) when examined in the non-CAMs should have higher relative energies compared to the ground state configurations. It is also reasonable to assume that the ***CAM*** states should be connected to the ground states through sequential series of minimal TM rotations (e.g., of $\pm 30°$) without crossing high energy barriers, that is, by avoiding high energy configurations.

3.3. Methods

Briefly, typical modeling of the TM regions of a GPCR and the mutants involves several main steps: sequence alignment to the selected template to define boundaries of TM helices; conformational calculations for individual TM helices; structural alignment of these helices to the selected templates; and final energy minimization with optimization of the side chain packing.

3.3.1. Force field

All energy calculations employ the ECEPP/2 force field with rigid valence geometry (Dunfield *et al.*, 1978; Nemethy *et al.*, 1983) and *trans*-conformations of Pro residues; residues Arg, Lys, Glu, and Asp are regarded as charged species. Energy calculations are routinely performed with the value of the macroscopic dielectric constant ε of 2 (the standard value for ECEPP corresponding to a protein environment).

3.3.2. Building TM regions

The procedure for building the TM bundle of a GPCR was described earlier (e.g., Nikiforovich *et al.*, 2006) and is described here in greater detail. The TM helical segments are typically found by sequence homology to the selected template (bovine rhodopsin or β_2AR); the first, middle, and last residues for each TM helix can be determined by alignment. Each individual helix can then be first subjected to energy minimization starting from the all-helical backbone conformations (i.e., the values of all dihedral angles ϕ and ψ are initially of $-60°$). Some limitations on the ϕ and ψ values ($-20° \geq \phi,\ \psi \geq -100°$) and the ω values ($150° \geq \omega \geq -150°$) are placed during energy minimization to mimic, to some extent, limitations on intrahelical mobility of TM segments immobilized in the membrane.

The resulting helical fragments are then aligned to the TM regions of the selected X-ray template (in the case of the C5aR, we used dark-adapted rhodopsin, PDB entry 1F88) by fitting the rms values calculated for C^{α} atoms. Further modeling consists of minimization of the sum of all intra- and interhelical interatomic energies in the multidimensional space of parameters assigned to each helix in the bundle. Those include the "global" parameters (related to movements of individual helices as rigid bodies, i.e., translations along the coordinate axes *X*, *Y*, *Z*, and rotations around these axes Tx, Ty, and Tz) and the "local" parameters (the dihedral angles of the side chains for all helices; the dihedral angles of the backbone are frozen).

For mutants, the corresponding residues are mutated without changing the dihedral angles of the backbone for the mutated residue.

The coordinate system for the global parameters is selected as follows: the long axial X coordinate axis for each TM helix is directed from the first to the last C^{α} atom; the Y axis is perpendicular to X and goes through the C^{α} atom of the "middle" residue of each helix; and the Z axis is built perpendicular to X and Y to maintain the right-handed coordinate system. The starting point of computational search for conformational states of the TM regions (the reference structure) is determined by global parameters corresponding to the X-ray structure of the template (all Tx rotations are assumed to be equal to 0°).

Typically, two types of energy calculations are performed. The "preliminary" energy calculations comprise energy minimization in the space of the global parameters along with optimization of spatial arrangements for each side chain at each convergence step of energy minimization by an algorithm described in the next subsection. "Full" energy calculations involve energy minimization not only within the space of global parameters, but also include the dihedral angles of the side chains. New repacking of spatial arrangements for each side chain at each convergence step is also performed at full energy calculations. Energy minimization typically proceeds to achieve a convergence criterion of $\Delta E < 1$ kcal/mol.

3.3.3. Optimizing positions of side chains

The algorithm developed earlier for optimization of spatial positions of side chains in energy calculations of proteins utilizes a stepwise grid search and consists of several steps (Nikiforovich *et al.*, 1991). First, initial values of n selected (or all) dihedral angles of the side chains (conventionally referred to as the χ angles) are arranged in a set of the θ_i^0 initial values, ($i = 1, \ldots, n$). Then the first dihedral angle, θ_1, possessing the initial value of $\theta_1{}^0$, is rotated with a chosen grid step, normally 30°, from −150° to 180°. All other angles are fixed in their θ_i^0 values. Rotation results in the energy profile where some angle value $\theta_1{}^{\min}$ corresponds to local energy minimum $E_{\min}(\theta_1)$. Then the θ_1 angle is fixed in the $\theta_1{}^{\min}$ value, and the procedure is repeated for each θ_i angle to θ_n; at the end of this run all θ_i^0 values became equal to $\theta_i^{\min}$. The second run starts again from θ_1, and so on, until all $\theta_i^{\min}$ remain unchanged, consistent with achieving the optimal values of θ_i angles.

This algorithm has been extensively used for optimizing the initial (prior to energy minimization) and final (after energy minimization) values of the dihedral angles χ_i of side chains and has been validated by successful design of many biologically active analogs of peptides (Nikiforovich, 1994). As an additional benefit, this algorithm produces the energy profiles along the χ_i angles in the final point of energy minimization revealing a "slice" of the multidimensional energy surface for each given χ_i. The algorithm is a path-dependent one, since its results may depend on the choice of the initial

θ_i^0 values; however, it is easily compensated by changing the order of the initial θ_i^0 angles and repeating the procedure. Typically, the pathway for optimization of the dihedral angles of the side chains involves first all χ_1 angles from TM1 to TM7, than all χ_2 angles from TM1 to TM7, then all χ_3 angles and so on until convergence.

3.3.4. Buildup procedure for rotational sampling

Our search for possible conformational transitions in the TM regions of GPCRs includes sampling of possible rotations of TM helices along their long X axes. This rotational sampling is performed according to a buildup procedure, thus allowing us to explore possible rotations of the seven TM helices with a grid step of 30° without actually sampling all $12^7 = 35{,}831{,}808$ configurations. The procedure was partly described earlier (Nikiforovich and Marshall, 2006).

The procedure is based on three main considerations. First, all available X-ray structures of the TM regions of GPCRs showed that the direct interactions between TM helices are within the five "triplets" of helices: TM1–TM2–TM7 (TM127), TM2–TM3–TM7 (TM237), TM3–TM4–TM5 (TM345), TM3–TM5–TM6 (TM356), and TM3–TM6–TM7 (TM367). Energy calculations for the triplets allow selection of configurations with relative energies not higher than that of the predefined energy cut-off. Then the selected configurations of the triplets can be combined in the packages of the four contacting helices, TM1237, TM2367, TM3567, and TM3456. After a new round of energy calculations and selection, the newly selected configurations can be combined into packages of five contacting helices and so on until the entire TM bundle is built. Depending on results of energy calculations for triplets, "quadruplets" and so on, one can choose the specific pattern of combining packages of helices into the larger packages to further reduce the amount of calculations. This sampling pattern is not path-dependent, since configurations discarded, say, at the level of any triplet, are not considered in further calculations anyway.

Second, in our particular case of considering CAMs, we are interested only in configurations of TM helices with relative energy values lower or at least comparable with those of the ground states of the receptors (initially defined as those with all TM rotations of 0°). This consideration determines the energy cut-offs for selection of low-energy configurations at any level of the buildup procedure.

Third, sampling of the TM regions can be also limited by selecting only rotations precluding direct exposure of the charged residues in the middle sections of each TM helix to the lipid core of the membrane. This structural feature was observed in all crystal structures of GPCRs present in the PDB. Likewise, it is reasonable to consider only TM rotations that orient the highly hydrophobic tryptophan residues in the middle sections of the TM helices toward the exterior of the TM bundles. Again, this was observed for all tryptophans in the crystal structures of GPCRs with a notable exception of the highly conserved $W^{6.48}$ that is oriented toward the interior of the bundle.

4. Modeling Structural Mechanisms of Constitutive Activity in C5aRs

This section briefly reports the main results of rotational sampling that modeled possible conformational changes occurring in the TM region of the C5aR upon constitutive activation. C5aR is an important receptor in the physiology of immune defense, inflammation, and human disease with a sequence of 350 amino acid residues sharing 22% sequence identity with rhodopsin. We have applied rotational sampling procedure to two CAMs of C5aR, NQ (I124N/L127Q) and F251A, and for two C5aRs with WT phenotype (i.e., C5aR itself) and the triple mutant NQ-N296A (I124N/L127Q/N296A) found by us earlier (Whistler *et al.*, 2002). The complete results of the study are to be published elsewhere.

4.1. Deducing the *CAM* state of C5aRs

First, the initial model of the ground state of the WT C5aR was built as described above. The TM helical segments in C5aR were determined by homology to bovine rhodopsin and human β2AR using the CLUSTAL-W algorithm. After energy minimization, individual TM helices were aligned to the template structure of the dark-adapted rhodopsin (the PDB entry 1F88) yielding the reference structure with all TM rotations of 0°. Then, the buildup rotational sampling was performed for all TM triplets in all four receptors, that is, WT, NQ, F251A, and NQ-N296A. Since it is not likely that relatively small molecular replacements leading to constitutive activity (as F251 → A251 in F251A or I124 → N124 and L127 → Q127 in NQ) would result in large scale reorientations in the TM helical bundle, we assumed that rotations of the TM helices corresponding to the constitutively activated states would not differ from the reference structure by more than ±90°.

At the next step, the configurations of the four-helical bundles, TM3567's, were obtained by combining the selected configurations of TM356 and TM367 for all four receptors. Full energy calculations of TM3567's revealed that energies of the reference configurations of TM3567's (the initial orientations of all TM helices of 0°) in all cases were significantly higher than those for the configurations with minimal energies. At the same time, some TM3567 configurations, which differed from the reference configuration only by minimal TM rotations of ±30°, possessed lower energies. This finding indicated that certain configurations with deviations from the reference structure by minimal TM rotations of $\pm 30^\circ$ can be collectively regarded as more likely models for the ground states of the corresponding C5aRs (e.g., configurations A_1 - A_7 in Table 21.2).

The five-member TM23567 bundles representing the core in the entire TM packages of GPCRs were subjected to full energy calculations. Selected TM23567 configurations were combined with the adjustable lowest-energy configurations of TM1 and TM4 obtained by full energy calculations of corresponding configurations of TM127's and TM345's, respectively, forming possible configurations of the entire seven-helical TM bundles. Full energy calculations revealed the final sets of configurations of the TM bundles. As an example, Table 21.2 lists all configurations of TM bundles of NQ found by rotational sampling and possessing energies not exceeding the minimal energy found for the probable ground states, (i.e., for configuration A_3 for NQ) by more than 20 kcal/mol. Reference configuration for rotational sampling (the 1F88 template) is denoted as ***Rh*** in Table 21.2; energy values (in kcal/mol) are rounded. Of note, the actual values of rotations around the long axes of TM helices obtained by energy

Table 21.2 Configurations corresponding to the low energy states of the NQ receptor

TM configuration	TM1	TM2	TM3	TM4	TM5	TM6	TM7	Energy (kcal/mol)	Connections
Rh	0	0	0	0	0	0	0	−513	A_1–A_7
A_1	0	330	30	30	330	30	30	−546	A_6/B
A_2	0	330	30	330	330	30	0	−552	$A_3/A_4/A_5/A_7/B$
A_3	0	330	30	330	330	30	30	−553	$A_2/A_4/A_5/A_7/B$
A_4	0	0	0	330	0	0	0	−540	$A_2/A_3/A_5/A_7$
A_5	0	330	0	330	330	30	30	−544	$A_2/A_3/A_4/A_7$
A_6	0	330	30	30	330	30	0	−538	A_1/B
A_7	0	330	0	330	330	30	0	−536	$A_2/A_3/A_4/A_5$
B	30	300	60	0	0	30	30	**−561**	$A_1/A_2/A_3/A_6/C/D$
C	**0**	**330**	**60**	30	**30**	**30**	**30**	**−557**	B/D
D	30	300	60	30	30	30	30	**−548**	B/C

calculations differed slightly from the round values corresponding to the starting TM configurations shown in Table 21.2. All listed configurations can be connected to each other and to the ground state configurations $\boldsymbol{A_i}$ by series of minimal TM rotations of $\pm 30°$; direct connections between each pair of configurations are shown in the last column of Table 21.2.

According to our main assumption, TM configurations associated with constitutive activity would be those with energies lower or at least comparable to those of the ground state structures and characteristic only to CAMs, NQ, and F251A, but not for WT or NQ-N296A. There was only one combination found by rotational sampling satisfying this requirement, namely that corresponding to configuration ***C*** for NQ (the corresponding rotations are shown in bold in Table 21.2). Therefore, this configuration may be regarded as the structure associated with constitutive activity of C5aR (the ***CAM*** state).

4.2. Structural features associated with constitutive activity

Major structural differences between the ground states ($\boldsymbol{A_i}$) and ***CAM*** configurations can be summarized by patterns of contacts between the side chains that may form hydrogen bonds, either directly, or through mediation of the water molecules (absent in our calculations). The contacts found by modeling in the various ground state configurations vary from receptor to receptor; however, there are some contacts that were observed most often for all four receptors. These interactions include hydrogen bonding between the β-carboxyl of D82 in TM2 and the β-amido group of N119 in TM3; between the hydroxyl of S85 in TM2 and the β-amido group of N119 in TM3; between the hydroxyl group of S131 in TM3 and the sulfhydryl group of C221 in TM5; and between the γ-carbonyl of Q259 and the N^{ε} hydrogen in W255, both in TM6. Figure 21.1A presents a sketch of the representative model of the ground state of WT showing the discussed contacts between the side chains.

Three persistent potential contacts through hydrogen bonding between side chains were observed for the ***CAM*** configuration found by rotational sampling for both NQ and F251A: contacts between the β-carboxyl of D82 in TM2 and the β-amido group of N296 in TM7; between the hydroxyl of S131 and the guanidine group of R134, both in TM3; and between the β-amido group of N296 in TM7 and the hydroxyl of Y300, both in TM7. Figure 21.1B illustrates the system of hydrogen bonding in the ***CAM*** configuration of the NQ receptor.

Comparison of Fig. 21.1A and B suggests that main conformational changes occurring in the TM regions of C5aR upon activation would break the contacts between N119 and S85 (and D82) due to rotation of TM3 and, at the same time, enhance the contact between D82 and N296. Accordingly, interactions involving N119 could stabilize the ground state of

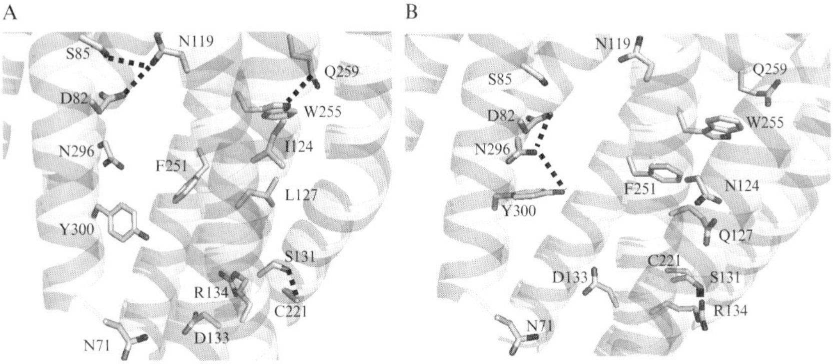

Figure 21.1 Sketch of the ground state of WT C5aR (A) and the activated state of NQ (B). The side chains are shown as sticks. The hydrogen bonding is shown as dashed lines. TM helices are shown as semitransparent cartoons. The sketches were prepared with use of PyMOL v0.99 software, www.pymol.org.

C5aR, and interactions involving N296 could stabilize the possible activated state. Additional stabilization of the ***CAM*** configurations may be also achieved by interaction N296–Y300. These findings provide a molecular mechanism for the fact that the N296A mutation suppresses the constitutive activity when substituted into the CAM NQ (Whistler *et al.*, 2002).

The modeling predicts several other notable conformational changes upon activation, namely, breaking the hydrogen bonding between S131 and C221 and movement of the side chain of R134 toward S131 resulting in the hydrogen bond between the two side chains. A hydrogen bond between W255 and Q259 represents another potential interaction broken in the activated state. For NQ, the side chain of F251, which corresponds to the *trans*-conformation (χ_1 angle *ca.* 180°) in the ***CAM*** configuration, is involved in stacking interactions with the W255 side chain (see Fig. 21.1B). As was shown by modeling, steric hindrance of the voluminous side chains of I124 and, especially, L127 in TM3 prevents the side chain of F251 from rotation toward the *trans*-conformation in some ground configurations of WT. At the same time, the side chains of N124 and Q127 in NQ and NQ-N296A are more flexible and accommodate the *trans*-conformation of the side chain of F251 even in the ground states.

4.3. Validation

Molecular modeling based on rotational TM sampling delineated several important elements in the structural mechanism of C5aR constitutive activation. First, it demonstrated that the activated state is stabilized by interactions involving the side chain of N296. Second, it was shown that interactions involving the side chain of N119 stabilize the ground state of the receptor by hydrogen bonding with S85 and D82. Third, the modeling demonstrated that interactions between S131 and either C221 or R134 stabilize, respectively, either the ground state or the activated state of the receptors. And, fourth, the activated state accommodates the *trans*-conformation the side chain of F251, while in the ground state of WT this conformation may be hindered by unfavorable interaction with the side chains of I124 and L127.

These results are consistent with the fact that all CAMs of C5aR found by random saturation mutagenesis (RSM) in yeast in the TM helices included mutations of either L127, or F251 (Baranski *et al.*, 1999; the only exception was *R306* mutant, where all mutations occurred in TM1; Geva *et al.*, 2000). Though NQ (I124N/L127Q) was the most pronounced CAMs both in the yeast and mammalian systems, some strong CAMs obtained in the RSM scans of TM3, such as *R35* and *R37* (Baranski *et al.*, 1999), did not feature the additional mutations of I124. On the other hand, single mutations of L127 (as in L127Q, Baranski *et al.*, 1999 or in L127A, Sen *et al.*, 2008) led to weaker CAMs. Abolishing the voluminous side chain of F251 in the F251A mutant yielded the strong CAM (Baranski *et al.*, 1999), whereas combining the mutations of L127 and F251 in NQ-F251N resulted in weaker basal activity than that of NQ (Sen *et al.*, 2008).

While the importance of possible contact between L127 and F251 for receptor activation could be suggested on the basis of the initial rhodopsin-like 3D model of C5aR in the resting state (configuration ***Rh***; Sen *et al.*, 2008), restoring of the WT phenotype in the NQ-N296A mutant did not have an adequate explanation other than general observation that N296 locates in the TM region of C5aR close to F251, which, in turn, is close to L127 (Hagemann *et al.*, 2006; Whistler *et al.*, 2002). Rotational sampling provided a plausible molecular mechanism for the role of N296A mutation in the NQ-N296A and highlighted a role for N296 in stabilization of the activated state of WT C5aR by the hydrogen bond between the side chains of D82 and N296. Additionally, according to our modeling, the activated state of the receptor is stabilized also by hydrogen bonding between N296 and Y300. Obviously, these hydrogen bonds cannot be maintained in NQ-N296A, or in the N296A mutant that was nonfunctional (Whistler *et al.*, 2002).

Results of rotational sampling emphasized also the importance of the hydrogen bond between the β-amido group of N119 and the hydroxyl of S85 and the β-carboxyl of D82 in maintaining the ground state of the receptor. The RSM technique found numerous mutants with replacements of N119 by hydroxyl-containing threonine, serine, or tyrosine as well as by isoleucine (*R25* mutant) or even proline (*R11*) were functional but did not display constitutive activity (Baranski *et al.*, 1999). On the other hand, mutants produced by RSM featured simultaneous multiple replacements of several residues, so effects of mutation of a specific residue can be masked by effects of other replacements. In this regard, very recent studies on site-directed mutagenesis of C5aR independently performed in our lab showed significant constitutive activity of the N119S mutant (Rana and Baranski, 2010); these experimental data are also consistent with the modeling results.

5. Conclusions and Perspectives

In the past decade, computational molecular modeling gained profound significance in studies of constitutive activity of GPCRs. In fact, practically all current papers related to structural aspects of CAMs employ various modeling approaches for rationalizing experimental results. Further development of this trend would most likely require progress in two main directions. First, it could be expected that the X-ray crystallography will produce direct structural evidence of the activated states of at least some CAMs. In conjunction with the experimental data on the ground states, it would provide molecular modeling with extremely useful information on the structural templates for both the starting and final points of the trajectory of conformational transition that could be modeled by computational approaches. Second, to model such transitions, computational approaches (and readily available computer resources) should be developed to an extent of modeling large scale molecular movements covering trajectories over milliseconds. Some recent efforts along this line look rather encouraging (Dror *et al.*, 2009; Grossfield *et al.*, 2007; Hurst *et al.*, 2010).

ACKNOWLEDGMENTS

This work was supported by NIH grants GM 71634 (GVN and TJB) and GM63720 (TJB).

REFERENCES

Altenbach, C., Kusnetzow, A. K., Ernst, O. P., Hofmann, K. P., and Hubbell, W. L. (2008). High-resolution distance mapping in rhodopsin reveals the pattern of helix movement due to activation. *Proc. Natl. Acad. Sci. USA* **105,** 7439–7444.

Angelova, K., Fanelli, F., and Puett, D. (2002). A model for constitutive lutropin receptor activation based on molecular simulation and engineered mutations in transmembrane helices 6 and 7. *J. Biol. Chem.* **277,** 32202–33213.

Ballesteros, J. A., and Weinstein, H. (1992). Analysis and refinement of criteria for predicting the structure and relative orientations of transmembranal helical domains. *Biophys. J.* **62,** 107–109.

Ballesteros, J. A., Jensen, A. D., Liapakis, G., Rasmussen, S. G., Shi, L., Gether, U., and Javitch, J. A. (2001). Activation of the beta 2-adrenergic receptor involves disruption of an ionic lock between the cytoplasmic ends of transmembrane segments 3 and 6. *J. Biol. Chem.* **276,** 29171–29177.

Baranski, T. J., Herzmark, P., Lichtarge, O., Gerber, B. O., Trueheart, J., Meng, E. C., Iiri, T., Sheikh, S. P., and Bourne, H. R. (1999). C5a receptor activation. Genetic identification of critical residues in four transmembrane helices. *J. Biol. Chem.* **274,** 15757–15765.

Befort, K., Zilliox, C., Filliol, D., Yue, S., and Kieffer, B. L. (1999). Constitutive activation of the delta opioid receptor by mutations in transmembrane domains III and VII. *J. Biol. Chem.* **274,** 18574–18581.

Biebermann, H., Schoneberg, T., Schulz, A., Krause, G., Gruters, A., Schultz, G., and Gudermann, T. (1998). A conserved tyrosine residue (Y601) in transmembrane domain 5 of the human thyrotropin receptor serves as a molecular switch to determine G-protein coupling. *FASEB J.* **12,** 1461–1471.

Borhan, B., Souto, M. L., Imai, H., Shichida, Y., and Nakanishi, K. (2000). Movement of retinal along the visual transduction path. *Science* **288,** 2209–2212.

Boucard, A. A., Roy, M., Beaulieu, M. E., Lavigne, P., Escher, E., Guillemette, G., and Leduc, R. (2003). Constitutive activation of the angiotensin II type 1 receptor alters the spatial proximity of transmembrane 7 to the ligand-binding pocket. *J. Biol. Chem.* **278,** 36628–36636.

Cherezov, V., Rosenbaum, D. M., Hanson, M. A., Rasmussen, S. G., Thian, F. S., Kobilka, T. S., Choi, H. J., Kuhn, P., Weis, W. I., Kobilka, B. K., and Stevens, R. C. (2007). High-resolution crystal structure of an engineered human beta2-adrenergic G protein-coupled receptor. *Science* **318,** 1258–1265.

Choi, G., Landin, J., Galan, J. F., Birge, R. R., Albert, A. D., and Yeagle, P. L. (2002). Structural studies of metarhodopsin II, the activated form of the G-protein coupled receptor, rhodopsin. *Biochemistry* **41,** 7318–7324.

Cotecchia, S., Rossier, O., Fanelli, F., Leonardi, A., and De Benedetti, P. G. (2000). The alpha 1a and alpha 1b-adrenergic receptor subtypes: Molecular mechanisms of receptor activation and of drug action. *Pharm. Acta Helv.* **74,** 173–179.

Cotecchia, S., Fanelli, F., and Costa, T. (2003). Constitutively active G protein-coupled receptor mutants: Implications on receptor function and drug action. *Assay Drug Dev. Technol.* **1,** 311–316.

Domazet, I., Holleran, B. J., Martin, S. S., Lavigne, P., Leduc, R., Escher, E., and Guillemette, G. (2009). The second transmembrane domain of the human type 1 angiotensin II receptor participates in the formation of the ligand binding pocket and undergoes integral pivoting movement during the process of receptor activation. *J. Biol. Chem.* **284,** 11922–11929.

Dror, R. O., Arlow, D. H., Borhani, D. W., Jensen, M. O., Piana, S., and Shaw, D. E. (2009). Identification of two distinct inactive conformations of the beta2-adrenergic receptor reconciles structural and biochemical observations. *Proc. Natl. Acad. Sci. USA* **106,** 4689–4694.

Dunfield, L. G., Burgess, A. W., and Scheraga, H. A. (1978). Energy parameters in polypeptides. 8. Empirical potential energy algorithm for the conformational analysis of large molecules. *J. Phys. Chem.* **82,** 2609–2616.

Farrens, D. L., Altenbach, C., Yang, K., Hubbell, W. L., and Khorana, H. G. (1996). Requirement of rigid-body motion of transmembrane helices for light activation of rhodopsin. *Science* **274,** 768–770.

Fowler, C. B., Pogozheva, I. D., Lomize, A. L., LeVine, H., 3rd, and Mosberg, H. I. (2004). Complex of an active mu-opioid receptor with a cyclic peptide agonist modeled from experimental constraints. *Biochemistry* **43,** 15796–15810.

Geva, A., Lassere, T. B., Lichtarge, O., Pollitt, S. K., and Baranski, T. J. (2000). Genetic mapping of the human C5a receptor: Identification of transmembrane amino acids critical for receptor function. *J. Biol. Chem.* **275,** 35393–35401.

Ghanouni, P., Steenhuis, J. J., Farrens, D. L., and Kobilka, B. K. (2001). Agonist-induced conformational changes in the G-protein-coupling domain of the β_2 adrenergic receptor. *Proc. Natl. Acad. Sci. USA* **98,** 5997–6002.

Govaerts, C., Lefort, A., Costagliola, S., Wodak, S. J., Ballesteros, J. A., Van Sande, J., Pardo, L., and Vassart, G. (2001). A conserved Asn in transmembrane helix 7 is an on/off switch in the activation of the thyrotropin receptor. *J. Biol. Chem.* **276,** 22991–22999.

Greasley, P. J., Fanelli, F., Rossier, O., Abuin, L., and Cotecchia, S. (2002). Mutagenesis and modelling of the alpha(1b)-adrenergic receptor highlight the role of the helix 3/helix 6 interface in receptor activation. *Mol. Pharmacol.* **61,** 1025–1032.

Groblewski, T., Maigret, B., Larguier, R., Lombard, C., Bonnafous, J.-C., and Marie, J. (1997). Mutation of Asn^{111} in the third transmembrane domain of the AT_{1A} angiotensin II receptor induces its constitutive activation. *J. Biol. Chem.* **272,** 1822–1826.

Grossfield, A., Feller, S. E., and Pitman, M. C. (2007). Convergence of molecular dynamics simulations of membrane proteins. *Proteins* **67,** 31–40.

Hagemann, I. S., Nikiforovich, G. V., and Baranski, T. J. (2006). Comparison of the retinitis pigmentosa mutations in rhodopsin with a functional map of the C5a receptor. *Vis. Res.* **46,** 4519–4531.

Huang, P., Visiers, I., Weinstein, H., and Liu-Chen, L. Y. (2002). The local environment at the cytoplasmic end of TM6 of the mu opioid receptor differs from those of rhodopsin and monoamine receptors: Introduction of an ionic lock between the cytoplasmic ends of helices 3 and 6 by a L6.30(275)E mutation inactivates the mu opioid receptor and reduces the constitutive activity of its T6.34(279)K mutant. *Biochemistry* **41,** 11972–11980.

Hubbell, W. L., Altenbach, C., and Khorana, H. G. (2003). Rhodopsin structure, dynamics and activation. *Adv. Prot. Chem.* **63,** 243–290.

Hurst, D. P., Grossfield, A., Lynch, D. L., Feller, S., Romo, T. D., Gawrisch, K., Pitman, M. C., and Reggio, P. H. (2010). A lipid pathway for ligand binding is necessary for a cannabinoid G protein-coupled receptor. *J. Biol. Chem.* **285,** 17954–17964.

Javitch, J. A., Fu, D., Liapakis, G., and Chen, J. (1997). Constitutive activation of the beta2 adrenergic receptor alters the orientation of its sixth membrane-spanning segment. *J. Biol. Chem.* **272,** 18546–18549.

Joubert, L., Claeysen, S., Sebben, M., Bessis, A.-S., Clark, R. D., Martin, R. S., Bockaert, J., and Dumius, A. (2002). A 5-HT4 receptor transmembrane network implicated in the activity of inverse agonists but not agonists. *J. Biol. Chem.* **277,** 25502–25511.

Marie, J., Koch, C., Pruneau, D., Paquet, J. L., Groblewski, T., Larguier, R., Lombard, C., Deslauriers, B., Maigret, B., and Bonnafous, J. C. (1999). Constitutive activation of the human bradykinin B2 receptor induced by mutations in transmembrane helices III and VI. *Mol. Pharmacol.* **55,** 92–101.

Marie, J., Richard, E., Pruneau, D., Paquet, J. L., Siatka, C., Larguier, R., Ponce, C., Vassault, P., Groblewski, T., Maigret, B., and Bonnafous, J. C. (2001). Control of conformational equilibria in the human B2 bradykinin receptor. Modeling of nonpeptidic ligand action and comparison to the rhodopsin structure. *J. Biol. Chem.* **276,** 41100–41111.

Martin, S. S., Boucard, A. A., Clement, M., Escher, E., Leduc, R., and Guillemette, G. (2004). Analysis of the third transmembrane domain of the human type 1 angiotensin II receptor by cysteine scanning mutagenesis. *J. Biol. Chem.* **279,** 51415–51423.

Mirzadegan, T., Benko, G., Filipek, S., and Palczewski, K. (2003). Sequence analyses of G-protein coupled receptors: Similarities to rhodopsin. *Biochemistry* **42,** 2759–2767.

Miura, S., and Karnik, S. S. (2002). Constitutive activation of angiotensin II type 1 receptor alters the orientation of transmembrane helix-2. *J. Biol. Chem.* **277,** 24299–24305.

Montanelli, L., Van Durme, J. J., Smits, G., Bonomi, M., Rodien, P., Devor, E. J., Moffat-Wilson, K., Pardo, L., Vassart, G., and Costagliola, S. (2004). Modulation of ligand selectivity associated with activation of the transmembrane region of the human follitropin receptor. *Mol. Endocrinol.* **18,** 2061–2073.

Mustafi, D., and Palczewski, K. (2009). Topology of class A G protein-coupled receptors: Insights gained from crystal structures of rhodopsins, adrenergic and adenosine receptors. *Mol. Pharmacol.* **75,** 1–12.

Nemethy, G., Pottle, M. S., and Scheraga, H. A. (1983). Energy parameters in polypeptides. 9. Updating of geometrical parameters, nonbonded interactions, and hydrogen bond interactions for the naturally occurring amino acids. *J. Phys. Chem.* **87,** 1883–1887.

Nikiforovich, G. V. (1994). Computational molecular modeling in peptide design. *Int. J. Pept. Protein Res.* **44,** 513–531.

Nikiforovich, G. V., and Marshall, G. R. (2003). 3D model for meta-II rhodopsin, an activated G-protein-coupled receptor. *Biochemistry* **42,** 9110–9120.

Nikiforovich, G. V., and Marshall, G. R. (2006). 3D modeling of the activated states of constitutively active mutants of rhodopsin. *Biochem. Biophys. Res. Commun.* **345,** 430–437.

Nikiforovich, G. V., Hruby, V. J., Prakash, O., and Gehrig, C. A. (1991). Topographical requirements for delta-selective opioid peptides. *Biopolymers* **31,** 941–955.

Nikiforovich, G. V., Mihalik, B., Catt, K. J., and Marshall, G. R. (2005). Molecular mechanisms of constitutive activity: Mutations at position 111 of the angiotensin AT_1 receptor. *J. Pept. Res.* **66,** 236–248.

Nikiforovich, G. V., Zhang, M., Yang, Q., Jagadeesh, G., Chen, H. C., Hunyady, L., Marshall, G. R., and Catt, K. J. (2006). Interactions between conserved residues in transmembrane helices 2 and 7 during angiotensin AT1 receptor activation. *Chem. Biol. Drug Des.* **68,** 239–249.

Parnot, C., Miserey-Lenkei, S., Bardin, S., Corvol, P., and Clauser, E. (2002). Lessons from constitutively active mutants of G protein-coupled receptors. *Trends Endocrin. Metabol.* **13,** 336–343.

Pogozheva, I. D., Chai, B. X., Lomize, A. L., Fong, T. M., Weinberg, D. H., Nargund, R. P., Mulholland, M. W., Gantz, I., and Mosberg, H. I. (2005). Interactions of human melanocortin 4 receptor with nonpeptide and peptide agonists. *Biochemistry* **44,** 11329–11341.

Prioleau, C., Visiers, I., Ebersole, B. J., Weinstein, H., and Sealfon, S. C. (2002). Conserved helix 7 tyrosine acts as a multistate conformational switch in the 5HT2C receptor. Identification of a novel "locked-on" phenotype and double revertant mutations. *J. Biol. Chem.* **277,** 36577–36584.

Proneth, B., Xiang, Z., Pogozheva, I. D., Litherland, S. A., Gorbatyuk, O. S., Shaw, A. M., Millard, W. J., Mosberg, H. I., and Haskell-Luevano, C. (2006). Molecular mechanism of the constitutive activation of the L250Q human melanocortin-4 receptor polymorphism. *Chem. Biol. Drug Des.* **67,** 215–229.

Rana, S. and Baranski, T. J. (2010) The third extracellular loop (EC3) - N terminus interaction is important for 7TM receptor function: implications for an activation microswitch region. *J. Biol. Chem.* doi:10.1074/jbc.M110.129213.

Rasmussen, S. G., Jensen, A. D., Liapakis, G., Ghanouni, P., Javitch, J. A., and Gether, U. (1999). Mutation of a highly conserved aspartic acid in the beta2 adrenergic receptor: Constitutive activation, structural instability, and conformational rearrangement of transmembrane segment 6. *Mol. Pharmacol.* **56,** 175–184.

Ringkananont, U., Van Durme, J., Montanelli, L., Ugrasbul, F., Yu, Y. M., Weiss, R. E., Refetoff, S., and Grasberger, H. (2006). Repulsive separation of the cytoplasmic ends of transmembrane helices 3 and 6 is linked to receptor activation in a novel thyrotropin receptor mutant (M626I). *Mol. Endocrinol.* **20,** 893–903.

Ruprecht, J. J., Mielke, T., Vogel, R., Villa, C., and Schertler, G. F. (2004). Electron crystallography reveals the structure of metarhodopsin I. *EMBO J.* **23,** 3609–3620.

Salom, D., Lodowski, D. T., Stenkamp, R. E., Le Trong, I., Golczak, M., Jastrzebska, B., Harris, T., Ballesteros, J. A., and Palczewski, K. (2006). Crystal structure of a photo-activated deprotonated intermediate of rhodopsin. *Proc. Natl. Acad. Sci. USA* **103,** 16123–16128.

Scheer, A., Fanelli, F., Costa, T., De Benedetti, P. G., and Cotecchia, S. (1997). The activation process of the alpha1B-adrenergic receptor: Potential role of protonation and hydrophobicity of a highly conserved aspartate. *Proc. Natl. Acad. Sci. USA* **94,** 808–813.

Scheerer, P., Park, J. H., Hildebrand, P. W., Kim, Y. J., Krauss, N., Choe, H. W., Hofmann, K. P., and Ernst, O. P. (2008). Crystal structure of opsin in its G-protein-interacting conformation. *Nature* **455,** 497–502.

Sen, S., Baranski, T. J., and Nikiforovich, G. V. (2008). Conformational movement of F251A contributes to the molecular mechanism of constitutive activation in the C5a receptor. *Chem. Biol. Drug Des.* **71,** 197–204.

Shi, L., Liapakis, G., Xu, R., Guarneri, F., Ballesteros, J. A., and Javitch, J. A. (2002). β_2 Adrenergic receptor activation. Modulation of the proline kink in transmembrane 6 by a rotamer toggle switch. *J. Biol. Chem.* **277,** 40989–40996.

Singh, R., Hurst, D. P., Barnett-Norris, J., Lynch, D. L., Reggio, P. H., and Guarnieri, F. (2002). Activation of the cannabioid CB1 receptor may involve a $W_{6.48}/F_{3.36}$ rotamer toggle switch. *J. Peptide Res.* **60,** 357–370.

Tan, K., Pogozheva, I. D., Yeo, G. S., Hadaschik, D., Keogh, J. M., Haskell-Leuvano, C., O'Rahilly, S., Mosberg, H. I., and Farooqi, I. S. (2009). Functional characterization and structural modeling of obesity associated mutations in the melanocortin 4 receptor. *Endocrinology* **150,** 114–125.

Tao, Y.-X., Huang, H., Wang, Z.-H., Yang, F., Williams, J. N., and Nikiforovich, G. V. (2010). Constitutive activity of neural melanocortin receptors. *Methods Enzymol.* **484,** 267–279.

Urizar, E., Claeysen, S., Deupi, X., Govaerts, C., Costagliola, S., Vassart, G., and Pardo, L. (2005). An activation switch in the rhodopsin family of G protein-coupled receptors: The thyrotropin receptor. *J. Biol. Chem.* **280,** 17135–17141.

Visiers, I., Ballesteros, J. A., and Weinstein, H. (2002). Three-dimensional representations of G protein-coupled receptor structures and mechanisms. *Methods Enzymol.* **343,** 329–371.

Whistler, J. L., Gerber, B. O., Meng, E., Baranski, T. J., von Zastrow, M., and Bourne, H. R. (2002). Constitutive activation and endocytosis of the complement factor 5a receptor: Evidence for multiple activated conformations of a G protein-coupled receptor. *Traffic* **3,** 866–877.

Ye, S., Zaitseva, E., Caltabiano, G., Schertler, G. F., Sakmar, T. P., Deupi, X., and Vogel, R. (2010). Tracking G-protein-coupled receptor activation using genetically encoded infrared probes. *Nature* **464,** 1386–1390.

Zhang, M., Mizrachi, D., Fanelli, F., and Segaloff, D. L. (2005). The formation of a salt bridge between helices 3 and 6 is responsible for the constitutive activity and lack of hormone responsiveness of the naturally occurring L457R mutation of the human lutropin receptor. *J. Biol. Chem.* **280,** 26169–26176.

CHAPTER TWENTY-TWO

TSH Receptor Monoclonal Antibodies with Agonist, Antagonist, and Inverse Agonist Activities

Jane Sanders, Ricardo Núñez Miguel, Jadwiga Furmaniak, *and* Bernard Rees Smith

Contents

Abstract

Autoantibodies in autoimmune thyroid disease (AITD) bind to the TSH receptor (TSHR) and can act as either agonists, mimicking the biological activity of TSH, or as antagonists inhibiting the action of TSH. Furthermore, some antibodies with antagonist activity can also inhibit the constitutive activity of the TSHR, that is, act as inverse agonists. The production of animal TSHR monoclonal antibodies (MAbs) with the characteristics of patient autoantibodies and the isolation of

FIRS Laboratories, RSR Ltd, Parc Ty Glas, Llanishen, Cardiff, United Kingdom

Methods in Enzymology, Volume 485
ISSN 0076-6879, DOI: 10.1016/S0076-6879(10)85022-0

human autoantibodies from patients with AITD has allowed us to analyze the interactions of these antibodies with the TSHR at the molecular level.

In the case of animal MAbs, advances such as DNA immunization allowed the production of the first MAbs which showed the characteristics of human TSHR autoantibodies (TRAbs). Mouse MAbs (TSMAbs 1–3) and a hamster MAb (MS-1) were obtained that acted as TSHR agonists with the ability to stimulate cyclic AMP production in CHO cells expressing the TSHR. In addition, a mouse TSHR MAb (MAb-B2) that had the ability to act as an antagonist of TRAbs and TSH was isolated and characterized. Also, a mouse TSHR MAb that showed TSH antagonist and TSHR inverse agonist activity (CS-17) was described.

Furthermore, a panel of human TRAbs has been obtained from the peripheral blood lymphocytes of patients with AITD and extensively characterized. These MAbs have all the characteristics of TRAbs and are active at ng/mL levels. To date, two human MAbs with TSHR agonist activity (M22 and K1–18), one human MAb with TSHR antagonist activity (K1–70) and one human MAb (5C9) with both TSHR antagonist and TSHR inverse agonist activity have been isolated.

Early experiments showed that the binding sites for TSH and for TRAbs with thyroid stimulating or blocking activities were located on the extracellular domain of the TSHR. Extensive studies using TSHRs with single amino acid mutations identified TSHR residues that were important for binding and biological activity of TSHR MAbs (human and animal) and TSH. The structures of several TSHR MAb Fab fragments were solved by X-ray crystallography and provided details of the topography of the antigen binding sites of antibodies with either agonist or antagonist activity. Furthermore stable complexes of the leucine-rich repeat domain (LRD) of the TSHR with a human MAb (M22) with agonist activity and with a human MAb (K1–70) with antagonist activity have been produced and their structures solved by X-ray crystallography at 2.55 and 1.9 Å resolution, respectively. Together these experiments have given detailed insights into the interactions of antibodies with different biological activities (agonist, antagonist, and inverse agonist) with the TSHR.

Although the nature of ligand binding to the TSHR is now understood in some detail, it is far from clear how these initial interactions lead to functional effects on activation or inactivation of the receptor.

1. Introduction

The TSH receptor (TSHR) is a major autoantigen in autoimmune thyroid disease (AITD) and TSHR autoantibodies (TRAbs) with thyroid stimulating (agonist) activity are responsible for the hyperthyroidism of Graves' disease (Rapoport *et al.*, 1998; Rees Smith *et al.*, 1988, 2007; Sanders *et al.*, 1997). However, in rare cases, TRAbs act as antagonists and prevent the binding and stimulating activity of TSH and this can cause hypothyroidism (McKenzie and Zakarija, 1992; Rapoport *et al.*, 1998;

Rees Smith *et al.*, 1988, 2007; Sanders *et al.*, 1997). Furthermore, a few of these TRAbs with TSH antagonist activity can inhibit the constitutive (basal) activity of the TSHR, that is, act as inverse agonists (Sanders *et al.*, 2008).

An important goal in understanding the different mechanisms involved in the activity of autoantibodies in patient sera that act as agonists, antagonists, and/or inverse agonists of the TSHR was to either produce animal monoclonal antibodies (MAbs) that have the characteristics of the autoantibodies found in patients sera or to isolate and immortalize human B cells expressing TRAbs from the lymphocytes of patients. The characteristics of patient serum TRAbs are: (1) high binding affinity for the TSHR and hence the ability to interact with the receptor and at very low concentrations (ng/mL) mimic the stimulating activity of TSH or block TSH-mediated stimulation of cyclic AMP production, (2) reactive with conformational epitopes on the TSHR, (3) inhibition of TSH binding to the TSHR by very low concentrations of TRAbs with either blocking (antagonist) or stimulating (agonist) activity.

Some of the mouse MAbs produced using purified TSHR preparations expressed in different systems were able to inhibit labeled TSH binding to the receptor but they did not have the characteristics of human TRAbs listed above (Costagliola *et al.*, 2002a; Davies *et al.*, 1998; Jeffreys *et al.*, 2002; Lenzner and Morgenthaler, 2003; Minich *et al.*, 2004; Oda *et al.*, 1998, 2000; Rapoport *et al.*, 1998; Shepherd *et al.*, 1999; Vlase *et al.*, 1998). However, these MAbs were useful in mapping three regions (amino acids (aa) 246–260, 277–296, and 381–385) of the TSHR extracellular domain important for TSH binding, although none of the MAbs reactive with these regions were strong agonists. In contrast, MAbs which bound to TSHR aa 381–385 were effective antagonists of TSH (Costagliola *et al.*, 2002a; Jeffreys *et al.*, 2002; Shepherd *et al.*, 1999) but not antagonists of TRAbs (Lenzner and Morgenthaler, 2003). Furthermore, they had no effect on the constitutive activity of the TSHR.

The conventional method of raising antibodies using immunization with various TSHR protein preparations can be compared to DNA immunization. The advantage of DNA immunization over the conventional method is that *in vivo* expression of correctly folded protein leads to production of antibodies recognizing conformational epitopes (Costagliola *et al.*, 1998; Hasan *et al.*, 1999).

DNA immunization resulted in the production of the first mouse TSHR MAbs (TSMAbs 1–3; Sanders *et al.*, 2002) with strong thyroid stimulating activity and the characteristics of patient serum autoantibodies in our laboratory and a hamster MAb (MS-1; Ando *et al.*, 2002) with similar characteristics in a different laboratory at about the same time. Subsequently more stimulating antibodies, with similar characteristics were produced in ours and other laboratories (Costagliola *et al.*, 2004; Gilbert *et al.*, 2006; Sanders *et al.*, 2004). In addition, a mouse MAb to the N-terminal region of the TSHR which did not inhibit labeled TSH binding to the TSHR but stimulated cyclic AMP production in TSHR transfected cells (i.e., not having the characteristics of patient serum TRAbs) was described (Costagliola *et al.*, 2002b, 2004).

Furthermore, a mouse MAb (MAb-B2) was obtained in our laboratory, also using DNA immunization that acted as a powerful antagonist of the TSHR stimulating activities of patient serum TRAbs and TSH thus resembling patient blocking type TRAbs (Sanders *et al.*, 2005). More recently, in a different laboratory, a mouse MAb with TSH antagonist activity and the ability to inhibit the constitutive activity of wild-type TSHR and TSHR with activating mutations was produced by immunization with adenovirus expressing the human TSHR A-subunit (aa 1–289; Chen *et al.*, 2007, 2008). However, no studies on the ability of this MAb to inhibit the stimulating activity of patient serum TRAbs were reported.

Since the discovery of thyroid stimulating autoantibodies over 50 years ago, many (albeit unsuccessful) efforts have been made to isolate and characterize TRAbs at the molecular level (McLachlan and Rapoport, 1996; Rees Smith *et al.*, 1988). Recently, though we have been successful in isolating four high affinity human MAbs to the TSHR, two with strong TSHR agonist activity (M22 and K1–18; Evans *et al.*, 2010; Rees Smith *et al.*, 2009; Sanders *et al.*, 2003, 2004), one with potent TSHR antagonist activity (K1–70; Evans *et al.*, 2010; Rees Smith *et al.*, 2009) and one with both TSHR antagonist and inverse agonist activity (5C9; Rees Smith *et al.*, 2007; Sanders *et al.*, 2008). M22 IgG, 5C9 IgG, K1–70 IgG, and K1–18 IgG were all far more potent than their donor serum IgGs (3000×, 3900×, 10,000×, and 10,000×, respectively).

These potency figures contrast markedly with those in previous reports of human MAbs reactive with the TSHR. For example, the blocking and the stimulating MAbs described by Valente *et al.* (1982) and Kohn *et al.* (1997) were of similar potency to the donor serum IgG. In addition, some of these MAbs reportedly reacted with gangliosides. Yoshida *et al.* (1988) have also reported the production of blocking and stimulating type TSHR human MAbs from patient lymphocytes. However, the MAbs described by Yoshida *et al.* (1988) were very different from 5C9, M22, K1–18, and K1–70, being far less potent and being able to react with reduced and denatured TSHR preparations in western blotting analysis. In addition, production of recombinant human MAbs to the TSHR with thyroid stimulating activity has been reported (Akamizu *et al.*, 1996, 1999) but these recombinant MAbs have affinities for the TSHR of about three orders of magnitude lower than 5C9, M22, K1–70, and K1–18. Furthermore, these recombinant TSHR MAbs did not inhibit labeled TSH binding to the TSHR. Also, blocking MAbs isolated from peripheral blood lymphocytes (Okuda *et al.*, 1994) and expressed as recombinant IgGs have been described (Moriyama *et al.*, 2003). However, they appeared to have low affinity for the TSHR, inhibited TSH binding weakly or not at all, and were weak blockers of TSH-mediated cyclic AMP production (e.g., 67% inhibition with 100 μg/mL IgG; Moriyama *et al.*, 2003). The isolation of B cell lines producing IgGs with weak TSH blocking activity has been reported from another laboratory, but those IgGs were unable to inhibit TSH binding to the receptor (Morgenthaler *et al.*, 1996).

We now describe the procedures used in our laboratory to produce both mouse and human MAbs to the TSHR that have the characteristics of patient serum autoantibodies. In particular, the human autoantibody 5C9 which has the ability to inhibit the constitutive activity of the TSHR as well as ligand-induced receptor activation. Also, how 5C9 compares with other TSHR MAbs is considered.

2. Production of Monoclonal Antibodies to the TSHR with the Characteristics of Patient Serum Autoantibodies

2.1. Production of mouse monoclonal antibodies to the TSHR by DNA immunization

The use of DNA immunization techniques in our laboratory resulted in the production of mouse MAbs to the TSHR with the characteristics of human autoantibodies. Briefly, 6- to 8-week-old NMRI (out bred) mice were injected intramuscularly with 100 μL of 10 mmol/L cardiotoxin 5 days before intramuscular immunization with 100 μg TSHR cDNA (pRC/CMV hTSHR; Oda *et al.*, 1998). TSHR DNA immunization was repeated at 3 week intervals for a total of five injections (Costagliola *et al.*, 1998; Hasan *et al.*, 1999). The mouse bleeds were tested for the presence and levels of TSHR antibodies using a ^{125}I-labeled TSH binding inhibition assay (RSR Ltd, Cardiff, UK). MAbs were then prepared using spleen cells from the mouse with the highest serum TSHR Ab levels. Isolated spleen cells were mixed with a mouse myeloma cell line (X63-Ag8.653; ECACC, Porton Down, UK) at a ratio of 1:2 and fused using 10% DMSO and 50% PEG (Sigma Aldrich, Poole, UK) according to previously described methods (de StGroth and Scheidegger, 1980). Cells were cultured in DMEM (supplemented with 20% fetal calf serum containing HAT to select for hybrids) and plated into 48-well plates. Supernatants from the cell cultures were then screened for TSHR antibodies (by inhibition of TSH binding) after 10–14 days of growth (Bolton *et al.*, 1999; Sanders *et al.*, 1999). The cells from the positive wells were recloned two times by limiting dilution at half a cell per well to obtain a pure clone.

Using this DNA immunization approach, we were successful in producing MAbs that acted as potent thyroid stimulators and inhibited TSH binding to the TSHR, that is, had similar features to the TRAbs found in patients with Graves' disease (Sanders *et al.*, 2002, 2004). Furthermore, using the same technique a mouse MAb which acted as a powerful antagonist of thyroid stimulating autoantibodies and TSH (MAb-B2; Sanders *et al.*, 2005) was produced which closely resembled patient serum TSHR antagonist type TRAbs. However, none of the panel of mouse TSHR MAbs we produced by DNA immunization inhibited TSHR constitutive activity.

2.2. Isolation and immortalization of human B cells expressing monoclonal autoantibodies to the TSHR

An important goal in thyroid research over several decades was the isolation of pure autoantibodies to the TSHR from patients with active AITD (McLachlan and Rapoport, 1996). This goal was finally achieved when the thyroid stimulating MAb M22 was produced (Sanders *et al.*, 2003, 2004). M22 is secreted by isolated, immortalized, and extensively recloned B cells obtained from the peripheral blood of a patient with Graves' disease and type 1 diabetes mellitus. The method used to produce M22 and all our human TSHR MAbs is described below.

Twenty milliliters of peripheral blood from the donor patient was mixed with an equal volume of RPMI media (Invitrogen, Paisley, UK) and gently layered onto Ficoll-paque (GE Healthcare, Little Chalfont, UK) and a gradient formed by centrifugation. The layer of peripheral blood lymphocytes was removed from the gradient, placed in RPMI media, washed, and counted. The cells were resuspended in IMDM media (Invitrogen), containing phytohaemagglutinin (Invitrogen) and 20% fetal calf serum then plated in 48-well plates (1×10^6 cells/well) on mouse macrophage feeder layers. Two-hundred microliters of cell culture supernatant containing Epstein Barr virus (EBV; ECACC) was added to each well and the cells incubated at 37 °C in 5% CO_2 in air and 95% humidity. Wells were screened for TSHR antibodies by inhibition of TSH binding assays after 10–14 days of culture (Bolton *et al.*, 1999; Sanders *et al.*, 1999). The cells from the positive wells were expanded and then fused with a mouse/human heterohybridoma cell line (K6H6/B5; ECACC) using the method described above for mouse spleen cells except ouabain (2 μmol/L) was used to kill unfused EBV immortalized B cells. Cells from TSHR antibody positive wells were then recloned four times by limiting dilution to obtain pure clones.

Using this technique we have been able to isolate human monoclonal autoantibodies to the TSHR with different activities (agonists, antagonist, and an inverse agonist). Two antibodies (M22 and K1–18; Evans *et al.*, 2010; Rees Smith *et al.*, 2009; Sanders *et al.*, 2003, 2004) with TSHR agonist activity are both potent stimulators of cyclic AMP production and strong inhibitors of TSH binding to the TSHR. The first human stimulating MAb M22 was obtained from the peripheral blood lymphocytes of a

19-year-old male with hyperthyroidism, high levels of TRAbs as measured by inhibition of TSH binding to TSHR coated tubes (400 U/L), and type 1 diabetes mellitus (Sanders *et al.*, 2003, 2004). The second stimulating autoantibody, K1–18 was obtained from the peripheral blood lymphocytes of a 54-year-old female patient with hypothyroidism and high levels of TRAbs (160 U/L) as measured by inhibition of TSH binding to TSHR coated tubes (Evans *et al.*, 2010; Rees Smith *et al.*, 2009). This patient had an 8-year history of AITD. She first presented with hyperthyroidism, was successfully treated for 3 years but then developed hypothyroidism. The patient's serum showed both stimulation of cyclic AMP production in TSHR transfected CHO cells and the ability to block the stimulation of cyclic AMP by TSH (i.e., had both TSHR agonist and antagonist activity). From the lymphocyte preparation used to isolate K1–18, we were also able to isolate the human MAb K1–70 with TSH antagonist activity (Evans *et al.*, 2010; Rees Smith *et al.*, 2009). K1–70 is a potent inhibitor of TSHR activation by TSH and TRAbs but has no effect on the constitutive activity of the TSHR. The observation that a stimulating MAb (K1–18) and a blocking MAb (K1–70) can be obtained from the same blood sample proves that antibodies with different activities can exist in patients at the same time.

Further to these studies, we have produced another human monoclonal autoantibody to the TSHR (5C9) which inhibits the constitutive activity of wild-type TSHR and TSHRs with activating mutations as well as blocking ligand-induced TSHR activation (Rees Smith *et al.*, 2007; Sanders *et al.*, 2008).

5C9 was isolated from the peripheral blood lymphocytes of a 27-year-old patient with postpartum hypothyroidism and high levels of TRAbs (260 U/L as measured by inhibition of TSH binding to TSHR coated tubes). The patient developed severe hypothyroidism (TSH levels 278 mU/L 6 months postpartum) after a normal pregnancy and delivery of a healthy child, and she was started on L-thyroxine. The patient's serum showed strong inhibition of TSH stimulated cyclic AMP production in TSHR transfected CHO cells and also showed cyclic AMP stimulating activity in CHO cells expressing the TSHR (i.e., had both agonist and antagonist activity).

3. Characterization of 5C9 a Human Autoantibody with TSH Antagonist and TSHR Inverse Agonist Activity

3.1. Inhibition of TSH-induced cyclic AMP stimulation

Purified IgG was obtained from culture supernatants by protein A affinity chromatography on Mabselect (GE Healthcare), characterized, and compared to the human MAb K1–70 and the mouse antibody MAb-B2, both with TSH antagonist activity.

The ability of 5C9 IgG and other preparations to inhibit the stimulating activity of porcine (p) TSH (RSR Ltd), and patient serum TRAbs was assessed using CHO cells expressing the TSHR as described previously (Sanders *et al.*, 2008). These studies were performed by comparing the stimulatory effect of TSH, or sera, in the absence or in the presence of 5C9 IgG, K1–70 IgG or MAb-B2 IgG. CHO cells expressing approximately 5 × 10^4 TSHRs per cell were seeded into 96-well plates at 3 × 10^4 cells/well, adapted into DMEM (Invitrogen) without fetal calf serum and then 50 μL test samples (5C9, K1–70, or MAb-B2) diluted in cyclic AMP assay buffer (NaCl free Hank's Buffered Salts solution containing 1 g/L glucose, 20 mmol/L HEPES, 222 mmol/L sucrose, 15 g/L bovine serum albumin, and 0.5 mmol/L 3 isobutyl-1-methyl-xanthine, pH 7.4) were added to the cell wells prior to addition of 50 μL pTSH or other thyroid stimulator (diluted as appropriate in cyclic AMP buffer) and incubated for 1 h at 37 °C. After removal of test solutions, cells were lysed and cyclic AMP concentrations in the lysates determined using a Correlate-EIA™ Direct Cyclic AMP Enzyme Immunoassay Kit (Assay Designs; Cambridge Bioscience, Cambridge, UK).

5C9, K1–70, and MAb-B2 all inhibited the stimulating activity of pTSH in CHO cells expressing the TSHR in a dose-dependent manner (Fig. 22.1). The dilution profiles of the two human MAbs 5C9 and K1–70 were similar and gave complete inhibition of TSH stimulation to background levels at 1 μg/mL while MAb-B2 was less effective giving

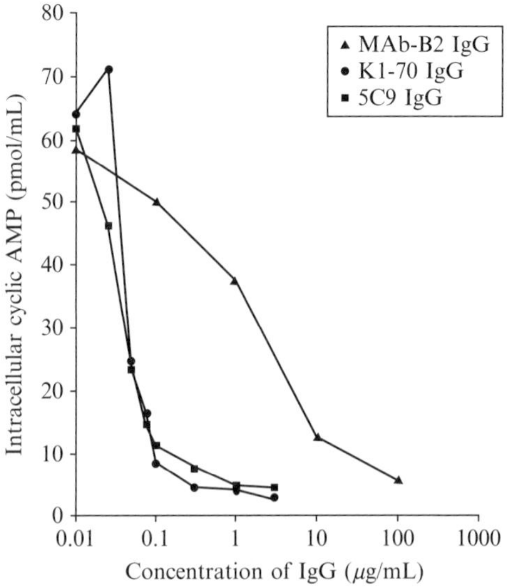

Figure 22.1 Inhibition of porcine TSH (3 ng/mL)-mediated stimulation of cyclic AMP production by TSHR monoclonal antibodies with antagonist activity. MAb-B2 IgG (▲); K1–70 IgG (●); 5C9 IgG (■).

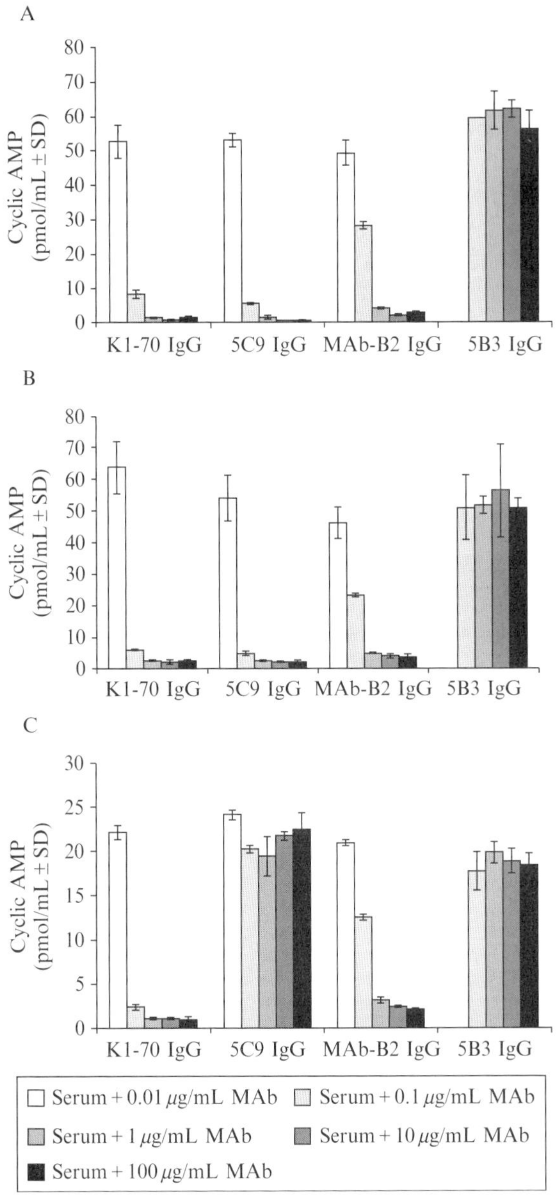

Figure 22.2 Blocking effects of human MAbs K1–70, 5C9, and mouse MAb-B2 on stimulation of cyclic AMP production by sera from three different patients with Graves' disease. (□) serum + 0.01 μg/mL MAb; □ serum + 0.1 μg/mL MAb; ■ serum + 1 μg/mL MAb; (■) serum + 10 μg/mL MAb; ■ serum + 100 μg/mL MAb. (A) Experiments carried out using serum T5. (B) Experiments carried out using serum T8. (C) Experiments carried out using serum T11. Sera diluted 1:10 (final concentration) in cyclic AMP assay buffer. Monoclonal antibody (MAb) IgG concentrations shown are final concentration. 5B3 is a human MAb to glutamic acid decarboxylase (Hayakawa *et al.*, 2002).

partial inhibition at 1 μg/mL and complete inhibition at 100 μg/mL. This may be due to the slightly lower TSHR affinity of MAb-B2 (2×10^{10} L/mol) compared to 5C9 (4×10^{10} L/mol) and K1–70 (4×10^{10} L/mol).

3.2. Inhibition of serum TSHR–autoantibody induced cyclic AMP stimulation

The ability of 5C9 IgG, K1–70 IgG, and MAb-B2 to inhibit the stimulating activity of patient serum TRAbs was assessed using CHO cells expressing the TSHR as described above.

Preliminary results from two separate studies indicated that in the presence of 5C9 IgG (100 μg/mL final concentration), the stimulating activity of 15/16 Graves' patient sera was reduced to near unstimulated (i.e., basal) levels (Sanders *et al.*, 2008). However, in the case of 1/16 sera, the stimulating activity was only reduced to about 50% by 5C9 IgG although the stimulating activity of all 16 sera was reduced to near basal levels in the presence of MAb-B2 (Sanders *et al.*, 2008). Furthermore, in a second series of experiments (Evans *et al.*, 2010), the stimulating activity of 14/15 Graves' sera was reduced completely by 5C9 but no effect was seen with 1/15 sera (T11). In contrast, both MAb-B2 and K1–70 were able to inhibit the stimulating activities to near basal levels in all 15 sera. The discrepant serum (T11) and two other sera (T5 and T8) were analyzed further (Fig. 22.2). In the case of sera T5 and T8, all three MAbs (5C9, K1–70, and MAb-B2) were able to inhibit stimulation of cyclic AMP in a dose-dependent manner. In contrast, with serum T11, no concentrations of 5C9 studied were effective inhibitors of stimulation, whereas K1–70 and MAb-B2 showed almost complete inhibition at 0.1 and 1 μg/mL, respectively.

Overall, in these two experiments 5C9 showed no effect on one out of the 31 sera tested (serum T11 in Fig. 22.2; Evans *et al.*, 2010; Sanders *et al.*, 2008) and was only able to reduce the stimulation of another serum by 50% (T3; Sanders *et al.*, 2008) while MAb-B2 and K1–70 were effective inhibitors of stimulation by all 31 sera studied. This indicates that although stimulating autoantibodies in different patient sera interact with the same region of the TSHR they do so in subtly different ways. Furthermore, this observation indicates that there are differences in the way 5C9, MAb-B2, and K1–70 bind to the TSHR even though MAb-B2 and K1–70 both inhibit binding of 5C9 to the TSHR effectively (Sanders *et al.*, 2008).

3.3. Inhibition of TSHR constitutive activity and effect on TSHR activating mutations

The methods used to introduce specific amino acid mutations into the TSHR sequence and transfection of mutated TSHR constructs into CHO cells using the Flp-In system (Invitrogen), have been described in detail previously (Sanders *et al.*, 2006). In addition to various TSHR amino acid mutations on the concave surface of the TSHR leucine-rich domain (LRD) (Sanders *et al.*, 2006), specific mutations of TSHR residues previously been reported to increase TSHR basal activity (S281I in the extracellular domain (Kopp *et al.*, 1997), I568T in extracellular loop 2 of the transmembrane domain (TMD; Parma *et al.*, 1995), and A623I in intracellular loop 3 of the TMD (Parma *et al.*, 1993)) were also produced. The effect of 5C9 on the constitutive activity of wild-type TSHR was studied using a CHO-K1 cell line expressing approximately 5×10^5 TSHRs per cell (Sanders *et al.*, 2008).

Our experiments revealed that 5C9 IgG was able to reduce the constitutive activity of wild-type TSHR and reduce the constitutive activity associated with all 3 TSHR activating mutations studied (S281I, I568T, and A623I; Rees Smith *et al.*, 2007; Sanders *et al.*, 2008). Although the constitutive activity of the wild-type TSHR was reduced in a dose-dependent manner by 5C9 IgG (71% inhibition at 1 μg/mL of 5C9), it was unaffected by MAb-B2 (0% inhibition over a concentration range of 0.001–10 μg/mL; Fig. 22.3A) and K1–70 (0% inhibition over a concentration range of 0.001–10 μg/mL; Rees Smith *et al.*, 2009).

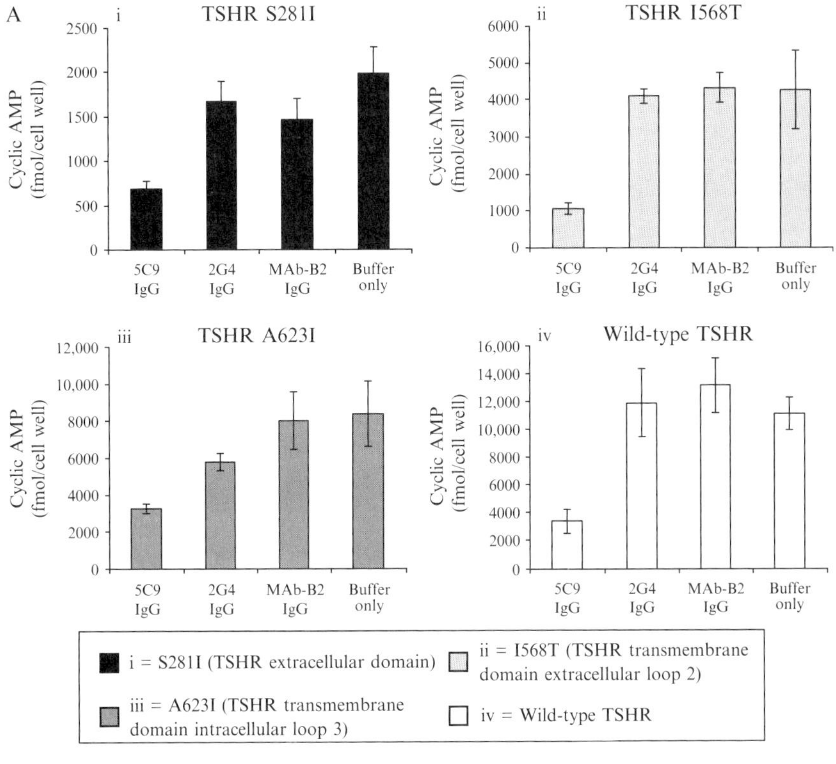

Figure 22.3 (Continued)

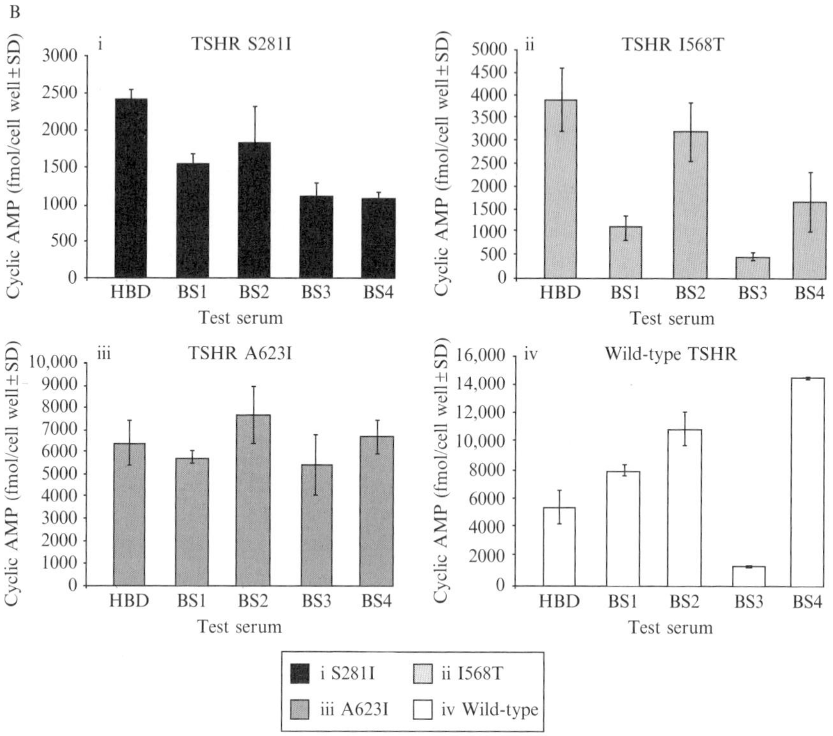

Figure 22.3 (A) Effect of 1 μg/mL of 5C9 IgG or MAb-B2 IgG on constitutive (i.e., basal) activity of wild-type TSHR and TSHRs containing activating mutations: (i) S281I (TSHR extracellular domain), (ii) I568T (TSHR transmembrane domain extracellular loop 2) , (iii) A623I (TSHR transmembrane domain intracellular loop 3) , (iv) wild-type TSHR . 2G4 is a negative control human autoantibody to thyroid peroxidase (TPO; Horimoto *et al.*, 1992). From Sanders *et al.* (2008). The publisher for this copyrighted material is Mary Ann Liebert, Inc. publishers and a license has been granted (number 2437100898401). (B) Effect of patient sera with TSH antagonist activity (BS1-4) on basal cyclic AMP activity of wild-type and TSHR containing activating mutations) (i) S281I , (ii) I568T , (iii) A623I (iv) wild-type . Healthy blood donor (HBD) sera were used as a negative control. The CHO cells transfected with wild-type TSHR in this experiment expressed approximately 500,000 receptors per cell (i.e., about 10× more than cells usually employed in cyclic AMP stimulation assays). Results shown are means of triplicate determinations ±SD.

In the presence of 5C9 IgG or MAb-B2 at 1 μg/mL TSHR S281I constitutive activity was inhibited by 60% and 4%, respectively (Fig. 22.3A). A similar response was observed in the case of the TSHR I568T activating mutation, the constitutive activity being inhibited by 74% with 1 μg/mL of 5C9 and unaffected (0% inhibition) by 1 μg/mL of MAb-B2 (Fig. 22.3A). In the case of TSHR A623I, the constitutive activity was reduced by 43%

by incubation with 1 μg/mL of 5C9 and unaffected by MAb-B2 (0% inhibition; Fig. 22.3A). Increasing the concentration of 5C9 IgG further (concentrations up to and including 100 μg/mL) showed no increase in the inhibition of cyclic AMP production in cells expressing the three activating mutations or wild-type TSHR (data not shown).

Overall therefore, the MAb 5C9 inhibited TSHR stimulation by TSH and TSHR as well as TSHR activity when no stimulating ligand was present (i.e., basal activity). This inhibiting effect on basal activity was observed with CHO cells transfected with wild-type TSHR or TSHR with different activating mutations (Rees Smith *et al.*, 2007; Sanders *et al.*, 2008). Consequently, 5C9 has at least two mechanisms of action: (a) preventing binding of activating ligand to the TSHR and (b) an effect on TSHR activation not dependent on activating ligand binding. The mechanism by which inhibition of TSHR basal activity occurs is not clear at present and the effect is not seen with the blocking type mouse MAb-B2 or human MAb K1–70. Recently, however, a mouse MAb (CS-17) to the TSHR has been described (Chen *et al.*, 2007, 2008) which has the ability to block both basal and TSH stimulated cyclic AMP production in COS-7 cells expressing wild-type TSHR and TSHRs with activating mutations. Both 5C9 and CS-17 were able to inhibit the constitutive activity of TSHR S281I, TSHR I568T, and TSHR A623I at 1 μg/mL; however, 5C9 appears to be more effective than CS-17 at low IgG concentrations.

In a second series of experiments, some patient sera with TSH antagonist activity were found to influence constitutive activity (Fig. 22.3B). However, individual sera showed different effects on the wild-type TSHR and on TSHRs with activating mutations. For example, BS3 was an effective inhibitor of the constitutive activity of the wild-type, I568T and S281I TSHRs (Fig. 22.3B). In contrast, BS4 inhibited the constitutive activity of I568T and S281I TSHRs but increased the basal activity of the wild-type TSHR (Fig. 22.3B). These differences in the effects of individual sera may be related in part at least to the presence of a mixture of blocking and stimulating autoantibodies and their different relative effects on wild-type TSHR and TSHRs with different activating mutations.

4. Effects of TSHR Mutations on the Activity of MAb 5C9

The effect of various TSHR amino acid mutations on the biological activity of TSHR MAbs was studied using CHO cells expressing wild-type or mutated TSHRs as described in detail previously (Sanders *et al.*, 2005, 2006, 2008). Flp-In-CHO cells expressing either wild-type or mutated TSHRs were seeded into 96-well plates and used to test the ability of 5C9 IgG, MAb-B2 IgG, and K1–70 IgG to block the stimulating activity of pTSH.

Out of all the TSHR mutations investigated, mutation of K129, D203, and F153 to alanine resulted in a complete loss of the ability of 5C9 IgG to block TSH stimulation of cyclic AMP production (Sanders *et al.*, 2008). Furthermore, TSHR mutations K183, E178, and E251 to alanine showed reduced 5C9 blocking activity. All these amino acids mutations are on the concave surface of the TSHR extracellular domain. A new series of TSHR mutations were constructed including amino acids from the TSHR cleavage domain (CD; amino acids 282–409), in particular, E297A, E297K, E303A, E303K, R312A, K313A, E325A, D382A, D382K, H384A, Y385A, and D386A. At least some of these TSHR CD residues have been implicated as being involved in the interaction of TSH with the TSHR and TSHR activation (Costagliola *et al.*, 2002a; Kosugi *et al.*, 1991; Mueller *et al.*, 2008, 2009). However, the ability of 5C9 to inhibit pTSH-induced stimulation of cyclic AMP production was not affected by any of the mutations in the TSHR CD (listed above) that we studied. The mouse MAb CS-17 (Chen *et al.*, 2007, 2008) which has TSH antagonist activity and inverse agonist activity has been reported to bind to amino acids on the convex surface of the TSHR (Y195 and S243) and Q235 on the concave surface. Consequently, TSHRs containing mutation of these amino acids (Y195A, S243A, and Q235A) and other potential candidates (Y148A, K201A, Y225A, and E247A) were produced and expressed in CHO cells in our laboratory. However, no effect on the ability of 5C9 to inhibit TSH-induced cyclic AMP was observed. Consequently our mutation studies only indicate that the human MAb 5C9 forms strong interactions with several amino acids on the concave surface of the TSHR. Also, as mutations in other parts of the TSHR (i.e., the CD and convex surface of the LRD) had no effect on the activity of 5C9 it is likely that 5C9 forms either no or weak interactions (no effect of mutations on 5C9 activity cannot discount interactions with these residues) with either the TSHR CD (implicated in TSH binding) or residues on the convex surface of the TSHR implicated in CS-17 binding (Chen *et al.*, 2008). Our experiments suggest that the interacting region for 5C9 on the TSHR may be different to that of the CS-17 MAb. However, the effects of Y195H on CS-17 were only detected on a background of S243P mutation and extensive mutation studies of the TSHR concave surface or CD on CS-17 activity have not been reported.

TSHR mutations that had an effect on the activity of 5C9 can be compared to TSHR mutations that affected the ability of MAb-B2 to inhibit TSH-induced cyclic AMP production in CHO-cells expressing the TSHR (D36R, R38A, K58A, R80A, R80D, Y82A, R109A, R109D, K129A, K129D, and F134A; Fig. 22.4, Table 22.1; Sanders *et al.*, 2005).

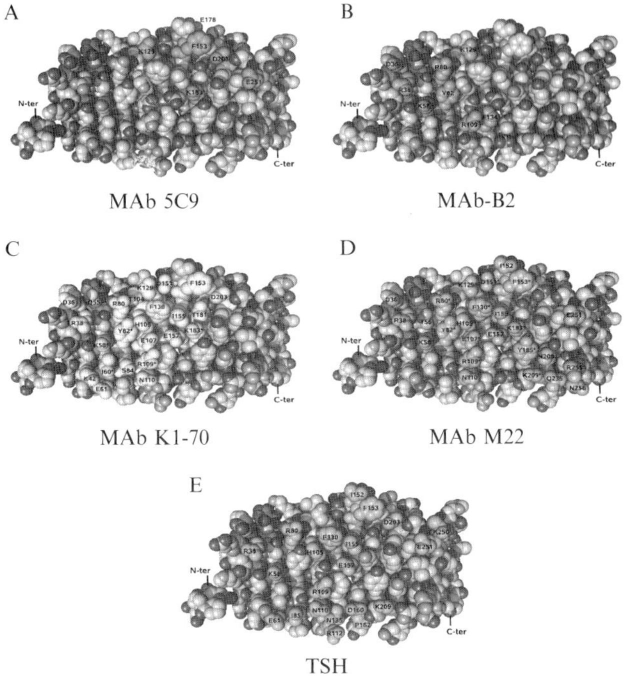

Figure 22.4 Space fill representation of the TSHR leucine-rich domain interactive surface (based on the crystal structure of the TSHR in complex with K1–70 solved at 1.9 Å resolution). Single amino acid mutations which affect inhibition of TSH-mediated stimulation of cyclic AMP production in TSHR transfected CHO cells in the case of: (A) MAb 5C9 shown in red (Table 22.1; antagonist and inverse agonist), (B) MAb-B2 shown in blue (Table 22.1; antagonist), (C) TSHR residues that interact with MAb K1–70 (antagonist) in the TSHR—K1–70 crystal structure are shown in yellow (⋆ denotes an effect on K1–70 activity in mutation experiments; see Table 22.1). (D) TSHR residues that interact with MAb M22 (agonist) in the TSHR—M22 crystal structure are shown in green (⋆ denotes an effect on M22 activity in mutation experiments; see Table 22.1). (E) TSHR residues that interact with TSH (agonist) in a TSH-TSHR comparative model are shown in orange (Table 22.1). (See Color Insert.)

Further studies on TSHR–autoantibody interactions were carried out using stable complexes of a part of the TSHR extracellular domain (aa 21–260; TSHR260) with the agonist MAb (M22) and with the antagonist MAb (K1–70) using X-ray crystallography. The crystal structures of (1) the thyroid stimulating human MAb M22 (Sanders *et al.*, 2007a) and (2) the K1–70 human MAb with TSH antagonist activity with TSHR260 (unpublished data) were solved at 2.55 and 1.9 Å resolution, respectively. These complexes also provide molecular level detail of the structure of the LRD of the TSHR (Sanders *et al.*, 2007a).

Table 22.1 TSHR residues important for TSH antagonist activity of 5C9, MAb-B2, and K1–70 and the agonist activity of M22 and TSH

TSHR residues important for TSH antagonist activity of 5C9 (mutation studies)	TSHR residues important for TSH antagonist activity of MAb-B2 (mutation studies)	TSHR residues that interact with the TSH antagonist K1–70 (data from the K1–70-TSHR crystal structure and mutation studies)	TSHR residues that interact with the TSHR agonist M22 (data from the M22-TSHR crystal structure and mutation studies)	TSHR residues that interact with TSH (data from the TSH-TSHR LRD comparative model)
	D36	*D36*	*D36*	
	R38	*R38*	*R38*	*R38*
		K42		
		Q55		
			T56	
	K58	*K58*★	*K58*	*K58*
		I60★		
		E61		*E61*
	R80	*R80*	*R80*★	*R80*
	Y82	*Y82*★	*Y82*★	
		S84		
				I85
		T104		
		H105	*H105*★	*H105*
		E107	*E107*★	
	R109	*R109*★	*R109*★	*R109*
		N110	*N110*	*N110*
				R112
K129	*K129*	*K129*	*K129*★	
		F130	*F130*★	*F130*
	F134			
				N135
		D151	*D151*★	

		I152	*I152*
F153	*F153*	*F153*★	*F153*
	I155	*I155*	*I155*
	E157	*E157*	*E157*
			D160
			P162
E178			
	T181		
K183	*K183*★	*K183*★	
		Y185★	
D203	*D203*		*D203*
		N208	
		K209★	*K209*
		Q235	
			K250
E251		*E251*	*E251*
		R255★	
		N256	

★ Denotes amino acids identified in the crystal structures, which when mutated affected the activity of the K1–70 or M22.

Amino acids depicted in italic are important for more than one antibody and/or TSH.

TSHR residues that interact with the antagonist K1–70 or the agonist M22 obtained from the crystal structures of their respective TSHR complexes and TSHR residues that interact with TSH in the comparative model include hydrogen bonds, salt bridges, nonhydrogen bonding polar interactions, hydrophobic contacts, van der Waals interactions (showing variation of accessible surface area on complexation >60 Å^2) and charge–charge interactions (strength $>6.0e-10$ N by residues or $>2.5e-10$ N by atoms).

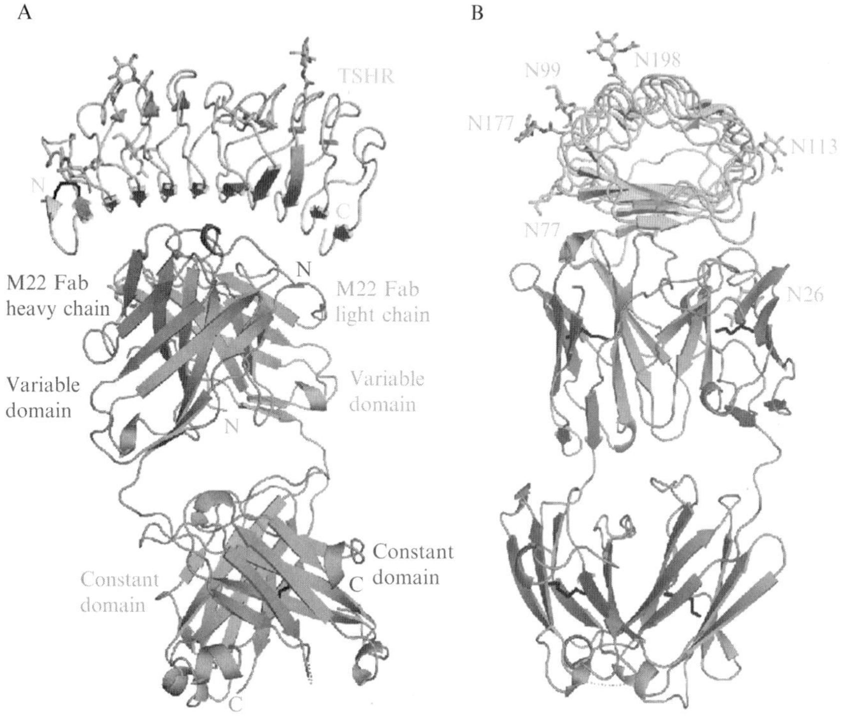

Figure 22.5 Crystal structure of TSHR260 in complex with thyroid stimulating monoclonal autoantibody (M22) Fab (solved at 2.55 Å resolution; Sanders *et al.*, 2007a). (A & B) Cartoon diagram of the complex shown in two aligned views related by a 90° rotation about the vertical axis. TSHR is in cyan, M22 light chain is in green, M22 heavy chain is in blue. The positions of amino-(N) and carboxy-(C) termini are indicated. The observed N-linked carbohydrates are shown in yellow and carbohydrate-bound asparagines residues are labeled. Disulphide bonds are in black. From Sanders *et al.* (2007a). The publisher for this copyrighted material is Mary Ann Liebert, Inc. publishers and a license has been granted (number 2437101285589). (See Color Insert.)

The TSHR has the shape of a slightly curved helical tube constructed from leucine-rich repeat motifs. It has opposed concave and convex surfaces, with an 11-stranded β-sheet located on the concave surface (10 parallel strands, one per repeat and an antiparallel strand at the N-terminus; Fig. 22.5). The inner surface of the tube is lined with hydrophobic residues. The closest homologue of TSHR is FSHR with which it shares 40.9% sequence identity (Sanders *et al.*, 1997). The crystal structure of the FSHR LRD in complex with FSH has been solved at 2.9 Å resolution (Fan and Hendrickson, 2005) and the root mean square deviation (rmsd) on C_α core atoms between the TSHR LRD and FSHR LRD structures is 1.1 Å indicating that the LRDs, of the two receptors, have almost identical structures (Sanders *et al.*, 2007a). Like that of FSHR the concave surface of the TSHR LRD is formed from an untwisted β-sheet (Fig. 22.5). The structure of the convex surface of the TSHR LRD presents eight small strands (two residues each) forming two 3-stranded β-sheets and one 2-stranded β-sheet (Fig. 22.5). There are no α helices in the TSHR LRD structure. All five (N77, N99, N113, N177, and N198) glycosylation sites are located on the convex surface of the TSHR LRD. A comparison of the TSHR structures in the complex with M22 or with K1–70 shows an rmsd on all C_α atoms between the structures of 0.51 Å thus confirming the solved structure of the TSHR LRD originally reported (Sanders *et al.*, 2007a). The N-terminal cysteines (C31 and C41) in the TSHR sequence are disulphide bonded in both complexes.

The crystal structures of the complexes show the interactions between the TSHR and M22 or the TSHR and K1–70 at the atomic level. There are strong interactions (hydrogen bonds or salt bridges) between the M22 or K1–70 heavy chains (HC) or light chains (LC) and the residues on the concave surface of the TSHR LRD. Many of the interactions observed in the crystal structure have been confirmed by experiments involving mutation of TSHR or MAb amino acids (Sanders *et al.*, 2006, 2007b).

Table 22.1 and Fig. 22.4 show a comparison of the TSHR residues that were identified as important for the activity of 5C9 and MAb-B2 by TSHR mutation experiments and the K1–70 and M22 TSHR contact residues confirmed by analysis of crystal structure data (residues that were identified as important for the activity of M22 and K1–70 by TSHR mutation experiments are marked with an "asterisk" in Table 22.1; Sanders *et al.*, 2006). The residues of TSH that interact with the TSHR LRD were obtained by comparative modeling (Núñez Miguel *et al.*, 2004, 2005, 2008, 2009) and are shown in Table 22.1 and Fig. 22.4 for comparison with the MAb–TSHR interactions. One residue TSHR K129 is important for interaction of all four MAbs (5C9, MAb-B2, K1–70, and M22) whether they have antagonist, inverse agonist, or agonist activity showing that even though the antibodies have different functional activities their binding sites on the concave surface of the TSHR overlap. The TSHR residues F153 and K183 are important for the activity of all three human MAbs (5C9, K1–70, and M22) but TSHR mutation experiments indicate that they are not important for activity of the mouse MAb-B2. In contrast, TSHR D203 is important for the activity of 5C9 and is an interacting amino acid in the TSHR-K1–70 complex but does not appear to be involved in MAb-B2 activity. Furthermore three residues, mutations of which affect 5C9 activity (F153, D203, and E251) also appear to be involved in the interactions between the TSHR and TSH. TSHR residues which are important for the biological activity of 5C9 are more C-terminal than the TSHR residues important for the activities of MAb-B2, K1–70, and M22 (Fig. 22.4).

In contrast, the antagonist MAb-B2 contact residues are located more N-terminal on the TSHR concave surface and seven of the eight residues identified as important for activity of MAb-B2 are also interacting residues in the TSHR complex with the antagonist K1–70 MAb. Identification of TSHR interacting residues using crystal structures of MAb–TSHR complexes allows a more detailed analysis compared to TSHR mutation experiments. For example, a mutation may not show any effect on the biological activity of an antibody if other interactions with the TSHR are not disrupted or are strong enough to compensate for the mutation. Remarkably, the crystal structures of the TSHR complexes with the agonist MAb M22 and the antagonist MAb K1–70 show an overlap of 16 TSHR residues (D36, R38, K58, R80, Y82, H105, E107, R109, N110, K129, F130, D151, F153, I155, E157, and K183) that interact with both MAbs. There is also an extensive overlap between the TSHR residues involved in antibodies binding to the TSHR and those predicted to be involved in TSH interaction with the receptor. Overall the data obtained from the crystal structures and mutation experiments suggests that although the binding sites of all four MAbs (5C9, MAb-B2, M22, and K1–70) overlap on the concave surface of the TSHR subtle differences in the contact amino acids most probably define the functional effects of the MAb on TSHR activity.

5. Structure of MAb 5C9 Fab

RNA was isolated from the hybridoma cell line expressing 5C9 IgG and both the HC and LC genes were cloned and sequenced (Sanders *et al.*, 2008). The available amino acid sequence was used to build a comparative model of 5C9 Fab using pdb-Id 2r8s as an initial template and then pdb-Id 1j05, 1sjx, 2or9, 2uzi, 1rz7, and 1mqk were used as templates to model the HC CDRs 1, 2, 3, and the LC CDRs 1, 2, and 3, respectively. A model of 5C9 was obtained using the program MODELLER (Sali and Blundell, 1989). Pdb-Id 2or9 was chosen to model the CDR 3 of the HC to allow construction of an internal disulfide bond. Model validation was carried out using PROCHECK (Laskowski *et al.*, 1993) and Verify3D (Luthy *et al.*, 1992). Ramachandran plot analysis for the 5C9 Fab model gave 96.8% of residues in allowed regions, 2.4% in marginal regions and 0.8% in disallowed regions. The structure of 5C9 was compared to the known crystal structures of MAb-B2, K1–70, and M22 (solved at 3.3 Å, Sanders *et al.*, 2005; 2.2 and 1.65 Å, Sanders *et al.*, 2004, respectively).

The structure of MAb 5C9 Fab is that of a typical Fab with a standard distribution of CDRs within the 5C9 antigen binding site. HC CDR 3 and LC CDR 3 are in central positions, HC CDR 1 and LC CDR 1 are further out, and HC CDR 2 and LC CDR 2 on the periphery. The crystal structures of MAb-B2 (antagonist), MAb K1–70 (antagonist), and MAb M22 (agonist) are also standard Fab structures. The HC CDR 3 of 5C9 is longer than the other three MAbs (18 amino acids long) and is expected to protrude outward from the V domain. It contains a three aa and a six aa insertion between the V/D and D/J, respectively, and contains the CX_4C motif of the D2-2 germline in which the two cysteine residues form an internal disulphide bond for loop stability.

Electrostatic potential surfaces of the antigen binding site were generated for all four MAbs (5C9, MAb-B2, K1–70, and M22) using PYMOL (Fig. 22.6; De Lano, 2002). The electrostatic potential surface of 5C9 is mainly positive with only a small negative patch centered on D61 (HC). 5C9 has few charged amino acids (five residues) and eight aromatic residues in its combining site therefore most of the combining surface has overall low electrostatic potential. Only three high positive potentials are observed centered on HC residues R64, R97, and R100G, and one (R24 LC) with a weak positive electrostatic potential. The aromatic residues are randomly distributed on the surface of the combining site except for two residues close to each other arising from amino acids inserted into HC CDR 3 (Fig. 22.6). It is possible that this aromatic patch may contact complementary aromatic residues in the TSHR LRD, two good candidates for such interaction could be TSHR Y195 or Y225. Y195 has previously been suggested as part of the epitope of CS-17 MAb with inverse agonistic activity (described above; Chen *et al.*, 2008).

In contrast, the surface of the antagonist mouse MAb-B2 antigen binding site is dominated by acidic patches in the center (Fig. 22.6). The predominant acidic patch corresponds to a deep cavity on the surface of the combining site centered on HC CDR 2 residues: E50, D52, and D54. In addition, acidic patches with low electrostatic potential are centered on D50 in LC CDR 2 and D101 in HC CDR 3. The basic patches with low electrostatic potential are represented by the LC CDR 2 K53 and the HC CDR 2 K62 and K64. It is of interest that the LC CDRs contributing to the antigen binding site of MAb-B2 are rich in serine residues (nine serine residues; 34.6% of total LC CDR 1–3 residues). In contrast, only two charged residues (D50 and K53 both in LC CDR 2) are from the LC CDRs. The aromatic amino acids (11 residues) present on the antigen binding surface of MAb-B2 are contributed mostly by the HC CDR 1–3.

The surface of the human MAb K1–70 (antagonist) antigen binding site (Fig. 22.6) is dominated by acidic patches on one side and basic patches on the other. The acidic patches are centered on D27B (CDR 1 LC), D50 (CDR 2 LC), D92 (CDR 3 LC), D31 (CDR 1 HC), D54 and D56 (CDR 2 HC), and D96 (CDR 3 HC). The basic patches are centered on residues K53 and R54 (CDR 2 LC), R94 (CDR 3 LC), R58 (CDR 2 HC), and R101 (CDR 3 HC). Outside the variable domain region K66 LC creates a basic patch. The surface of K1–70 Fab also contains aromatic residues from

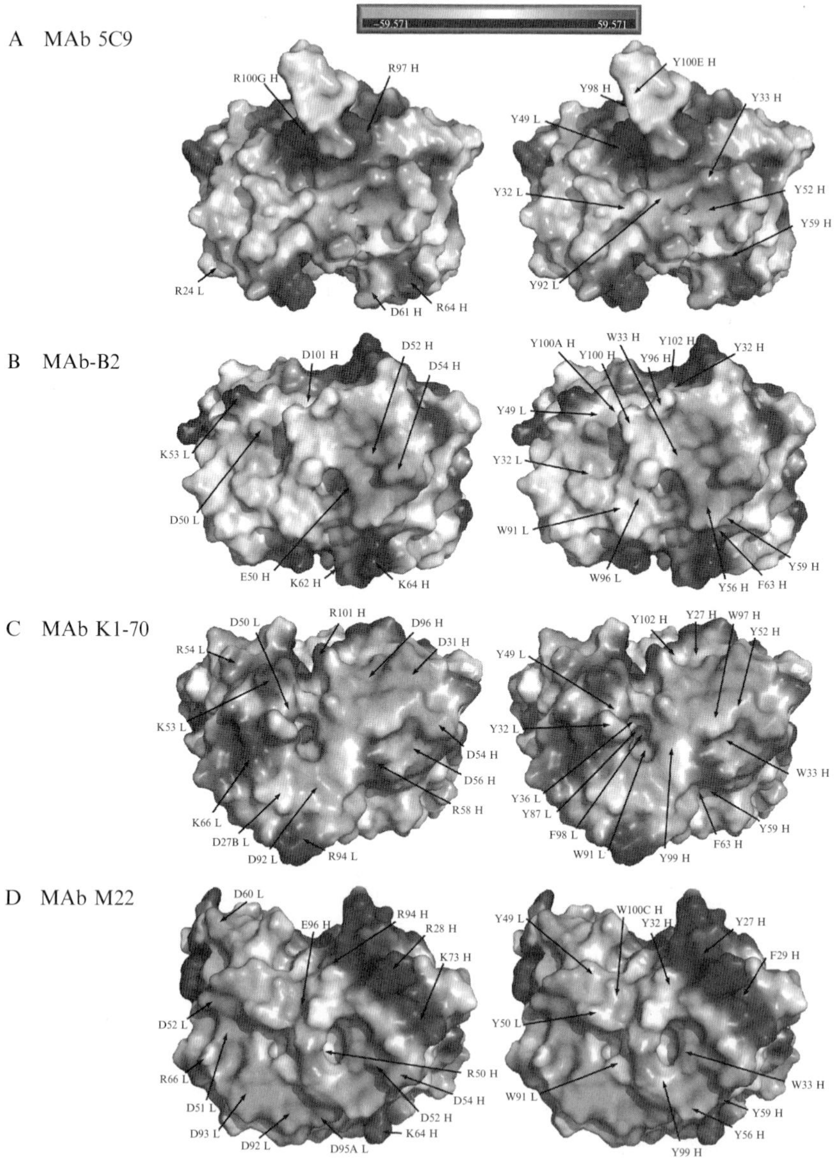

Figure 22.6 The electrostatic surface potential of the antigen binding regions of: (A) Human MAb 5C9 (antagonist and inverse agonist; derived from the comparative model of 5C9 structure), (B) mouse MAb-B2 (antagonist; derived from the crystal structure solved at 3.3 Å resolution), (C) human MAb K1–70 (antagonist; derived from the crystal structure solved at 2.2 Å resolution), (D) human MAb M22 (agonist; derived from the crystal structure solved at 1.65 Å resolution). Position of the charged residues is indicated in the figures on the left-hand side of the panel and the position of aromatic residues is indicated in the figures on the right-hand side of the panel. Acidic patches are shown in red and basic patches in blue (See Color Insert).

both the variable regions and residues located close to them (14 residues). The surface of the combining site is highly irregular with a cavity at the center surrounded mostly by aromatic residues and D50 LC. The central cavity is also lined by aromatic residues.

In comparison, the surface of the agonist human MAb M22 antigen binding site is highly charged with acidic patches on one side and basic patches on the other side (Fig. 22.6). The acidic patches are centered on D51 LC and D52 (CDR 2 LC); D92 LC, D93, and D95A (CDR 3 LC); D52 HC and D54 (CDR 2 HC); and E96 (CDR 3 HC). The basic patches are centered on HC residues R28 and R50 (CDR 2 HC) and R94. The surface of the variable region of M22 is irregular and the hypervariable regions of the HC protrude further from the framework than the hypervariable regions of the LC.

The electrostatic potential distribution on the antigen binding surfaces of M22 and K1–70 is similar; however, the negatively charged residues are contributed from the LC in the case of M22 and from the HC in the case of K1–70. Conversely, the positively charged amino acids on the M22 surface are predominantly from M22 HC and on the K1–70 surface from K1–70 LC.

Overall the electrostatic potential surface of 5C9 is less charged than the other three MAbs and the HC CDR 3 protrudes further out of the combining surface and these differences in the topography of its antigen binding site may be related to the inverse agonist activity of 5C9.

6. Conclusions

Analysis of TSHR MAb–TSHR interactions has provided valuable insights into how antibodies with different biological functions bind to the TSHR. For example, mutation of the TSHR at R255 to an oppositely charged residue (e.g., R255D) has a marked effect on the stimulating activity of both human and mouse TSHR MAbs and patient serum stimulating autoantibodies (Sanders *et al.*, 2006). However, the TSHR R255D mutation had no effect on the stimulating activity of TSH itself or the antagonist activity of TSHR blocking antibodies (MAb-B2, K1–70, and 5C9) or the antagonist activity of patient serum blocking autoantibodies. In the TSHR-M22 crystal structure TSHR R255 forms strong interactions with three residues of the agonist MAb M22 (Sanders *et al.*, 2007a). In contrast, there are no interactions between the TSHR R255 and antagonist MAb K1–70 in the crystal structure as K1–70 binds to the central and N-terminal region of the TSHR LRD only. These observations indicate that TSHR R255 is critical for TSHR stimulation by autoantibodies with agonist activity but not for the activity of autoantibodies with antagonist or inverse agonist properties. Observations of this type may be helpful in developing specific inhibitors of thyroid stimulating antibodies.

Recent experimental and crystallographic studies provide compelling evidence that the major binding sites of TSHR antibodies with different biological activities and the binding site of the hormone TSH are localized on the concave surface of the TSHR LRD and the contact residues for all these ligands show considerable overlap. Furthermore the surface areas of the interacting interfaces in the complexes of TSHR with agonist or antagonist MAbs (crystal structures) or TSH (comparative model) are similar (Núñez Miguel *et al.*, 2004, 2005, 2008, 2009; Sanders *et al.*, 2007a). However, some differences in the binding characteristics of antibodies of different activities to the TSHR have emerged from these studies. It can be observed that the stimulating antibody M22 (and TSH) form interactions with all 10 leucine-rich repeats (LRR) on the concave surface of the TSHR LRD while the blocking antibodies tend to bind to the central and the N-terminal part of this surface (there are no interactions with the 9th and the 10th LRR in the TSHR-K1–70 complex; Sanders *et al.*, 2007a,b; Núñez Miguel *et al.*, 2004, 2005, 2008, 2009). The way the human MAb 5C9, which has antagonist and inverse agonist activity, positions itself on the TSHR differs from the other antibodies. In particular, 5C9 interactions are orientated to the central and the C-terminal regions of the concave surface of the LRD and may reach to its edge and over to the convex surface of the structure. It may be that these differences in the binding arrangements of 5C9 with the TSHR are related to its inverse agonist activity.

The actual mechanisms involved in TSHR activation by TSH and by TRAbs or how TSHR antibodies cause inverse agonism are not known at present. However, the first step in these processes, that is, ligand binding to the TSHR, is now understood in some detail. Also, analysis of TSHR TMD structural models together with extensive mutations of TMD amino acids has provided some insight into the signaling process within the TMD (Kleinau and Krause, 2009). However, how activation of the TSHR TMD is triggered following ligand binding to the LRD and how inverse agonist type TSHR MAbs are able to affect constitutive activity is far from clear at present. It seems likely that ligand-induced changes in the relative positions or movements of three TSHR domains (LRD, CD, and TMD) are an important part of receptor activation and future structural studies should provide key insights into this process.

ACKNOWLEDGMENT

We thank Carol James for expert preparation of the manuscript.

REFERENCES

Akamizu, T., Matsuda, F., Okuda, J., Li, H., Kanda, H., Watanabe, T., Honjo, T., and Mori, T. (1996). Molecular analysis of stimulatory anti-thyrotropin receptor antibodies (TSAbs) involved in Graves' disease. Isolation and reconstruction of antibody genes, and production of monoclonal TSAbs. *J. Immunol.* **157,** 3148–3152.

Akamizu, T., Moriyama, K., Miura, M., Saijo, M., Matsuda, F., and Nakao, K. (1999). Characterization of recombinant monoclonal antithyrotropin receptor antibodies (TSHRAbs) derived from lymphocytes of patients with Graves' disease: Epitope and binding study of two stimulatory TSHRAbs. *Endocrinology* **140,** 1594–1601.

Ando, T., Latif, R., Pritsker, A., Moran, T., Nagayama, Y., and Davies, T. F. (2002). A monoclonal thyroid-stimulating antibody. *J. Clin. Invest.* **110,** 1667–1674.

Bolton, J., Sanders, J., Oda, Y., Chapman, C., Konno, R., Furmaniak, J., and Rees Smith, B. (1999). Measurement of thyroid-stimulating hormone receptor autoantibodies by ELISA. *Clin. Chem.* **45,** 2285–2287.

Chen, C. R., McLachlan, S. M., and Rapoport, B. (2007). Suppression of thyrotropin receptor constitutive activity by a monoclonal antibody with inverse agonist activity. *Endocrinology* **148,** 2375–2382.

Chen, C. R., McLachlan, S. M., and Rapoport, B. (2008). Identification of key amino acid residues in a thyrotropin receptor monoclonal antibody epitope provides insight into its inverse agonist and antagonist properties. *Endocrinology* **149,** 3427–3434.

Costagliola, S., Rodien, P., Many, M. C., Ludgate, M., and Vassart, G. (1998). Genetic immunization against the human thyrotropin receptor causes thyroiditis and allows production of monoclonal antibodies recognizing the native receptor. *J. Immunol.* **160,** 1458–1465.

Costagliola, S., Panneels, V., Bonomi, M., Koch, J., Many, M. C., Smits, G., and Vassart, G. (2002a). Tyrosine sulfation is required for agonist recognition by glycoprotein hormone receptors. *EMBO J.* **21,** 504–513.

Costagliola, S., Franssen, J. D., Bonomi, M., Urizar, E., Willnich, M., Bergmann, A., and Vassart, G. (2002b). Generation of a mouse monoclonal TSH receptor antibody with stimulating activity. *Biochem. Biophys. Res. Commun.* **299,** 891–896.

Costagliola, S., Bonomi, M., Morgenthaler, N. G., Van Durme, J., Panneels, V., Refetoff, S., and Vassart, G. (2004). Delineation of the discontinuous-conformational epitope of a monoclonal antibody displaying full in vitro and in vivo thyrotropin activity. *Mol. Endocrinol.* **18,** 3020–3034.

Davies, T. F., Bobovnikova, Y., Weiss, M., Vlase, H., Moran, T., and Graves, P. N. (1998). Development and characterization of monoclonal antibodies specific for the murine thyrotropin receptor. *Thyroid* **8,** 693–701.

De Lano, W. L. (2002). The Pymol Molecular Graphics System. DeLano Scientific, San Carlos, CA http://pymol.sourceforge.net/.

de StGroth, S. F., and Scheidegger, D. (1980). Production of monoclonal antibodies: Strategy and tactics. *J. Immunol. Methods* **35,** 1–21.

Evans, M., Sanders, J., Tagami, T., Sanders, P., Young, S., Roberts, E., Wilmot, J., Hu, X., Kabelis, K., Clark, J., Holl, S., Richards, T., *et al.* (2010). Monoclonal autoantibodies to the TSH receptor, one with stimulating activity and one with blocking activity, obtained from the same blood sample. *Clin. Endocrinol.* **73,** 404–412.

Fan, Q. R., and Hendrickson, W. A. (2005). Structure of human follicle-stimulating hormone in complex with its receptor. *Nature* **433,** 269–277.

Gilbert, J. A., Gianoukakis, A. G., Salehi, S., Moorhead, J., Rao, P. V., Khan, M. Z., McGregor, A. M., Smith, T. J., and Banga, J. P. (2006). Monoclonal pathogenic antibodies to the thyroid-stimulating hormone receptor in Graves' disease with potent thyroid-stimulating activity but differential blocking activity activate multiple signaling pathways. *J. Immunol.* **176,** 5084–5092.

Hasan, U. A., Abai, A. M., Harper, D. R., Wren, B. W., and Morrow, W. J. (1999). Nucleic acid immunization: Concepts and techniques associated with third generation vaccines. *J. Immunol. Methods* **229,** 1–22.

Hayakawa, N., Premawardhana, L. D. K. E., Powell, M., Masuda, M., Arnold, C., Sanders, J., Evans, M., Chen, S., Jaume, J. C., Baekkeskov, S., Rees Smith, B., and Furmaniak, J. (2002). Isolation and characterization of human monoclonal autoantibodies to glutamic acid decarboxylase. *Autoimmunity* **35,** 343–355.

Horimoto, M., Petersen, V. B., Pegg, C. A. S., Fukuma, N., Wabayashi, N., Kiso, Y., Furmaniak, J., and Rees Smith, B. (1992). Production and characterization of a human monoclonal thyroid peroxidase autoantibody. *Autoimmunity* **14,** 1–7.

Jeffreys, J., Depraetere, H., Sanders, J., Oda, Y., Evans, M., Kiddie, A., Richards, T., Furmaniak, J., and Rees Smith, B. (2002). Characterization of the thyrotropin binding pocket. *Thyroid* **12,** 1051–1061.

Kleinau, G., and Krause, G. (2009). Thyrotropin and homologous glycoprotein hormone receptors: Structural and functional aspects of extracellular signaling mechanisms. *Endocr. Rev.* **30,** 133–151.

Kohn, L. D., Suzuki, K., Hoffman, W. H., Tombaccini, D., Marcocci, C., Shimojo, N., Watanabe, Y., Amino, N., Cho, B. Y., Kohno, Y., Hirai, A., and Tahara, K. (1997). Characterization of monoclonal thyroid-stimulating and thyrotropin binding-inhibiting autoantibodies from a Hashimoto's patient whose children had intrauterine and neonatal thyroid disease. *J. Clin. Endocrinol. Metab.* **82,** 3998–4009.

Kopp, P., Muirhead, S., Jourdain, N., Gu, W.-X., Jameson, J. L., and Rodd, C. (1997). Congenital hyperthyroidism caused by solitary toxic adenoma harboring a novel somatic mutation (serine281 to isoleucine) in the extracellular domain of the thyrotropin receptor. *J. Clin. Invest.* **100,** 1634–1639.

Kosugi, S., Ban, T., Akamizu, T., and Kohn, L. D. (1991). Site-directed mutagenesis of a portion of the extracellular domain of the rat thyrotropin receptor important in autoimmune thyroid disease and nonhomologous with gonadotropin receptors. Relationship of functional and immunogenic domains. *J. Biol. Chem.* **266,** 19413–19418.

Laskowski, R. A., MacArthur, M. W., Moss, D. S., and Thornton, J. M. (1993). PROCHECK—A program to check the stereochemical quality of protein structures. *J. Appl. Crystallogr.* **26,** 283–291.

Lenzner, C., and Morgenthaler, N. G. (2003). The effect of thyrotropin-receptor blocking antibodies on stimulating autoantibodies from patients with Graves' disease. *Thyroid* **13,** 1153–1161.

Luthy, R., Bowie, J. U., and Eisenberg, D. (1992). Assessment of protein models with 3-dimensional profiles. *Nature* **356,** 83–85.

McKenzie, J. M., and Zakarija, M. (1992). Fetal and neonatal hyperthyroidism and hypothyroidism due to maternal TSH receptor antibodies. *Thyroid* **2,** 155–159.

McLachlan, S. M., and Rapoport, B. (1996). Monoclonal, human autoantibodies to the TSH receptor—The Holy Grail and why are we looking for it? *J. Clin. Endocrinol. Metab.* **81,** 3152–3154.

Minich, W. B., Lenzner, C., and Morgenthaler, N. G. (2004). Antibodies to TSH-receptor in thyroid autoimmune disease interact with monoclonal antibodies whose epitopes are broadly distributed on the receptor. *Clin. Exp. Immunol.* **136,** 129–136.

Morgenthaler, N. G., Kim, M. R., Tremble, J., Huang, G. C., Richter, W., Gupta, M., Scherbaum, W. A., McGregor, A. M., and Banga, J. P. (1996). Human immunoglobulin G autoantibodies to the thyrotropin receptor from Epstein-Barr virus-transformed B lymphocytes: Characterization by immunoprecipitation with recombinant antigen and biological activity. *J. Clin. Endocrinol. Metab.* **81,** 3155–3161.

Moriyama, K., Okuda, J., Saijo, M., Hattori, Y., Kanamoto, N., Hataya, Y., Matsuda, F., Mori, T., Nakao, K., and Akamizu, T. (2003). Recombinant monoclonal thyrotropin-stimulation blocking antibody (TSBAb) established from peripheral lymphocytes of a hypothyroid patient with primary myxedema. *J. Endocrinol. Invest.* **26,** 1076–1080.

Mueller, S., Kleinau, G., Jaeschke, H., Paschke, R., and Krause, G. (2008). Extended hormone binding site of the human thyroid stimulating hormone receptor: Distinctive acidic residues in the hinge region are involved in bovine thyroid stimulating hormone binding and receptor activation. *J. Biol. Chem.* **283,** 18048–18055.

Mueller, S., Kleinau, G., Szkudlinski, M. W., Jaeschke, H., Krause, G., and Paschke, R. (2009). The superagonistic activity of bovine thyroid-stimulating hormone (TSH) and the human TR1401 TSH analog is determined by specific amino acids in the hinge region of the human TSH receptor. *J. Biol. Chem.* **284,** 16317–16324.

Núñez Miguel, R., Sanders, J., Jeffreys, J., Depraetere, H., Evans, M., Richards, T., Blundell, T. L., Rees Smith, B., and Furmaniak, J. (2004). Analysis of the thyrotropin receptor–thyrotropin interaction by comparative modeling. *Thyroid* **14,** 991–1011.

Núñez Miguel, R., Sanders, J., Blundell, T. L., Rees Smith, B., and Furmaniak, J. (2005). Comparative modeling of the thyrotropin receptor. *Thyroid* **15,** 746–747.

Núñez Miguel, R., Sanders, J., Chirgadze, D. Y., Blundell, T. L., Furmaniak, J., and Rees Smith, B. (2008). FSH and TSH binding to their respective receptors: Similarities, differences and implication for glycoprotein hormone specificity. *J. Mol. Endocrinol.* **41,** 145–164.

Núñez Miguel, R., Sanders, J., Chirgadze, D. Y., Furmaniak, J., and Rees Smith, B. (2009). Thyroid stimulating autoantibody M22 mimics TSH binding to the TSH receptor leucine rich domain: A comparative structural study of protein–protein interactions. *J. Mol. Endocrinol.* **42,** 381–395.

Oda, Y., Sanders, J., Roberts, S., Maruyama, M., Kato, R., Perez, M., Petersen, V. B., Wedlock, N., Furmaniak, J., and Rees Smith, B. (1998). Binding characteristics of antibodies to the TSH receptor. *J. Mol. Endocrinol.* **20,** 233–244.

Oda, Y., Sanders, J., Evans, M., Kiddie, A., Munkley, A., James, C., Richards, T., Wills, J., Furmaniak, J., and Rees Smith, B. (2000). Epitope analysis of the human thyrotropin (TSH) receptor using monoclonal antibodies. *Thyroid* **10,** 1051–1059.

Okuda, J., Akamizu, T., Sugawa, H., Matsuda, F., Hua, L., and Mori, T. (1994). Preparation and characterization of monoclonal antithyrotropin receptor antibodies obtained from peripheral lymphocytes of hypothyroid patients with primary myxedema. *J. Clin. Endocrinol. Metab.* **79,** 1600–1604.

Parma, J., Duprez, L., Van Sande, J., Cochaux, P., Gervy, C., Mockel, J., Dumont, J., and Vassart, G. (1993). Somatic mutations in the thyrotropin receptor gene cause hyperfunctioning thyroid adenomas. *Nature* **365,** 649–651.

Parma, J., Van Sande, J., Swillens, S., Tonacchera, M., Dumont, J., and Vassart, G. (1995). Somatic mutations causing constitutive activity of the thyrotropin receptor are the major cause of hyperfunctioning thyroid adenomas: Identification of additional mutations activating both the cyclic adenosine 3′,5′-monophosphate and inositol phosphate-Ca^{2+} cascades. *Mol. Endocrinol.* **9,** 725–733.

Rapoport, B., Chazenbalk, G. D., Jaume, J. C., and McLachlan, S. M. (1998). The thyrotropin (TSH) receptor: Interaction with TSH and autoantibodies. *Endocr. Rev.* **19,** 673–716.

Rees Smith, B., McLachlan, S. M., and Furmaniak, J. (1988). Autoantibodies to the thyrotropin receptor. *Endocr. Rev.* **9,** 106–121.

Rees Smith, B., Sanders, J., and Furmaniak, J. (2007). TSH receptor antibodies. *Thyroid* **17,** 923–938.

Rees Smith, B., Sanders, J., Evans, M., Tagami, T., and Furmaniak, J. (2009). TSH receptor–autoantibody interactions. *Horm. Metab. Res.* **41,** 448–455.

Sali, A., and Blundell, T. L. (1989). Definition of general topological equivalence in protein structures. *J. Mol. Biol.* **212,** 403–428.

Sanders, J., Oda, Y., Roberts, S. A., Maruyama, M., Furmaniak, J., and Rees Smith, B. (1997). Understanding the thyrotropin receptor function–structure relationship. *Baillières Clin. Endocrinol. Metab.* **11,** 451–479.

Sanders, J., Oda, Y., Roberts, S., Kiddie, A., Richards, T., Bolton, J., McGrath, V., Walters, S., Jaskólski, D., Furmaniak, J., and Rees Smith, B. (1999). The interaction of TSH receptor autoantibodies with ^{125}I-labelled TSH receptor. *J. Clin. Endocrinol. Metab.* **84,** 797–802.

Sanders, J., Jeffreys, J., Depraetere, H., Richards, T., Evans, M., Kiddie, A., Brereton, K., Groenen, M., Oda, Y., Furmaniak, J., and Rees Smith, B. (2002). Thyroid-stimulating monoclonal antibodies. *Thyroid* **12,** 1043–1050.

Sanders, J., Evans, M., Premawardhana, L. D. K. E., Depraetere, H., Jeffreys, J., Richards, T., Furmaniak, J., and Rees Smith, B. (2003). Human monoclonal thyroid stimulating autoantibody. *Lancet* **362,** 126–128.

Sanders, J., Jeffreys, J., Depraetere, H., Evans, M., Richards, T., Kiddie, A., Brereton, K., Premawardhana, L. D., Chirgadze, D. Y., Núñez Miguel, R., Blundell, T. L., Furmaniak, J., *et al.* (2004). Characteristics of a human monoclonal autoantibody to the thyrotropin receptor: Sequence structure and function. *Thyroid* **14,** 560–570.

Sanders, J., Allen, F., Jeffreys, J., Bolton, J., Richards, T., Depraetere, H., Nakatake, N., Evans, M., Kiddie, A., Premawardhana, L. D. K. E., Chirgadze, D. Y., Miguel, R. N., *et al.* (2005). Characteristics of a monoclonal antibody to the thyrotropin receptor that acts as a powerful thyroid-stimulating autoantibody antagonist. *Thyroid* **15,** 672–682.

Sanders, J., Bolton, J., Sanders, P., Jeffreys, J., Nakatake, N., Richards, T., Evans, M., Kiddie, A., Summerhayes, S., Roberts, E., Miguel, R. N., Furmaniak, J., *et al.* (2006). Effects of TSH receptor mutations on binding and biological activity of monoclonal antibodies and TSH. *Thyroid* **16,** 1195–1206.

Sanders, J., Chirgadze, D. Y., Sanders, P., Baker, S., Sullivan, A., Bhardwaja, A., Bolton, J., Reeve, M., Nakatake, N., Evans, M., Richards, T., Powell, M., *et al.* (2007a). Crystal structure of the TSH receptor in complex with a thyroid-stimulating autoantibody. *Thyroid* **17,** 395–410.

Sanders, J., Núñez Miguel, R., Bolton, J., Bhardwaja, A., Sanders, P., Nakatake, N., Evans, M., Furmaniak, J., and Rees Smith, B. (2007b). Molecular interactions between the TSH receptor and a thyroid-stimulating monoclonal autoantibody. *Thyroid* **17,** 699–706.

Sanders, J., Evans, M., Betterle, C., Sanders, P., Bhardwaja, A., Young, S., Roberts, E., Wilmot, J., Richards, T., Kiddie, A., Small, K., Platt, H., *et al.* (2008). A human monoclonal autoantibody to the thyrotropin receptor with thyroid-stimulating blocking activity. *Thyroid* **18,** 735–746.

Shepherd, P. S., Da Costa, C. R., Cridland, J. C., Gilmore, K. S., and Johnstone, A. P. (1999). Identification of an important thyrotrophin binding site on the human thyrotrophin receptor using monoclonal antibodies. *Mol. Cell. Endocrinol.* **149,** 197–206.

Valente, W. A., Vitti, P., Yavin, Z., Yavin, E., Rotella, C. M., Grollman, E. F., Toccafondi, R. S., and Kohn, L. D. (1982). Monoclonal antibodies to the thyrotropin receptor: Stimulating and blocking antibodies derived from the lymphocytes of patients with Graves' disease. *Proc. Natl. Acad. Sci. USA* **79,** 6680–6684.

Vlase, H., Weiss, M., Graves, P. N., and Davies, T. F. (1998). Characterization of the murine immune response to the murine TSH receptor ectodomain: Induction of hypothyroidism and TSH receptor antibodies. *Clin. Exp. Immunol.* **113,** 111–118.

Yoshida, T., Ichikawa, Y., Ito, K., and Homma, M. (1988). Monoclonal antibodies to the thyrotropin receptor bind to a 56-kDa subunit of the thyrotropin receptor and show heterogeneous bioactivities. *J. Biol. Chem.* **263,** 16341–16347.

CHAPTER TWENTY-THREE

Current Standards, Variations, and Pitfalls for the Determination of Constitutive TSHR Activity *In Vitro*

Sandra Mueller, Holger Jaeschke, *and* Ralf Paschke

Contents

Abstract

Constitutively activating mutations of the TSHR are the major cause for non-autoimmune hyperthyroidism, which is based on ligand independent, permanent receptor activation. Several reports have highlighted the difficulties to determine whether a TSHR mutation is constitutively active or not especially for borderline cases with only a slight increase of the basal cAMP activity. Current methods to precisely classify such mutants as constitutively active or not, are limited. In some cases, *in vitro* characterization of TSHR mutants has led to false positive conclusions regarding constitutive TSHR activity and subsequently the molecular origin of hyperthyroidism. For characterization of constitutive TSHR activity, a particular point to consider is that basal receptor activity tightly correlates with the receptor number expressed on the cell surface. Therefore,

Department for Internal Medicine, Neurology and Dermatology; Clinic for Endocrinology and Nephrology, University of Leipzig, Leipzig, Germany

Methods in Enzymology, Volume 485
ISSN 0076-6879, DOI: 10.1016/S0076-6879(10)85023-2

a comparison of the receptors basal activity in relation to the wild type is only possible with determination of the receptor cell surface expression. Thus, the experimental approaches to determine constitutive TSHR activity should consider the receptor's cell surface expression.

We here provide a description of three methods for the determination of constitutive TSHR activity: (A) the evaluation of constitutive TSHR activity under conditions of equal receptor expression; (B) computation of the specific constitutive activity; and (C) the linear regression analysis (LRA). To date, LRA is the best experimental approach to characterize the mutant's basal activity as a function of TSHR cell surface expression. This approach utilizes a parallel measurement of basal cAMP values and receptor cell surface expression and therefore provides a more reliable decision with respect to the presence or absence of constitutive activity.

1. Introduction

Ligand independent, basal receptor signaling, also referred to as constitutive receptor activity, has not only been described for glycoprotein hormone receptors (GPHRs) but also for numerous G-protein-coupled receptors (Liebmann, 2004; Rosenbaum *et al.*, 2009; Seifert and Wenzel-Seifert, 2002; Smit *et al.*, 2007). Compared to the other GPHRs, follicle stimulating hormone receptor (FSHR) and luteinizing hormone/choriogonadotropin receptor (LHCGR), the thyroid-stimulating hormone receptor (TSHR) is much more susceptible to constitutive activity by receptor mutations (Cetani *et al.*, 1996; Dufau, 1998; Parma *et al.*, 1993; Simoni *et al.*, 1997; Szkudlinski *et al.*, 2002). Such constitutively activating mutations (CAMs) of the TSHR are the major cause for nonautoimmune hyperthyroidism, which is based on ligand independent, permanent activation of the $G_{\alpha}s$- (cAMP), and in some cases, also the $G_{\alpha}q$-(IP)-mediated signaling pathway (Duprez *et al.*, 1998; Parma *et al.*, 1994).

In vitro determination of the constitutive activity of mutant TSHRs especially for borderline cases with only a slight increase of the basal cAMP activity is difficult (Mueller *et al.*, 2009). This is supported by divergent *in vitro* data for the level of constitutive activity of the same mutation, which were published for several constitutively activating TSHR mutations (Kleinau *et al.*, 2007), although the same cell system was used (Table 23.1). Such divergent *in vitro* results may be related to differences in assay conditions like different cell count, transfected DNA amount, transfection reagents, which can influence transfection efficiency, as well as different cAMP (3′-5′-cyclic adenosine monophosphate) measuring systems.

It is very likely that mutants with pronounced constitutive activity lead to consistent results and interpretation regarding their constitutive activity

Table 23.1 Divergent increases of the basal activity for selected TSHR mutations described in different studies using the COS-7 cell system

TSHR mutation	Constitutive activity fold over wt TSHR	References
S281L	2.7–5.4	Claeysen *et al.* (2002), Jaeschke *et al.* (2006), Kleinau *et al.* (2007)
S281N	2.2–5.3	Duprez *et al.* (1997), Gruters *et al.* (1998), Ho *et al.* (2001), Jaeschke *et al.* (2006), Kleinau *et al.* (2007), Neumann *et al.* (2005a)
I486F	4.0–8.2	Claeysen *et al.* (2002), Kleinau *et al.* (2007), Neumann *et al.* (2005a), Parma *et al.* (1995)
L629F	2.1–6.5	Fuhrer *et al.* (1997b), Kleinau *et al.* (2007), Parma *et al.* (1997), Tonacchera *et al.* (1998), Wonerow *et al.* (2000)
I630L	2.5–6.4	Holzapfel *et al.* (1997), Kleinau *et al.* (2007), Tonacchera *et al.* (1998), Wonerow *et al.* (2000)
V656F	4.6–7.1	Fuhrer *et al.* (1997a), Kleinau *et al.* (2007), Neumann *et al.* (2005a), Wonerow *et al.* (2000)

despite variable data in different publications. In contrast, divergent and ambiguous results could well lead to variable classifications regarding constitutive activity for TSHR mutants with only a slight increase of the basal activity. Indeed, determination of basal mutant TSHR activity of such borderline cases with only a slightly increased level of basal cAMP activity previously led to different classification regarding constitutive activation (Table 23.2) if analyzed in different cell–vector systems (Jaeschke *et al.*, 2010; Mueller *et al.*, 2009; Neumann *et al.*, 2001; Niepomniszcze *et al.*, 2006; Russo *et al.*, 1999, 2000; Tonacchera *et al.*, 1996).

In detail, TSHR mutation N670S previously classified as CAM (Neumann *et al.*, 2001; Tonacchera *et al.*, 1996) showed constitutive activity when tested in HEK_{GT} (human embryonic kidney 293 grip tight cells) but not in COS-7 cells using the same assay system (Mueller *et al.*, 2009). Moreover, reexamination of the basal cAMP activity of mutation R310C after transient expression in COS-7 and also in HEK 293 cells did not show constitutive activity for R310C (Mueller *et al.*, 2009) in contrast to the previous characterization in CHO cells stably expressing the mutated receptor construct (Russo *et al.*, 2000). Such differences regarding the basal cAMP level using stable versus transient expression were also be observed for TSHR mutations L677V and T620I (Table 23.2). The previous classification of R310C (Russo *et al.*, 2000),

Table 23.2 Divergent classification of selected TSHR mutations regarding constitutive receptor activity based on different functional *in vitro* characterization and cell systems

TSHR mutation	Basal cAMP accumulation fold over wt TSHR	Assay type	Cell system	Constitutive activity	References
R310C	2.0	Basal cAMP	Stable CHO cells	Yes	Russo *et al.* (2000)
	1.1	LRA[a]	Transient transfection in COS-7	No	Mueller *et al.* (2009)
	0.5	LRA[a]	Transient transfection in HEK_{GT}	No	Mueller *et al.* (2009)
N670S	1.7	Basal cAMP and SCA[b]	Transient transfection in COS-7	Yes	Tonacchera *et al.* (1996)
	2.4	Basal cAMP	Transient transfection in COS-7	Yes	Neumann *et al.* (2001)
	1.1	LRA[a]	Transient transfection in COS-7	No	Mueller *et al.* (2009)
	2.5	LRA[a]	Transient transfection in HEK_{GT}	Yes	Mueller *et al.* (2009)
T620I	3.0–5.0	Basal cAMP	Stable 3T3-Vill cells	Yes	Niepomniszcze *et al.* (2006)
	0.4	LRA[a]	Transient transfection in COS-7	No	Jaeschke *et al.* (2010)
L677V	3.4	Basal cAMP	Stable CHO cells	Yes	Russo *et al.* (1999)
	1.4	LRA[a]	Transient transfection in COS-7	No	Jaeschke *et al.* (2010)

[a] Linear regression analysis of constitutive activity.
[b] Specific constitutive activity.

T620I (Niepomniszcze *et al.*, 2006), and L677V (Russo *et al.*, 1999) as CAMs is most likely due to their characterization in cell lines stably expressing the mutated receptor construct without assessing the respective receptor numbers per cell. However, the basal activity of GPCRs tightly correlates in a linear relationship with the number of receptors expressed on the cell surface (Vlaeminck-Guillem *et al.*, 2002). Therefore, the basal cAMP value without the consideration of the cell surface expression is only a preliminary indication for the constitutive activity of TSHR mutations. A comparison of the receptor's basal activity in relation to the wild type (wt) is only possible with determination of the receptor's cell surface expression. Thus, the experimental approaches to determine constitutive TSHR activity should consider the receptor's cell surface expression. However, the *in vitro* methods to precisely classify mutants with only a slight increase of the basal activity as constitutively active or not are limited. We here provide a description of the best current approaches to characterize constitutive TSHR activity (Table 23.3).

2. Detection of Constitutive TSHR Activity

There is the possiblility to express the TSHR mutants and wt at the same level to determine the basal cAMP activity at comparable cell surface expressions (Neumann *et al.*, 2001; Table 23.3). The TSHR cell surface expression can be measured by fluorescence activated cell sorting (FACS) using an appropriate anti-TSHR antibody or by the labeled ligand TSH (thyroid stimulating hormone) binding which results in the determination of Bmax (maximum number of binding sites) which should correlate with the respective FACS value assuming a one site binding model. Alternatively, the specific constitutive activity (SCA) can be computed based on the mutant's basal TSHR activity and cell surface expression resulting in a constitutive activity value normalized to the cell surface expression (Agretti *et al.*, 2003; Duprez *et al.*, 1997; Govaerts *et al.*, 2001; Tonacchera *et al.*, 1996; Urizar *et al.*, 2005; Vlaeminck-Guillem *et al.*, 2002; Wonerow *et al.*, 2000). Furthermore, the constitutive receptor activity (not only for the TSHR) can be determined by linear regression analysis (LRA), which combines measuring basal cAMP activity and receptor cell surface expression measured by FACS or Bmax in parallel (Ballesteros *et al.*, 2001; De Leener *et al.*, 2006; Hjorth *et al.*, 1998; Jaeschke *et al.*, 2006; Kleinau *et al.*, 2008; Montanelli *et al.*, 2004; Mueller *et al.*, 2009; Van Sande *et al.*, 1995; Vlaeminck-Guillem *et al.*, 2002). The resulting slopes are to date the best parameter to determine whether a receptor is constitutively active or not.

Table 23.3 Previously published methods to determine constitutive TSHR activity by *in vitro* characterization considering receptor cell surface expression

Method	Short description	References
Measuring basal cAMP using the same receptor expression level	– Reducing the cell surface expression of the wt and mutant TSHR to the level of the mutant with the lowest cell surface expression by cotransfection with plasmids containing, for example, another receptor – Measuring basal cAMP values and comparison to the wt TSHR	Neumann *et al.* (2001)
Determination of the specific constitutive activity (SCA)	– Computation of the specific constitutive activity of TSHR mutants based on the basal cAMP activity value normalized to the cell surface expression measured by FACS analysis or TSH binding (Bmax) – Comparison of the mutant cAMP value versus wt TSHR	Agretti *et al.* (2003), Govaerts *et al.* (2001), Tonacchera *et al.* (1996), Urizar *et al.* (2005), Vlaeminck-Guillem *et al.* (2002), Wonerow *et al.* (2000)
Linear regression analysis (LRA)	– Determination of basal cAMP activity and receptor cell surface expression (measured by FACS or Bmax) in parallel using increasing DNA amounts for transfection – Plotting of the basal activity as a function of receptor expression – The resulting slope represents the constitutive activity level in comparison to the wt TSHR	Jaeschke *et al.* (2006), Kleinau *et al.* (2008), Mueller *et al.* (2009), Van Sande *et al.* (1995), Vlaeminck-Guillem *et al.* (2002)

2.1. Evaluation of constitutive TSHR activity setting up an equal receptor expression

For comparison of the constitutive activity level of wt and several TSHR mutants with different cell surface expression levels, the basal cAMP activity has to be determined at comparable expression values. Therefore, the cell surface expression of all mutants and wt is reduced to the level of the mutant with the lowest expression (Neumann *et al.*, 2001). The TSHR constructs expressed at a higher level have to be cotransfected at first with various amounts of plasmids containing, for example, another receptor like the vasopressin receptor (Neumann *et al.*, 2001) to decrease the expression of the respective mutant with higher expression to the level of the one with the lowest expression. After setting up the cotransfected plasmid amount for each mutant and the wt, the basal cAMP values can be measured and compared to the wt TSHR to evaluate the constitutive activity level of each mutant. However, using another GPCR like the vasopressin receptor for cotransfection may lead to an interference and subsequently to a falsification of the measured intracellular signal, which can influence the evaluation of TSHR constitutive activity. Therefore, the better option is to use a non-GPCR cDNA construct.

2.2. Computation of the specific constitutive activity

The constitutive activity of TSHR mutants can also be computed based on the basal cAMP or IP value, which is normalized to the cell surface expression measured by TSH binding (Bmax) (Tonacchera *et al.*, 1996) or FACS analysis (Agretti *et al.*, 2003; Govaerts *et al.*, 2001; Tonacchera *et al.*, 1996; Urizar *et al.*, 2005; Vlaeminck-Guillem *et al.*, 2002; Wonerow *et al.*, 2000). Therefore, the transfection efficiency for each construct has to be kept constant for a given batch of cells. To compute the SCA, the basal cAMP (or IP) as well as the cell surface expression via Bmax or FACS analysis have to be determined. The SCA value of each mutant is calculated by the formula shown in Fig. 23.1. In detail, for the evaluation of the mutant's SCA, the basal cAMP

$$SCA_{mutant} = (cAMP_{mutant} - cAMP_{vector}) / (MFU_{mutant} - MFU_{vector})$$

$$SCA_{wt\_normalized} = SCA_{mutant} / SCA_{wt}$$

Figure 23.1 Computation of specific constitutive activity of TSHR mutants adapted from Agretti *et al.* (2003), Govaerts *et al.* (2001), Urizar *et al.* (2005), Vlaeminck-Guillem *et al.* (2002). SCA, specific constitutive activity; cAMP, basal cAMP value; MFU, mean fluorescence unit measured by FACS analysis. For comparison of the constitutive activity of TSHR mutants with each other, the SCA value should be normalized to the wt TSHR. SCA can also be performed for IP instead of cAMP.

$$cAMP_{mutant\_TE} = [cAMP_{mutant} - cAMP_{vector} \times (1 - TE)] / cAMP_{vector} \times TE$$

$$RCA_{mutant} = cAMP_{mutant\_TE} / cAMP_{wt\_TE}$$

$$Expr_{mutant\_wt} = (Expr_{mutant} - Expr_{vector}) / (Expr_{wt} - Expr_{vector})$$

$$SCA_{mutant} = RCA_{mutant} / Expr_{mutant\_wt}$$

Figure 23.2 Formula for evaluation of specific constitutive activity (SCA) of TSHR mutants considering transfection efficiency and receptor expression adapted to Duprez *et al.* (1997), Wonerow *et al.* (2000). $cAMP_{mutant\ or\ vector}$: mutant's or empty vector's basal cAMP value; $cAMP_{mutant\_TE}$ is the mutant's basal cAMP accumulation considering transfection efficiency; TE: transfection efficiency, defined as the fraction of cells having taken and amplified the vector-based constructs; RCA_{mutant}: relative constitutive activity of the mutant receptor over the wt TSHR; Expr: expression; $Expr_{mutant\_wt}$: expression of the mutant receptor relative to the wt; SCA_{mutant}: mutant's SCA.

value of receptor transfected cells minus the basal cAMP of empty vector transfected cells has to be divided by the fluorescence value of receptor transfected cells minus the fluorescence value of empty vector transfected cells. This formula can also be used for calculation of SCA using Bmax instead of FACS (Tonacchera *et al.*, 1996) or IP instead of cAMP. For comparison of the mutant SCA values with the wt TSHR, they have to be normalized to the SCA value of the wt TSHR leading to the determination of the increase in the SCA over that of the wt TSHR.

Another computation method for SCA (Fig. 23.2) considers apart from the TSHR cell surface expression also the transfection efficiency both measured by FACS analysis (Duprez *et al.*, 1997; Wonerow *et al.*, 2000). More precisely, the basal cAMP accumulation of the mutant considering transfection efficiency ($cAMP_{mutant\_TE}$) is thus the ratio between the cAMP level in cells expressing the TSHR mutant ($cAMP_{mutant}$), over the basal cAMP in the same number of cells transfected with the empty vector ($cAMP_{vector}$). The mutant's relative constitutivity (RCA_{mutant}) is obtained by $cAMP_{mutant\_TE}$ in comparison to the wt, whereas the mutant's specific constitutive activity (SCA_{mutant}) is expressed as the ratio of RCA_{mutant} and realtive mutant receptor expression ($Expr_{mutant\_wt}$). The comparison of the SCA values calculated for each TSHR mutant determines the fold increase in constitutivity of each mutant over the wt TSHR if their expression levels are identical.

2.3. Determination of constitutive TSHR activity by linear regression analysis

The currently best experimental approach to characterize the basal TSHR activity independently from the receptor's cell surface expression is the determination of the LRA (Jaeschke *et al.*, 2006; Kleinau *et al.*, 2008;

Mueller *et al.*, 2009; Van Sande *et al.*, 1995; Vlaeminck-Guillem *et al.*, 2002). LRA combines parallel measuring of basal cAMP signaling and cell surface expression in the same experiment and therefore allows a more reliable decision with respect to the presence or absence of constitutive activity. For LRA, cells (see Section 3) are transfected with increasing DNA amounts of TSHR mutants and wt, respectively (see Section 3), in a parallel approach for the determination of basal cAMP activity as well as cell surface expression measured by FACS analysis (see Section 3) or TSH binding (see Section 3).

Using the linear regression function of Graph Pad Prism for Windows, LRA is determined by plotting the basal cAMP accumulation minus the empty vector value (*y*-axis) as a function of receptor expression minus the empty vector value (*x*-axis), resulting in a linear increase of basal cAMP accumulation for each mutant or wt TSHR, represented as a slope (Fig. 23.3). The slope of the TSHR mutants is calculated by the equation: slope = delta Y/delta X. Based on this analysis, a TSHR mutation is constitutively active if its slope is significantly increased compared to the wt. For direct comparison of the mutants with each other, the slope of the wt TSHR is set at 1 due to the presence of constitutive wt TSHR activity (Cetani *et al.*, 1996) and the slopes of the mutants are calculated according to this. However, the limitations of this method to precisely classify a TSHR mutation with only a slight increase in basal cAMP activity as consitutively active or not have previously been shown (Mueller *et al.*, 2009).

3. Methods and Required Materials for LRA

3.1. Cell culture and transient expression of mutant TSHRs

Based on our experience, we recommend the use of COS-7 cells as the most prominent cell system for transient transfection to functionally characterize TSHR mutations *in vitro*. COS-7 cells are cultured in Dulbecco's modified Eagle's Medium (DMEM) supplemented with 10% fetal bovine serum, 100 U/ml penicillin, and 100 μg/ml streptomycin (e.g., Gibco Life technologies, Paisley, UK) at 37 °C in a humidified 5% CO_2 incubator and are transiently transfected with plasmids (e.g., pcDNA) containing wt or mutant TSHRs using, for example, the GeneJammer® Transfection Reagent (Stratagene, Amsterdam, NL) according to the manufacturer's instructions. COS-7 cells are transfected in 24-well plates (0.5×10^5 cells per well) with increasing concentrations of wt and mutant TSHR constructs (e.g., 50, 100, 200, 300, 400, and 500 ng DNA per well) for bTSH binding or FACS analysis and cAMP accumulation in parallel. To transfect a constant DNA amount of 500 ng, empty vector plasmid DNA was added to the mutant or wt DNA. However, it has been demonstrated that the

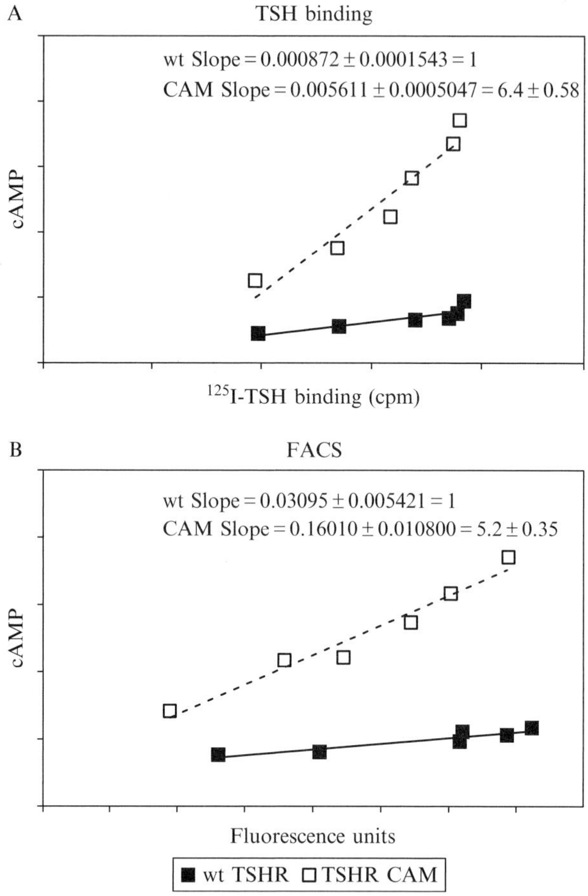

Figure 23.3 Representative example of linear regression analysis (LRA) of constitutive activity as a function of TSHR expression investigated using COS-7 cells and transient transfection with increasing amounts of DNA encoding the wt TSHR or the TSHR constitutively activating mutation (CAM). Constitutive activity of the TSHR CAM is shown by a steep increment of the slope, which is expressed as basal cAMP formation (*y*-axis) as a function of receptor expression determined by (A) ^{125}I-bTSH binding or (B) FACS analysis (*x*-axis). Slopes are calculated by LRA using the equation: slope = delta *Y*/delta *X* of Graph Pad Prism 4.0 for Windows. The slope of TSHR wt is set at 1 due to the presence of constitutive wt TSHR activity and the slope of the TSHR CAM is calculated according to this.

characterization of TSHR mutations regarding their constitutive activity should consider the influence of the *in vitro* cell and/or vector system on the basal cAMP accumulation (Mueller *et al.*, 2009). In detail, the study showed that the *in vitro* characterization by LRA of TSHR mutation with only a slight increase in basal cAMP activity may lead to different classifications regarding constitutive activity when tested in different combinations of the cell–vector system (Mueller *et al.*, 2009). In contrast, mutants with pronounced constitutive activity most likely lead to consistent results and interpretation regarding their constitutive activity independent from the cell–vector system.

Results of previous *in vitro* classification of TSHR mutants obtained with stable transfection of cells differ regarding constitutive activity from transient transfection experiments (Table 23.2). This is most likely due to differences in receptor number on the cell surface between mutants and wt TSHR using stable cell clones. Thus, using stable transfected cells for the determination of constitutive TSHR activity necessarily requires the determination of the respective TSHR copy number for comparison of the expression and basal cAMP level with the wt.

3.2. Determination of cell surface expression

The cell surface expression value necessary for evaluation of the constitutive activity of TSHR mutants can be determined by (i) FACS analysis or (ii) bTSH binding.

3.2.1. ^{125}I-bTSH (bovine thyroid stimulating hormone) binding assay

For radioligand binding assays, 48 h after transfection, cells are washed once with modified Hank's buffer (5.36 m*M* KCl, 0.44 m*M* KH_2PO_4, 0.41 m*M* $MgSO_4$, 0.33 m*M* Na_2HPO_4, 5.55 m*M* glucose) supplemented with 1.3 m*M* $CaCl_2$, 280 m*M* sucrose, 0.2% bovine serum albumin (BSA), and 2.5% milk powder. Cells are incubated in the same buffer in the presence of 140,000–180,000 cpm of ^{125}I-TSH (e.g., ^{125}I-bovine TSH: Brahms, Hennigsdorf, Germany; specific activity: approx. 50 μCi/μg) supplemented with 5 mU/ml nonlabeled TSH (e.g., bTSH: Sigma Chemical Co., St. Louis, MO) for 4 h. To avoid or minimize effects of internalization caused by incubation at room temperature, cells should be incubated at 4 °C. The empty plasmid vector is used as a control for nonspecific binding. Afterward, cells have to be washed with the same ice-cold buffer, solubilized with 500 μl 1 N NaOH, and radioactivity is measured in a gamma counter.

To exclude alterations of the affinity for TSH by mutations and possible effects of internalization LRA should be determined using FACS analysis instead of ^{125}I-bTSH binding.

However, for both LRA methods it cannot be excluded that TSHR mutations could alter both TSH binding and the recognition site of the anti-TSHR antibody. Discrepancies in cell surface expression determined by FACS and TSH binding have previously been observed for some TSHR mutants (Arseven *et al.*, 2000; Biebermann *et al.*, 1998; Claus *et al.*, 2005; Gozu *et al.*, 2005; Mueller *et al.*, 2006; Neumann *et al.*, 2001, 2005b).

Therefore, we recommend comparison of cell surface expression measured by FACS analysis and ^{125}I-TSH binding for each mutant.

3.2.2. FACS analysis

Cells are detached from the dishes 48 h after transfection using 1 m*M* PBS and transferred in Falcon 2054 tubes. Cells are washed once with PBS containing 0.1% BSA and incubated for 1 h with an appropriate primary anti-TSHR monoclonal antibody (e.g., a 1:400 dilution of the antibody 2C11: MAK 1281, Linaris, Wertheim-Bettingen; 10 μg/ml) in the same buffer at 4 °C. Cells are washed twice and incubated at 4 °C for 1 h in the dark with an appropriate secondary antibody (e.g., a 1:400 dilution of Alexa Fluor 488-conjugated goat anti-mouse IgG: Invitrogen, Oregon). Cells are washed twice and fixed with 1% paraformaldehyde before FACS analysis (e.g., FACscan Becton Dickinson and Co., Franklin Lakes, NJ). Receptor expression is determined by the mean fluorescence intensity (MFI). The percentage of signal positive cells corresponds to transfection efficiency, which should be approximately 50–60% of viable cells for each mutant in each experiment because it was shown that the detection of constitutive activity for TSHR mutants also requires a certain level of transfection efficiency (Mueller *et al.*, 2009).

3.3. cAMP accumulation assay

Measurement of basal cAMP accumulation is performed 48 h after transfection. Cells are incubated with serum-free DMEM without antibiotics containing 1 m*M* 3-isobutyl-1-methylxanthine (IBMX; e.g., Sigma Chemical Co.) for 1 h at 37 °C in a humified 5% CO_2 incubator. Subsequently, the reaction should be terminated by aspiration of the medium and cells have to be washed once with ice-cold PBS and then lysed by addition of 300 μl/well 0.1 N HCl for 30 min. After collecting supernatants into 1.5 ml tubes and drying for 24 h at 54 °C, cAMP content of the cell extracts is determined using for instance the cAMP AlphaScreen^{TM} Assay (PerkinElmer^{TM} Life Sciences, Zaventem, Belgium) according to the manufacturer's instructions.

4. Conclusion

Constitutive activity of the TSHR can be determined by the three described methods. However, evaluation of constitutive activity under conditions of an equal receptor expression using additional cotransfection of a functional cDNA (e.g., GPCRs, Arrestin) may lead to interference of the cotransfected protein with intracellular cell signaling or cell cycle mechanisms and therefore may result in changes of the measured constitutive activity level. Moreover, determination of SCA is a fairly crude (computation) procedure to determine constitutive receptor activity which lacks a parallel approach for the determination of basal cAMP and cell surface expression and therefore equal conditions like the same cell count or the same transfection efficiency. In contrast, such equal experimental conditions are used by LRA, which is therefore the current best approach to characterize the TSHR mutant's basal activity *in vitro*. Moreover, LRA can also be used for other GPCRs. This was previously shown for, for example, the FSHR (De Leener *et al.*, 2006; Montanelli *et al.*, 2004), β2-adrenergic receptor (Ballesteros *et al.*, 2001), 5-hydroxytryptamine-4 receptors (Pellissier *et al.*, 2009), and the glucagon receptor (Hjorth *et al.*, 1998). However, determination of the constitutive receptor activity by the three described methods depends on the availability of specific anti-receptor antibodies, which in many cases requires a relatively large extracellular receptor domain necessary for immunoglobulin binding at the intact cell surface or alternatively a ligand, which can be labeled for binding studies. For other receptors like GPCR orphan receptors, the only current option for the evaluation of constitutive activity is the determination of basal G-protein signaling as a function of increasing amounts of transfected DNA (Eggerickx *et al.*, 1995).

ACKNOWLEDGMENT

The research relevant to this chapter was supported by grants from the Deutsche Forschungsgemeinschaft (DFG/Pa 423/14-1, DFG/Ja 1927/1-1 and DFG/Pa 423/15-2).

REFERENCES

Agretti, P., De Marco, G., Collecchi, P., Chiovato, L., Vitti, P., Pinchera, A., and Tonacchera, M. (2003). Proper targeting and activity of a nonfunctioning thyroid-stimulating hormone receptor (TSHr) combining an inactivating and activating TSHr mutation in one receptor. *Eur. J. Biochem.* **270,** 3839–3847.

Arseven, O. K., Wilkes, W. P., Jameson, J. L., and Kopp, P. (2000). Substitutions of tyrosine 601 in the human thyrotropin receptor result in increase or loss of basal activation of the cyclic adenosine monophosphate pathway and disrupt coupling to Gq/11. *Thyroid* **10,** 3–10.

Ballesteros, J. A., Jensen, A. D., Liapakis, G., Rasmussen, S. G., Shi, L., Gether, U., and Javitch, J. A. (2001). Activation of the beta 2-adrenergic receptor involves disruption of an ionic lock between the cytoplasmic ends of transmembrane segments 3 and 6. *J. Biol. Chem.* **276,** 29171–29177.

Biebermann, H., Schoneberg, T., Schulz, A., Krause, G., Gruters, A., Schultz, G., and Gudermann, T. (1998). A conserved tyrosine residue (Y601) in transmembrane domain 5 of the human thyrotropin receptor serves as a molecular switch to determine G-protein coupling. *FASEB J.* **12,** 1461–1471.

Cetani, F., Tonacchera, M., and Vassart, G. (1996). Differential effects of NaCl concentration on the constitutive activity of the thyrotropin and the luteinizing hormone/chorionic gonadotropin receptors. *FEBS Lett.* **378,** 27–31.

Claeysen, S., Govaerts, C., Lefort, A., Van Sande, J., Costagliola, S., Pardo, L., and Vassart, G. (2002). A conserved Asn in TM7 of the thyrotropin receptor is a common requirement for activation by both mutations and its natural agonist. *FEBS Lett.* **517,** 195–200.

Claus, M., Jaeschke, H., Kleinau, G., Neumann, S., Krause, G., and Paschke, R. (2005). A hydrophobic cluster in the center of the third extracellular loop is important for thyrotropin receptor signaling. *Mol. Endocrinol.* **146,** 5197–5203.

De Leener, A., Montanelli, L., Van Durme, J., Chae, H., Smits, G., Vassart, G., and Costagliola, S. (2006). Presence and absence of follicle-stimulating hormone receptor mutations provide some insights into spontaneous ovarian hyperstimulation syndrome physiopathology. *J. Clin. Endocrinol. Metab.* **91,** 555–562.

Dufau, M. L. (1998). The luteinizing hormone receptor. *Annu. Rev. Physiol.* **60,** 461–496.

Duprez, L., Parma, J., Costagliola, S., Hermans, J., Van Sande, J., Dumont, J. E., and Vassart, G. (1997). Constitutive activation of the TSH receptor by spontaneous mutations affecting the N-terminal extracellular domain. *FEBS Lett.* **409,** 469–474.

Duprez, L., Parma, J., Van Sande, J., Rodien, P., Dumont, J. E., Vassart, G., and Abramowicz, M. (1998). TSH receptor mutations and thyroid disease. *Trends Endocrinol. Metab.* **9,** 133–140.

Eggerickx, D., Denef, J. F., Labbe, O., Hayashi, Y., Refetoff, S., Vassart, G., Parmentier, M., and Libert, F. (1995). Molecular cloning of an orphan G-protein-coupled receptor that constitutively activates adenylate cyclase. *Biochem. J.* **309** (Pt. 3), 837–843.

Fuhrer, D., Holzapfel, H. P., Wonerow, P., Scherbaum, W. A., and Paschke, R. (1997a). Somatic mutations in the thyrotropin receptor gene and not in the Gs alpha protein gene in 31 toxic thyroid nodules. *J. Clin. Endocrinol. Metab.* **82,** 3885–3891.

Fuhrer, D., Wonerow, P., Willgerodt, H., and Paschke, R. (1997b). Identification of a new thyrotropin receptor germline mutation (Leu629Phe) in a family with neonatal onset of autosomal dominant nonautoimmune hyperthyroidism. *J. Clin. Endocrinol. Metab.* **82,** 4234–4238.

Govaerts, C., Lefort, A., Costagliola, S., Wodak, S. J., Ballesteros, J. A., Van Sande, J., Pardo, L., and Vassart, G. (2001). A conserved Asn in transmembrane helix 7 is an on/off switch in the activation of the thyrotropin receptor. *J. Biol. Chem.* **276,** 22991–22999.

Gozu, H., Avsar, M., Bircan, R., Claus, M., Sahin, S., Sezgin, O., Deyneli, O., Paschke, R., Cirakoglu, B., and Akalin, S. (2005). Two novel mutations in the sixth transmembrane segment of the thyrotropin receptor gene causing hyperfunctioning thyroid nodules. *Thyroid* **15,** 389–397.

Gruters, A., Schoneberg, T., Biebermann, H., Krude, H., Krohn, H. P., Dralle, H., and Gudermann, T. (1998). Severe congenital hyperthyroidism caused by a germ-line neo mutation in the extracellular portion of the thyrotropin receptor. *J. Clin. Endocrinol. Metab.* **83,** 1431–1436.

Hjorth, S. A., Orskov, C., and Schwartz, T. W. (1998). Constitutive activity of glucagon receptor mutants. *Mol. Endocrinol.* **12,** 78–86.

Ho, S. C., Van Sande, J., Lefort, A., Vassart, G., and Costagliola, S. (2001). Effects of mutations involving the highly conserved S281HCC motif in the extracellular domain of the thyrotropin (TSH) receptor on TSH binding and constitutive activity. *Mol. Endocrinol.* **142,** 2760–2767.

Holzapfel, H. P., Fuhrer, D., Wonerow, P., Weinland, G., Scherbaum, W. A., and Paschke, R. (1997). Identification of constitutively activating somatic thyrotropin receptor mutations in a subset of toxic multinodular goiters. *J. Clin. Endocrinol. Metab.* **82,** 4229–4233.

Jaeschke, H., Neumann, S., Kleinau, G., Mueller, S., Claus, M., Krause, G., and Paschke, R. (2006). An aromatic environment in the vicinity of serine 281 is a structural requirement for thyrotropin receptor function. *Mol. Endocrinol.* **147,** 1753–1760.

Jaeschke, H., Mueller, S., Eszlinger, M., and Paschke, R. (2010). Lack of in vitro constitutive activity for three previously reported TSH-receptor (TSHR)-mutations identified in patiens with nonautoimmune hyperthyroidism. *Clin. Endocrinol.* (Oxf) in press.

Kleinau, G., Brehm, M., Wiedemann, U., Labudde, D., Leser, U., and Krause, G. (2007). Implications for molecular mechanisms of glycoprotein hormone receptors using a new sequence–structure–function analysis resource. *Mol. Endocrinol.* **21,** 574–580.

Kleinau, G., Jaeschke, H., Mueller, S., Raaka, B. M., Neumann, S., Paschke, R., and Krause, G. (2008). Evidence for cooperative signal triggering at the extracellular loops of the TSH receptor. *FASEB J.* **22,** 2798–2808.

Liebmann, C. (2004). G protein-coupled receptors and their signaling pathways: Classical therapeutical targets susceptible to novel therapeutic concepts. *Curr. Pharm. Des.* **10,** 1937–1958.

Montanelli, L., Van Durme, J. J., Smits, G., Bonomi, M., Rodien, P., Devor, E. J., Moffat-Wilson, K., Pardo, L., Vassart, G., and Costagliola, S. (2004). Modulation of ligand selectivity associated with activation of the transmembrane region of the human follitropin receptor. *Mol. Endocrinol.* **18,** 2061–2073.

Mueller, S., Kleinau, G., Jaeschke, H., Neumann, S., Krause, G., and Paschke, R. (2006). Significance of ectodomain cysteine boxes 2 and 3 for the activation mechanism of the thyroid-stimulating hormone receptor. *J. Biol. Chem.* **281,** 31638–31646.

Mueller, S., Gozu, H. I., Bircan, R., Jaeschke, H., Eszlinger, M., Lueblinghoff, J., Krohn, K., and Paschke, R. (2009). Cases of borderline in vitro constitutive thyrotropin receptor activity: How to decide whether a thyrotropin receptor mutation is constitutively active or not? *Thyroid* **19,** 765–773.

Neumann, S., Krause, G., Chey, S., and Paschke, R. (2001). A free carboxylate oxygen in the side chain of position 674 in transmembrane domain 7 is necessary for TSH receptor activation. *Mol. Endocrinol.* **15,** 1294–1305.

Neumann, S., Claus, M., and Paschke, R. (2005a). Interactions between the extracellular domain and the extracellular loops as well as the 6th transmembrane domain are necessary for TSH receptor activation. *Eur. J. Endocrinol.* **152,** 625–634.

Neumann, S., Krause, G., Claus, M., and Paschke, R. (2005b). Structural determinants for g protein activation and selectivity in the second intracellular loop of the thyrotropin receptor. *Mol. Endocrinol.* **146,** 477–485.

Niepomniszcze, H., Suarez, H., Pitoia, F., Pignatta, A., Danilowicz, K., Manavela, M., Elsner, B., and Bruno, O. D. (2006). Follicular carcinoma presenting as autonomous functioning thyroid nodule and containing an activating mutation of the TSH receptor (T620I) and a mutation of the Ki-RAS (G12C) genes. *Thyroid* **16,** 497–503.

Parma, J., Duprez, L., Van Sande, J., Cochaux, P., Gervy, C., Mockel, J., Dumont, J., and Vassart, G. (1993). Somatic mutations in the thyrotropin receptor gene cause hyperfunctioning thyroid adenomas. *Nature* **365,** 649–651.

Parma, J., Duprez, L., Van Sande, J., Paschke, R., Tonacchera, M., Dumont, J., and Vassart, G. (1994). Constitutively active receptors as a disease-causing mechanism. *Mol. Cell. Endocrinol.* **100,** 159–162.

Parma, J., Van Sande, J., Swillens, S., Tonacchera, M., Dumont, J., and Vassart, G. (1995). Somatic mutations causing constitutive activity of the thyrotropin receptor are the major cause of hyperfunctioning thyroid adenomas: Identification of additional mutations activating both the cyclic adenosine 3′,5′-monophosphate and inositol phosphate-Ca2+ cascades. *Mol. Endocrinol.* **9,** 725–733.

Parma, J., Duprez, L., Van Sande, J., Hermans, J., Rocmans, P., Van Vliet, G., Costagliola, S., Rodien, P., Dumont, J. E., and Vassart, G. (1997). Diversity and prevalence of somatic mutations in the thyrotropin receptor and Gs alpha genes as a cause of toxic thyroid adenomas. *J. Clin. Endocrinol. Metab.* **82,** 2695–2701.

Pellissier, L. P., Sallander, J., Campillo, M., Gaven, F., Queffeulou, E., Pillot, M., Dumuis, A., Claeysen, S., Bockaert, J., and Pardo, L. (2009). Conformational toggle switches implicated in basal constitutive and agonist-induced activated states of 5-hydroxytryptamine-4 receptors. *Mol. Pharmacol.* **75,** 982–990.

Rosenbaum, D. M., Rasmussen, S. G., and Kobilka, B. K. (2009). The structure and function of G-protein-coupled receptors. *Nature* **459,** 356–363.

Russo, D., Wong, M. G., Costante, G., Chiefari, E., Treseler, P. A., Arturi, F., Filetti, S., and Clark, O. H. (1999). A Val 677 activating mutation of the thyrotropin receptor in a Hurthle cell thyroid carcinoma associated with thyrotoxicosis. *Thyroid* **9,** 13–17.

Russo, D., Betterle, C., Arturi, F., Chiefari, E., Girelli, M. E., and Filetti, S. (2000). A novel mutation in the thyrotropin (TSH) receptor gene causing loss of TSH binding but constitutive receptor activation in a family with resistance to TSH. *J. Clin. Endocrinol. Metab.* **85,** 4238–4242.

Seifert, R., and Wenzel-Seifert, K. (2002). Constitutive activity of G-protein-coupled receptors: Cause of disease and common property of wild-type receptors. *Naunyn Schmiedebergs Arch. Pharmacol.* **366,** 381–416.

Simoni, M., Gromoll, J., and Nieschlag, E. (1997). The follicle-stimulating hormone receptor: Biochemistry, molecular biology, physiology, and pathophysiology. *Endocr. Rev.* **18,** 739–773.

Smit, M. J., Vischer, H. F., Bakker, R. A., Jongejan, A., Timmerman, H., Pardo, L., and Leurs, R. (2007). Pharmacogenomic and structural analysis of constitutive g protein-coupled receptor activity. *Annu. Rev. Pharmacol. Toxicol.* **47,** 53–87.

Szkudlinski, M. W., Fremont, V., Ronin, C., and Weintraub, B. D. (2002). Thyroid-stimulating hormone and thyroid-stimulating hormone receptor structure–function relationships. *Physiol. Rev.* **82,** 473–502.

Tonacchera, M., Van Sande, J., Cetani, F., Swillens, S., Schvartz, C., Winiszewski, P., Portmann, L., Dumont, J. E., Vassart, G., and Parma, J. (1996). Functional characteristics of three new germline mutations of the thyrotropin receptor gene causing autosomal dominant toxic thyroid hyperplasia. *J. Clin. Endocrinol. Metab.* **81,** 547–554.

Tonacchera, M., Chiovato, L., Pinchera, A., Agretti, P., Fiore, E., Cetani, F., Rocchi, R., Viacava, P., Miccoli, P., and Vitti, P. (1998). Hyperfunctioning thyroid nodules in toxic multinodular goiter share activating thyrotropin receptor mutations with solitary toxic adenoma. *J. Clin. Endocrinol. Metab.* **83,** 492–498.

Urizar, E., Claeysen, S., Deupi, X., Govaerts, C., Costagliola, S., Vassart, G., and Pardo, L. (2005). An activation switch in the rhodopsin family of G protein coupled receptors: The thyrotropin receptor. *J. Biol. Chem.* **280**(17), 17135–17141.

Van Sande, J., Swillens, S., Gerard, C., Allgeier, A., Massart, C., Vassart, G., and Dumont, J. E. (1995). In Chinese hamster ovary K1 cells dog and human thyrotropin receptors activate both the cyclic AMP and the phosphatidylinositol 4, 5-bisphosphate cascades in the presence of thyrotropin and the cyclic AMP cascade in its absence. *Eur. J. Biochem.* **229,** 338–343.

Vlaeminck-Guillem, V., Ho, S. C., Rodien, P., Vassart, G., and Costagliola, S. (2002). Activation of the cAMP pathway by the TSH receptor involves switching of the ectodomain from a tethered inverse agonist to an agonist. *Mol. Endocrinol.* **16,** 736–746.

Wonerow, P., Chey, S., Fuhrer, D., Holzapfel, H. P., and Paschke, R. (2000). Functional characterization of five constitutively activating thyrotrophin receptor mutations. *Clin. Endocrinol. (Oxf)* **53,** 461–468.

CHAPTER TWENTY-FOUR

Toward the Rational Design of Constitutively Active KCa3.1 Mutant Channels

Line Garneau, Hélène Klein, Lucie Parent, *and* Rémy Sauvé

Contents

Abstract

The Ca^{2+} activated potassium channel of intermediate conductance KCa3.1 is now emerging as a therapeutic target for a large variety of health disorders. KCa3.1 is a tetrameric membrane protein with each subunit formed of six transmembrane helices (S1–S6). Ca^{2+} sensitivity is conferred by the Ca^{2+} binding protein calmodulin (CaM), with the CaM C-lobe constitutively bound to an intracellular domain of the channel C-terminus, located proximal to the membrane and connected to the S6 transmembrane segment. Patch clamp single channel recordings have demonstrated that binding of Ca^{2+} to CaM allows the channel to transit dose dependently from a nonconducting to an

Department of Physiology, Groupe d'étude des protéines membranaires, Université de Montréal, Montreal, Canada

Methods in Enzymology, Volume 485
ISSN 0076-6879, DOI: 10.1016/S0076-6879(10)85024-4

ion-conducting configuration. Here we present a general strategy to generate KCa3.1 mutant channels that remain in an ion-conducting state in the absence of Ca^{2+}. Our strategy is first based on the production of a 3D model of the channel pore region, followed by SCAM experiments to confirm that residues along each of the channel S6 transmembrane helix form the channel pore lumen as predicted. In a simple model, constitutive activity can be obtained by removing the steric hindrances inside the channel pore susceptible to prevent ion flow when the channel is in the closed configuration. Using charged MTS reagents and Ag^{+} ions as probes acting on Cys residues engineered in the pore lumen, we found that the S6 transmembrane helices of KCa3.1 cannot form a pore constriction tight enough to prevent ion flow for channels in the closed state. These observations ruled out experimental strategies where constitutive activity would be generated by producing a "leaky" closed channel. A more successful approach consisted however in perturbing the channel open/closed state equilibrium free energy. In particular, we found that substituting the hydrophobic residue V282 in S6 by hydrophilic amino acids could lock the channel in an open-like state, resulting in channels that were ion conducting in the absence of Ca^{2+}.

1. Introduction

1.1. KCa channels

Calcium signaling cascades play a prominent role in a large variety of cellular processes. Calcium activated potassium channels constitute in this regard key effectors during Ca^{2+} signaling as increases in cystosolic Ca^{2+} concentration cause an enhanced channel activity resulting in a variation of the membrane potential in both excitable and nonexcitable cells. Based on their single channel conductance, genetic relationship and mechanisms of Ca^{2+} activation, the eight KCa channels identified so far form two well defined groups. The first group contains the KCa1.1, KCa4.1, KCa4.2, and KCa5.1 channels characterized by rather large unitary conductance (> 100 pS). The best studied channel of this group is KCa1.1 (Maxi KCa), which is both voltage and Ca^{2+} sensitive. Ca^{2+} sensitivity in this case is linked to specific domains of the channel structure involved in Ca^{2+} binding. The second group refers to the three small conductance channels, KCa2.1, KCa2.2, and KCa2.3 and to the intermediate conductance KCa3.1 channel. These channels are voltage insensitive, and Ca^{2+} sensitivity is conferred by the Ca^{2+}-binding protein calmodulin (CaM) constitutively bound in C-terminus to a membrane-proximal region that is highly conserved among these channels (Khanna *et al.*, 1999). KCa3.1 and KCa2.x present, however, distinct pharmacological profiles with KCa3.1 and KCa2.x, respectively, inhibited by TRAM-34 (Wulff *et al.*, 2000) and apamin (Kohler *et al.*, 1996).

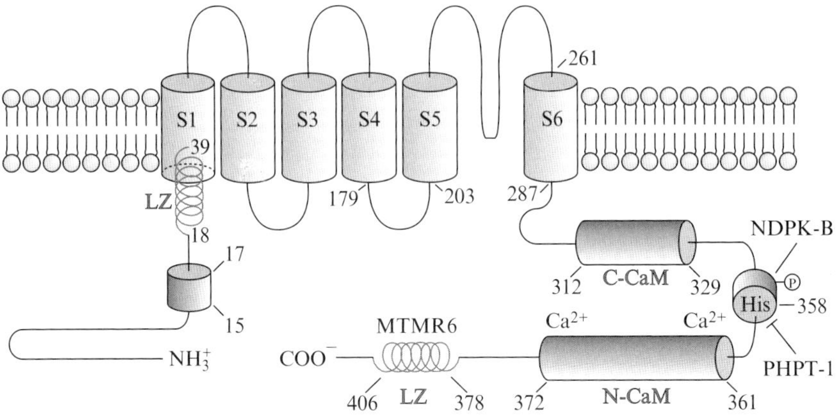

Figure 24.1 Membrane topology of KCa3.1. KCa3.1 is a tetrameric protein with each monomer organized in six transmembrane segments plus a pore region between segments 5 and 6. The channel Ca^{2+} sensitivity is conferred by calmodulin, with the CaM C-lobe (C-CaM) constitutively bound to the 312–329 segment of the channel C-terminal region. A stretch of 11 amino acids from 361 to 372 is responsible for the Ca^{2+}-dependent binding of the CaM N-lobe (N-CaM) to the channel. The channel regulation by ATP is mediated in part by NDPK-B and PHPT-1 which phosphorylates and dephosphorylates H358, while the presence of a coil–coil segment at the channel C-terminal end is involved in the binding of the MTMR6 phosphatase. We have shown that AMPK-γ1-subunit can interact with a domain extending from 380 to 400 in KCa3.1 C-terminus (Klein *et al.*, 2009). Finally, KCa3.1 contains a 15-RKR-17 motif in N-terminus required for ATP regulation, and two leucine zipper motifs (LZ) in N- and C-termini respectively, critical for channel assembly and trafficking.

Cloning and functional expression of the KCa3.1 channel have revealed that KCa3.1 is a tetrameric membrane protein with each subunit organized in six transmembrane segments S1–S6 with a pore motif between segments 5 (S5) and 6 (S6) (Fig. 24.1). Each subunit contains in C-terminus a Ca^{2+}-dependent and a Ca^{2+}-independent CaM binding domain, for an overall stoichiometry of four CaM per channel (Khanna *et al.*, 1999). Recent studies have also shown that internal ATP stimulates the human KCa3.1 channel activity *via* phosphorylation by the nucleoside diphosphate kinase NDPK-B of a histidine residue at position 358 located within the channel C-terminal region (Srivastava *et al.*, 2006b). The stimulatory action of ATP requires Ca^{2+} with ATP leading to an apparent increase in Ca^{2+} sensitivity (Srivastava *et al.*, 2006a). Work from our laboratory has also provided the first evidence that the C-terminal region of KCa3.1 is interacting with the γ1-subunit of the metabolic sensing kinase AMPK (Klein *et al.*, 2009), with KCa3.1 activity decreasing in response to AMPK stimulation. Altogether these data support a model where KCa3.1 is part of a multiprotein complex involving several protein kinases.

1.2. KCa3.1 physiological role in health and diseases

There is now strong evidence that the KCa3.1 channel plays a prominent role in a large variety of physiological events. For instance, the KCa3.1 blocker TRAM-34 has been documented to suppress acute immune reactions involving memory B and T cells (Wulff *et al.*, 2004). KCa3.1 is also well known to constitute a major determinant to the endothelium-dependent control of vascular tone. Data from several laboratories have established that KCa3.1 activation is an obligatory step to the EDHF (endothelium-derived hyperpolarizing factor) vasodilation process, with TRAM-34 and apamin causing an inhibition of the EDHF-induced vasorelaxation (Feletou and Vanhoutte, 2007). A strong upregulation of KCa3.1 was observed in the mitogenesis of rat fibroblast, vascular smooth muscle cells (Grgic *et al.*, 2005), and cancer cell lines, indicating that KCa3.1 represents an important regulator of cell proliferation (Grgic *et al.*, 2005; Jager *et al.*, 2004; Ouadid-Ahidouch *et al.*, 2004). Notably, the use of the KCa3.1 inhibitor TRAM-34 was found to prevent restenosis, an effect directly related to an upregulated expression of KCa3.1 in coronary artery vascular smooth muscle cells following balloon catheter injury (Kohler *et al.*, 2003). Recent data also demonstrated that KCa3.1 is involved in renal fibroblast proliferation and fibrogenesis (Grgic *et al.*, 2009) suggesting that KCa3.1 may represent a therapeutic target for the treatment of fibrotic kidney disease. Finally, increasing evidence argues for a prominent role of KCa3.1 in Cl^- secreting epithelial cells by maintaining an electrochemical gradient favorable to Cl^- and Na^+ transepithelial transport. The strong coupling between basolateral K^+ channel activation and apical Cl^- secretion has led to the proposal of using K^+ channel openers such as DCEBIO (Singh *et al.*, 2001) or 4-chloro-benzo[*F*]isoquinoline (CBIQ; Szkotak *et al.*, 2004) as therapeutic agents to correct fluid secretion in epithelia presenting ion transport defects as found in cystic fibrosis. In contrast, blockage of KCa3.1 might be beneficial in treating pathological conditions characterized by an excessive fluid secretion. In this regard, the use of the KCa3.1 blocker clotrimazole was documented to normalize salt and water transport in secretory diarrhea (Rufo *et al.*, 1997). Altogether, these data support KCa3.1 as promising therapeutic target for a large variety of health disorders.

1.3. Rationale for the design of constitutively active KCa3.1 channels

Investigating the KCa3.1 structures involved in transducing Ca^{2+} binding into channel opening is essential, not only to understand how the channel actually works, but also to determine how channel activity can be affected by blockers or potentiators. Critical to this process is the identification of residues that when mutated lead to channels that are ion conducting in the absence of Ca^{2+}. These residues may either critically contribute to the energy balance between the channel open/closed configurations, or play a pivotal role in coupling the channel activation gate to the conformational change triggered by the binding of Ca^{2+} to the CaM/KCa3.1 complex. Identification of these residues thus requires to localize the channel activation gate and/or the structures responsible to maintain the gate in an open state. A study of the residues involved in constitutive activity may also offer the possibility to generate KCa3.1 mutants susceptible to be used in high-throughput screening (HTPS) assays. Testing drugs by fluorescence-based membrane potential measurements represents a valid screening strategy as long as the changes in membrane potential accurately reflect the activity of the channel of interest. In this regard, the study of KCa3.1 is particularly challenging. First, the maximum open probability of KCa3.1 rarely exceeds 0.2 in saturating Ca^{2+} conditions, so that changes in membrane potential due to KCa3.1 inhibition in resting Ca^{2+} conditions might not be important enough to be detectable through fluorescence measurements. Second, HTPS cannot exclude indirect effects of the drugs on proteins involved in KCa3.1 regulation. For instance, drugs that will affect intracellular Ca^{2+} homeostasis or regulate the activity of protein kinases such as AMPK or NDPK-B are expected to modify the KCa3.1 contribution to membrane potential independently of a direct action on the channel itself. To circumvent this problem, one may consider the use of a KCa3.1 channel constitutively active in zero Ca^{2+}. Besides being insensitive to intracellular Ca^{2+} fluctuations, this channel would also be independent of ATP-based regulatory mechanisms. An automated selection of channel inhibitors could under these conditions truly reflect an effect of the drug on the channel itself, and not on some secondary mechanisms. To be truly applicable, however, this approach requires that the structure of the channel constitutively active be representative of the wild-type channel in the open configuration. Here we present the global strategy that was implemented to produce KCa3.1 mutant channels that showed constitutive activation properties.

2. Production of Constitutively Active KCa3.1 Mutant Channels

2.1. Production of a model structure of the pore region

The rational design of a constitutively active KCa3.1 mutant first requires to identify some of the channel key structural features. Unfortunately, the 3D structure of KCa3.1 is currently unknown. A 3D representation of KCa3.1 pore region could, however, be generated by applying a comparative modeling approach where segments of the channel amino acid sequence were translated into 3D structural data (Baker and Sali, 2001; Sali and Blundell, 1993). Comparative modeling, or homology modeling, is usually based on a number of steps: (1) identifying a known structure (the template structure) with an amino acid sequence homologous to the sequence of the protein to be modeled (the target sequence); (2) aligning the template sequence and secondary structure to the target amino acid sequence; (3) building model structures using a specialized software such as MODELLER; and (4) testing the model. Identification of suitable template structures can be obtained using the computational tools provided by specialized servers such as SAM-T08 (http://compbio.soe.ucsc.edu/SAM_T08/T08-query.html) and/or T-TASSER (http://zhanglab.ccmb.med.umich.edu/). This procedure is usually complemented by a comparative sequence analysis with T-COFFEE and MUSCLE as to fine tune the sequence alignments that will serve as input files for MODELLER. Using such a procedure for KCa3.1, we identified the Kv1.2 (Long *et al.*, 2005; PDB:2a79) and MlotiK1 (Clayton *et al.*, 2008; PDB:3BEH) potassium channel crystal structures as suitable templates to generate a 3D representation of the open/closed KCa3.1 pore region (Fig. 24.2). MlotiK1 is a voltage insensitive internal ligand-gated channel with six transmembrane helices per monomer. The MlotiK1structure is representative of a ligand-gated channel in the closed configuration, whereas the voltage-gated Kv1.2 channel structure is more representative of a channel in the open state. Performing automated homology modeling with MODELLER V9.3 (Sali and Blundell, 1993) requires the initial production of a large number of models for each template. We routinely set the number of models to 150, and the best models are selected based on the value of the objective function (roughly related to the energy of the model) provided by MODELLER, together with the RMS deviation computed for the Cα of the model structure backbone relative to the template. The overall structural quality of the top five model 3D structures can be checked by PROCHECK (Laskowski *et al.*, 1993) so that aberrant structural features (rotation angles, bond lengths, etc.) can be user adjusted during the procedure. When applied to KCa3.1, the models derived from Kv1.2 and MlotiK1 predicted that the residues V275, T278, A279, V282, and A286 of the S6 transmembrane segment should be lining the channel pore with residues C276, C277, L280, and L281 oriented opposite to the pore lumen (Fig. 24.2). An analysis of the model generated from the MlotiK1 structure indicated furthermore that the narrowest segment of the conduction pathway should be located at the level of the C-terminal end of the four S6 transmembrane helices (A282–A286). This region is predicted to line with the bundle crossing domain of the inverted tepee structure originally described for the KcsA channel (Doyle *et al.*, 1998). More importantly, the bundle crossing region has been proposed to play a prominent role in controlling ion flow in several K^+ channels. For instance,

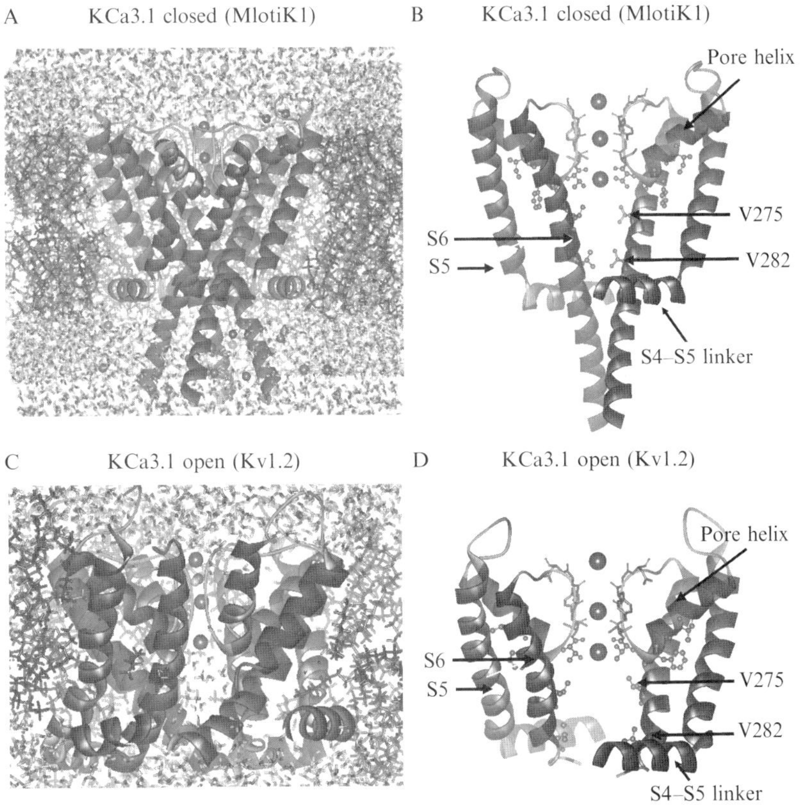

Figure 24.2 Model structures of KCa3.1. A and C. Global side view representations of the closed (A) and open (C) KCa3.1 embedded in a lipid bilayer (green) separating two electrolyte solutions containing 150 m*M* KCl. The closed KCa3.1 3D representation was obtained by homology modeling using the MlotiK1 structure (PDB:3BEH) as template, whereas the open representation is based on the Kv1.2 channel structure (PDB:2a79). Both models include the S4–S5 linker plus the S5–S6 pore region of the channel. (B and D) Detailed view of the pore region with only two monomers represented for clarity. Turns colored in blue refer to residues predicted to be facing the pore (V275, T278, A279, V282, and A286). These predictions were confirmed in three SCAM analyses produced by our laboratory. Templates selected with I-TASSER and SAM-T08 servers. CPK representation of K^+ ions in purple. Representation by DS Visualizer. (See Color Insert.)

a major contribution of the bundle crossing region to channel gating was hypothesized from the KirBac1.1 channel structure, where semiconserved phenylalanine residues at the C-terminal end of the inner helices (S6 in KCa3.1) appeared to clash, suggesting that they form a hydrophobic gate capable to block permeation of K^+ ions (Kuo *et al.*, 2003). Similarly,

Cd^{2+}-dependent block experiments on the voltage activated *Shaker* channel have revealed that the residue V474 equivalent to V282 in KCa3.1, likely contributes to the formation of a very tight constriction site in the Shaker pore (Webster *et al.*, 2004). It follows that constitutive activation could in principle be obtained by reducing steric hindrance at the level of the bundle crossing region, so that ions would still diffuse through the channel pore despite the channel being in a closed configuration. This is the "leaky" closed state hypothesis for constitutive activity.

2.2. Experimental identification of the residues forming the channel pore

The first step to determine whether relieving steric hindrance at the bundle crossing region in KCa3.1 can lead to the formation of an ion-conducting channel in zero Ca^{2+} is to establish experimentally the nature of the residues predicted to be lining the channel pore. Predictions derived from homology modeling can formally be tested using a method termed substituted-cysteine accessibility method (SCAM) which allows site-selective modification of accessible cysteine in a protein. Each residue of the S6 transmembrane segment was mutated one at a time to a cysteine (Cys) and the effects on channel activity of small, charged, sulfhydryl-specific reagents measured in patch clamp experiments. Cysteine residues engineered along the S6 segment can either be located in a water-accessible environment or buried into the protein. Since the reaction of sulfhydryl-specific reagents such as MTS (methanethiosulfonate) is 10^9 slower for the protonated than deprotonated form of Cys (Karlin and Akabas, 1998), one expects that only Cys residues exposed to a water filled channel pore will be modified by MTS compared to Cys exposed to a nonaqueous environment. In addition, the irreversible binding of a charged sulfhydryl reagent to a Cys facing the channel pore is likely to inhibit ion flow due to electrostatic interactions and steric hindrance as well. $MTSET^+$ ([2-(trimethylammonium)ethyl] methanethiosulfonate) is a specific sulfhydryl reagent predicted to fit into a cylinder of 2.9 radius and 9 Å in length when undergoing an all-*trans* configuration (without the hydrated shell). Because it is positively charged at neutral pH and poorly liposoluble, this molecule is generally considered to be an excellent probe to investigate the structural features of cationic channel pore. The smaller positively charged MTS reagent $MTSEA^+$ (aminoethyl methanethiosulfonate: 2.4 Å radius × 9 Å long) has also been extensively used in SCAM type experiments, but with a $pK_a > 8.5$, 6% of the molecules remains on the average in a neutral form. A neutral molecule could in principal diffuse through the membrane and modify Cys residues located transmembrane to the side of application. This problem can be circumscribed by adding exogenous Cys (1 m*M*) to the solution bathing the membrane surface opposite to the side of $MTSEA^+$

application (Holmgren *et al.*, 1996). It is recommended finally to use MTS reagents at concentrations lower than 1 mM, to avoid nonspecific channel inhibition coming from overcrowding the channel pore. However, low MTS concentrations (10 μM) often result in modification time constants of the order of minutes, thus requiring highly stable current recordings with no detectable rundown.

The experimental procedure underlying this type of SCAM experiments consists essentially in patch clamp recordings of KCa3.1 channels in the inside-out configuration where the MTS reagent is applied internally by means of a fast solution change system (RSC-160, BioLogic, Grenoble, France). We found that, for most applications, an exchange time less than 30 ms was sufficient to get a reliable measurement of the current change initiated following MTS application. By fitting the time course of the current variation to a single exponential function, the modification rate of the target Cys by a given reagent can be estimated using: modification rate (M^{-1} s^{-1}) = 1/(MTS concentration (in M) × time constant of the current change (in s)). The bath and patch pipette solutions usually contained (in mM) 200 K_2SO_4, 1.8 $MgCl_2$, 0.025 $CaCl_2$, 25 HEPES, buffered at pH 7.4 with KOH. A high K^+ concentration maximize the signal to noise ratio in unitary current recordings and the use of sulfate salts enables to chelate contaminant divalent cations such as Ba^{2+} (maximum free Ba^{2+} concentration: 0.5 nM in 200 mM K_2SO_4) which may otherwise block KCa3.1 by interacting directly with the channel selectivity filter. A sulfate salt also minimizes the contribution coming from endogenous Ca^{2+}-activated Cl^- channels in experiments where KCa3.1 is expressed in *Xenopus* oocytes. Site-directed mutagenesis of KCa3.1 is routinely performed using the QuickChange Site-Directed Mutagenesis kit (Stratagene). Oocytes (stage V or VI) used for channel expression are obtained from *Xenopus laevis* frogs anaesthetized with 3-aminobenzoic acid ethyl ester. The follicular layer is removed by incubating the oocytes in a Ca^{2+}-free Barth's solution containing collagenase (1.6 mg/ml; Sigma-Aldrich) for 60 min. Defolliculated oocytes are stored at 18 °C in Barth's solution supplemented with 5% horse serum, 2.5 mM Na-pyruvate, 100 U/ml penicillin, 0.1 mg/ml kanamycin, and 0.1 mg/ml streptomycin. The Barth's solution contains (in mM) 88 NaCl, 3 KCl, 0.82 $MgSO_4$, 0.41 $CaCl_2$, 0.33 $Ca(NO_3)_2$, and 5 HEPES (pH 7.6). Oocytes are patched 3–5 days after injection of 0.1-1 ng of the cDNA coding for KCa3.1. Prior to patch clamping, defolliculated oocytes are bathed in a hyperosmotic solution containing (in mM) 250 KCl, 1 $MgSO_4$, 1 EGTA, 50 sucrose, and 10 HEPES buffered at pH 7.4 with KOH. The vitelline membrane is then peeled off using fine forceps, and the oocyte is transferred to a superfusion chamber for patch clamp measurements.

This procedure was used to map the pore structure of the open KCa3.1 channel. We found that $MTSET^+$ applied internally caused a total inhibition of the V275C, T278C, and V282C mutants and partially blocked the A279C and V284C channels. In contrast, $MTSET^+$ initiated a strong channel activation of the A283C and A286C mutant channels. These results are in agreement with the proposed model obtained by homology where V275, T278, A279, and V282 are lining the channel pore in the open state while supporting a bundle crossing region of KCa3.1 extending from V282 to A286 (Klein *et al.*, 2007).

2.3. Testing the current leak hypothesis

Central to the current leak hypothesis is the proposal that in the closed state, the channel bundle crossing region should form a seal tight enough to prevent ion flow through the channel pore. Removing steric constraints at this site should consequently lead to the formation of a channel that is ion conducting in the absence of Ca^{2+} (constitutive activity). The contribution of the S6 segment in the bundle crossing region to the regulation of ion flow can formally be tested by measuring to what extent sulfhydryl-specific reagents of various sizes can access a Cys residue engineered deep in the channel pore when the channel is in the closed configuration. The choice of the probe is crucial in this type of experiments. Because of its size, $MTSET^+$ is not truly representative of a K^+ ion in solution. In fact, the bundle crossing region could be impermeable to $MTSET^+$ for the closed channel, but still be permeable to K^+ ions under identical conditions. $MTSEA^+$ is of smaller size than $MTSET^+$ (4.8 Å diameter compared to 5.8 Å) but this reagent may lead to false positive results as it can exist in a neutral protonated form and thus access a target Cys deep in the channel pore by diffusing directly through the membrane. Better results can be obtained using Ag^+ as thiol modifying agent. As mentioned by Lu and Miller (1995), Ag^+ constitutes an excellent probe to study K^+ channels as both ions are very similar in size with a van der Waals radius of 1.52 Å for K^+ and 1.29 Å for Ag^+ (Marcus, 1988; Shannon, 1976). Ag^+ is, however, highly reactive, so that Ag^+ solutions need to be prepared by adding $AgNO_3$ to strictly Cl^- free solutions. In addition, the free Ag^+ concentration needs to be stabilized using a chelating agent such as EDTA at high concentrations. We found that 60 m*M* EDTA gave reproducible results. As the diffusion limited modification rate of Cys by Ag^+ is of the order of 10^8 M^{-1} s^{-1}, Ag^+ should be used at nanomolar concentrations to yield time constant of Cys modification within the seconds range. The free Ag^+ concentration can be calculated with programs such as Eqcal (Biosoft, Cambridge, UK). Finally, the target Cys should be engineered deep in the channel pore, well above the bundle crossing region, close to the selectivity filter. Our model of the KCa3.1 region predicts that the Val at 275 should be located in proximity of the selectivity filter, facing the channel central cavity. In accordance with this model, we found that the V275C mutant in the open state can be specifically blocked through the irreversible binding of sulfhydryl reagents such as $MTSEA^{+}$ or $MTSET^{+}$ (Klein *et al.*, 2007).

This experimental procedure was applied to our study of the KCa3.1 bundle crossing region. In these experiments, EDTA, EDTA + Ag^{+}, EDTA solutions were applied repetitively in 1.5 s pulses each separated by a test perfusion with a solution containing 25 μ*M* Ca^{2+}. Modification rates were calculated from the current inhibition curve obtained in response to cumulative Ag^{+} applications. Our results showed that Cys residues engineered inside the channel central cavity at position 275 were readily accessible to Ag^{+} applied internally (7 n*M*) with the channel in the closed configuration (see Fig. 24.3). In contrast, larger molecules such as $MTSET^{+}$ (5.8 Å diameter) showed a 10^{4}-fold difference in accessibility between the channel closed and open configurations (Fig. 24.6; Klein *et al.*, 2007). These observations demonstrate that the bundle crossing region for the closed KCa3.1 channel is permeable to K^{+} ions, but impermeable to larger molecules such as $MTSET^{+}$. This proposal is in line with Cys mutagenesis data from cyclic nucleotide-gated channels suggesting that the inner helices may form a constriction at the C-terminal end of the channel pore tight enough to restrict the accessibility of reagents larger than K^{+} to the channel cavity, but nonobstructive to K^{+} ion flow (Flynn and Zagotta, 2001, 2003; Xiao *et al.*, 2003). Similar conclusions supported by SCAM data were also

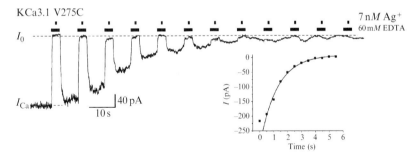

Figure 24.3 Evidence against the leaky channel model for constitutive activation. Inside-out patch clamp recording illustrating the action of Ag^{+} (7 n*M*) on the closed V275C channel. Ag^{+} was applied for 0.5 s during a 3-s perfusion period with a Ca^{2+} free solution containing (in m*M*): 150 KMES, 60 EDTA, 10 HEPES, pH 7.3, with 25 μ*M* $AgNO_3$ for a free Ag^{+} concentration of 7 n*M*. The inhibitory effect of a 0.5s Ag^{+} application was estimated from the current intensity recorded after replacing the zero Ca^{2+} solution by a solution containing 25 μ*M* Ca^{2+} (test current). The time dependent variation of the test currents obtained from the repetitive application of Ag^{+} at 0.2 Hz is illustrated in the inserted panel on the right. These results argue against the presence along the channel pore of steric constraints that would impair K^{+} ion flow when the channel is in the closed configuration. Originally published in Garneau *et al.* (2009). © The American Society for Biochemistry and Molecular Biology.

derived from studies performed on the KCa2.2 and Kir2.1 channels (Bruening-Wright *et al.*, 2002; Xiao *et al.*, 2003). It follows that constitutively activated KCa3.1 channels are not likely to be generated by simply removing steric constraints at the bundle crossing site, since this site does not constitute an active barrier controlling K^+ ion flow in this case.

2.4. Constitutive activity and the energetics of the S6 transmembrane helix

The exact molecular mechanism underlying KCa3.1 opening in response to Ca^{2+} binding to the CaM/KCa3.1 complex remains to be elucidated. This point is crucial to the design of a constitutively active KCa3.1 channel. Important structural information pertinent to channel gating was however obtained through the crystallization of CaM bound to the rat KCa2.2-CaM binding domain in the presence of Ca^{2+} (Schumacher *et al.*, 2001). These results suggested a large-scale conformational rearrangement taking place in the presence of Ca^{2+} where the N-lobe of CaM binds to a segment in the C-terminus of an adjacent monomer, resulting in a dimerization of contiguous subunits within the channel structure. This rearrangement would in turn lead to a rotation/translation of the associated S6 transmembrane helix and to the opening of the ion-conducting pore (Maylie *et al.*, 2004; Schumacher *et al.*, 2001, 2004; Wissmann *et al.*, 2002). Added to our Ag^+ ion-based results suggesting an active channel gate at the level of selectivity filter region, this model argues for a crucial role of S6 movements to the KCa3.1 opening process. In fact, the difference in free energy between the channel closed and open state is likely to be governed by the changes in free energy coming from S6 residues making different contacts with their surrounding milieu when the channel is in the open compared to the closed state configuration.

To determine to what extent the interactions between the side chain of a residue along S6 with its atomic environment contributes to the difference in free energy between the channel open and closed state, one can minimize these interactions by replacing the residue of interest by a Gly (Garneau *et al.*, 2009). As constitutively active channel mutants will be ion conducting in zero Ca^{2+}, it is also essential to establish a reference current level corresponding to a non ion-conducting channel. This problem can be circumvented by performing Gly substitutions on the V275C channel, so that $MTSEA^+$ or $MTSET^+$ could be used as specific irreversible blockers of the channel pore. Under these conditions, constitutive activity can be estimated by measuring the ratio $R = (I_{EGTA} - I_0)/(I_{Ca} - I_0)$, where I_{EGTA} and I_{Ca} refer, respectively, to the current measured in zero and 25 μM internal free Ca^{2+}, and I_0 the current level following channel block by $MTSEA^+$. For non constitutively active channels, we expect $I_{EGTA} = I_0$ for $R = 0$ as the current level in zero Ca^{2+} will correspond to the background current obtained following a total block of the channel pore. Values of $R > 0$ would be indicative of a channel where $I_{EGTA} > I_0$, thus indicating that the current level obtained in the absence of agonist does not correspond to the current level expected for a nonconducting channel. Using this paradigm, we found that the V275C-V282G and V275C-A279G mutants led to R values equal to 1.0 ± 0.02 ($n = 3$) and 0.2 ± 0.02 ($n = 3$), respectively, indicating that V275C-V282G or V275C-A279G were ion conducting in zero Ca^{2+} conditions (Fig. 24.4). Other double mutant channels generated for residues in S6 extending from C276 to A286 did not show constitutive activation properties with R values of 0 (Fig. 24.4D).

A change in the channel open/closed equilibrium energy leading to constitutive activity may result either from a stabilization of the channel open configuration, and/or a destabilization of the channel closed configuration. A straightforward approach useful to discriminate between these two mechanisms consists in measuring the channel mean open and closed time at the single channel level. In conditions where N independent channels are present, the channel mean open time $\langle t_o \rangle$ can formally be obtained from $\langle t_o \rangle = \langle t_o^{(r)} \rangle [\mathrm{Po}(N - r) + r(1 - \mathrm{Po})] / (1 - \mathrm{Po})$ where $\langle t_o^{(r)} \rangle$ is the mean open time when r channels are simultaneously open and Po the channel open probability. Po can be calculated from current amplitude histograms assuming that the probability of having 'r' channels open simultaneously among N, Po(r), obeys a binomial distribution ($\mathrm{Po}(r) = N!/[(N - r)!r!]\ \mathrm{Po}^r(1 - \mathrm{Po})^{N-r}$). We found through a single channel analysis of the V275C-A279G mutant that the channel open probability increases from 0.5 at $[Ca^{2+}]_i < 0.1$ nM to 0.97 in 25 μM Ca^{2+}, an effect essentially attributable to a 15- to 30-fold increase and 2- to 3-fold decrease of the channel mean open and closed times, respectively. Such a behavior contrasts with the effect of Ca^{2+} on the wild-type KCa3.1 channel where increasing the internal Ca^{2+} concentration causes a strong reduction of the channel mean closed time with the channel mean open time remaining rather unchanged. This analysis thus supports a model whereby constitutive activity is obtained by essentially locking the channel in an open configuration (Garneau *et al.*, 2009).

Without a detailed description of the microenvironment surrounding the V282 and/or A279 residues in the closed and open state, their respective contribution to the free energy barriers governing the closed to open state transitions is doomed to remain largely undetermined. Hydrophobic interactions have been, however, identified as key determinants to the open/closed equilibrium energy in several ion channels. For instance, constitutively active channels were generated by replacing through yeast screening analysis, the residue V188 (V282 in KCa3.1) in GIRK2 (Yi *et al.*, 2001) by more hydrophilic amino acids. As discussed by Karplus (1997), a quantification in terms of generic hydrophobic effects of the energy difference governing dynamic changes in protein conformations is probably valid for hydrophobic residues only.

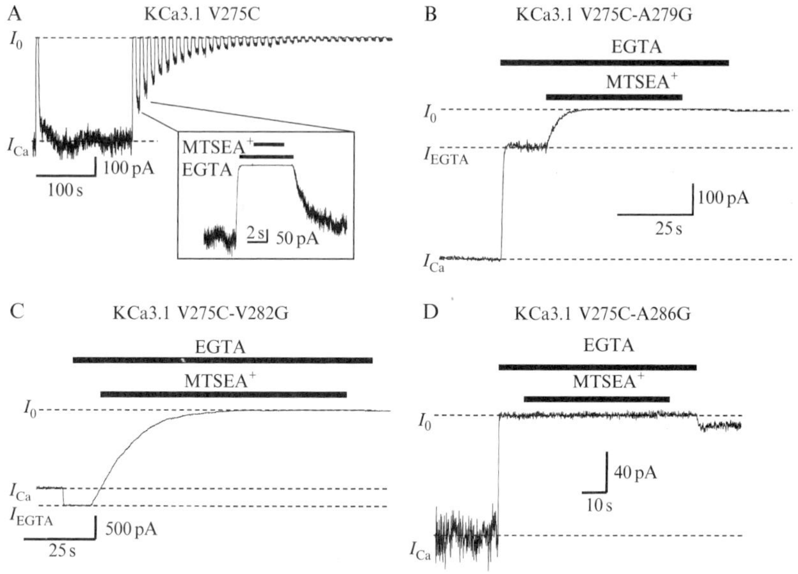

Figure 24.4 Glycine scan along the S6 segment below the Gly hinge at 274. Inside-out recordings of inward currents illustrating the effect of MTSEA$^+$ on the channel mutants. Recordings obtained in symmetrical 200 mM K_2SO_4 conditions at $Vm = -60$ mV. The symbol I_{Ca} refers to the current level in 25 μM Ca^{2+}, I_{EGTA} to the current level in 1 mM EGTA, and I_0 to the current level following inhibition by MTSEA$^+$. Under these experimental conditions, Ca^{2+}-activated K^+ current are represented as inwardly directed currents relative to the zero current level I_0. (A) Control experiment illustrating the effect of internal MTSEA$^+$ (1 mM) application on the closed V275C channel. MTSEA$^+$ was applied for 3 s during 5.5 s pulses in zero Ca^{2+} at a frequency of 0.1 Hz. The accessibility to MTSEA$^+$ of cysteines in the channel cavity at position 275 for the closed V275C channel was estimated from the time constant of inhibition of the test inward currents measured in 25 μM internal Ca^{2+} at the end of each pulse. The successive applications of MTSEA$^+$ in zero Ca^{2+} did not result in this case in a gradual decrease of the I_0 current, but the clear inhibition of the test currents in 25 μM Ca^{2+} confirms binding of MTSEA$^+$ to the cysteines at 275. (B) Effect of MTSEA$^+$ applied internally on the V275C–A279G double mutant. This current record shows that internal application of MTSEA$^+$ (1 mM) in zero Ca^{2+} resulted in a strong inward current inhibition for a modification rate of 297 ± 5 M^{-1} s^{-1} ($n = 3$), in support of an ion-conducting conformation in zero Ca^{2+}. (C) Response of V275C–V282G to MTSEA$^+$. As seen, the addition of MTSEA$^+$ (1 mM) in zero Ca^{2+} caused a strong inward current inhibition for a modification rate of 100 ± 15 $M^{-1}s^{-1}$ ($n = 13$) confirming that the V275C–V282G mutant was conducting in the absence of Ca^{2+}. (D) An identical perfusion protocol applied to the V275C–A286G mutant failed to provide evidence of an MTSEA$^+$-induced inhibition of I_0, in accordance with V275C–A286G being nonconductive in the absence of Ca^{2+}. The observation that channel activity could not be recovered by the addition of Ca^{2+} following MTSEA$^+$ exposure confirmed that MTSEA$^+$ had access to the cysteines in the channel cavity in zero Ca^{2+}. Originally published in Garneau *et al.* (2009). © The American Society for Biochemistry and Molecular Biology.

The energetics associated to the transfer of a polar or charged residue from a water to a hydrophobic environment require the introduction of additional factors such as H-bound, dipole–dipole and/or Coulombic interactions, which critically depend on the residue microenvironment. Such structural factors remain still ill defined for KCa3.1. If, however, hydrophobic effects constitute the dominant force driving V282 and/or A279 from an aqueous to a nonaqueous environment upon channel closure, there should be a strong correlation between constitutive activity and the free energy of solvation at these particular sites. Selecting residues differing in hydrophobicity relative to Val and/or Ala largely depends on the hydrophobicity scale chosen. In the pure hydrophobicity scale proposed by Karplus, polar atoms of residues are ignored, so that hydrophobic and environment dependent contributions to hydrophobic effects are separated (Karplus, 1997). This scale predicts that the substitution of V282 (3.38 kcal/mol) by residues such as Ala (2.15 kcal/mol), Gln (1.63 kcal/mol), Ser (1.4 kcal/mol), and Gly (1.18 kcal/mol) should lead to a reduced contribution of hydrophobic interactions to the KCa3.1 equilibrium energy, while the substitutions Ile (3.38 kcal/mol) or Leu (4.10 kcal/mol) would be equivalent to Val in terms of hydrophobic effects. By systematically varying the hydrophobicity of the residue at position 282, we found a strong correlation between free energy of solvation and constitutive activity, with residues characterized by a hydrophobic energy for side chain burial less than 1.2 kcal/mol compared to Val more likely to lead to constitutively active channels (Garneau *et al.*, 2009; see Fig. 24.5). This observation truly supports a strategy whereby constitutively active KCa3.1 channels can be generated by decreasing the hydrophobicity of the residues at position 282 in S6, thus stabilizing the channel open configuration. The same approach applied to the A279 residue failed, however, to produce A279 mutants constitutively active, despite A279G being active in zero Ca^{2+}. This observation points toward either the presence of a side chain at 279 causing systematically a destabilization of the channel open state, or to an effect directly linked to the special structural properties of the Gly residue, likely to be due to an increase in flexibility of the S6 segment at this site.

2.5. Are constitutively active KCa3.1 representative of the channel open configuration?

The question remains to what extent the constitutively active V275C-A279G and V275C-V282G mutant channels are representative of the wild-type KCa3.1 in the open state. Formally, the pore structure of the constitutively active V275C-A279G and V275C-V282G channels can be probed by measuring the accessibility of MTS reagents to cysteine residues engineered at position 275. We showed in a previous work that the modification rate of the V275C mutant by $MTSEA^+$ (4.8 Å diameter) is poorly state dependent, in contrast to $MTSET^+$ which yielded modification rates at least 10^3-fold slower

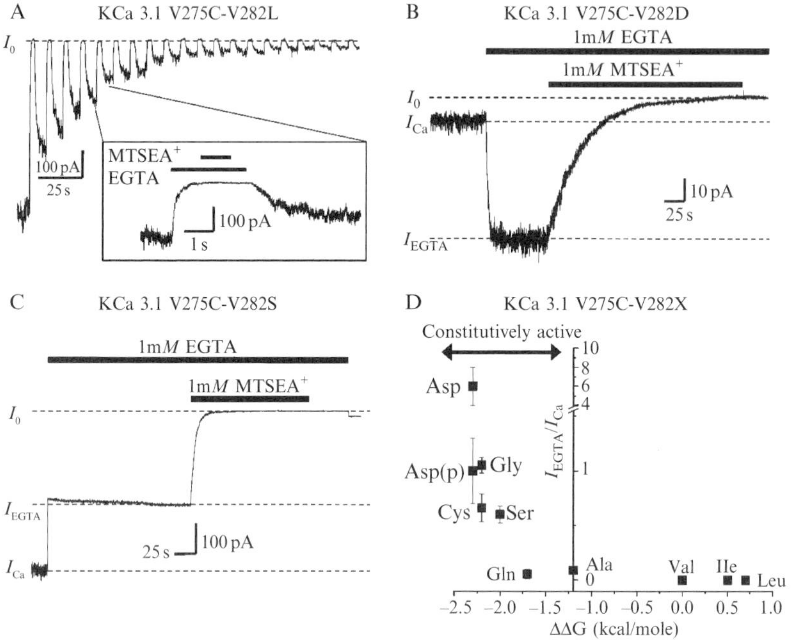

Figure 24.5 Inside-out current recordings illustrating the effect on constitutive activation of mutating V282 by residues of different size and/or hydrophobicity. Inward current recordings performed in symmetrical 200 m*M* K_2SO_4 conditions at $Vm = -60$ mV. The symbols I_{Ca}, I_{EGTA} and I_0, refer to the current recorded at saturating 25 μ*M* Ca^{2+}, zero Ca^{2+} or following inhibition with 1 m*M* $MTSEA^+$. Ca^{2+}-activated K^+ currents are represented as inward currents relative to the zero current level I_0. The substitution V282L (A) did not result in a constitutively active channel as demonstrated by the absence of current variations in zero Ca^{2+} ($I_0 = I_{EGTA}$) despite a progressive block of the test inward currents in 25 μ*M* Ca^{2+} by $MTSEA^+$. Application of $MTSEA^+$ in zero Ca^{2+} (EGTA) caused, however, a clear current inhibition with the V275C–V282D (B) and V275C–V282S (C) channel mutants indicating that the V282D and V282S mutations successfully led to constitutive activation. Notably, in contrast to V275C–V282S, the inward current measured with the V275C–V282D mutant was higher in zero Ca^{2+} (I_{EGTA}) than in 25 μ*M* Ca^{2+} conditions (I_{Ca}). These results are summarized in (D) illustrating the correlation between constitutive activation and hydrophobic energy for side chain burial. Energies were taken from Karplus (1997) and expressed relative to Val. This analysis suggests that residues with a hydrophobic energy for side chain burial less than 1.2 kcal/mol compared to Val are more likely to lead to constitutively active channels. Asp (p) refers to the predicted I_{EGTA}/I_{Ca} for V275C–V282D when the channel open probability in 25 μ*M* is corrected for the blocking action of Ca^{2+} due to Ca^{2+} binding to the channel selectivity filter. Originally published in Garneau *et al.* (2009). © The American Society for Biochemistry and Molecular Biology.

in zero than in saturating Ca^{2+} conditions (Klein *et al.*, 2007). The accessibility of cysteine residues at 275 to $MTSET^+$ can thus be used to assess the conformational state of the open and closed KCa3.1 pore structures. If the S6 segment of the V275C-A279G and V275C-V282G mutants moves minimally in response of Ca^{2+} binding to the CaM/KCa3.1 complex, we expect little difference in $MTSET^{+}$ accessibility for cysteines at 275 with and without Ca^{2+}. Conversely, if the binding of Ca^{2+} to the CaM/KCa3.1 complex still induces a movement of the S6 segment, we expect the modification rate by $MTSET^{+}$ of the cysteine residue at 275 to be Ca^{2+} dependent despite little effect on channel open probability. Finally, if the rates of modification by $MTSET^{+}$ of the V275C-A279G and V275C-V282G channels in zero Ca^{2+} correspond to the rates measured for the open V275C mutant, we will conclude that the pore structure of the constitutive active state of these double mutants is structurally equivalent to the KCa3.1 open configuration. The results of these experiments are summarized in Fig. 24.6. Modification rates were calculated as previously described. Significant differences ($p < 0.05$) were seen between the modification rates by $MTSET^{+}$ measured with and without Ca^{2+} for the V275C channel and for the two constitutively active mutant channels. This observation indicates that despite

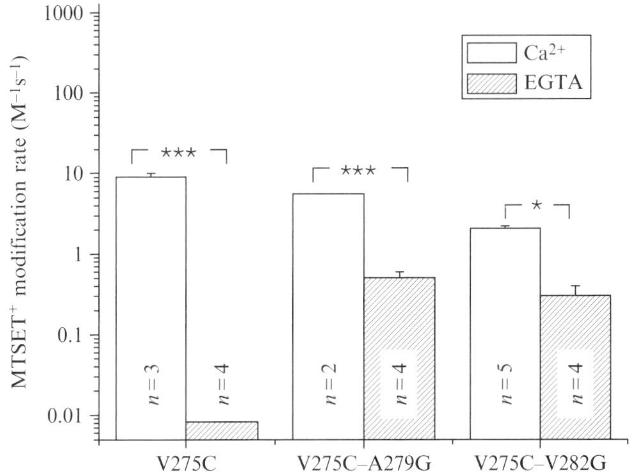

Figure 24.6 Bar graph illustrating the state dependent accessibility to $MTSET^{+}$ of cysteine residues at position 275. *p* values of less than 0.05 and 0.0005 are represented as * and ***, respectively. The modification rates measured for the V275C channel and the two V275C–V282G and V275C–A279G mutants differed significantly ($p < 0.05$) with and without Ca^{2+}. However, whereas the modification rates measured in zero and 25 μM Ca^{2+} for the V275C channel differed by 10^3- to 10^4-fold, this difference is reduced to less than 10-fold for the constitutively active channels. This effect is attributable to the modification rates in zero Ca^{2+}, which appeared 50–100 times faster for the constitutively active mutants compared to the V275C control channel. These observations suggest that the pore structure of the V275C–V282G and V275C–A279G mutants in zero Ca^{2+} better approximates the V275C open than closed configurations. Originally published in Garneau *et al.* (2009). © The American Society for Biochemistry and Molecular Biology.

constitutive activity both mutant channels remained Ca^{2+} sensitive. However, whereas the modification rates by $MTSET^+$ measured in zero and 25 μ*M* Ca^{2+} for the V275C channel differed by 10^3- to 10^4-fold, this difference was reduced to less than 10-fold for the constitutively active channels, an effect attributable to the modification rates in zero Ca^{2+} which appeared 50–100 times faster for the constitutively active mutants compared to the V275C control channel. These results strongly suggest that mutating the residues at positions 279 and 282 into Gly affects the channel geometry in zero Ca^{2+} so that molecules such as $MTSET^+$ with a van der Waals diameter of 5.8 Å have now a greater access to the channel cavity comparatively to the closed V275C mutant. Finally, our results showed that the accessibility of $MTSET^+$ to V275C in 25 μ*M* Ca^{2+} is significantly slower ($p < 0.05$) for the V275C-V282G mutant relative to V275C, an indication that the geometry of the constitutively active state is not totally equivalent to the wild-type open state. Altogether these observations argue for a structure of the constitutively active KCa3.1 that is relatively close to the channel open state without, however, being totally equivalent.

3. Concluding Remarks

We have presented a general strategy by which constitutively active KCa3.1 channels can be generated. Our strategy includes (1) the production of a 3D model of the channel pore region coupled to SCAM experiments to identify residues in the S6 transmembrane segment facing the channel pore, (2) a Gly scan of the S6 segment to determine to what extent the interactions between the side chains of key residues along S6 with their surrounding milieu contribute to the open/closed KCa3.1 equilibrium energy, and (3) a perturbation of the channel open/closed equilibrium by varying the contribution of hydrophobic effects to the energy balance leading to constitutive activity. The procedure described in step 1 is essential to establish if constitutive activity cannot be generated by simply removing steric constraints along the channel pore so that the channel would be ion conducting in the closed state. This mechanism does not seem applicable to KCa3.1. In contrast, perturbing the hydrophobic energy of the S6 segment by substituting the Val at position 282 by more hydrophilic amino acids was sufficient to lock the channel in an open-like state. The production of constitutively active mutants thus appears to be tightly linked to the energy balance between the channel open and closed states. In this regard, the results obtained with KCa3.1 are representative of a general mechanism already documented for the Shaker, GIRK2, and KIR3.1/KIR3.4 channels, where constitutive activity is strongly correlated to the energetics of the channel S6 transmembrane segment.

REFERENCES

Baker, D., and Sali, A. (2001). Protein structure prediction and structural genomics. *Science* **294,** 93–96.

Bruening-Wright, A., Schumacher, M. A., Adelman, J. P., and Maylie, J. (2002). Localization of the activation gate for small conductance Ca2+-activated K+ channels. *J. Neurosci.* **22,** 6499–6506.

Clayton, G. M., Altieri, S., Heginbotham, L., Unger, V. M., and Morais-Cabral, J. H. (2008). Structure of the transmembrane regions of a bacterial cyclic nucleotide-regulated channel. *Proc. Natl. Acad. Sci. USA* **105,** 1511–1515.

Doyle, D. A., Cabral, J. M., Pfuetzner, R. A., Kuo, A., Gulbis, J. M., Cohen, S. L., Chait, B. T., and MacKinnon, R. (1998). The structure of the potassium channel: Molecular basis of K^+ conduction and selectivity. *Science* **280,** 69–77.

Feletou, M., and Vanhoutte, P. M. (2007). Endothelium-dependent hyperpolarizations: Past beliefs and present facts. *Ann. Med.* **39,** 495–516.

Flynn, G. E., and Zagotta, W. N. (2001). Conformational changes in S6 coupled to the opening of cyclic nucleotide-gated channels. *Neuron* **30,** 689–698.

Flynn, G. E., and Zagotta, W. N. (2003). A cysteine scan of the inner vestibule of cyclic nucleotide-gated channels reveals architecture and rearrangement of the pore. *J. Gen. Physiol.* **121,** 563–582.

Garneau, L., Klein, H., Banderali, U., Longpre-Lauzon, A., Parent, L., and Sauve, R. (2009). Hydrophobic interactions as key determinants to the KCa3.1 channel closed configuration: An analysis of KCa3.1 mutants constitutively active in zero Ca2+. *J. Biol. Chem.* **284,** 389–403.

Grgic, I., Eichler, I., Heinau, P., Si, H., Brakemeier, S., Hoyer, J., and Kohler, R. (2005). Selective blockade of the intermediate-conductance Ca2+-activated K+ channel suppresses proliferation of microvascular and macrovascular endothelial cells and angiogenesis in vivo. *Arterioscler. Thromb. Vasc. Biol.* **25,** 704–709.

Grgic, I., Kiss, E., Kaistha, B. P., Busch, C., Kloss, M., Sautter, J., Muller, A., Kaistha, A., Schmidt, C., Raman, G., Wulff, H., Strutz, F., *et al.* (2009). Renal fibrosis is attenuated by targeted disruption of KCa3.1 potassium channels. *Proc. Natl. Acad. Sci. USA* **106,** 14518–14523.

Holmgren, M., Liu, Y., Xu, Y., and Yellen, G. (1996). On the use of thiol-modifying agents to determine channel topology. *Neuropharmacology* **35,** 797–804.

Jager, H., Dreker, T., Buck, A., Giehl, K., Gress, T., and Grissmer, S. (2004). Blockage of intermediate-conductance Ca2+-activated K+ channels inhibit human pancreatic cancer cell growth in vitro. *Mol. Pharmacol.* **65,** 630–638.

Karlin, A., and Akabas, M. H. (1998). Substituted-cysteine accessibility method. *Methods Enzymol.* **293,** 123–145.

Karplus, P. A. (1997). Hydrophobicity regained. *Protein Sci.* **6,** 1302–1307.

Khanna, R., Chang, M. C., Joiner, W. J., Kaczmarek, L. K., and Schlichter, L. C. (1999). hSK4/hIK1, a calmodulin-binding KCa channel in human T lymphocytes. Roles in proliferation and volume regulation. *J. Biol. Chem.* **274,** 14838–14849.

Klein, H., Garneau, L., Banderali, U., Simoes, M., Parent, L., and Sauvé, R. (2007). Structural determinants of the closed KCa3.1 channel pore in relation to channel gating: Results from a substituted cysteine accessibility analysis. *J. Gen. Physiol.* **129,** 299–315.

Klein, H., Garneau, L., Trinh, N. T., Prive, A., Dionne, F., Goupil, E., Thuringer, D., Parent, L., Brochiero, E., and Sauve, R. (2009). Inhibition of the KCa3.1 channels by AMP-activated protein kinase in human airway epithelial cells. *Am. J. Physiol. Cell Physiol.* **284,** 389–403.

Kohler, m., Hirschberg, B., Bond, C. T., Kinzie, J. M., Marrion, N. V., Maylie, J., and Adelman, J. P. (1996). Small-conductance, calcium-activated potassium channels from mammalian brain. *Science* **273,** 1709–1714.

Kohler, R., Wulff, H., Eichler, I., Kneifel, M., Neumann, D., Knorr, A., Grgic, I., Kampfe, D., Si, H., Wibawa, J., Real, R., Borner, K., *et al.* (2003). Blockade of the intermediate-conductance calcium-activated potassium channel as a new therapeutic strategy for restenosis. *Circulation* **108,** 1119–1125.

Kuo, A., Gulbis, J. M., Antcliff, J. F., Rahman, T., Lowe, E. D., Zimmer, J., Cuthbertson, J., Ashcroft, F. M., Ezaki, T., and Doyle, D. A. (2003). Crystal structure of the potassium channel KirBac1.1 in the closed state. *Science* **300,** 1922–1926.

Laskowski, R. A., MacArthur, M. W., Moss, D. S., and Thornton, J. M. (1993). PROCHECK: A program to check the stereochemical quality of protein structures. *J. Appl. Crystallogr.* **26,** 283–291.

Long, S. B., Campbell, E. B., and MacKinnon, R. (2005). Crystal structure of a mammalian voltage-dependent Shaker family K+ channel. *Science* **309,** 897–903.

Lu, Q., and Miller, C. (1995). Silver as a probe of pore-forming residues in a potassium channel. *Science* **268,** 304–307.

Marcus, Y. (1988). Ionic radii in aqueous solutions. *Chem. Rev.* **88,** 1475–1498.

Maylie, J., Bond, C. T., Herson, P. S., Lee, W. S., and Adelman, J. P. (2004). Small conductance Ca2+-activated K+ channels and calmodulin. *J. Physiol.* **554,** 255–261.

Ouadid-Ahidouch, H., Roudbaraki, M., Delcourt, P., Ahidouch, A., Joury, N., and Prevarskaya, N. (2004). Functional and molecular identification of intermediate-conductance Ca(2+)-activated K(+) channels in breast cancer cells: Association with cell cycle progression. *Am. J. Physiol. Cell Physiol.* **287,** C125–C134.

Rufo, P. A., Merlin, D., Riegler, M., Ferguson-Maltzman, M. H., Dickinson, B. L., Brugnara, C., Alper, S. L., and Lencer, W. I. (1997). The antifungal antibiotic, clotrimazole, inhibits chloride secretion by human intestinal T84 cells via blockade of distinct basolateral K+ conductances. Demonstration of efficacy in intact rabbit colon and in an in vivo mouse model of cholera. *J. Clin. Invest.* **100,** 3111–3120.

Sali, A., and Blundell, T. L. (1993). Comparative protein modelling by satisfaction of spatial restraints. *J. Mol. Biol.* **234,** 779–815.

Schumacher, M. A., Rivard, A. F., Bachinger, H. P., and Adelman, J. P. (2001). Structure of the gating domain of a Ca2+-activated K+ channel complexed with Ca2+/calmodulin. *Nature* **410,** 1120–1124.

Schumacher, M. A., Crum, M., and Miller, M. C. (2004). Crystal structures of apocalmodulin and an apocalmodulin/SK potassium channel gating domain complex. *Structure (Camb.)* **12,** 849–860.

Shannon, R. D. (1976). Revised effective ionic radii and systematic studies of interatomic distances in halides and chalcogenides. *Acta Crystallogr.* **A32,** 751–767.

Singh, S., Syme, C. A., Singh, A. K., Devor, D. C., Bridges, R. J., Lambert, L. C., DeLuca, A., and Frizzell, R. A. (2001). Benzimidazolone activators of chloride secretion: Potential therapeutics for cystic fibrosis and chronic obstructive pulmonary disease. Bicarbonate and chloride secretion in Calu-3 human airway epithelial cells. *J. Gen. Physiol.* **296,** 600–611.

Srivastava, S., Choudhury, P., Li, Z., Liu, G., Nadkarni, V., Ko, K., Coetzee, W. A., and Skolnik, E. Y. (2006a). Phosphatidylinositol 3-phosphate indirectly activates KCa3.1 via 14 amino acids in the carboxy terminus of KCa3.1. *Mol. Biol. Cell* **17,** 146–154.

Srivastava, S., Li, Z., Ko, K., Choudhury, P., Albaqumi, M., Johnson, A. K., Yan, Y., Backer, J. M., Unutmaz, D., Coetzee, W. A., and Skolnik, E. Y. (2006b). Histidine phosphorylation of the potassium channel KCa3.1 by nucleoside diphosphate kinase B is required for activation of KCa3.1 and CD4 T cells. *Mol. Cell* **24,** 665–675.

Szkotak, A. J., Murthy, M., MacVinish, L. J., Duszyk, M., and Cuthbert, A. W. (2004). 4-Chloro-benzo[F]isoquinoline (CBIQ) activates CFTR chloride channels and KCNN4 potassium channels in Calu-3 human airway epithelial cells. *Br. J. Pharmacol.* **142,** 531–542.

Webster, S. M., del Camino, D., Dekker, J. P., and Yellen, G. (2004). Intracellular gate opening in shaker K+ channels defined by high-affinity metal bridges. *Nature* **428,** 864–868.

Wissmann, R., Bildl, W., Neumann, H., Rivard, A. F., Klocker, N., Weitz, D., Schulte, U., Adelman, J. P., Bentrop, D., and Fakler, B. (2002). A helical region in the C-terminus of small-conductance Ca2+-activated K+ channels controls assembly with apo-calmodulin. *J. Biol. Chem.* **277,** 4558–4564.

Wulff, H., Miller, M. J., Hansel, W., Grissmer, S., Cahalan, M. D., and Chandy, K. G. (2000). Design of a potent and selective inhibitor of the intermediate-conductance Ca2+-activated K+ channel, IKCa1: A potential immunosuppressant. *Proc. Natl. Acad. Sci. USA* **97,** 8151–8156.

Wulff, H., Knaus, H. G., Pennington, M., and Chandy, K. G. (2004). K+ channel expression during B cell differentiation: Implications for immunomodulation and autoimmunity. *J. Immunol.* **173,** 776–786.

Xiao, J., Zhen, X. G., and Yang, J. (2003). Localization of PIP2 activation gate in inward rectifier K+ channels. *Nat. Neurosci.* **6,** 811–818.

Yi, B. A., Lin, Y. F., Jan, Y. N., and Jan, L. Y. (2001). Yeast screen for constitutively active mutant G protein-activated potassium channels. *Neuron* **29,** 657–667.

CHAPTER TWENTY-FIVE

Fusion Proteins as Model Systems for the Analysis of Constitutive GPCR Activity

Erich H. Schneider* *and* Roland Seifert†

Contents

Abstract

In many cases, the coexpression of GPCRs with G-proteins and/or regulators of G-protein signaling (RGS-proteins) allows a successful reconstitution of high-affinity agonist binding and functional responses. However, in some cases, coexpressed GPCRs and G-proteins interact inefficiently, resulting in weak [^{35}S]GTPγS- and

* Laboratory of Molecular Immunology, NIAID/NIH, Bethesda, Maryland, USA
† Institute of Pharmacology, Medical School of Hannover, Hannover, Germany

Methods in Enzymology, Volume 485
ISSN 0076-6879, DOI: 10.1016/S0076-6879(10)85025-6

steady-state GTPase assay signals. This may be, for example, caused by a rapid dissociation of the G-protein from the plasma membrane, as has been reported for $G\alpha_S$. Moreover, for a detailed characterization of GPCR/G-protein interactions, it may be required to work with a defined GPCR/G-protein stoichiometry and to avoid cross-interaction with endogenous G-proteins. Cross-talk to endogenous G-proteins has been shown to play a role in some mammalian expression systems. These problems can be addressed by the generation of GPCR–Gα fusion proteins and their expression in Sf9 insect cells. When the C-terminus of the receptor is fused to the N-terminus of the G-protein, a 1:1 stoichiometry of both proteins is achieved. In addition, the close proximity of GPCR and G-protein in fusion proteins leads to enhanced interaction efficiency, resulting in increased functional signals. This approach can also be extended to fusion proteins of GPCRs with RGS-proteins, specifically when steady-state GTP hydrolysis is used as read-out. GPCR–RGS fusion proteins optimize the interaction of RGS-proteins with coexpressed Gα subunits, since the location of the RGS-protein is close to the site of receptor-mediated G-protein activation. Moreover, in contrast to coexpression systems, GPCR–Gα and GPCR–RGS fusion proteins provide a possibility to imitate physiologically occurring interactions, for example, the precoupling of receptors and G-proteins or the formation of complexes between GPCRs, G-proteins and RGS-proteins (transducisomes). In this chapter, we describe the technique for the generation of fusion proteins and show the application of this approach for the characterization of constitutively active receptors.

1. Introduction

In GPCR/G-protein coexpression systems, it is sometimes difficult to perform sensitive functional assays. As previously reported for the β_2AR/$G\alpha_s$ interaction (Seifert *et al.*, 1999b), specifically $G\alpha_s$ proteins tend to rapidly dissociate from the membrane, precluding a continuous stimulation by the agonist-activated receptor. This results only in weak [^{35}S]GTPγS and steady-state GTPase signals (Seifert and Wieland, 2005). However, when the C-terminus of the receptor is fused to the N-terminus of the G-protein (Fig. 25.1A), both interaction partners are in close proximity. In the case of the β_2AR, this considerably increased the functional signal and restored high-affinity agonist binding (Bertin *et al.*, 1994; Seifert *et al.*, 1998a). Depending on the cellular environment, the close proximity of receptor and G-protein can reduce cross-interaction with other G-protein types present in the cell. However, to completely avoid such cross-activation of other G-proteins, use of Sf9 cells as an expression system is recommended (Seifert *et al.*, 1999c). By photoaffinity-labeling of G-protein α-subunits with [γ-^{32}P]GTP azidoanilide, cross-talk of the G_i-protein-coupled formyl peptide receptor (FPR) to endogenous G-proteins was excluded (Wenzel-Seifert *et al.*, 1999), whereas such cross-talk is a serious problem in

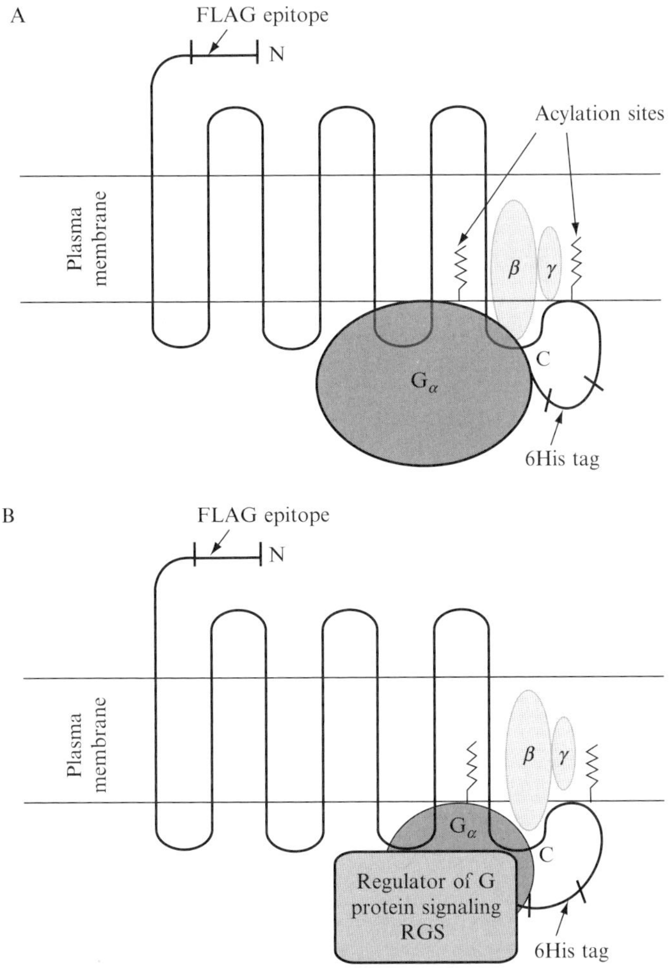

Figure 25.1 Structure of N-terminally FLAG-tagged GPCR–Gα and GPCR–RGS fusion proteins. (A) Schematic representation of a GPCR–Gα fusion protein. The C-terminus of the GPCR is fused to the N-terminus of the Gα subunit, using a 6-His tag as a linker. The receptor as well as the Gα subunit may be anchored in the membrane via acyl residues. The fusion results in close physical proximity of both proteins, enhancing the interaction efficiency of Gα with the corresponding site at the GPCR, for example, the third intracellular loop. In addition, the Gβγ complex is depicted that forms the heterotrimeric G-protein with Gα. (B) Schematic representation of a GPCR–RGS fusion protein. The N-terminus of the regulator of G-protein signaling is fused to the C-terminus of the FLAG-tagged receptor by a 6-His linker. The fusion protein brings the RGS-protein into close proximity to the coexpressed Gα subunit. In addition, the Gβγ complex is depicted.

mammalian expression systems (Burt *et al.*, 1998; Sautel and Milligan, 1998). In order to compensate for this problem, pertussis toxin-insensitive G-protein α-subunits have to be fused to GPCRs and expressed in pertussis toxin-treated cells (Burt *et al.*, 1998; Sautel and Milligan, 1998). Unfortunately, those mutants may themselves affect receptor/G-protein coupling substantially (Carr *et al.*, 1998). Again, these experimental complications argue for preferentially using Sf9 insect cells as expression system for GPCR–Gα fusion proteins, particularly when G_i/G_o-proteins are considered.

An additional advantage of fusion proteins is the 1:1 stoichiometry of receptor and G-protein, which allows a more detailed characterization of receptor/G-protein activation in terms of pharmacological receptor profile (specifically for agonists and inverse agonists) and the kinetics of G-protein activation. Moreover, G-protein coupling specificity can easily be determined by comparison of fusion proteins consisting of a specific GPCR and different Gα subunits. This was, for example, described for the FPR (Wenzel-Seifert *et al.*, 1999).

The fusion protein approach is not confined to GPCR–Gα fusion proteins. Also a regulator of G-protein signaling (RGS-protein), which stimulates the intrinsic GTPase activity of Gα subunits, can be fused to a GPCR (Fig. 25.1B), yielding a fully functional protein (Bahia *et al.*, 2003). This increases the probability of complex formation between the RGS-protein and the receptor-activated Gα subunit and enhances the effect of RGS-proteins on steady-state GTPase activity (Schneider and Seifert, 2009). An implication of a stimulatory effect of RGS-proteins on receptor-mediated GTP hydrolysis is that GTP hydrolysis becomes the rate-limiting step of the G-protein cycle. This could be due to a relative paucity of G-proteins relative to receptors as has been reported for the visual rhodopsin/transducin system (Burns and Pugh, 2009).

In this chapter, we explain in detail, how fusion protein cDNAs are generated. We compare the pharmacological behavior of fusion proteins with the properties of the corresponding coexpression systems and show the advantages of the fusion protein approach to addressing specific problems of GPCR signal transduction research.

2. Expression of Fusion Proteins: hH_4R–$G\alpha_{i2}$ and hH_4R–GAIP as Paradigms

2.1. Molecular biology: Generation of fusion proteins

Fusion proteins are generated by the method of overlap extension PCR, which is shown in Fig. 25.2. In the following, we describe the use of this method to generate DNA encoding the hH_4R–$G\alpha_{i2}$ fusion protein. The hexahistidine tagged C-terminus of FLAG-hH_4R was fused to the

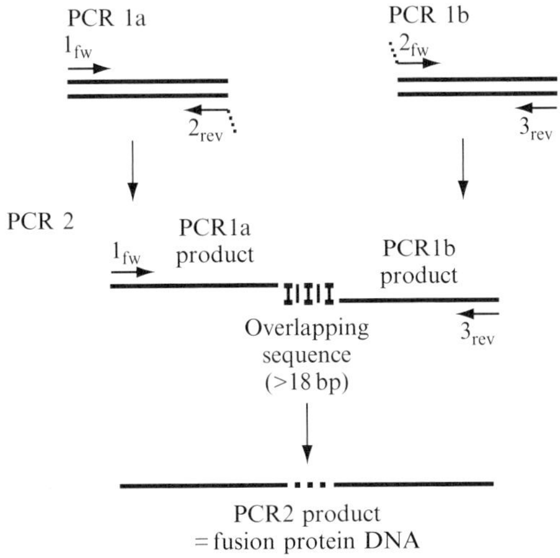

Figure 25.2 Overlap extension PCR to generate a fusion protein gene. The two fragments to be fused are synthesized in the two PCR steps 1a and 1b. The design of the 2_{rev} and 2_{fw} primer allows the generation of an overlapping linker sequence (at least 18 bp, e.g., a 6-His tag) that is shared by the products from PCR 1a and 1b. The primers 1_{fw} and 3_{rev} may serve to introduce restriction sites, additional tags or a 3′ stop codon. The products from PCR 1a and 1b are fused in PCR2, using the primers 1fw and 3rev. The fusion becomes possible, because the products from PCR 1a and 1b can hybridize *via* their overlapping sequence (in case of the proteins described in this chapter at least 6-His tag). The PCR product is then cut by restriction enzymes and cloned into a target plasmid.

N-terminus of $G\alpha_{i2}$. Using the hexahistidine tag as a linker between receptor and G-protein has several advantages. First, the fusion protein is better protected against proteolysis by divalent cation-dependent proteases. Second, it is, in principle, possible to purify the fusion protein by nickel chromatography and third, the fusion protein can be easily detected by immunoblotting, using an anti-His_6 antibody. The length of the hexahistidine tag is appropriate to allow an efficient interaction between receptor and G-protein. The fusion of FLAG-hH_4R-His_6 with $G\alpha_{i2}$ is performed in three steps (Fig. 25.2).

Four primers are required. Primer 1fw encodes bp 526–561 of the hH_4R sequence (5′-GCC ATC ACA TCA TTC TTG GAA TTC GTG ATC CCA GTC-3′) and contains an *Eco*RI restriction site, which occurs only once in hH_4R, but not in $G\alpha_{i2}$ and the intended fusion protein sequence. Primer 3rev encodes the C-terminus of $G\alpha_{i2}$ (5′-GGT CGA CTC TAG AGG TCA GAA GAG GCC ACA GTC-3′). Both primers contain specific restriction sites (1fw: *Eco*RI; 3rev: *Xba*I) that allow further processing of the product, for example, cloning of the fusion protein DNA in a target plasmid. It is also important to include a stop codon in primer 3rev. Primer

2fw (5′-CAC CAT CAT CAC CAT CAC ATG GGC TGC ACC GTG AGC-3′) encodes the 18 bp sequence of the His_6 DNA and additionally an 18 bp part of the $G\alpha_{i2}$ N-terminus. Primer 2rev is complementary to 2fw (5′-GCT CAC GGT GCA GCC CAT GTG ATG GTG ATG ATG GTG-3′).

In PCR 1a the sequence between primer 1fw, which contains the *Eco*RI site of the hH_4R, and the 2rev fusion primer is amplified using the pGEM-3Z-SF-H_4R-His_6 plasmid as template (Fig. 25.2). In PCR 1b the $G\alpha_{i2}$ sequence between the 2fw fusion primer and the antisense primer 3rev is amplified, using the pGEM-3Z-SF-β_2AR-His_6-$G\alpha_{i2}$ plasmid as template. The product of PCR 1a is a part of the hH_4R sequence with a hexahistidine sequence and a part of the $G\alpha_{i2}$ N-terminus on the 3′ end. PCR 1b yields the $G\alpha_{i2}$ sequence with a hexahistidine DNA at the 5′ end. In PCR 2 the products of PCR 1a and 1b were used as templates together with the primers 1fw and 3rev. Since both templates can hybridize in the region of the hexahistidine sequence and the $G\alpha_{i2}$ N-terminus, PCR2 yields the fusion protein DNA, encoding a part of the hH_4R (beginning at the *Eco*RI site), followed by a hexahistidine tag and the $G\alpha_{i2}$ sequence with an *Xba*I site 3′ of the stop codon (Fig. 25.2).

This fragment is then digested with *Eco*RI and *Xba*I and cloned into a pGEM-3Z-SF-hH_4R-His_6 plasmid, which was digested before with the same enzymes to obtain the full-length fusion protein DNA sequence. Alternatively, it is also possible to synthesize the complete FLAG-tagged hH_4R-His_6-$G\alpha_{i2}$ DNA by overlap extension PCR instead of starting at the *Eco*RI site of hH_4R. However, then the product, which has to be generated in PCR2 is markedly larger, which may reduce product yield and increases the probability of point mutations. In any case, confirmation of the PCR-amplified cDNA sequences is mandatory.

The same approach can be applied to synthesize any fusion protein, for example, also hH_4R–RGS4 and hH_4R–GAIP (Schneider and Seifert, 2009). It is only important that the sequence, which is shared by the PCR1a and PCR1b product, is sufficiently long to produce a stable DNA hybridization in PCR2. Normally, this is achieved by sequences of at least 18 bp.

2.2. Protein expression in Sf9 cell membranes

The methods of Sf9 cell culture, baculovirus generation and membrane preparation are described in Chapter 28. If not indicated otherwise, in the following, hH_4R–$G\alpha_{i2}$ is coexpressed with $G\beta_1\gamma_2$. The hH_4R–RGS4- as well as the hH_4R–GAIP fusion protein is always coexpressed with $G\alpha_{i2}$ and $G\beta_1\gamma_2$. These fusion protein systems are compared with the nonfused hH_4R, which is coexpressed with $G\alpha_{i2}$ and $G\beta_1\gamma_2$.

The immunoblot shown in Fig. 25.3 demonstrates that hH_4R–$G\alpha_{i2}$ and the fusion proteins of hH_4R with RGS4 and GAIP are readily expressed in

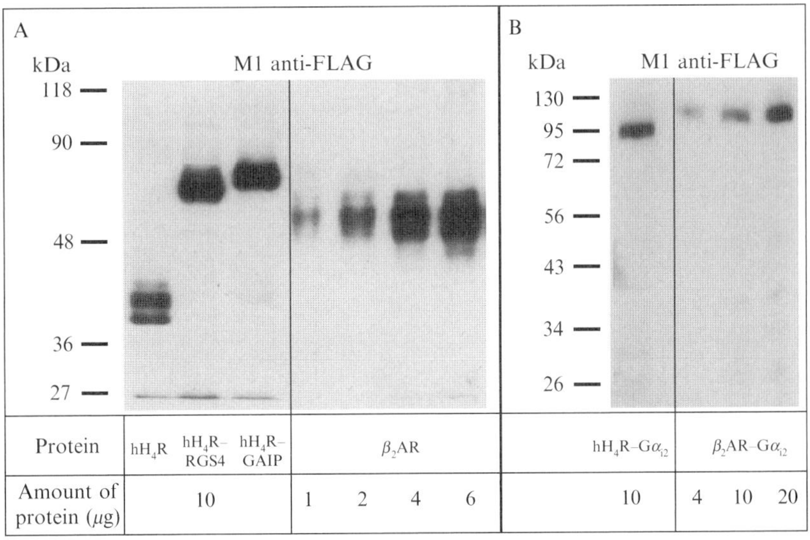

Figure 25.3 Immunoblotting of proteins, expressed in Sf9 cell membranes. The data were taken from previous publications: (A) Adapted from Schneider and Seifert (2009). (B) Adapted with permission of Schneider *et al.* (2009). Copyright 2009 American Chemical Society. (A) Lanes 1–3: Staining of FLAG-epitope tagged hH_4R (lane 1), hH_4R–RGS4 (lane 2), and hH_4R–GAIP (lane 3) (all coexpressed with $G\alpha_{i2}$ and $G\beta_1\gamma_2$) with the M1 anti-FLAG antibody. Lanes 4–7: Increasing concentrations of a standard membrane, expressing 7.5 pmol/mg of FLAG-tagged β_2AR (determined by saturation binding with [^{3}H]dihydroalprenolol). Comparison of the β_2AR band intensities with the signals in lanes 1–3 yielded estimated expression levels of 1.8 pmol/mg (hH_4R), 3.1 pmol/mg (hH_4R–RGS4), and 3.0 pmol/mg (hH_4R–GAIP). (B) Lane 1: FLAG-tagged hH_4R–$G\alpha_{i2}$ fusion protein (coexpressed with $G\beta_1\gamma_2$), stained with the M1 anti-FLAG antibody. Lanes 2–4: Increasing concentrations of a standard membrane, expressing 3.5 pmol/mg of FLAG-tagged β_2AR–$G\alpha_{i2}$ fusion protein (determined by saturation binding with [^{3}H]dihydroalprenolol). Comparison of the band intensities with the signal in lane 1 (hH_4R–$G\alpha_{i2}$) yielded 7 pmol/mg.

Sf9 cell membranes. Interestingly, the fusion proteins are always expressed at a higher level than wild-type hH_4R. A comparison of the hH_4R–, hH_4R–RGS4, and hH_4R–GAIP signals (Fig. 25.3A, lanes 1–3) with different concentrations of a FLAG-β_2AR expressing standard membrane (7.5 pmol/mg, determined by [^{3}H]dihydroalprenolol saturation binding; Fig. 25.3A, lanes 4–7) reveals expression levels of 3.1 pmol/mg for hH_4R–RGS4 and 3.0 pmol/mg for hH_4R-GAIP, but only 1.8 pmol/mg for the wild-type hH_4R. The expression level of hH_4R–$G\alpha_{i2}$ (Fig. 25.3B, lane 1) was ~7.0 pmol/mg, as determined with a FLAG-β_2AR-$G\alpha_{i2}$ standard membrane (3.5 pmol/mg, determined by [^{3}H]dihydroalprenolol binding; Fig. 25.3B, lanes 2–4). Possibly, the $G\alpha_{i2}$ subunit fused to hH_4R enhances correct folding of the receptor protein by a chaperone-like effect or simply stabilizes the receptor protein in the cell membrane. The higher expression levels of hH_4R–RGS4 and hH_4R–GAIP may also be caused by a chaperone-like effect of the RGS-protein part or by an increased recruitment of $G\alpha$ subunits to the membrane, which helps stabilizing the receptor protein. A third explanation could be that $G\alpha_{i2}$ or RGS-proteins fused to hH_4R may simply prevent the receptor from proteolytic degradation. As previously reported (Schneider and Seifert, 2009; Schneider *et al.*, 2009), these differences in expression level between fusion proteins and nonfused receptor were also found in radioligand binding studies with [^{3}H]histamine ([^{3}H] HA).

2.3. Assays for the characterization of GPCR/$G\alpha$/RGS fusion proteins

High-affinity agonist binding assays for the determination of ternary complex formation, steady-state GTPase assays for the measurement of intrinsic GTPase activity of $G\alpha$, as well as [^{35}S]GTPγS binding assays for the detection of G-protein activation have been described in much detail in Chapter 28. These protocols can be applied without any change to the fusion protein systems described in this chapter. The same protein amounts per sample can be used for characterization of both coexpression- and fusion protein systems.

Measurement of adenylyl cyclase (AC) activity is less important in case of $G\alpha_i$-coupled receptors, since the inhibitory effect of most $G\alpha_i$-coupled receptors on AC is only low, even after prestimulation of the system with forskolin. This is particularly true for the Sf9 insect cell expression system (Seifert *et al.*, 2002). Thus, AC assays do not play a role for the characterization of the constitutive activity of hH_4R, which is used in this chapter as a paradigm for a receptor fused to $G\alpha_i$ or RGS-proteins. However, AC assays are highly important for the investigation of $G\alpha_s$-coupled receptors like the β_2AR. A detailed description of the AC assay has been provided in Seifert and Wieland (2005).

3. Investigation of GPCR Constitutive Activity with Fusion Proteins: H_4R as Paradigm

3.1. [^{3}H]histamine binding

High-affinity agonist binding with [^{3}H]HA revealed that the pharmacological properties of hH_4R–$G\alpha_{i2}$ are very similar to those of the nonfused hH_4R. Neither the K_D value of [^{3}H]HA, nor the K_i values of HA (histamine) and THIO (thioperamide) did significantly differ between hH_4R and hH_4R–$G\alpha_{i2}$ (Schneider *et al.*, 2009). As shown in Fig. 25.4, hH_4R–$G\alpha_{i2}$ also showed

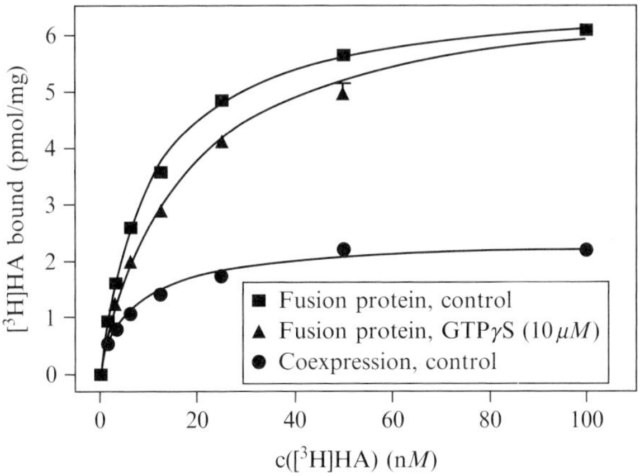

Figure 25.4 High-affinity agonist binding with the $hH_4R–G\alpha_{i2}$ fusion protein (coexpressed with $G\beta_1\gamma_2$). Saturation binding was performed with increasing concentrations of [^{3}H]HA, using membranes expressing $hH_4R–G\alpha_{i2}$ (+$G\beta_1\gamma_2$) under control conditions (■) and in the presence of GTPγS (10 μ*M*, ▲). For comparison, a saturation binding under control conditions, determined with the coexpression system (hH_4R + $G\alpha_{i2}$ + $G\beta_1\gamma_2$), is shown (●). The data shown are from one representative experiment in triplicates, mean ± S.E.M.

a G-protein independent high-affinity state, which could not be abrogated by 10 μ*M* of GTPγS. The only one difference is the B_{max} value, which is markedly higher in case of $hH_4R–G\alpha_{i2}$ compared to hH_4R. This is shown in Fig. 25.4 by the [^{3}H]HA saturation curves for the coexpression system (hH_4R + $G\alpha_{i2}$ + $G\beta_1\gamma_2$) and the fusion protein ($hH_4R–G\alpha_{i2}$ + $G\beta_1\gamma_2$). Interestingly, in contrast to the nonfused hH_4R coexpressed with $G\alpha_{i2}$ and $G\beta_1\gamma_2$ (cf. chapter 28), it is not possible to further increase the expression level of $hH_4R–G\alpha_{i2}$ (coexpressed with $G\beta_1\gamma_2$) by addition of HA or THIO to the cell culture (Schneider *et al.*, 2009). This indicates that $G\alpha_{i2}$ fused to hH_4R may already maximally stabilize the receptor, precluding any further stabilizing effects of the hH_4R ligands.

Also RGS4 and GAIP caused almost no alteration in receptor pharmacology. A comparison of the standard hH_4R ligands histamine, imetit, immepip, R-α-methylhistamine, 5-methylhistamine, iodophenpropit, thioperamide, and JNJ-7777120 (1-[(5-chloro-1H-indol-2-yl)carbonyl]-4-methyl-piperazine) revealed no difference in binding affinities, when hH_4R was compared with hH_4R–GAIP (both coexpressed with $G\alpha_{i2}$ and $G\beta_1\gamma_2$; Schneider and Seifert, 2009). Only hH_4R–RGS4 (coexpressed with $G\alpha_{i2}$ and $G\beta_1\gamma_2$) showed significantly lower binding affinities of histamine and JNJ-7777120, compared to hH_4R (+$G\alpha_{i2}$ + $G\beta_1\gamma_2$; Schneider and Seifert, 2009). A similar effect of RGS4 was previously shown for exogenously added purified RGS4. The potency of UK14304 (5-bromo-6-(2-imidazolin-2-ylamino) quinoxaline) to stimulate GTPase activity of α_{2A}AR-Val351-Gα_{o1} (a fusion protein with a pertussis toxin-resistant G_i-protein mutant) expressed in COS-7 cells was decreased in the presence of 100 n*M* of recombinant RGS4 (Cavalli *et al.*, 2000).

The example of hH_4R–RGS4 indicates that in some cases a fusion protein may slightly pharmacologically differ from the nonfused protein. Thus, if a fusion protein is used for the characterization of receptor ligands, it should first be carefully characterized and compared with the nonfused receptor.

3.2. GTPase assay

The nonfused hH_4R as well as hH_4R–Gα_{i2}, hH_4R–RGS4, and hH_4R–GAIP was characterized in steady-state GTPase enzyme kinetic studies. The effects of HA (agonist) and THIO (inverse agonist) were determined under standard conditions in the presence of 100 n*M* of GTP. Every assay contained 100 m*M* of sodium chloride to suppress a part of the constitutive activity. This increases the agonist-induced signal without considerably influencing the inverse agonist-caused inhibition of GTPase activity. Similarly to the nonfused hH_4R, also hH_4R–RGS4, hH_4R–GAIP, and hH_4R–Gα_{i2} show pronounced sodium insensitivity of the active state (data not shown). A detailed description of the GTPase assay is provided in chapter 28 of this volume.

3.2.1. The hH_4R–Gα_{i2} fusion protein

As indicated by the first triplet of bars in Fig. 25.5A/B, the HA-induced GTPase signal in the coexpression system (hH_4R + Gα_{i2} + G$\beta_1\gamma_2$) is relatively low, hardly 30% above baseline level. Also THIO reduces the baseline activity only by ~30%, compared to control conditions.

Under these conditions it is very difficult to obtain a relative signal that is sufficiently high to allow the accurate characterization of hH_4R ligands in steady-state GTPase assays. As previously demonstrated for the β_2AR (Seifert *et al.*, 1998a), fusion of a GPCR to its cognate Gα subunit can considerably improve signal intensity in steady-state GTPase and in [^{35}S]GTPγS binding assays. Thus, this approach was also applied to the hH_4R. In fact, as shown in Fig. 25.5A, expression of hH_4R–Gα_{i2} (+G$\beta_1\gamma_2$) led to a pronounced increase of absolute GTPase activity. However, the signal intensity is increased to the same extent under control conditions as well as in the presence of histamine or thioperamide, yielding the same relative signal as in the coexpression system. This is shown in Fig. 25.5B, where the GTPase activities are converted to percentage of baseline activity (second triplet). This suggests that the low relative agonist-induced signal of hH_4R is not so much caused by an inefficient G-protein activation, but rather by the high constitutive activity, which constitutes at least 50% of the total ligand-regulated

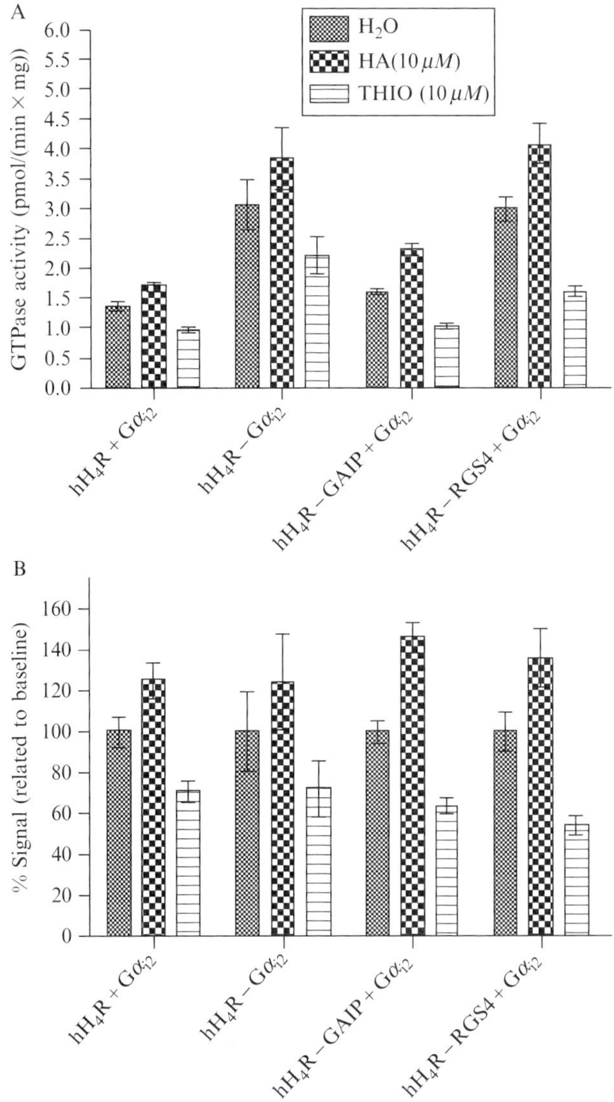

Figure 25.5 Comparison of the standard coexpression system ($hH_4R + G\alpha_{i2} + G\beta_1\gamma_2$) with various fusion protein systems in steady-state GTPase assays (standard conditions, in the presence of 100 n*M* of GTP and 100 m*M* of NaCl). All expression systems shown in this figure additionally contain $G\beta_1\gamma_2$. (A) Absolute GTPase activities under control conditions (first bar of each triplet), in the agonist-stimulated system (HA, 10 μ*M*; second bar) and in the inverse agonist-inhibited system (THIO, 10 μ*M*; third bar). (B) Relative GTPase activities, related to control conditions (set to 100 %, first bar of each triplet), in the presence of HA (10 μ*M*, second bar) and THIO (10 μ*M*, third bar). Data are from four to eight independent experiments with at least three different membranes in triplicates, mean ± S.E.M.

GTPase activity. It should be noted that experiments with coexpression systems (hH_4R + $G\alpha_{i2}$ vs. $G\alpha_{i2}$ alone) revealed that THIO is only a partial inverse agonist, substantially underestimating the total H_4R-regulated GTPase activity (Schneider *et al.*, 2009).

The hH_4R–$G\alpha_{i2}$ fusion protein was also characterized in steady-state GTPase enzyme kinetic experiments. In Fig. 25.6A the percentage of total ligand-regulated GTPase activity in the presence of HA (10 μ*M*) and THIO (10 μ*M*) in the fusion protein system (hH_4R–$G\alpha_{i2}$ + $G\beta_1\gamma_2$) is compared with the corresponding data from the coexpression system (hH_4R + $G\alpha_{i2}$ + $G\beta_1\gamma_2$, shown by the dashed curves). The curves were generated by subtraction of the data under control conditions from the curves recorded in the presence of HA and THIO. This results in a positive curve representing the HA-stimulated GTPase activity and in a negative signal showing the THIO-inhibited constitutive activity. The percentage of total ligand-regulated GTPase activity was calculated by relating all data to $\Delta(V_{\text{max, HA}} - V_{\text{max, THIO}})$. Interestingly, the fusion of hH_4R and $G\alpha_{i2}$ resulted in a decrease of the HA-induced signal from 60% (coexpression system) to 45% (fusion protein system) and the constitutive activity was increased by ~15% to reach a level of 55% (Schneider *et al.*, 2009). At first glance, this is contradictory to the aforementioned finding that the relative HA-induced signal is not changed in case of the fusion protein, compared to the coexpression system (Fig. 25.5A/B). However, the results in Fig. 25.5 were obtained in the presence of 100 n*M* of GTP, which is the standard condition for the characterization of ligands in the steady-state GTPase assay. By contrast, the percentages of HA- and THIO-induced signal in Fig. 25.6A were calculated from GTPase activities occurring at a GTP concentration of 1500 n*M* or higher. Obviously, there is some improvement of hH_4R–$G\alpha_{i2}$ interaction in the fusion protein, which, however, becomes visible only in the presence of high GTP concentrations. Moreover, as recently reported (Schneider and Seifert, 2009), the K_M value of GTP at $G\alpha_{i2}$ was decreased in case of hH_4R–$G\alpha_{i2}$, compared to the coexpression system, under both HA- and THIO conditions. Although this effect did not reach significance, it may indicate that fusion of hH_4R to $G\alpha_{i2}$ improves receptor/G-protein interaction, resulting in an increased GTP affinity of $G\alpha$ in step 3 of the G-protein cycle, where the ternary complex is disrupted by binding of a guanine nucleotide (cf. Fig. 28.2 in Chapter 28).

3.2.2. Fusion of hH_4R with the RGS-proteins RGS4 or GAIP (RGS19)

3.2.2.1. Absolute GTPase activity, relative HA- and THIO-induced signals and K_M values A second approach to increase the signal in steady-state GTPase assays is the expression of a GPCR–RGS fusion protein. RGS-proteins are classified in eight subfamilies, showing high structural diversity. A common feature is the RGS domain, which consists of 120 amino acids

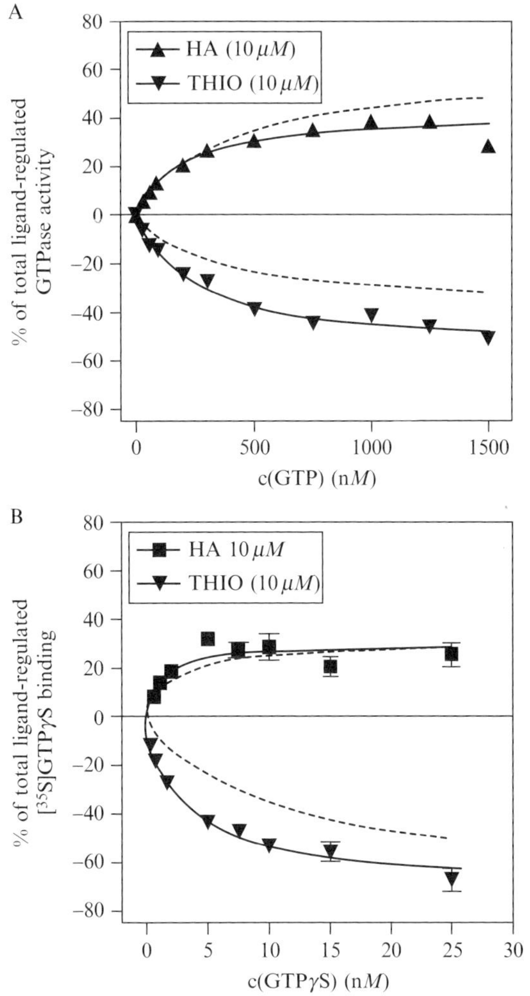

Figure 25.6 Functional analysis of the $hH_4R–G\alpha_{i2}$ fusion protein in steady-state GTPase enzyme kinetics and [^{35}S]GTPγS saturation binding. (A) Steady-state GTPase assay enzyme kinetics of the fusion protein system ($hH_4R–G\alpha_{i2} + G\beta_1\gamma_2$) in the presence of HA (10 μ*M*, ▲) and THIO (10 μ*M*, ▼). The control curve was subtracted from the curves for the HA-stimulated and THIO-inhibited system, yielding a positive signal for the agonist-stimulated GTPase activity and a negative signal for the inverse agonist-inhibited system. The percentage of total ligand-regulated GTPase activity was calculated by relating all GTPase activities to $\Delta(V_{max,\ HA} - V_{max,\ THIO})$. (B) [^{35}S]GTPγS saturation binding, determined with the fusion protein system ($hH_4R–G\alpha_{i2} + G\beta_1\gamma_2$) in the presence of HA (10 μ*M*, ■) and THIO (10 μ*M*, ▼). The control curve was subtracted from the curves for the HA-stimulated and THIO-inhibited system, yielding a positive signal for the agonist-stimulated [^{35}S]GTPγS binding and a negative signal for the inverse agonist-inhibited system. The percentage of total ligand-regulated GTPase activity was calculated by relating all data to $\Delta(B_{max,HA} - B_{max,\ THIO})$. For comparison, in (A) and (B) the corresponding data of the coexpression system ($hH_4R + G\alpha_{i2} + G\beta_1\gamma_2$) are shown as dashed curves. Data are from four to six independent experiments with two to three replicates, mean ± S.E.M. Adapted with permission of Schneider *et al.* (2009). Copyright 2009 American Chemical Society.

and interacts with $G\alpha$ subunits. This increases the intrinsic GTPase activity of $G\alpha$ (Willars, 2006). Thus, RGS-proteins can help enhancing signal intensity in steady-state GTPase assays. As previously reported (Schneider and Seifert, 2009) and also shown in Chapter 28, coexpression of hH_4R ($+G\alpha_{i2} + G\beta_1\gamma_2$) with the RGS-proteins RGS4 and RGS19 (=GAIP; $G\alpha$ interacting protein) did not lead to a significant increase of baseline GTPase activity or HA induced signal. However, when the fusion proteins hH_4R–RGS4 and hH_4R–GAIP are coexpressed with $G\alpha_{i2}$ and $G\beta_1\gamma_2$, pronounced effects on $G\alpha$ activity become visible. Interestingly, GAIP and RGS4 differentially alter the hH_4R-induced GTPase signal. Figure 25.5A shows that hH_4R–RGS4 leads to a general increase of GTPase activity under control conditions as well as in the presence of HA (10 μM) and of THIO (10 μM). By contrast, hH_4R–GAIP results in an almost unaltered baseline (control conditions) and a selective enhancement of the agonist-induced (HA, 10 μM) signal. As shown by Fig. 25.5B, both hH_4R–RGS4 and hH_4R–GAIP (coexpressed with $G\alpha_{i2}$ and $G\beta_1\gamma_2$) show an increased relative effect of HA and THIO, compared to the hH_4R, coexpressed with $G\alpha_{i2}$ and $G\beta_1\gamma_2$). However, the relative HA-stimulated GTPase activity is by ~10% higher in case of hH_4R–GAIP, compared with hH_4R–RGS4.

At first glance, it is hardly conceivable why RGS4 and GAIP should differentially affect hH_4R-stimulated GTPase activity. However, this may be explained by differing G-protein selectivity of the two RGS-proteins. RGS4 accelerates the GTPase activity of both $G\alpha_i$- (Berman *et al.*, 1996; Cavalli *et al.*, 2000) and $G\alpha_q$ proteins (Xu *et al.*, 1999), whereas GAIP prefers the $G\alpha_i$ subtype (Wieland and Mittmann, 2003). However, it is reported that GAIP interacts rather weakly with $G\alpha_{i2}$ (De Vries *et al.*, 1995). In fact, according to the corresponding entry (P49795) in the database UniProt (2009), GAIP binds to $G\alpha_{i/o}$ proteins with the order of preference $G\alpha_{i3} > G\alpha_{i1} > G\alpha_o \gg G\alpha_{i2}$. Thus, the stimulatory effect of GAIP on $G\alpha_{i2}$ GTPase activity may become visible only at very high concentrations of activated GTP-bound $G\alpha_{i2}$, for example, when hH_4R is activated by HA. By contrast, RGS4 may stimulate $G\alpha_{i2}$ more potently than GAIP, resulting in a GTPase activating effect, which is already visible in the constitutively active system. Thus, different affinities of GAIP and RGS4 for $G\alpha_{i2}$ may explain why hH_4R–GAIP enhances the relative HA-stimulated GTPase activity more effectively than hH_4R–RGS4. Finally, this also shows that constitutive activity can be markedly influenced by the type of RGS-protein available to the receptor/G-protein complex.

Despite the only minor effects of coexpressed RGS-proteins on absolute and relative hH_4R-induced GTPase activities, the K_M values in enzyme kinetic experiments were clearly altered (Schneider and Seifert, 2009). The $G\alpha$ GTPase K_M value of the HA (10 μM) stimulated system was significantly increased from 462 ± 181 nM ($hH_4R + G\alpha_{i2} + G\beta_1\gamma_2$) to

882 ± 172 n*M* ($hH_4R + G\alpha_{i2} + G\beta_1\gamma_2$ + GAIP). In the case of the hH_4R–GAIP fusion protein (coexpressed with $G\alpha_{i2}$ and $G\beta_1\gamma_2$) the K_M value (HA, 10 μ*M*) was slightly but not significantly reduced, compared to the system with coexpressed GAIP (Schneider and Seifert, 2009). A rather speculative explanation for the larger K_M effect of the coexpression system may be the fact that coexpressed GAIP can be anchored to the membrane *via* a palmitoylated cysteine string motif. This may result in more membrane-bound GAIP protein in the coexpression system, compared to the hH_4R–GAIP fusion protein system, where the number of GAIP molecules always equals the number of receptor molecules. The highest K_M value (970 ± 145 n*M*) was determined for the hH_4R–RGS4 fusion protein (coexpressed with $G\alpha_{i2}$ and $G\beta_1\gamma_2$), again in the presence of 10 μ*M* of HA (Schneider and Seifert, 2009). An increase of GTPase K_M values in the presence of RGS-proteins was already previously reported for $G\alpha_{o1}$ and $G\alpha_{i2}$, stimulated by the α_{2A} adrenoceptor (Cavalli *et al.*, 2000). Interestingly, no significant effect of RGS-proteins on the Gα GTPase K_M value was observed in the THIO-inhibited system (Schneider and Seifert, 2009). Seemingly, the $G\alpha_{i2}$ conformation induced by the constitutively active hH_4R is quite different from that in the presence of agonist (HA, 10 μ*M*) and is less susceptible to the RGS-protein-caused K_M effect. Another explanation may be that the concentration of activated Gα subunits in the constitutively active system is simply too low for a strongly visible RGS-protein effect.

3.2.2.2. G-protein selectivity of the hH_4R–GAIP fusion protein If GAIP shows different potencies at interacting with the various $G\alpha_{i/o}$ subunits, it is to be expected that fusion of hH_4R to GAIP also changes the $G\alpha_i$ subtype selectivity. Thus, before hH_4R–GAIP is used to characterize receptor ligands in functional assays, it should be determined whether the fused RGS-protein affects the interaction of hH_4R with $G\alpha_{i/o}$ proteins. Figure 25.7A shows the G-protein selectivity profile of hH_4R, expressed alone or coexpressed with $G\beta_1\gamma_2$ and $G\alpha_{i1}$, $G\alpha_{i2}$, $G\alpha_{i3}$, or $G\alpha_o$. In Fig. 25.7B, the same combinations are investigated, but with hH_4R–GAIP instead of the nonfused hH_4R. The fusion of hH_4R to GAIP only slightly affects the baseline activity in all of the compared systems, but leads to a pronounced enhancement of HA-stimulated GTPase activity. The selective improvement of the HA-induced signal is most pronounced with $G\alpha_{i1}$ and $G\alpha_{i2}$, but also occurs in the presence of $G\alpha_{i3}$ and $G\alpha_o$. Obviously, the threshold of G-protein activation that has to be exceeded to make the RGS-protein effect visible does not exclusively exist in the $G\alpha_{i2}$-containing system. Moreover, the G-protein coupling profile of hH_4R–GAIP is very similar to that of the nonfused hH_4R, showing a preference for $G\alpha_{i2}$ (specifically with regard to the absolute HA-induced signal), compared to $G\alpha_{i1}$ and $G\alpha_{i3}$ and only a

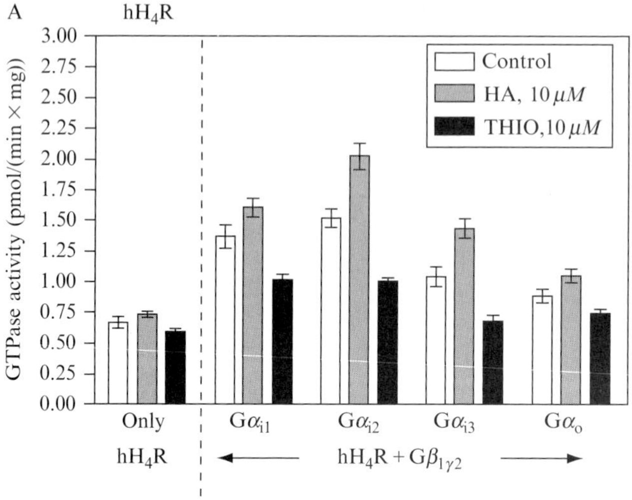

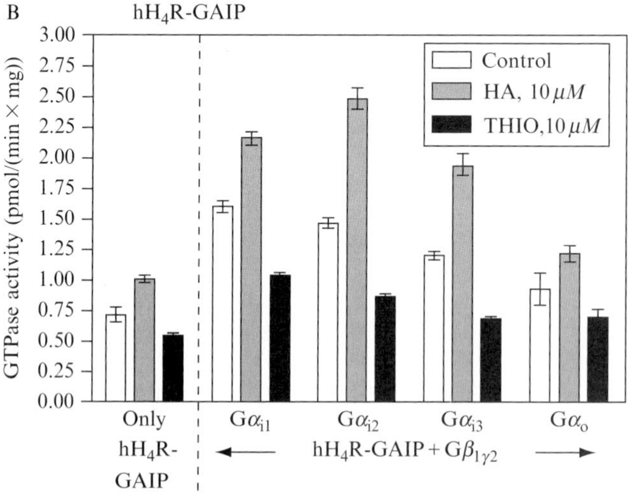

Figure 25.7 G-protein selectivity of hH_4R and hH_4R–GAIP, determined in the steady-state GTPase assay. (A) G-protein selectivity of hH_4R, determined in the steady-state GTPase assay. The hH_4R was expressed in the absence of mammalian G-proteins or in the presence of $G\beta_1\gamma_2$, combined with $G\alpha_{i1}$, $G\alpha_{i2}$, $G\alpha_{i3}$, or $G\alpha_o$. The first bar of each triplet shows control conditions (H_2O), the second bar indicates the HA-induced signal (10 μM of HA) and the third bar depicts the THIO-inhibited signal (10 μM of THIO). All assays were performed in the presence of 100 nM of GTP and 100 mM of NaCl. (B) G-protein selectivity of hH_4R–GAIP, determined in the steady-state GTPase assay. The data were determined in analogy to the results shown in panel (A), but with membranes expressing hH_4R–GAIP. Data are from two independent experiments with three to four replicates, mean ± S.E.M. (A) Adapted with permission of Schneider *et al.* (2009). Copyright 2009 American Chemical Society. (B) Adapted from Schneider and Seifert (2009).

very weak interaction with $G\alpha_o$. Interestingly, the hH_4R–GAIP fusion protein seems to unmask an interaction of hH_4R with endogenous insect cell G-proteins, because, in contrast to the coexpression system ($hH_4R + G\alpha_{i2} + G\beta_1\gamma_2$), a small but pronounced HA-induced signal becomes visible (Fig. 25.7B, first bar triplet). In summary, we can conclude that the G-protein specificity of hH_4R–GAIP is governed by the properties of the receptor rather than by the GAIP part.

3.3. [^{35}S]GTPγS binding

A detailed protocol of the GTPγS binding assay is provided in chapter 28 of this volume. Figure 25.6B shows the results from [^{35}S]GTPγS saturation binding with the hH_4R–$G\alpha_{i2}$ fusion protein (continuous line), compared with the data from the coexpression system (dashed line). Similar to the results from steady-state GTPase assays (Fig. 25.6A), the fusion of hH_4R to $G\alpha_{i2}$ causes an increase of thioperamide-sensitive constitutive activity. However, the difference in constitutive activity between hH_4R ($+G\alpha_{i2} + G\beta_1\gamma_2$) and hH_4R–$G\alpha_{i2}$ ($+G\beta_1\gamma_2$) exists only at lower [^{35}S]GTPγS concentrations. At higher concentrations of [^{35}S]GTPγS, the curves representing constitutive GTPγS binding for hH_4R–$G\alpha_{i2}$ and the coexpression system converge at ~70% of total ligand-regulated [^{35}S]GTPγS binding. The difference in constitutive activity is actually caused by a largely increased [^{35}S]GTPγS affinity of hH_4R–$G\alpha_{i2}$, compared to coexpressed $G\alpha_{i2}$. The apparent K_D value of [^{35}S]GTPγS in the THIO-inhibited system was 3.6 ± 0.7 nM at hH_4R–$G\alpha_{i2}$, but 10.0 ± 4.0 nM in the coexpression system (Schneider *et al.*, 2009). Also the K_D value in the presence of HA (10 μM) was reduced in the fusion protein system (0.9 ± 0.2 nM), compared to the coexpression system (3.4 ± 1.3 nM) (Schneider *et al.*, 2009). Similar to the steady-state GTPase data discussed in the preceding section, this indicates again that the interaction of hH_4R with $G\alpha_{i2}$ is more efficient in the fusion protein than in the nonfused hH_4R.

The hH_4R–$G\alpha_{i2}$ fusion protein provides also a useful tool to characterize receptor/G-protein coupling in a system with defined stoichiometry. This is demonstrated by the coupling-factor, which is the ratio of the amount of total ligand-regulated [^{35}S]GTPγS binding to the B_{max} value from radioligand binding. As previously reported (Schneider *et al.*, 2009), the nonfused hH_4R shows catalytic signaling with a coupling factor of ~5. By contrast, hH_4R–$G\alpha_{i2}$ shows a coupling factor of ~1 (1.2 ± 0.6), indicating a 1:1 stoichiometry of receptor and G-protein. In addition, a coupling factor of 1 shows that there is no cross-interaction of hH_4R–$G\alpha_{i2}$ with insect cell G-proteins. This finding is in accordance with data obtained for the FPR and β_2AR (Wenzel-Seifert and Seifert, 2000; Wenzel-Seifert *et al.*, 1999).

4. Application of the Fusion Protein Approach to other GPCRs

4.1. Fusion proteins of the β_2-adrenoceptor (β_2AR) with $G\alpha$ subunits

The first GPCR/$G\alpha$ fusion protein described in literature was β_2AR–$G_s\alpha$, expressed in S49 cyc-cells, a cell line that lacks $G_s\alpha$ expression (Bertin *et al.*, 1994). In the following years, β_2AR–$G_s\alpha$ became one of the most intensively studied GPCR/$G\alpha$ fusion protein systems. Expression of β_2AR–$G\alpha_s$ in Sf9 cells resulted in enhanced efficiency of the β_2AR–$G_s\alpha$ interaction, as indicated by increased high-affinity agonist binding and higher absolute agonist-induced AC-activity (Seifert *et al.*, 1998a). Interestingly, the steady-state GTPase- and [^{35}S]GTPγS binding signals were more enhanced than AC activity. This can be explained by limited availability of AC that does not allow a full transduction of the G-protein activation to the level of cAMP synthesis, despite the efficient stimulation of GDP/GTP exchange. The limited availability of AC molecules leads to an apparently increased constitutive GPCR activity in AC assays compared to GTPase assays, the latter not being dependent on AC molecules (Seifert *et al.*, 1998a). This is also the reason, why membranes with rather low expression levels of GPCR–$G\alpha_s$ fusion proteins are recommended for AC assays (Seifert and Wieland, 2005). Taken together, β_2AR–$G\alpha_s$ can be successfully used to study the modulation of $G\alpha_s$-function by β_2AR agonists and inverse agonists, specifically at the level of the AC assay (Seifert *et al.*, 1999a) and in the GTPγS binding assay at late time points (Wenzel-Seifert and Seifert, 2000). A detailed description of the AC assay has been provided in Seifert and Wieland (2005). A typical inverse agonist that was used to characterize constitutive activity of β_2AR–$Gs\alpha_s$ in GTPγS binding- (Wenzel-Seifert and Seifert, 2000) and AC (Seifert *et al.*, 1999a) assays is ICI-118551 (erythro-DL-1(7-methylindan-4-yloxy)-3-isopropylaminobutan-2-ol).

For the $G\alpha_s$ isoforms $Gs\alpha_S$, $Gs\alpha_L$, and $G\alpha_{olf}$, it was demonstrated that the fusion protein approach is much more sensitive at dissecting differences between G-proteins than the coexpression system (Gille and Seifert, 2003). For example, β_2AR fusion proteins with $Gs\alpha_L$, $Gs\alpha_S$, $G\alpha_{i2}$, $G\alpha_{i3}$, $G\alpha_q$, or $G\alpha_{16}$ were compared in Sf9 cells. Radioligand binding and [^{35}S]GTPγS binding assays demonstrated a differential interaction of β_2AR with G_q, G_s, and G_i-proteins. Moreover, ligand-specific conformations that differentially activate various G-proteins were shown. It was demonstrated that the constitutive activity of the β_2AR–$G\alpha$ fusion proteins depends on the G-protein fusion partner. The inverse agonist ICI-118551 showed the highest efficacy, when acting at β_2AR–$Gs\alpha_L$. However, when β_2AR was fused to $Gs\alpha_S$, $G\alpha_{i2}$, $Gi\alpha_3$, $G\alpha_{16}$, and $G\alpha_q$, there was only a minor effect of

ICI-118551, indicating a very low and almost absent constitutive activity with those G-proteins. (Wenzel-Seifert and Seifert, 2000). However, it should be mentioned in this context that the expression of functional $G\alpha_q$ in Sf9 cells is hampered by yet unknown factors (Houston *et al.*, 2002).

The fusion protein approach can also provide insights into the influence of $G\alpha$ subunits on constitutive receptor activity. A comparison of β_2AR–$Gs\alpha_s$ and β_2AR–$Gs\alpha_L$ (β_2AR fused to the short and long $Gs\alpha$ splice variant) showed that $Gs\alpha_L$ conferred constitutive activity to β_2AR, indicating that unoccupied β_2AR induces GDP dissociation more effectively at $Gs\alpha_L$ than at $Gs\alpha_S$ (Seifert *et al.*, 1998b). Taken together, these data show that constitutive activity is influenced by both, the tendency of the receptor to spontaneously adopt the active state and the GDP binding affinity of the $G\alpha$ subunit.

Although fusion proteins are very useful as tools to investigate signal transduction and constitutive activity, the results have to be interpreted with caution. Sometimes, the close proximity of G-protein and receptor in a fusion protein can also result in surprising effects. For example, the FPR–$G\alpha_{i2}$ fusion protein reduced AC-activity as expected, but β_2AR–$G\alpha_{i2}$ activated AC. Obviously, the β_2AR–$G\alpha_{i2}$ fusion protein brings $G\alpha_{i2}$ in close proximity to the $G\alpha_s$ binding site of AC, causing AC activation (Seifert *et al.*, 2002).

4.2. Fusion proteins of the histamine H_2 and H_3 receptor with $G\alpha$ subunits

The influence of the $G\alpha_s$ isoform on constitutive activity of the human histamine H_2 receptor (hH_2R) was investigated by expressing the corresponding fusion proteins in Sf9 cells. In contrast to the corresponding fusion proteins with β_1- and β_2AR (Wenzel-Seifert *et al.*, 2002), the constitutive activity of hH_2R–$Gs\alpha_S$ and hH_2R–$Gs\alpha_L$ was similar. Thus, the GDP affinity of G-proteins does not generally determine constitutive activity of receptors (Wenzel-Seifert *et al.*, 2001). Famotidine is a typical inverse hH_2R agonist, which reduces constitutive activity of both hH_2R-$Gs\alpha_S$ and hH_2R-$Gs\alpha_L$ (Wenzel-Seifert *et al.*, 2001). H_2R–$Gs\alpha_S$ fusion proteins were also used for medicinal chemistry studies. For example, new H_2R agonists were pharmacologically characterized in steady-state GTPase assays, using hH_2R–$Gs\alpha_S$ or the guinea pig H_2R–$Gs\alpha_s$ fusion protein (gpH_2R–$Gs\alpha_S$) (Kraus *et al.*, 2009).

The human histamine H_3R was coexpressed with $G\alpha_{i1-3}$ or $G\alpha_{o1}$ (always combined with $G\beta_1\gamma_2$) in Sf9 cells. Moreover, hH_3R–$G\alpha_{i2}$ and hH_3R–$G\alpha_{o1}$ were expressed in combination with $G\beta_1\gamma_2$ (Schnell *et al.*, 2009). All hH_3R/G-protein combinations showed comparably high constitutive activity in steady-state GTPase assays. The pharmacological profiles of various imidazole-containing ligands were independent of the coexpressed $G\alpha$ subunit. Similar results were obtained with the fusion proteins

(Schnell *et al.*, 2009). Examples for inverse H_3R agonists are thioperamide (which is also an inverse H_4R agonist) and ciproxyfan (Schnell *et al.*, 2009).

4.3. Formyl peptide receptor: $G\alpha$ fusion proteins

FPR–$G\alpha_{i1}$, –$G\alpha_{i2}$, and –$G\alpha_{i3}$ fusion proteins (combined with $G\beta_1\gamma_2$) were expressed in Sf9 cells and compared with the corresponding coexpression systems to determine stoichiometry and specificity of FPR–$G\alpha_i$ interaction (Wenzel-Seifert *et al.*, 1999). As indicated by high-affinity agonist binding, steady-state GTPase assays, [^{35}S]GTPγS binding and photolabeling of $G\alpha$, the FPR–$G\alpha$ fusion proteins showed highly efficient G-protein coupling and no insect cell G-proteins were cross-activated. The coupling efficiency of FPR was similar with $G\alpha_{i1}$, $G\alpha_{i2}$, and $G\alpha_{i3}$. FPR stimulates G-proteins in a linear manner with 1:1 stoichiometry, even when coexpressed with the $G\alpha$-subunit. Surprisingly, a low-affinity FPR state, which couples efficiently to G-proteins, was found in this system (Wenzel-Seifert *et al.*, 1999). A highly efficient inverse agonist that can be used to characterize FPR constitutive activity is cyclosporin H (Wenzel-Seifert and Seifert, 1993).

4.4. Other GPCR–$G\alpha$ fusion proteins

In a very detailed study all muscarinic receptor subtypes (M_1R–M_5R) were fused to several $G\alpha$ subunits ($G\alpha_{i1}$, $G\alpha_{i2}$, $G\alpha_s$, $G\alpha_{11}$, $G\alpha_{16}$, and chimera of $G\alpha$-subunits; Guo *et al.*, 2001). M_2R–$G\alpha_{i1}$ and M_4R–$G\alpha_{i1}$ turned out to be useful tools to screen agonists and antagonists in [^{35}S]GTPγS binding assays. M_2R–$G\alpha_{i1}$ expressed in Sf9 cells still has M_2R-like pharmacology in radioligand- and [^{35}S]GTPγS binding studies and shows efficient signaling. A typical M_2R-inverse agonist is pirenzepine (Daeffler *et al.*, 1999). A constitutively active $D_{2S}R$–$G\alpha_o$ fusion protein was expressed in Sf9 cells in order to characterize D_2R antagonists in radioligand binding and [^{35}S]GTPγS binding assays (Gazi *et al.*, 2003). The majority of the investigated compounds were inverse agonists, which shows again the usefulness of the fusion protein approach for the investigation of constitutive activity and inverse agonism.

ACKNOWLEDGMENTS

We thank Drs Katharina Wenzel-Seifert and David Schnell for fruitful collaboration and stimulating discussions and Mrs A. Seefeld and G. Wilberg for expert technical assistance. This work was supported by the Graduate Training Program GRK 760: (Medicinal Chemistry: Molecular Recognition—Ligand–Receptor-Interactions) and GRK 1441 (Regulation of allergic inflammation in lung and skin) of the Deutsche Forschungsgemeinschaft. This work was supported by the European Cooperation in the Field of Scientific and Technical Research (COST) action #BM0806 ("Recent advances in histamine receptor H_4R research"), funded by the European Comission (7th European Framework Program).

REFERENCES

Bahia, D. S., Sartania, N., Ward, R. J., Cavalli, A., Jones, T. L., Druey, K. M., and Milligan, G. (2003). Concerted stimulation and deactivation of pertussis toxin-sensitive G proteins by chimeric G protein-coupled receptor-regulator of G protein signaling 4 fusion proteins: Analysis of the contribution of palmitoylated cysteine residues to the GAP activity of RGS4. *J. Neurochem.* **85,** 1289–1298.

Berman, D. M., Kozasa, T., and Gilman, A. G. (1996). The GTPase-activating protein RGS4 stabilizes the transition state for nucleotide hydrolysis. *J. Biol. Chem.* **271,** 27209–27212.

Bertin, B., Freissmuth, M., Jockers, R., Strosberg, A. D., and Marullo, S. (1994). Cellular signaling by an agonist-activated receptor/$G_s\alpha$ fusion protein. *Proc. Natl. Acad. Sci. USA* **91,** 8827–8831.

Burns, M. E., and Pugh, E. N., Jr. (2009). RGS9 concentration matters in rod phototransduction. *Biophys. J.* **97,** 1538–1547.

Burt, A. R., Sautel, M., Wilson, M. A., Rees, S., Wise, A., and Milligan, G. (1998). Agonist occupation of an α_{2A}-adrenoreceptor-$G_{i1}\alpha$ fusion protein results in activation of both receptor-linked and endogenous G_i proteins. Comparisons of their contributions to GTPase activity and signal transduction and analysis of receptor-G protein activation stoichiometry. *J. Biol. Chem.* **273,** 10367–10375.

Carr, I. C., Burt, A. R., Jackson, V. N., Wright, J., Wise, A., Rees, S., and Milligan, G. (1998). Quantitative analysis of a cysteine351glycine mutation in the G protein $G_{i1}\alpha$: Effect on α_{2A}-adrenoceptor-$G_{i1}\alpha$ fusion protein activation. *FEBS Lett.* **428,** 17–22.

Cavalli, A., Druey, K. M., and Milligan, G. (2000). The regulator of G protein signaling RGS4 selectively enhances α_{2A}-adrenoreceptor stimulation of the GTPase activity of $G_{o1}\alpha$ and $G_{i2}\alpha$. *J. Biol. Chem.* **275,** 23693–23699.

Daeffler, L., Schmidlin, F., Gies, J. P., and Landry, Y. (1999). Inverse agonist activity of pirenzepine at M_2 muscarinic acetylcholine receptors. *Br. J. Pharmacol.* **126,** 1246–1252.

De Vries, L., Mousli, M., Wurmser, A., and Farquhar, M. G. (1995). GAIP, a protein that specifically interacts with the trimeric G protein $G\alpha_{i3}$, is a member of a protein family with a highly conserved core domain. *Proc. Natl. Acad. Sci. USA* **92,** 11916–11920.

Gazi, L., Wurch, T., Lopez-Giménez, J. F., Pauwels, P. J., and Strange, P. G. (2003). Pharmacological analysis of a dopamine D_{2Short}:$G\alpha_o$ fusion protein expressed in Sf9 cells. *FEBS Lett.* **545,** 155–160.

Gille, A., and Seifert, R. (2003). Co-expression of the β_2-adrenoceptor and dopamine D_1-receptor with $G_s\alpha$ proteins in Sf9 insect cells: Limitations in comparison with fusion proteins. *Biochim. Biophys. Acta* **1613,** 101–114.

Guo, Z. D., Suga, H., Okamura, M., Takeda, S., and Haga, T. (2001). Receptor-$G\alpha$ fusion proteins as a tool for ligand screening. *Life Sci.* **68,** 2319–2327.

Houston, C., Wenzel-Seifert, K., Bürckstümmer, T., and Seifert, R. (2002). The human histamine H_2-receptor couples more efficiently to Sf9 insect cell G_s-proteins than to insect cell G_q-proteins: Limitations of Sf9 cells for the analysis of receptor/G_q-protein coupling. *J. Neurochem.* **80,** 678–696.

Kraus, A., Ghorai, P., Birnkammer, T., Schnell, D., Elz, S., Seifert, R., Dove, S., Bernhardt, G., and Buschauer, A. (2009). N^G-acylated aminothiazolylpropylguanidines as potent and selective histamine H_2 receptor agonists. *ChemMedChem* **4,** 232–240.

Sautel, M., and Milligan, G. (1998). Loss of activation of G_s but not G_i following expression of an α_{2A}-adrenoceptor-$G_{i1}\alpha$ fusion protein. *FEBS Lett.* **436,** 46–50.

Schneider, E. H., and Seifert, R. (2009). Histamine H_4 receptor–RGS fusion proteins expressed in Sf9 insect cells: A sensitive and reliable approach for the functional characterization of histamine H_4 receptor ligands. *Biochem. Pharmacol.* **78,** 607–616.

Schneider, E. H., Schnell, D., Papa, D., and Seifert, R. (2009). High constitutive activity and a G-protein-independent high-affinity state of the human histamine H_4-receptor. *Biochemistry* **48,** 1424–1438.

Schnell, D., Burleigh, K., Trick, J., and Seifert, R. (2009). No evidence for functional selectivity of proxyfan at the human histamine H_3-receptor coupled to defined G_i/G_o protein heterotrimers. *J. Pharmacol. Exp. Ther.* **332,** 996–1005.

Seifert, R., and Wieland, T. (2005). G Protein-Coupled Receptors as Drug Targets: Analysis of Activation and Constitutive Activity. Wiley-VCH, Weinheim.

Seifert, R., Lee, T. W., Lam, V. T., and Kobilka, B. K. (1998a). Reconstitution of β_2-adrenoceptor-GTP-binding-protein interaction in Sf9 cells–high coupling efficiency in a β_2-adrenoceptor-Gsα fusion protein. *Eur. J. Biochem.* **255,** 369–382.

Seifert, R., Wenzel-Seifert, K., Lee, T. W., Gether, U., Sanders-Bush, E., and Kobilka, B. K. (1998b). Different effects of $G_s\alpha$ splice variants on β_2-adrenoreceptor-mediated signaling. The β_2-adrenoreceptor coupled to the long splice variant of $G_s\alpha$ has properties of a constitutively active receptor. *J. Biol. Chem.* **273,** 5109–5116.

Seifert, R., Gether, U., Wenzel-Seifert, K., and Kobilka, B. K. (1999a). Effects of guanine, inosine, and xanthine nucleotides on β_2-adrenergic receptor/G_s interactions: Evidence for multiple receptor conformations. *Mol. Pharmacol.* **56,** 348–358.

Seifert, R., Wenzel-Seifert, K., Gether, U., Lam, V. T., and Kobilka, B. K. (1999b). Examining the efficiency of receptor/G-protein coupling with a cleavable β_2-adrenoceptor-$G_{s\alpha}$ fusion protein. *Eur. J. Biochem.* **260,** 661–666.

Seifert, R., Wenzel-Seifert, K., and Kobilka, B. K. (1999c). GPCR-Gα fusion proteins: Molecular analysis of receptor-G-protein coupling. *Trends Pharmacol. Sci.* **20,** 383–389.

Seifert, R., Wenzel-Seifert, K., Arthur, J. M., Jose, P. O., and Kobilka, B. K. (2002). Efficient adenylyl cyclase activation by a β_2-adrenoceptor-$G_i\alpha_2$ fusion protein. *Biochem. Biophys. Res. Commun.* **298,** 824–828.

UniProt (2009). The Universal Protein Resource (UniProt) 2009. *Nucleic Acids Res.* **37,** D169–D174.

Wenzel-Seifert, K., and Seifert, R. (1993). Cyclosporin H is a potent and selective formyl peptide receptor antagonist. Comparison with N-t-butoxycarbonyl-L-phenylalanyl-L-leucyl-L-phenylalanyl-L- leucyl-L-phenylalanine and cyclosporins A, B, C, D, and E. *J. Immunol.* **150,** 4591–4599.

Wenzel-Seifert, K., and Seifert, R. (2000). Molecular analysis of β_2-adrenoceptor coupling to G_s-, G_i-, and G_q-proteins. *Mol. Pharmacol.* **58,** 954–966.

Wenzel-Seifert, K., Arthur, J. M., Liu, H. Y., and Seifert, R. (1999). Quantitative analysis of formyl peptide receptor coupling to $G_i\alpha_1$, $G_i\alpha_2$, and $G_i\alpha_3$. *J. Biol. Chem.* **274,** 33259–33266.

Wenzel-Seifert, K., Kelley, M. T., Buschauer, A., and Seifert, R. (2001). Similar apparent constitutive activity of human histamine H_2-receptor fused to long and short splice variants of $G_{s\alpha}$. *J. Pharmacol. Exp. Ther.* **299,** 1013–1020.

Wenzel-Seifert, K., Liu, H. Y., and Seifert, R. (2002). Similarities and differences in the coupling of human β_1- and β_2-adrenoceptors to $G_{s\alpha}$ splice variants. *Biochem. Pharmacol.* **64,** 9–20.

Wieland, T., and Mittmann, C. (2003). Regulators of G-protein signalling: Multifunctional proteins with impact on signalling in the cardiovascular system. *Pharmacol. Ther.* **97,** 95–115.

Willars, G. B. (2006). Mammalian RGS proteins: Multifunctional regulators of cellular signalling. *Semin. Cell Dev. Biol.* **17,** 363–376.

Xu, X., Zeng, W., Popov, S., Berman, D. M., Davignon, I., Yu, K., Yowe, D., Offermanns, S., Muallem, S., and Wilkie, T. M. (1999). RGS proteins determine signaling specificity of G_q-coupled receptors. *J. Biol. Chem.* **274,** 3549–3556.

CHAPTER TWENTY-SIX

Screening for Novel Constitutively Active CXCR2 Mutants and Their Cellular Effects

Giljun Park,* Tom Masi,* Chang K. Choi,† Heejung Kim,* Jeffrey M. Becker,* *and* Tim E. Sparer*

Contents

Abstract

Chemokines play an important role in inflammatory, developmental, and homeostatic processes. Deregulation of this system results in various diseases including tumorigenesis and cancer metastasis. Deregulation can occur when constitutively active mutant (CAM) chemokine receptors are locked in the "on" position. This can lead to cellular transformation/tumorigenesis.

The CXC chemokine receptor 2 (CXCR2) is a G-protein-coupled receptor (GPCR) expressed on neutrophils, some monocytes, endothelial cells, and some epithelial cells. CXCR2 activation with CXC chemokines induces leukocyte migration, trafficking, leukocyte degranulation, cellular differentiation, and angiogenesis. Activation of CXCR2 can lead to cellular transformation. We hypothesized that CAM CXCR2s may play a role in cancer development. In order to identify CXCR2 CAMs, potential mutant CXCR2 receptors were screened using a modified *Saccharomyces cerevisiae* high-throughput system. *S. cerevisiae* has been

* The University of Tennessee, Department of Microbiology, Knoxville, Tennessee, USA
† Michigan Technological University, Department of Mechanical Engineering-Engineering Mechanics, Houghton, Michigan, USA

Methods in Enzymology, Volume 485
ISSN 0076-6879, DOI: 10.1016/S0076-6879(10)85026-8

used successfully to identify GPCR/G-protein interactions and autocrine selection for peptide agonists. The CXCR2 CAMs identified from this screen were characterized in mammalian cells. Their ability to transform cells *in vitro* was shown using foci formation, soft-agar growth, impedance measurement assays, and *in vivo* tumor growth following hind flank inoculation into mice. Signaling pathways contributing to cellular transformation were identified using luciferase reporter assays. Studying constitutively active GPCRs is an approach to "capturing" pluridimensional GPCRs in a "locked" activation state. In order to address the residues necessary for CXCR2 activation, we used *S. cerevisiae* for screening novel CAMs and characterized them using mammalian reporter assays.

1. Introduction

Although there are many contributing factors to tumorigenesis, chemokines and their receptors can contribute to tumorigenesis and metastasis (Muller *et al.*, 2001). Chemokines function normally as small chemoattractant cytokines that are involved in inflammatory, developmental, and homeostatic processes (Zlotnik, 2006). Deregulation of this system is associated with the development of cancers and metastasis. One example of the disruption of the chemokine pathway is the constitutively active mutant (CAM) receptors (Kakinuma and Hwang, 2006; Muller *et al.*, 2001). These CAMs cause an alteration in the structural constraints of the receptor, locking it into an active conformation. This conformational change induces ligand-independent activation called constitutive activity (Gether and Kobilka, 1998). CAM GPCRs are associated with the development of a variety of human diseases (Leurs *et al.*, 1998). The best example of a constitutively active chemokine receptor causing cancer is the Kaposi sarcoma (KS) herpesvirus chemokine receptor, ORF74, which is a viral homolog for human CXCR2 (Gershengorn *et al.*, 1998). This receptor has a "VRY" motif instead of the normal "DRY" motif between the third transmembrane domain and second intracellular loop and has been shown to be constitutively active. Stable lines of NIH3T3 fibroblasts expressing ORF74 lead to cellular transformation (Burger *et al.*, 1999; Rosenkilde *et al.*, 1999, 2001). These transfectants are tumorigenic in nude mice and a transgenic mouse expressing ORF74 developed tumors resembling KS lesions (Guo *et al.*, 2003).

CXC chemokine receptor 2 (CXCR2) is a G-protein-coupled receptor (GPCR; Kristiansen, 2004) expressed on neutrophils, some monocytes, endothelial cells, and some epithelial cells and has homology with ORF74 (Horton *et al.*, 2007). Activation of CXCR2 with CXC chemokines leads to leukocyte migration, trafficking, leukocyte degranulation, cell differentiation, and chemokine-mediated angiogenesis (Ahuja and Murphy, 1996; Baggiolini *et al.*, 1994).

Utilizing the activation of the natural pheromone response pathway, *S. cerevisiae* has been successfully used to identify GPCR/G-proteins interactions and select peptide agonists (Erickson *et al.*, 1998; King *et al.*, 1990; Klein *et al.*, 1998; Price *et al.*, 1995). Recently, allosteric peptide agonists for CXCR4 were identified using this system. The pheromone responsive pathway was modified to include hybrid G alpha proteins to interact with CXCR4, elimination of growth arrest genes, and an auxotrophic marker under the control of a pheromone responsive element. With these modifications a library of CXCR4 ligands was screened to identify novel agonists (Sachpatzidis *et al.*, 2003).

To identify CXCR2 CAMs, the open reading frame (ORF) of CXCR2 was randomly mutated by PCR and expressed in this genetically engineered yeast. Using a simple selection for growth on media lacking histidine in the presence of 3-amino-1,2,4-triazole (3-AT), we identified novel CXCR2 CAMs. Constitutive activation was confirmed with the induction of β-galactosidase under the control of a pheromone responsive element. In order to characterize the CXCR2 CAMs in the mammalian system and to link them to cellular transformation, the first step in tumorigenesis, we generated stable transfectants using CXCR2 CAMs and measured foci formation, anchorage-independent growth in soft agar, impedance, and tumor formation *in vivo*. To address the downstream constitutively active signal transduction pathways, immunoblotting and flowcytometry were used to identify specific phosphorylated signaling pathways. Luciferase reporter constructs with NF-κB responsive elements were used to identify important signaling pathways that could contribute to cellular transformation. Here, we describe detailed approaches to successful identification and cellular characterization of CXCR2 CAMs.

2. Establishment of a Yeast System to Identify CXCR2 CAMs

In this part, we describe screening procedures for CXCR2 CAMs in two genetically modified yeast strains, CY1141 and CY12946. CY1141 expresses a hybrid G alpha subunit, which encodes the first 33 N-terminal residues of human $G\alpha_{i2}$ replaced with the first 41 N-terminal residues of the endogenous yeast GPA1 protein. Also CY12946 contains a modified G alpha subunit, which expresses the last five C-terminal residues of yeast GPA1 replaced with human $G\alpha_{i2}$. These genetically modified G proteins allow coupling to the mammalian receptor, which induces yeast pheromone response signaling. To identify CXCR2 CAMs, wild-type (WT)

CXCR2 was randomly mutated by error-prone PCR and the PCR products were cloned into the yeast expression vector, p426GPD. This plasmid contains an origin of replication, a glyceraldehyde-3-phosphate dehydrogenase (GPD) promoter driving expression of the inserted ORF and a uracil (URA) selectable marker. Transformation into these yeast strains allowed the selection of CXCR2 CAMs that activate the signaling pathway leading to the activation of the pheromone responsive element and the gene for histidine auxotrophy. Yeast transformants expressing a CXCR2 CAM will grow on selective growth medium lacking both uracil (MLU) and histidine in the presence of 3-AT. A β-galactosidase assay was used to measure the degree of receptor activation, via *FUS1-lacZ* activation.

2.1. Receptor expression in *S. cerevisiae* strains CY1141 and CY12946

To identify CXCR2 CAMs, *S. cerevisiae* genetically modified strains are necessary in order to allow G-protein activation of the yeast pheromone responsive signaling pathway. Yeast strain CY1141 (*MATα FUS1p-HIS3 can1 far1-1442 gpa1 (41)-Gααi2 his3 leu2 lys2 ste14::trp1::LYS2 ste3-1156 tbt1-1 trp1 ura3*) contains an integrated copy of a hybrid Gα subunit ($GPA1_{(41)}G\alpha_{i2}$) in which the N-terminal 33 residues of human $G\alpha_{i2}$ are substituted with the 41 N-terminal residues of GPA1 (Klein *et al.*, 1998). For strain CY12946 (*MAT2 FUS1p-HIS3 GPA1Gααi2(5) can1 far1-1442 his3 leu2 lys2 sst2-2 ste14::trp1::LYS2 ste3-1156 tbt1-1 trp1 ura3*) also contains a chimeric G protein $GPA1G\alpha_{i2(5)}$ in which the five residues of C-terminal GPA1 were replaced with the C-terminal five residues of human $G\alpha_{i2}$ (Zhang *et al.*, 2002). Both strains were initially used because it was unclear which hybrid G protein would couple with CXCR2. Immunofluorescence and subcellular fractionation were used to demonstrate proper expression and cellular localization.

2.1.1. Required materials

- *S. cerevisiae strains*: CY1141 and CY12946 (From Dr Broach, Princeton, NJ)
- *Plasmids*: p426GPD (ATCC, #87361)
- *Growth medium*: YPD is a basic enriched media to culture CY1141 and CY12946. It contains 1% of BactoYeast extract (Difco), 2% of Bacto-Peptone (Difco), and 2% dextrose.
- *Selection medium*: MLU contains 1× Yeast Nitrogen Base (YNB) without amino acids plus ammonium sulfate (Difco), 2% glucose, 1% casamino acids (Difco). When selecting for histidine and/or leucine auxotrophies in combination with uracil, the casamino acids are deleted from the media because it contains both histidine and leucine.

- *Reagents for yeast transformation*: 1.0 *M* lithium acetate (LiAc) and 50% (w/v) polyethylene glycol (PEG) 3350 are sterilized with 0.2 and 0.45 μm pore size filtration respectively and stored at room temperature. Single-strand DNA (ssDNA, salmon sperm DNA, Sigma-Aldrich) is boiled for 10 min, cooled down on ice, and frozen at −20 °C.
- *Reagents for yeast immunofluorescence*: Potassium phosphate (KPi, 0.1 *M*, pH 6.5), SHA (1 *M* Sorbitol, 0.1 *M* NaHEPES (pH 7.5), 5 m*M* NaN_3), 4% formaldehyde, β-mercaptoethanol, and yeast cell wall lytic enzyme (Fisher, *Cat. no. BP2683-25*). WT buffer: 1% fat free dry milk, 0.5 mg/ml BSA, 150 m*M* NaCl, 50 m*M* HEPES (pH 7.5), 0.1% Tween 20, and 1 m*M* NaN_3. Anti-CXCR2 antibody (E-2, SantaCruz Biotechnology). Hoechst dye (1 μg/ml, Molecular Probe) for nuclear staining is used to identify individual cells.
- *Reagents for subcellular fractionation*: Sorbitol buffer (10 m*M* Tris, pH 7.6, 0.8 *M* Sorbitol, 10 m*M* NaN_3, 10 m*M* KF, 1 m*M* EDTA, pH 8.0) and sucrose buffer (10 m*M* Tris, pH 7.6, 1 m*M* EDTA, 10% (w/v) sucrose).
- *Immunoblotting*: NuPAGE 12% Bis–Tris SDS-polyacrylamide gel (Invitrogen), PVDF membrane (Invitrogen), Primary antibodies: anti-FLAG M2 antibody (Eastman Kodak Co.), antihuman CXCR2 (sc-7304-SantaCruz Biotechnology) Secondary antibody: anti-mouse HRP-conjugated secondary Ab (eBioscience). Blocking media: 1% dried milk in phosphate buffered saline (PBS)/0.1% Tween 20. Wash media: PBS/0.1% Tween 20. ECL development kit (Amersham). 2× Laemmli sample buffer: 4% SDS, 20% glycerol, 10% 2-mercaptoethanol, 0.004% bromphenol blue, and 0.125 μ*M* Tris–HCl, pH approx. 6.8.
- *Devices*: Fluorescence Microscope (Olympus BX50), Ultracentrifuge (Beckman, TL-100), SDS PAGE gel apparatus with power supply (Hoeffer), and wet transfer blotter (Invitrogen).

2.1.2. Lithium acetate method for yeast transformation

Maintain and culture yeast strains on YPD medium. Inoculate a fresh colony into 5 ml YPD and incubate with shaking overnight at 30 °C. Using the overnight culture inoculate 50 ml of YPD in a sterile flask to a final density of 5×10^6 cells/ml. Incubate the culture as above until a cell concentration of 2×10^7 cell/ml (usually 3–5 h) is achieved. Centrifuge the cells at 4000 rpm for 5 min and discard the supernatant. Wash the cells once in 25 ml sterile water, centrifuge, and then resuspend in 1 ml of 100 m*M* LiAc and transfer to a 1.5-ml microcentrifuge tube. Centrifuge the cells at top speed in a microcentrifuge for 15 s and remove the LiAc. Resuspend the cells in 400 μl of 100 m*M* LiAc. For each plasmid to be transformed, aliquot 50 μl of cells into a clean microfuge tube. Boil the ssDNA for 5 min and keep on ice until needed. Centrifuge each aliquot of cells and the discard the supernatant. In the following order, carefully add each of the following reagents: 240 μl of 50% PEG, 36 μl of 1.0 *M* LiAc, 25 μl of ssDNA, 5 μl (0.1–10 μg) of cloned DNA (e.g., WT CXCR2 cloned into p426 GPD) and 45 μl of sterile water. Vortex the mixture until the cell pellet is completely resuspended (about 1–2 min). Include a negative control reaction without cloned DNA. Incubate the mixture at 30 °C for 30 min and heat shock at 42 °C for 25 min. Centrifuge the heat-shocked cells at top speed in a microcentrifuge for 2 min and then remove the supernatant. Resuspend the cells in 1 ml of sterile water and plate 200 μl on selective media lacking uracil and incubate at 30 °C. Transformed yeast colonies will be seen in 2–3 days.

2.1.3. Yeast immunofluorescence analysis

Fix 1×10^8 cultured yeast cells in 0.1 *M* KPi (pH 6.5) and 4% formaldehyde for at least 1 h then centrifuge and resuspend again in fix (as above) for at least 12 h but less than 24 h. After fixation, resuspend the cells in 5 ml of SHA (1 *M* Sorbitol, 0.1 *M* NaHEPES, pH 7.5, 5 m*M* NaN_3) and store at 4 °C for up to 2 weeks. To remove the cell wall, incubate 500 μl of the resuspended cells in 1 ml of SHA, 0.2% of β-mercaptoethanol, and 10 mg/ml of yeast cell wall lytic enzyme (Fisher, *BP2683-25*) at 30 °C for 1.5 h. After incubation, centrifuge the cells and resuspend in 1% SDS, wash with 1 ml of SHA, and resuspend with SHA (~100 μl). After spinning down the cells (microfuge, max speed for 1 min), incubate the cells with a specific antihuman CXCR2 antibody (sc-7304, SantaCruz Biotechnology) (1:100 dilution in WT buffer: 1% fat free dry milk, 0.5 mg/ml of BSA, 150 m*M* of NaCl, 50 m*M* HEPES, pH 7.5, 0.1% Tween 20, and 1 m*M* NaN_3) overnight at 4 °C. Wash the cells five times with WT buffer, and then add an anti-mouse secondary antibody FITC (sc-2099, SantaCruz Biotechnology) (1:1000 dilution in WT buffer). After incubation at room temperature for 30 min, wash the cells five times with WT buffer and stain with 100 μl of Hoechst dye (1 μg/ml, Molecular Probe). Wash the cells three times in sterile water and store at 4 °C in the dark. Take photomicrographs using an Olympus BX50 microscope and photograph with pictureframe version 2.3 (Olympus; Fig. 26.1).

2.1.4. Subcellular fractionation

Harvest 1×10^7 cells grown in selective MLU (1 min, max speed in microfuge), wash once with 1 ml Sorbitol buffer (10 m*M* Tris, pH 7.6, 0.8 *M* Sorbitol, 10 m*M* NaN_3, 10 m*M* KF, 1 m*M* EDTA, pH 8.0), centrifuge, and remove the supernatant. Wash again with 1 ml sucrose buffer (10 m*M* Tris, pH 7.6, 1 m*M* EDTA, 10% (w/v) sucrose) and resuspend the cells in 1 ml sucrose buffer containing protease inhibitors (10 μg/ml phenylmethylsulfonyl fluoride, 2μg/ml leupeptin, and 2 μg/ml pepstatin A). Mechanically disrupt the cells with glass beads and centrifuge at 300×*g* for 5 min to remove any unlysed cells. Mix 0.5 ml of supernatant

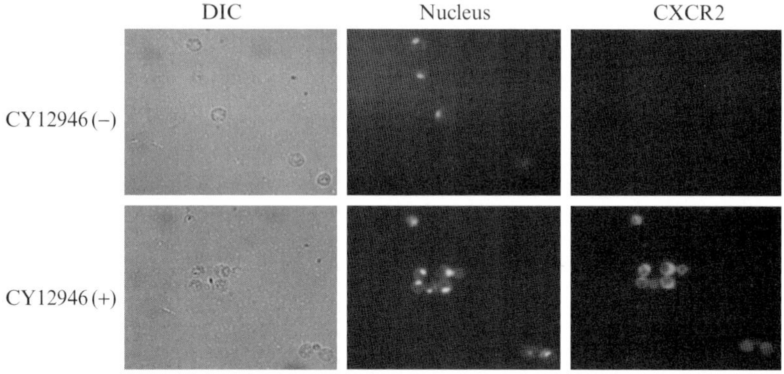

Figure 26.1 Human CXCR2 expression in transformed yeast strains. Spheroplasts were generated from CXCR2 untransformed (−) and transformed (+) yeast strain, CY12946. Cells were fixed and stained with an anti-CXCR2 antibody (E-2, SantaCruz Biotechnology) and a nuclear stain (Hoechst, Molecular Probe). Cells were observed and photographed using an Olympus BX50 fluorescence microscope (100× magnification) equipped with a CCD camera and analyzed using pictureframe version 2.3 (Olympus).

with 0.5 ml of 50% (w/v) sucrose in 10 m*M* Tris (pH 7.6), 1 m*M* EDTA and layer on top of a 4-ml of 30–60% linear sucrose gradient prepared in 10 m*M* Tris (pH 7.6) and 1 m*M* EDTA. For the gradient separation, overlay the cells on the sucrose cushion and centrifuge at 150,000×*g* for 20 h in an ultracentrifuge (Beckman, TL-100). Collect fractions (≈250 μl) from bottom of the gradient by inserting a long blunt-end needle into the bottom of the tube. Make sure that the needle rests on the bottom of the ultracentrifuge tube. Dilute 10 μl of each fraction 1:1 in 2× sample buffer. Warm the samples for 10 min at 37 °C and then load on a NuPAGE 12% Bis–Tris SDS-polyacrylamide gel (Invitrogen). Other gels can be substituted but this manufactured gel gave the best-looking gel for publication. Transfer the proteins via wet transfer onto a PVDF membrane according to the manufacturer's instructions (Invitrogen). After transfer, block the membrane in 1% nonfat dried milk in PBS/0.1% Tween 20. The yeast pheromone receptor, Ste2p-FLAG, which BJS21 strain expresses, was used as a positive control for plasma membrane expression of the receptor. After blocking, probe the membrane for Ste2p-FLAG expression with an anti-FLAG M2 antibody (1:25,000 dilution, Eastman Kodak Co.) or antihuman CXCR2 antibody (1:10,000 dilution, sc-7304, SantaCruz Biotechnology), which are incubated at 4 °C overnight. Make all antibody dilutions in wash buffer. Wash twice with PBS/0.1% Tween 20, then add the anti-mouse HRP-conjugated secondary Ab (1:10,000 dilution, eBioscience) for 30 min. Wash blots 4× with PBS/0.1% Tween 20 and then detect with ECL (Amersham; Fig. 26.2). Once localization and expression are confirmed, proceed to the mutagenesis step.

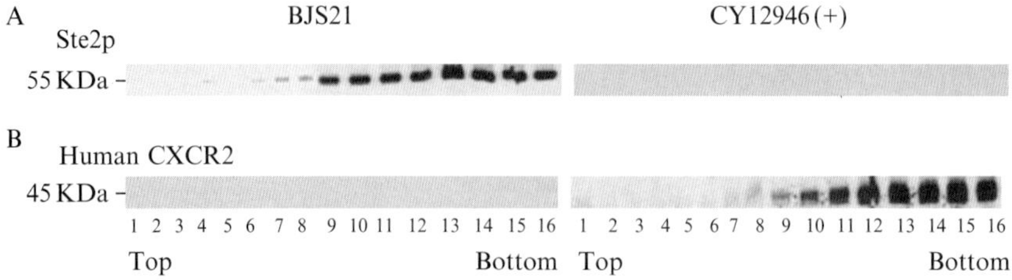

Figure 26.2 CXCR2 is expressed in the CY12946 strain and localizes to the plasma membrane. Total membrane preparations from control strain, BJS21 that expresses Ste2p (Son *et al.*, 2004), and CXCR2 transformed CY12946 (+) strain were run over a 4-ml sucrose gradient (30–60%) and 250-μl fractions were collected from the bottom to the top of the tube. Ten microliters of each fraction was run on the SDS PAGE gel and immunoblotted with either (A) anti-FLAG antibody or (B) CXCR2 (E-2, SantaCruz Biotechnology) followed by anti-mouse HRP antibody. Blots were developed with ECL (Amersham).

2.2. Identification of CXCR2 CAMs in the yeast strains

CXCR2 CAMs are identified in the *FUS1-HIS3* yeast strains when they grow on modified MLU lacking histidine and in the presence of 3-AT, which is used to suppress endogenous *HIS* expression. The CAMs activate the chimeric G proteins resulting in activation of the pheromone response pathway and subsequent activation of the *FUS1* promoter upregulating the *HIS* auxotrophy gene. In order to confirm and quantify the degree of constitutive activation, a plasmid containing β-galactosidase (*lacZ*) under the control of the *FUS1* promoter was cotransformed into these strains. The plasmid *FUS1-lacZ* (pMD1325) allows expression of *lacZ* using the same pheromone response pathway. This plasmid also contains a *LEU2* marker for selecting transformants (Sommers *et al.*, 2000). Once the transformants were created, the amount of β-galactosidase produced was quantified using a standard kit.

2.2.1. Required materials

- *Plasmids*: pMD1325 (a gift from Dr Dumont, University of Rochester, School of Medicine and Dentistry)
- *Reagents for error-prone PCR*: Primers forward and reverse for the CXCR2 ORF were synthesized to include unique restriction sites (forward with *Hin*dIII 5′-CCC*AAGCTT*ATGGAAGATTTTAACATGGAGAGTG-3′ and reverse with *Xba*I 5′-GC*TCTAGATTA*GAGAGTAGTGGAAGTG TGC-3′). Low-fidelity Taq polyerase (Fisher BioReagents) is preferred for error-prone PCR to high-fidelity Taq polymerase. In addition dATP, dGTP, dTTP, dCTP, and manganese chloride were added separately to

the reaction so that their concentrations could be adjusted in order to increase the error rate of the PCR.
- β-*Galactosidase assay*: Basal activity of β-galactosidase was measured using a yeast β-galactosidase assay kit (Pierce) according to the manufacturer's instruction. The amount of β-galactosidase is expressed as Miller units.
- *Cloning of PCR* products into p426GPD: Restriction enzymes (HindIII and XbaI) (NEB). β-Agarase (NEB). *3-AT* (MP Biomedicals) is used to suppress endogenous *HIS* expression.
- *Devices*: Gel electrophoresis apparatus (Owl). Thermocycler (Eppendorf)

2.2.2. Random mutagenesis

Perform random mutagenesis by using error-prone PCR (Matsumura and Ellington, 2002). Synthesize primers up and downstream of the CXCR2 ORF to include unique restriction sites for cloning into the expression vector (*Hin*dIII and *Xba*I). Alter three parameters in order to increase the mutagenic potential of the PCR. First, use low-fidelity Taq polymerase (Fisher BioReagents). Second, adjust the final concentration of dTTP and dCTP to 0.8 m*M* while keeping the remaining nucleotides at 0.25 m*M* (Matsumura and Ellington, 2002). Alteration of the ratio of nucleotides increases the likelihood of misincorporations during product synthesis. Finally, add manganese chloride to a final concentration of 500 n*M*. Manganese chloride concentrations lower than 500 n*M* are less mutagenic. Thermocycler settings: hot start: 94 °C 1 min; cycle: boil: 94 °C 1 min, anneal: 67 °C 1 min, extension: 68 °C 1 min 15 s for 50 cycles. Final extension: 68 °C 7 min, hold: 4 °C. The PCR product is cleaned up using PCR clean up kit according to the manufacturer's instructions (Qiagen). Digest the PCR products from the random mutagenesis step with Hin*dIII* and Xba*I* (NEB), and gel extracted using β agarase (NEB) and ligate into the p426GPD yeast expression vector, which is also digested and gel extracted. Select yeast colonies (about 200) on media lacking uracil and histidine and containing 3-AT. Extract plasmid encoding the CXCR2 CAMs using a yeast plasmid miniprep kit (Zymoprep™, ZYMO Research). Then transform the extracted yeast plasmid into *Escherichia coli* (MAX Efficiency® DH5 α™ Competent Cells, Invitrogen) following manufacturer's instructions. This step is necessary in order to acquire enough DNA for sequencing. Purify the CXCR2 CAMs/p426GPD plasmid using a miniprep kit (Promega) and sequence using primers flanking the multiple cloning site.

2.2.3. β-Galactosidase assay

This assay is used to confirm that the identified CXCR2 CAMs are truly constitutively active and quantify their signaling. Plasmid pMD1325 (a gift from Dr Dumont) contains a *FUS1-lacZ* that is induced when the receptor is activated (Sommers *et al.*, 2000). β-Galactosidase assay was performed as previously described (Kim *et al.*, 2009). Briefly, grow transformed yeast strains CY1141 and/or CY12946 (with the p426GPD expressing CXCR2 CAMs and pMD1325) overnight in selective growth media lacking uracil and leucine at 30 °C until they reach 5×10^6 cells/ml. Wash with sterile water and grow at 30 °C in selective media for one doubling based on a hemocytometer count. Measure basal activity of β-galactosidase, using a yeast β-galactosidase assay kit (Pierce) according to the manufacturer's instructions. This allows the comparison of β-galactosidase production from yeast expressing WT CXCR2 containing strains to the activity of each mutant.

3. Establishment of a Mammalian System to Characterize CXCR2 CAMs

In this part, we describe the experimental approaches to the identification of CAM-induced cellular transformation. In general, there are five characteristics of transformed cells: immortalization, growth in reduced serum, anchorage-independent growth, loss of contact inhibition, and tumor formation in nude mice (Alberts *et al.*, 2008; Bais *et al.*, 1998; McGlennen *et al.*, 1992). Mutations in cell cycle control genes (Cordon-Cardo, 1995), growth factors (Inoue and Nukiwa, 2005), transcription factors (Latchman, 1996), and degradation pathways (Guardavaccaro and Pagano, 2004) can lead to cellular transformation. To characterize CXCR2 CAM-transformed cells, electrical impedance, foci formations, and soft-agar growth assays were applied after the establishment of stable cell lines expressing the different CAMs. Tumor growth in nude mice was also used as an assessment of transformative ability. To delve further into how CXCR2 CAMs alter signal transduction pathways, luciferase assays for the activation of specific transcription factors were carried out. Recently, impedance measurements were used to discern transformed cells from normal cells (Park *et al.*, 2009). These microimpedance measurements were used to quantitatively examine the proliferation and the morphological changes associated with cellular transformation.

3.1. Establishment of stable cell lines expressing CXCR2 CAMs

NIH3T3 cells, established from NIH Swiss mouse embryo culture (Jainchill *et al.*, 1969), have been used historically for identifying transforming events. To identify fully transformed cells due to CXCR2 CAMs, we generated clonal, stable NIH3T3 cell lines. These lines were then used for analysis in the various assays for assessing the degree of transformation.

3.1.1. Required materials

- *Plasmid*: pcDNA3.1 (Invitrogen), pcDNA3.1+GFP
- *Transfection reagent*: Lipofectamine 2000™ (Invitrogen)
- *Cell line*: NIH3T3 (ATCC)
- *Growth medium*: DMEM with 10% bovine calf serum (Hyclone)
- *Neomycin*: G418 Sulfate (Cellgro)
- Trypsin–EDTA solution (Lonza)
- Ice-cold PBS (Hyclone)
- PE-conjugated human CXCR2 specific antibody (R&D Systems, *FAB331P*) for FACs analysis
- 37% paraformaldehyde (Fisher)
- 1% goat serum (Fisher Scientific) in PBS
- *Devices*: Microscope and hemocytometer, FACalibur (BD Bioscience)
- *Disposable material*: Sterile cloning disc (3 mm, Scienceware Inc), 6-well plate (Corning), and/or 100 (or 60) mm cell culture dishes (BD Falcon), and cell culture flasks (150 cm^2, BD Falcon).

3.1.2. Transfection of NIH3T3 cells to establish stable cell lines expressing CXCR2 CAMs

In a 6-well dish, transfect 2 μg of DNA and 4 μl of lipofectamine 2000 (Invitrogen) into 5×10^6 NIH3T3 cells. The transfection efficiency was optimized by varying DNA and lipofectamine concentrations with an indicator plasmid such as pcDNA3.1 expressing GFP. This is based on the recommendations of manufacturer's protocol. About 24–48 posttransfection, change the growth medium with medium supplemented with 0.8 g/L of G418 sulfate. Replace the medium with fresh selection media and wash with ice-cold PBS every 3 days for 14–21 days. This is the period required for the selection of stably transfected NIH3T3 cells. Untransfected NIH3T3 cells will detach and will be washed away during the media exchanges. Once all of the cells have died in the untransfected control, keep the growing cells in selective medium until colonies develop. Once colonies have formed, remove media from the plate and circle colonies with a colored pen on the bottom of the plate. Choose 5–10 colonies per construct to check for expression. Place a cloning disc (Scienceware) that has been soaked in trypsin–EDTA solution (Lonza), over the top of the colony for 2–3 min. Place the disc containing the stable clone into a well of a 24-well plate containing selection media. Change the media every day until the cells have achieved 80–90% confluency. Trypsinize the confluent well of cells and transfer to a 6-well dish and maintain in selection media until 80–90% confluency. Expand the cells into a T-25 cell culture flask and analyze for CXCR2 expression. Verify the surface expression of the receptor for each of the clonal population using immunostaining followed by flow cytometric analysis using a FACs machine (Calibur, BD Biosciences). Trypsinize 1×10^6 stably transfected cells, wash with ice-cold PBS and then block with 1% goat serum in PBS. Incubate cells with a PE-conjugated human CXCR2 specific antibody (R&D Systems, FAB331P) for 30 min at room temperature. Wash cells once with PBS and fix with 4% paraformaldehyde and analyze with flowcytometry. After verifying expression, expand cells and store in 10% DMSO in media in the freezer (− 140 °C) (liquid nitrogen is adequate) until further analysis. This is step is important as NIH3T3 cells will spontaneously transform upon passages above 10–15.

3.2. Characterization of CXCR2 CAMs in mammalian cells

To characterize the constitutive activity of the different CXCR2 CAMs, NIH3T3 stable cell lines were assessed for anchorage-independent growth and the loss of contact inhibition (McGlennen *et al.*, 1992). Additionally, impedance measurements allowed quantitative measurements of cellular proliferation and morphological changes associated with cellular transformation. This electrochemical technique has been applied to biological studies that include cellular barrier function, attachment, spreading, and adhesion (Burns *et al.*, 2000; Choi *et al.*, 2007; Gainor *et al.*, 2001; Kataoka *et al.*, 2002). To characterize the specific genes that may be involved in these morphological changes, luciferase reporter assay was utilized. For example, Cannon and Cesarman (2004) demonstrated that KSHV ORF74 constitutively activates transcription of AP-1, NF-κB, CREB, and NFAT-responsive promoters using luciferase assays. The ultimate test of whether these CXCR2 CAMs lead to tumorgenicity was the implantation of cells into the hind flank of nude mice and measuring tumor formation *in vivo*.

3.2.1. Required materials

- *Cell line*: NIH3T3 (ATCC) and HEK293 (ATCC)
- *Transfection reagent*: Lipofectamine 2000^{TM} (Invitrogen)
- *Soft-agar*: Low melt agarose (Boston Bioproduct)
- Six to eight week old athymic *nu/nu* mice (Jackson Laboratory)
- Crystal violet solution (25% ethanol, 1% formaldehyde, 0.125% NaCl, 0.25% crystal violet)
- *p*-Iodonitrotetrazolium violet (Sigma-Aldrich)
- CXCR2 antagonist (SB225002 (Calbiochem))
- Dual-Glo^{TM} Luciferase Assay System (Promega)
- *ImageJ* version 1.43 (Rasband, W. S., ImageJ, U. S. National Institutes of Health, Bethesda, MD, USA, http://rsb.info.nih.gov/ij/, 1997–2009)
- LabVIEW (National Instruments Corp.)
- *Devices*: Image analyzer (LAS 4000, Fujifilm Co.), Gold electrode array (Applied Biophysics), and SR830 lock-in amplifier (Stanford Research Instruments)

- *Disposable material*: Tissue culture grade 6-well dishes (Corning) and 50 ml centrifuge tubes (Corning)

3.2.2. Foci formation assay

Seed 100 stably transfected NIH3T3 cells on top of 2 × 10^5 untransfected NIH3T3 cells in normal growth medium with CXCR2 antagonist, SB225002 (Calbiochem) added to a final concentration of 1 μ*M*. This is added in order to prevent growth of the WT CXCR2 transfectants, which may respond to CXC chemokines present in the serum. Change the growth media every 2 or 3 days. After 5–7 days the cells should start to form colonies. 10–14 days after initial the seeding, fix the cells in 70% ethanol and stain with crystal violet solution (minimum of 1 h). If the cells grow too fast and begin to detach from the plate, stop the culture earlier and fix and stain. To enumerate the number of colonies, photograph the well and analyze the images using *ImageJ* version 1.43 (Rasband, W. S., ImageJ, U. S. National Institutes of Health, Bethesda, MD, USA, http://rsb.info.nih.gov/ij/, 1997–2009).

3.2.3. Soft-agar growth assay

Blend a total of 1 × 10^3 stably transfected NIH3T3 cells in prewarmed (37 °C) 0.4% soft-agar containing regular medium and pour the mixture on top of 0.8% agar in a 6-well plate. Feed cells every 3 days with five drops of normal 10% FBS growth media supplemented with 1uM CXCR2 antagonist (SB225002) per well. After incubation for 3 weeks, add 500 μl of 0.5 mg/ml *p*-iodonitrotetrazolium violet and incubate for overnight at 37 °C. Photograph wells and visualize colonies under an image analyzer (LAS 4000, Fujifilm Co.). Count and measure the number and size of colonies using *ImageJ*.

3.2.4. Electrical impedance measurements for cellular proliferation

Figure 26.3 shows diagrams of the three key factors for cellular microimpedance measurements (Fig. 26.3A and B) and images of what those measurements represent (Fig. 26.3C). Impedance is the frequency-dependent opposition of a conductor to the flow of an alternating electrical current. An alternating current is used for bioelectrical impedance measurements because it penetrates the cells at low levels of voltage and amperage. The corresponding experimental setup consists of lock-in amplifier, data acquisition board, computer, and electrodes. Cellular microimpedance data are analyzed using dynamic biophysical analysis in LabVIEW (National Instruments Corp.). An ac 1 Vrms reference signal via a series 1 MΩ resistor is provided as a reference voltage. A National Instruments SCXI-1127 switch is employed to connect the various working electrodes with the counter electrode of each array. The source voltage generator resistance (Rs) was 50 and the input impedance equivalent to a parallel resistor (Rv) and capacitor (Cv) combination of 10 MΩ and 10 pF, respectively.

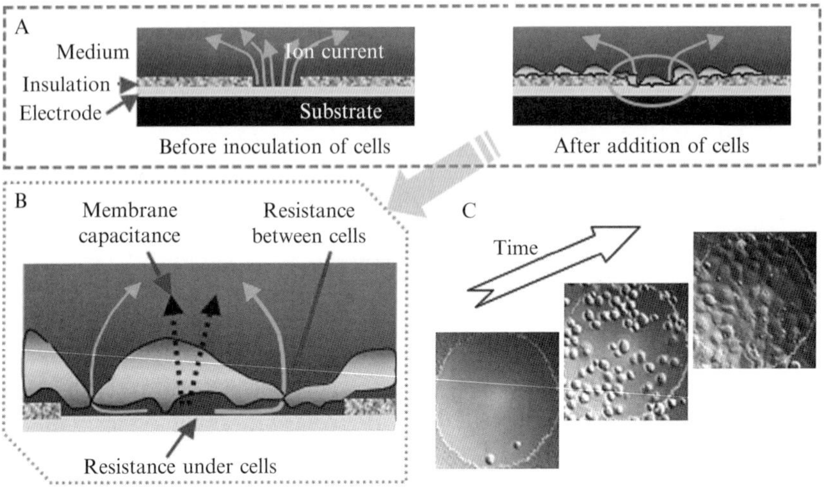

Figure 26.3 Overview of cellular microimpedance measurement. Impedance measurements reflect the electrical resistance and reactance across a cellular membrane and through a cell monolayer. (A) Schematic drawings show current flows before and after inoculation of cells. Each chamber contains a substrate with an electrode, a layer of conductive material, an insulation layer, and a small chamber containing the cellular growth medium. As cells attach, there is a modification of the current flow. (B) This drawing shows the three important parameters in impedance measurements, which characterize the cellular barrier (1) the current flow under the cells, (2) the current flow between the cells, and (3) the capacitively coupled current through the cellular membranes. (C) Cellular growth images on 250-μm diameter electrodes. These photomicrographs show cellular changes in their morphology, their adherence to each other, and to the substrate overtime.

During the cellular microimpedance scans, data are acquired at a rate of 32 Hz for 2 s using a 30-ms filter time constant and 12-dB/decade roll off. Averages and standard deviation estimates are obtained from the 64 sampled data points over the 2-s time intervals. During the experiments, cultures are maintained at 37 °C and 5% CO_2. Before data acquisition, naked scans are used to optimize the sensitivity, check for any electrode debris or defects and, most importantly, as a reference level for normalization. After repeated careful examination to select pertinent a cell density, a total of 3×10^4 transfectants and untransfected NIH3T3 cells are inoculated in 400 μl normal growth medium onto the electrode. One well contains no cells to provide a control.

3.2.5. Electrical impedance measurements for the foci formation assay

The impedance apparatus, as described above, is also employed to dynamically examine colony-like foci formation of CXCR2 CAM transfectants. 400 μl of normal growth medium containing 6×10^4 untransfected NIH3T3 cells is filled in each well to produce a cellular base layer to provide growth factors. A total of 3×10^4 transfectants or untransfected controls are seeded on top of the untransfected NIH3T3 layer. Seeding is performed at the saturated growth time point (approximately 27 h) with the addition of 100 μl of normal growth medium. Impedance measurements continue without any interruption during the seeding process for another 61 h to produce a second set of scans. Time-dependent changes in the normalized resistance and reactance can be obtained in real time. The bioelectrical impedance measurement can provide a noninvasive, sensitive, and quick way of dynamically examining cellular physiological changes.

3.2.6. Luciferase reporter assays

To measure the basal level of NF-κB, transiently cotransfect HEK293 cells (4.0×10^5) in a 6-well plate, with 2.0 μg of 3× MHC-Luc (Sherrill *et al.*, 2009). This follows the same protocol that was used to establish the stable transfectants. 0.5 μg of pRL-CMV (the empty vector) as an internal control and 0.5 μg of CXCR2 constructs with LipofectamineTM 2000 (Invitrogen). At 24 h posttransfection, synchronize the transfected cells with serum starvation and then harvest. Assay the cells with Dual-Glo™ Luciferase Assay System following manufacturer's instructions (Promega).

3.2.7. Tumor formation *in vivo*

Six to eight week old athymic *nu/nu* mice (Jackson Laboratory) are anesthetized using isoflourane in a chamber. Once they have been sedated, they are removed and their nose placed into a 50-ml conical tube that has a gauze soaked with isoflourane at the base. Lay the mouse down and rub the flank with an alcohol pad. Subcutaneously inject NIH3T3 transfectants (2×10^5/mice in 100 μl of PBS) into the flank. This protocol was reviewed and approved by the Institutional Animal Care and Concerns Committee of the University of Tennessee. Once tumors begin to appear (from 3 to 4 weeks), measure daily using a vernier caliper (Fisher Scientific 12-125-1). If one member in the group achieves a tumor size of 1.5 cm, all members of the group were euthanized. Resect, measure (height × length × width) with a caliper, and place tumors into 10% buffered formalin. Cut thin sections and stain with hematoxylin and eosin to characterize the types of cells that have infiltrated and the morphology of the tumors themselves. We have used a commercial company for this step, but any pathology lab should have the ability to process these samples.

REFERENCES

Ahuja, S. K., and Murphy, P. M. (1996). The CXC chemokines growth-regulated oncogene (GRO) alpha, GRObeta, GROgamma, neutrophil-activating peptide-2, and epithelial cell-derived neutrophil-activating peptide-78 are potent agonists for the type B, but not the type A, human interleukin-8 receptor. *J. Biol. Chem.* **271,** 20545–20550.

Alberts, B., *et al.* (2008). Molecular Biology of the Cell. Garland Science, New York.

Baggiolini, M., *et al.* (1994). Interleukin-8 and related chemotactic cytokines–CXC and CC chemokines. *Adv. Immunol.* **55,** 97–179.

Bais, C., *et al.* (1998). G-protein-coupled receptor of Kaposi's sarcoma-associated herpesvirus is a viral oncogene and angiogenesis activator. *Nature* **391,** 86–89.

Burger, M., *et al.* (1999). Point mutation causing constitutive signaling of CXCR2 leads to transforming activity similar to Kaposi's sarcoma herpesvirus-G protein-coupled receptor. *J. Immunol.* **163,** 2017–2022.

Burns, A. R., *et al.* (2000). Analysis of tight junctions during neutrophil transendothelial migration. *J. Cell Sci.* **113** (Pt 1), 45–57.

Cannon, M. L., and Cesarman, E. (2004). The KSHV G protein-coupled receptor signals via multiple pathways to induce transcription factor activation in primary effusion lymphoma cells. *Oncogene* **23,** 514–523.

Choi, C. K., *et al.* (2007). An endothelial cell compatible biosensor fabricated using optically thin indium tin oxide silicon nitride electrodes. *Biosens. Bioelectron.* **22,** 2585–2590.

Cordon-Cardo, C. (1995). Mutations of cell cycle regulators. Biological and clinical implications for human neoplasia. *Am. J. Pathol.* **147,** 545–560.

Erickson, J. R., *et al.* (1998). Edg-2/Vzg-1 couples to the yeast pheromone response pathway selectively in response to lysophosphatidic acid. *J. Biol. Chem.* **273,** 1506–1510.

Gainor, J. P., *et al.* (2001). Platelet-conditioned medium increases endothelial electrical resistance independently of cAMP/PKA and cGMP/PKG. *Am. J. Physiol. Heart Circ. Physiol.* **281,** H1992–H2001.

Gershengorn, M. C., *et al.* (1998). Chemokines activate Kaposi's sarcoma-associated herpesvirus G protein-coupled receptor in mammalian cells in culture. *J. Clin. Invest.* **102,** 1469–1472.

Gether, U., and Kobilka, B. K. (1998). G protein-coupled receptors. II. Mechanism of agonist activation. *J. Biol. Chem.* **273,** 17979–17982.

Guardavaccaro, D., and Pagano, M. (2004). Oncogenic aberrations of cullin-dependent ubiquitin ligases. *Oncogene* **23,** 2037–2049.

Guo, H. G., *et al.* (2003). Kaposi's sarcoma-like tumors in a human herpesvirus 8 ORF74 transgenic mouse. *J. Virol.* **77,** 2631–2639.

Horton, L. W., *et al.* (2007). Opposing roles of murine duffy antigen receptor for chemokine and murine CXC chemokine receptor-2 receptors in murine melanoma tumor growth. *Cancer Res.* **67,** 9791–9799.

Inoue, A., and Nukiwa, T. (2005). Gene mutations in lung cancer: Promising predictive factors for the success of molecular therapy. *PLoS Med.* **2,** e13.

Jainchill, J. L., *et al.* (1969). Murine sarcoma and leukemia viruses: Assay using clonal lines of contact-inhibited mouse cells. *J. Virol.* **4,** 549–553.

Kakinuma, T., and Hwang, S. T. (2006). Chemokines, chemokine receptors, and cancer metastasis. *J. Leukoc. Biol.* **79,** 639–651.

Kataoka, N., *et al.* (2002). Measurements of endothelial cell-to-cell and cell-to-substrate gaps and micromechanical properties of endothelial cells during monocyte adhesion. *Proc. Natl. Acad. Sci. USA* **99,** 15638–15643.

Kim, H., *et al.* (2009). Identification of specific transmembrane residues and ligand-induced interface changes involved in homo-dimer formation of a yeast G protein-coupled receptor. *Biochemistry* **48,** 10976–10987.

King, K., *et al.* (1990). Control of yeast mating signal transduction by a mammalian beta 2-adrenergic receptor and Gs alpha subunit. *Science* **250,** 121–123.

Klein, C., *et al.* (1998). Identification of surrogate agonists for the human FPRL-1 receptor by autocrine selection in yeast. *Nat. Biotechnol.* **16,** 1334–1337.

Kristiansen, K. (2004). Molecular mechanisms of ligand binding, signaling, and regulation within the superfamily of G-protein-coupled receptors: Molecular modeling and mutagenesis approaches to receptor structure and function. *Pharmacol. Ther.* **103,** 21–80.

Latchman, D. S. (1996). Transcription-factor mutations and disease. *N. Engl. J. Med.* **334,** 28–33.

Leurs, R., *et al.* (1998). Agonist-independent regulation of constitutively active G-protein-coupled receptors. *Trends Biochem. Sci.* **23,** 418–422.

Matsumura, I., and Ellington, A. D. (2002). Mutagenic polymerase chain reaction of protein-coding genes for in vitro evolution. *Methods Mol. Biol.* **182,** 259–267.

McGlennen, R. C., *et al.* (1992). Cellular transformation by a unique isolate of human papillomavirus type 11. *Cancer Res.* **52,** 5872–5878.

Muller, A., *et al.* (2001). Involvement of chemokine receptors in breast cancer metastasis. *Nature* **410,** 50–56.

Park, G., *et al.* (2009). Electrical impedance measurements predict cellular transformation. *Cell Biol. Int.* **33,** 429–433.

Price, L. A., *et al.* (1995). Functional coupling of a mammalian somatostatin receptor to the yeast pheromone response pathway. *Mol. Cell. Biol.* **15,** 6188–6195.

Rosenkilde, M. M., *et al.* (1999). Agonists and inverse agonists for the herpesvirus 8-encoded constitutively active seven-transmembrane oncogene product, ORF-74. *J. Biol. Chem.* **274,** 956–961.

Rosenkilde, M. M., *et al.* (2001). Virally encoded 7TM receptors. *Oncogene* **20,** 1582–1593.

Sachpatzidis, A., *et al.* (2003). Identification of allosteric peptide agonists of CXCR4. *J. Biol. Chem.* **278,** 896–907.

Sherrill, J. D., *et al.* (2009). Activation of intracellular signaling pathways by the murine cytomegalovirus G protein-coupled receptor M33 occurs via PLC-beta/PKC-dependent and -independent mechanisms. *J. Virol.* **83,** 8141–8152.

Sommers, C. M., *et al.* (2000). A limited spectrum of mutations causes constitutive activation of the yeast alpha-factor receptor. *Biochemistry* **39,** 6898–6909.

Son, C. D., *et al.* (2004). Identification of ligand binding regions of the *Saccharomyces cerevisiae* alpha-factor pheromone receptor by photoaffinity cross-linking. *Biochemistry* **43,** 13193–13203.

Zhang, W. B., *et al.* (2002). A point mutation that confers constitutive activity to CXCR4 reveals that T140 is an inverse agonist and that AMD3100 and ALX40-4C are weak partial agonists. *J. Biol. Chem.* **277,** 24515–24521.

Zlotnik, A. (2006). Chemokines and cancer. *Int. J. Cancer* **119,** 2026–2029.

CHAPTER TWENTY-SEVEN

A Method for Parallel Solid-Phase Synthesis of Iodinated Analogs of the Cannabinoid Receptor Type I (CB_1) Inverse Agonist Rimonabant

Alan C. Spivey* *and* Chih-Chung Tseng†

Contents

* Department of Chemistry, Imperial College, London, United Kingdom
† Chemical Synthesis Laboratory@Biopolis, Institute of Chemical and Engineering Sciences, Singapore

Methods in Enzymology, Volume 485
ISSN 0076-6879, DOI: 10.1016/S0076-6879(10)85027-X

Abstract

Rimonabant (acomplia) is a 1,5-diarylpyrazole derivative that acts as a type 1 cannabinoid receptor (CB_1) inverse agonist. Here, we overview the role of this type of molecule in regulation of these receptors and their potential as starting points for the development of molecular probes to image the central nervous system (CNS). We then describe a novel protocol for the solid-phase parallel chemical synthesis of iodinated rimonabant analogs using germanium-functionalized, cross-linked polystyrene as the solid-support (or "resin"). The method allows for rapid derivatization at the key C-3 position of rimonabant from a common resin-bound precursor. The desired iodinated analogs are then obtained by *ipso*-iododegermylative cleavage from the resin using sodium iodide/*N*-chlorosuccinimide (NCS) in a fashion that ought to be readily adapted to the rapid preparation of isotopically labeled iodine derivatives for molecular imaging of CNS activity by positron emission tomography (PET, using ^{124}I) and single photon emission computerized tomography (SPECT, using ^{123}I) techniques. Toward this goal, we also show that the NCS-derived succinimide by-product that is released from the resin concomitantly with the potential imaging probe molecules can be readily and selectively removed by treatment of the crude soluble product mixture with a solid-supported hydrazide scavenger resin.

1. Introduction

1.1. Cannabinoid receptors

Cannabinoid receptors obtained their name as a result of their response to cannabinoids, for example, Δ^9-tetrahydrocannabinol (Δ^9-THC) from *Cannabis sativa* (marijuana) and synthetic analogs. Cannabinoid receptors are members of the endocannabinoid system and are key mediators of many psychological processes (Wotjak, 2005). These receptors belong to the rhodopsin family of G protein-coupled receptors (GPCRs). Two subtypes of receptors have been identified: type 1 (CB_1) and type 2 (CB_2). The CB_1 receptor abounds in the brain and central nervous system (CNS), whereas the CB_2 is found mainly in the immune system (Howlett *et al.*, 2002).

As deduced by complementary DNA encoding (Matsuda *et al.*, 1990) and molecular cloning (Gerard *et al.*, 1991), human CB_1 receptors are encoded by an amino sequence of 472 residues. As for other proteins belonging to the rhodopsin family, the CB_1 receptor has seven transmembrane helical (TMH) domains of which the TMH-4 and -5 domains form the high-affinity ligand binding site (Shire *et al.*, 1996).

Studies of the physiological roles of the CB_1 receptor have revealed strong correlations with inhibition of adenylyl cyclase, pain control, regulation of ion channels, modulation of energy intake, and various other signal transduction pathways (Howlett, 2004; Pagotto *et al.*, 2006). Among these functions, arguably the most promising feature of this receptor's regulatory profile from a therapeutic-potential standpoint, is (or at least was, see below) its role in energy regulation and metabolism. In particular, animal models revealed that CB_1 knockout mice were leaner than wild-type species (Di Marzo *et al.*, 2001). Blocking the CB_1 receptor potentially therefore provides a therapeutic strategy for treatment of obesity and metabolic syndrome as well as drug additions and addiction to smoking (Di Marzo *et al.*, 2001).

1.2. Antagonists and inverse agonists of CB_1 receptors

A 1,5-diarylpyrazole-based ligand, *SR141716A*, later known as rimonabant (acomplia), was released in 1994 by *Sanofi-Aventis* to target the CB_1 receptor. It displayed nanomolar affinity for this receptor and just micromolar affinity for the CB_2 receptor making it the first efficient, competitive agent for selective blocking of CB_1 receptors (Rinaldi-Carmona *et al.*, 1994). Rimonabant was first categorized as a CB_1 receptor antagonist; however, after comprehensive pharmacological studies taking into account the intrinsic activity of the endocannabinoid system, it was recategorized as an inverse agonist (Pertwee, 2005). An analog, *AM251*, which has an iodine-substituent in place of a chlorine atom in the 5-phenyl ring of rimonabant, has similar inhibitory activity and selectivity for CB_1 whilst also providing an opportunity to introduce a radioactive-iodine isotope (^{123}I or ^{124}I) to allow visualization of the CNS by position emission tomography (PET) or single photon emission computed tomography (SPECT; Howlett, 2004; Pagotto *et al.*, 2006). A ring-constrained analog *NESS0327* was reported as having femtomolar affinity for the CB_1 receptor *in vitro*; however, no inhibition was shown when administered orally (Murineddu *et al.*, 2005; Stoit *et al.*, 2002). The poor biodistribution of this compound *in vivo* doubtless contributed to this outcome. Extensive studies have also been reported toward the development of new CB_1 ligands by changing the central core of rimonabant to imidazole, triazole, and phenyl units (e.g., *O-1803*) and by *de novo* design to give novel scaffolds such as found in *CP-945598* (*Pfizer*), *MK-0364* (taranabant, *Merck*), and *AVE-1625* (*Sanofi-Aventis*; Scheme 27.1).

Clinical trials of rimonabant for treatment of obesity suggested that it was a promising and effective medicine and so the compound was approved by the European Medicines Agency for sale in Europe as acomplia in 2006. Although only minor side-effects were reported during the clinic trials, several serious psychiatric disorder cases emerged once the drug was available on prescription. As a consequence, this drug was withdrawn from the market in 2008 and many other ongoing projects targeting the

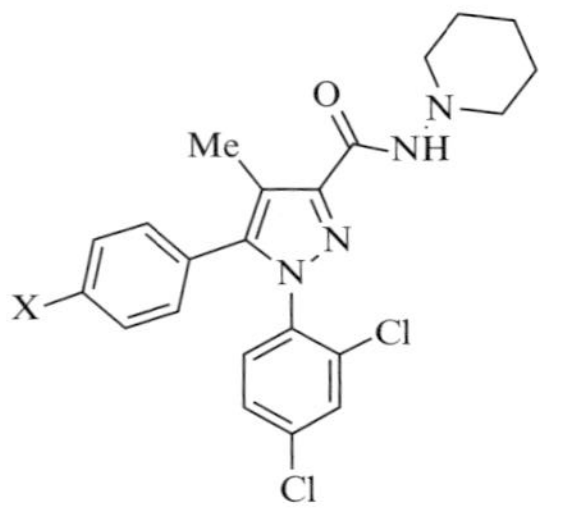

SR141716A X = Cl (rimonabant) $hCB_1 = 5.6\,nM$, $hCB_2 > 1\,\mu M$
AM251 X = I $hCB_1 = 7.5\,nM$, $hCB_2 = 2.3\,\mu M$

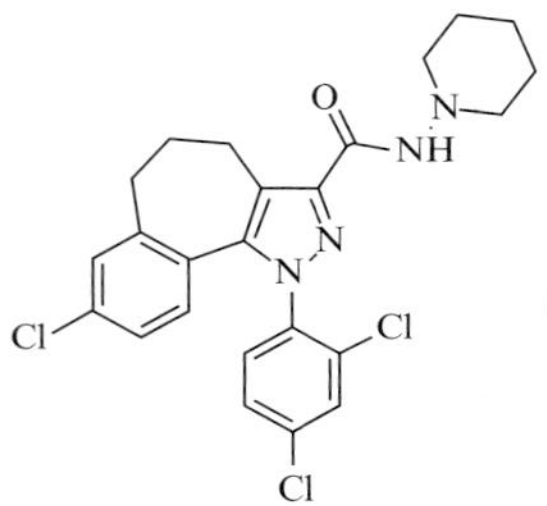

NESS0327 $hCB_1 = 350\,fM$ (*in vitro*) $hCB_2 = 442\,nM$

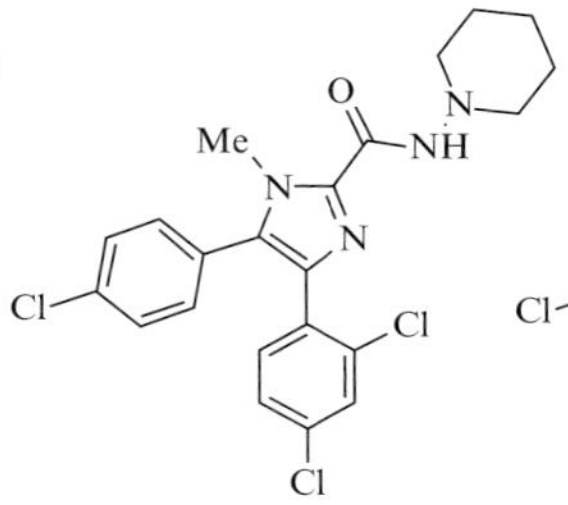

$hCB_1 = 4\,nM$
$hCB_2 = 300\,nM$

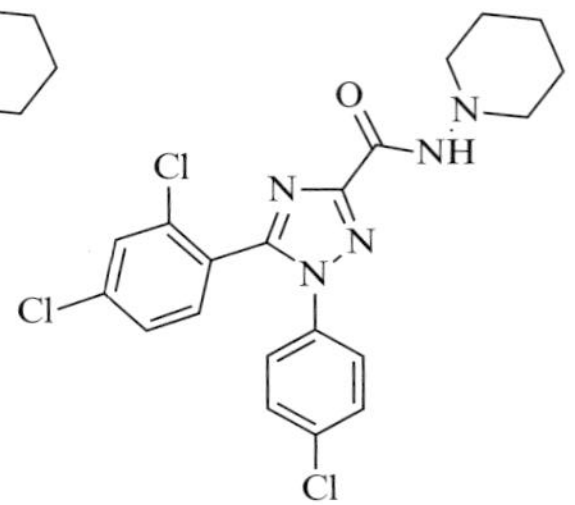

$hCB_1 = 356\,nM$
$hCB_2 = 3562\,nM$

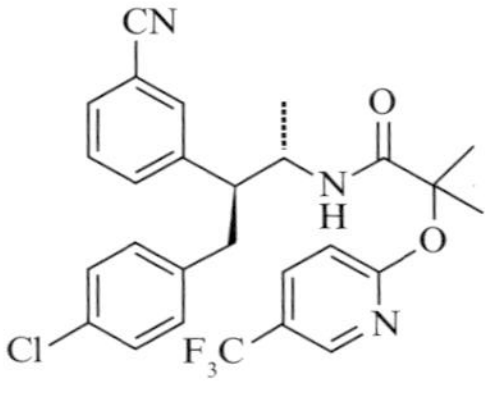

MK-0634 (taranabant) $hCB_1 = 0.3\,nM$ $hCB_2 = 284\,nM$

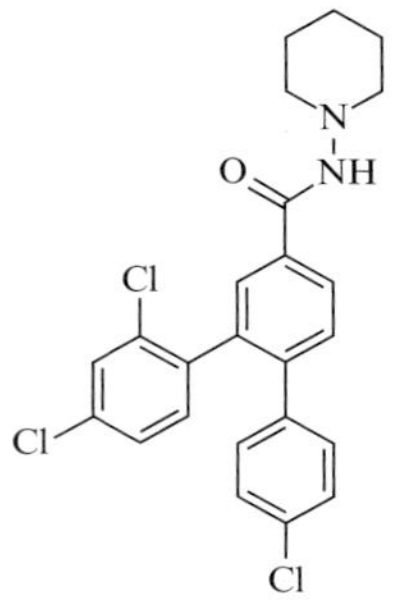

O-1803 $rCB_1 = 113\,nM$

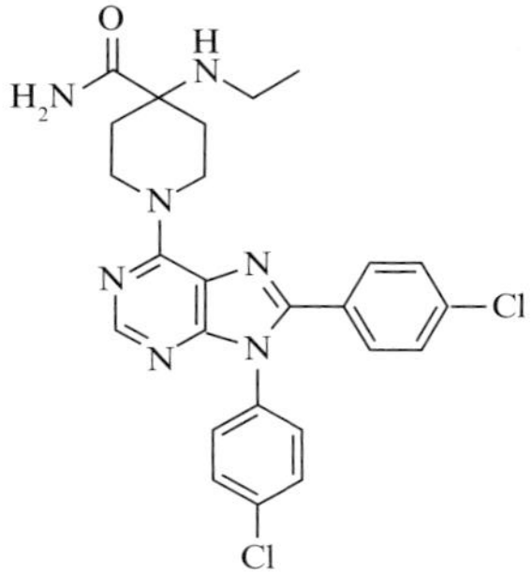

CP-945598 $hCB_1 = 0.7\,nM$

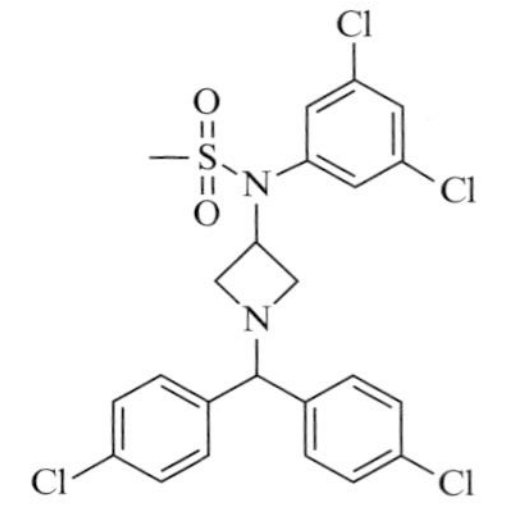

AVE-1625 $hCB_1 = 0.16–0.44\,nM$

Scheme 27.1 "Drug-like" potential CB_1 inhibitors.

CB_1 receptor as a strategy for developing antiobesity drugs were also apparently terminated (Jones, 2008).

1.3. Conformational analysis of rimonabant and computational modeling of its binding features with the CB_1 receptor

Computational docking analyses of rimonabant and related structures in the CB_1 receptor binding site have revealed that its gross orientation is set by an essential salt bridge between the carbonyl moiety and Lys192/Asp366 (Hurst *et al.*, 2002). π–π Stacking interactions between the two substituted phenyl groups and the aromatic-rich TMH 3–6 regions of the CB_1 receptor also constitute an important anchor for this binding mode, as are lipophilic interactions between the piperidinyl group and a hydrophobic cavity comprising Val196, Phe170, Leu387, and Met384. Notwithstanding these constraints, several distinct conformational minima are apparently energetically accessible (Scheme 27.2; Lange and Kruse, 2005).

Conformations in which torsional angle τ_1 (about the bond between the pyrazole ring and the acyl hydrazine carbonyl) enforces an s-*trans* pose and the torsional angle τ_2 (about the N–N bond of the hydrazine unit) is in one of four particular values appear to be favored. The torsional angles τ_3 and τ_4 control the conformation of the two aryl groups, and these vary depending on the substituents on these rings (Thomas *et al.*, 2006).

It is not clear what exact molecular mechanisms cause the adverse psychiatric side effects. However, the endocannabinoid system is known to be involved in our ability to manage stress and anxiety and so the onset of depression as a result of disruption of these pathways is perhaps unsurprising.

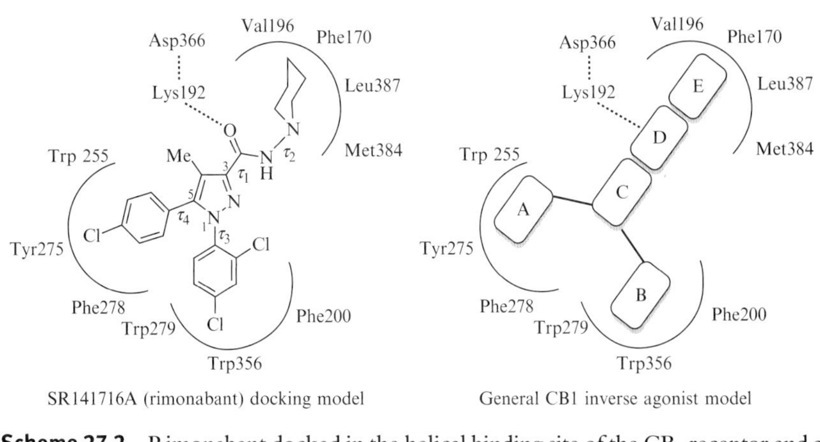

Scheme 27.2 Rimonabant docked in the helical binding site of the CB_1 receptor and a general inverse agonist-receptor interaction model (Lange and Kruse, 2005).

Elaboration of the psychiatric functions controlled by the endocannabinoid system will be necessary if the effects caused by blocking the receptor are to be fully understood and safe modulators are to be developed.

2. Concepts for Molecular Imaging

Molecular imaging refers to a raft of multidisciplinary technologies encompassing medicine, chemistry, pharmacology, and nuclear medicine which allow noninvasive monitoring and visualization of living biological systems in real time. Molecular imaging generally involves administering specific biomarkers or radiolabeled probes into living organisms for characterization and quantification of biological processes or for pathological diagnosis at the cellular or subcellular level by tomographic image generation. PET and SPECT in particular have had a significant impact on our ability to monitor a plethora of medical phenomena. Currently, the repertoire of conditions that can be illuminated by these techniques appears to be primarily limited by our ability to develop high-affinity radiolabeled ligands to target selectively the required tissue or pathological area.

The probe molecules for PET and SPECT imaging are normally positron decay nuclei labeled pharmaceuticals. Ideally, the introduced radiolabel should have no significant effect on the physicochemical and pharmacokinetic properties of the molecule. The characteristics of radioactive nuclei routinely used in this type of molecular imaging are listed in Table 27.1.

These radioisotopes are all β^+ decay or γ-ray emitting nuclei with short radioactive half-lives ($t_{1/2}$). Their short radioactive half-lives impose stringent restrictions on the synthetic methods used for labeling. Only a very limited repertoire of labeled precursors are available and these have to be incorporated into the target probe molecule via rapid and efficient chemical transformations, ideally in a manner that is operationally simple, minimizes waste generation and facilitates rapid purification to high levels of purity.

3. Conventional Methods for Preparation of Radiolabeled Pharmaceuticals for Imaging the CB_1 Receptor

The scaffold of rimonabant constitutes a promising starting point for the development of radiolabeled molecular probes for *in vivo* studies of the brain. Various studies have been reported in this area and can be categorized into two groups according to the method by which the radioactive-iodine atom is introduced.

Table 27.1 Characteristics of commonly used radioactive nuclei in PET and SPECT scanning

Nucleus	$t_{1/2}$	Maximum energy (MeV)	Mode of decay	Examples of clinical applications
^{18}F	109 min	0.64	β^+ (97%) EC (3%)[a]	Presynaptic dopaminergic system; Parkinson's disease (PET)
^{11}C	20.3 min	0.97	β^+ (99%)	Dopamine D_2/D_3 receptors (PET)
^{13}N	10 min	1.20	β^+ (100%)	Glioblastoma; tumor (PET)
^{124}I	4.2 d	2.14	β^+ (25%) EC (75%)	Tumor hypoxia tracer (PET)
^{15}O	2 min	1.74	β^+ (100%)	Solid tumor blood perfusion (PET)
^{99m}Tc	6 h	140 (89%)[a]	γ-Ray	Brain, kidneys (SPECT)
^{123}I	13.2 h	159 (83%)[a]	γ-Ray	Dopaminergic system (SPECT)
^{201}Tl	73 h	167 (10%)[a], 80 (20%)[a], 71 (47%)[a]	γ-Ray	Bronchioloalveolar cell carcinoma (SPECT)
^{111m}In	2.83 d	245 (94%)[a]	γ-Ray	Somatostain receptors (SPECT)

[a] γ-Ray energy (keV).

3.1. Radiolabeling of the rimonabant scaffold by nucleophilic substitution

Derivatization of the rimonabant core to introduce a radiolabel must of course take into account the above-discussed binding models with the CB_1 receptor and established structure–activity relationship (SAR) data in order to have maximum chance of retaining potency. Modifications to the substrate that concomitantly enhance affinity and biodistribution are also desirable. As a consequence, modification of rimonabant at the 4-methyl, 5-aryl, and 3-carboxamide moieties has received the most attention.

Mathews has prepared [^{18}F] and [^{11}C]-labeled analogs by nucleophilic substitution reactions at the 4-methyl and 5-aryl units, respectively (Mathews *et al.*, 2000, 2002). Galety has described replacing the 3-acyl hydrazine motif with a [^{18}F]-4-fluorophenyl amide unit. The [^{18}F]-4-fluorophenylaniline was shown to be a particularly promising derivative and this was prepared by a S_NAr reaction of dinitrobenzene with [^{18}F]-fluoride anion (Scheme 27.3; Lia *et al.*, 2005).

Pharmacological data for these tracer candidates revealed good inhibition of the CB_1 receptor in their unlabeled form. Moreover, the [^{18}F] and [^{11}C]-labeled probes provided high-quality scanning images, although the short half-life of these radionuclei restricts the range of potential applications for these tracers.

3.2. Radiolabeling of rimonabant by *ipso*-iododestannylation

As indicated above, the iodine analog of rimonabant *AM251* has similar binding characteristics as the parent drug. Radiolabeled [^{123}I]-*AM251* has been prepared and has been evaluated in animal models for SPECT scanning. Another candidate, [^{123}I]-*AM281* was also prepared for this purpose and was the first SPECT agent for visualizing the human cannabinoid system (Lan *et al.*, 1999a). The method used to synthesize these aryl iodide CB_1 ligands was to effect *ipso*-iododestannylation of the corresponding aryltributylstannane precursors by treatment with a mixture of radiolabeled KI and chloroamine-T or with radiolabeled molecular iodine. Arylstannanes are generally prepared by palladium(0)-catalyzed coupling of the corresponding aryl bromides with hexabutyldistannane. [I^{124}]-*AM281* for PET imaging was also prepared in this fashion (Scheme 27.4; Berdinga *et al.*, 2006).

Due to the involvement of palladium complexes and organostannanes in these processes, tedious postreaction purification procedures are critical and necessary to remove these potentially harmful heavy metals prior to administration of the labeled analogs for *in vivo* trials. The time required for

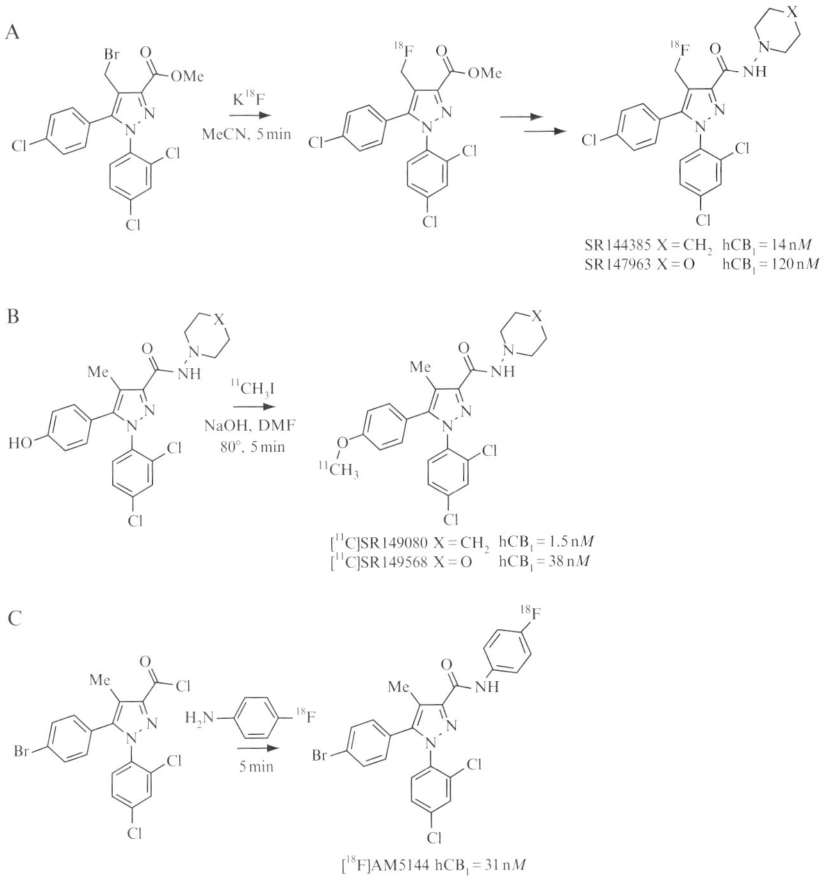

Scheme 27.3 Mathews' (A, B) and Galety's approaches (C) to the preparation of radiolabeled rimonabant analogs for PET scanning (Lia *et al.*, 2005; Mathews *et al.*, 2000, 2002).

this purification inevitably degrades the quality of the resulting tomographic images due to attenuation of the radioactivity. Additionally, preparation of these analogs for screening as potential imaging agents relies on the development of individual, bespoke synthetic procedures in solution for each new candidate tracer which is expensive and time-consuming (Lan *et al.*, 1999b). The method described below was developed to address these deficiencies in the current state-of-the-art for preparing radioiodinated aryliodide-based imaging agents (Hockley *et al.*, 2009; Spivey *et al.*, 2007a)

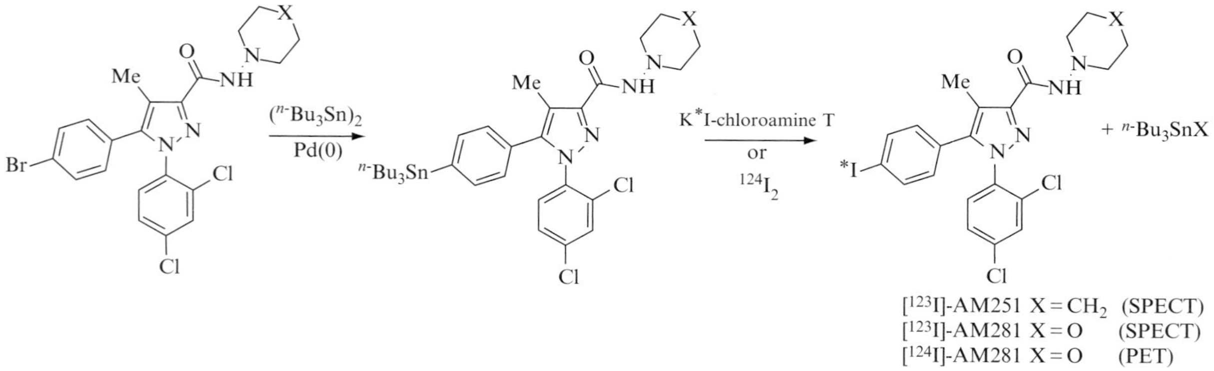

Scheme 27.4 Preparation of radioiodinated molecular probes via *ipso*-iododestannylation (Berdinga *et al.*, 2006).

4. Parallel Solid-Phase Synthesis of Iodinated Analogs of the CB_1 Receptor Inverse Agonist Rimonabant

We envisaged that by employing a rimonabant-derived labeling precursor bound to an insoluble polymeric solid-support ("resin") via a trialkylgermane linker at the 4-position of the phenyl ring attached to C-5 of the pyrazole core it would be possible to effect *ipso*-iododegermylation to introduce a radiolabeled iodine atom is a fashion analogous to that currently achieved by *ipso*-iododestannylation in solution. This innovation would avoid the need for the radiochemist to handle tin-containing materials and also the requirement for any postreaction purification. Ge-containing compounds display a very favorable toxicity profile compared to their Sn congeners (Asai and Miracle, 1980; Goodman, 1988; Patai, 1995) and the only molecules released into solution should be those of the labeled product which should therefore be obtained in a relatively pure form simply by filtering off the insoluble resin. Moreover, unlike arylstannanes, arylgermanes are robust toward a wide range of conditions for multistep chemical synthesis and so afford the possibility to make a range of derivatives by on-resin synthetic diversification of an advanced precursor prior to final radio-iodinative release; a potentially valuable feature during preclinical imaging agent development. Experimental SAR data, particularly from previous attempts to develop effective PET ligands based on the rimonabant core, have shown that the nature of the C-3 hydrazide group has a strong effect on their receptor affinity and biodistribution profiles (Donohue *et al.*, 2006; Jagerovic *et al.*, 2008; Lan *et al.*, 1999b; Thomas *et al.*, 2005; Tobishi *et al.*, 2007). We therefore selected to target this position for variation. The envisaged approach for this parallel, solid-phase synthesis (SPS) of an array of iodinated analogs of rimonabant having various acyl units at the 3-position is outlined below (Scheme 27.5).

A method that provides proof-of-concept for this approach, albeit using nonradiolabeled iodine, is described below.

4.1. Solid-phase synthesis of iodinated rimonabant analogs

The choice of solid-support employed for SPS is a crucial parameter in determining its success. Our initial investigations to develop efficient conditions for *ipso*-iododegermylation of model aryltrialkygermanes in solution demonstrated that a polar reaction medium is essential for efficient, clean, and fast reactions (Spivey *et al.*, 2009). Consequently, HypoGel®-200 resin is employed for the SPS method. This resin comprises polyethyleneglycol (PEG) grafted 1% 1,4-divinylbenzene (DVB) cross-linked polystyrene beads and displays superior

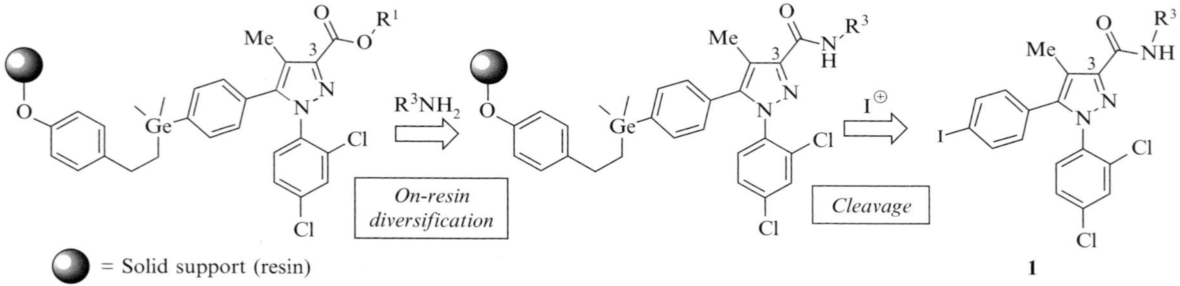

1

Scheme 27.5 Proposed SPS approach to iodinated rimonabant analogs 1.

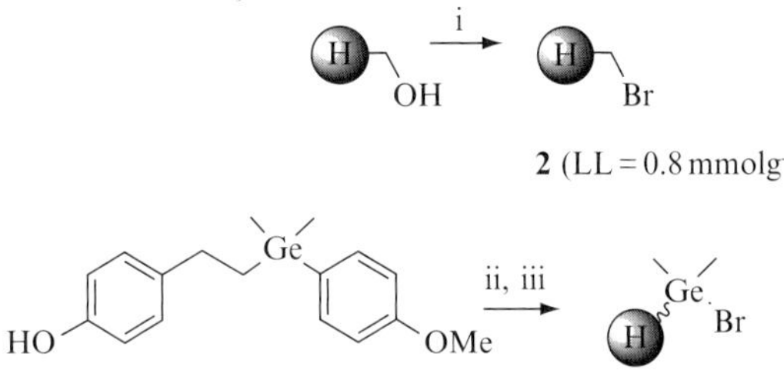

2 (LL = 0.8 mmolg^{-1})

3

4 (LL = 0.75 mmolg^{-1})

Scheme 27.6 Preparation of germyl functionalized resin 4.
Reagents and conditions: (i) PPh_3, CBr_4, CH_2Cl_2, 0 °C–RT, 24 h (ii) **3**, K_2CO_3, DMF, 80 °C, 24 h (iii) Conc. HBr, CH_2Cl_2, RT, 16 h.

swelling behavior in hydrophilic reaction media relative to alternative non-grafted 1% DVB cross-linked polystyrene resins (i.e., Merrifield type).

Commerically available HypoGel®-OH (n = 5, 110–150 μm diameter beads, 0.8 mmol g^{-1}) is first converted to HypoGel®-Br **2** (0.8 mmol g^{-1}) quantitatively by Appel bromination using a combination of PPh_3 and CBr_4 in CH_2Cl_2. The preformed germyl linker **3** is then introduced by Williamson etherification and the resulting resin converted into the germyl bromide resin **4** by treatment with 48% HBr in CH_2Cl_2 (Scheme 27.6; Turner *et al.*, 2007).

Further elaboration of the germyl bromide resin **4** is achieved by its reaction with lithiated *tert*-butyl diarylpyrazolyl ester **5** to give resin **6.** This functionalized resin can serve as a pivotal intermediate for the parallel preparation of C-3 functionalized derivatives by on-resin "amidation" of the *tert*-butyl ester moiety of resin **6** with hydrazines and amines facilitated by lithium bis(trimethylsilyl)amide (LHMDS) in THF at RT. The method is exemplified by the preparation of an array five selected diarylpyrazolyl resins **7a–e** shown below (Scheme 27.7).

Cleavage of the model candidate tracers from the resin with concomitant introduction of iodine onto the C-5 aryl ring at the position where the germanium linker has been attached is achieved by *ipso*-iododegermylation. This requires a combination of *N*-chlorosuccinimide (NCS) and NaI in acidic media. The Ge resins **7a–e** are swollen in individual reaction cartridges with CH_2Cl_2 in the presence of NaI, prior to addition of a solution of NCS in TFA-HOAc at RT. The resin beads are then filtered off and the desired iodinated rimonabant analogs **1a–e** remain in the filtrate solution (Table 27.2; Spivey *et al.*, 2009).

In these examples, the *ipso*-iododegermylative cleavage of aryl pyrazolyl germane resins **7a–d** was complete in 1 h (entries 1–4) and the 4-fluoroanilinyl functionalized pyrazolyl resin **7e** in 6 h (entry 5). Given that the radioactive half-lives of [^{124}I] and [^{123}I] are 4.2 d and 13.2 h, respectively, these reaction times are not incompatible with translation to a radiolabeling method.

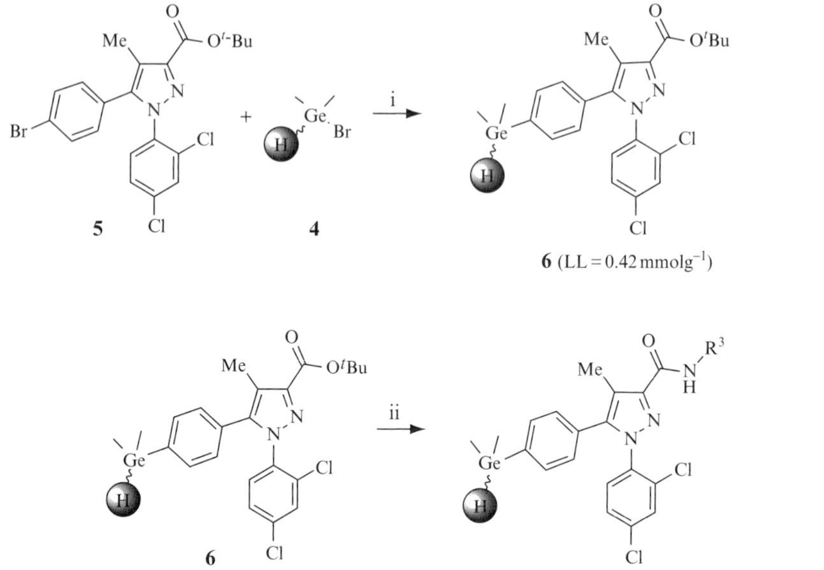

7a R^3 = morpholin-4-yl (LL = 0.42 mmol g^{-1})
7b R^3 = piperidin-1-yl (LL = 0.42 mmol g^{-1})
7c R^3 = *N'*,*N'*-dimethylamino (LL = 0.42 mmol g^{-1})
7d R^3 = cyclohexyl (LL = 0.38 mmol g^{-1})
7e R^3 = 4-fluoroanilinyl (LL = 0.35 mmol g^{-1})

Scheme 27.7 Parallel SPS of C-3 functionalized pyrazole resins **7a–e**. Reagents and conditions (i) ^{t}BuLi, THF, −78 °C–RT, 16 h (ii) R^3NH_2, LHMDS, THF, RT, 16 h.

However, cognizant that more rapid reaction kinetics would improve radiochemical yields, resin **7b** was also exposed to the iodination conditions for just 5 min; this reaction furnished iodinated aryl pyrazole *AM251* (**1b**) in 8% after column chromatography (entry 6). Moreover, resubjecting resin **1b** used in this experiment to the iodination conditions for a further 5 min gave an additional 5% of the desired product (entry 7). These results suggest that a short-exposure protocol could probably be developed for radiolabeling and still provide sufficient quantities of radioiodinated tracer for high-resolution imaging whilst also offering the prospect of utilizing the resin in a reusable cartridge format for employment in several sequential labeling runs.[1]

[1] *NB.* The method has not, as yet, been performed using a radioactive iodide salt for actual imaging probe labeling and so this is conjecture at this point in time. A significant difference between a typical radiolabeling scenario and the experiments described in this method is that radioactive reagents are invariably employed at low concentrations, and under these conditions adsorption of the labeled products by resin beads or glass apparatus may be significant. We envisage that microfluidic techniques may hold the key to optimizing the cleavage protocol and succinimide scavenging to achieve sufficiently highly reproducibility for clinical use.

Table 27.2 Parallel preparation of iodinated C-3 functionalized rimonabant analogs

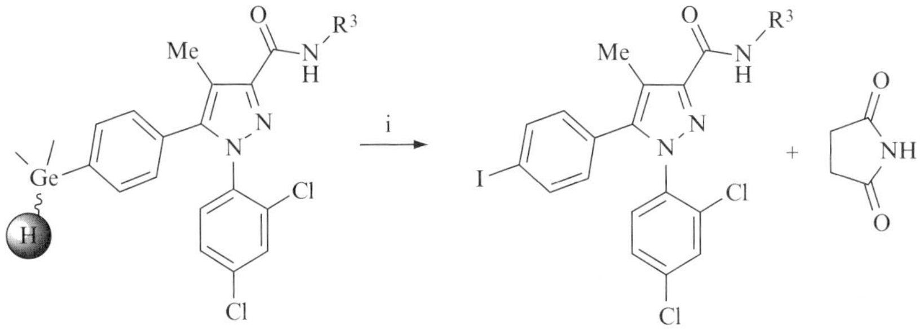

7a	R^3 = morpholin-4-yl	AM281(**1a**)
7b	R^3 = piperidin-1-yl	AM251(**1b**)
7c	R^3 = *N′*,*N′*-dimethylamino	**1c**
7d	R^3 = cyclohexyl	**1d**
7e	R^3 = 4-fluoroanilinyl	**1e**

Entry	Substrate	Time (h)	Product	Yield (%)
1	**a**	1	*AM281* (**1a**)	63
2	**b**	1	*AM251* (**1b**)	60
3	**c**	1	**1c**	60
4	**d**	1	**1d**	55
5	**e**	6	**1e**	42
6	**b**	5 min	*AM251* (**1b**)	8 (1st run)
7	**b**	5 min	*AM251* (**1b**)	5 (2nd run)

Reagents and conditions: (i) NaI, NCS, TFA-HOAc (1:3, v/v), RT.

Analysis of the crude iodinated aryl pyrazole *AM251* (**1b**) directly after filtration by means of HPLC (UV detector set at 254 nm) reveals that it has a purity of just 75% owing to the presence of succinimide, a by-product of the sodium iodide "activating" reagent NCS. Clearly, it is crucial to remove this by-product prior to injection for *in vivo* evaluation, and so an efficient and rapid process to achieve this has been developed. Thus, a nucleophilic scavenger resin is deployed to effect *in situ* removal of the succinimide. Hydrazine-functionalized *Si*-propyl hydrazine resin **8,** which is prepared by treatment of commercially available *Si*-propyl bromide solid-support (40–

63 μm diameter beads, 1.43 mmol g^{-1}) with propylhydrazine hydrate and K_2CO_3 in DMF, is particularly effective for this purpose.

Thus, a freshly prepared crude product mixture of iodinated aryl pyrazole *AM251* (**1b**) following *ipso*-iododegermylation is immediately treated with the *Si*-propyl hydrazine beads **8** (1.43 mmol g^{-1}) in refluxing EtOH for 1 h. The suspended beads are then filtered off to leave a solution of the desired iodinated pyrazole *AM251* (**1b**) in 96% purity (as determined by HPLC[2]; Scheme 27.8).

The success of this one-pot, sequential, functionalizative-cleavage then purification protocol augurs well for the development of an efficient procedure to prepare arrays of radiolabeled rimonabant analogs for PET and SPECT imaging. The method should also be amenable to adaption for preparation of arrays of a wide range of other aryl iodide-based imaging probes given the chemical stability of the arylgermane linker toward a diverse range of reaction diversification conditions (Spivey *et al.*, 2007b).

5. Materials and Methods

5.1. General directions

Solvents and reagents: All solvents were redistilled before use. Reagents and solvents were used as commercially supplied unless otherwise stated and handled in accordance with COSHH or local regulations.

HypoGel®-OH: Commercially supplied from Fluka.

NaI: Commercially supplied from Lancaster.

N-Chlorosuccinimide: Rapidly recrystallized from glacial acetic acid and washed with cold H_2O followed by drying *in vacuo*.

Si-propyl bromide: Commercially supplied from Crawford Scientific, UK.

CBr_4: Sublimed under vacuum 0.1 mmHg at 120 °C immediately before use.

Chromatography: Flash chromatography (FC) was carried out on silica gel (BDH Silica gel for FC) according to the method described by Still *et al.* (1978). TLC was performed on aluminium-backed silica gel plates (Merck Silica gel 60 F_{254}) which were developed with UV fluorescence (254 and 365 nm) and $KMnO_4$(aq)/Δ.

^1H NMR spectra: These were recorded at 400 or 500 MHz on Bruker-DRX 400, AV-400, or AV-500 instruments, respectively. Chemical shifts (δ_H) are given in parts per million (ppm) as referenced to the appropriate

[2] The HPLC chromatogram for pure **AM251**(**1b**) shows three distinct peaks. The relative intensity of these peaks varies depending on the pH of the eluent system used. We presume that these peaks correspond to rotamers about τ_1 and/or τ_2 (Clayden *et al.*, 2009).

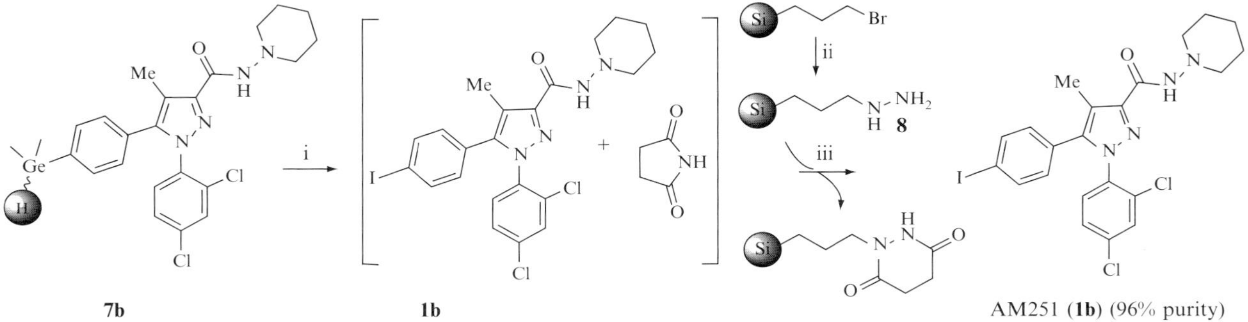

Scheme 27.8 Sequential preparation/purification of iodinated rimonabant analog **AM251** using a solid-supported hydrazine **8** to scavenge succinimide.

Reagents and conditions (i) NaI, NCS, TFA-HOAc (1:3, v/v), RT, 1 h; (ii) NH_2NH_2 hydrate, K_2CO_3, DMF, 80 °C, 24 h; (iii) EtOH, 80 °C, 1 h.

residual solvent peak. Broad signals are assigned as "br." All spectra were compared with reference spectra of authentic products and reagents.

^{13}C NMR spectra: These were recorded at 100.6 MHz on Bruker AV-400 instrument or at 125.1 MHz on a Bruker AV-500 instrument. Chemical shifts (δ_C) are given in ppm as referenced to $CHCl_3$, and are assigned as s, d, t, and q, for C, CH, CH_2, and CH_3, respectively.

Mass spectra: Low-resolution and high-resolution spectra were recorded on a VG Prospec spectrometer, with molecular ions and major peaks being reported. Intensities are given as percentages of the base peak. HRMS values are valid to ± 5 ppm.

GC/MS: Analyses were carried out using a Trio-1 with HP 5890 gas chromatograph.

Melting points: Analyses were carried out using a Khofler hot stage and are uncorrected.

5.2. Synthesis of diarylpyrazole core and Ge-functionalized solid-supported resins

5.2.1. 1-(2,4-Dichlorophenyl)-5-(4-bromophenyl)-4-methyl-1H-pyrazole-3-carboxylic acid ethyl ester

The title material was prepared as described by Lan *et al.* (1999b). 4-Bromopropiophenone (4.5 g, 21.1 mmol), a solution of LHMDS (26.4 mL, 26.4 mmol, 1 *M* in hexanes), diethyl oxalate (3.85 g, 26.4 mmol, 3.58 mL), and 2,4-dichlorophenyl hydrazine hydrochloride (4.96 g, 23.2 mmol) were used. Purification by FC (hexane/EtOAc, 90/10) gave *1-(2,4-dichlorophenyl)-5-(4-bromophenyl)-4-methyl-1H-pyrazole-3-carboxylic acid ethyl ester* (8.44 g, 88%) as an orange solid.

5.2.2. 5-(4-Bromophenyl)-1-(2,4-dichlorophenyl)-4-methyl-1*H*-pyrazole-3-carboxylic acid

To a solution of 1-(2,4-dichlorophenyl)-5-(4-bromophenyl)-4-methyl-1H-pyrazole-3-carboxylic acid ethyl ester (1.02 g, 2.246 mmol) in THF (15 mL) was added NaOH (2 g, 50 mmol) and H_2O (4 mL). The reaction mixture was refluxed at 70 °C for 6 h and extracted with Et_2O (3×50 mL). The combined organic extracts were dried over Na_2SO_4 and the solvent removed *in vacuo*. The residue was purified by recrystallization to give *5-(4-bromophenyl)-1-(2,4-dichlorophenyl)-4-methyl-1H-pyrazole-3-carboxylic acid* as a white solid (0.947 g, 99%); ^{1}H NMR (400 MHz, $CDCl_3$): δ 2.35 (s, 3H, CH_3), 7.01 (d, $J = 8.4$ Hz, 2H, ArH), 7.30 (dd, $J = 8.5$, 2.0 Hz, 1H, ArH), 7.34 (d, $J = 8.5$ Hz, 1H, ArH), 7.41 (d, $J = 2.0$ Hz, 1H, ArH), 7.47 (d, $J = 8.4$ Hz, 2H, ArH); ^{13}C NMR (100.7 MHz, $CDCl_3$): δ 9.6 (q), 119.6 (s), 123.5 (s), 127.2 (s), 127.9 (d), 130.2 (d), 130.5 (d), 131.1 (2d), 131.9 (2d), 132.8 (s), 135.6 (s), 136.3 (s), 142.1 (s), 143.4 (s), 166.0 (s); *m/z* (CI^+) (rel. intensity) 452 [M(^{79}Br)H^+,

100], 339 (60), 297 (20); HRMS (CI^+) calc'd for $C_{17}H_{12}{}^{79}Br^{35}Cl_2N_2O_2$ $(MH)^+$ 424.9459, found 424.9447 (Δ −2.9 ppm).

5.2.3. 5-(4-Bromophenyl)-1-(2,4-dichlorophenyl)-4-methyl-1*H*-pyrazole-3-carboylic acid *tert*-butyl ester 5

According to the method of Armstrong (Armstrong *et al.*, 1988), to a solution of *5-(4-bromophenyl)-1-(2,4-dichlorophenyl)-4-methyl-1H-pyrazole-3-carboxylic acid* (0.947 g, 2.223 mmol) in CH_2Cl_2 (10 mL) was added *tert*-butyl 2,2,2-trichloroacetimidate (TBTA, 1.215 g, 5.56 mmol) in cyclohexane (10 mL) and BF_3-Et_2O (27.3 μL, 0.223 mmol) for 16 h at RT. The resulting mixture was extracted with Et_2O (3 × 50 mL) and the combined organic extracts were dried over Na_2SO_4 and the solvent removed *in vacuo*. The residue was purified by FC (hexane/EtOAc, 10/90) to give *pyrazole carboxylic acid tert-butyl ester* **5** as a colorless oil. ^1H NMR (400 MHz, $CDCl_3$): δ 1.63 (s, 9H, 3 × CH_3), 2.28 (s, 3H, CH_3), 6.98 (d, J = 8.4 Hz, 2H, ArH), 7.29 (dd, J = 8.5, 2.0 Hz, 1H, ArH), 7.35 (d, J = 8.5 Hz, 1H, ArH), 7.36 (d, J = 2.0 Hz, 1H, ArH), 7.44 (d, J = 8.4 Hz, 2H, ArH); ^{13}C (100.7 MHz, $CDCl_3$): δ 10.0 (q), 20.3 (3q), 81.8 (s), 118.3 (s), 123.1 (s), 127.3 (s), 127.7 (d), 130.1 (d), 130.7 (d), 131.1 (2d), 131.8 (2d), 132.9 (s), 135.8 (s), 136.0 (s), 142.7 (s), 144.5 (s), 162.0 (s); IR ν_{max} (neat): 3010, 2915, 1710, 1490, 1430, 1385, 1360, 1300, 1250, 1190, 1135, 1100, 1055, 1020, 965, 870 cm^{-1}; m/z (CI^+) (rel. intensity) 481 [M(^{79}Br)H^+, 80], 427 (55), 367 (50), 311 (30); HRMS (CI^+) calc'd for $C_{21}H_{20}{}^{79}Br^{35}Cl_2N_2O_2$ $(MH)^+$ 481.0085, found 481.0089 (Δ 0.8 ppm).

5.2.4. HypoGel®-Br 2

To a suspension of HypoGel®-200-OH (n = 5, 110–150 μm diameter beads, 20.0 g, LL = 0.8 mmol g^{-1}, 16.0 mmol) in CH_2Cl_2 at 0 °C was added triphenylphosphine (8.34 g, 32 mmol) and carbon tetrabromide (21.22 g, 64 mmol; Turner *et al.*, 2007). The resulting yellow suspension was warmed to RT and stirred for 24 h. The solvent was then removed by filtration and the resin was washed with H_2O (3 × 100 mL), MeOH (3 × 100 mL), THF (3 × 100 mL), and CH_2Cl_2 (3 × 100 mL). The resulting resin as dried *in vacuo* at 50 °C for 16 h to furnish *bromo resin* **2** as yellow grains (21.0 g, LL = 0.8 mmol g^{-1} was determined by the weight increase of the resin); IR ν_{max} (KBr): 3060, 3020, 2890, 1944, 1875, 1805, 1750, 1600, 1490, 1455, 1355, 1295, 1105, 980, 890 cm^{-1}.

5.2.5. HypoGel®-dimethylgermyl bromide 4

A suspension of HypoGel®-Br **2** (5.0 g, LL = 0.8 mmol g^{-1}, 4 mmol), K_2CO_3 (2.76 g, 20 mmol), TBAI (1.48 g, 4 mmol), and *dimethylgermyl phenol* **3** (3.31 g, 10 mmol), which was prepared as described by Turner *et al.* (2007), in DMF (15 mL) was heated to 80 °C for 24 h. The solvent was removed by filtration and the resin was then washed with H_2O

(3 × 20 mL), MeOH (3 × 20 mL), THF (3 × 20 mL), and CH_2Cl_2 (3 × 20 mL). The resulting *dimethyl-4-anisolyl germyl resin* was swollen in CH_2Cl_2 (20 mL) and conc. HBr (2.0 mL, 48 wt%) added. The reaction mixture was stirred for 16 h at RT followed by filtration to remove the solvent. The resin was successively rinsed with H_2O (3 × 20 mL), MeOH (3 × 20 mL), THF (3 × 20 mL), and CH_2Cl_2 (3 × 20 mL) and then dried *in vacuo* at 50 °C for 16 h to give *dimethylgermyl bromide resin* **4** as light brown granules (5.84 g, LL = 0.75 mmol g^{-1} as determined by the mass of anisole by-product recovered). IR ν_{max} (KBr): 3050, 3010, 2930, 1950, 1840, 1600, 1450, 1110, 960, 850 cm^{-1}.

5.3. Synthesis of surrogate resin 6 and general procedure for the preparation of HypoGel®-aryl pyrazolyl germanes 7

5.3.1. HypoGel®-dimethylgermyl-pyrazole-*tert*-butyl ester 6

To a suspension of HypoGel®-dimethylgermyl bromide **4** (5.0 g, LL = 0.75 mmol g^{-1}, 3.75 mmol) and pyrazole carboxylic acid *tert*-butyl ester **5** (3.62 g, 7.5 mmol) in THF (20 mL) was cooled to −78 °C and a solution of ^{t}BuLi (6.03 mL, 8.44 mmol, 1.4 *M* in pentane) added dropwise (Scheme 4.3). The reaction mixture was stirred at this temperature for 2 h, warmed to RT, and stirred for further 14 h. A solution of 1.0 *M* NH_4Cl was added until no effervescence occurred, and then the solvent was removed by filtration and the resin was then washed with H_2O (3 × 20 mL), MeOH (3 × 20 mL), THF (3 × 20 mL), and CH_2Cl_2 (3 × 20 mL) to afford *dimethylgermyl-pyrazole-tert-butyl ester resin* **6** as brown granules (5.66 g, 55% conversion was determined by the weight increase of the resin and the mass of recovered debrominated pyrazole carboxylic acid *tert*-butyl ester, LL = 0.42 mmol g^{-1}); IR ν_{max} (KBr): 3050, 3010, 2920, 1940, 1830, 1670, 1650, 1600, 1505, 1455, 1242, 1100, 980, 860 cm^{-1}.

5.3.2. HypoGel®-dimethylgermyl-pyrazole carboxylic acid morpholin-4-ylamide 7a

To a suspension of HypoGel®-dimethylgermyl-pyrazole-*tert*-butyl ester **6** (0.5 g, 0.21 mmol, LL = 0.42 mmol g^{-1}) and *N*-aminomorpholine (107.3 mg, 101 μL, 1.05 mmol) in THF (5 mL) was added a solution of LHNDS (1.16 μL, 1.16 mmol, 1.0 *M* in THF) dropwise and the resulting suspension stirred for 16 h at RT. The solvent was removed by filtration and the resin was then washed with H_2O (3 × 20 mL), MeOH (3 × 20 mL), THF (3 × 20 mL), and CH_2Cl_2 (3 × 20 mL) to afford *pyrazole morpholin-4-ylamide resin* **7a** as brown grains [0.5 g, LL = 0.42 mmol g^{-1}, as determined by swelling a sample of the resin (30.0 mg) in CH_2Cl_2 and adding of conc. HBr (1.0 mL, 48 wt%) for 1 h at RT to achieve the protonolysis of the resin to give corresponding protonated *1-(2,4-dichlorophenyl)-4-methyl-5-phenyl-1H-pyrazole-3-carboxylic acid morpholin-4-ylamide* (5.4 mg, 12.52 mmol)]. IR

(KBr): 3450 (brs), 3020, 1945, 1800, 1670, 1655, 1600, 1560, 1450, 1350, 1300, 1245, 1195, 1100, 860, 755 cm^{-1}

5.3.3. HypoGel®-dimethylgermyl-pyrazole carboxylic acid piperidin-1-ylamide 7b

Using the *general procedure*, HypoGel®-dimethylgermyl-pyrazole-*tert*-butyl ester **6** (0.45 g, 0.189 mmol, LL = 0.42 mmol g^{-1}), *N*-aminopiperdine (94.7 mg, 102 μL, 0.945 mmol), and a solution of LHMDS (1.04 mL, 1.04 mmol, 1.0 *M* in hexanes) were employed. The solvent was removed by filtration and the resin was washed successively with H_2O (3 × 20 mL), MeOH (3 × 20 mL), THF (3 × 20 mL), and CH_2Cl_2 (3 × 20 mL) to give *pyrazole piperidin-1-ylamide resin* **7b** as brown grains (0.45 g, LL = 0.42 mmol g^{-1}). IR (KBr): 3420 (brs), 3020, 1940, 1875, 1805, 1690, 1600, 1500, 1490, 1350, 1300, 1245, 1200, 1100, 890, 755 cm^{-1}.

5.3.4. HypoGel®-dimethylgermyl-pyrazole carboxylic acid *N′,N′*-dimethylhydrazide **7c**

Using the *general procedure*, HypoGel®-dimethylgermyl-pyrazole-*tert*-butyl ester **6** (0.5 g, 0.21 mmol, LL = 0.42 mmol g^{-1}), *N′,N′*-dimethylhydrazine (63.1 mg, 80 μL, 1.05 mmol), and a solution of LHMDS (1.16 μL, 1.16 mmol, 1.0 *M* in hexanes) were employed. The solvent was removed by filtration and the resin was washed successively with H_2O (3 × 20 mL), MeOH (3 × 20 mL), THF (3 × 20 mL), and CH_2Cl_2 (3 × 20 mL) to afford *pyrazole N′,N′-dimethylhydrazide resin* **7c** as brown grains (0.5 g, LL = 0.42 mmol g^{-1}). IR (KBr): 3450 (brs), 3020, 2920, 1940, 1870, 1800, 1670, 1600, 1500, 1490, 1455, 1350, 1245, 1110, 900, 755 cm^{-1}.

5.3.5. HypoGel®-dimethylgermyl-pyrazole carboxylic acid cyclohexylamide 7d

Using the *general procedure*, HypoGel®-dimethylgermyl-pyrazole-*tert*-butyl ester **6** (0.5 g, 0.21 mmol, LL = 0.42 mmol g^{-1}), 4-fluoroaniline (104 mg, 120 μL, 1.05 mmol), and a solution of LHMDS (1.16 μL, 1.16 mmol, 1.0 *M* in hexanes) were employed. The solvent was removed by filtration and the resin was washed successively by H_2O (3 × 20 mL), MeOH (3 × 20 mL), THF (3 × 20 mL), and CH_2Cl_2 (3 × 20 mL) to give *pyrazole cyclohexylamide resin* **7d** as brown grains (0.45 g, LL = 0.38 mmol g^{-1}). IR (KBr): 3420 (brs), 3050, 3010, 2920, 1945, 1875, 1800, 1650, 1600, 1500, 1455, 1350, 1245, 1195, 1100, 765 cm^{-1}.

5.3.6. HypoGel®-dimethylgermyl-pyrazole carboxylic acid 4-fluorophenylamide 7e

Using the *general procedure*, HypoGel®-dimethylgermyl-pyrazole-*tert*-butyl ester **6** (0.5 g, 0.21 mmol, LL = 0.42 mmol g^{-1}), cyclohexylamine (116.7 mg, 100 μL, 1.05 mmol), and a solution of LHMDS (1.16 μL,

1.16 mmol, 1.0 *M* in hexanes) were employed. The solvent was removed by filtration and the resin was washed successively with H_2O (3 × 20 mL), MeOH (3 × 20 mL), THF (3 × 20 mL), and CH_2Cl_2 (3 × 20 mL) to afford *pyrazole 4-fluorophenylamide resin* **7e** as dark red grains (0.43 g, LL = 0.35 mmol g^{-1}). IR (KBr): 3380 (brs), 3045, 3015, 2920, 1945, 1870, 1805, 1645, 1600, 1500, 1455, 1350, 1300, 1245, 1100, 785 cm^{-1}.

5.4. Typical procedure for *ipso*-iododegermylation

5.4.1. 1-(2,4-Dichlorophenyl)-5-(4-iodophenyl)-4-methyl-1*H*-pyrazole-3-carboxylic acid morpholin-4-ylamide AM281 (1a)

Entry 1

Pyrazole morpholin-4-ylamide resin **7a** (52 mg, 0.022 μmol, LL = 0.42 mmol g^{-1}) and NaI (16.4 mg, 0.11 mmol) were suspended in CH_2Cl_2 (0.5 mL) in a cartridge and a solution of NCS (11.7 mg, 0.087 mmol) in TFA-HOAc (5 mL, 1:3, v/v) was added (Table 27.1). The reaction mixture was shaken at RT for the indicated time. The solvent was removed by filtration and the resin washed with H_2O (3 × 10 mL) and CH_2Cl_2 (3 × 10 mL). Solid Na_2SO_5 was then added to the collected organic filtrate until a colorless solution formed. The pH of this solution was adjusted to pH 8.0 with K_2CO_3. The mixture was extracted with CH_2Cl_2 (3 × 10 mL) and combined organic extracts were dried over Na_2SO_4 and the solvent removed *in vacuo*. The residue was purified by FC (hexane/EtOAc, 50/50) to afford *iodinated pyrazole-3-carboxylic acid morpholin-4-ylamide AM281* (**1a**) as a white solid (7.7 mg, 63%).

5.4.2. 1-(2,4-Dichloro-phenyl)-5-(4-iodophenyl)-4-methyl-1*H*-pyrazole-3-carboxylic acid piperidin-1-ylamide AM251 (1b)

Entry 2

Using the *typical procedure* described for entry 1, pyrazole piperidin-1-ylamide resin **7b** (55.2 mg, 0.023 mmol, LL = 0.42 mmol g^{-1}), NaI (17.2 mg, 0.115 mmol), and NCS (12.3 mg, 0.092 mmol) were employed (Lan *et al.*, 1996). The collected organic filtrate was analyzed by HPLC prior to purification by FC (hexane/EtOAc, 50/50) to give *iodinated pyrazole-3-carboxylic acid piperidin-1-ylamide AM251* (**1b**) (7.7 mg, 60%) as a pale yellow solid. ^1H NMR (400 MHz, $CDCl_3$): δ 1.34–1.47 (m, 2H, CH_2), 1.67–1.77 (m, 4H, 2×CH_2), 2.35 (s, 3H, CH_3), 2.76–2.92 [m, 4H, $N(CH_2)_2$], 6.84 (d, *J* = 8.4 Hz, 2H, ArH), 7.27–7.30 (m, 2H, ArH), 7.42 (d, *J* = 1.8 Hz, 2H, ArH), 7.64 (d,

$J = 8.4$ Hz, 3H, NH + ArH); ^{13}C (100.7 MHz, $CDCl_3$): δ 9.3 (q), 23.3 (t), 25.4 (2t), 57.1 (2t), 95.0 (s), 118.2 (s), 127.9 (d), 128.2 (s), 130.3 (d), 130.5 (d), 131.1 (2d), 132.9 (s), 135.8 (s), 135.9 (s), 137.7 (2d), 143.0 (s), 144.4 (s), 160.0 (s); HRMS (ESI) calc'd for $C_{22}H_{22}Cl_2IN_4O$ $(MH)^+$ 555.0215, found 555.0227 (Δ 2.2 ppm).

5.4.3. 1-(2,4-Dichlorophenyl)-5-(4-iodophenyl)-4-methyl-1*H*-pyrazole-3-carboxylic acid *N*′,*N*′-dimethylhydrazide **1c**

Entry 3

Using the *typical procedure* described for entry 1, *N*′,*N*′-dimethylhydrazide resin **7c** (43 mg, 0.018 mmol, LL = 0.42 mmol g^{-1}), NaI (13.5 mg, 0.09 mmol), and NCS (9.6 mg, 0.072 mmol) were employed. Purification by FC (EtOAc/Hexane, 50/50) gave *iodinated pyrazole-3-carboxylic acid N*′,*N*′*-dimethylhydrazide* **1c** as a white solid (5.5 mg, 60%). ^{1}H NMR (400 MHz, $CDCl_3$): δ 2.33 (s, 3H, CH_3), 2.67 [s, 6H, $N(CH_3)_2$], 6.82 (d, $J = 8.3$ Hz, 2H, ArH), 7.25–7.28 (m, 2H, ArH), 7.39 (d, $J = 1.4$ Hz, 1H, ArH), 7.62 (d, $J = 8.3$ Hz, 3H, NH + ArH); ^{13}C (100.7 MHz, $CDCl_3$): δ 9.3 (q), 47.5 (2q), 95.0 (s), 118.0 (s), 127.8 (d), 128.0 (s), 130.2 (d), 130.4 (d), 131.0 (2d), 132.8 (s), 135.7 (s), 135.9 (s), 137.6 (2d), 142.9 (s), 144.1 (s), 160.2 (s); HRMS (ESI) calc'd for $C_{19}H_{18}Cl_2IN_4O$ $(MH)^+$ 514.9902, found 514.9905 (Δ 0.6 ppm).

5.4.4. 1-(2,4-Dichlorophenyl)-5-(4-iodophenyl)-4-methyl-1*H*-pyrazole-3-carboxylic acid cyclohexylamide (**1d**)

Entry 4

Using the *typical procedure* described in entry 1, pyrazole cyclohexylamide resin **7d** (65 mg, 0.0247 mmol, LL = 0.35 mmol g^{-1}), NaI (18.5 mg, 0.124 mmol), and NCS (13.2 mg, 0.099 mmol) were employed. Purification by FC (hexane/EtOAc, 50/50) gave *iodinated pyrazole-3-carboxylic acid cyclohexylamide* **1d** as a white solid (7.5 mg, 55%). ^{1}H NMR (400 MHz, $CDCl_3$): δ 1.19–1.30 (m, 3H, 3×CH), 1.35–1.45 (m, 2H, CH_2), 1.60–1.66 (m, 1H, CH), 1.72–1.77 (m, 2H, CH_2), 1.98–2.01 (m, 2H, CH_2), 2.36 (s, 3H, CH_3), 3.89–3.99 (m, 1H, CH), 6.85 (d, $J = 8.4$ Hz, 3H, NH + ArH), 7.29 (s, 2H, ArH), 7.42 (s, 1H, ArH), 7.64 (d, $J = 8.4$ Hz, 2H, ArH); ^{13}C (100.7 MHz, $CDCl_3$): δ 9.4 (q), 25.0 (2t), 25.6 (t), 33.2 (2t), 47.9 (d), 94.9 (s), 117.7 (s), 127.9 (d), 128.3 (s), 130.3 (d), 130.6 (d), 131.1 (2d + s), 132.9 (s), 135.9 (s), 137.7 (2d), 143.1 (s), 145.2 (s), 161.7 (s); HRMS (ESI) calc'd for $C_{23}H_{23}Cl_2IN_3O$ $(MH)^+$ 554.0263, found 554.0269 (Δ 1.1 ppm).

5.4.5. 1-(2,4-Dichlorophenyl)-5-(4-iodophenyl)-4-methyl-1*H*-pyrazole-3-carboxylic acid 4-fluorophenylamide (**1e**)

Entry 5

Using the *typical procedure* described for entry 1, pyrazole (4-fluorophenyl) amide resin **7e** (102 mg, 0.0357 mmol, LL = 0.35 mmol g^{-1}), NaI (26.8 mg, 0.179 mmol), and NCS (20.0 mg, 0.15 mmol) were employed. Purification by FC (hexane/EtOAc, 50/50) gave *iodinated pyrazole-3-carboxylic acid (4-fluoro-phenyl)amide* **1e** as a white solid (8.5 mg, 42%). ^{1}H NMR (400 MHz, $CDCl_3$): δ 2.42 (s, 3H, CH_3), 6.89 (d, J = 8.3 Hz, 2H, ArH), 7.00–7.07 (m, 2H, ArH), 7.29–7.34 (m, 2H, ArH), 7.45 (d, J = 1.8 Hz, 1H, ArH), 7.60–7.65 (m, 2H, ArH), 7.67 (d, J = 8.3 Hz, 2H, ArH), 8.76 (s, 1H, NH); ^{13}C (100.7 MHz, $CDCl_3$): δ 9.5 (q), 95.2 (s), 114.9 (s), 115.6 (2d, $^2J_{C-F}$ = 22.5 Hz), 118.2 (s), 121.4 (2d, $^3J_{C-F}$ = 8.3 Hz), 124.9 (s), 127.9 (d), 130.4 (d), 130.5 (d), 131.1 (2d), 133.0 (s), 133.8 (s), 135.7 (s), 136.2 (s), 137.8 (2d), 144.2 (d, $^1J_{C-F}$ = 109.2 Hz), 160.4 (s); ^{19}F NMR (367 MHz, $CDCl_3$): δ −118.1 (s, 1F); HRMS (ESI) calc'd for $C_{23}H_{16}Cl_2IN_3O$ $(MH)^+$ 565.9699, found 565.9693 (Δ −1.1 ppm).

5.4.6. 1-(2,4-Dichlorophenyl)-5-(4-iodophenyl)-4-methyl-1*H*-pyrazole-3-carboxylic acid piperidin-1-ylamide AM251 (**1b**)

Entry 6

Using the *typical procedure* described for entry 1, pyrazole piperidin-1-ylamide resin **7b** (102.1 mg, 0.0429 mmol, LL = 0.42 mmol g^{-1}), NaI (32.1 mg, 0.214 mmol), and NCS (22.9 mg, 0.172 mmol) were employed. The reaction mixture was shaken for 5 min at RT followed by filtration to remove the resin. Purification by PLC (hexane/EtOAc, 50/50) gave *iodinated pyrazole-3-carboxylic acid piperidin-1-ylamide AM251* (**1b**) (1.9 mg, 8%) as a pale yellow solid. Analytical data as described previously.

Entry 7

The recovered pyrazole piperidin-1-ylamide resin **7b** from *entry 6* was resubjected to the identical conditions at RT for 5 min. The filtration liquid was collected and purified by PLC (EtOAc/hexane, 50/50) to give *iodinated pyrazole-3-carboxylic acid piperidin-1-ylamide AM251* (**1b**) (1.1 mg, 5%). Analytical data as described previously.

5.5. Sequential *ipso*-iododegermylation/succinimide scavenging protocol

5.5.1. *Si*-propyl hydrazine resin 8

To a suspension of *Si*-propyl bromide solid-support (0.50 g, 40–63 μm diameter beads, LL = 1.43 mmol g^{-1}, 0.715 mmol) in DMF (10 mL) was added K_2CO_3 (0.98 g, 7.1 mmol), NaI (1.07 g, 7.1 mmol), and hydrazine hydrate (230 mg, 7.1 mmol, 223 μL). The mixture was warmed to 80 °C for 24 h (Scheme 4.4). The solvent was then removed by filtration and the resin was washed with H_2O (3 × 100 mL), EtOH (3 × 100 mL), THF (3 × 100 mL), and CH_2Cl_2 (3 × 100 mL). The resulting resin was dried *in vacuo* at RT for 16 h to furnish *Si-hydrazine resin* **8** as white particles (0.466 g, LL = 1.43 mmol g^{-1}, as determined by the weight loss of the resin); IR (KBr): 3450 (brs), 3300 (brs), 2950, 1630, 1405, 1110 cm^{-1}.

Using the *typical iodination procedure* described above, pyrazole piperidin-1-ylamide resin **7b** (25.1 mg, 0.0105 mmol, LL = 0.42 mmol g^{-1}), NaI (7.9 mg, 0.0527 mmol), and NCS (5.6 mg, 0.042 mmol) were employed. The collected organic filtrate was dried *in vacuo* and resuspended in EtOH (5 mL) in the presence the *Si*-propyl hydrazine resin **8** (150 mg, 0.215 mmol). The amount of succinimide scavenging resin **8** was employed based on the theoretical amount of NCS expected to remain. The mixture was then refluxed at 80 C for 1 h. The resin was then filtered off and the filtrate analyzed by HPLC. The chromatogram revealed a significant improvement in purity (96%) of the desired product **1b**, relative to that obtained without the use of the succinimide scavenger resin. HPLC parameters: Column: XTerra RP-8, 4.6 mm × 150 mm, 5 μm; mobile phase: 0.1% TFA/MeCN: H_2O = 70:30; flow rate: 1 mL/min; detection: UV diode array at 250 nm; injection volume: 5 μL.

REFERENCES

Armstrong, A., Brackenrldge, I., Jackson, R. P. W., & Rirk, J. M. (1988). A New Method for the Preparation of Tertiary Butyl Ethers and Esters. *Tetrahedron Lett.*, 2483–2486.

Asai, K., and Miracle, C. (1980). Organic Germanium. Japan Publications Inc., Tokyo.

Berdinga, G., Schneiderb, U., Gielowa, P., Buchertc, R., Donnerstagd, F., Brandaue, W., Knappa, W. H., Emrichb, H. M., and Müller-Vahlb, K. (2006). Feasibility of central cannabinoid CB1 receptor imaging with [^{124}I] AM281 PET demonstrated in a schizophrenic patient. *Psychiatry Res. Neuroim.* **147,** 249–256.

Clayden, J., Moran, W. J., Edwards, P. J., and LaPlante, S. R. (2009). The challenge of atropisomerism in drug discovery. *Angew. Chem. Int. Ed.* **48,** 6398–6401.

Di Marzo, V., Goparaju, S. K., Wang, L., Liu, J., Batkai, S., Jarai, Z., Fezza, F., Miura, G. I., Palmiter, R. D., Sugiura, T., and Kunos, G. (2001). Leptin-Regulated endocannabinoids are involved in maintaining food intake. *Nature* **410,** 822–825.

Donohue, S. R., Halldinb, C., and Pike, V. W. (2006). Synthesis and structure–activity relationships (SARs) of 1, 5-diarylpyrazole cannabinoid type-1 (CB1) receptor ligands for potential use in molecular imaging. *Bioorg. Med. Chem.* **14,** 3712–3720.

Gerard, C. M., Mollreau, C., Vassart, G., and Parmentier, M. (1991). Molecular cloning of human cannabinoid receptor. *Biochem. J.* **279,** 129–134.

Goodman, S. (1988). Therapeutic effects of organic germanium. *Med. Hypotheses* **26,** 207–215.

Hockley, B. G., Scott, P. J. H., and Kilbourn, M. R. (2009). Solid-phase radiochemistry. *In* "Linker Strategies in Solid-Phase Organic Synthesis," (P. J. H. Scott, ed.).Wiley, Chichester.

Howlett, A. C. (2004). Efficiency in CB1 receptors-mediated signal transduction. *Brit. J. Pharmacol.* **142,** 1209–1218.

Howlett, A. C., Barth, F., Bonner, T. I., Cabral, G., Casellas, P., Devane, W. A., Felder, C. C., Herkenham, M., Mackie, K., Martin, B. R., Mechoulam, R., and Pertwee, R. G. (2002). Classification of cannabinoid receptors. *Pharmacol. Rev.* **54,** 161–202.

Hurst, D. P., Lynch, D. L., Barnett-Norris, J., Hyatt, S., Seltzman, H., Zhong, M., Song, Z.-H., Nie, J.-J., Lewis, D., and Reggio, P. H. (2002). Rimonabant interaction with Lys 3.28 (192) is crucial for its inverse agonism at CB1. *Mol. Pharmacol.* **62,** 1274–1287.

Jagerovic, N., Fernandez-Fernandez, C., and Goya, P. (2008). CB_1 cannabinoid antagonists: Structure–activity relationships and potential therapeutic applications. *Curr. Top. Med. Chem.* **8,** 205–230.

Jones, D. (2008). End of the line for CB1 as an anti-obseity agent? *Nat. Rev. Drug Discov.* **7,** 961–962.

Lan, R., Gatleyz, S. J., and Makriyannisl, A. (1996). Preparation of Iodine-123 labeled AM251: A potential SPECT radioligand for the brain cannabinoid CB1 receptor. *J. Labelled Comp. Radiopharm.* **38,** 875–881.

Lan, R., Fan, P., Gatley, S. J., Volkow, N. D., Fernado, S. R., Pertwee, R., and Makriyannis, A. (1999a). [123I]-AM281. *AAPS PharmSci.* **1**Article 4.

Lan, R., Liu, Q., Fan, P., Lin, S., Fernando, S. R., McCallion, D., Pertwee, R., and Makriyannis, A. (1999b). Structure–activity relationships of pyrazole derivatives as cannabionod receptors antagonists. *J. Med. Chem.* **42,** 769–776.

Lange, J. H. M., and Kruse, C. G. (2005). Medical chem strategies to CB1 receptor antagonists. *Drug Discov. Today* **10,** 693–702.

Lia, Z., Gifforda, A., Liub, Q., Thotapallyb, R., Dinga, Y.-S., Makriyannisb, A., and Gatley, S. J. (2005). Candidate PET radioligands for cannabinoid CB1 receptors: [^{18}F] AM5144 and related pyrazole compounds. *Nucl. Med. Biol.* **32,** 361–366.

Mathews, W. B., Scheffel, U., Finley, P., Ravert, H. T., Frank, R. A., Rinaldi-Carmona, M., Barth, F., and Dannals, R. F. (2000). Biodistribution of [^{18}F] SR144385 and [^{18}F] SR147963: Selective radioligands for positron emission tomographic studies of brain cannabinoid receptors. *Nucl. Med. Biol.* **27,** 757–762.

Mathews, W. B., Scheffel, U., Rauseo, P. A., Ravert, H. T., Frank, R. A., Ellames, G. J., Herbert, J. M., Barth, F., Rinaldi-Carmona, M., and Dannals, R. F. (2002). Carbon-11 labelled rimonabant. *Nucl. Med. Biol.* **29,** 671–677.

Matsuda, L. A., Lolait, S. J., Brownstein, M. J., Young, A. C., and Bonner, T. I. (1990). Structure of cannabinoid receptors. *Nature* **346,** 561–564.

Murineddu, G., Ruiu, S., Loriga, G., Manca, I., Lazzari, P., Reali, R., Pani, L., Toma, L., and Pinna, G. A. (2005). Tricyclic pyrazoles. 3. Synthesis, biological evaluation, and molecular modeling of analogues of the cannabinoid antagonist 8-chloro-1-(2′4′-dichlorophenyl)-N-piperidin-1-yl-1, 4, 5, 6- tetrahydrobenzo[6, 7]cyclohepta[1, 2-c] pyrazole-3-carboxamide. *J. Med. Chem.* **48,** 7351–7362.

Pagotto, U., Marsicano, G., Cota, D., Lutz, B., and Pasquali, R. (2006). The emerging role of the endocannabinoid system in endocrine regulation and energy balance. *Endocr. Rev.* **27,** 73–100.

Patai, S. (1995). The Chemistry of Organic Germanium. Tin and Lead Compounds. John Wiley & Sons Ltd., Chichester, England.

Pertwee, R. G. (2005). Inverse agonism and neutral antagonism at cannabionid CB1 receptors. *Life Sci.* **76,** 1307–1324.

Rinaldi-Carmona, M., Barth, F., Hkaulme, M., Shireb, D., Calandrab, B., Congy, C., Martinez, S., Maruani, J., Neliat, G., Caputb, D., Ferrarab, P., Soubrie, P., *et al.* (1994). SR1417 16A, a potent and selective antagonist of the brain cannabinoid receptor. *FEBS Lett.* **350,** 240–244.

Shire, D., Calandra, B., Delpech, M., Dumont, X., Kaghad, M., Fur, L. G., Caput, D., and Ferrara, P. (1996). Structural features of CB1 receptor. *J. Biol. Chem.* **271,** 6941–6946.

Spivey, A. C., Martin, L. J., Noban, C., Jones, T. C., Ellames, G. J., and Kohler, A. D. (2007a). Opportunities for isotopic labelling via phase-tagged synthesis with organogermanium linkers. *J. Label. Comp. Radiopharm.* **50,** 281.

Spivey, A. C., Tseng, C.-C., Hannah, J. P., Gripton, C. J. G., de Fraine, P., Parr, N. J., and Scicinski, J. J. (2007b). Light-fluorous safety-catch arylgermanes—Exceptionally robust, photochemically activated precursors for biaryl synthesis by Pd(0) catalysed cross-coupling. *Chem. Commun.* 2926–2928.

Spivey, A. C., Tseng, C.-C., Jones, T. C., Kohler, A. D., and Ellames, G. J. (2009). A method for parallel solid-phase synthesis of iodinated analogues of the CB1 receptor inverse agonist rimonabant. *Org. Lett.* **11,** 4760–4763.

Still, W. C., Kahn, M., and Mitra, A. (1978). Rapid chromatographic technique for preparative seperations with moderate resolution. *J. Org. Chem.* **43,** 2923–2924.

Stoit, A. R., Lange, J. H. M., den Hartog, A. P., Ronken, E., Tipker, K., van Stuivenberg, H. H., Dijksman, J. A. R., Wals, H. C., and Kruse, C. G. (2002). Design, synthesis and biological activity of rigid CB1 ligands. *Chem. Pharm. Bull.* **50,** 1109–1113.

Thomas, B. F., Francisco, M. E. Y., Seltzman, H. H., Thomas, J. B., Fix, S. E., Schulz, A.-K., Gilliam, A. F., Pertwee, R. G., and Stevenson, L. A. (2005). Synthesis of long-chain amide analogs of the cannabinoid CB1 receptor antagonist N-(piperidinyl)-5-(4-chlorophenyl)-1-(2, 4-dichlorophenyl)-4-methyl-1H-pyrazole-3-carboxamide (SR141716) with unique binding selectivities and pharmacological activities. *Bioorg. Med. Chem.* **13,** 5463–5474.

Thomas, B., Zhang, Y., Brackeen, M., Page, K., Mascarella, S., and Seltzman, H. (2006). Conformational characteristics of the interaction of SR141716A with the CB1 cannabinoid receptor as determined through the use of conformationally constrained analogs. *AAPS J.* **8,** E665–E671.

Tobishi, S., Sasada, T., Nojiri, Y., Yamamoto, F., Mukai, T., Ishwata, K., and Maeda, M. (2007). Methoxy- and fluorine-substituted analogs of O-1302: Synthesis and in vitro binding affinity for the CB1 cannabinoid receptor. *Chem. Pharm. Bull.* **55,** 1213–1217.

Turner, D. J., Anemian, R., Mackie, P. R., Cupertino, D. C., Yeates, S. G., Turner, M. L., and Spivey, A. C. (2007). Towards a general solid phase approach for the iterative synthesis of conjugated oligomers using a germanium based linker—First solid phase synthesis of an oligo-(triarylamine). *Org. Biomol. Chem.* **5,** 1752–1763.

Wotjak, C. T. (2005). Role of endogenous cannabinoids in cognition and emotionality. *Mini Rev. Med. Chem.* **5,** 659–670.

CHAPTER TWENTY-EIGHT

Coexpression Systems as Models for the Analysis of Constitutive GPCR Activity

Erich H. Schneider* *and* Roland Seifert†

Contents

Abstract

The investigation of constitutive activity of GPCRs in transfected mammalian cells is often hampered by the presence of other constitutively active receptors that generate a high background signal. This impairs the measurement of constitutive activity and of inverse agonistic effects, both of which often occur in a relatively small signal range. Moreover, constitutive activity of a GPCR depends on the interacting G-protein. Since the commonly used mammalian cells contain a set of several different G-protein types, it is very difficult to investigate the influence of specific Gα and Gβγ subunits on constitutive activity in more detail in these expression systems. Here, we show that the Sf9 cell/baculovirus expression system provides excellent conditions for the characterization of constitutively active GPCRs. Sf9 cells express a restricted set of G-protein subtypes that show only a limited capability of interacting with mammalian GPCRs. Moreover, the Sf9

* Laboratory of Molecular Immunology, NIAID/NIH, Bethesda, Maryland, USA
† Institute of Pharmacology, Medical School of Hannover, Hannover, Germany

Methods in Enzymology, Volume 485
ISSN 0076-6879, DOI: 10.1016/S0076-6879(10)85028-1

cell/baculovirus expression system allows the combined expression of up to four different proteins encoded by the respective genetically modified baculoviruses. Using the highly constitutively active human histamine H_4R (hH_4R) as a paradigm, we demonstrate how the coexpression of hH_4R with different signaling proteins (Gα, Gβγ, and RGS-proteins) in combination with sensitive functional assays (high-affinity agonist binding and steady-state GTPase- and GTPγS-binding assays) allows in-depth studies of constitutive activity. The preparation of Sf9 cell membranes, coexpressing hH_4R and various additional proteins, is described in detail as well as the procedures of the different functional assays. Moreover, we show that coexpression of GPCRs with signal transduction components in Sf9 cells can also be applied to the characterization of other constitutively active receptors, for example, the formyl peptide receptor and β_2-adrenoceptor.

1. Introduction

According to the two-state model of receptor activation (Fig. 28.1), a GPCR can occur in an active state R* (stabilized by agonists and G-proteins) and an inactive state R (stabilized by inverse agonists and sodium cations). Both states are in equilibrium, and the balance between R and R* depends on the receptor type as well as on its environment (Seifert and Wenzel-Seifert, 2002). It is now

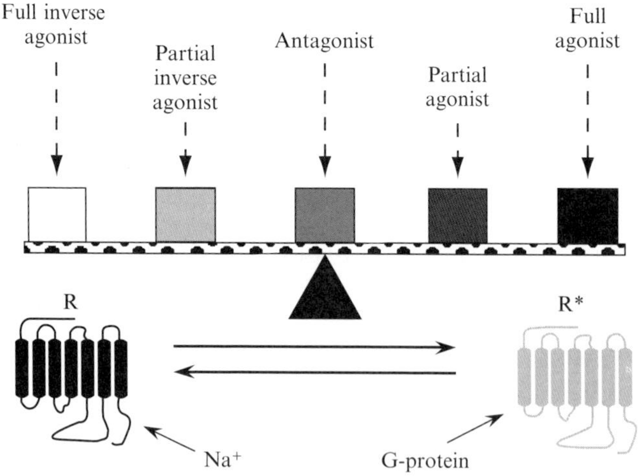

Figure 28.1 Two-state model of receptor activation. As shown by the scales, G-protein-coupled receptors exist in equilibrium between an inactive state R (left side) and an active state R* (right side). Agonists and G-proteins stabilize the R* state. Constitutive activity is defined as spontaneous adoption of the R* state in the absence of an agonist. Full inverse agonists and Na^+ stabilize the R state and therefore reduce constitutive activity. Partial inverse agonists and partial agonists shift the equilibrium only partially toward R- or R*-state, respectively. Neutral antagonists bind to the receptor without altering the R/R* balance.

commonly accepted that GPCRs can be constitutively active, which means that they spontaneously adopt the active state, even in the absence of an agonist.

The investigation of constitutive activity of a specific GPCR in mammalian cells may be hampered by other endogenously expressed constitutively active receptors that produce a high background signal in functional assays. This was, for example, observed for HL-60 promyelocytic cells, where cyclosporin H did not reduce basal Gα_i activity, although it is an inverse agonist at the highly constitutively active formyl peptide receptor (FPR), which is expressed by this cell line (Wenzel-Seifert and Seifert, 1993; Wenzel-Seifert *et al.*, 1998). Moreover, it is difficult to determine the influence of specific G-proteins on constitutive activity (Seifert and Wenzel-Seifert, 2002), since mammalian cells express many G-protein subtypes at the same time.

Sf9 insect cells do not express constitutively active receptors and possess only a limited set of Gα-subunits that in many cases do not interact with mammalian proteins (Schneider and Seifert, 2010). By coinfection with different baculoviruses that encode the proteins of interest, GPCRs can be coexpressed with up to three other proteins (e.g., components of the signal transduction pathway) in any possible combination. Thus, Sf9 cells provide a unique system for the investigation of constitutive activity in an environment with low background and defined G-protein composition. In this chapter, we use the human histamine H_4 receptor (hH_4R) as a paradigm to demonstrate how a highly constitutively active GPCR can be coexpressed with signal transduction proteins and characterized in high-affinity agonist binding, [^{35}S]GTPγS binding, and steady-state GTPase assays. For all assays, we provide detailed protocols that are universally applicable for the characterization of constitutive activity of any GPCR expressed in Sf9 cells.

2. Coexpression of GPCRs and Signaling Proteins: hH_4R as Paradigm

2.1. Sf9 cell culture and baculovirus generation

2.1.1. Devices and disposables

- Incubator with integrated shaker (e.g., Innova 40R from New Brunswick Scientific, Edison, NJ), capable of maintaining a temperature of 28 °C (no CO_2 supplementation) and continuous shaking of 150 rpm.
- 25 cm^2 cell culture flasks for preparing the P1 virus generation.
- Erlenmeyer flasks (100–1000 mL), glass (reusable), or plastic (disposable) with vented caps (e.g., polycarbonate Erlenmeyer flasks with flat cap from Corning Lifesciences, Lowell, MA).
- Aluminum-foil wrapped glass bottles for baculovirus storage.
- Laboratory centrifuge, capable of centrifuging 50 mL tubes at a speed of up to 3000 rpm

2.1.2. Buffers and reagents

- Cell culture medium, for example, Sf900 II serum-free medium (Invitrogen, Carlsbad, CA, USA); alternatively, also the Insect XPRESS medium (Lonza, Walkersville, MD) can be used. We have not noticed performance differences between both media.
- The medium should be supplemented with 5% fetal bovine serum (FBS; PAA laboratories, Pasching, Austria) and 100 μg/mL gentamicin (Lonza, Walkersville, MD). We have not noticed performance differences between various types of FBS. Although Sf9 cells do grow in the absence of serum, the addition of serum enhances growth and protein expression.
- BaculoGold™ transfection kit from Invitrogen or comparable products.
- pVL-1392 plasmid, containing the gene of interest.

2.1.3. Instructions for Sf9 cell culture and baculovirus generation

In contrast to mammalian cell lines, Sf9 cell cultures do not require CO_2 supplementation and are maintained at 28 °C. Many media available on the market are serum-free formulations, for example, the Sf900 II medium (Invitrogen, Carlsbad, CA). However, supplementation of the media with 5% (v/v) of FBS is beneficial for cell growth and protein production. To prevent bacterial contamination, 0.1 mg/mL of gentamicin may be added. Sf9 insect cells can be cultured as a monolayer or in suspension. Normally, monolayer cultures are kept as a "backup culture" that can be used to restore suspension cultures, when a contamination with bacteria, fungus, or baculovirus occurs. Suspension cultures are maintained in disposable Erlenmeyer flasks at a density of $1.0–6.0 \times 10^6$ cells/mL. We have made the experience that maintaining Sf9 cells in reusable glass flasks, although less expensive, results in rather frequent bacterial and fungal contamination and, therefore, cannot be recommended. However, for terminal culture of Sf9 cells after baculovirus infection (routinely 48 h), glass Erlenmeyer flasks can be used without particular risk of bacterial or fungal contamination. Continuous shaking at ~150 rpm ensures optimal oxygenation of the cells. Normally, the volume of a suspension culture in shaker flasks should not exceed 40% of the flask volume. This ensures optimal mixing of the medium and oxygenation. Sf9 cell suspension cultures should be passaged three times weekly and seeded at a density of $\sim 1 \times 10^6$ cells/mL. It is very important to centrifuge the cells and resuspend them in fresh medium at each passage. If the cells are diluted only with new medium, cell viability and protein expression may be impaired. Another critical point is contamination of noninfected Sf9 cell cultures with baculoviruses, for example, by medium spilled from an infected culture. Baculoviruses can even cross the sterile filter of cell culture flasks. It is recommended that noninfected and infected cultures be kept in different incubators. Accidental baculoviral contamination of an Sf9 cell culture can be easily recognized by typical signs of infection, for example, deformation of cell shape, reduced growth, and finally the death of the whole cell population.

For baculovirus generation, in our hands the BaculoGold^{TM} transfection kit from Invitrogen yielded the best results. The gene encoding the protein of interest has to be cloned into the multiple cloning site of a pVL-1392 expression plasmid. Sf9 cells are cotransfected with the pVL-1392 plasmid and linearized baculovirus DNA to achieve homologous recombination of both components and production of functional baculoviruses encoding the protein of interest. Here, we provide a simple protocol for baculovirus generation using the BaculoGold kit, which works reliably in most cases.

Adherent Sf9 cells (4×10^6 cells in a 25-cm^2 flask) are cotransfected with the pVL-1392 plasmid encoding the gene of interest and the linearized baculovirus DNA according to the manufacturer's instructions. After 7 days in culture, a 1:100 dilution of the supernatant fluid (P1) is used to infect another cell culture (P2, density 2×10^6 cells/mL), which is again maintained for 7 days. Within this time the death of virtually the entire cell population should occur, which is a sign of efficient virus propagation (Seifert and Wieland, 2005). This P2 culture is centrifuged for 10 min at 3000 rpm. The supernatant fluid should be stored in the dark and is used in a 1:20 dilution to infect another Sf9 cell culture (P3) with 3.0×10^6 cells/mL The P3 supernatant is isolated after 48 h, when the cells show signs of infections but are still intact. The P3 supernatant fluid is normally used to infect Sf9 cells for membrane preparations. The P3 virus suspension can be stored at 4 °C for up to 5 years without significant loss of potency to induce protein expression (Seifert and Wieland, 2005). Only very rarely we have observed contamination of virus stocks by fungus. However, after prolonged storage, virus stocks should be carefully shaken since virus particles sediment to the bottom. Baculoviruses are light-sensitive. Thus, storage vessels should be wrapped with aluminum foil. We recommend that backups of the P3 viruses be kept at −80 °C. Routinely, we use 1:100 dilutions of P3 virus stocks for protein expression. In case of decreased protein expression, we generate a P4 virus (7 day culture of a culture seeded at 3.0×10^6 cells/mL with a 1:20 dilution of the P3 virus).

2.2. Preparation of membranes expressing the protein of interest

2.2.1. Devices and disposables

- Dounce homogenizers for preparation of the Sf9 cell homogenate
- Plastic transfer pipettes
- Ultracentrifuge
- UV/VIS photometer for determination of protein concentration
- Standard cell culture laboratory centrifuge

2.2.2. Buffers and reagents

- Cell culture medium (cf. Section 2.1)
- PBS buffer for washing the infected Sf9 cells: NaCl (137 m*M*), KCl (2.6 m*M*), $MgCl_2$ (0.5 m*M*), $CaCl_2$ (0.9 m*M*), KH_2PO_4 (1.5 m*M*), and Na_2HPO_4 (0.8 m*M*), pH 7.4
- Lysis buffer with protease inhibitors: 10 m*M* Tris–HCl, pH 7.4, with EDTA (1 m*M*), PMSF (0.2 m*M*), benzamidine (10 μg/mL), and leupeptin (10 μg/mL)
- Binding buffer: 12.5 m*M* $MgCl_2$, 1 m*M* EDTA in 75 m*M* Tris–HCl, pH 7.4
- Bio-Rad DC protein assay kit (Bio-Rad, Hercules, CA, USA) or an equivalent kit for determination of protein concentrations

2.2.3. Membrane preparation protocol

To express the proteins of interest, an Sf9 cell culture (3×10^6 cells/mL, 50–125 mL) is infected with P3 virus suspension, yielding a final virus dilution of 1:100 (1 mL of supernatant to 100 mL of cell suspension). Normally, virus titer determination is not required. The level of protein expression in Sf9 cells is more dependent on the quality of the Sf9 cells (well-fed cells provided with fresh medium three times per week) than on the virus titer. In fact, attempts to regulate protein expression by virus titer are poorly predictable. To ensure optimal protein yields, the cells should be suspended in fresh medium prior to an infection.

It is possible to combine up to four different baculoviruses, encoding receptors and components of the intracellular signal transduction cascade. For the experiments described in this chapter, hH_4R- and $G\beta_1\gamma_2$-encoding baculoviruses are combined with baculoviruses containing the gene for $G\alpha_{i1-3}$ or $G\alpha_o$. In quadruple infections also baculoviruses encoding the regulators of G-protein signaling RGS4 or GAIP were added. RGS proteins accelerate GTP hydrolysis which can be the rate-limiting step in some systems (Kleemann *et al.*, 2008; Schneider and Seifert, 2009; Fig. 28.2). To avoid cross-contamination, ideally only one baculovirus stock should be processed under the sterile hood at a specific time. Moreover, when switching from one baculovirus stock to another, the sterile bench as well as the pipettes has to be carefully decontaminated.

After infection, the cells are cultured under standard conditions (28 °C, no CO_2 supplementation, 150 rpm) and normally harvested for membrane preparation after ~48 h. We have made the experience that there is little difference in protein expression in a 42–50-h window. A typical 100 mL culture of baculovirus-infected Sf9 cells yields approximately 35–55 mg of membrane protein. If incubation is performed for too long a time, cell lysis may occur, leading to proteolysis of proteins, including the GPCR to be investigated.

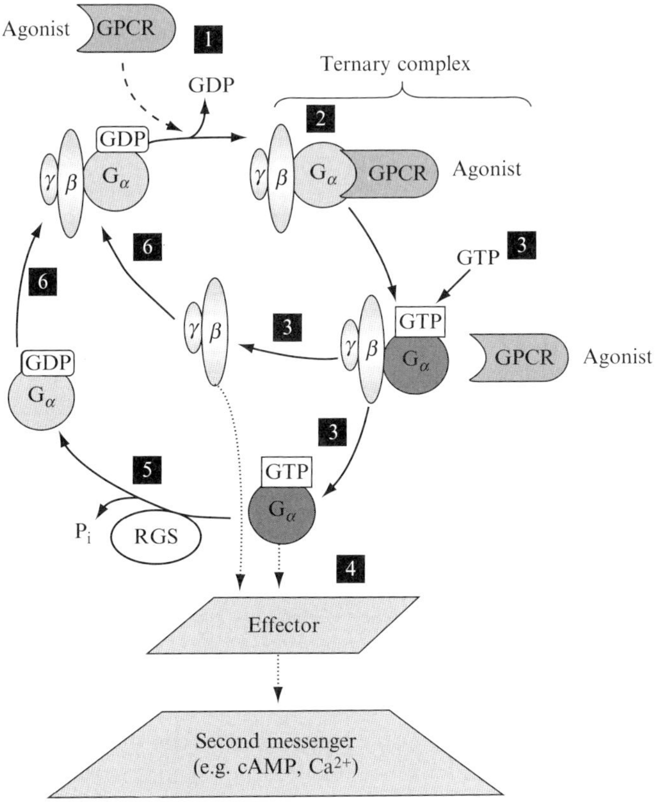

Figure 28.2 G-protein cycle. An agonist-activated or a constitutively active GPCR interacts with the heterotrimeric G-protein GDP-GαGβγ. This promotes the release of GDP (step 1), yielding the so-called ternary complex, which consists of agonist-occupied receptor, Gα-subunit, and Gβγ-subunit (step 2). The ternary complex shows high agonist binding affinity and can be characterized in high-affinity agonist binding assays. Binding of guanine nucleotides like GTP disrupts the ternary complex (step 3) and leads to the release of Gα-GTP (active form of Gα) and Gβγ (step 3). Step 3 is addressed by [^{35}S]GTPγS binding assays. Gα-GTP and Gβγ can activate effector molecules (step 4), leading to the generation of second messengers. Gα-GTP inactivates itself by its intrinsic GTPase activity, which cleaves GTP (step 5), yielding Gα-GDP and inorganic phosphate (P_i). Step 5 is addressed by steady-state GTPase assays and can be accelerated by regulators of G-protein signaling (=RGS). In step 6, Gα-GDP and Gβγ reassociate to a heterotrimeric G-protein, which is then available for a new G-protein cycle.

All steps of the membrane preparation are performed at 4 °C. One hundred milliliters of the infected cell suspension is centrifuged (10 min at 1000×*g*) and washed once with 50 mL of PBS (composition given in section 2.2.2). Thereafter, cells are lysed in 15 mL of lysis buffer by 25 strokes in a Dounce homogenizer and the homogenate is centrifuged for 5 min at 500×*g*. The supernatant suspension, which contains the membrane fraction, is carefully removed using plastic transfer pipettes and kept for further processing. The pellet contains nuclei and unbroken cells and is discarded. The borderline between pellet and supernatant is diffuse, and it is recommended that too much of the supernatant fluid be not removed to avoid contamination with components from the pellet. The isolated supernatant is centrifuged for 20 min at 40,000×*g* and the resultant membrane pellet is resuspended in 20 mL of lysis buffer. After another centrifugation step (20 min at 40,000×*g*), the membranes are suspended in binding buffer. At this point in the protocol, it is possible to concentrate the membranes by choosing a smaller volume of binding buffer. We recommend that membranes are suspended in such a way that a final protein concentration of 1.0–1.5 mg/mL is achieved. To achieve this protein concentration, as a rule of thumb, membranes from a 100-mL culture should be suspended in 30 mL of BB. In this step, the membrane is homogenized 15 times, using a 10-20 mL syringe with a 20-gauge needle. After determination of protein concentration with the Bio-Rad DC protein assay kit (Bio-Rad, Hercules, CA), the suspension is aliquoted and can be stored at − 80 °C for at least 4 years (Seifert and Wieland, 2005). The protein concentration may also be determined after thawing of the frozen homogenate, which reflects more accurately the condition of the membrane used in later assays. There can be some loss of protein due to adsorption during the freeze/thaw cycle.

When the receptor is tagged, for example, with a FLAG epitope, it is easily possible to detect protein expression with the M1 anti-FLAG antibody on a Western blot. A raw estimation of receptor expression is performed by comparing the band intensity of the receptor with the signal of a reference GPCR protein, which was quantified in radioligand binding assays. Figure 28.3A (first lane) shows the signal of the FLAG-tagged hH_4R, coexpressed with $G\alpha_{i2}$ and $G\beta_1\gamma_2$ in Sf9 cell membranes. For comparison, a dilution series of FLAG-β_2AR-expressing membrane (7.5 pmol/mg, determined by [^{3}H]dihydroalprenolol binding) was used (Fig. 28.3A, lanes 2–4), resulting in an estimated hH_4R expression level of about 2 pmol/mg. The expression level of GPCRs in Sf9 cells depends on the age and concentration of the virus stock, on the health state of the cells, but also on the type of protein to be expressed. In the literature, for some receptors very high expression levels are reported, for example, up to 65 pmol/mg for the human substance P receptor (Nishimura *et al.*, 1998).

The expression of $G\alpha$ subunits can also reach high levels. The first lane in Fig. 28.3B shows the band of $G\alpha_{i2}$ in membranes coexpressing hH_4R, $G\alpha_{i2}$, and $G\beta_1\gamma_2$, stained with a $G\alpha_{i1/2}$ antibody. Comparison with a dilution series of purified $G\alpha_{i2}$ protein (Fig. 28.3B, lanes 2–4) yielded an estimated expression level of 330 pmol/mg. This more than 100-fold excess of G-protein compared to hH_4R ensures that hH_4R signal transduction is not compromised by limited G-protein availability. As already mentioned above, it is even possible to coexpress five different proteins in Sf9 cells. Figure 28.3C shows the regulators of

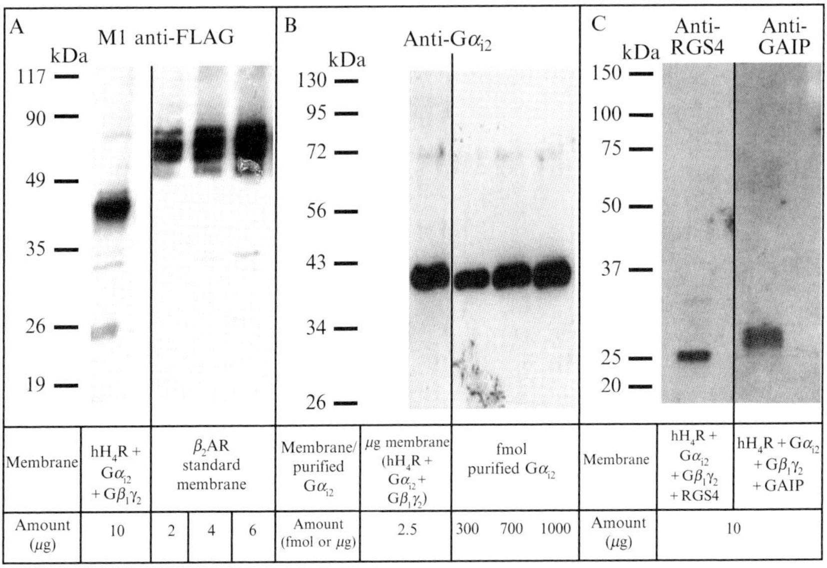

Figure 28.3 Immunoblotting of proteins, expressed in Sf9 cell membranes. The data were taken from previous publications. (A, B). Adapted with permission of Schneider *et al.* (2009). Copyright 2009 American Chemical Society; (C) adapted from Schneider and Seifert (2009). (A) Lane 1: Staining of FLAG-epitope tagged hH_4R (coexpressed with $G\alpha_{i2}$ and $G\beta_1\gamma_2$) with the M1 anti-FLAG antibody. Lane 2–4: Increasing concentrations of a standard membrane, expressing 7.5 pmol/mg of FLAG-tagged β_2AR (determined by saturation binding with [^{3}H]dihydroalprenolol). Comparison of the β_2AR band intensities with the hH_4R signal yielded an estimated hH_4R expression level of 2 pmol/mg membrane. (B) Lane 1: $G\alpha_{i2}$ (coexpressed with hH_4R and $G\beta_1\gamma_2$), stained with an anti-$G\alpha_{i2}$ antibody. For protein quantification, a dilution series of purified $G\alpha_{i2}$ protein standard was applied to lanes 2–4. Quantification of the signal in lane 1 yields an expression level of approximately 330 pmol/mg membrane. (C) The regulators of G-protein signaling RGS4 (lane 1) and GAIP (lane 2), coexpressed with hH_4R, $G\alpha_{i2}$, and $G\beta_1\gamma_2$ in a quadruple infection, were detected with specific antibodies.

G-protein signaling RGS4 and GAIP (for a more detailed discussion cf. Section 3.3.3), expressed in membranes that additionally contain hH_4R, $G\alpha_{i2}$, and $G\beta_1\gamma_2$.

3. Investigation of GPCR Constitutive Activity with Coexpression Systems

In this section, we provide the detailed protocols for high-affinity agonist binding assays and steady-state GTPase- and [^{35}S]GTPγS binding experiments.

3.1. High-affinity agonist binding with [^{3}H]histamine ([^{3}H]HA)

3.1.1. Required materials

3.1.1.1. Devices and disposables

- M48 Brandel Harvester (48 wells, Biomedical Research and Development Laboratories Inc., Gaithersburg, MD) or an alternative vacuum filtration device and appropriate GF/C or GF/B filters. Both filters work equally well.
- β-Scintillation counter for determination of bound radioactivity
- Excentric platform shaker with antiskid rubber mat (e.g., Innova 2050 from New Brunswick Scientific, Edison, NJ) and flexible strings to attach binding racks safely on platform. Each rack is covered with alumina foil to avoid spillage of tube content. Up to four racks can be safely placed on the platform mentioned above.
- Five-milliliter reaction tubes for preparation of the binding samples
- Tabletop centrifuge with cooling function and capable of up to 15,000 g

3.1.1.2. Buffers and reagents

- Appropriate [^{3}H]HA solutions. For the data shown in this chapter, we used [^{3}H]HA with specific activities in the range of 14–18 Ci/mmol (1 mCi/mL) from Perkin Elmer, Boston, MA. Tenfold concentrated stock solutions were prepared in binding buffer. It is possible to store these stock solutions for several days at 4 °C.
- Tenfold concentrated solutions of test compounds (A 100-μ*M* thioperamide (THIO) solution was used for nonspecific binding)
- Binding buffer: 12.5 m*M* $MgCl_2$, 1 m*M* EDTA in 75 m*M* Tris–HCl, pH 7.4
- Polyethyleneimine solution (0.3%, m/v)
- Scintillator for β-counting (e.g., Rotiszint eco plus from Carl Roth GmbH, Karlsruhe, Germany)
- Membrane expressing the proteins of interest (e.g., $hH_4R + G\alpha_{i2} + G\beta_1\gamma_2$)

3.1.2. Assay procedure

3.1.2.1. Competition bindings The radioligand binding assay described in the following protocol is performed with a sample volume of 250 μL. Prior to the experiment, the membrane should be centrifuged for 10 min at 4 °C and 14,000×*g* and resuspended in binding buffer. The membrane is homogenized 30 times, using a standard insulin syringe with a 24- or 27-gauge needle. This procedure has the purpose of eliminating, as far as possible, endogenous guanine nucleotides that could interfere with assays. For competition bindings, 25 μL of the (10-fold concentrated) test compound is added first, followed by 25 μL of the (10-fold concentrated) radioligand solution. For hH_4R competition binding experiments, ideally 10 n*M* of [^{3}H]HA and five appropriate concentrations between 1 n*M* and 10 μ*M* of the test compound (final concentrations) should be used. For nonspecific binding ($cpm_{nonspecific}$), the radioligand is completely displaced from the receptor by THIO (10 μ*M* final concentration), which is used instead of the test compound. Thereafter, 150 μL of binding buffer is added. The binding buffer is supplemented with bovine serum albumin (BSA) to yield a final concentration of 0.2% (m/v) of BSA in the sample. The binding reaction is started by addition of 50 μL of Sf9 cell membranes suspended in binding buffer. The amount of membrane in the samples should not be too high, to avoid binding of more than 10% of the totally added radioligand. Moreover, for each experiment the totally added radioactivity (cpm_{added}) has to be determined by adding 25 μL of the [^{3}H]HA solution directly to a scintillation vial.

3.1.2.2. Saturation bindings When K_d and B_{max} values are determined in a saturation binding experiment, the samples are prepared essentially as described for the competition binding experiments. Instead of one single [^{3}H]HA concentration, several concentrations in the range between 1 and 100 n*M* (final) are used. Total binding is determined with samples containing only [^{3}H]HA in binding buffer supplemented with 0.2% (m/v) BSA. Ideally, for each of the [^{3}H]HA concentrations, nonspecific binding is determined by additional preparation of samples containing a 1000-fold excess of THIO or another unlabeled ligand with sufficiently high affinity.

3.1.2.3. Sample processing After addition of the membranes, the samples are mixed and incubated for 60 min at 25 °C under continuous shaking at 200–250 rpm. Separation of bound and free radioligand is achieved by filtration through GF/C filters (Schleicher and Schuell, Dassel, Germany). It is absolutely essential to soak the filters ~10 min prior to filtration with a solution containing 0.3% (m/v) polyethyleneimine, since otherwise nonspecific filter-binding of [^{3}H]HA is exceedingly high and masks any specific [^{3}H]HA binding. After that, the filters are washed three times with 2 mL of ice-cold binding buffer (4 °C) and the binding reaction is stopped by filtering the samples. The sample tubes and filters should be washed twice with ice-cold binding buffer to ensure complete transfer of the samples and to remove background radioactivity from the filter. The filters are then directly transferred to scintillation vials and soaked in scintillation liquid for ~4 h (or overnight). Finally, the filter-bound radioactivity is determined by liquid scintillation counting. The counts per minute (cpm) are transformed to absolute amounts of bound radioligand by the following equation (cpm_{total} is the amount of radioactivity of the samples, prior to subtraction of nonspecific binding):

$$\frac{\mathrm{pmol}}{\mathrm{mg}} = \frac{\left(\mathrm{cpm}_{\mathrm{total}} - \mathrm{cpm}_{\mathrm{nonspecific}}\right) \cdot \mathrm{pmol}\left(\left[^{3}\mathrm{H}\right]\mathrm{HA}\right)/\mathrm{tube}}{\mathrm{cpm}_{\mathrm{added}} \cdot \mathrm{mg}_{\mathrm{protein}}/\mathrm{tube}}$$

3.1.3. Interpretation of the results

3.1.3.1. Disruption of the ternary complex with guanine nucleotides According to the G-protein cycle (Fig. 28.2), an agonist-stimulated GPCR promotes the dissociation of GDP from the cognate Gα subunit (step 1) and forms the "ternary complex," consisting of the GPCR, the agonist, and the heterotrimeric G-protein (step 2). In the ternary complex, the receptor exhibits the highest affinity for an agonist. Consequently, this affinity is reduced, when the complex is disrupted, for example, by binding of GTP to the Gα subunit (step 3). Thus, addition of guanine nucleotides to binding assay samples with an agonistic radioligand should lead to a dramatic decrease of radioligand binding affinity. This is, for example, observed for the FPR, which is highly constitutively active when coexpressed with $G\alpha_{i2}$ and $G\beta_1\gamma_2$ in Sf9 cells (Wenzel-Seifert *et al.*, 1998). Addition of 10 μ*M* of GTPγS leads to a disruption of the ternary complex and to a reduction of binding affinity of the agonist [^{3}H]fMLF to FPR (Fig. 28.4B). At the same time, B_{max} is reduced, since the

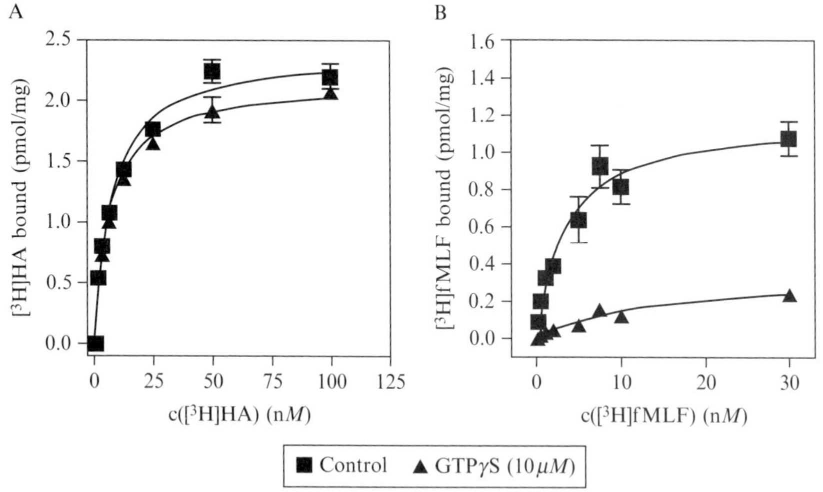

Figure 28.4 Saturation binding assays with agonistic radioligands at the hH_4R (A) and the formyl peptide receptor (FPR) (B), both coexpressed with $G\alpha_{i2}$ and $G\beta_1\gamma_2$. The data were taken from previous publications. (A) Adapted with permission of Schneider *et al.* (2009). Copyright 2009 American Chemical Society. (B) This research was originally published in Wenzel-Seifert *et al.* (1998). Copyright by the American Society for Biochemistry and Molecular Biology. (A) Saturation binding with [^{3}H]HA and membranes coexpressing hH_4R, $G\alpha_{i2}$, and $G\beta_1\gamma_2$. Binding was recorded in the presence (▲) and absence (■) of GTPγS (10 μ*M*) and shows that hH_4R can adopt a G-protein-independent high-affinity state. (B) Saturation binding with [^{3}H]fMLF and membranes coexpressing FPR-26, $G\alpha_{i2}$, and $G\beta_1\gamma_2$. Binding was recorded in the presence (▲) and absence (■) of GTPγS (10 μ*M*), showing a pronounced GTPγS sensitivity of the high-affinity state. Data in (A) and (B) are from one representative experiment with two to three replicates, mean ± S.E.M.

concentrations of radioligand are not high enough to label the newly formed low-affinity binding sites. Interestingly, the hH_4R behaves completely differently. Addition of 10 μM of GTPγS does not reduce the binding affinity of the agonist [^{3}H]HA (Fig. 28.4A). Even in the presence of 100 μM of GTPγS or in the complete absence of G-proteins (data not shown), no significant reduction of agonist binding affinity occurs (Schneider *et al.*, 2009). This demonstrates that, in contrast to other GPCRs, hH_4R shows a G-protein independent high-affinity state, which may also explain the extraordinarily high constitutive activity of this receptor. The profoundly different behavior of the two highly constitutively active receptors FPR and hH_4R in high-affinity agonist binding also demonstrates that the relation between constitutive activity and high-affinity receptor states is not yet completely understood and obviously more complex than suggested by the two-state model or the G-protein cycle.

3.1.3.2. Conformational instability of constitutively active receptors High constitutive activity is a result of a reduced energy barrier between active and inactive state of a receptor, leading to an increased conformational flexibility of the protein. This suggests that constitutively active receptors show a reduced conformational stability, undergoing denaturation more easily than GPCRs with low constitutive activity. This phenomenon can be investigated in two ways. First, receptor instability can be recognized by reduced expression levels in Sf9 cells and by an increased B_{max} value in the presence of stabilizing factors (e.g., ligands). Second, receptors (purified or together with coexpressed G-proteins in membranes) can be incubated at 37 °C and the B_{max} value of a radioligand is determined at specific time points.

Both experimental setups revealed that hH_4R in fact shows high conformational instability. As reported previously (Schneider *et al.*, 2009), addition of the agonist HA (10 μM) or the inverse agonist THIO (1 μM) to Sf9 cell cultures (coinfected with baculoviruses encoding hH_4R, $G\alpha_{i2}$, and $G\beta_1\gamma_2$) resulted in a more than 70% increase of the B_{max} value in [^{3}H]HA binding. Surprisingly, this effect was not observed, when hH_4R receptor protein was quantified by Western blots. This indicates that the increase of the B_{max} value is due to a conformational stabilization of already membrane-inserted GPCR, but not caused by an increased hH_4R protein production (Schneider *et al.*, 2009). When membranes coexpressing hH_4R, $G\alpha_{i2}$, and $G\beta_1\gamma_2$ were incubated for 45 min at 37 °C, [^{3}H]HA binding declined by more than 60%. Addition of HA to the samples stabilized the receptor. The inverse agonist THIO was even able to reconstitute binding sites that were inactivated at 37 °C (Schneider *et al.*, 2009). This shows that inactivation of a conformationally instable and constitutively active GPCR is not necessarily a "one-way street," but some ligands may be able to reconstitute the original agonist-binding conformation.

Conformational instability was already earlier reported for the constitutively active β_2AR mutant β_2AR_{CAM} (Gether *et al.*, 1997). Both wtβ_2AR and β_2AR_{CAM} were purified by nickel chromatography and incubated at 37 °C. Radioligand binding assays with [^{3}H]dihydroalprenolol showed that the half-life of β_2AR_{CAM} was by 75% shortened compared to wtβ_2AR (12.3 vs. 49.9 min). Interestingly, receptor stability was considerably increased in the presence of both the agonist isoproterenol and the inverse agonist ICI 118551 (erythro-DL-1(7-methylindan-4-yloxy)-3-isopropylaminobutan-2-ol)

This was also observed in Sf9 cells infected with β_2AR- and β_2AR_{CAM}-encoding baculoviruses. The expression of β_2AR_{CAM} was markedly reduced, compared to wtβ_2AR. However, when the infected cells were cultured in the presence of isoproterenol or ICI 118551, expression of β_2AR_{CAM}, but not of wtβ_2AR, was markedly increased. This demonstrates again that the increased constitutive activity of β_2AR_{CAM} reduces receptor stability and that the receptor is stabilized by ligands, irrespective of the quality of action (Gether *et al.*, 1997).

3.2. [^{35}S]GTPγS binding

In this assay, step 3 in the G-protein cycle (Fig. 28.2) is addressed. The radioactive guanine nucleotide [^{35}S]GTPγS is added to membranes, coexpressing a GPCR with Gα and G$\beta\gamma$. [^{35}S]GTPγS displaces GDP from the Gα subunit, but, in contrast to GTP, it cannot be hydrolyzed by the intrinsic GTPase activity of Gα. Thus, step 5 of the G-protein cycle cannot take place and Gα-[^{35}S]GTPγS complexes are enriched in the membrane with increasing incubation time. After filtration, this membrane-bound radioactivity can be determined by scintillation counting. It should be noted that, in contrast to the steady-state GTPase assay, the relative intensity of agonist- and inverse agonist-induced signals depends on the incubation time. Stimulation of Gα by a GPCR results in an increased affinity for [^{35}S]GTPγS and a reduction of GDP affinity. This is reflected by an increased binding after a specific incubation time or in an accelerated binding kinetics of GTPγS. Since the GDP affinity of Gα subunits is relatively low, GDP has to be added to the samples in a final concentration between 0.1 and 10 μM (in most cases 1 μM). This ensures that the Gα subunits are "preloaded" with GDP, reduces constitutive [^{35}S]GTPγS binding and increases the sensitivity of the assay.

The [^{35}S]GTPγS binding assay is readily suited to analyze constitutively active Gα_s- and Gα_i-coupled GPCRs. However, it should be noted that the signals for Gα_s-coupled receptors (e.g., β_2AR) are very low in coexpression systems. In this case the signal-to-background ratio can be considerably improved by using GPCR–Gα_s fusion proteins, as described in the corresponding chapter of this volume. In general, Gα_q proteins show only a low affinity for GTPγS and the intrinsic guanine nucleotide exchange occurs only slowly. Thus, for the analysis of constitutive activity of Gα_q-

coupled receptors, other approaches like reconstitution with purified squid retinal $G\alpha_q$ protein (Hartman and Northup, 1996) or a binding/immunoprecipitation assay (Barr *et al.*, 1997) have been described. We do not have experience with those assays in our laboratory. In the following, we focus on the application of [^{35}]GTPγS binding assays for the analysis of hH_4R constitutive activity and provide protocols for three different types of [^{35}S]GTPγS assays, suited for a sample volume of 250 μL.

3.2.1. Required materials

3.2.1.1. Devices and disposables The same as in Section 3.1 (high-affinity agonist binding assays).

3.2.1.2. Buffers and reagents

- [^{35}S]GTPγS stock solution. We obtained the solution (1250 Ci/mmol, 12.5 mCi/mL, 80 μL) from Perkin Elmer (Boston, MA). The stock solution was diluted with 720 μL of an appropriate buffer (e.g., 10 m*M* Tricine buffer, pH 7.6, supplemented with 1 m*M* of dithiothreitol) and aliquots of 50 μL were stored on dry ice for up to 4 months. For the individual experiment, the required amount of radioactivity was diluted with water to yield a 10-fold concentrated stock solution. Radioactive decay (factor = 0.9921/day) was considered in the calculations.
- Tenfold concentrated solutions of test compounds
- Tenfold concentrated GDP solution (between 1 n*M* and 100 μ*M*)
- Binding buffer: 12.5 m*M* $MgCl_2$, 1 m*M* EDTA in 75 m*M* Tris–HCl, pH 7.4
- Scintillator for β-counting
- Membrane expressing the proteins of interest (e.g., $hH_4R + G\alpha_{i2} + G\beta_1\gamma_2$)

3.2.2. Assay procedure

3.2.2.1. Saturation bindings This type of [^{35}S]GTPγS binding assays determines the affinity of [^{35}S]GTPγS to Gα under different conditions. A saturation binding curve is recorded with increasing concentrations of [^{35}S]GTPγS under control conditions, in the presence of an agonist and in the presence of an inverse agonist. All samples should contain a final concentration of 1 μ*M* of GDP and 0.05% (m/v) of BSA. Prior to the experiment, membranes are centrifuged (10 min at 4 °C and 13,000 rpm) and resuspended in binding buffer (containing 1.25 μ*M* of GDP and 0.0625% (m/v) of BSA). The membrane is homogenized 30 times, using a standard insulin syringe with a 24- or 27-gauge needle. First, 25 μL of a 10-fold concentrated ligand solution (HA, THIO, or water for control conditions) are added to the tubes, followed by 150 μL of binding buffer, containing GDP (1.25 μ*M*) and BSA (0.0625%, m/v). Then, 25 μL of a 10-fold concentrated [^{35}S]GTPγS solution (in H_2O) are added. Normally, in a

hH_4R saturation binding, we use final [^{35}S]GTPγS concentrations between 0.2 and 25 nM. To avoid large amounts of radioactivity, the [^{35}S]GTPγS can be diluted with nonlabeled GTPγS. However, this also reduces signal intensity and assay sensitivity. In each experiment, nonspecific [^{35}S]GTPγS binding ($cpm_{nonspecific}$) should be determined with samples containing 10 μM of cold GTPγS. Normally, nonspecific binding is in the range between 0.2% and 0.4% of total binding. Moreover, the activity of a defined volume of [^{35}S]GTPγS stock solution (directly added to the scintillation vial) should be determined to calculate the absolute amount of radioactivity added to each individual sample (cpm_{added}).

Finally, the binding is initiated by addition of 50 μL of the membrane suspension, yielding a final membrane concentration of 5–20 μg per sample. In membranes coexpressing hH_4R, $G\alpha_{i2}$, and $G\beta_1\gamma_2$, it may be necessary to reduce constitutive activity by stabilizing the inactive state (cf. Fig. 28.1) with a final concentration of 100 mM of NaCl, specifically when agonist-induced signals are to be investigated. Incubations are performed for 240 min at 25 °C and shaking at 250 rpm.

3.2.2.2. Time course studies For time course studies, the composition of the samples is essentially the same as described for the saturation binding assays, but with a [^{35}S]GTPγS concentration between 0.4 and 1 nM. The sample volume is upscaled (1700 μL for nine time points). The membrane concentration is 0.03 μg/μL. NaCl at concentrations between 25 and 150 mM can be added to samples to reduce constitutive GPCR activity. The effect of NaCl on constitutive activity varies substantially among various GPCRs, FPR being very sensitive (Wenzel-Seifert *et al.*, 1998) and H_4R being very insensitive (Schneider *et al.*, 2009). The samples are started by addition of membrane suspension and incubated at 25 °C, shaking at 250 rpm. At defined time points, 175 μL aliquots, corresponding to 5.25 μg of membrane protein are taken from the total sample volume and filtered by a Millipore 1225 vacuum sampling manifold (Millipore, Bedford, MA), followed by three washes with 2 mL of binding buffer (4 °C). Nonspecific [^{35}S]GTPγS binding is determined in the presence of 10 μM of cold GTPγS.

3.2.2.3. Determination of Gα-GDP affinity For this type of analysis, the samples are prepared as described for the saturation binding assays, but with a [^{35}S]GTPγS concentration between 0.4 and 1 nM. Instead of a final GDP concentration of 1 μM, increasing concentrations of GDP are used in a range from 0.1 nM to 100 μM. In this way, a displacement curve is generated, where GDP competes with [^{35}S]GTPγS for the Gα-subunit. An incubation time of 240 min is required for the hH_4R, coexpressed with $G\alpha_{i2}$ and $G\beta_1\gamma_2$.

3.2.2.4. Characterization of ligands The [^{35}S]GTPγS binding assay can also be used to functionally characterize agonists, neutral antagonists and inverse agonists. To perform this kind of analysis, the samples are essentially prepared as described for the saturation bindings, but with a [^{35}S]GTPγS concentration between 0.4 and 1 n*M*. If required, 100 m*M* of NaCl (final) may be added to the sample to reduce constitutive activity. It should be noted that, in contrast to steady-state GTPase assays, the effect of ligands can depend on the incubation time. The effect of inverse agonists is increasing with incubation time (Seifert and Wieland, 2005). To avoid interexperimental variations, agonist and inverse agonist efficacies and potencies should be determined at equilibrium conditions. This is very difficult to achieve for the hH_4R, since the association kinetics of [^{35}S]GTPγS is so slow (Fig. 28.5D) and it is necessary to incubate the samples for up to 240 min (Schneider *et al.*, 2009).

3.2.2.5. Sample processing Free and bound [^{35}S]GTPγS are separated by filtration through GF/C filters using a Brandel Harvester (for the time course studies, single samples are stopped separately using a Millipore 1225 vacuum sampling manifold). In contrast to the [^{3}H]HA binding assay, the filters do not have to be presoaked with polyethyleneimine. After filtration, the filters are washed three times with 2 mL of binding buffer (4 °C). Prior to scintillation counting, the filters have to be incubated in scintillation liquid for 2–3 h (or overnight).

The cpm are transformed to absolute amounts of bound radioligand by the following equation (cpm_{total} is the amount of radioactivity in the samples, prior to subtraction on nonspecific binding):

$$\frac{\text{pmol}}{\text{mg}} = \frac{\left(\text{cpm}_{\text{total}} - \text{cpm}_{\text{non-specific}}\right) \cdot \text{pmol}\left(\left[^{35}\text{S}\right]\text{GTP}\gamma\text{S}\right)/\text{tube}}{\text{cpm}_{\text{added}} \cdot \text{mg}_{\text{protein}}/\text{tube}}$$

3.2.3. Interpretation of the results

[^{35}S]GTPγS assays that determine the activity of hH_4R agonists and inverse agonists yield in general results similar to those of the corresponding steady-state GTPase assays. The only difference is the dependency of ligand efficacies on incubation time, which was already mentioned above. Thus, in the following, we focus mainly on the interpretation of saturation bindings and binding kinetics.

Figure 28.5A and B show the results of [^{35}S]GTPγS saturation bindings. Similarly as described below for the steady-state GTPase enzyme kinetic studies (Fig. 28.6A and B), the control curve (without ligand, only solvent) was subtracted from the curves determined in the presence of the agonist

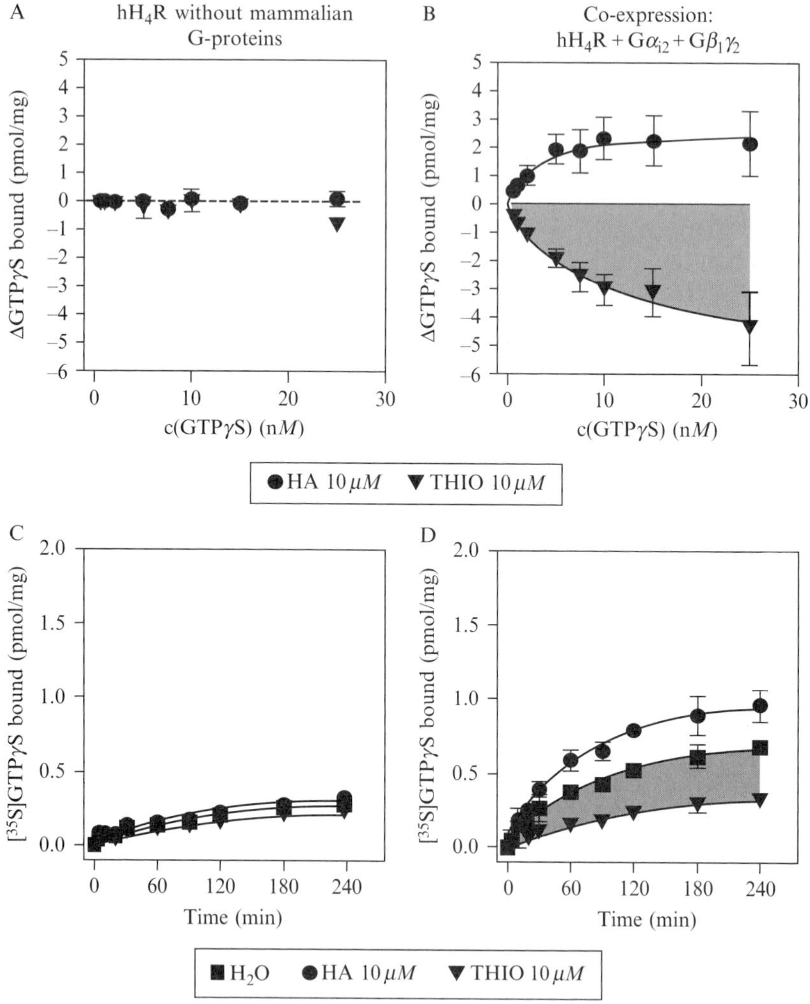

Figure 28.5 [^{35}S]GTPγS saturation binding and binding kinetics to membranes expressing hH_4R with or without mammalian G-proteins, determined in the presence of 100 m*M* of NaCl. Figure adapted with permission of Schneider *et al.* (2009). Copyright 2009 American Chemical Society. (A, B) [^{35}S]GTPγS saturation binding was determined with membranes expressing hH_4R without mammalian G-proteins (A) and coexpressing hH_4R with mammalian $G\alpha_{i2}$ and $G\beta_1\gamma_2$ (B), always in the presence of saturating concentrations (10 μ*M*) of HA (●) or THIO (▼). To generate the curves in panels (A) and (B), saturation binding was first determined under control conditions (H_2O) and in the presence of HA and THIO. Then, the control curve (representing the constitutive activity) was subtracted from the other two curves, resulting in a positive curve for HA-induced [^{35}S]GTPγS binding and a negative one for THIO-inhibited constitutive activity. The THIO-inhibited constitutive activity is highlighted as a gray area. (C, D) [^{35}S]GTPγS binding kinetics were determined with membranes expressing hH_4R without mammalian G-proteins (C) and coexpressing hH_4R with mammalian $G\alpha_{i2}$ and $G\beta_1\gamma_2$ (D), always in the presence of saturating concentrations (10 μ*M*) of HA (●) or THIO (▼) and under control conditions (H_2O, ■).

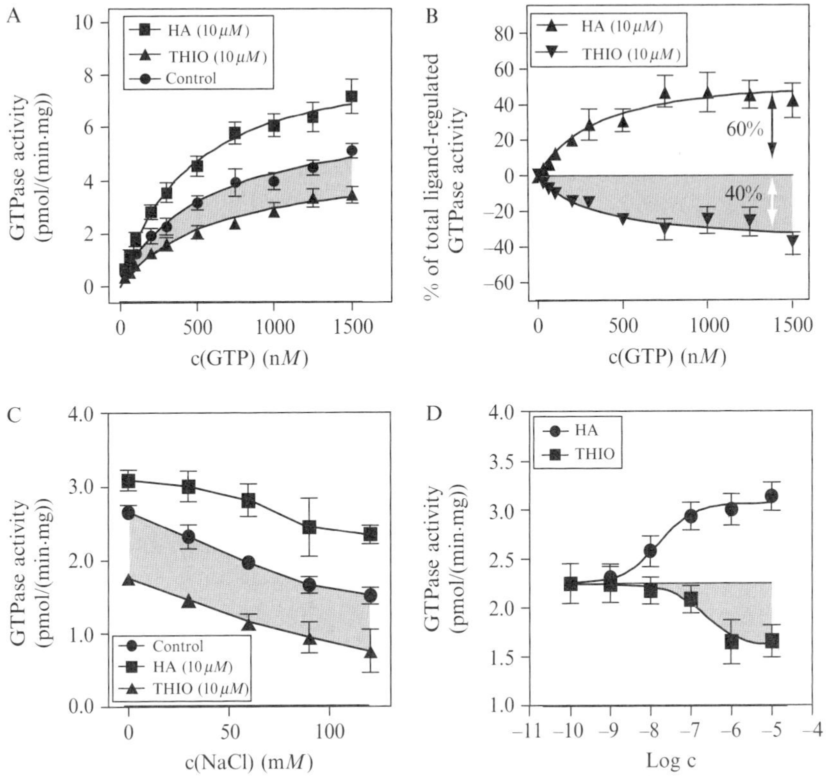

Figure 28.6 The hH_4R-modulated $G\alpha_{i2}$ steady-state GTPase enzyme activity in membranes coexpressing hH_4R with mammalian $G\alpha_{i2}$ and $G\beta_1\gamma_2$. (A–C) Adapted with permission of Schneider *et al.* (2009). Copyright 2009 American Chemical Society. The areas highlighted in gray in panels A–D show the inverse agonist-inhibited constitutive activity of hH_4R. (A, B) Michaelis–Menten kinetics with increasing concentrations of the substrate GTP, determined in the presence of 100 m*M* of NaCl. (A) First, data were recorded under control conditions (H_2O, ●) and in the presence of saturating concentrations (10 μ*M*) of HA (■) and THIO (▲). (B) Then, the control curve was subtracted from the HA- and THIO curve, yielding a positive curve for HA-induced GTPase activity (▲) and a negative curve for THIO-inhibited constitutive activity (▼). All values were transformed to % of total ligand-regulated GTPase activity ($V_{max(HA)} - V_{max(THIO)}$). (C) Influence of increasing NaCl concentrations on the steady-state GTPase activity of $G\alpha_{i2}$ under control conditions (●) and in the presence of saturating (10 μ*M*) concentrations of HA (■) and THIO (▲). All data were recorded in the presence of 100 n*M* of GTP. (D) Concentration–effect curves of the full agonist HA (●) and the inverse agonist THIO (■), recorded in the steady-state GTPase assay in the presence of 100 n*M* of GTP and 100 m*M* of NaCl.

HA and the inverse agonist THIO. The result is a positive saturation curve showing the stimulatory effect of HA and a negative saturation curve that shows the amount of constitutive activity inhibited by THIO.

The difference between these two curves corresponds to the total ligand-regulated [^{35}S]GTPγS binding. As demonstrated by Fig. 28.5A, the interaction of hH_4R with endogenous G-proteins produces no significant signal, not even at higher concentrations of GTPγS. This shows that Sf9 cells are a "clean" system for most Gα_i-coupled receptors, allowing the coexpression of specific Gα_i subunits and the investigation of their influence on constitutive receptor activity.

However, when coexpressed with mammalian Gα_{i2} and G$\beta_1\gamma_2$, the hH_4R shows highly effective G-protein coupling (Fig. 28.5B). The agonist HA and the inverse agonist THIO induce significant effects. The gray area in Fig. 28.5B demonstrates that the THIO-inhibited constitutive activity is about 70% of the total ligand-regulated activity, which is surprising, when it is considered that all results in Fig. 28.5 were obtained in the presence of 100 m*M* of NaCl. This shows that the constitutive activity of hH_4R is highly resistant to NaCl and confirms the steady-state GTPase data shown in Fig. 28.6C (discussion in Section 3.3.3).

The results of [^{35}S]GTPγS saturation binding assays can be used for the determination of the "coupling factor" of a GPCR, that is, the number of G-proteins, which can be activated by one single GPCR molecule. To determine the coupling factor, it is required to divide the B_{max} value from a radioligand binding assay (cf. Section 3.1) by the total ligand-regulated [^{35}S]GTPγS binding. The coupling factor of hH_4R is ~5, which means that one receptor molecule can stimulate up to five Gα-subunits (Schneider *et al.*, 2009). This is also termed catalytic G-protein activation. By contrast, one FPR molecule activates only one Gα subunit, which is referred to as linear G-protein activation (Wenzel-Seifert *et al.*, 1999).

The behavior of hH_4R in [^{35}S]GTPγS binding kinetics is shown in Fig. 28.5C and D. Again, the signal in a system expressing only hH_4R in the absence of coexpressed mammalian G-proteins is very low (Fig. 28.5C), indicating an absent or unproductive interaction with endogenous insect cell Gα_i-like proteins. As soon as mammalian Gα_{i2} and G$\beta_1\gamma_2$ are coexpressed with hH_4R, a very large [^{35}S]GTPγS binding signal is caused by the constitutive activity of hH_4R and additionally increased by the agonist HA. A typical characteristic of hH_4R signaling is the slow kinetics of Gα_{i2}-stimulation. Even after 240 min of incubation equilibrium is hardly reached. THIO causes a clear reduction of [^{35}S]GTPγS binding and the contribution of THIO-inhibited constitutive activity is again highlighted by the gray area in Fig. 28.5D. The results show that high constitutive activity of a GPCR is not necessarily connected with a fast Gα_i activation kinetic, but both receptor properties can coexist independently of each other.

When [^{35}S]GTPγS is displaced by GDP, the GDP affinity is expected to be reduced in the presence of agonists, since the activated receptor promotes GDP release in step 1 of the G-protein cycle (Fig. 28.2). Conversely, inverse agonists can increase GDP affinity by inhibiting the GDP release promoted by the constitutively activated receptor.

3.3. Steady-state GTPase assay

Steady-state GTPase assays determine the hydrolysis of Gα-associated GTP by the intrinsic GTPase activity of the Gα subunit (Fig. 28.2, step 5). This is achieved by adding radiolabeled [γ-^{32}P]GTP to the membrane samples. The amount of radioactive inorganic phosphate, which is released by the GTPase activity of Gα, is directly proportional to the extent of receptor activation, provided that steady-state conditions are maintained. In the following, we give a detailed description of the experimental protocol and explain how this functional assay can be applied for the characterization of a constitutively active GPCR like the hH_4R.

3.3.1. Required materials

3.3.1.1. Devices and disposables

- Heating block, capable of maintaining a temperature of 25 °C
- β-Scintillation counter for determination of bound radioactivity
- Tabletop centrifuge with cooling function and capable of up to 15,000×*g*
- 1.5 mL Eppendorf tubes for preparation of the binding samples

3.3.1.2. Buffers and reagents

- *[γ-^{32}P]GTP solution* The stock solution that was obtained by an enzymatic labeling procedure (5–10 mCi in 130 μL) was stored on dry ice in a lead container. The labeling procedure has been described elsewhere (Walseth and Johnson, 1979). Portions of this stock solution were diluted with Tricine buffer (10 m*M*, pH 7.6) to yield a final radioactive concentration of 2 μCi/μL. This dilution was used for the experiments. The volume of radioactive solution needed for a single experiment was calculated under consideration of radioactive decay (factor = 0.9526/day). All [γ-^{32}P]GTP containing solutions have to be stored at −80 °C to ensure stability. Under these conditions, the limiting factor is the radioactive decay and not chemical instability. In our hands, one batch of freshly labeled [γ-^{32}P]GTP can be used for ~4 weeks. For GTPase assays, also commercial [γ-^{32}P]GTP preparations can be used. However, we have made the experience that chemical stability of the [γ-^{32}P]GTP prepared in-house was much better than in the commercial preparations. This is a crucial point since chemical instability results in higher blank values and hence, decreased signal-to-noise ratio. In several cases, we even experienced chemical decomposition of commercial [γ-^{32}P]GTP during the relatively short GTPase assay, resulting in a continued increase in [^{32}P]$_i$ harvested. Such a systematic error compromises data interpretation and data interpretation severely. Accordingly, if commercial [γ-^{32}P]GTP is used, we strongly recommend strict storage of stock solutions at −80 °C, dilution of [γ-^{32}P]GTP in Tricine buffer or reaction mixture as well as blank values at the beginning, in the middle and at the end of each sample series, since reactions are not started at the same time but in 10–15 s intervals.

- Tenfold concentrated solutions of test compounds (GTP 10 m*M* for blank value)
- Tris–HCl buffer, pH 7.4, 10 m*M* (for suspension of the membrane) and 20 m*M* (for dilution of [γ-^{32}P]GTP solution)
- Reaction mixture, twofold concentrated (100 m*M* Tris, 0.2 m*M* AppNHp, 200 n*M* GTP, 200 μ*M* ATP, 10 m*M* $MgCl_2$, 0.2 m*M* EDTA, 2.4 m*M* creatine phosphate, 20 μg/mL creatine kinase and 0.4% (m/v) BSA); if necessary, NaCl may be added to suppress constitutive activity
- Chilled slurry (4 °C) consisting of 5% (m/v) activated charcoal in 50 m*M* NaH_2PO_4, pH 2.0
- Membrane expressing the proteins of interest (e.g., $hH_4R + G\alpha_{i2} + G\beta_1\gamma_2$)
- Scintillator for β-counting

3.3.2. Assay procedure

After thawing the membranes, they are centrifuged and resuspended in 10 m*M* Tris–HCl, pH 7.4. The membrane is homogenized 30 times, using a standard insulin syringe with a 24- or 27-gauge needle. Steady-state GTPase assays are normally performed with a sample volume of 100 μL. Samples are prepared by adding 10 μL of a 10-fold concentrated ligand solution to the assay tubes, followed by 50 μL of twofold concentrated reaction mixture (cf. section 3.3.1.2) and 20 μL of membrane suspension (normally around 500 μg/mL, yielding a final concentration of 10 μg/sample). The hH_4R shows extraordinarily high constitutive activity, leading to a very small signal-to-background ratio for agonist-induced signals. Thus, for the characterization of agonists or unknown ligands it may be beneficial to suppress a part of the constitutive activity by sodium chloride, since sodium ions stabilize the inactive state of many GPCRs (Fig. 28.1). For hH_4R, 100 m*M* of sodium chloride are recommended, because the constitutive activity of this receptor is surprisingly resistant to the effect of sodium ions. In the presence of 100 m*M* of sodium chloride, both agonists and inverse agonists induce acceptable signal intensities. If it is required, the sodium chloride is added to the reaction mixture, which is described in Section 3.3.1.2. The samples (80 μL/tube) are incubated for 2 min at 25 °C. The reaction is started by addition of 20 μL of [γ-^{32}P]GTP (0.1 μCi) per tube to yield a final sample volume of 100 μL, containing the ligands, [γ-^{32}P]GTP, AppNHp (0.1 m*M*), GTP (100 n*M*), ATP (100 μ*M*), $MgCl_2$ (5 m*M*), EDTA (0.1 m*M*), creatine phosphate (1.2 m*M*), creatine kinase (10 μg/mL), and BSA (0.2%, m/v). AppNHp is a nucleotide, which is routinely added to inhibit nucleotidases in the membrane. It can be omitted from steady-state GTPase assays with Sf9 cells without compromising assay sensitivity. However, when steady-state GTPase assays are performed with mammalian cell membranes that contain relatively high nucleotidase activities, AppNHp should be added to the samples. Otherwise substrate depletion may occur.

The total amount of radioactivity added to each tube should be determined by directly adding the corresponding volume of [γ-^{32}P]GTP stock solution or of a defined dilution to a scintillation vial (= cpm_{added}). All stock and work dilutions of [γ-^{32}P]GTP are prepared in 20 m*M* Tris–HCl, pH 7.4. The reactions are normally conducted for 20 min at 25 °C and the samples are started in 10–15 s intervals. This allows a maximum of 120 samples to be processed in one experiment. If the 20 min incubation time is exceeded for some reasons, it has to be ensured that the reaction is still in steady-state and the substrate is not depleted. Otherwise the signal does not increase any more linearly with time, but reaches saturation. This may reduce or even eliminate the differences between the signals induced by the constitutively active-, agonist-stimulated-, or inverse agonist-inhibited receptor. The experimental conditions should always be chosen in a way that not more than 10% of the total amount of [γ-^{32}P]GTP added is converted to $^{32}P_i$.

To terminate the reaction after 20 min, 900 μL of a chilled slurry (4 °C) consisting of 5% (m/v) activated charcoal in 50 m*M* NaH_2PO_4, pH 2.0, are added per tube. We strongly recommend cutting off the end of the pipet tip in order to avoid obstruction by the viscous charcoal slurry. Otherwise, variations in the pipetted volume can result. The charcoal absorbs nucleotides but not $^{32}P_i$. The charcoal-quenched reaction mixtures are then centrifuged for 7 min at room temperature at 15,000×*g*. The supernatant contains $^{32}P_i$, which was produced by G-protein mediated cleavage of GTP. Six hundred microliters of the supernatant fluid is added to water-filled scintillation vials for the determination of $^{32}P_i$ by liquid scintillation counting. Since ^{32}P produces Čerenkov radiation in water, no organic scintillator is needed. It is extremely important that the supernatant is not contaminated with charcoal from the pellet. The charcoal, which has bound the excess of nonprocessed radioactive substrate, would produce outliers with increased signal intensity.

One critical point in the steady-state GTPase assay is spontaneous degradation of the relatively instable [γ-^{32}P]GTP. The stock solutions should be continuously stored at −80 °C. Storage at higher temperatures causes accelerated degradation, producing free $^{32}P_i$, which leads to a high signal background. Spontaneous [γ-^{32}P]GTP degradation can be determined with tubes containing all of the above-described components plus 1 m*M* of unlabeled GTP that, by competition with [γ-^{32}P]GTP, prevents [γ-^{32}P]GTP hydrolysis by enzymatic activities present in Sf9 membranes. Normally, spontaneous [γ-^{32}P]GTP degradation should not exceed ~1% of the total amount of radioactivity added, using 20 m*M* Tris–HCl, pH 7.4, as solvent for [γ-^{32}P]GTP.

Sometimes, spontaneous degradation of [γ-^{32}P]GTP can even occur during the experiment, which is indicated by increasing baseline values. Thus, it is recommended to determine several baseline values within each sample series. This provides an internal reference for the other samples measured in temporal proximity to the baseline values. In our experience, stability of [γ-^{32}P]GTP can be increased by diluting it with reaction mixture instead of Tris-buffer. In this case, the corresponding volume of the reaction mixture, which is added to the tubes, has to be replaced by Tris-buffer (20 m*M*, pH 7.4) to maintain the correct final concentrations of all components. In our experience, the lowest nonenzymatic [γ-^{32}P]GTP degradation is achieved when the nucleotide is synthesized in-house by the procedure of Walseth and Johnson. The procedure has been described in detail elsewhere (Walseth and Johnson, 1979). With this procedure, we routinely obtained blank values of ~0.1% of the total radioactivity. We terminated the reactions by adding chilled absolute ethanol to the reaction sample so that a final ethanol concentration of 50% (v/v) is achieved. From this primary stock solution, secondary stock solutions were prepared as described in section 3.3.1.2.

To determine the Michaelis–Menten constant of the Gα subunit, the samples are prepared with a higher amount of [γ-^{32}P]GTP (0.4–0.5 μCi/tube). Unlabeled GTP is added in increasing concentrations from 0 to 1500 n*M*. Due to the displacement of [γ-^{32}P]GTP from the Gα subunit, the signal-to-noise ratio of the GTPase signal is reduced by unlabeled GTP. Therefore, unlabeled GTP can only be used at concentrations higher than 1.5 μM, when a higher percentage of radioactively labeled GTP is used.The steady-state GTPase activity (pmol/mg/min) is calculated as follows:

$$\frac{\text{pmol}}{\text{mg}\cdot\text{min}} = \frac{(\text{cpm}_{\text{total}} - \text{cpm}_{\text{GTP}}) \cdot \text{pmol}_{\text{GTP,unlabeled}}/\text{tube} \cdot 1.67}{\text{cpm}_{\text{added}} \cdot \text{min}_{\text{incubation}} \cdot \text{mg}_{\text{protein}}/\text{tube}}$$

$\text{Cpm}_{\text{total}}$ is the radioactivity determined in the respective sample, cpm_{GTP} is the blank value determined in the presence of 1 m*M* of GTP, and $\text{pmol}_{\text{GTP,unlabeled}}$ is the amount of unlabeled GTP present in the 100 μL reaction volume. The concentration of radioactive [^{32}P]GTP is negligible. The factor 1.67 accounts for the fact that only 600 μL out of 1 mL are counted. The total amount of radioactivity in the sample is expressed as $\text{cpm}_{\text{added}}$, incubation time ($\text{min}_{\text{incubation}}$) is normally 20 min.

3.3.3. Applications of the GTPase assay and interpretation of the results

3.3.3.1. Determination of K_M- and V_{max}-values Increasing GTP concentrations enhance GTPase activity. In case of Gα_i, stimulated by hH_4R, this effect reaches saturation (V_{max}) at substrate concentrations higher than 1500 μM (Fig. 28.6A). Under steady-state conditions, such curves are comparable with saturation curves in radioligand binding assays. The K_M value is the GTP concentration at which the GTPase activity is 50% of V_{max} and at which 50% of all available Gα subunits are occupied by the substrate. Figure 28.6A shows Michaelis–Menten kinetic curves recorded with membranes coexpressing hH_4R, Gα_{i2}, and G$\beta_1\gamma_2$ in the absence of ligands as well as in the presence of the agonist HA (10 μM) or the inverse agonist THIO (10 μM). The THIO-sensitive part of the constitutive receptor activity is shown as gray area. Similar to the method explained in Section 3.2.3 for the [^{35}S]GTPγS assay, the control curve can be subtracted from the curves recorded in the presence of HA or THIO. Relating the resulting ΔV_{max} values to the total ligand-regulated GTPase activity ($V_{max\,(HA)} - V_{max(THIO)}$) yields Fig. 28.6B. Now the percentage of constitutive activity, which is eliminated by the inverse agonist, can be calculated. Figure 28.6B reveals that in case of hH_4R (coexpressed with Gα_{i2} and G$\beta_1\gamma_2$) the THIO-sensitive constitutive activity amounts to 40% of the total ligand-regulated GTPase activity (gray area).

In analogy to the calculation of the coupling factor from [^{35}S]GTPγS saturation binding data (cf. Section 3.2.3), the so-called turnover number can be calculated from the steady-state GTPase data. The ratio of B_{max} from the binding assay and ΔV_{max} from the steady-state GTPase enzyme kinetics is $\sim$4 min^{-1}, which means that one hH_4R molecule can stimulate the Gα_{i2}-catalyzed hydrolysis of up to four GTP molecules per minute (Schneider *et al.*, 2009).

3.3.3.2. Effect of sodium on constitutive activity As shown in the diagram of the two-state model of GPCR activation (Fig. 28.1), sodium ions can act like a full inverse agonist and stabilize the inactive state of a receptor. In case of FPR (coexpressed in Sf9 cells with Gα_{i2} and G$\beta_1\gamma_2$), 50 mM of NaCl are sufficient to completely eliminate constitutive activity (Seifert and Wenzel-Seifert, 2001b). By contrast, the potency of NaCl at reducing constitutive activity of hH_4R is much weaker. Even at NaCl concentrations higher than 100 mM, agonist-independent activity is preserved (gray area in Fig. 28.6C). The Na^+ insensitivity of the R state of the hH_4R supports the conclusion drawn from the radioligand binding assays in the presence and absence of GTPγS, namely that there is a G-protein independent high-affinity state of hH_4R. Moreover, the differences in NaCl sensitivity of hH_4R and hFPR suggest that the constitutive activity of these two receptors is caused by distinct mechanisms.

The hH_4R inverse agonist THIO was long considered a full inverse agonist, since it reduces constitutive hH_4R activity in reporter gene assays with hH_4R-transfected SK-N-MC cells to the same level as a treatment with pertussis toxin (PTX; Lim *et al.*, 2005). However, Fig. 28.6C shows that, even in the presence of saturating concentrations of THIO, sodium chloride is still able to cause a decrease in constitutive hH_4R activity.

This indicates that THIO is not a full inverse agonist. In fact, some hH_4R ligands with a higher inverse agonistic efficacy than THIO have already been identified (Deml *et al.*, 2009; Smits *et al.*, 2008).

3.3.3.3. Functional characterization of GPCR ligands Figure 28.6C shows that in the absence of sodium chloride the effect of saturating concentrations of the full agonist HA is only very low, which precludes the characterization of agonistic ligands in concentration–response curves. However, after addition of 100 m*M* of sodium chloride, the THIO-sensitive constitutive activity and the HA-induced GTPase activity are comparably high (Fig. 28.6B and C). Under these conditions, it becomes possible to characterize both agonists and inverse agonists in steady-state GTPase assays. Figure 28.6D shows the effect of increasing concentrations of the model compounds HA and THIO on hH_4R-stimulated $G\alpha_{i2}$ GTPase activity in the presence of 100 m*M* of sodium chloride. Again, the THIO-sensitive constitutive activity is highlighted in gray. It should be noted that the data in Fig. 28.6D were obtained in the presence of 100 n*M* of GTP. Figure 28.6B shows that the signal intensity of both agonist-stimulated and inverse agonist-inhibited GTPase activity can be enhanced by increasing the GTP concentration. However, one has to bear in mind that this may also cause a rightward shift of concentration–effect curves and an increase of EC_{50} values. Thus, the GTP-concentration should be kept constant, when the results of different steady-state GTPase assays have to be compared. In a recent publication, the inverse agonistic activity of a series of hH_4R antagonists structurally related to JNJ-7777120 was determined in steady-state GTPase assays with membranes coexpressing $hH_4R + G\alpha_{i2} + G\beta_1\gamma_2$ (Schneider *et al.*, 2010).

3.3.3.4. Determination of G-protein specificity of hH_4R Coexpression of GPCRs with signal transduction proteins in Sf9 cells provides the unique possibility to flexibly combine receptors with different $G\alpha$ and $G\beta\gamma$ subunits. In standard mammalian expression systems, it is very difficult to investigate GPCR–$G\alpha$ coupling selectivity, specifically when discrimination between different $G\alpha_i$ isoforms is intended. Normally, this is only possible by inactivating endogenous $G\alpha_i$ with PTX and coexpressing the GPCR with the respective PTX-resistant $G\alpha_i$ mutant. By contrast, Sf9 cells express only one $G\alpha_i$ isoform, which does not interact with mammalian GPCRs like FPR or hH_4R (Quehenberger *et al.*, 1992; Schneider *et al.*, 2009). Thus, the pharmacology of these receptors can be characterized in the presence of specific coexpressed $G\alpha$-subunits. Figure 28.7 shows that the constitutive activity of hH_4R is the highest in the presence of $G\alpha_{i1}$ and $G\alpha_{i2}$, followed by $G\alpha_{i3}$ and $G\alpha_o$. If hH_4R is expressed in the absence of mammalian G-proteins, only a very low, if any, HA-stimulated or THIO-inhibited signal can be detected.

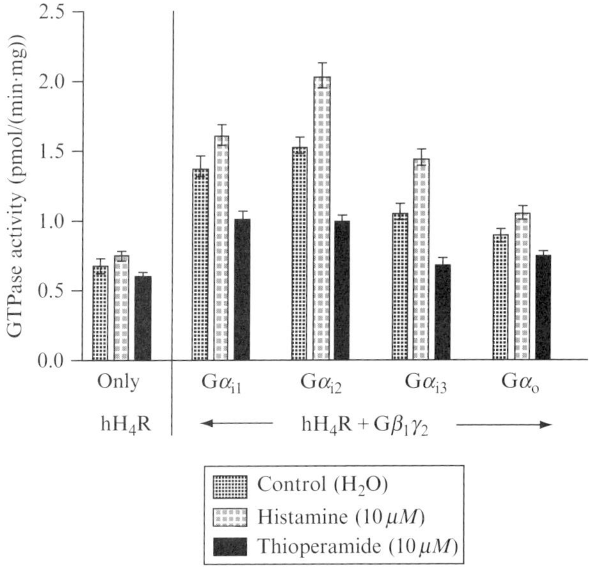

Figure 28.7 G-protein selectivity of hH_4R, determined in the steady-state GTPase assay. The hH_4R was expressed in the absence of mammalian G-proteins or in the presence of $G\beta_1\gamma_2$, combined with $G\alpha_{i1}$, $G\alpha_{i2}$, $G\alpha_{i3}$, or $G\alpha_o$. The first bar of each triplet shows control conditions (H_2O), the second bar indicates the HA-induced signal (10 μ*M* of HA), and the third bar depicts the THIO-inhibited signal (10 μ*M* of THIO). All assays were performed in the presence of 100 n*M* of GTP and 100 m*M* of NaCl. Data are from two independent experiments with three to four replicates, mean ± S.E.M. Adapted with permission of Schneider *et al.* (2009). Copyright 2009 American Chemical Society.

3.3.3.5. Coexpression of RGS proteins to enhance the steady-state GTPase signal RGS proteins (regulators of G-protein signaling) are classified in eight subfamilies, showing high structural diversity. A common feature is the RGS domain, which consists of 120 amino acids and interacts with Gα subunits. This increases the intrinsic GTPase activity of Gα (Willars, 2006). Thus, coexpressed RGS-proteins can help to enhance signal intensity in steady-state GTPase assays (Fig. 28.2, the RGS protein is acting in step 5 of the G-protein cycle). This is specifically useful for receptors, which stimulate Gα only sluggishly or show such a high constitutive activity that the relative agonist-induced signal is only very low. Since both is the case for hH_4R, we coexpressed the RGS-proteins RGS4 or GAIP together with hH_4R, $G\alpha_{i2}$, and $G\beta_1\gamma_2$ by performing quadruple infections of Sf9 cells. The corresponding Western blot in Fig. 28.3C shows the presence of RGS4 and GAIP in these membranes. However, as recently reported (Schneider and Seifert, 2009), the baseline GTPase activity was not significantly increased in membranes coexpressing hH_4R, G-proteins, and RGS4 or GAIP, compared to the RGS-free membranes. Only in the coexpression system containing RGS4 the relative THIO-induced signal was significantly increased. However, as described in chapter 25, in case of hH_4R, RGS-proteins are much more effective, when they are fused to the C-terminus of the receptor (Schneider and Seifert, 2009).

A more successful application of a coexpression system, containing GPCR and RGS-proteins was reported for the histamine H_1R (Houston *et al.*, 2002). When only the hH_1R or gpH_1R protein is expressed in Sf9 cells, there is no productive interaction with endogenous $G\alpha_q$ proteins. However, after coexpression of RGS4 or GAIP with gpH_1R or hH_1R, an interaction of H_1R with endogenous $G\alpha_q$ in insect cells was unmasked. This resulted in agonist-induced signals of up to 174% (gpH_1R + RGS4; Houston *et al.*, 2002).

However, it should always be considered that RGS proteins do not only stimulate the GTPase activity of $G\alpha$ subunits, but may form signaling complexes consisting of agonist, GPCR, G-protein, and RGS-protein (Benians *et al.*, 2005). Moreover, it is reported that some RGS-proteins, for example, RGS4, can directly interact with GPCRs (Zeng *et al.*, 1998). Thus, RGS-proteins act in a very complex way. If they are used to increase the sensitivity of a test system in the GTPase assay, it has to be ensured that they do not change the pharmacological properties of the receptor.

3.4. Adenylyl cyclase assay (AC-assay)

Measurement of adenylyl cyclase (AC) activity is less important in case of $G\alpha_i$-coupled receptors, since the inhibitory effect of most $G\alpha_i$-coupled receptors on AC is only low, even after prestimulation of the system with forskolin. This is particularly true for the Sf9 insect cell expression system (Seifert *et al.*, 2002). Thus, AC assays do not play a role for the characterization of the constitutive activity of hH_4R, which is used in this chapter as a paradigm for a receptor fused to $G\alpha_i$ or RGS-proteins. However, AC assays are highly important for the investigation of $G\alpha_s$-coupled receptors like the β_2AR. A detailed description of the AC assay has been provided in Seifert and Wieland (2005).

4. Application to Other Receptors

In this chapter, we focused on the investigation of constitutive activity of the hH_4R, coexpressed with $G\alpha_{i2}$ and $G\beta_1\gamma_2$ in Sf9 cells. We also discussed the usefulness of RGS proteins to enhance the steady-state GTPase signal in some systems, for example, by unmasking an interaction between the hH_1R and endogenous insect cell G-proteins. In the following, we briefly discuss that such Sf9 cell coexpression systems were also successfully used to investigate the constitutive activity of other GPCRs.

One of the constitutively active receptors that were most extensively studied in Sf9 cell coexpression systems is the human FPR. The FPR expressed alone in Sf9 cells is not coupling to endogenous G-proteins (Quehenberger *et al.*, 1992). Coexpression with only $G\alpha_{i2}$ or $G\beta_1\gamma_2$ did not improve signal intensity. However, once coexpressed with both $G\alpha_{i2}$ and $G\beta_1\gamma_2$, FPR shows extraordinarily high constitutive activity in [^{35}S]GTPγS assays. Moreover, high-affinity agonist binding was sensitive to GTPgS. (cf. Fig. 28.4B; Wenzel-Seifert *et al.*, 1998). In contrast to hH_4R, for which the [^{35}S]GTPγS binding kinetics can be hardly saturated, even after 240 min of incubation, the maximum signal of FPR (coexpressed with $G\alpha_{i2}$ and $G\beta_1\gamma_2$) is already reached after 60 min.

In addition to the studies performed with the wild-type FPR, also FPR mutations were coexpressed with $G\alpha_{i2}$ and $G\beta_1\gamma_2$ and characterized in Sf9 cells. FPR-F110S and -C126W, two naturally occurring FPR mutants associated with juvenile periodontitis showed almost complete (F110S) or complete (C126W) loss of $G\alpha_i$–protein coupling (Seifert and Wenzel-Seifert, 2001a). The role of N-terminal FPR glycosylation for signaling was investigated with mutants, where the asparagine residues in positions 4, 10, and 179 or only in positions 4 and 10 were exchanged against glutamine residues (Wenzel-Seifert and Seifert, 2003a). The isoforms FPR-98 (L101, N192, A346) and FPR-G6 (V101, K192, A346) were coexpressed with $G\alpha_{i2}$ and $G\beta_1\gamma_2$ in Sf9 cells and compared with FPR-26 (V101, N192, E346) in high-affinity agonist binding and [^{35}S]GTPγS binding assays (Wenzel-Seifert and Seifert, 2003b). Among all FPR isoforms examined, FPR-26 possesses the highest constitutive activity.

Cyclosporin H acts as an inverse agonist at FPR coexpressed with $G\alpha_{i2}$ and $G\beta_1\gamma_2$ in Sf9 cells (Wenzel-Seifert *et al.*, 1998), but does not reduce basal $G\alpha_i$-protein activity in FPR expressing HL-60 cells (Wenzel-Seifert and Seifert, 1993). In fact, HL-60 cells also express complement C5a-, leukotriene B4 (LTB_4)-, and platelet-activating factor (PAF) receptors that may compensate for the inhibition of FPR constitutive activity. By separately expressing C5aR, LTB_4R, and PAFR (always together with $G\alpha_{i2}$ and $G\beta_1\gamma_2$) in Sf9 cells, pronounced constitutive activity of C5aR was unmasked (Seifert and Wenzel-Seifert, 2001b). FPR and C5aR showed higher constitutive activity than LTB_4R and PAFR and NaCl suppressed basal G-protein activity with different potencies. No structural instability of these constitutively active receptors was observed (Seifert and Wenzel-Seifert, 2001b).

Another example for a GPCR, of which the constitutive activity was characterized in a coexpression system in Sf9 cells, is the chemokine receptor CXCR4. In the presence of $G\alpha_{i2}$ and $G\beta_1\gamma_2$, no evidence for constitutive activity of CXCR4 was found. This indicates that the constitutive activity of GPCRs in the Sf9 cell coexpression system is in fact an individual feature of the investigated receptors and not only caused by the high protein expression levels.

ACKNOWLEDGMENTS

We thank Drs Katharina Wenzel-Seifert and David Schnell for fruitful collaboration and stimulating discussions and Mrs A. Seefeld and G. Wilberg for expert technical assistance. This work was supported by the Graduate Training Program GRK 760: (Medicinal Chemistry: Molecular Recognition—Ligand–Receptor-Interactions) and GRK 1441 (Regulation of allergic inflammation in lung and skin) of the Deutsche Forschungsgemeinschaft. This work was supported by the European Cooperation in the Field of Scientific and Technical Research (COST) action #BM0806 ("Recent advances in histamine receptor H_4R research"), funded by the European Comission (7th European Framework Program).

REFERENCES

Barr, A. J., Brass, L. F., and Manning, D. R. (1997). Reconstitution of receptors and GTP-binding regulatory proteins (G proteins) in Sf9 cells. A direct evaluation of selectivity in receptor G protein coupling. *J. Biol. Chem.* **272,** 2223–2229.

Benians, A., Nobles, M., Hosny, S., and Tinker, A. (2005). Regulators of G-protein signaling form a quaternary complex with the agonist, receptor, and G-protein. A novel explanation for the acceleration of signaling activation kinetics. *J. Biol. Chem.* **280,** 13383–13394.

Deml, K. F., Beermann, S., Neumann, D., Strasser, A., and Seifert, R. (2009). Interactions of histamine H_1-receptor agonists and antagonists with the human histamine H_4-receptor. *Mol. Pharmacol.* **76,** 1019–1030.

Gether, U., Ballesteros, J. A., Seifert, R., Sanders-Bush, E., Weinstein, H., and Kobilka, B. K. (1997). Structural instability of a constitutively active G protein-coupled receptor. Agonist-independent activation due to conformational flexibility. *J. Biol. Chem.* **272,** 2587–2590.

Hartman, J. I., and Northup, J. K. (1996). Functional reconstitution in situ of 5-hydroxytryptamine$_{2c}$ ($5HT_{2c}$) receptors with α_q and inverse agonism of $5HT_{2c}$ receptor antagonists. *J. Biol. Chem.* **271,** 22591–22597.

Houston, C., Wenzel-Seifert, K., Bürckstümmer, T., and Seifert, R. (2002). The human histamine H_2-receptor couples more efficiently to Sf9 insect cell G_s-proteins than to insect cell G_q-proteins: Limitations of Sf9 cells for the analysis of receptor/G_q-protein coupling. *J. Neurochem.* **80,** 678–696.

Kleemann, P., Papa, D., Vigil-Cruz, S., and Seifert, R. (2008). Functional reconstitution of the human chemokine receptor CXCR4 with G_i/G_o-proteins in Sf9 insect cells. *Naunyn Schmiedebergs Arch. Pharmacol.* **378,** 261–274.

Lim, H. D., van Rijn, R. M., Ling, P., Bakker, R. A., Thurmond, R. L., and Leurs, R. (2005). Evaluation of histamine H_1-, H_2-, and H_3-receptor ligands at the human histamine H_4 receptor: Identification of 4-methylhistamine as the first potent and selective H_4 receptor agonist. *J. Pharmacol. Exp. Ther.* **314,** 1310–1321.

Nishimura, K., Frederick, J., and Kwatra, M. M. (1998). Human substance P receptor expressed in Sf9 cells couples with multiple endogenous G proteins. *J. Recept. Signal Transduct. Res.* **18,** 51–65.

Quehenberger, O., Prossnitz, E. R., Cochrane, C. G., and Ye, R. D. (1992). Absence of G_i proteins in the Sf9 insect cell. Characterization of the uncoupled recombinant N-formyl peptide receptor. *J. Biol. Chem.* **267,** 19757–19760.

Schneider, E. H., and Seifert, R. (2009). Histamine H_4 receptor–RGS fusion proteins expressed in Sf9 insect cells: A sensitive and reliable approach for the functional characterization of histamine H_4 receptor ligands. *Biochem. Pharmacol.* **78,** 607–616.

Schneider, E. H., and Seifert, R. (2010). Sf9 cells: A versatile model system to investigate the pharmacological properties of G protein-coupled receptors. *Pharmacol. Ther.* [Epub ahead of print] doi:10.1016/j.pharmthera.2010.07.005.

Schneider, E. H., Schnell, D., Papa, D., and Seifert, R. (2009). High constitutive activity and a G-protein-independent high-affinity state of the human histamine H_4-receptor. *Biochemistry* **48,** 1424–1438.

Schneider, E. H., Strasser, A., Thurmond, R. L., and Seifert, R. (2010). Structural requirements for inverse agonism and neutral antagonism of indole-, benzimidazole- and thienopyrrole-derived histamine H_4R ligands. *J. Pharmacol. Exp. Ther.* **334,** 513–521.

Seifert, R., and Wenzel-Seifert, K. (2001a). Defective G_i protein coupling in two formyl peptide receptor mutants associated with localized juvenile periodontitis. *J. Biol. Chem.* **276,** 42043–42049.

Seifert, R., and Wenzel-Seifert, K. (2001b). Unmasking different constitutive activity of four chemoattractant receptors using Na^+ as universal stabilizer of the inactive (R) state. *Recept. Channels* **7,** 357–369.

Seifert, R., and Wenzel-Seifert, K. (2002). Constitutive activity of G-protein-coupled receptors: Cause of disease and common property of wild-type receptors. *Naunyn Schmiedebergs Arch. Pharmacol.* **366,** 381–416.

Seifert, R., and Wieland, T. (2005). G Protein-Coupled Receptors as Drug Targets: Analysis of Activation and Constitutive Activity. Wiley-VCH, Weinheim.

Seifert, R., Wenzel-Seifert, K., Arthur, J. M., Jose, P. O., and Kobilka, B. K. (2002). Efficient adenylyl cyclase activation by a β_2-adrenoceptor–$G_i\alpha_2$ fusion protein. *Biochem. Biophys. Res. Commun.* **298,** 824–828.

Smits, R. A., de Esch, I. J., Zuiderveld, O. P., Broeker, J., Sansuk, K., Guaita, E., Coruzzi, G., Adami, M., Haaksma, E., and Leurs, R. (2008). Discovery of quinazolines as histamine H_4 receptor inverse agonists using a scaffold hopping approach. *J. Med. Chem.* **51,** 7855–7865.

Walseth, T. F., and Johnson, R. A. (1979). The enzymatic preparation of [α-^{32}P]nucleoside triphosphates, cyclic [^{32}P] AMP, and cyclic [^{32}P] GMP. *Biochim. Biophys. Acta* **562,** 11–31.

Wenzel-Seifert, K., and Seifert, R. (1993). Cyclosporin H is a potent and selective formyl peptide receptor antagonist. Comparison with N-t-butoxycarbonyl-L-phenylalanyl-L-leucyl-L-phenylalanyl-L-leucyl-L-phenylalanine and cyclosporins A, B, C, D, and E. *J. Immunol.* **150,** 4591–4599.

Wenzel-Seifert, K., and Seifert, R. (2003a). Critical role of N-terminal N-glycosylation for proper folding of the human formyl peptide receptor. *Biochem. Biophys. Res. Commun.* **301,** 693–698.

Wenzel-Seifert, K., and Seifert, R. (2003b). Functional differences between human formyl peptide receptor isoforms 26, 98, and G6. *Naunyn Schmiedebergs Arch. Pharmacol.* **367,** 509–515.

Wenzel-Seifert, K., Hurt, C. M., and Seifert, R. (1998). High constitutive activity of the human formyl peptide receptor. *J. Biol. Chem.* **273,** 24181–24189.

Wenzel-Seifert, K., Arthur, J. M., Liu, H. Y., and Seifert, R. (1999). Quantitative analysis of formyl peptide receptor coupling to $G_i\alpha_1$, $G_i\alpha_2$, and $G_i\alpha_3$. *J. Biol. Chem.* **274,** 33259–33266.

Willars, G. B. (2006). Mammalian RGS proteins: Multifunctional regulators of cellular signalling. *Semin. Cell Dev. Biol.* **17,** 363–376.

Zeng, W., Xu, X., Popov, S., Mukhopadhyay, S., Chidiac, P., Swistok, J., Danho, W., Yagaloff, K. A., Fisher, S. L., Ross, E. M., Muallem, S., and Wilkie, T. M. (1998). The N-terminal domain of RGS4 confers receptor-selective inhibition of G protein signaling. *J. Biol. Chem.* **273,** 34687–34690.

CHAPTER TWENTY-NINE

Modeling and Simulation of Inverse Agonism Dynamics

L. J. Bridge

Contents

Abstract

With the recent discovery and increased recognition of constitutive activity of G-protein coupled receptors (GPCRs) and inverse agonists have come a number of important questions. The signaling mechanisms underlying inverse agonist effects on constitutively active systems need to be elucidated qualitatively. Furthermore, quantitative analysis is needed to support experimental observations, characterize the pharmacology of the ligands and systems of interest, and to provide numerical predictions of dynamic physiological responses to inverse agonists in an effort toward drug design. Here, we review the concept of inverse agonism and describe the application of mathematical and computational techniques to models of inverse agonists in GPCR systems. Numerical simulation results for active G-protein levels demonstrate a variety of dynamic features including inhibition of agonist-induced peak–plateau responses, undershoots, multiple time scales, and both surmountable and insurmountable inverse agonism.

Centre for Mathematical Medicine and Biology, School of Mathematical Sciences, University of Nottingham, United Kingdom

Methods in Enzymology, Volume 485
ISSN 0076-6879, DOI: 10.1016/S0076-6879(10)85029-3

1. Introduction

In the past two decades, there has been increasing recognition and study of the capacity of G-protein coupled receptor (GPCR) systems to exhibit constitutive activity. This concept of constitutive activity refers to the ability of a receptor to produce a response in the absence of an agonist. Adopting a typical two-state (Bridge *et al.*, 2010) or multiple-state (Leff *et al.*, 1997) model of receptor activation whereby a receptor may exist in an active or inactive confirmation, it is useful here for us to immediately give the following definitions:

- An *agonist* is a ligand that binds with higher affinity to active receptors over inactive ones. Equally, it is a ligand that, when bound to receptor, increases the propensity for receptor activation.
- An *antagonist*, or *neutral antagonist*, is a ligand that binds with equal affinity to both active receptors and inactive ones. Equally, it is a ligand that, when bound to receptor, does not affect the propensity for receptor activation.
- An *inverse agonist* is a ligand that binds with higher affinity to inactive receptors over active ones. Equally, it is a ligand that, when bound to receptor, decreases the propensity for receptor activation.

With these definitions in place, we see that the basal activity in a ligand-free system is thus constitutive activity, and an agonist will induce a response that is above basal. If we add the loose definition of efficacy as "the extent of functional change imparted to receptor" (Gilchrist, 2007), then we see that the term "negative efficacy ligand" found in the literature simply means an inverse agonist.

The concept of an inverse agonist had been neglected for many years, despite being implicit (or even parametrically explicit) in widely adopted receptor theory, and many ligands previously assumed to be antagonists of certain receptors were later found to be inverse agonists (Costa and Cotecchia, 2005). If an antagonist competes with an agonist for receptor binding sites, then the effect on agonist-induced signaling is typically a decrease in response. Such a decrease in response is not sufficient, however, to classify an inhibitory competing ligand; with the increased understanding of constitutively active systems has come the reclassification of many GPCR-targeting therapeutic drugs as inverse agonists, as also noted in other reviews (Bond and IJzerman, 2006; Parra and Bond, 2007; Strange, 2002). It is the effect of the inhibitory ligand on the agonist-independent activity that is key. If, upon introduction of a ligand, the basal ligand-free response is noticeably reduced, then the ligand is an inverse agonist (Strange, 2002). In Negus (2006), assays to detect basal activity and identify inverse agonist action are discussed, together with the need to distinguish between basal activity levels ("tones") which are due to constitutive receptor activity, endogenous agonists, or both. The mathematical models which we present later may be applied to any of these situations in order to quantify inverse agonist effects.

With the increased awareness of the concept of inverse agonism has come great effort to detect the effects it may have on signal transduction. In theory, any preparation used to measure agonist-induced responses can also be used to study inhibition by an inverse agonist, assuming some measurable level of constitutive activity which is clear above background noise (Chidiac, 2002). Whatever the level at which inverse agonist action is assessed (receptor, G-protein, or downstream), if the basal response is small relative to the noise in the system, then the inverse agonist-induced change in response will be difficult to detect. Thus, with little or no constitutive activity, an inverse agonist will behave as an antagonist (Gilchrist, 2007). As such, a sensible strategy for *in vitro* experiments designed to detect inverse agonism is to maximize receptor expression in order to increase constitutive activity. Total receptor and G-protein levels are crucial parameters in effecting measurable basal responses (Chidiac, 2002; Parra and Bond, 2007), and mathematical modeling has been used to explain this in detail (Woodroffe *et al.*, 2010a).

In addition to studies aimed at understanding the fundamentals of inverse agonist mechanism, and also its molecular basis (Vilardaga *et al.*, 2005), researchers are naturally questioning the clinical and therapeutic significance of these drugs. Association of constitutive activity with certain diseases suggests that negative efficacy might well be an important drug property (Kenakin, 2005). Antagonist and inverse agonist properties are drug design considerations when (i) the targeted disease is being produced by a mutation of the receptor, resulting in increased constitutive activity and (ii) when the disease is due to excess endogenous agonist. In (i), an antagonist is of no use if there is no hormone to block, and an inverse agonist is required, whereas an antagonist may be an effective treatment in case (ii) (Bond and IJzerman, 2006). Despite this intuitive theoretical argument, the potential therapeutic differences between inverse agonists and antagonists and the impact of this theory on the drug discovery process have remained unclear. Both Parra and Bond (2007) and Strange (2002) ask whether there is really any clinical difference between the two ligand types, while Bond and IJzerman (2006) suggest that perhaps the primary question in the field, now that constitutive activity and inverse agonism have been experimentally established, is whether these ideas are physiologically relevant after all.

Both qualitative and quantitative questions thus remain concerning the fundamentals of inverse agonist signaling and its physiological significance, and mathematical modeling and scientific computing can provide powerful tools toward answering these questions. Some preliminary models have been presented for quantifying the effects of inverse agonism. In Chidiac

(2002), a two-state receptor model is described, including receptor desensitization. As is common in cell signaling and pharmacological modeling studies, the analysis is at equilibrium; this is also true of the extended cubic ternary complex model analysis in Negus (2006), the simplified model in Leurs *et al.* (2000), and the inverse agonism mechanism and consitutively active mutant models in Strange (2002).

Steady-state results of course provide only a subset of the information which one can gain from a dynamic model. Where time-course data are available from GPCR signaling experiments, the use of dynamic models will aid attempts to understand the signaling kinetics which lead to dynamic responses of interest, and provide an insight into the mechanisms involved. Timing of ligand-induced GPCR-mediated responses is modeled in Shea *et al.* (2000), and more recent studies have employed numerical and asymptotic methods to identify key reaction sequences over the multiple time scales of G-protein activation responses. In particular, the influence of system parameters including receptor concentration and precoupling rates on purely constitutive activity is analyzed in Woodroffe *et al.* (2010a), and agonist-induced "peak–plateau" responses are studied in Woodroffe *et al.* (2010b). Some existing explanations of antagonist behavior are discussed in Lew *et al.* (2000), together with the need for a full kinetic model. This need has been addressed in Bridge *et al.* (2010), with the development of a model for agonist–antagonist competition, where the inhibitory effect of the antagonist on agonist-induced dynamic responses is characterized as surmountable or insurmountable under various parameter regimes.

In this work, we show how the models and numerical methods developed in Woodroffe *et al.* (2010b) and Bridge *et al.* (2010) may be applied to study the effects of inverse agonism. In Strange (2002), the idea of a *spectrum of efficacy*, characterizing drugs from inverse agonist, through antagonist and partial agonist, to full agonist is discussed. For most GPCRs this spectrum is exhibited, and we can imagine a model containing a parametric description of ligand types. The measure of efficacy for our model, ζ, is defined in Section 2.1.1, and it is this parameter that we vary in order to examine inverse agonism rather than just agonism and antagonism.

This chapter now proceeds with a discussion of the temporal characteristics and types of dynamic responses that mathematical modeling predicts for inverse agonist action for the case of a single ligand inducing a change in response from the basal (constitutively active) level, and in the case of two ligands in competition for GPCR receptor sites. Our goal here is to be illustrative rather than exhaustive; while the results here are not intended to be numerically accurate for any particular ligand–receptor system, we aim to demonstrate the power of mathematical modeling and computation as a tool for the analysis of a variety of response features, and to provide the reader with a framework for the study of inverse agonism dynamics.

2. Single-Ligand Analysis

2.1. Model formulation

The formulation and analysis which follows is based upon the cubic ternary complex framework which has been widely adopted by authors including (de Lean *et al.*, 1980; Kenakin, 2001; Shea *et al.*, 2000). First, we consider the problem of a ligand L binding to a GPCR, which may also be bound by a G-protein G. The receptor is assumed to be able to exist either in an inactive state R or in an active one $R^\star$. We now describe the development of the single-ligand model of Woodroffe *et al.* (2010b) for all possible receptor states. We label G-protein bound inactive receptor RG, ligand-bound, G-protein bound active receptor $LR^\star G$, and so on. Downstream signaling is assumed to continue in the presence of active G-protein α subunits with GTP attached. Dissociation of the G-protein into α and β_γ subunits may take place from either of the complexes $R^\star G$ (the active precoupled complex which dissociates to give a signal in a constitutively active system) or $LR^\star G$. The dissociation and exchange of α_{GDP} for α_{GTP} is treated as a single step, and it is the concentration of α_{GTP} that is taken as the response of interest. The full ternary complex model thus consists of eight receptor states and the G-protein cycle, as shown schematically in Fig. 29.1.

The constants K_L, K_P, and K_G are equilibrium rate constants, and the factors μ, ν, and ζ are microaffinity constants, included in a way that ensures thermodynamic detailed balance.

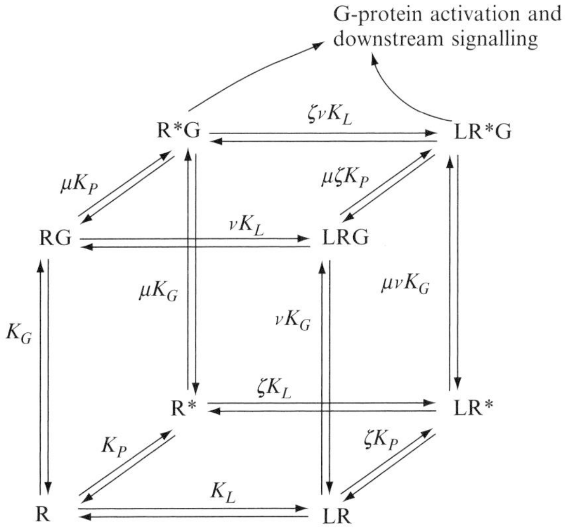

Figure 29.1 Cubic ternary complex model with eight receptor states.

2.1.1. The "efficacy" parameter ζ

The parameter ζ measures the preference of the ligand for active over inactive receptor. Equally, it measures the relative propensity for receptor activation when the receptor is ligand bound rather than free. As such, the value of ζ controls whether the ligand L is an agonist, antagonist, or inverse agonist, with $\zeta > 1$ signifying an agonist, $\zeta = 1$ an antagonist, and $0 < \zeta < 1$ an inverse agonist. The dynamics of α_{GTP} signals in response to agonist and antagonist stimuli have been studied in detail (Bridge *et al.*, 2010; Woodroffe *et al.*, 2010b), and we use these methods here to show how inverse agonist effects may be analyzed numerically in a similar fashion. The magnitude of ζ is clearly a measure of efficacy or "strength" of the ligand, and we pay particular attention to varying this parameter.

It is worth noting here that considering a continuous spectrum of efficacy, it is perhaps unlikely that we would ever find a ligand with exactly a desired numerical value of ζ. It is not surprising to read that precisely neutral antagonists are likely to be rare (Milligan, 2003).

2.1.2. Reaction list

The next step toward formulating a mathematical description of the signaling system dynamics is to list the reactions involved. Both forward and backward rates are included in all the reactions, and further constants $\theta_x (x = \mu, \nu, \zeta, \nu\mu, \zeta\mu, \zeta\nu)$ are included to account for the possibility that the thermodynamic effects may be proportionately greater in different reactions. This is discussed further in Woodroffe *et al.* (2010b). The reactions are as follows:

$$R \underset{k_{\mathrm{deact}}}{\overset{k_{\mathrm{act}}}{\rightleftharpoons}} R^*, L + R \underset{k_{l-}}{\overset{k_{l+}}{\rightleftharpoons}} LR, LR \underset{\zeta_- k_{\mathrm{deact}}}{\overset{\zeta_+ k_{\mathrm{act}}}{\rightleftharpoons}} LR^*, L + R^* \underset{\theta_\zeta \zeta_- k_{l-}}{\overset{\theta_\zeta \zeta_+ k_{l+}}{\rightleftharpoons}} LR^*, \quad (29.1a)$$

$$R + G \underset{k_{g-}}{\overset{k_{g+}}{\rightleftharpoons}} RG, R^* + G \underset{\theta_\mu \mu_- k_{g-}}{\overset{\theta_\mu \mu_+ k_{g+}}{\rightleftharpoons}} R^*G, RG \underset{\mu_- k_{\mathrm{deact}}}{\overset{\mu_+ k_{\mathrm{act}}}{\rightleftharpoons}} R^*G, \quad (29.1b)$$

$$LR + G \underset{\theta_\nu \nu_- k_{g-}}{\overset{\theta_\nu \nu_+ k_{g+}}{\rightleftharpoons}} LRG, LR^* + G \underset{\theta_{\nu\mu} \mu_- \nu_- k_{g-}}{\overset{\theta_{\nu\mu} \mu_+ \nu_+ k_{g+}}{\rightleftharpoons}} LR^*G, LRG \underset{\theta_{\zeta\mu} \zeta_- \mu_- k_{\mathrm{deact}}}{\overset{\theta_{\zeta\mu} \zeta_+ \mu_+ k_{\mathrm{act}}}{\rightleftharpoons}} LR^*G, \quad (29.1c)$$

$$L + RG \underset{\nu_- k_{l-}}{\overset{\nu_+ k_{l+}}{\rightleftharpoons}} LRG, L + R^*G \underset{\theta_{\zeta\mu} \zeta_- \mu_- k_{l-}}{\overset{\theta_{\zeta\mu} \zeta_+ \mu_+ k_{l+}}{\rightleftharpoons}} LR^*G, R^*G \underset{k_{\mathrm{GTP}-}}{\overset{k_{\mathrm{GTP}+}}{\rightleftharpoons}} R^* + \alpha_{\mathrm{GTP}} + \beta\gamma, \quad (29.1d)$$

$$LR^*G \underset{\theta_a \nu_+ k_{\mathrm{GTP}-}}{\overset{\theta_a \nu_- k_{\mathrm{GTP}+}}{\rightleftharpoons}} LR^* + \alpha_{\mathrm{GTP}} + \beta\gamma, \alpha_{\mathrm{GTP}} \underset{k_{gd-}}{\overset{k_{gd+}}{\rightleftharpoons}} \alpha_{\mathrm{GDP}}, \alpha_{\mathrm{GDP}} + \beta\gamma \underset{k_{RA-}}{\overset{k_{RA+}}{\rightleftharpoons}} G. \quad (29.1e)$$

While included above for completeness, the reverse reaction in the dissociation of $R^{\star}G$ complexes is assumed negligible (Woodroffe *et al.*, 2010a,b) and hence we shall take $k_{\mathrm{GTP}-} = 0$. The equilibrium constants and rate constants are related as follows:

$$K_P = \frac{k_{\mathrm{act}}}{k_{\mathrm{deact}}}, K_L = \frac{k_{l+}}{k_{l-}}, K_G = \frac{k_{g+}}{k_{g-}}, K_{Gd} = \frac{k_{gd+}}{k_{gd-}}, \tag{29.2a}$$

$$K_{RA} = \frac{k_{RA+}}{k_{RA-}}, \mu = \frac{\mu_+}{\mu_-}, \nu = \frac{\nu_+}{\nu_-}, \zeta = \frac{\zeta_+}{\zeta_-}. \tag{29.2b}$$

2.1.3. Mathematical model

The concentrations of the 12 species involved are governed by the following system of ordinary differential equations (ODEs), where the law of mass action has been applied to the reactions listed in Eqs. (29.1) to find derivatives with respect to time t:

$$\frac{\mathrm{d}[R]}{\mathrm{d}t} = k_{\mathrm{deact}}[R^*] - k_{\mathrm{act}}[R] + k_{l-}[LR] - k_{l+}[L][R] + k_{g-}[RG] - k_{g+}[R][G], \tag{29.3a}$$

$$\begin{aligned}\frac{\mathrm{d}[R^*]}{\mathrm{d}t} &= k_{\mathrm{act}}[R] - k_{\mathrm{deact}}[R^*] + \theta_{\zeta}\zeta_- k_{l-}[LR^*] \\ &\quad - \theta_{\zeta}\zeta_+ k_{l+}[L][R^*] + k_{\mathrm{GTP}+}[R^*G],\end{aligned} \tag{29.3b}$$

$$\begin{aligned}\frac{\mathrm{d}[LR]}{\mathrm{d}t} &= k_{l+}[L][R] - k_{l-}[LR] + \zeta_- k_{\mathrm{deact}}[LR^*] \\ &\quad -\zeta_+ k_{\mathrm{act}}[LR] + \theta_{\nu}\nu_- k_{g-}[LRG] - \theta_{\nu}\nu_+ k_{g+}[LR][G],\end{aligned} \tag{29.3c}$$

$$\begin{aligned}\frac{\mathrm{d}[LR^*]}{\mathrm{d}t} &= \zeta_+ k_{\mathrm{act}}[LR] - \zeta_- k_{\mathrm{deact}}[LR^*] + \theta_{\zeta}\zeta_+ k_{l+}[L][R^*] \\ &\quad - \theta_{\zeta}\zeta_- k_{l-}[LR^*] + \theta_{\mu\nu}\mu_-\nu_- k_{g-}[LR^*G] \\ &\quad - \theta_{\mu\nu}\mu_+\nu_+ k_{g+}[LR^*][G] + \theta_a\nu_- k_{\mathrm{GTP}+}[LR^*G],\end{aligned} \tag{29.3d}$$

$$\begin{aligned}\frac{\mathrm{d}[RG]}{\mathrm{d}t} &= k_{g+}[R][G] - k_{g-}[RG] + \mu_- k_{\mathrm{deact}}[R^*G] \\ &\quad -\mu_+ k_{\mathrm{act}}[RG] + \nu_- k_{l-}[LRG] - \nu_+ k_{l+}[L][RG],\end{aligned} \tag{29.3e}$$

$$\begin{aligned}\frac{\mathrm{d}[R^*G]}{\mathrm{d}t} &= \theta_{\mu}\mu_+ k_{g+}[R^*][G] - \theta_{\mu}\mu_- k_{g-}[R^*G] + \mu_+ k_{\mathrm{act}}[RG] \\ &\quad - \mu_- k_{\mathrm{deact}}[R^*G] + \theta_{\zeta\nu}\zeta_-\nu_- k_{l-}[LR^*G] \\ &\quad - \theta_{\zeta\nu}\zeta_+\nu_+ k_{l+}[L][R^*G] - k_{\mathrm{GTP}+}[R^*G],\end{aligned} \tag{29.3f}$$

$$\frac{\mathrm{d}[LRG]}{\mathrm{d}t} = \theta_v v_+ k_{g+}[LR][G] - \theta_v v_- k_{g-}[LRG] + \theta_{\zeta\mu}\zeta_-\mu_- k_{\mathrm{deact}}[LR^*G] - \theta_{\zeta\mu}\zeta_+\mu_+ k_{\mathrm{act}}[LRG] + v_+ k_{l+}[L][RG] - v_- k_{l-}[LRG], \tag{29.3g}$$

$$\frac{\mathrm{d}[LR^*G]}{\mathrm{d}t} = \theta_{\mu v}\mu_+ v_+ k_{g+}[LR^*][G] - \theta_{\mu v}\mu_- v_- k_{g-}[LR^*G] + \theta_{\zeta\mu}\zeta_+\mu_+ k_{\mathrm{act}}[LRG] - \theta_{\zeta\mu}\zeta_-\mu_- k_{\mathrm{deact}}[LR^*G] + \theta_{\zeta v}\zeta_+ v_+ k_{l+}[L][R^*G] - \theta_{\zeta v}\zeta_- v_- k_{l-}[LR^*G] - \theta_a v_- k_{\mathrm{GTP}+}[LR^*G], \tag{29.3h}$$

$$\frac{\mathrm{d}[\alpha_{\mathrm{GTP}}]}{\mathrm{d}t} = \theta_a v_- k_{\mathrm{GTP}+}[LR^*G] - \theta_a v_+ k_{\mathrm{GTP}-}[LR*][\alpha_{\mathrm{GTP}}][\beta\gamma] + k_{gd-}[\alpha_{\mathrm{GDP}}] - k_{gd+}[\alpha_{\mathrm{GTP}}] + k_{\mathrm{GTP}+}[R^*G], \tag{29.3i}$$

$$\frac{\mathrm{d}[\beta\gamma]}{\mathrm{d}t} = \theta_a v_- k_{\mathrm{GTP}+}[LR^*G] + k_{RA-}[G] - k_{RA+}[\alpha_{\mathrm{GDP}}][\beta\gamma] + k_{\mathrm{GTP}+}[R^*G], \tag{29.3j}$$

$$\frac{\mathrm{d}[\alpha_{\mathrm{GDP}}]}{\mathrm{d}t} = k_{gd+}[\alpha_{\mathrm{GTP}}] - k_{gd-}[\alpha_{\mathrm{GDP}}] + k_{RA-}[G] - k_{RA+}[\alpha_{\mathrm{GDP}}][\beta\gamma], \tag{29.3k}$$

$$\frac{\mathrm{d}[G]}{\mathrm{d}t} = k_{g-}[RG] - k_{g+}[R][G] + \theta_\mu \mu_- k_{g-}[R^*G] - \theta_\mu \mu_+ k_{g+}[R^*][G] + \theta_v v_- k_{g-}[LRG] - \theta_v v_+ k_{g+}[LR][G] + \theta_{\mu v}, \mu_- v_- k_{g-}[LR^*G] - \theta_{\mu v}\mu_+ v_+ k_{g+}[LR^*][G] + k_{RA+}[\alpha_{\mathrm{GDP}}][\beta\gamma] - k_{RA-}[G]. \tag{29.3l}$$

In order to explore the dynamics of Eq. (29.3) in response to a ligand dose $[L]$, we solve the system subject to initial conditions which are the drug-free ($[L] = 0$) levels of each species.

2.2. Numerical solution method

Consider a system with total receptor concentration R_{tot} and total G-protein concentration G_{tot}. To find the basal values of the 12 species' concentrations, we could attempt to set the left-hand sides of Eq. (29.3) equal to zero, and solve for the 12 variables analytically. In general, we would not expect this to be possible due to the nonlinearity in the system, and instead we choose to integrate the system numerically.

ODE solvers are widely available, both as code modules for programmers and as routines embedded in software packages and tools such as MATLAB® (The MathWorks, Inc.). Many other packages specific to systems and computational biology modeling have also recently been developed. All computations shown here have used the MATLAB® stiff ODE solver ode15s, which is ideal for solutions that evolve over various time scales.

The drug-free basal (constitutive) concentrations are found by solving Eq. (29.3) subject to an arbitrary initial condition with $[R] = R_{tot}$ and $[G] = G_{tot}$, until a steady-state has been reached. The steady-state values are then taken as the new initial conditions at time $t = 0$, when the ligand L is introduced, and the system is integrated again over a specified time interval. The ODE solver output is a time course for each of the species' concentrations, and we now proceed to describe particular features of the α_{GTP} dynamics in particular in response to varying ligand concentration $[L]$ and "efficacy" ζ. Unless otherwise stated, the results shown are for parameter values taken from Tables 29.1 and 29.2, and time t is in seconds. We mostly use the on/off rates for the antagonists given in Table 29.2, for both our antagonist and inverse agonist computations.

2.3. Numerical results

In this section, we show numerical results (for α_{GTP} levels) which illustrate both the variety of computations that we can perform and the range of dynamic features of responses to inverse agonism that our model can predict. These results are therefore intended to be illustrative of the methods rather than exhaustive or accurate predictions for any particular receptors or ligands.

2.3.1. Time courses

In Fig. 29.2(A), we plot time courses showing the different effects of agonist, antagonist, and inverse agonist. The ligand concentrations are such that the fold/basal level is low, but that the distinguishing features are clear. After the addition of an agonist, the α_{GTP} level rises to a peak on a time scale on the order of 100 s, and then decreases toward a plateau steady-state value. The mechanism for this behavior is described in Woodroffe *et al.* (2010b). There is no change in response upon addition of antagonist, which does not encourage receptor activation. The addition of an inverse agonist results in a decrease in response from the basal level. For the parameters used here, the response tends to a steady-state value which represents a reduction in constitutive activity of about a half, and this steady-state is approached over a timescale on the order of 10 min.

In Fig. 29.2(B), we show the effect of varying ζ from $\zeta \gg 1$ (strong agonist) to $\zeta \ll 1$ (strong inverse agonist). We see the agonist-induced

Table 29.1 Key quantities, taken from Bridge *et al.* (2010)

Parameter	Meaning	Value	Units
k_{act}	Receptor activation rate	1	s^{-1}
k_{deact}	Receptor deactivation rate	1×10^3	s^{-1}
k_{g-}	G-protein unbinding rate	0.1	$M^{-1}\ s^{-1}$
k_{g+}	G-protein binding rate	1×10^8	s^{-1}
k_{gd-}	α_{GTP} activation rate	1×10^{-4}	s^{-1}
k_{gd+}	α_{GTP} deactivation rate	0.01	s^{-1}
$k_{\mathrm{RA}-}$	G-protein dissociation rate	1.3×10^{-3}	s^{-1}
$k_{\mathrm{RA}+}$	G-protein reassociation rate	7×10^5	$M^{-1}\ s^{-1}$
$k_{\mathrm{GTP}+}$	$R^{\star}G$ dissociation rate	1	s^{-1}
μ	Preference of G for $R^{\star}$ over R	2	–
ν	Preference of L for RG over R	1	–
$\zeta + I$	Effect of ζ_I on forward reaction	1	–
$\theta_{\bullet}$	Thermodynamic scalings	1	–
ζ	Preference of L for $R^{\star}$ over R	1	–
ν_A	Preference of A for RG over R	1	–
ν_I	Preference of I for RG over R	1	–
ζ_{A}	Preference of A for $R^{\star}$ over R	1	–
ζ_I	Preference of I for $R^{\star}$ over R	1	–
$\mu+$	Effect of μ on forward reaction	1	–
$\mu-$	Effect of μ on backward reaction	0.5	–
$\nu+$	Effect of ν on forward reaction	1	–
$\nu-$	Effect of ν on backward reaction	1	–
$\zeta+$	Effect of ζ on forward reaction	1	–
$\zeta-$	Effect of ζ on backward reaction	1	–
ν_{+I}	Effect of ν_I on forward reaction	1	–
ν_{-I}	Effect of ν_I on backward reaction	1	–
ζ_{-I}	Effect of ζ_I on backward reaction	1	–
R_{tot}, G_{tot}	Total R, G concentration	4.15×10^{-10}	M

response change from a peak–plateau nature to a smaller plateau-only signal, and then the saturating inverse agonist effect.

2.3.2. The effect of inverse agonist strength and concentration on dynamics

In Fig. 29.3, we zoom in on time courses for inverse agonist only, and show the effects of inverse agonist strength ζ and concentration $[L]$ on the α_{GTP} response. Either increasing $[L]$ or decreasing ζ (<1), while keeping all other parameters fixed, has the effect of lowering the response. Also, as $[L]$ becomes very large or as ζ becomes very small, we see the approach to a limiting time course.

Table 29.2 Ligand binding rates

Agonist rates from Sykes *et al.* (2009)			Antagonist rates from Dowling and Charlton (2006)		
Agonist	k_{l+} ($M^{-1}s^{-1}$)	k_{l-} (s^{-1})	Antagonist	k_{l+} ($M^{-1}s^{-1}$)	k_{l-} (s^{-1})
Acetylcholine	3.4×10^3	0.11	Atropine	2.5×10^7	4.5×10^{-3}
Carbachol	6.9×10^2	0.14	Clidinium	1.8×10^6	4.8×10^{-4}
Methacholine	2.0×10^3	0.14	Ipratropium	8.0×10^6	1.2×10^{-3}
Oxotremorine-M	2.9×10^3	0.14	NMS	7.5×10^6	2.8×10^{-4}
Bethanachol	8.0×10^2	0.26	Tiotropium	2.7×10^6	2.5×10^{-5}
Oxotremorine	9.4×10^4	0.31			
Pilocarpine	1.0×10^4	0.35			

To summarize the effect of $[L]$ and ζ on the responses, we show in Fig. 29.4 plots of concentration–response relations for varying ζ. These relations will, of course, change dynamically, and we show results taken at $t = 60$ s (usually before a peak in the case of agonist) and at $t = 1000$ s (when the system is near to its steady-state). The larger response is seen at $t = 60$s, while α_{GTP} is on a sharp rise before peaking, and we see the usual shape of an agonist-induced dose–response curve, with a decreasing maximum value as ζ is decreased. At $t = 1000$ s, the response is smaller, and it is easier to see the decreasing dose–response curves for ligands which are inverse agonists ($\zeta < 1$).

We can use our models and numerical methods to examine a variety of features of the simulated α_{GTP} response. Closer inspection of the time courses in Fig. 29.3(B) reveals nonmonotonic responses for $\zeta = 0.5$ and $\zeta = 0.3$. There is an "undershoot" in the time course for weak inverse agonists, a feature that we do not see for the stronger inverse agonists. While the effect is small in this case, it is possible that the undershoot effect would be larger for different parameter sets. Defining the undershoot as the difference between the lowest α_{GTP} value and the steady-state value, we can see in Fig. 29.5 that the undershoot has a maximum magnitude of about 1% of the basal response. While this does not seem particularly significant, it is interesting that the effect is confined to a narrow region of (ζ, L)-space, which extends as the off-rate is increased. It is yet to be shown whether this undershoot effect appears in practice, but the mathematical tools are in place to investigate further.

3. Two Competing Ligands

Many drugs that had previously been thought to be antagonists have subsequently been shown to be inverse agonists. Clearly a model that allows analysis of the effects of antagonist versus inverse agonist parameters in

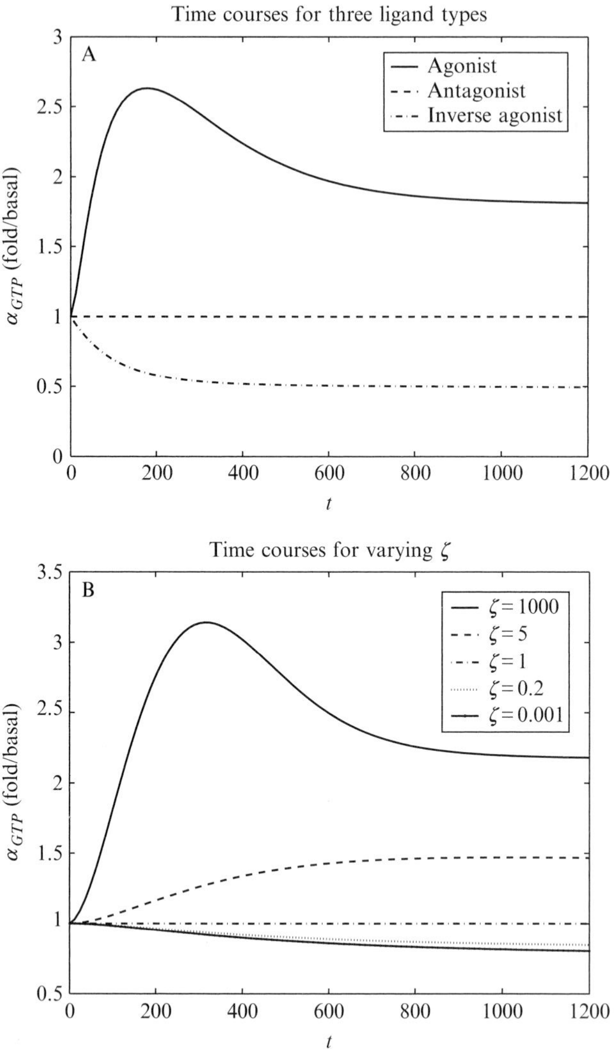

Figure 29.2 Time courses after the addition of agonist, antagonist, and inverse agonist. (A) For agonist, we take carbachol on/off rates ($k_{lb\pm}$), and $\zeta = 1000$, while for inverse agonist, we take NMS on/off rates and $\zeta = 0.001$. (B) We set the ligand on/off rates $k_{lb+} = 10^4\ M^{-1}\ s^{-1}$, $k_{lb} = 10^{-3}\ s^{-1}$.

systems where antagonism may previously have been assumed is a valuable tool toward understanding and predicting inverse agonism behavior. In particular, we can build on the model of the previous section by considering a system in which an inverse agonist competes with an agonist, to demonstrate the inhibitory effect on the agonist-induced response.

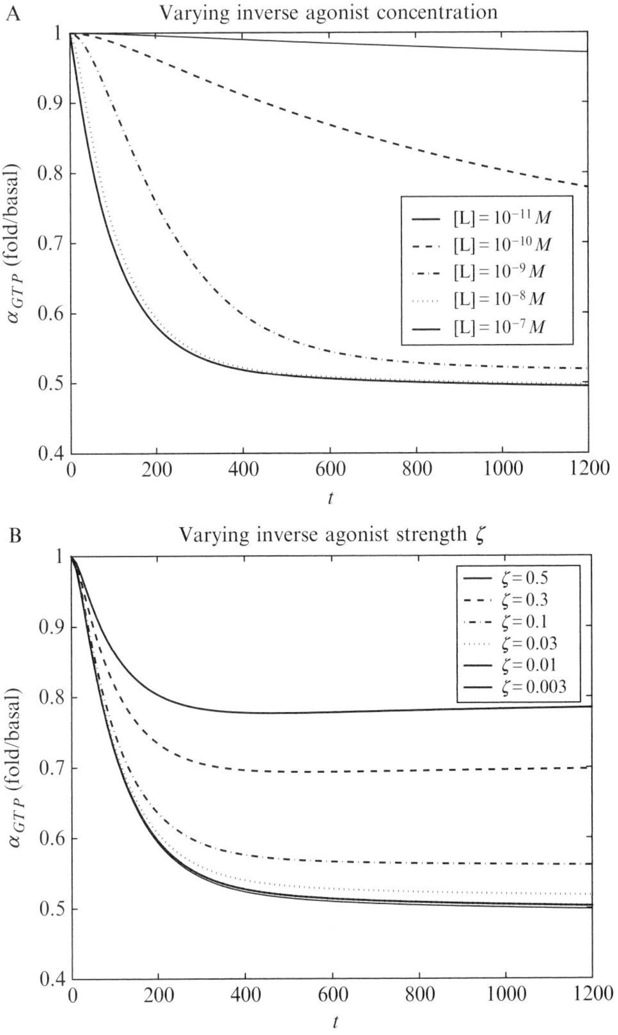

Figure 29.3 Time courses after the addition of inverse agonist. (A) Varying concentration $[L]$ of a strong inverse agonist with $\zeta = 0.001$, using NMS on/off rates. (B) Varying ζ for a fixed concentration $[L] = 10^{-8}$ M of an inverse agonist with NMS on/off rates.

The model of Bridge *et al.* (2010) for the dynamics of α_{GTP} in a system with agonist–antagonist competition provides an ideal framework for such a study, and we now proceed to apply this model to systems with an agonist A with $\zeta_A > 1$ competing with an inverse agonist $\zeta_I < 1$.

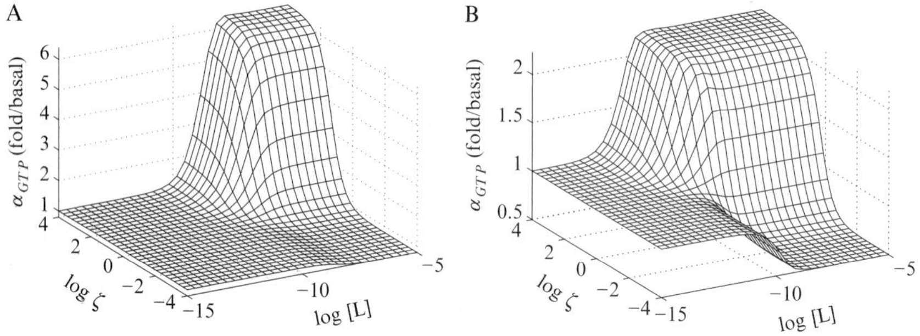

Figure 29.4 Concentration–response relations with varying ζ, for a ligand with NMS on/off rates. (A) Response at $t = 60$ s. (B) Response at $t = 1000$ s.

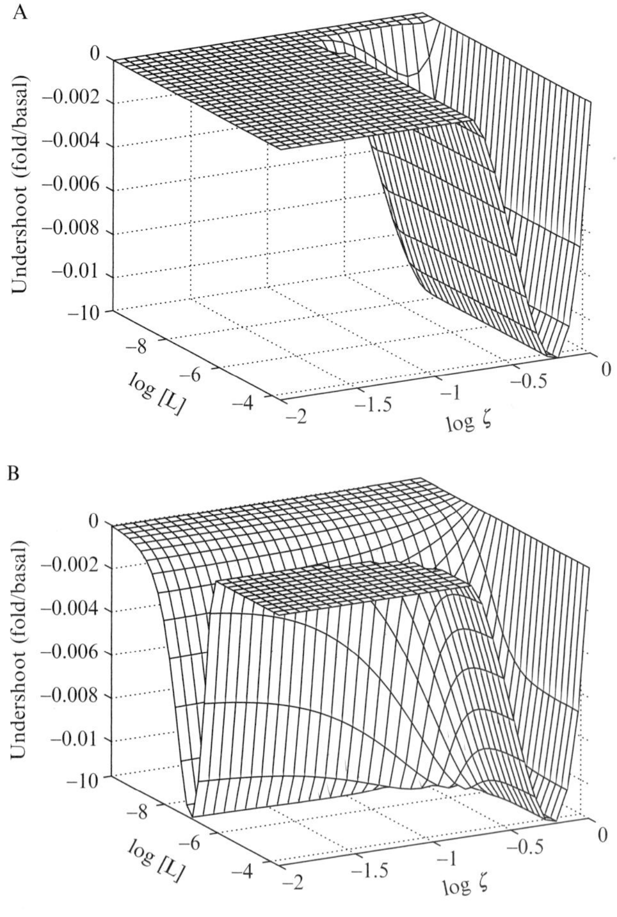

Figure 29.5 Undershoot in α_{GTP}, varying ζ and $[L]$, for a ligand with NMS on rate. (A) Using NMS off rate. (B) Using very fast off rate $k_{lb-} = 1\ s^{-1}$.

3.1. Model formulation

A schematic of the two-ligand system is shown in Fig. 29.6. The equilibrium constants are as before, but now subscripts A and I refer to ligands A and I, which are taken to be agonist and inverse agonist, respectively.

The reaction list this time reads as follows:

$$R \underset{k_{\text{deact}}}{\overset{k_{\text{act}}}{\rightleftharpoons}} R^*, A + R \underset{k_{a-}}{\overset{k_{a+}}{\rightleftharpoons}} AR, AR \underset{\zeta_{-a}k_{\text{deact}}}{\overset{\zeta_{+a}k_{\text{act}}}{\rightleftharpoons}} AR*, A + R^* \underset{\theta_{\zeta a}\zeta_{-a}k_{a-}}{\overset{\theta_{\zeta a}\zeta_{+a}k_{a+}}{\rightleftharpoons}} AR^*, \tag{29.4a}$$

$$R + G \underset{k_{g-}}{\overset{k_{g+}}{\rightleftharpoons}} RG, R^* + G \underset{\theta_\mu \mu_- k_{g-}}{\overset{\theta_\mu \mu_+ k_{g+}}{\rightleftharpoons}} R^*G, RG \underset{\mu_- k_{\text{deact}}}{\overset{\mu_+ k_{\text{act}}}{\rightleftharpoons}} R^*G, \tag{29.4b}$$

$$\begin{aligned} &AR + G \underset{\theta_{va}v_{-a}k_{g-}}{\overset{\theta_{va}v_{+a}k_{g+}}{\rightleftharpoons}} ARG, AR^* + G \underset{\theta_{v\mu a}\mu_- v_{-a}k_{g-}}{\overset{\theta_{v\mu a}\mu_+ v_{+a}k_{g+}}{\rightleftharpoons}} AR^*G, \\ &ARG \underset{\theta_{\zeta\mu a}\zeta_{-a}\mu_- k_{\text{deact}}}{\overset{\theta_{\zeta\mu a}\zeta_{+a}\mu_+ k_{\text{act}}}{\rightleftharpoons}} AR^*G, \end{aligned} \tag{29.4c}$$

$$\begin{aligned} &A + RG \underset{v_{-a}k_{a-}}{\overset{v_{+a}k_{a+}}{\rightleftharpoons}} ARG, A + R^*G \underset{\theta_{\zeta\mu a}\zeta_{-a}\mu_- k_{a-}}{\overset{\theta_{\zeta\mu a}\zeta_{+a}\mu_+ k_{a+}}{\rightleftharpoons}} AR^*G, \\ &R^*G \underset{k_{\text{GTP}-}}{\overset{k_{\text{GTP}+}}{\rightleftharpoons}} R^* + \alpha_{\text{GTP}} + \beta\gamma, \end{aligned} \tag{29.4d}$$

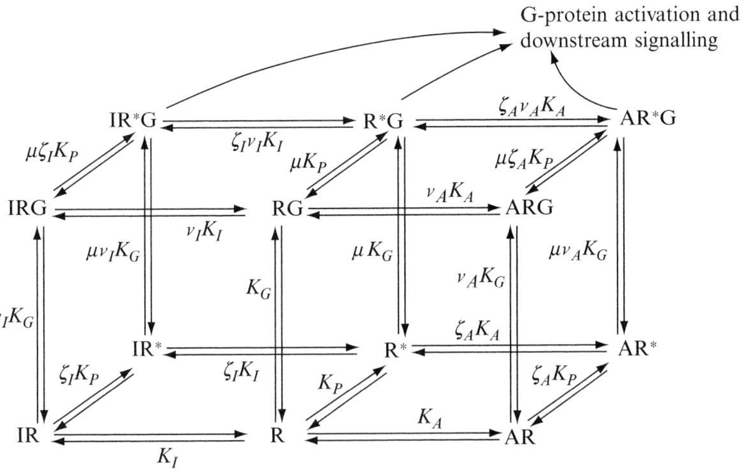

Figure 29.6 Extended cubic ternary complex for 12 receptor states, showing equilibrium rate constants. The two competing ligands A and I in this study are agonist and inhibitor (inverse agonist or antagonist).

$$AR^*G \underset{\theta_a\nu_{+a}k_{\mathrm{GTP}-}}{\overset{\theta_a\nu_{-a}k_{\mathrm{GTP}+}}{\rightleftharpoons}} AR^* + \alpha_{\mathrm{GTP}} + \beta\gamma, \alpha_{\mathrm{GTP}} \underset{k_{gd-}}{\overset{k_{gd+}}{\rightleftharpoons}} \alpha_{\mathrm{GDP}}, \alpha_{\mathrm{GDP}} + \beta\gamma \underset{k_{RA-}}{\overset{k_{RA+}}{\rightleftharpoons}} G, \tag{29.4e}$$

$$I + R \underset{k_{i-}}{\overset{k_{i+}}{\rightleftharpoons}} IR, IR \underset{\zeta_{-i}k_{\mathrm{deact}}}{\overset{\zeta_{+i}k_{\mathrm{act}}}{\rightleftharpoons}} IR^*, IR^*G \underset{\theta_i\nu_{+i}k_{\mathrm{GTP}-}}{\overset{\theta_i\nu_{-i}k_{\mathrm{GTP}}}{\rightleftharpoons}} IR^* + \alpha_{\mathrm{GTP}} + \beta\gamma, \tag{29.4f}$$

$$IR + G \underset{\theta_{\nu i}\nu_{-i}k_{g-}}{\overset{\theta_{\nu i}\nu_{+i}k_{g+}}{\rightleftharpoons}} IRG, IR^* + G \underset{\theta_{\nu\mu i}\mu_{-}\nu_{-i}k_{g-}}{\overset{\theta_{\nu\mu i}\mu_{+}\nu_{+i}k_{g+}}{\rightleftharpoons}} IR^*G, IRG \underset{\theta_{\zeta\mu i}\zeta_{-i}\mu_{-}k_{\mathrm{deact}}}{\overset{\theta_{\zeta\mu i}\zeta_{+i}\mu_{+}k_{\mathrm{act}}}{\rightleftharpoons}} IR^*G, \tag{29.4g}$$

$$I + RG \underset{\nu_{-i}k_{i-}}{\overset{\nu_{+i}k_{i+}}{\rightleftharpoons}} IRG, I + R^*G \underset{\theta_{\zeta\mu i}\zeta_{-i}\mu_{-}k_{i-}}{\overset{\theta_{\zeta\mu i}\zeta_{+i}\mu_{+}k_{i+}}{\rightleftharpoons}} IR^*G, I + R^* \underset{\theta_{\zeta i}\zeta_{-i}k_{i-}}{\overset{\theta_{\zeta i}\zeta_{+i}k_{i+}}{\rightleftharpoons}} IR^*. \tag{29.4h}$$

Again, $k_{\mathrm{GTP}-}$ is taken to be zero, and now subscripts a and A refer to agonist, and i and I to inverse agonist. The mathematical model for this two-ligand problem is a system of ODEs for the time derivatives of concentrations of the 16 species.

3.2. Numerical results

Time courses and dose–response relations in this section are again generated by integrating the governing equations numerically, and we pay particular attention to how adding inverse agonist to a system affects the peak–plateau features of an agonist-induced response.

3.2.1. Time courses: Simultaneous addition and pretreatment, varying [*I*]

In Fig. 29.7, we show time courses, after the introduction of an agonist to a system with some constitutive basal activity. The bottom row contains time courses where the system also includes varying concentrations of inverse agonist I, which are introduced simultaneously with the agonist (C) or 30 min prior to the agonist (D). For comparison, the top row contains the corresponding time courses for varying concentrations of antagonist. In (B), in common with the pretreatment results of Bridge *et al.* (2010), we see that increasing antagonist concentration results in a decreased response, and a loss of the sharp peak. If instead an inverse agonist is used, the basal (time $t = 0$) level is even lower, as is the response at all later times. The strong inverse agonist, as we would expect, has a greater inhibitory effect than the antagonist. When the antagonist and agonist are simultaneously added (A), the response over short times is greater than that for the pretreated case, and the sharp peak is still apparent. This is also the case for the inverse agonist

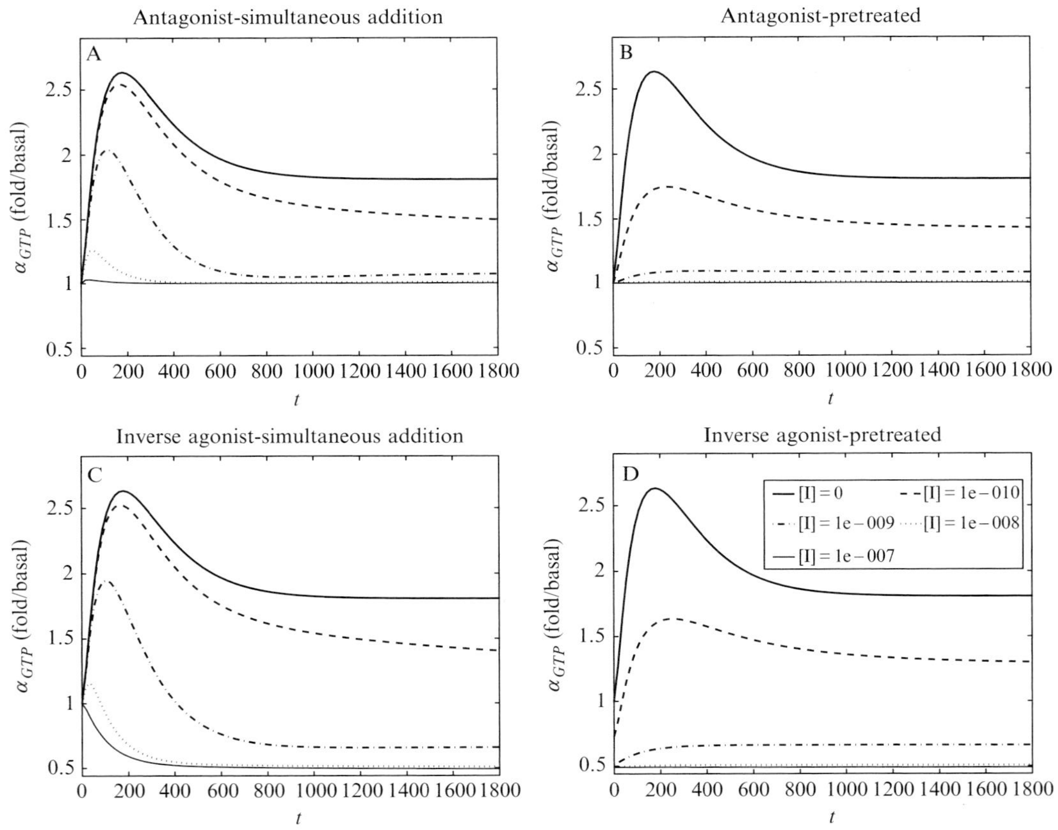

Figure 29.7 Time courses of α_{GTP} after the addition of agonist with $[A] = 10^{-5}$ M. Agonist on/off rates are for carbachol and $\zeta_A = 1000$; antagonist and inverse agonist on/off rates are for NMS. (A) Simultaneous addition of antagonist. (B) 30 min pretreatment with antagonist. (C) Simultaneous addition of inverse agonist with $\zeta_I = 0.001$. (D) 30 min pretreatment with inverse agonist with $\zeta_I = 0.001$.

(C), where the sharp peak is retained, except for the highest inverse concentration where agonist action is not sufficient to produce any noticeable response above basal. With a smaller agonist concentration (see Fig. 29.8), inverse agonism is the dominant effect, and the α_{GTP} response is lower still.

3.2.2. Time courses: Effect of varying inverse agonist efficacy ζ_I

In Fig. 29.9, we examine the effect of the inverse agonist "efficacy" or strength ζ_I, and also the "stickiness," measured by the off-rate k_{Ib-I}, in the cases of simultaneous addition of agonist and inverse agonist and pretreatment with inverse agonist. As ζ_I decreases, as in the case of increasing [I] with fixed ζ_I, the inverse agonist acts to lower the response and a limiting response is approached. In the case of simultaneous addition, this decrease in ζ_I also results in an earlier peak. The pretreated system again gives a lower signal, which, for the stickier (more slowly dissociating) inverse agonist, suppresses the peak response altogether (D).

3.2.3. Dose–response relations: Surmountable and insurmountable inverse agonism

As in Bridge *et al.* (2010) and Woodroffe *et al.* (2010b), it is convenient to summarize the agonist-induced [α_{GTP}] dynamics by plotting dose–response curves or surfaces for the peak (maximum) and plateau (eventual steady-state) concentrations. Such plots are shown in Fig. 29.10, for a system pretreated with an inverse agonist which is sticky (A) and less sticky (B). The plateau responses in both cases display what we refer to as *surmountable inverse agonism*, whereby the agonist-induced response level which decreases as [I] increases can be recovered by increasing [A]. The peak response displays regions where the response is surmountable, but also regions (high [A] and low [I]) where a drop in response caused by increasing [I] cannot be nullified by increasing [A]; we say that the peak is subject to *partial insurmountable inverse agonism*. For the less sticky inverse agonist, we observe the appropriate shift in the I-curves, and the magnitude of the insurmountable drop in peak response is decreased.

In Fig. 29.11, we show corresponding dose–response relations for the case of simultaneous addition of A and I. Now we see surmountable inverse agonism in all regions of (A, I)-space, for both peak and plateau responses. Clearly there are rich competition dynamics leading to the differences in the observed types of inverse agonist effect in the different cases, and these certainly warrant further study. The solution of the governing system of ODEs gives time courses for each of the species involved, and examination of these may give a clearer picture of the key reaction steps and signaling dynamics. The concentrations for a subset of the species involved in a typical simulation are shown in Fig. 29.12. Displaying time on a log scale allows us to identify dominant effects over various time scales, and such

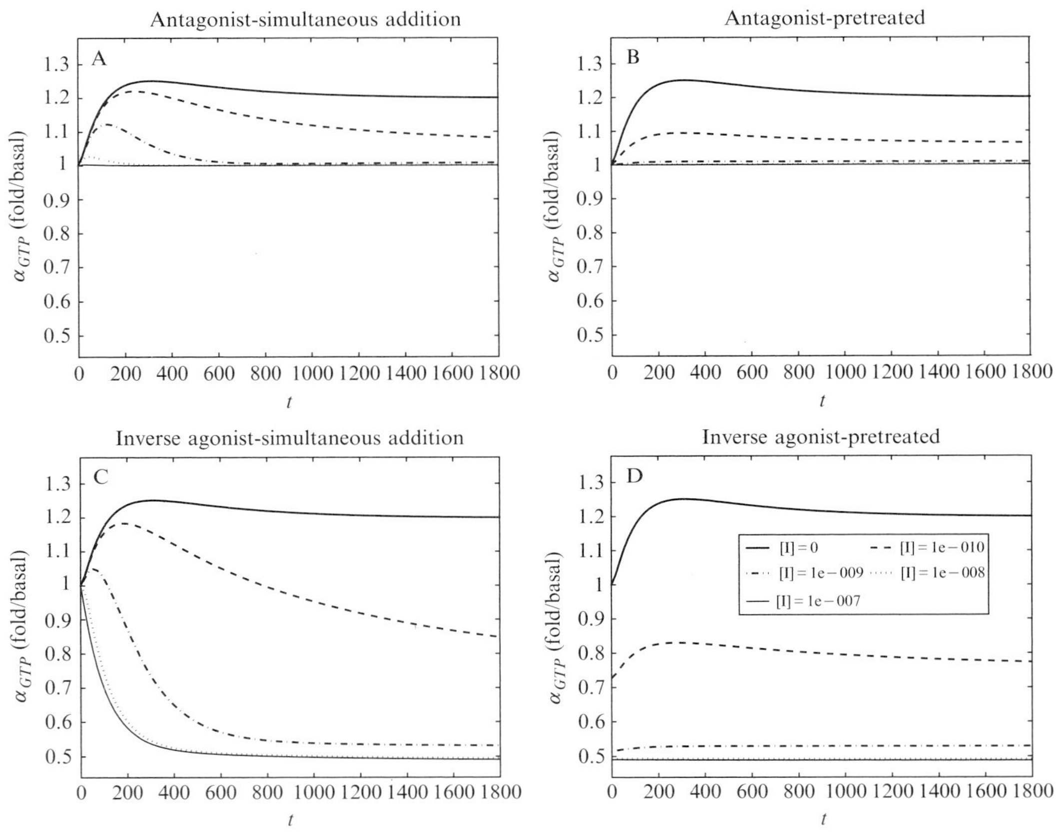

Figure 29.8 Time courses as in Fig. 29.7, with smaller agonist concentration $[A] = 10^{-6}\ M$.

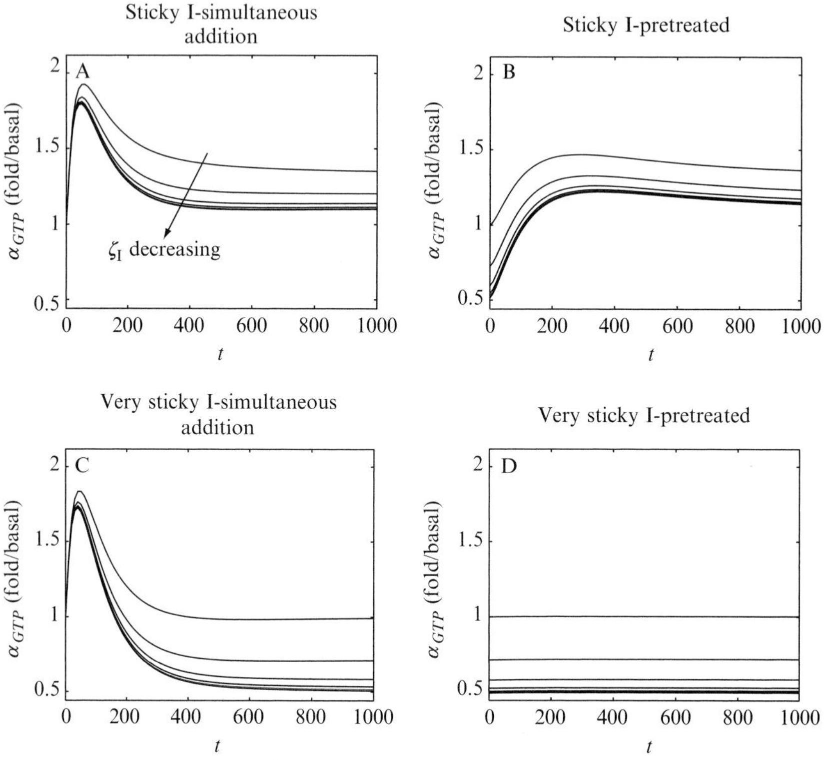

Figure 29.9 Time courses of α_{GTP} after the addition of agonist with $[A] = 10^{-5}$ M. Agonist on/off rates are for carbachol and $\zeta_A = 1000$; inverse agonist on rates is for atropine, with varying inverse agonist efficacy ζ_I. Moderately "sticky" inverse agonist has $k_{Ib-I} = 4.5 \times 10^{-3}$ s^{-1}, while very sticky inverse agonist has $k_{Ib-I} = 4.5 \times 10^{-5}\,s^{-1}$. Inverse agonist concentration is $[I] = 5 \times 10^{-9}$ M. (A) Simultaneous addition of moderately "sticky" inverse agonist. (B) 30 min pretreatment with moderately "sticky" inverse agonist. (C) Simultaneous addition of very sticky inverse agonist. (D) 30 min pretreatment with very sticky inverse agonist.

plots can therefore be valuable in clarifying the complex dynamics. These ideas can be formalized through the application of asymptotic methods as in Bridge *et al.* (2010).

4. Discussion

In this chapter, we have presented mathematical models for dynamics of α_{GTP} signals in response to inverse agonists at GPCRs, and described the numerical techniques for their solution. Modeling transient signaling, rather

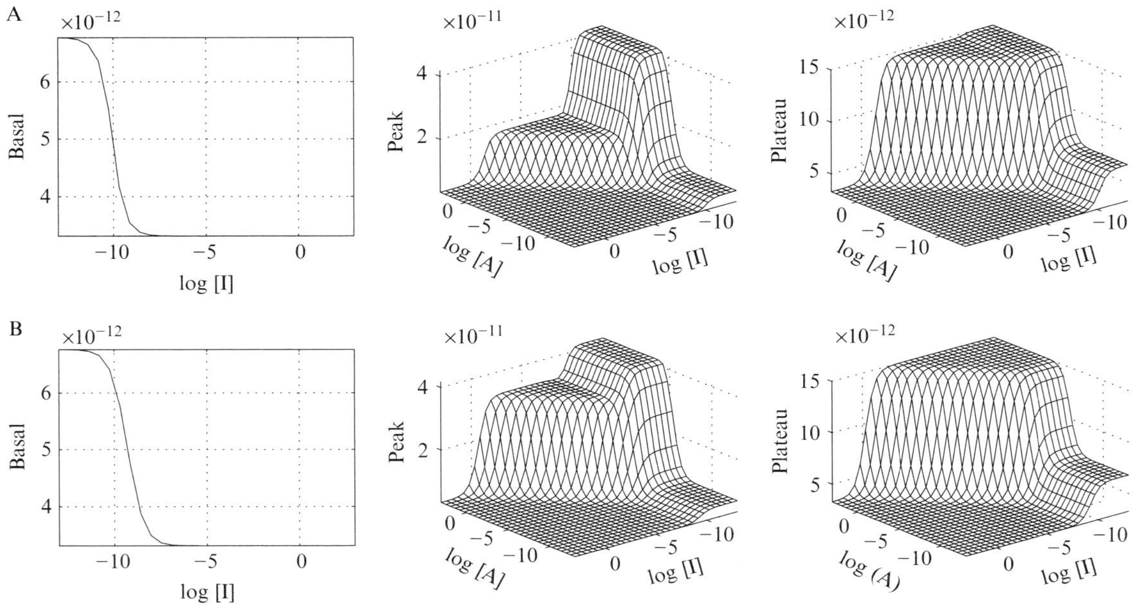

Figure 29.10 Concentration–response (α_{GTP}) relations for 30 min pretreatment with strong inverse agonist followed by addition of agonist. We take agonist on/off rates as those for acetylcholine, and inverse agonist on rate as that for NMS. (A) $k_{lb-I} = 2.8 \times 10^{-4}\ s^{-1}$, (B) $k_{lb-I} = 2.8 \times 10^{-3}\ s^{-1}$.

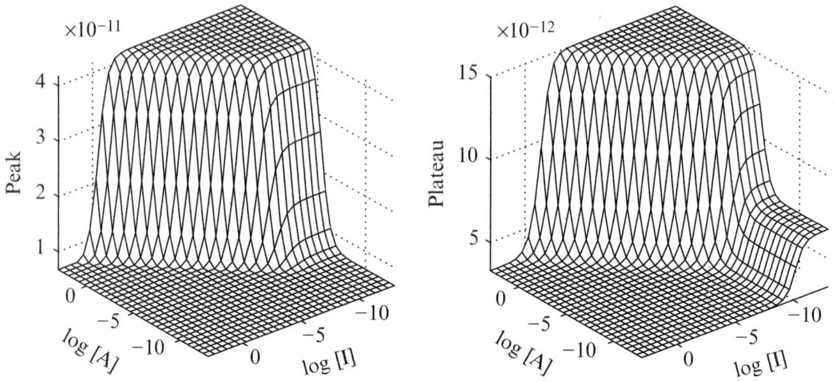

Figure 29.11 Concentration–response (α_{GTP}) relations for simultaneous addition of strong inverse agonist and agonist. We take agonist on/off rates as those for acetylcholine, and inverse agonist on/off rates as those for NMS.

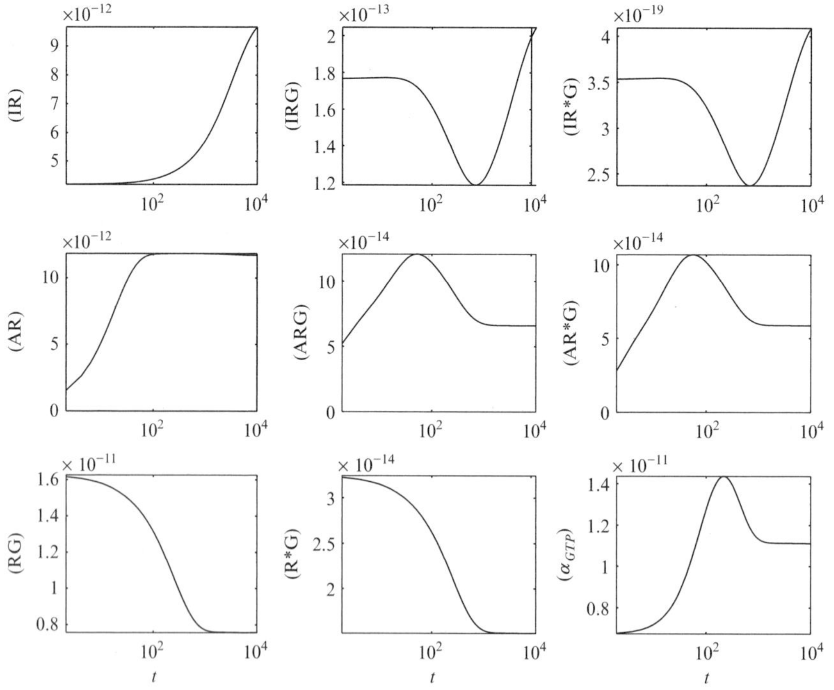

Figure 29.12 Time courses for various species after the addition of agonist to an inverse agonist pretreated system.

than simply equilibrium levels, can predict dynamic response features including inhibition, undershoots, multiple time scales, and both surmountable and insurmountable inverse agonism. The analysis of such responses can highlight differences between antagonist and inverse agonist action and provide insight into the mechanisms of inverse agonism.

The modeling and numerical computation here is largely in the spirit of Woodroffe *et al.* (2010b) and Bridge *et al.* (2010), and numerical results are focused on inverse agonist effects on signals in constitutively active systems, and comparisons between inverse agonist and antagonist effects on exogenous agonist-induced responses. While these results reveal some novel dynamic behavior, the story is far from complete. Constitutive activity can be effected by generating a nonsignaling complex (such as *IRG* in our model). Since inverse agonist efficacy is widely discussed in the literature, our main focus here has been on varying the parameter ζ. A reduction in basal activity could also lead from a repression of receptor-G-protein association (Costa and Cotecchia, 2005; Strange, 2002). This suggests that ν may also be an important parameter to investigate. Of course, the other parameters in the system all play a role; we would expect, for example, that R_{tot} and G_{tot} influence basal levels of $[\alpha_{GTP}]$. Efforts toward a better understanding of inverse agonism dynamics would benefit from the application of parameter sensitivity analysis and asymptotic methods to our models, in order to identify the importance of certain parameters in certain regimes and the time scales over which the solution evolves. Also, as noted throughout, apparent basal activity *in vivo* may be due to either constitutive activity, endogenous agonists, or both. The simulation technique we describe here could be applied to systems where an inverse agonist is introduced to a system initially at equilibrium under the action of an endogenous agonist.

We conclude that it is clear that mathematical modeling has a role to play in understanding the effect of inverse agonists on GPCR signaling dynamics. The framework and techniques presented here will hopefully provide a platform upon which to build qualitative and quantitative answers to some of the outstanding questions in the field.

Appendix A. Parameter Values

The values taken for the various parameters in our model are shown in Tables 29.1 and 29.2.

Acknowledgments

The author gratefully acknowledges valuable discussions with Lauren May, Joëlle Alcock, John King, Markus Owen, and Steve Hill, and financial support from BBSRC and a Wellcome Trust Value in People award.

REFERENCES

Bond, R. A., and IJzerman, A. P. (2006). Recent developments in constitutive receptor activity and inverse agonism, and their potential for GPCR drug discovery. *TiPS* **27**(2), 92–96.

Bridge, L. J., King, J. R., Hill, S. J., and Owen, M. R. (2010). Mathematical modelling of signalling in a two-ligand G-protein coupled receptor system: Agonist–antagonist competition. *Math. Biosci.* **223,** 115–132.

Chidiac, P. (2002). Considerations in the evaluation of inverse agonism and protean agonism at G-protein-coupled receptors. *Methods Enzymol.* **343,** 3–16.

Costa, T., and Cotecchia, S. (2005). Historical review: Negative efficacy and constitutive activity of G-protein-coupled receptors. *TiPS* **26**(12), 618–624.

de Lean, A., Stadel, J. M., and Lefkowitz, R. J. (1980). A ternary complex model explains the agonist-specific binding properties of the adenylate cyclase coupled β-adrenergic receptor. *J. Biol. Chem.* **255,** 7108–7117.

Dowling, M. R., and Charlton, S. J. (2006). Quantifying the association and dissociation rates of unlabelled antagonists at the muscarinic M_3 receptor. *Br. J. Pharm.* **148,** 927–937.

Gilchrist, A. (2007). Modulating G-protein-coupled receptors: From traditional pharmacology to allosterics. *TiPS* **28**(8), 431–437.

Kenakin, T. (2001). Inverse, protean and ligand-selective agonism: Matters of receptor confirmation. *FASEB J.* **15,** 598–611.

Kenakin, T. (2005). The physiological significance of constitutive receptor activity. *TiPS* **26** (12), 603–605.

Leff, P., Scaramellini, C., Law, C., and McKechnie, K. (1997). A three-state receptor model of agonist action. *TiPS* **18,** 355–362.

Leurs, R., Pena, M. S. R., Bakker, R. A., Alewijnse, A. E., and Timmerman, H. (2000). Constitutive activity of G-protein coupled receptors and drug action. *Pharm. Acta Helv.* **74,** 327–331.

Lew, M. J., Ziegas, J., and Christopoulos, A. (2000). Dynamic mechanisms of non-classical antagonism by competitive AT_1 receptor antagonists. *TiPS* **21**(10), 376–381.

Milligan, G. (2003). Constitutive activity and inverse agonists of G protein-coupled receptors: A current perspective. *Mol. Pharmacol.* **64,** 1271–1276.

Negus, S. S. (2006). Some implications of receptor theory for in vivo assessment of agonists, antagonists and inverse agonists. *Biochem. Pharmacol.* **71,** 1663–1670.

Parra, S., and Bond, R. A. (2007). Inverse agonism: From curiosity to accepted dogma, but is it clinically relevant? *Curr. Opin. Pharmacol.* **7,** 146–150.

Shea, L. D., Neubig, R. R., and Linderman, J. J. (2000). Timing is everything—The role of kinetics in G-protein activation. *Life Sci.* **68,** 647–658.

Strange, P. G. (2002). Mechanisms of inverse agonism at G-protein coupled receptors. *TiPS* **23**(2), 89–95.

Sykes, D. A., Dowling, M. R., and Charlton, S. J. (2009). Exploring the mechanism of agonist efficacy: A relationship between efficacy and agonist dissociation rate from the muscarinic M_3 receptor. *Mol. Pharmacol.* **76**(3), 543–551.

Vilardaga, J.-P., Steinmeyer, R., Harms, G. S., and Lohse, M. J. (2005). Molecular basis of inverse agonism in a G protein-coupled receptor. *Nat. Chem. Biol.* **1**(1), 25–28.

Woodroffe, P. J., Bridge, L. J., King, J. R., Chen, C. Y., and Hill, S. J. (2010a). Modelling of the activation of G-protein coupled receptors: Drug free constitutive receptor activity. *J. Math. Biol.* **60,** 313–346.

Woodroffe, P. J., Bridge, L. J., King, J. R., and Hill, S. J. (2010b). Modelling the activation of G-protein coupled receptors by a single drug. *Math. Biosci.* **219**(1), 32–55.

CHAPTER THIRTY

Design and Use of Constitutively Active STAT5 Constructs

Michael A. Farrar*,†

Contents

Abstract

The transcription factor *S*ignal *T*randucer and *A*ctivator of *T*ranscription 5 (STAT5) plays an important role in numerous biological processes including, but not limited to, (i) homeostasis of hematopoietic stem cells, (ii) development of essentially all blood cell lineages, (iii) growth hormone effects, (iv) differentiation of mammary epithelium, and (v) central nervous system control of metabolism. Two key tools for deciphering STAT5 biology have involved the use of mice in which the *Stat5a* and *Stat5b* genes can be conditionally deleted (*Stat5*$^{FL/FL}$ mice) and the development of systems in which STAT5a or STAT5b is rendered constitutively active. In this chapter, the distinct mechanisms that have been developed to render STAT5 constitutively active and their use in probing biological processes are discussed.

1. Introduction

The STAT family of transcription factors consists of seven distinct members that have wide-ranging biological activities (Leonard and O'Shea, 1998). The transcription factor STAT5, which belongs to the STAT family of transcription factors, exists in two isoforms called STAT5a and STAT5b. These two proteins exhibit 95% sequence identity and are encoded by two

* Center for Immunology, Masonic Cancer Center, University of Minnesota, Minneapolis, USA
† Department of Laboratory Medicine and Pathology, University of Minnesota, Minneapolis, USA

Methods in Enzymology, Volume 485
ISSN 0076-6879, DOI: 10.1016/S0076-6879(10)85030-X

closely linked genes that are separated by less than 14,000 base pairs in the genome (Copeland *et al.*, 1995; Cui *et al.*, 2004; Lin *et al.*, 1996). The entire *Stat5a* and *Stat5b* gene locus spans approximately 100,000 base pairs on murine chromosome 11 or human chromosome 17. Regulation of *Stat5a* and *Stat5b* transcription remains poorly studied but both genes appear to be relatively ubiquitously expressed. Given the high sequence conservation between STAT5a and STAT5b, it is perhaps not surprising that they appear to be relatively functionally interchangeable. Mice lacking either STAT5a or STAT5b have relatively mild defects in prolactin-dependent mammary differentiation or sexually dimorphic growth hormone-dependent effects, respectively (Liu *et al.*, 1997; Udy *et al.*, 1997). It remains unclear whether this reflects true differences in function between STAT5a and STAT5b, or differences in relative expression of STAT5a versus STAT5b in different cell types. In contrast, mice lacking both *Stat5a* and *Stat5b* exhibit a perinatal lethal phenotype and exhibit multiple defects including anemia and virtual absence of B and T lymphocytes (Cui *et al.*, 2004; Yao *et al.*, 2006).

The STAT5a and STAT5b proteins consist of an N-terminal domain that is involved in promoting STAT5 dimerization, a DNA-binding domain, an SH2 domain, a C-terminal regulatory tyrosine residue, and a C-terminal transactivation domain (Fig. 30.1). The N-terminal domain plays an important role in directing the generation of specific STAT dimers—thus the N-terminal domain of STAT1 interacts specifically with the N-terminal domain of other STAT1 proteins; likewise, the N-terminal domain of STAT5a interacts uniquely with the N-terminal domain of other STAT5a proteins while the N-terminal domain of STAT5b interacts uniquely with the N-terminal domain of other STAT5b proteins (Ota *et al.*, 2004). This arrangement suggests that STAT5a and STAT5b may preferentially form homodimers with each other upon activation, although heterodimers of STAT5a and STAT5b have been suggested to form as well. The DNA-binding domain interacts with a conserved DNA-binding sequence consisting of TTCXXXGAA (Darnell, 1997). An important feature of STAT5 is that it can bind to DNA as a tetramer to closely spaced (6–7 base pair separation) STAT consensus binding sites—the binding specificity when interacting with such tetramer sequences is much less stringent (John *et al.*, 1999). The SH2 domain of STATs is what drives the initial interaction of STAT proteins with phosphorylated tyrosine residues in the cytoplasmic tails of cytokine receptors that activate specific STATs (Greenlund *et al.*, 1994). This allows for cytokine receptor-associated kinases (most typically belonging to the Jak family) to phosphorylate either Tyr-694 (STAT5a) or Tyr-699 (STAT5b). These phosphorylated tyrosine residues then interact with neighboring STAT5 proteins via reciprocal phospho-tyrosine–SH2 interactions that result in release of STAT5 from the activated receptor complex and the rearrangement of STAT5 as a head to tail homodimer (Fig. 30.2). This new homodimeric form of STAT5

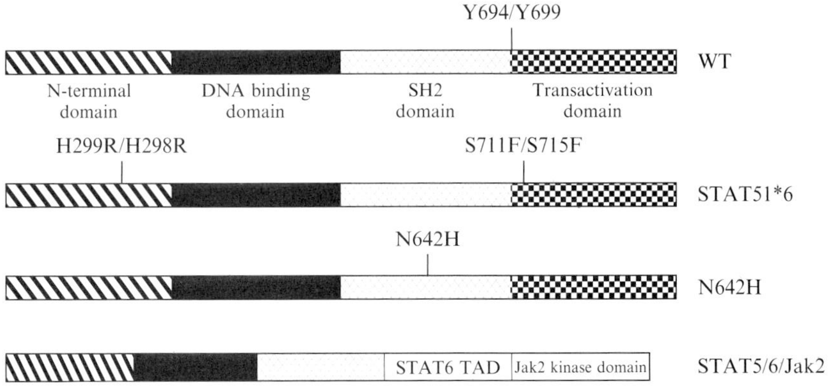

Figure 30.1 Schematic outline of the various domains of STAT5. The top construct depicts WT STAT5 and includes an N-terminal domain involved in promoting STAT5 homodimerization, a DNA-binding domain, SH2 domain, and C-terminal transactivation domain. The relative positions of Tyr-694 (STAT5a) or Tyr-699 (STAT5b), the residue which is phosphorylated and promotes the formation of a distinct STAT5 homodimer and hence activation, are shown. Schematic descriptions of the three distinct constitutively active forms of STAT5 that have been developed so far are also shown. The relative positions of point mutations that promote constitutive STAT5 activation (H299R/H298R, N642H, and S711F/S715F) are indicated; the mutation listed first at each position refers to locations in STAT5a while the second mutation refers to that location in STAT5b. The bottom construct consists of the first 750 amino acids of ovine STAT5 fused to the transactivation domain from STAT6 and the kinase domain from Jak2. The respective names of the constructs are listed on the right.

can then translocate to the nucleus where it binds to STAT5-regulted genes. Once in the nucleus, STAT5 can also interact with either coactivators such as CBP (Pfitzner *et al.*, 1998) or corepressors such as *S*ilencing-*M*ediator-of-*R*etinoid-and-*T*hyroid-receptor (SMRT; Nakajima *et al.*, 2001) to positively or negatively influence gene transcription.

2. Design of Constitutively Active STAT5 Constructs

To study the role that STAT5 plays in biological systems, several groups generated different constitutively active versions of STAT5. The first such construct was developed by Groner and colleagues and involved a fusion protein consisting of the first 750 amino acids of ovine STAT5 (and thus lacking the transactivation domain), linked to amino acids 677–847 of human STAT6 (encoding the more active transactivation of STAT6) and then amino acids 757–1129 encoding the kinase domain of murine Jak2 (Berchtold *et al.*, 1997; Fig. 30.1). This construct was efficiently

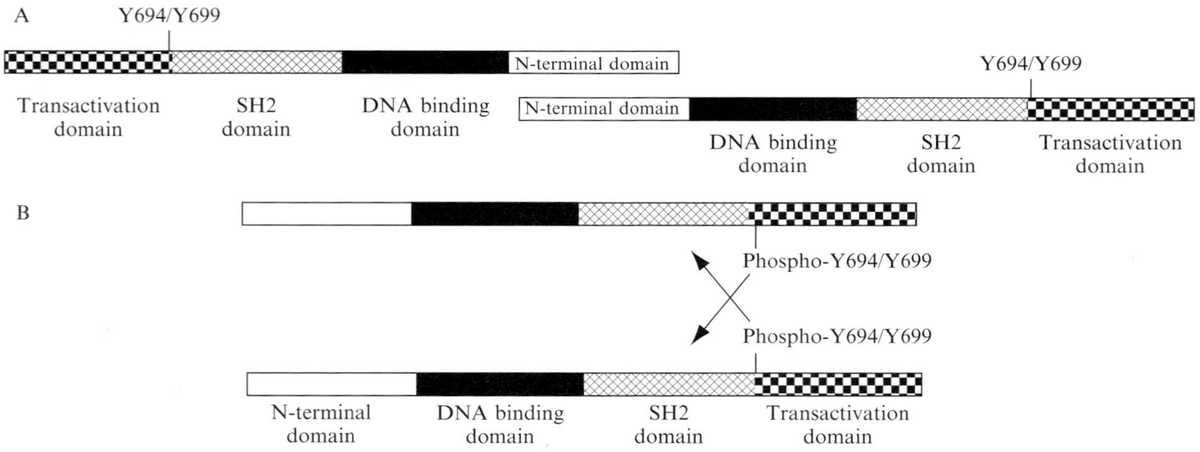

Figure 30.2 (A) Prior to activation STAT5a and STAT5b exist as homodimers organized in an antiparallel head to head fashion that is mediated via interactions between the N-terminal domains. (B) Phosphorylation of Tyr-694/Tyr-699 results in binding of this phosphorylated tyrosine residue to the SH2 domain of a neighboring STAT5 molecule resulting in the generation of a novel parallel homodimeric configuration that can translocate to the nucleus, bind DNA, and activate gene transcription.

phosphorylated on tyrosine, presumably by the linked Jak2 kinase domain, and resulted in cytokine-independent transcription of a number of previously defined STAT5 target genes.

The second approach taken to generate a constitutively active form of STAT5 involved random PCR-directed mutagenesis of *Stat5a* by Kitamura and colleagues (Onishi *et al.*, 1998). A library of such mutated *Stat5a* constructs was then cloned into a retrovirus that was used to infect BAF cells, a cell line that requires IL-3-dependent STAT5 activation for proliferation and survival. This resulted in a limited number of BAF cell clones that could proliferate in the absence of IL-3. Sequencing of the *Stat5a* constructs from these clones led to the identification of two recurring point mutations—Histidine-299 → Arginine and Serine-711 → Phenylalanine (Fig. 30.1). Introduction of just these two point mutants into otherwise wild-type (WT) STAT5a constructs also allowed for IL-3-independent growth of BAF cells, indicating that these two mutations alone lead to constitutive activation of STAT5. This construct is typically referred to as STAT5A1*6. Subsequent work demonstrated that these two mutations alone were sufficient to induce constitutive phosphorylation of the STAT5A1*6 construct on Tyr-694 and that this tyrosine phosphorylation step is required for STAT5A1*6 activity. The latter result indicates that the STAT5A1*6 construct still requires tyrosine phosphorylation and is thus perhaps best thought of as a modestly constitutively active form of STAT5 that can be hyperactivated by cytokine stimulation. Introduction of the homologous mutations in STAT5b (His298 → Arg and Ser715 → Phe) resulted in similar constitutive phosphorylation of the homologous Tyr-699 residue and hence constitutive activation of STAT5b.

To try and address the role of these two point mutations in rendering STAT5a constitutively active, BAF cells were transfected with either the H299R or S711F mutant constructs alone. The H299R construct alone was not tyrosine phosphorylated and did not exhibit detectable binding to STAT5 consensus sequences when tested in electrophoretic mobility shift assays. In contrast, both the double mutant and the S711F single mutant exhibited a low level of constitutive tyrosine phosphorylation and appeared to form STAT5 homodimers capable of binding STAT5 consensus binding sites in electrophoretic mobility shift assays. However, only the double mutant was capable of restoring IL-3-independent growth of transduced BAF cells. The exact mechanism by which these mutations act remains to be defined. Work by Ihle and colleagues demonstrated that histidine-299 is required for interaction of STAT5a with the corepressor molecule SMRT receptor (Nakajima *et al.*, 2001). Thus, these results suggested that weaker interaction with a presumably inhibitory molecule (SMRT) might account for the effect of this mutation. Subsequent work, discussed later in this chapter, has shed doubt on this interpretation. The role of the S711F mutation also remains obscure. It has been suggested that mutation of serine-711 may enhance STAT5 dimerization due to its proximity to tyrosine-694, although no clear mechanism to account for such an effect has been described. Alternatively, mutation of serine-711 may prevent proteolytic cleavage of the C-terminus of STAT5, a mechanism that normally functions to limit STAT5 activation (Azam *et al.*, 1997). Neither of these suggestions has been explored experimentally so the mechanism by which the S711F point mutation renders STAT5 constitutively active remains unknown.

The STAT5A1*6 and STAT5B1*6 constructs have been used in many *in vitro* and *in vivo* studies. A critical observation made by Moriggl *et al.* (2005) was that *in vivo* only the S711F mutation was critical for rendering STAT5 constitutively active and the H299R mutation in fact rendered STAT5a inactive. This discovery was made in experiments in which the STAT5A1*6 construct was reintroduced into cells in which the endogenous *Stat5a* and *Stat5b* genes were either present or had been deleted. These transduced cells were then reintroduced into lethally irradiated WT host mice. All mice that received cells containing *both* the endogenous *Stat5a* and *Stat5b* genes *and* the STAT5A1*6 construct developed multilineage leukemia; in contrast, none of the mice that received cells in which the STAT5A1*6 construct was present but the endogenous *Stat5a/b* genes were deleted developed leukemia. Moreover, the STAT5A1*6 construct failed to restore STAT5-dependent blood cell formation in cells lacking the endogenous STAT5 genes. Additional work by this same group established that transduction of either STAT5 sufficient or deficient bone marrow with the S711F mutant but not the H299R mutant resulted in cell transformation. Moreover, the H299R mutant was incapable of restoring T cell proliferation in mice lacking the endogenous *Stat5a/b* genes while the S711F mutant could restore T cell proliferation in this assay. Thus, these studies made four very important points. First, the STAT5A1*6 (and presumably STAT5B1*6) construct is functional only if it can heterodimerize with the WT endogenous form of either STAT5a or STAT5b. Second, for *in vivo* studies the only mutation that appears to be critical to rendering STAT5 constitutively active is the S711F mutation. Third, the H299R mutation renders STAT5 inactive unless it can heterodimerize with WT STAT5. Fourth, the specific mutation of serine-711 to phenylalanine is critical to activation as similar constructs in which serine-711 is mutated to alanine do not render STAT5 constitutively active.

The last constitutively active form of STAT5 that has been described is one in which asparagine-642, located in the SH2 domain of STAT5, is mutated to histidine (Fig. 30.1). This construct was also discovered by Kitamura and colleagues as part of their random PCR mutagenesis screen to identify constitutively active mutants of STAT5 (Ariyoshi *et al.*, 2000). Like the STAT5A1*6 construct, the N642H mutation results in constitutive phosphorylation of Tyr-694; mutation of Y694 to phenylalanine prevents activation by the N642H mutation. Thus, this construct also requires prior phosphorylation by an undefined kinase. The mechanism of action remains undefined but may involve stochastic low-level phosphorylation of STAT5 by kinases that is not effectively reversed by phosphatases due to the introduced mutation. Interestingly, when comparing the sequence of STAT5's SH2 domain with that of many *other* SH2 domains, the amino acid corresponding to position 642 is typically a histidine; the presence of a histidine at this position is thought to stabilize the interaction of the SH2 domain with the corresponding phosphorylated tyrosine residue. Thus, this mutation reverts STAT5s SH2 domain back to the consensus SH2 sequence, which may stabilize the homodimeric active form of STAT and thereby may prevent dephosphorylation, and hence inactivation of STAT5. Finally, the N642H mutant appears to be more strongly activated than the STAT5A1★6 construct as assessed by tyrosine phosphorylation of STAT5 N642H constructs (Ariyoshi *et al.*, 2000). However, it is important to point out that only a few studies have actually studied this constitutively active form of STAT5; thus, how well it will function in cell types other than BAF cells has not been extensively tested.

3. Use of Constitutively Active STAT5 Constructs

Constitutively active forms of STAT5 have been used in many *in vitro* systems to study the effects of STAT5 on specific biological functions. For example, Bissell and colleagues used STAT5A1★6 and STAT5B1★6 constructs to explore the role of STAT5 in regulating transcription of the *β-casein* gene in mammary epithelial cells (Xu *et al.*, 2009). Previous work by this group had established that prolonged prolactin-dependent signals led to acetylation of histone H4 at the *β-casein* promoter and that this effect required the presence of laminin-111. To determine whether the role of laminin-111 was simply to prolong STAT5 activation, epithelial cells were infected with retroviruses expressing either STAT5A1★6 or STAT5B1★6. In this setting STAT5 did not appear to be activated in the absence of prolactin signaling, but prolactin-dependent STAT5 activation was greatly prolonged. Using this system Bissell and colleagues demonstrated that the primary function of laminin-111 was to ensure prolonged STAT5 activation and that in the presence of activated STAT5 constructs laminin-111-dependent signals were no longer required for prolactin-dependent *β-casein* transcription. Thus, in this experimental setup the STAT5A1★6 constructs do not act as constitutively active constructs, since they still appear to require cytokine signaling for measureable activity, but rather function as a hyperactive form of STAT5 that allows one to probe the effect of short-term versus prolonged STAT5 signaling on gene transcription.

The latter study points out the importance of tracking STAT5 phosphorylation in cells that have been transduced with constitutively active forms of STAT5. This can be done using conventional Western blotting approaches for phospho-STAT5 using phospho-STAT5 specific antibodies (we have successfully used antiphospho-STAT5 antibodies from both Santa Cruz and Cell Signaling to detect Tyr-694/Tyr-699 phosphorylated forms of STAT5). In circumstances where cell numbers are more limited, such as early lymphoid progenitor populations, one can use flow cytometry to characterize phospho-STAT5 levels and STAT5 activation. A protocol based on work by Teague and colleagues (Van De Wiele *et al.*, 2004) that works well is outlined here.

Reagents

- Antiphospho-STAT5-PE (we use Becton Dickinson, catalog #612567, Clone 47; Teague and colleagues found that antiphospho-STAT5 antibodies from Cell Signaling also worked well)
- Isotype control PE (eBioscience catalog#12-4719-73, anti-mouse IgG1-PE)
- CALTAG Fix and Perm Kit #GAS-004

1. Surface staining – 2–10 × 10^6 cells per condition, use FITC, APC, Cascade Blue, or Alexa 700-conjugated monoclonal antibodies (methanol in step 7 quenches PE, PerCP, and PE-Cy conjugates; methanol fixation is critical to obtaining good signals as formaldehyde fixation does not work well in our hands). This step allows one to identify the cell population of interest. If one is using a homogenous tissue culture cell line and do not need to stain for additional cell surface markers, then this step can be omitted.
2. Wash the cells twice with serum-free media (SFM).
3. Resuspend the cells in 0.5 ml of warm (37 °C) SFM, and preincubate cells for 20 min in a 37 °C water bath.
4. IL-7 cytokine stimulation. Make a 50-ng/ml working dilution of IL-7 in SFM, and add 0.5 ml of this working dilution to the appropriate tubes of cells that were preincubated at 37 °C. For the "no cytokine" controls add 0.5 ml of SFM. Incubate all the tubes at 37 °C for 20 min. This step is for use with cells that express the IL-7 receptor (e.g., progenitor B cells or T cells). If one is interested in assessing signaling downstream of another receptor, substitute the appropriate cytokine; for example, this protocol works equally well for assessing STAT5 phosphorylation in response to IL-2 (Burchill *et al.*, 2007).
5. Wash the cells once in SFM, and remove all traces of supernatant.
6. Resuspend the cell pellet in 200 μl of Reagent A (Fixation Medium) from the Caltag Fix and Perm Kit and incubate at 37 °C for 15 min.

7. Add 1 ml of ice-cold 100% methanol while vortexing gently.
8. Incubate the cells on ice for 30 min or overnight at 4 °C in the dark (good stopping point). Transfer the cells to Eppendorf tubes after incubation.
9. Centrifuge for 15 min at 25 °C (room temperature) at 10,000 rpm. (Alternatively after step 7, the cells can be transferred to 12 × 75 5 ml polystyrene round bottom FACS tubes. All subsequent washes and incubations can be done in these tubes. In this case, cells should be centrifuged at 1500 rpm in a table-top centrifuge.)
10. Wash the cells twice in 1 ml of 1× phosphate buffered saline (PBS)/5% fetal bovine serum (1 × PBS/5% FBS), and centrifuge as in step 9.
11. Intracellular staining—resuspend each cell pellet in 80 μl of Reagent B (Perm Buffer) from the Caltag Fix and Perm Kit and add 20 μl PE-conjugated anti-pSTAT5 or the PE isotype control.
12. Incubate for 1 h at room temperature in the dark.
13. Wash the cells twice in 1 ml 1× PBS/5% FBS, and centrifuge as in step 9.
14. Resuspend in 1× PBS/5% FBS and analyze by flow cytometry.

We have used this approach both for tissue culture cells and for primary cells stained directly *ex vivo* (Will *et al.*, 2006). In the latter case, using transgenic mice expressing the STAT5B1★6 construct in lymphocytes we were able to document low-level constitutive activation in pro-B and pre-B cells and hyperactivation following stimulation with IL-7 (Fig. 30.3).

Several groups including my own have also used constitutively active forms of STAT5 to assess STAT5 function in whole animal models. For example, we have generated mice expressing the STAT5B1★6 construct (referred to as STAT5b-CA in our papers) throughout B and T lymphocyte development. These studies provide an important example of how expression levels affect the outcome of such studies. Most of our transgenic founder lines expressed superphysiological amounts of the constitutively active STAT5 construct and all those transgenic founders died by 4–6 weeks of birth due to massive expansion of progenitor B cells and CD8+ T cells. In contrast, two founder lines expressed levels of STAT5b1★6 that were ~50% that of the endogenous *Stat5a* and *Stat5b* genes (Burchill *et al.*, 2003). These mice survived and exhibited no outwardly obvious phenotype. We did observe constitutive activation of STAT5 as determined by measuring phospho-STAT5 levels in both B and T lymphocytes (Fig. 3 and data in Burchill *et al.*, 2003). In addition, lymphocyte development was perturbed in these mice as they exhibited increased numbers of progenitor B cells, CD8+ memory T cells, CD4+ regulatory T cells, γδ T cells, and NK T cells. Importantly, these results were essentially the exact reciprocal of findings obtained later on with mice in which the *Stat5a/b* genes had been deleted in B and T lymphocytes (Burchill *et al.*, 2007; Yao *et al.*,

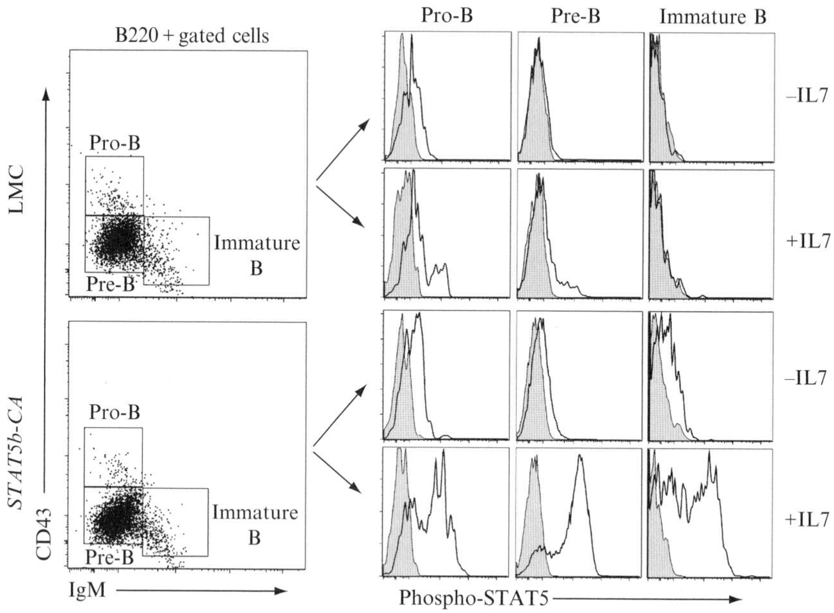

Figure 30.3 Phospho-STAT5 expression in pro-B, pre-B, and immature-B cell subsets. Bone marrow from 5- to 10-week old *STAT5b-CA* and littermate control (LMC) mice was harvested, stimulated with or without IL-7, and stained with antibodies for CD43, IgM, B220, and phospho-STAT5. $B220^+$ cells were gated on and CD43 and IgM markers were used to distinguish pro-B ($CD43^+IgM^-$), pre-B ($CD43^-IgM^-$), and immature-B ($CD43^-IgM^+$) cell subsets (left panels). Phospho-STAT5 expression was then examined in the histograms to the right. Shaded, gray histograms represent isotype controls while black outlined histograms represent staining with phospho-STAT5 specific antibodies. These results are representative of five separate experiments (16 *STAT5b-CA* and 10 LMC mice). This figure is reproduced from Will *et al.* (2006); Copyright 2006. The American Association of Immunologists, Inc.

2006). Finally, we used these STAT5b-CA mice to probe the role of STAT5 in IL-7 and IL-2-dependent B cell and T cell development. By crossing STAT5b-CA mice to *Il7r−/−* or *Il2rβ−/−* mice, we were able to establish a key role for STAT5 in the development of progenitor B cells and regulatory T cells downstream of IL-7 and IL-2, respectively (Burchill *et al.*, 2007, 2008; Goetz *et al.*, 2004). Subsequent studies using mice in which both STAT5a and STAT5b were selectively deleted in B cells or T cells confirmed the key role of STAT5 in IL-7-dependent B cell development and IL-2-dependent regulatory T cell differentiation. Thus, a constitutively active form of STAT5b, when present at physiologically appropriate levels, very accurately predicted the function of STAT5 in lymphocyte differentiation. These results validate the use of constitutively active forms of STAT5 but also underscore the importance of measuring the amount of constitutively active STAT5 protein present and the level of constitutive STAT5 activation (as assessed by tyrosine phosphorylation of STAT5).

4. Concerns with the Use of Constitutively Active STAT5 Constructs

There are a number of factors that need to be considered when designing experiments using constitutively active forms of STAT5. The first issue is what type of construct one should use. No studies exist in which the three distinct constitutively active forms of STAT5 have been compared in similar experimental systems; however, a comparison of the activity of the N642H, H299R/S711F, and S711F mutant forms of STAT5 has been carried out in zebrafish (the zebrafish mutations are actually N646H, H298R/N714F, and N714F; Lewis *et al.*, 2006). In these studies the N646H form of zebrafish STAT5 (referred to as zSTAT5 hereafter) exhibited slightly higher levels of STAT5 activation when compared with the H2989R/N714F mutant as assessed by measuring phosphorylation of zSTAT5 equivalent of Tyr-694. Capped mRNA for these constructs was injected in zebrafish embryos to explore the function of these constructs. Both the N646H and H298R/N714F zSTAT5 mRNAs resulted in a clear increase in myeloid, erythroid, and B lineage cells, although interestingly the effect was more pronounced with the H298R/N714F mutant. In contrast, the phenotype of the N714F STAT5 mutant was no different from that observed following injection of WT zSTAT5. Thus, these results suggest that both the mammalian N642H and H299R/S711F STAT5 mutants are likely to be effective constitutively active forms of STAT5, although the H299R/S711F construct may be slightly more active.

The STAT5/6/Jak2 fusion as well as the STAT1*6 and N642H forms of STAT5 have all been used in cell lines and depending on the level of expression lead to constitutive activation of STAT5 as assessed by STAT5A phosphorylation of Tyr-694 (or STAT5B phosphorylation of Tyr-699). The STAT5A1*6 or STAT5B1*6 constructs have been used most frequently and thus we know somewhat more about how they can be used. A potential advantage of these constructs is that they are perhaps less contrived than the STAT5/STAT6/Jak2 fusion construct. For example, the STAT51*6 constructs and STAT5 N642H mutant have the benefit of including the appropriate STAT5 transactivation domain, which may affect function. However, it is important to point out that the STAT51*6 constructs will not work in cells that lack endogenous STAT5 expression (Moriggl *et al.*, 2005). In those cases, either the STAT5/6/Jak2 or N642H construct would be a better choice. Alternatively, it may be worth generating the STAT5A S711F mutant, which appears to function as a constitutively active transcription factor when expressed in cells of the hematopoietic lineage *in vivo*. An important caveat is that the S711F mutant may be a weaker constitutively active form of STAT5 that may not function in all systems (e.g., the zebrafish studies described above). Finally, in the few experiments in which STAT5A1*6 and STAT5B1*6 have been compared side by side no differences were observed (Onishi *et al.*, 1998; Xu *et al.*, 2009). Thus, at this point there is no reason to expect that the use of either constitutively active STAT5a or STAT5b will lead to distinct results.

A second issue to consider is how to deliver these constructs. *In vitro* studies in tissue culture cell lines have typically used standard retroviral infection approaches. This works well and gives a range of constitutively active STAT5 expression that may be useful for some studies. *In vivo* studies have relied on either retroviral infection or transgenesis. An important caveat is that care must be taken when using retroviral transduction as high-level expression of constitutively active forms of STAT5 can lead to cell transformation. For example, transducing hematopoietic stem cells with constitutively active forms of STAT5 invariably leads to mixed multilineage leukemia when these cells are reintroduced into host mice (Antov *et al.*, 2003; Moriggl *et al.*, 2005). In contrast, several different transgenic lines exist in which the activated form of STAT5 has been expressed in distinct cell lineages. These include the lymphocyte specific form of STAT5B1*6 generated by my lab as well as a similar transgenic mouse generated by Ghysdael and colleagues in which a mutant form (Ser-711 → Phe) of STAT5A was expressed throughout lymphocyte development (Burchill *et al.*, 2003; Joliot *et al.*, 2006). The phenotype of these two distinct transgenic mouse strains appears to be quite similar, although the H298R/S715F STAT5b transgenic may exhibit a slightly stronger phenotype; whether the latter result is due to subtle differences in transgene expression levels or the fact that one mouse includes both the H298R and S711F mutations while the other contains only the S711F mutation remains unclear. Nevertheless, despite differences in the actual constructs (H298R; S7115F STAT5b vs. S711F STAT5a) both of these transgenics are likely to be useful for studies exploring the role of STAT5 signaling in lymphocyte development. Finally, a constitutively active form of STAT5 (the STAT5/6/Jak2 fusion version) has also been expressed as a transgene in the mammary gland driven by regulatory elements form the β-lactoglobulin gene (Iavnilovitch *et al.*, 2002). This construct resulted in mammary cancer in multiparous female mice and if available may be useful for studies of STAT5 during mammary differentiation and transformation. It is important to point out that in both this transgenic mouse and the two lymphocyte specific strains described above, the level of constitutively active STAT5 protein expression ranged from 50% to 100% of the endogenous STAT5 levels.

In conclusion, a number of distinct STAT5 constructs have been described. When these constructs are expressed at physiologically relevant levels, they have provided important insights into the mechanism underlying STAT5-dependent cellular differentiation, proliferation, and survival both *in vitro* and *in vivo*.

ACKNOWLEDGMENTS

I thank Dr Lynn Heltemes Harris for critical review of this manuscript. I am currently supported by a Leukemia and Lymphoma Society Scholar award.

REFERENCES

Antov, A., *et al.* (2003). Essential role for STAT5 signaling in CD25+CD4+ regulatory T cell homeostasis and maintenance of self-tolerance. *J. Immunol.* **171,** 3435–3441.

Ariyoshi, K., *et al.* (2000). Constitutive activation of STAT5 by a point mutation in the SH2 domain. *J. Biol. Chem.* **275,** 24407–24413.

Azam, M., *et al.* (1997). Functionally distinct isoforms of STAT5 are generated by protein processing. *Immunity* **6,** 691–701.

Berchtold, S., *et al.* (1997). Cytokine receptor-independent, constitutively active variants of STAT5. *J. Biol. Chem.* **272,** 30237–30243.

Burchill, M. A., *et al.* (2003). Distinct effects of STAT5 activation on CD4+ and CD8+ T cell homeostasis: Development of CD4+CD25+ regulatory T cells versus CD8+ memory T cells. *J. Immunol.* **171,** 5853–5864.

Burchill, M. A., *et al.* (2007). IL-2 receptor beta-dependent STAT5 activation is required for the development of Foxp3+ regulatory T cells. *J. Immunol.* **178,** 280–290.

Burchill, M. A., *et al.* (2008). Linked T cell receptor and cytokine signaling govern the development of the regulatory T cell repertoire. *Immunity* **28,** 112–121.

Copeland, N. G., *et al.* (1995). Distribution of the mammalian Stat gene family in mouse chromosomes. *Genomics* **29,** 225–228.

Cui, Y., *et al.* (2004). Inactivation of Stat5 in mouse mammary epithelieum during pregnancy reveals distinct functions in cell proliferation, survival, and differentiation. *Mol. Cell. Biol.* **24,** 8037–8047.

Darnell, J. E., Jr. (1997). STATs and gene regulation. *Science* **277,** 1630–1635.

Goetz, C. A., *et al.* (2004). STAT5 activation underlies IL7 receptor-dependent B cell development. *J. Immunol.* **172,** 4770–4778.

Greenlund, A. C., *et al.* (1994). Ligand-induced IFN gamma receptor tyrosine phosphorylation couples the receptor to its signal transduction system (p91). *EMBO J.* **13,** 1591–1600.

Iavnilovitch, E., *et al.* (2002). Overexpression and forced activation of stat5 in mammary gland of transgenic mice promotes cellular proliferation, enhances differentiation, and delays postlactational apoptosis. *Mol. Cancer Res.* **1,** 32–47.

John, S., *et al.* (1999). The significance of tetramerization in promoter recruitment by Stat5. *Mol. Cell. Biol.* **19,** 1910–1918.

Joliot, V., *et al.* (2006). Constitutive STAT5 activation specifically cooperates with the loss of p53 function in B-cell lymphomagenesis. *Oncogene* **25,** 4573–4584.

Leonard, W. J., and O'Shea, J. J. (1998). Jaks and STATs: Biological implications. *Annu. Rev. Immunol.* **16,** 293–322.

Lewis, R. S., *et al.* (2006). Constitutive activation of zebrafish Stat5 expands hematopoietic cell populations in vivo. *Exp. Hematol.* **34,** 179–187.

Lin, J. X., *et al.* (1996). Cloning of human Stat5B. Reconstitution of interleukin-2-induced Stat5A and Stat5B DNA binding activity in COS-7 cells. *J. Biol. Chem.* **271,** 10738–10744.

Liu, X., *et al.* (1997). Stat5a is mandatory for adult mammary gland development and lactogenesis. *Genes Dev.* **11,** 179–186.

Moriggl, R., *et al.* (2005). Stat5 tetramer formation is associated with leukemogenesis. *Cancer Cell* **7,** 87–99.

Nakajima, H., *et al.* (2001). Functional interaction of STAT5 and nuclear receptor co-repressor SMRT: Implications in negative regulation of STAT5-dependent transcription. *EMBO J.* **20,** 6836–6844.

Onishi, M., *et al.* (1998). Identification and characterization of a constitutively active STAT5 mutant that promotes cell proliferation. *Mol. Cell. Biol.* **18,** 3871–3879.

Ota, N., *et al.* (2004). N-domain-dependent nonphosphorylated STAT4 dimers required for cytokine-driven activation. *Nat. Immunol.* **5,** 208–215.

Pfitzner, E., *et al.* (1998). p300/CREB-binding protein enhances the prolactin-mediated transcriptional induction through direct interaction with the transactivation domain of Stat5, but does not participate in the Stat5-mediated suppression of the glucocorticoid response. *Mol. Endocrinol.* **12,** 1582–1593.

Udy, G. B., *et al.* (1997). Requirement of STAT5b for sexual dimorphism of body growth rates and liver gene expression. *Proc. Natl. Acad. Sci. USA* **94,** 7239–7244.

Van De Wiele, C. J., *et al.* (2004). Thymocytes between the beta-selection and positive selection checkpoints are nonresponsive to IL-7 as assessed by STAT-5 phosphorylation. *J. Immunol.* **172,** 4235–4244.

Will, W. M., *et al.* (2006). Attenuation of IL-7 receptor signaling is not required for allelic exclusion. *J. Immunol.* **176,** 3350–3355.

Xu, R., *et al.* (2009). Sustained activation of STAT5 is essential for chromatin remodeling and maintenance of mammary-specific function. *J. Cell Biol.* **184,** 57–66.

Yao, Z., *et al.* (2006). Stat5a/b are essential for normal lymphoid development and differentiation. *Proc. Natl. Acad. Sci. USA* **103,** 1000–1005.

CHAPTER THIRTY-ONE

In Vitro and *In Vivo* Assays of Protein Kinase CK2 Activity

Renaud Prudent, Céline F. Sautel, Virginie Moucadel, Béatrice Laudet, Odile Filhol, *and* Claude Cochet

Contents

Abstract

Protein kinase CK2 (formerly casein kinase 2) is recognized as a central component in the control of the cellular homeostasis; however, much remains unknown regarding its regulation and its implication in cellular transformation and carcinogenesis. Moreover, study of CK2 function and regulation in a cellular context is complicated by the dynamic multisubunit architecture of this protein kinase. Although a number of robust techniques are available to assay CK2 activity *in vitro*, there is a demand for sensitive and specific assays to evaluate its activity in living cells. We hereby provide a detailed description of several assays for monitoring the CK2 activity and its subunit interaction in living cells. The guidelines presented herein should enable researchers in the field to establish strategies for cellular screenings of CK2 inhibitors.

INSERM, U873, CEA, iRTSV/LTS, Université Joseph Fourier, Grenoble, France

Methods in Enzymology, Volume 485
ISSN 0076-6879, DOI: 10.1016/S0076-6879(10)85031-1

1. Introduction

Protein phosphorylation is a recurrent theme in biology and is recognized as a major regulatory device that controls nearly all aspects of living cells. In eukaryotic cells, this reversible protein modification is catalyzed by protein kinases that are central components of signaling pathways regulating homeostasis. Extensive studies have established protein kinase CK2 (formerly casein kinase 2) as a mastermind enzyme, regulating cross-talk among multiple signaling pathways critical to cell differentiation, proliferation, and survival (Filhol *et al.*, 2004; Litchfield, 2003). Importantly, experimental evidence linking increased expression and activity of CK2 to human cancers underscores the relevance of CK2 biology to cellular transformation and carcinogenesis (Duncan and Litchfield, 2008; Ruzzene and Pinna, 2009). CK2 is endowed with the peculiar molecular architecture of a multisubunit protein kinase exhibiting a quaternary structure composed of two catalytic subunits, CK2α and CK2α′, and two regulatory, CK2β, subunits (Pinna, 2002). In addition, the crystal structure of this complex as well as live cell imaging studies (Filhol *et al.*, 2003; Niefind *et al.*, 2001) suggested that CK2 subunits can coexist in the cell without forming the holoenzyme complex. As a signaling protein, CK2 can be targeted to different cellular compartments in response to various stress stimuli and the dynamic association of its subunits together with its substrate-dependent subcellular targeting are likely points of regulation (Filhol and Cochet, 2009; Filhol *et al.*, 2004). Due to the critical regulatory role that this kinase plays in cell fate determination, there is an increase in activity geared at the development of CK2-specific molecular tools.

In this chapter, we describe methodologies to evaluate the catalytic activity of endogenous CK2 in living cells and to assay *in vitro* and *in vivo* the CK2 subunit interaction.

2. Monitoring of CK2 Catalytic Activity in Living Cells

We setup a cell-based assay that allows the accurate evaluation of endogenous CK2 activity in cell cultures in response to potential CK2 inhibitory compounds. The assay uses a stable cell line expressing a reporter plasmid (*pEYFPc1-SβS*) consisting of several repeats of a CK2 peptide substrate fused to EYFP (Prudent *et al.*, 2008).

2.1. Reagents

HeLa cells grown in Dulbecco's modified eagle's medium (DMEM; Invitrogen Life Technologies, Inc.), supplemented with 10% (v/v) fetal calf serum (FBS, BioWest).

mAb anti-GFP (Roche, Cat. # 1814460)

Goat anti-mouse-HRP secondary antibody (Sigma, Cat. # A4416)

ECL plus Western blotting detection system (GE Healthcare)

4,5,6,7-Tetrabromo-1-benzotriazole (TBB) (Calbiochem, Cat. # 218697)

The plasmid (*pEYFPc1-SβS*) is designed by adding six CK2 consensus phosphorylation sites at the C terminus of the YFP protein (pEYFPc1-SβS). pEYFPc1-SβS was obtained from pEYFPc1-CK2β (Filhol *et al.*, 2003) after two subsequent rounds of mutations with the Quickchange-Site Directed mutagenesis kit (Stratagene) using for mutagenesis 1: 5′-GCTCAAGCTTCGGATTCTGAAGACGACGATACCGCGG-GCCCG-3′ and 5′-CGGGCCCGCGGTATCGTCGTCTTCAGAA-TCCGAAGCTTGAGC-3′ and for mutagenesis 2: 5′-GCTCTGAG-GAGGTGTCCGAGGTCGACTGGTTCTGAGGGCTCCGT-3′ and 5′-CGG GCCCGCGGTATCGTCGTCTTCAGAATCCGAAGCTT-GAGC-3′.

2.2. Procedure

2.2.1. Generation of a stable cell line

For engineering of the stable cell line expressing the CK2 activity reporter, pEYFPc1-SβS is amplified using: 5′- CGCAATTGCGCCACCATGGT-GAGCAAG and CCCTAGGGCTCAGAACCAGTCGACCTCGG-3′ and subcloned in Lentiviral vector (pLVX DsRed MCS+, Invitrogen) using *Mfe*I and *Avr*II restriction sites. Lentiviral production is conducted according to manufacturer's recommendation (Clontech).

To accomplish incorporation into lentiviral particles, pLVX vector, pSPAX.2 packaging vector encoding HIV Gag, Pol, and Rev proteins are used in combination with pMD2.G (encoding VSV-G). In brief, the DNA–Lipofectamine 2000 complexes are added to a suspension of 293T cells (5×10^6 in 10 mL of serum-supplemented medium) and incubated overnight. The medium containing the DNA–Lipofectamine 2000 complexes is then replaced with complete DMEM containing 10% FCS, and the virus-containing supernatants are harvested 48 h posttransfection. Virus stocks are filtrated 0.45 μm, aliquoted into 1.5 mL tubes, and freezed at −80 °C (Mangeot *et al.*, 2002). For infection, HeLa cells are plated into 24-well plates (5×10^4 in 500 μL of serum-supplemented DMEM). The day later, adherent cells are incubated with pEYFPc1-SβS lentiviral particles (1–5 MOI) diluted in 250 μL of serum-supplemented medium containing 8 μg/μL polybrene (Sigma). After 4 h, 500 μL of medium are added to cultures and transduction is maintained for 16 h before to wash cells and change the medium. For stable transduction, puromycin selection starts 36 h postinfection (at the concentration of 1 μg/mL) and is maintained during all cell culture.

2.2.2. Endogenous CK2 activity assay

HeLa cells expressing the CK2 activity reporter (pLVX-EYFPc1-SβS) are plated at 2 × 10^6 cells/well in 6-well plates. One day later, cells are incubated for 24 or 48 h with fresh medium containing the compounds to be tested. At the end of incubation, cells are washed with PBS, and directly lysed in the wells for 30 min at 4 °C in 100 μL of lysis buffer (50 m*M* Tris–HCl, pH 7.4, 0.15 *M* NaCl, 2 m*M* EDTA, 1% Triton X-100, phosphatase inhibitor cocktail 1 and 2 (1/100, P2850, and P5726, Sigma) and protease inhibitor cocktail (1/1000, P8340, Sigma). Cell lysates are collected and clarified by centrifugation at 12,000 rpm at 4 °C during 30 min. Protein concentration in the supernatant (whole cell extract) is determined using the BCA kit according to the manufacturer's recommendations (Pierce). Fifty micrograms of proteins are loaded with a mix of 2% glycerol and 0.01% bromophenol blue and separated on 12% native-polyacrylamide gel in migration buffer (250 m*M* Tris, 2 *M* Glycine). After electrotransfer on a PVDF membrane, the blot is blocked with 1% BSA and incubated for 2 h at 4 °C with the mAb anti-GFP at 1/2000 dilution. After four washes in TBS-T 0.1% Tween 20, the membrane is incubated for 1 h at 20 °C with a goat anti-mouse-HRP secondary antibody at 1/10,000 dilution and GFP is revealed with the ECL plus Western blotting detection system according to the manufacturer's recommendations. Figure 31.1 shows the effect of TBB, a commercially available CK2 inhibitor on the endogenous CK2 activity of HeLa cells in culture.

Comments: As some chemical compounds may have toxic effects on cells, to ensure reproductibility of the assay, it is advised that the cell layer be carefully washed before lysing the cells. The assay can also be performed on cells transiently transfected with the pEYFPc1-SβS reporter plasmid. However, we found that to ensure reproductibility between experiments, the use of a stable cell line expressing the CK2 activity reporter is preferable.

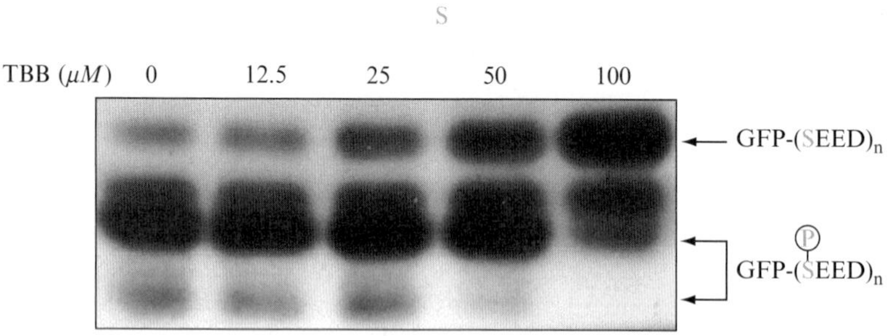

Figure 31.1 Inhibition of CK2 activity in living cells. HeLa cells expressing a chimeric CK2 activity YFP-based reporter were incubated during 24 h with different concentrations of TBB or an equivalent amount of DMSO as a control. Cell extracts were then analyzed by native electrophoresis and YFP was revealed by immunoblotting. The CK2-dependent phosphorylated S or nonphosphorylated P S forms of the CK2 peptide substrate reporter are indicated by arrows.

3. Assays of CK2 Subunit Interaction

Since the free CK2α subunit and the holoenzyme have distinct, though overlapping, substrate specificity profiles, it could be anticipated that such a balance is a crucial point of regulation of many cellular processes governed by this multifunctional enzyme. Thus, analyzing the CK2 subunit interaction may lead to the discovery of substrate-selective CK2 inhibitors and provides possibilities to pharmacologically test the importance of this interaction in tumor growth and viral infections.

3.1. *In vitro* CK2α–CK2β interaction assays

A solid-phase interaction assay was developed in which plate-bound MPB-CK2β binds soluble [^{35}S]methionine-labeled CK2α (Fig. 31.2; Laudet *et al.*, 2007).

Below are listed the reagents and procedure for the assay.

3.1.1. Reagents

Sulfo-NHS-LC-LC-biotin [sulfosuccinimidyl-6′-(biotinamido)-6-hexanamido hexanoate] which has a spacer arm of 30.5 Å length (Pierce # 21338).

Biotinylated MBP-CK2β (5 μg/mL in 50 m*M* Tris–HCl, pH 7.5; 0.4 *M* NaCl)

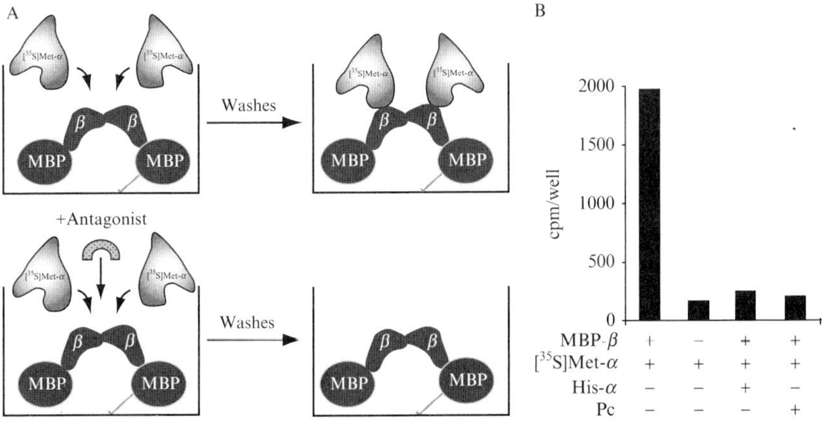

Figure 31.2 *In vitro* CK2 subunit interaction assay. (A) Scheme of the assay. (B) Plate-bound MPB-CK2β binds soluble [^{35}S]methionine-labeled CK2α which is displaced by 0.5 μ*M* of unlabeled His-CK2α (His-α) or 60 μ*M* of CK2β-derived cyclic peptide (Pc; Laudet *et al.*, 2007).

[^{35}S]methionine-labeled CK2α (10^6 cpm/mL in 50 m*M* Tris–HCl, pH 7.5; 0.15 *M* NaCl; 0.1 m*M* dithiothreitol)
Unlabeled CK2α (0.3 mg/mL in 50 m*M* Tris–HCl, pH 7.5; 0.15 *M* NaCl; 0.1 m*M* dithiothreitol)
Reacti-Bind streptavidin-coated high-binding-capacity 96-well plates (Pierce # 15500).
TNT® T7 Coupled Reticulocyte Lysate System (Promega Cat. # L4610).
Biosciences Redivue TM L-[^{35}S]methionine (1000 Ci/mmol) from Amersham Biosciences Cat. #AG1094).
Buffer A: 50 m*M* Tris–HCl, pH 7.5, 0.4 *M* NaCl
Buffer B: 50 m*M* Tris–HCl, pH 7.2, 0.15 *M* NaCl, and 0.05% Tween 20
Buffer C: 50 m*M* Tris–HCl, pH 7.2, 0.15 *M* NaCl and 3% BSA

3.1.2. Procedures

3.1.2.1. Recombinant proteins Expression and purification of chicken recombinant MBP-CK2β are performed as described previously (Chantalat *et al.*, 1999; Leroy *et al.*, 1999). Human recombinant CK2β subunit is expressed in *Escherichia coli* and purified to homogeneity as previously described (Heriche *et al.*, 1997).

3.1.2.2. Biotinylation of MBP-CK2β Prepare a freshly made solution of Sulfo-NHS-LC-LC-biotin (10 m*M*) in H_2O_2. Sulfo-NHS-LC-LC-biotin reacts with primary amines. Incubate MBP-CK2β (20 μ*M*) in PBS containing 0.4 *M* NaCl for 2 h at 4 °C with a 20-fold molar excess of Sulfo-NHS-LC-LC-biotin. At the end of the incubation, nonreacted biotin reagent is removed by size-exclusion chromatography on a PD-10 Desalting column equilibrated in Buffer A.

3.1.2.3.* In vitro *synthesis of [^{35}S]methionine-labeled CK2α Five micrograms of pSG5-CK2α plasmid is mixed with 200 μL of TNT®T7 Quick Master Mix and 220 μCi of [^{35}S]methionine in a final volume of 250 μl. After 2 h at 30 °C, the mixture is gel-filtrated on a BioSpin P6 column (Biorad) equilibrated in 50 m*M* Tris–HCl (pH 7.4), 150 m*M* NaCl. Aliquots of [^{35}S]methionine-labeled CK2α are stored at −20 °C.

3.1.2.4. Binding assay Reacti-Bind streptavidin-coated 96-well plates are coated with 50 μl of 5 μg/mL of biotinylated MBP-CK2β for 1 h at room temperature (RT). After three washes with Buffer B, the wells are blocked with 100 μl of Buffer C for 1 h at RT. After three washes with Buffer B, competing molecules or unlabeled CK2α along with [^{35}S]methionine-labeled CK2α (10^5 cpm) are added to each well in 50 μl of Buffer A.

The plates are incubated for 1 h at RT, and, after three washes with buffer B, each well is scapped with 2 × 100 μl of 10% SDS and the radioactivity determined using a scintillation counter. Positive controls (100% competition) are determined with a 10-fold molar excess of untagged CK2α. Negative controls are performed in the absence of plate-bound MPB-CK2β.

Comments: We found that 5 μg/mL of biotinylated MBP-CK2β (250 ng/well) is a saturating concentration for the Reacti-Bind streptavidin-coated well plates.

The assay is robust exhibiting a Z' factor >0.5 (Zhang *et al.*, 1999) and can be adapted for robotic screenings of chemical or peptide libraries. This assay can be also performed using GFP-CK2α-tagged protein instead of [^{35}S]methionine-labeled CK2α. However, we found that in this case, the recorded signal can be biased by the intrinsic autofluorescence of chemical molecules.

4. Visualization of CK2α–CK2β Interaction in Living Cells

4.1. Bimolecular fluorescence complementation

Two principal methods have been used to visualize the localization of protein interactions in living cells. Fluorescence resonance energy transfer (FRET) analysis is based on changes in the fluorescence intensities and lifetimes of two fluorophores that are brought sufficiently close together. In contrast, Bimolecular fluorescence complementation (BiFC) analysis is based on the association between two nonfluorescent fragments of a fluorescent protein when they are brought in proximity to each other by an interaction between proteins fused to the fragments (Kerpolla, 2008). This makes BiFC analysis potentially more sensitive and avoids interference from changes in fluorescence intensity or lifetime caused by cellular conditions unrelated to protein interactions.

4.1.1. Reagents

The BiFC assay is technically straightforward and can be performed using standard molecular biology and cell culture reagents and a regular fluorescence microscope.

Full-length chicken CK2α in pSG5 plasmid (Heriche *et al.*, 1997)

pEYFPc1 vector purchased from Clontech Laboratories

*Eco*RI and *Bam*HI restriction enzymes are from Invitrogen, and *Sac*I *Age*I from new England Biolabs.

HeLa cells grown in DMEM supplemented with 10% FBS.

LipofectamineTM 2000 reagent (Invitrogen)

Mouse anti-GFP antibody (Abcam Cat. # 1218-100)

Goat anti-mouse IgG-conjugated Cy3 (indocarbocyanine) (Molecular probes)
Hoechst 33342 (Sigma)
Vectashield mounting medium (Vector Labs, H-1000)

4.1.2. Devices and materials

Lab-Tek II chambers® (Nalge Nunc Cat. # 154534)

Fluorescence microscope. We use a Zeiss Axiovert 200 microscope equipped with a 40 × 1.3 Plan-Neofluar® objective.

4.1.3. Procedure

4.1.3.1. BiFC plasmid constructs Full-length chicken CK2α sequence is cloned into the pEYFPc1 vector using *Eco*RI restriction sites, then oriented with *Bam*HI digestion to generate the pEYFPc1-CK2α construct. Full-length mouse CK2β sequence is cloned into the same pEYFPc1 vector using *Bam*HI restriction sites, and then oriented with *Eco*RI digestion to generate the pEYFPc1-CK2β construct. Subsequently, the EYFP (enhanced yellow fluorescent protein) sequence is replaced by the 1–154 or the 155–238 fragments of EYFP. These fragments are then PCR amplified and inserted using *Sac*I-*Age*I to generate BiFC plasmid constructs. The resulting N- and C-terminal YFP-CK2 fusions are referred to as: YN-CK2α, YN-CK2β, YC-CK2α, YC-CK2β, respectively. Primers are for YN-CK2α or YN-CK2β: 5′G CTA CCG GTC GCC ACC ATG GTG AGC AAG 3′ and 5′ CAC AAC GTC TAT ATC ATG CGA GCT CAG GCT TCG AAT TCT GC 3′, respectively. Primers are for YC-CK2α or YC-CK2β: 5′ CCG TCA GAT CCG CTC GCG CTA CCG GTC ATG GCC GAC AAG CAG AAG AAC GGC 3′ and 5′ CG AAG CTT GAG CTC GAG ATC TGA GTC CGG 3′, respectively. All the constructs are verified by sequencing.

4.1.3.2. Cell transfection One day before transfection, HeLa cells are seeded in Lab-Tek chambers and transfected with the different BiFC plasmid constructs using Lipofectamine 2000 reagent according to the manufacturer's instructions. Four hours after transfection, cells are washed with PBS and incubated with fresh medium for 24 h at 37 °C and then switched to 30 °C for 2 h to promote fluorophore maturation. Cells are then fixed with 4% paraformaldehyde (PFA) for 10 min, permeabilized with 0.5% Triton X-100 for 10 min, preincubated with 5% goat serum for 30 min at RT, and incubated with the mouse anti-GFP as primary antibody for 1 h at RT, washed with PBS and incubated with goat anti-mouse IgG-conjugated Cy3 for 45 min at RT in the dark. Nuclei are stained by 2 μg/mL Hoechst 33342 and coverslips are mounted with mounting medium and examined using a fluorescence microscope. The results are expressed as % of cells that are above the fluorescence threshold observed in cells expressing only YN-CK2α or YN-CK2β. Because the amount of expressed proteins is critical to the interpretation of results, the expression levels of the different chimeras in transfected cells are also analyzed by Western blotting using anti-CK2α or anti-CK2β antibodies as previously described (Filhol *et al.*, 2003; Martel *et al.*, 2006).

Immunofluorescence images of a typical BiFC assay are shown in Fig. 31.3. Cells transfected with full-length EYFP fused to CK2α (panel A) exhibit a strong fluorescence after immunostaining of EYFP. As expected, cells separately transfected with YN-CK2α or YN-CK2β expression vectors did not show any immunostaining (panels B and C). Cotransfection of YN-CK2α and YN-CK2β yields fluorescence complementation through reconstitution of EYFP in a significant number of cells (panel D). In contrast, a weak complementation is observed with a CK2β-YF double mutant (panel E) confirming that amino acids Tyr188 and Phe190 in CK2β are key hotspots for CK2α binding in living cells (Laudet *et al.*, 2007).

Comments: Preliminary experiments have shown that a fraction of YN-CK2α or YC-CK2α and YN-CK2β or YC-CK2β could interact with

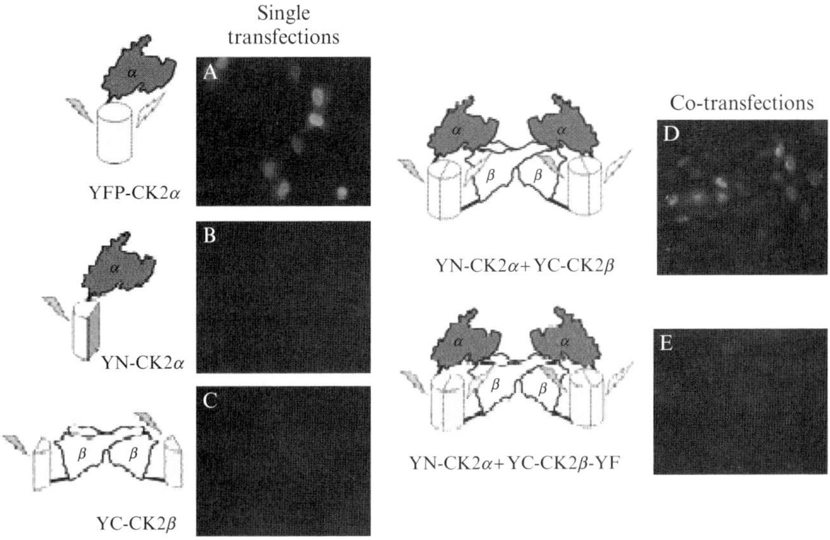

Figure 31.3 Immunofluorescence images of a typical BiFC assay. Immunostaining of HeLa cells transfected with plasmids expressing YFP fused to CK2α, or YFP fragments (YN or YC) fused to CK2α (panels A–C) or cotransfected with YFP fragment (YN) fused to CK2α and YFP fragment (YC) fused to CK2β (panel D) or YFP fragment (YN) fused to CK2α and YFP fragment (YC) fused to Y188A/F190A CK2β double mutant (CK2β-YF) (E). The schematic diagrams on the left of the images represent the experimental strategies used. (See Color Insert.)

endogenous CK2β and CK2α, trapping a part of YFP devoted to fluorescence complementation (Kerppola, 2008). Therefore, to enhance the signal, we advise to perform immunostaining of YFP by indirect immunofluorescence using the mouse anti-GFP antibody (Abcam Cat. # 1218-100), which recognizes only full-length YFP, but neither N-ter (1–154) nor C-ter (155–238) EYFP. Based on our experience, under similar transfection conditions, polyclonal anti-GFP antibody (Abcam Cat. # 290-50) did not discriminate between full-length and EYFP half-molecules. Also, analysis by immunoblotting of the expression of interacting proteins in the transfected cells is a necessary step before to draw any conclusion.

4.2. Visualizing endogenous CK2 subunit interaction

The *in situ* Proximity Ligation Assay (*in situ* PLA) described by Soderberg *et al.* (2006 # 1368) provides a sensitive means to detect and visualize individual protein–protein interactions *in situ*. In this technique, individual protein interactions are made visible through DNA amplification. Importantly, *in situ* PLA allows point detection of individual endogenous proteins, which reveals both their subcellular locations and the frequency of occurrence. This opens new opportunities to accurately quantify protein interaction in unmodified single cells. CK2 holoenzyme complex detection in living cells can be performed using an adaptation of the *in situ* PLA (Duolink *in situ* PLA). The assay can also be used to explore the ability of chemical compounds to impact on the formation of the CK2 holoenzyme in living cells.

4.2.1. Principle of *in situ* PLA assay

The assay involves five steps: (1) fixed and permeabilized adherent cells are incubated with two primary antibodies (one of mouse origin and one of rabbit origin) that bind to the interacting proteins to be detected; (2) two secondary antibodies (PLA probes), one anti-mouse Ig (PLA probe MINUS), and one anti-rabbit Ig (PLA probe PLUS), conjugated with oligonucleotides are added and bind to their respective primary antibody; (3) two oligonucleotides are added together with a ligase to generate a circular oligonucleotide when the PLA probes are in close vicinity; (4) after addition of nucleotides together with a polymerase, the oligonucleotide arm of one of the PLA probes acts as a primer for a rolling circle amplification (RCA) reaction, generating a concatemeric product; (5) fluorescently labeled oligonucleotides will hybridize to the RCA products generating visible fluorescent dots that are easily analyzed by fluorescence microscopy.

4.2.2. Reagents and equipments

Polyclonal anti-CK2α antibody (Laramas *et al.*, 2007) and monoclonal anti-CK2β antibody (unpublished) that have been validated in IHC.
Duolink *in situ* PLA assay kit (Cat. # 74601, Olink Bioscience)
96-well plates (PerkinElmer, Cat. # 6005182)
Hela cells cultivated in DMEM (Invitrogen Life Technologies, Inc.) supplemented with 10% (v/v) FBS (BioWest)
Fluorescence microscope equipped with excitation/emission filters compatible with Texas Red or Cy3 fluorophores.
Ca^{2+} and Mg^{2+}-free PBS 1×
4% PFA
PBS 0.5% Triton X-100
TBS 0.05%Tween 20 (TBS-T)
SSC 2×: 300 m*M* NaCl, 30 m*M* sodium citrate, pH 7

4.2.3. Procedure

CK2 holoenzyme complex detection in living cells can be performed using the Duolink *in situ* PLA assay kit (Olink). HeLa cells plated into 96-well plates at 2×10^4 cells/well are incubated for 24 h in the absence or presence of chemical compounds. Then, they are fixed for 20 min at 37 °C with 0.1 mL PFA, double-washed with 0.2 mL PBS and permeabilized for 20 min at RT with PBS 0.5% Triton X-100. After two washes, the samples are incubated for 2 h at RT with 0.1 mL of the 1× Duolink blocking solution before overnight incubation with a polyclonal anti-CK2α antibody (1/4000) and a monoclonal anti-CK2β antibody (1/400). The next day, the CK2α–CK2β interactions can be visualized using the Duolink *in situ* PLA assay kit following the manufacturer's instructions. The samples are double-washed with TBS-T (5 min) and incubated for 1 h at 37 °C with two secondary antibodies (PLA probes) diluted in 1× Duolink antibody diluent solution. After two washes with TBS-T (10 min), the samples are incubated for 30 min at 37 °C with 40 μl of 1× hybridization solution followed by two washes under gentle agitation with TBS-T (2 min) before adding the Ligase solution for 30 min at 37 °C. After two washes under gentle agitation (2 min), the Duolink amplification solution is added for 2.5 h at 37 °C. After two washes with TBS-T (4 min) under gentle agitation, samples are incubated with the detection solution for 1.5 h at 37 °C in the dark. After sequential washes (2× SSC, 5 min; 1× SSC, 5 min; 0.2× SSC, 5 min) 0.02× SSC containing 2 mg/ml Hoechst 33258 and Phalloïdin Alexa Fluor 488 (Invitrogen A12379), 1/200 for 15 min. Then, the samples are covered with 0.2 mL of 0.02× SSC. Image capture based on 1*z*-plane, are carried out with an In Cell Analyzer 1000 (GE Healthcare) with a 20× objective. The PLA detection reaction products are seen as bright fluorescent dots called blobs, allowing the quantitative effect of inhibitory molecules to be digitally recorded by computer-assisted image analysis. For example, fluorescent blobs can be enumerated using the Blob-Finder software recommanded by Olink (http://www.cb.uu.se/~amin/Blob-Finder/). As illustrated in Fig. 31.4, negligible specific PLA signal was observed in cells when either of the primary antibody against one of the CK2 subunits was omitted, or when PLA probe was omitted highlighting the strict dependence on dual recognition of CK2α–CK2β complexes to yield a PLA signal.

Comments: The assay requires to use two primary antibodies raised in different species and binding to the targets under the same conditions (fixation, buffer, etc.). Incubation of the cells with the primary CK2 antibodies is a crucial step: very specific antibodies must be used and it is important to find the correct dilutions for each antibody to avoid background signal. Molecules interfering with the CK2α–CK2β interaction may lead to decreased numbers of blobs and/or smaller blobs. Care must

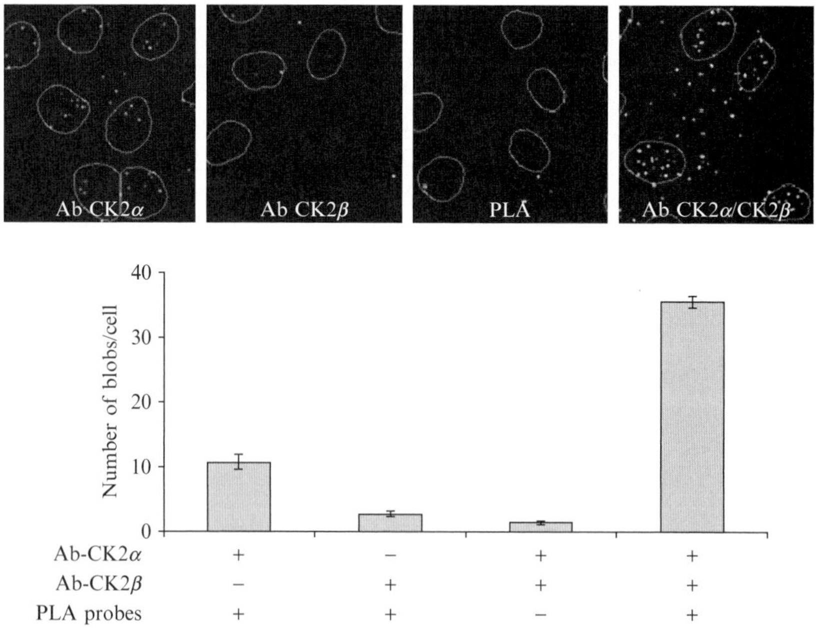

Figure 31.4 Direct observation of individual endogenous CK2 complexes *in situ* by proximity ligation. The upper panel shows a number of discrete fluorescent dots ("PLA signals" or "Blobs") in various locations of the studied cells. The pictures show the raw images based on 1*z*-plane with the nuclei delimited in green. (For interpretation of the references to color in this figure legend, the reader is referred to the Web version of this chapter.)

be taken with aggregated autofluorescent compounds that can be posted by the Blob-Finder software as true blobs.

Comparison of blobs number per nuclei using Blob-Finder software or blobs number per cell using In Cell Analyzer 1000 Workstation 3.5 after phalloidin staining yields similar results demonstrating that the procedure gives only weak background signals.

ACKNOWLEDGMENTS

The research relevant to this chapter was supported by the Institut National de la Santé et de la Recherche Médicale (INSERM), the Commissariat à l'Energie atomique (CEA), the Ligue Nationale Contre le Cancer (équipe labellisée 2007), the Institut National du Cancer (Grant No. 57), and the Agence Nationale de la Recherche (PCV08_324733). The authors thank Laurence Lafanechère and Emmanuelle Soleilhac for computer-assisted PLA signal acquisition and Alexandre Deshiere for technical support in lentivirus production.

REFERENCES

Chantalat, L., *et al.* (1999). Crystallization and preliminary X-ray diffraction analysis of the regulatory subunit of human protein kinase CK2. *Acta Crystallogr. D Biol. Crystallogr.* **55,** 895–897.

Duncan, J. S., and Litchfield, D. W. (2008). Too much of a good thing: The role of protein kinase CK2 in tumorigenesis and prospects for therapeutic inhibition of CK2. *Biochim. Biophys. Acta* **1784,** 33–47.

Filhol, O., and Cochet, C. (2009). Cellular functions of protein kinase CK2: A dynamic affair. *Cell. Mol. Life Sci.* **66,** 1830–1839.

Filhol, O., *et al.* (2003). Live-cell fluorescence imaging reveals the dynamics of protein kinase CK2 individual subunits. *Mol. Cell. Biol.* **23,** 975–987.

Filhol, O., *et al.* (2004). Protein kinase CK2: A new view of an old molecular complex. *EMBO Rep.* **5,** 351–355.

Heriche, J. K., *et al.* (1997). Regulation of protein phosphatase 2A by direct interaction with casein kinase 2alpha. *Science* **276,** 952–955.

Kerppola, T. K. (2008). Bimolecular fluorescence complementation : visualization of molecular interactions in living cells. *Methods Cell. Biol.* **85,** 431–470.

Laramas, M., *et al.* (2007). Nuclear localization of protein kinase CK2 catalytic subunit (CK2alpha) is associated with poor prognostic factors in human prostate cancer. *Eur. J. Cancer* **43,** 928–934.

Laudet, B., *et al.* (2007). Structure-based design of small peptide inhibitors of protein kinase CK2 subunit interaction. *Biochem. J.* **408,** 363–373.

Leroy, D., *et al.* (1999). Mutations in the C-terminal domain of topoisomerase II affect meiotic function and interaction with the casein kinase 2 beta subunit. *Mol. Cell. Biochem.* **191,** 85–95.

Litchfield, D. W. (2003). Protein kinase CK2: Structure, regulation and role in cellular decisions of life and death. *Biochem. J.* **369,** 1–15.

Mangeot, P. E., *et al.* (2002). High levels of transduction of human dendritic cells with optimized SIV vectors. *Mol. Ther.* **5,** 283–290.

Martel, V., *et al.* (2006). p53-dependent inhibition of mammalian cell survival by a genetically selected peptide aptamer that targets the regulatory subunit of protein kinase CK2. *Oncogene* **25,** 7343–7353.

Niefind, K., *et al.* (2001). Crystal structure of human protein kinase CK2: Insights into basic properties of the CK2 holoenzyme. *EMBO J.* **20,** 5320–5331.

Pinna, L. A. (2002). Protein kinase CK2: A challenge to canons. *J. Cell Sci.* **115,** 3873–3878.

Prudent, R., *et al.* (2008). Salicylaldehyde derivatives as new protein kinase CK2 inhibitors. *Biochim. Biophys. Acta* **1780**(12), 1412–1420.

Ruzzene, M., and Pinna, L. A. (2009). Addiction to protein kinase CK2: A common denominator of diverse cancer cells? *Biochim. Biophys. Acta* **1804**(3), 499–504.

Zhang, J. H., *et al.* (1999). A simple statistical parameter for use in evaluation and validation of high throughput screening assays. *J. Biomol. Screen.* **4,** 67–73.

Author Index

C

D

G

H

I

J

K

M

T

U

V

Z

Subject Index

D

E

F

G

H

W

Y

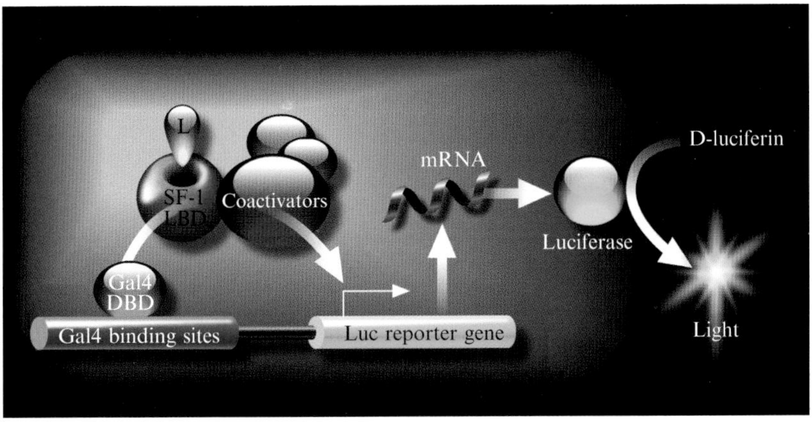

Mabrouka Doghman *et al.*, Figure 1.2 The Gal4 system. Mammalian cells are cotransfected with a reporter plasmid harboring Gal4 binding sites and another plasmid to express a Gal4 DBD–SF-1 LBD fusion protein. This protein binds to the Gal4 binding sites, interacts with coactivator complexes and increases the expression of the luciferase reporter gene. This enzyme generates light from the D-luciferin substrate. L, ligand.

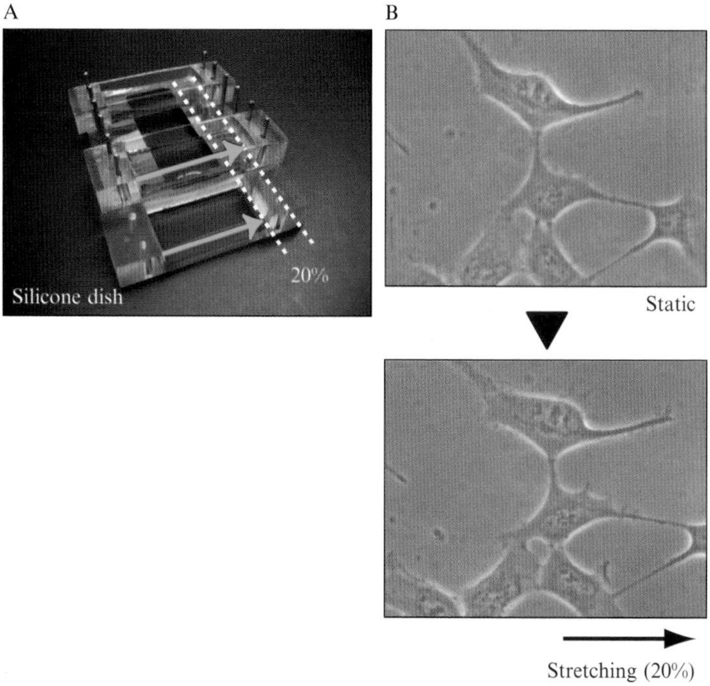

Hiroshi Akazawa *et al.*, Figure 2.1 Passive stretch of cells cultured on extensible silicone dishes (A) Cells were plated on collagen-coated silicone rubber dishes, and silicone dishes were passively stretched longitudinally by 20%. (B) HEK293cells before and after stretching were shown as phase-contrast images. Figure is adopted from the work of Yasuda *et al.* (2008).

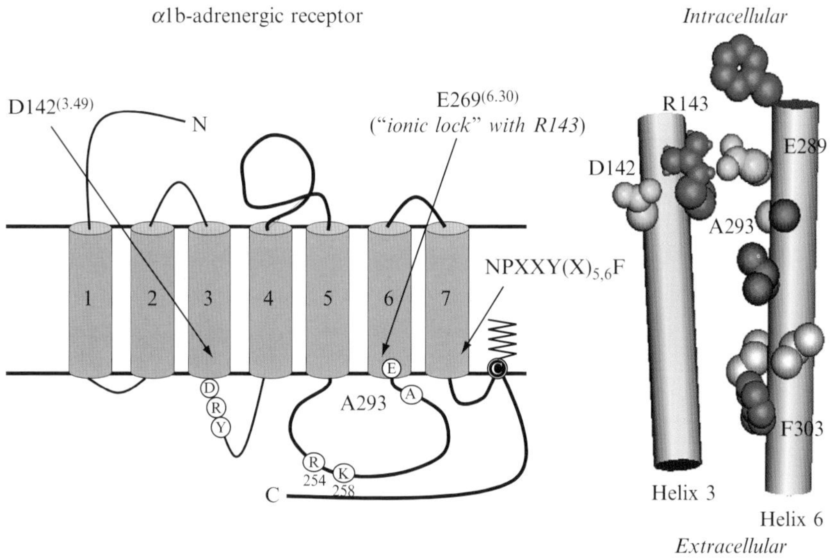

Susanna Cotecchia, Figure 7.1 The α_{1b}-adrenergic receptor (AR). (Left) Topographical model of the receptor displaying key amino acids involved in receptor function mentioned in the text. (Right) Relative position of helices 3 and 6 in the homology model of the wild-type α_{1b}-AR. Comparative modeling and molecular dynamic simulations were performed as described in Greasley *et al.* (2002). The view displays the amino acids of helices 3 and 6 involved in receptor activation. Van der Waals spheres, whose radius has been reduced by 40%, depict each side chain. The color indicates the effect of mutations: white/no effect, green/constitutively activating, red/impairing receptor signaling, violet/either impairing or constitutively activating depending upon the substituent amino acid.

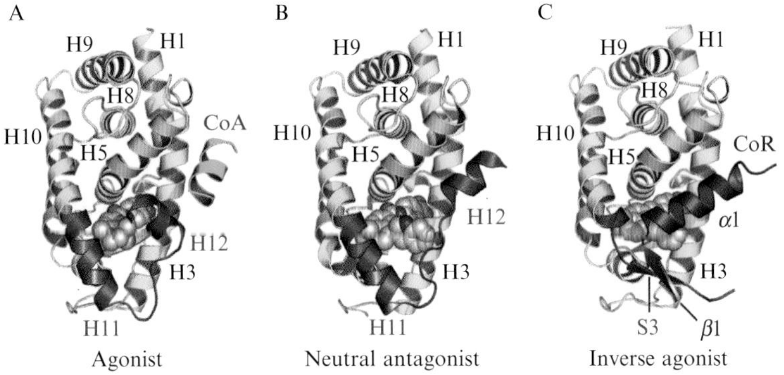

William Bourguet *et al.*, Figure 10.1 Crystal structures of RARα LBD in various functional states. (A) The active form of the receptor is induced by agonist binding and allows coactivator (CoA, green) binding. (B) In contrast, binding of neutral antagonists provokes a displacement of H12 that prevents CoA recruitment and maintains the receptor in an inactive state. (C) The repressive state can be obtained in the presence of an inverse agonist that stabilizes the interaction with corepressors (CoR, violet). The C-terminal portion of RARα LBD that is subjected to important conformational changes is highlighted in red.

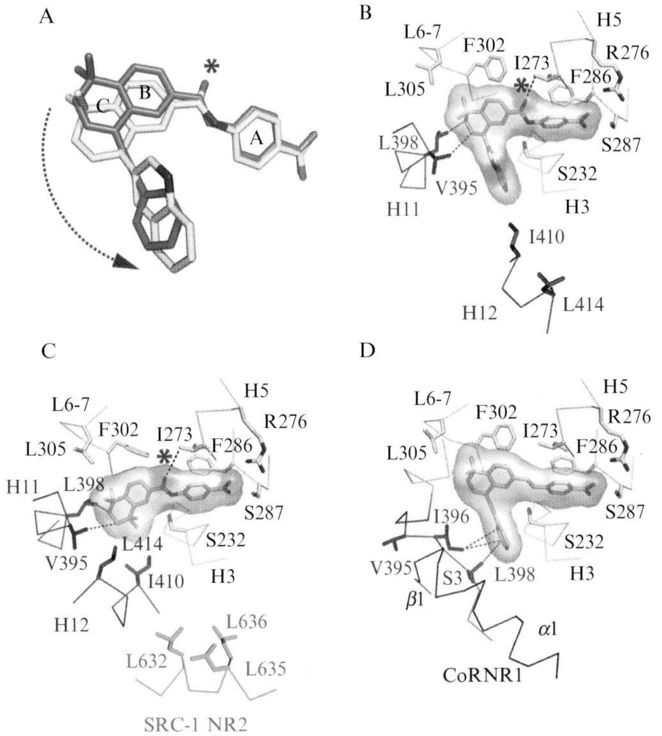

William Bourguet *et al.*, Figure 10.2 Structural basis of antagonism and inverse agonism in RAR. (A) The antagonist BMS614 [6] (yellow) and the inverse agonist BMS493 [4a] (blue) adopt different positions in RAR LBP. The asterisk indicates a steric clash between the carbonyl moiety of BMS614 [6] and Ile273, which induces a repositioning (red arrow) of the antagonist relative to BMS493 [4a]. (B) The RAR LBP with antagonist BMS614 [6] bound. Stabilization of helix H11 through van der Waals interactions with V395 and L398 is indicated by red dotted lines. (C) The RAR LBP with agonist Am580 [3] bound. Stabilization of helix H11 through van der Waals interactions with V395 and L398 is indicated by red dotted lines. (D) The RAR LBP with the bound inverse agonist BMS493 [4a]. Stabilization of β-strand S3 is indicated by red dotted lines. The color code is identical to that used in Figure 10.1.

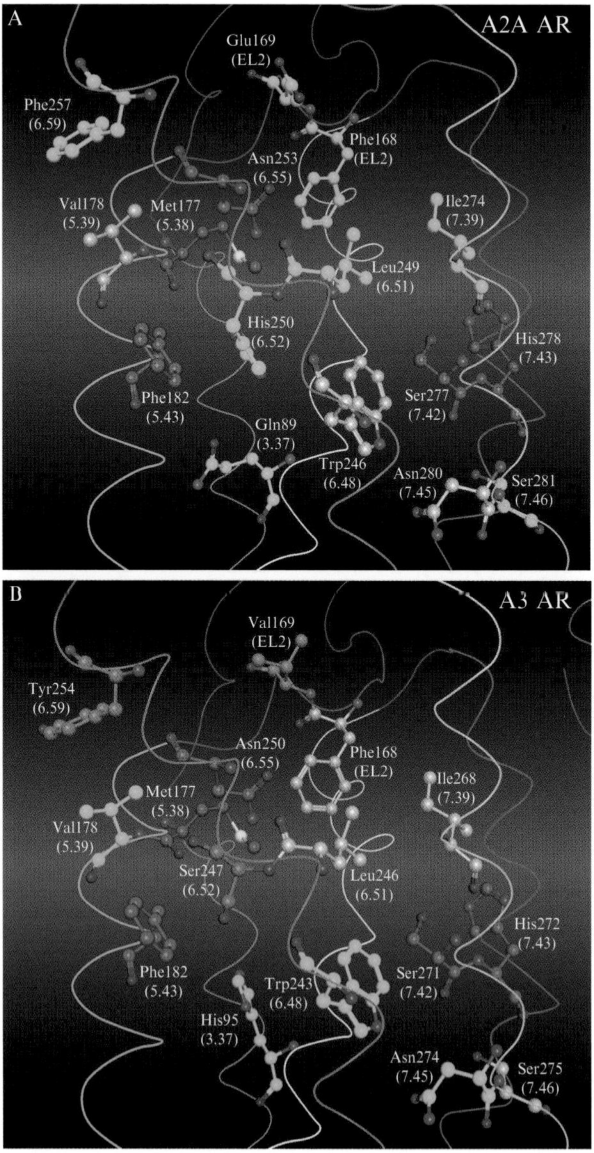

Silvia Paoletta *et al.*, Figure 13.4 Comparison of available mutagenesis data for amino acids affecting antagonists binding on (A) hA_{2A} and (B) hA_3 adenosine receptor subtypes. Amino acids color legend: *light green*: mutagenesis data available for one subtype; *dark green*: mutagenesis data available for both subtypes; *light pink*: conserved residues whose mutagenesis data are available only for the other subtype; *dark pink*: not conserved residues whose mutagenesis data are available only for the other subtype.

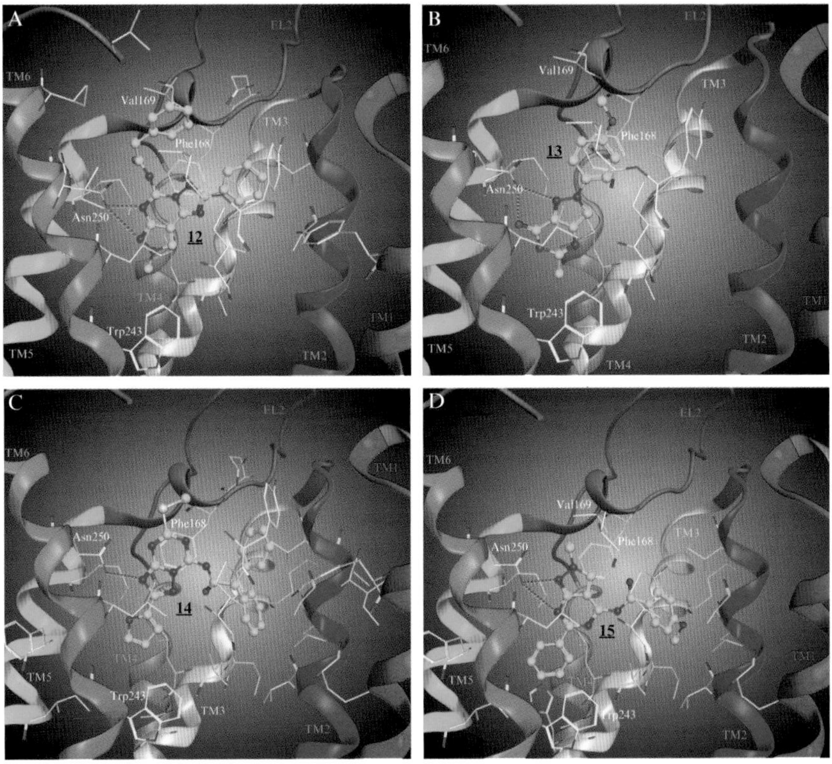

Silvia Paoletta *et al.*, Figure 13.5 Hypothetical binding modes, obtained after docking simulations inside the hA_3AR binding site, of (A) compound 12; (B) compound 13; (C) compound 14; (D) compound 15. Poses are viewed from the membrane side facing TM6, TM7, and TM1. The view of TM7 is voluntarily omitted. Side chains of some amino acids important for ligand recognition and H-bonding interactions are highlighted. Hydrogen atoms are not displayed.

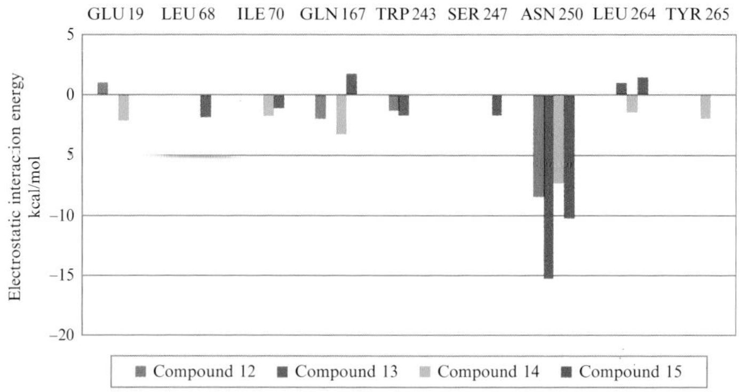

Silvia Paoletta *et al.*, Figure 13.6 Electrostatic interaction energy (in kcal/mol) between compounds 12, 13, 14, and **15** and particular hA_3AR residues involved in ligand recognition.

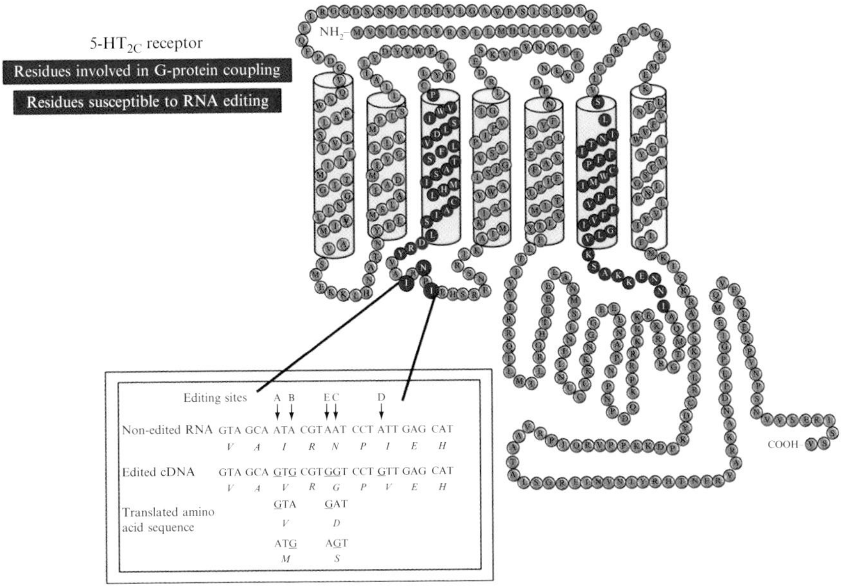

Maria Fe Lanfranco *et al.*, Figure 18.1 Schematic representation of the 5-HT_{2C}R (adapted from Julius *et al.*, 1988, 1989). Predicted amino acid residues linked to G-protein coupling are denoted in blue (Roth *et al.*, 1998; Wess, 1998) and amino acid residues at which RNA editing occurs are denoted in red (Niswender *et al.*, 1998). RNA, cDNA nucleotide, and predicted amino acid sequences for the nonedited and fully edited transcripts are shown with the editing sites indicated by arrows [↓] (Berg *et al.*, 2008b; Herrick-Davis *et al.*, 2005; Sanders-Bush *et al.*, 2003).

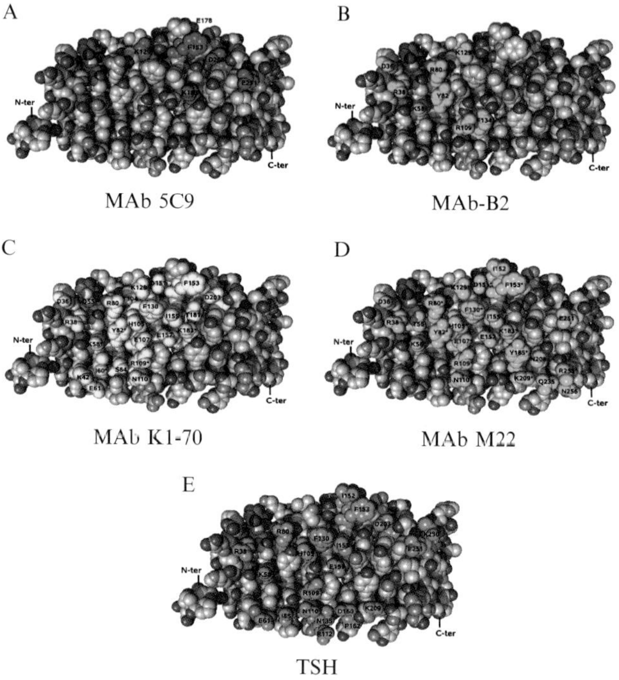

Jane Sanders *et al.*, Figure 22.4 Space fill representation of the TSHR leucine-rich domain interactive surface (based on the crystal structure of the TSHR in complex with K1–70 solved at 1.9 Å resolution). Single amino acid mutations which affect inhibition of TSH-mediated stimulation of cyclic AMP production in TSHR transfected CHO cells in the case of: (A) MAb 5C9 shown in red (Table 22.1; antagonist and inverse agonist), (B) MAb-B2 shown in blue (Table 22.1; antagonist), (C) TSHR residues that interact with MAb K1–70 (antagonist) in the TSHR—K1–70 crystal structure are shown in yellow (⋆ denotes an effect on K1–70 activity in mutation experiments; see Table 22.1), (D) TSHR residues that interact with MAb M22 (agonist) in the TSHR—M22 crystal structure are shown in green (⋆ denotes an effect on M22 activity in mutation experiments; see Table 22.1), (E) TSHR residues that interact with TSH (agonist) in a TSH-TSHR comparative model are shown in orange (Table 22.1).

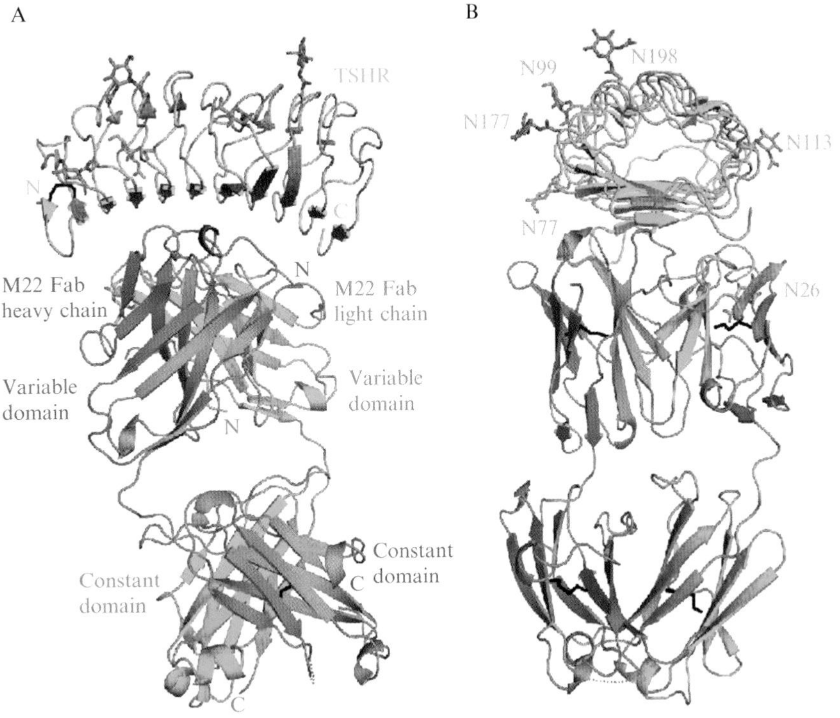

Jane Sanders *et al.*, Figure 22.5 Crystal structure of TSHR260 in complex with thyroid stimulating monoclonal autoantibody (M22) Fab (solved at 2.55 Å resolution; Sanders *et al.*, 2007a). (A & B) Cartoon diagram of the complex shown in two aligned views related by a 90° rotation about the vertical axis. TSHR is in cyan, M22 light chain is in green, M22 heavy chain is in blue. The positions of amino-(N) and carboxy-(C) termini are indicated. The observed N-linked carbohydrates are shown in yellow and carbohydrate-bound asparagines residues are labeled. Disulphide bonds are in black. From Sanders *et al.* (2007a). The publisher for this copyrighted material is Mary Ann Liebert, Inc. publishers and a license has been granted (number 2437101285589).

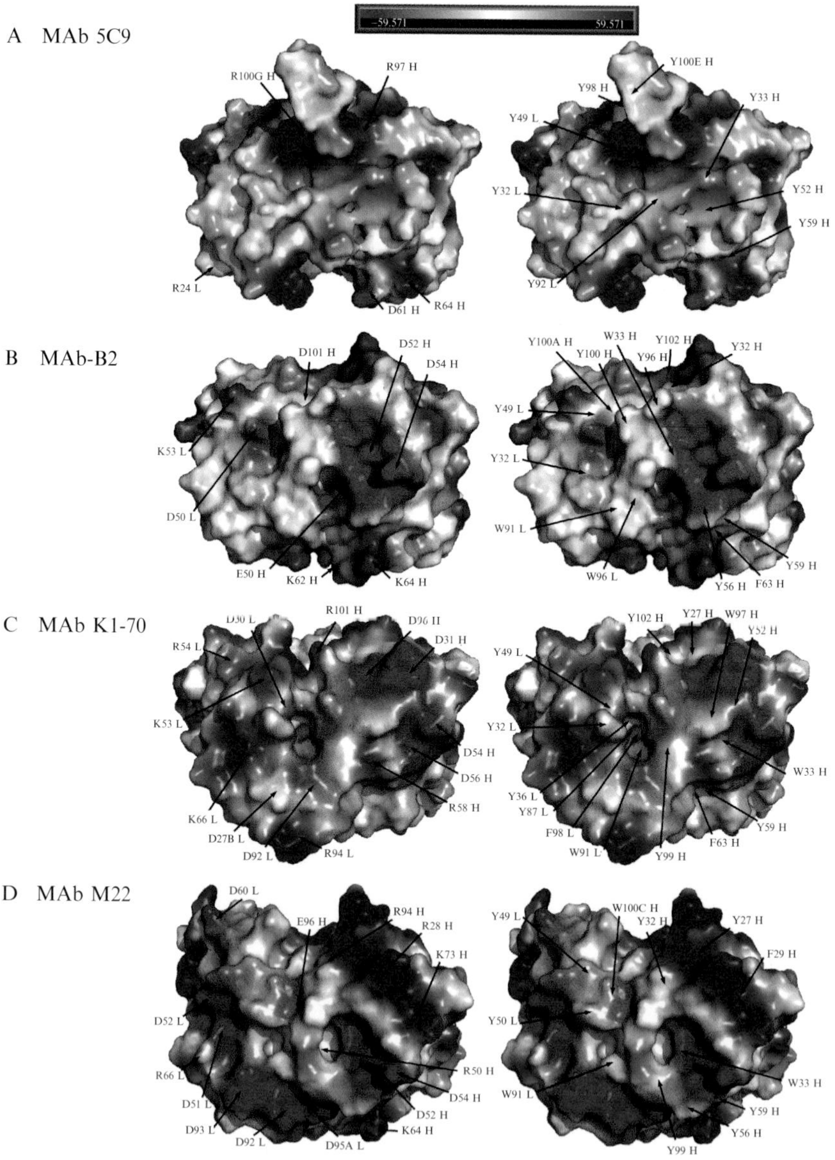

Jane Sanders *et al.*, Figure 22.6 The electrostatic surface potential of the antigen binding regions of: (A) Human MAb 5C9 (antagonist and inverse agonist; derived from the comparative model of 5C9 structure), (B) mouse MAb-B2 (antagonist; derived from the crystal structure solved at 3.3 Å resolution), (C) human MAb K1–70 (antagonist; derived from the crystal structure solved at 2.2 Å resolution), (D) human MAb M22 (agonist; derived from the crystal structure solved at 1.65 Å resolution). Position of the charged residues is indicated in the figures on the left-hand side of the panel and the position of aromatic residues is indicated in the figures on the right-hand side of the panel. Acidic patches are shown in red and basic patches in blue.

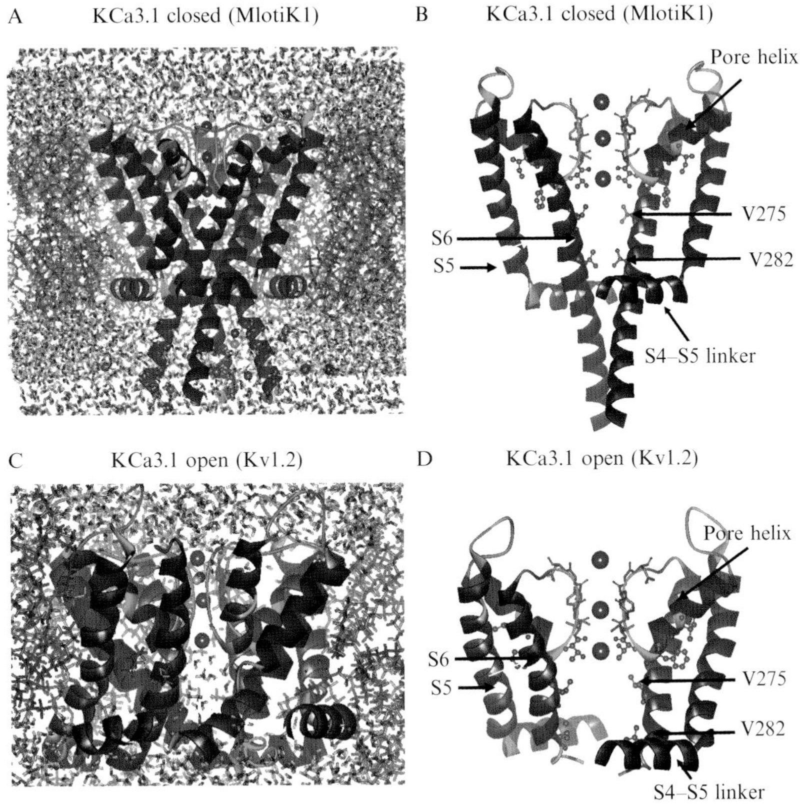

Line Garneau *et al.*, Figure 24.2 Model structures of KCa3.1. A and C. Global side view representations of the closed (A) and open (C) KCa3.1 embedded in a lipid bilayer (green) separating two electrolyte solutions containing 150 m*M* KCl. The closed KCa3.1 3D representation was obtained by homology modeling using the MlotiK1 structure (PDB:3BEH) as template, whereas the open representation is based on the Kv1.2 channel structure (PDB:2a79). Both models include the S4–S5 linker plus the S5–S6 pore region of the channel. (B and D) Detailed view of the pore region with only two monomers represented for clarity. Turns colored in blue refer to residues predicted to be facing the pore (V275, T278, A279, V282, and A286). These predictions were confirmed in three SCAM analyses produced by our laboratory. Templates selected with I-TASSER and SAM-T08 servers. CPK representation of K^+ ions in purple. Representation by DS Visualizer.

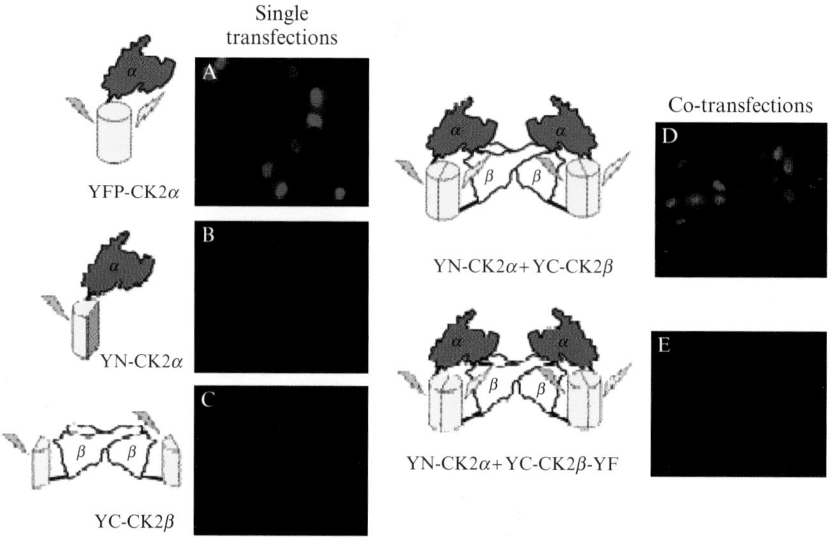

Renaud Prudent *et al.*, Figure 31.3 Immunofluorescence images of a typical BiFC assay. Immunostaining of HeLa cells transfected with plasmids expressing YFP fused to CK2α, or YFP fragments (YN or YC) fused to CK2α (panels A–C) or cotransfected with YFP fragment (YN) fused to CK2α and YFP fragment (YC) fused to CK2β (panel D) or YFP fragment (YN) fused to CK2α and YFP fragment (YC) fused to Y188A/F190A CK2β double mutant (CK2β-YF) (E). The schematic diagrams on the left of the images represent the experimental strategies used.